Water and Wastewater Technology

Sixth Edition

Mark J. Hammer

Mark J. Hammer, Jr.

PEARSON

Prentice
Hall

Upper Saddle River, New Jersey
Columbus, Ohio

Library of Congress Cataloging-in-Publication Data

Hammer, Mark J.,
   Water and wastewater technology / Mark J. Hammer, Mark J. Hammer, Jr. — 6th ed.
    p. cm.
   ISBN 0-13-174542-5
   1. Water-supply engineering. 2. Sewage disposal. I. Hammer, Mark J., II. Title.
   TD345.H25 2008
   628.1—dc22

                                        2007013442

Editor in Chief: Vernon Anthony
Acquisitions Editor: Eric Krassow
Associate Managing Editor: Christine Buckendahl
Editorial Assistant: Nancy Kesterson
Production Coordination: GGS Book Services
Production Editor: Holly Shufeldt
Design Coordinator: Diane Ernsberger
Cover Designer: Jason Moore
Production Manager: Deidra Schwartz
Director of Marketing: David Gesell
Executive Marketing Manager: Derril Trakalo
Marketing Assistant: Les Roberts

This book was set in Trump Mediaeval and Gill Sans by GGS Book Services. It was printed and bound by Edwards Brothers. The cover was printed by Phoenix Color Corp.

Pearson Education Ltd.
Pearson Education Singapore Pte. Ltd.
Pearson Education Canada, Ltd.
Pearson Education—Japan

Pearson Education Australia Pty. Limited
Pearson Education North Asia Ltd.
Pearson Educación de Mexico, S.A. de C.V.
Pearson Education Malaysia Pte. Ltd.

10 9 8 7 6 5 4 3 2 1
ISBN 13: 978-0-13-174542-1
ISBN 10: 0-13-174542-5

*In memory of our
parents and grandparents
Bertha (Grundahl) Hammer
and Herbert Hammer*

PREFACE

This book provides comprehensive coverage of the fundamental principles and current practices in water processing, water distribution, wastewater collection, wastewater treatment, sludge processing, advanced wastewater treatment, and water reuse. The objective is to transfer knowledge of these subjects to persons throughout the world who are interested in continuing their study in sanitary technology and engineering, and to persons interested in operation and maintenance of water and wastewater facilities.

This sixth edition updates the subject matter, illustrations, and problems to incorporate new concepts and issues related to the water environment. In this edition, the revision of text and addition of new problems are in the chapters on water processing, wastewater processing, advanced wastewater treatment, and water reuse.

Based on our experience in education, we believe our readers benefit from a review of the disciplines that have specific applications in water supply and wastewater management. Therefore, the introductory chapters cover fundamentals of chemistry, biology, hydraulics, and hydrology applicable to sanitary studies. The subject of water quality is also introduced to understand the reasons for the selection of processes in water and wastewater treatment.

The approach to presenting water supply and wastewater management is traditional, covering water distribution, processing, and operation of systems separated from wastewater collection, treatment, and operation. We have carefully integrated the subject matter in each area so readers can clearly understand the interrelationships between individual unit operations and integration of systems as a whole. Advanced wastewater treatment and water reuse, the subjects of the final chapter, are increasingly important in many regions for pollution control and beneficial use of water resources.

The extensive use of illustrations is intended to increase the understanding of concepts and show modern equipment and facilities. Also, numerous sample calculations assist the reader in the application of equations, charts, and tabulated data. Answers are given for some of the homework problems, mainly to help persons interested in individual study.

A discussion of the book's contents is given in Chapter 1, "Introduction."

SUPPLEMENTS

To access supplementary materials online, instructors need to request an instructor access code. Go to **www.prenhall.com**, click the **Instructor Resource Center** link, and then click **Register Today** for an instructor access code. Within 48 hours after registering you will receive a confirming e-mail including an instructor access code. Once you have received your code, go to the site and log on for full instructions on downloading the materials you wish to use.

ACKNOWLEDGMENTS

Particular thanks are due to Joseph R. V. Flora, University of South Carolina; Enos C. Inniss, University of Texas at San Antonio; Blaine Stong, Horry-Georgetown Technical College; and Robert Sharp, Manhattan College, for their assistance with the sixth edition text review.

We hope this book will be of benefit to both present and future colleagues who teach, study, and practice in the area of the water environment.

Mark J. Hammer
Mark J. Hammer, Jr.

CONTENTS

Contents

CHAPTER 1

Introduction

The hydrologic cycle describes the movement of water in nature. Evaporation from the ocean is carried over land areas by maritime air masses. Vapor from inland waters and transpiration from plants add to atmospheric moisture that eventually precipitates as rain or snow. Rainfall may percolate into the ground, join surface watercourses, be taken up by plants, or reevaporate. Groundwater and surface flows drain toward the ocean for recycling.

Humans intervene in the hydrologic cycle, generating artificial water cycles (Figure 1–1). Some communities withdraw groundwater for public supply, but the majority rely on surface sources. After processing, water is distributed to households and industries. Wastewater is collected in a sewer system and transported to a plant for treatment prior to disposal. Conventional methods provide only partial recovery of the original water quality. Dilution into a surface watercourse and purification by nature yield additional quality improvement. However, the next city downstream is likely to withdraw the water for a municipal supply before complete rejuvenation. This city in turn treats and disposes of its wastewater by dilution. This process of withdrawal and return by successive municipalities in a river basin results in indirect water reuse.

During dry weather, maintaining minimum flow in many small rivers relies on the return of upstream wastewater discharges. Thus, an artificial water cycle within the natural hydrologic scheme involves (a) surface-water withdrawal, processing, and distribution; (b) wastewater collection, treatment, and disposal back to surface water by dilution; (c) natural purification in a river; and (d) repetition of this scheme by cities downstream.

Discharge of conventionally treated wastewaters to lakes, reservoirs, and estuaries, which act like lakes, accelerates eutrophication. The resulting deterioration of water quality interferes with indirect reuse for public supply and water-based recreational activities. Consequently, advanced wastewater treatment by either mechanical plants or land disposal techniques has been introduced into the artificial water cycle, involving inland lakes and reservoirs.

Figure 1–1

Integration of natural and human-generated water cycles.

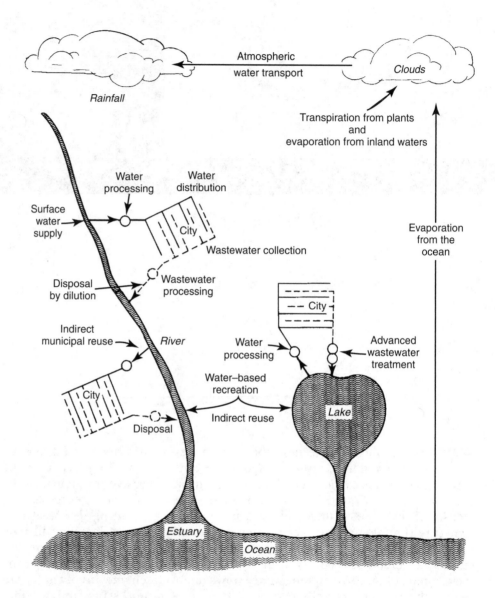

Installation of advanced treatment systems that reclaim wastewater to nearly its original quality has encouraged several cities to consider direct reuse of water for industrial processing, recreational lakes, irrigation, groundwater recharge, and other applications. However, direct return for a potable water supply is not being encouraged because of potential health hazards from viruses and traces of toxic substances that are difficult to detect and may not be removed in water reclamation. Another problem is the buildup of dissolved salts that can be removed only by costly demineralization processes. Nevertheless, with the increase in demand for fresh water, direct water reuse by some metropolitan areas may be realistic in the future.

The basic sciences of chemistry, biology, hydraulics, and hydrology are the foundation for understanding water supply and pollution control. Chemical principles find their greatest application in water processing, while wastewater treatment relies on biological systems. Knowledge of hydraulics is the key to water distribution and wastewater collection. Quality of the water environment is the focus of the indirect water reuse cycle. As illustrated in Figure 1–2, chemistry, hydrology, biology, water processing, and wastewater treatment converge to give a perception of

Figure 1–2

Flow diagram relating basic science areas to the disciplines of water and wastewater technology and the integration of various aspects of water supply and pollution control.

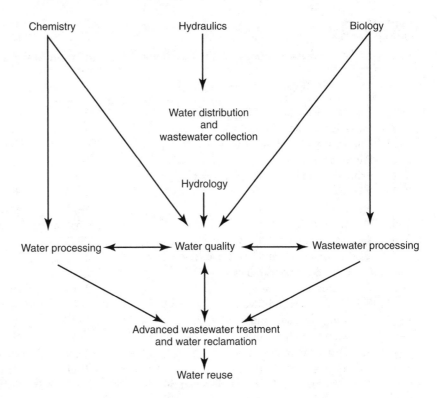

water quality. Finally, groundwater recharge and water reuse are provided through the technology of advanced wastewater treatment and water reclamation.

In this book, the fundamental principles of science are integrated with different aspects of sanitary technology. Table 1–1 lists prerequisite readings included in this book for various subject areas. The purpose of this table is to help the student correlate information from the text on a particular subject of interest. For example, prior to studying water quality in Chapter 5, a student should review chemistry fundamentals, biology of microorganisms, waterborne diseases, and the hydrology of rivers and lakes. Prerequisite readings for water distribution systems are selections from Chapter 4. The instructor who is preparing a lecture sequence on this topic may prefer to integrate the principles of hydraulics with the descriptive material from the chapter on water distribution. The advantage is that students can read about real pipe networks while solving simplified pipe flow problems. Conversely, when water processing is taught, it may be most advantageous for the student to review the portions on

chemistry listed in Table 1–1 prior to doing the reading assignments in Chapter 7.

Wastewater flows and characteristics are discussed separately in Chapter 9, since this information is needed for both collection systems and wastewater processing. The listing under collection systems also includes applied hydraulics. Conventional municipal wastewater treatment relies on biological processing, and therefore an understanding of living systems is indispensable. The chapters on operation of systems are not intended to give all-inclusive subject coverage. Since this book deals primarily with art and practice, a great deal of information relative to operation is presented throughout the book's descriptive material.

Advanced wastewater treatment incorporates both biological unit operations and chemical processes that are similar to those applied in water treatment. Therefore, to understand the concepts described in Chapter 13, the reader must have adequate knowledge about the treatment of both wastewater and water. Reuse of reclaimed water employs the most recent treatment technology and requires a comprehensive understanding of both treatment processes and water quality.

TABLE 1–1

Listing of Prerequisite Readings for Various Subject Areas in Water and Wastewater Technology

WATER QUALITY, CHAPTER 5	
2–1	Elements, Radicals, and Compounds
2–9	Organic Compounds
2–11	Laboratory Chemical Analyses
3–1	Bacteria and Fungi
3–2	Protozoa and Multicellular Animals
3–3	Viruses
3–4	Algae
3–5	Waterborne Diseases
3–6	Testing for Enteric Viruses
3–7	Testing for *Giardia* and *Cryptosporidium*
3–8	Coliform Bacteria as Indicator Organisms
4–12	Flow in Streams and Rivers
4–13	Hydrology of Lakes and Reservoirs
4–14	Groundwater Hydrology
5–1	Safe Drinking Water Act
5–4	Clean Water Act
5–5	National Pollutant Discharge Elimination System (NPDES)

WATER DISTRIBUTION SYSTEMS, CHAPTER 6	
4–1	Water Pressure
4–2	Pressure-Velocity-Head Relationships
4–3	Flow in Pipes Under Pressure
4–4	Centrifugal Pump Characteristics
4–5	System Characteristics
4–6	Equivalent Pipes
4–7	Computer Analysis of Pipe Networks
4–9	Flow Measurement in Pipes

WATER PROCESSING, CHAPTER 7	
2–1	Elements, Radicals, and Compounds
2–2	Chemical Water Analysis
2–3	Hydrogen Ion Concentration and pH
2–4	Chemical Equilibria
2–5	Chemical Kinetics
2–6	Gas Solubility
2–7	Alkalinity
2–8	Colloids and Coagulation
2–11	Laboratory Chemical Analyses
4–1	Water Pressure
4–2	Pressure-Velocity-Head Relationships
4–9	Flow Measurement in Pipes

WASTEWATER COLLECTION SYSTEMS, CHAPTER 10	
4–1	Water Pressure
4–2	Pressure-Velocity-Head Relationships
4–4	Centrifugal Pump Characteristics
4–5	System Characteristics
4–8	Gravity Flow in Circular Pipes
4–10	Flow Measurement in Open Channels
4–11	Amount of Storm Runoff
Chapter 9	Wastewater Flows and Characteristics

WASTEWATER PROCESSING, CHAPTER 11	
2–9	Organic Compounds
2–10	Organic Matter in Wastewater
3–1	Bacteria and Fungi
3–2	Protozoa and Multicellular Animals
3–4	Algae
3–10	Biochemical Oxygen Demand
3–11	Biological Treatment Systems
3–12	Biological Kinetics
4–1	Water Pressure
4–2	Pressure-Velocity-Head Relationships
4–10	Flow Measurement in Open Channels
Chapter 9	Wastewater Flows and Characteristics

OPERATION OF WATERWORKS, CHAPTER 8	
Chapter 5	Water Quality
Chapter 6	Water Distribution Systems
Chapter 7	Water Processing

OPERATION OF WASTEWATER SYSTEMS, CHAPTER 12	
Chapter 5	Water Quality
Chapter 9	Wastewater Flows and Characteristics
Chapter 10	Wastewater Collection Systems
Chapter 11	Wastewater Processing

ADVANCED WASTEWATER TREATMENT, CHAPTER 13
Wastewater Processing Plus Prerequisites
Water Processing Plus Prerequisites

WATER REUSE, CHAPTER 14
Water Quality Plus Prerequisites
Wastewater Processing Plus Prerequisites
Water Processing Plus Prerequisites
Advanced Wastewater Treatment

Chemistry

This chapter provides basic information about chemistry as it applies to water and wastewater technology. Selected data are compiled and presented as an introduction to the chapters dealing with water quality, pollution, and chemical-treatment processes. For example, the characteristics of common elements, radicals, and inorganic compounds are tabulated. The usual method for presenting chemical water analysis is described, since it is not normally presented in general chemistry textbooks. Sections on chemical reactions, alkalinity, and coagulation emphasize important aspects of applied water chemistry. Since organic chemistry traditionally has not been included in introductory courses, persons practicing in water supply and pollution control are not generally exposed to this area of chemistry. For this reason an introduction on the nomenclature of organic compounds and a brief description of the organic matter in wastewater are provided. Finally, the importance, technique, and equipment used in selected laboratory analyses are discussed. Water-quality parameters and their characteristics can be understood better when testing procedures are known.

2–1   ELEMENTS, RADICALS, AND COMPOUNDS

The fundamental chemical identities that form all substances are referred to as elements. Each element differs from any other in weight, size, and chemical properties. Names of elements common to water and wastewater technology along with their symbols, atomic weights, common valence, and equivalent weights are given in Table 2–1. Symbols for elements are used in writing chemical formulas and equations.

Atomic weight is the weight of an element relative to that of hydrogen, which has an atomic weight of unity. This weight, expressed in grams, is called one gram atomic weight of the element. For example, one gram atomic weight of aluminum (Al) is 27.0 grams. Equivalent or combining weight of an element is equal to atomic weight divided by the valence.

TABLE 2–1

Basic Information on Common Elements

NAME	SYMBOL	ATOMIC WEIGHT	COMMON VALENCE	EQUIVALENT WEIGHT[a]
Aluminum	Al	27.0	3+	9.0
Arsenic	As	74.9	3+	25.0
Barium	Ba	137.3	2+	68.7
Boron	B	10.8	3+	3.6
Bromine	Br	79.9	1−	79.9
Cadmium	Cd	112.4	2+	56.2
Calcium	Ca	40.1	2+	20.0
Carbon	C	12.0	4−	
Chlorine	Cl	35.5	1−	35.5
Chromium	Cr	52.0	3+	17.3
			6+	
Copper	Cu	63.5	2+	31.8
Fluorine	F	19.0	1−	19.0
Hydrogen	H	1.0	1+	1.0
Iodine	I	126.9	1−	126.9
Iron	Fe	55.8	2+	27.9
			3+	
Lead	Pb	207.2	2+	103.6
Magnesium	Mg	24.3	2+	12.2
Manganese	Mn	54.9	2+	27.5
			4+	
			7+	
Mercury	Hg	200.6	2+	100.3
Nickel	Ni	58.7	2+	29.4
Nitrogen	N	14.0	3−	
			5+	
Oxygen	O	16.0	2−	8.0
Phosphorus	P	31.0	5+	6.0
Potassium	K	39.1	1+	39.1
Selenium	Se	79.0	6+	13.1
Silicon	Si	28.1	4+	6.5
Silver	Ag	107.9	1+	107.9
Sodium	Na	23.0	1+	23.0
Sulfur	S	32.1	2−	16.0
Zinc	Zn	65.4	2+	32.7

[a]Equivalent weight (combining weight) equals atomic weight divided by valence.

Some elements appear in nature as gases, for example, hydrogen, oxygen, and nitrogen; mercury appears as a liquid; others appear as pure solids, for instance, carbon, sulfur, phosphorus, calcium, copper, and zinc; and many occur in chemical combination with each other in compounds. Atoms of one element unite with those of another in a definite ratio defined by their valence. Valence is the combining power of an element based on that of the hydrogen atom, which has an assigned value of 1. Thus, an element with a valence of 2+ can replace two hydrogen atoms in a compound, or in the case of 2− can react with two hydrogen atoms. Sodium has a valence of 1+, and chlorine has a valence of 1−; therefore, one sodium atom combines with one chlorine atom to form sodium chloride (NaCl), common salt. Nitrogen at a valence of 3− can combine with three hydrogen atoms to form ammonia gas (NH_3). The weight of a compound, equal to the sum of the weights of the combined elements, is referred to as molecular weight, or simply mole. The molecular weight of NaCl is 58.4 grams; one mole of ammonia gas is 17.0 grams.

Certain groupings of atoms act together as a unit in a large number of different molecules. These, referred to as radicals, are given special names, such as the hydroxyl group (OH^-). The most common radicals in ionized form are listed in Table 2–2. Radicals themselves are not compounds, but join with other elements to form compounds. Data on inorganic compounds common to water and wastewater chemistry are given in Table 2–3. The proper name, formula, and molecular weight are included for all of the chemicals listed. Popular names, for example, alum for aluminum sulfate, are included in brackets. For chemicals used in water treatment, one common use is given; many have other applications not included. Equivalent weights for compounds and hypothetical combinations—for example—$Ca(HCO_3)_2$, involved in treatment are provided.

■ EXAMPLE 2–1

Calculate the molecular and equivalent weights of ferric sulfate.

Solution

The formula from Table 2–3 is $Fe_2(SO_4)_3$.

Using atomic weight data from Table 2–1,

TABLE 2–2

Common Radicals Encountered in Water

Name	Formula	Molecular Weight	Electrical Charge	Equivalent Weight
Ammonium	NH_4^+	18.0	$1+$	18.0
Hydroxyl	OH^-	17.0	$1-$	17.0
Bicarbonate	HCO_3^-	61.0	$1-$	61.0
Carbonate	$CO_3^=$	60.0	$2-$	30.0
Orthophosphate	$PO_4^{\equiv}$	95.0	$3-$	31.7
Orthophosphate, monohydrogen	$HPO_4^=$	96.0	$2-$	48.0
Orthophosphate, dihydrogen	$H_2PO_4^-$	97.0	$1-$	97.0
Bisulfate	HSO_4^-	97.0	$1-$	97.0
Sulfate	$SO_4^=$	96.0	$2-$	48.0
Bisulfite	HSO_3^-	81.0	$1-$	81.0
Sulfite	$SO_3^=$	80.0	$2-$	40.0
Nitrite	NO_2^-	46.0	$1-$	46.0
Nitrate	NO_3^-	62.0	$1-$	62.0
Hypochlorite	OCl^-	51.5	$1-$	51.5

Fe	$2 \times 55.8 =$	111.6
S	$3 \times 32.1 =$	96.3
O	$12 \times 16.0 =$	$\underline{192.0}$

Molecular weight = 399.9 or 400 grams

The ferric (oxide iron) atom has a valence of $3+$, thus a compound with two ferric atoms has a total electrical charge of $6+$. (Three sulfate radicals have a total of $6-$ charges.)

$$\text{Equivalent weight} = \frac{\text{molecular weight}}{\text{electrical charge}}$$

$$= \frac{400}{6}$$

$$= 66.7 \text{ grams per equivalent weight}$$

■ ■ ■

2–2   CHEMICAL WATER ANALYSIS

When placed in water, inorganic compounds dissociate into electrically charged atoms and radicals referred to as ions. This breakdown of substances into their constituent ions is called ionization. An ion is represented by the chemical symbol of the element, or radical, followed by superscript $^+$ or $^-$ signs to indicate the number of unit charges on the ion. Consider the following: sodium, Na^+; chloride, Cl^-; aluminum, Al^{+++}; ammonium, NH_4^+; and sulfate, $SO_4^=$.

Laboratory tests on water, such as those outlined in Section 2–11, determine concentrations of particular ions in solution. Test results are normally expressed as weight of the element or radical in milligrams per liter of water, abbreviated as mg/l. Some books use the term parts per million (ppm), which is for practical purposes identical in meaning to mg/l, since 1 liter of water weighs 1,000,000 milligrams. In other words, 1 mg per liter (mg/l) equals 1 mg in 1,000,000 mg, which is the same as 1 part by weight in 1 million parts by weight (1 ppm). The concentration of a substance in solution can also be expressed in milliequivalents per liter (meq/l), representing the combining weight of the ion, radical, or compound.

TABLE 2–3

Basic Information on Common Inorganic Chemicals

Name	Formula	Common Usage	Molecular Weight	Equivalent Weight
Activated carbon	C	Taste and odor control	12.0	n.a.[a]
Aluminum sulfate (filter alum)	$Al_2(SO_4)_3 \cdot 14.3\ H_2O$	Coagulation	600	100
Aluminum hydroxide	$Al(OH)_3$	(Hypothetical combination)	78.0	26.0
Ammonia	NH_3	Chloramine disinfection	17.0	n.a.
Ammonium fluosilicate	$(NH_4)_2SiF_6$	Fluoridation	178	n.a.
Ammonium sulfate	$(NH_4)_2SO_4$	Coagulation	132	66.1
Calcium bicarbonate	$Ca(HCO_3)_2$	(Hypothetical combination)	162	81.0
Calcium carbonate	$CaCO_3$	Corrosion control	100	50.0
Calcium fluoride	CaF_2	Fluoridation	78.1	n.a.
Calcium hydroxide	$Ca(OH)_2$	Softening	74.1	37.0
Calcium hypochlorite	$Ca(OCl)_2 \cdot 2H_2O$	Disinfection	179	n.a.
Calcium oxide (lime)	CaO	Softening	56.1	28.0
Carbon dioxide	CO_2	Recarbonation	44.0	22.0
Chlorine	Cl_2	Disinfection	71.0	n.a.
Chlorine dioxide	ClO_2	Taste and odor control	67.0	n.a.
Copper sulfate	$CuSO_4$	Algae control	160	79.8
Ferric chloride	$FeCl_3$	Coagulation	162	54.1
Ferric hydroxide	$Fe(OH)_3$	(Hypothetical combination)	107	35.6
Ferric sulfate	$Fe_2(SO_4)_3$	Coagulation	400	66.7
Ferrous sulfate (copperas)	$FeSO_4 \cdot 7H_2O$	Coagulation	278	139
Fluorosilicic acid	H_2SiF_6	Fluoridation	144	n.a.
Hydrochloric acid	HCl	n.a.	36.5	36.5
Magnesium hydroxide	$Mg(OH)_2$	Defluoridation	58.3	29.2
Oxygen	O_2	Aeration	32.0	16.0
Ozone	O_3	Disinfection	48.0	n.a.
Potassium permanganate	$KMnO_4$	Oxidation	158	n.a.
Sodium aluminate	$NaAlO_2$	Coagulation	82.0	n.a.
Sodium bicarbonate (baking soda)	$NaHCO_3$	pH adjustment	84.0	84.0
Sodium carbonate (soda ash)	Na_2CO_3	Softening	106	53.0
Sodium chloride (common salt)	$NaCl$	Ion-exchanger regeneration	58.4	58.4
Sodium fluoride	NaF	Fluoridation	42.0	n.a.
Sodium hexametaphosphate	$(NaPO_3)_n$	Corrosion control	n.a.	n.a.
Sodium hydroxide	$NaOH$	pH adjustment	40.0	40.0
Sodium hypochlorite	$NaOCl$	Disinfection	74.4	n.a.
Sodium silicate	Na_4SiO_4	Coagulation aid	184	n.a.
Sodium fluorosilicate	Na_2SiF_6	Fluoridation	188	n.a.
Sodium thiosulfate	$Na_2S_2O_3$	Dechlorination	158	n.a.
Sulphur dioxide	SO_2	Dechlorination	64.1	n.a.
Sulfuric acid	H_2SO_4	pH adjustment	98.1	49.0
Water	H_2O	n.a.	18.01	n.a.

[a]n.a. = not applicable.

Milliequivalents can be calculated from milligrams per liter by

$$meq/l = mg/l \times \frac{valence}{atomic\ weight}$$

$$= \frac{mg/l}{equivalent\ weight} \qquad (2\text{--}1)$$

In the case of a radical or compound the equation reads

$$meq/l = mg/l \times \frac{electrical\ charge}{molecular\ weight}$$

$$= \frac{mg/l}{equivalent\ weight} \qquad (2\text{--}2)$$

Equivalent weights for selected elements, radicals, and inorganic compounds are given in Tables 2–1, 2–2, and 2–3, respectively.

A typical chemical water analysis is in Table 2–4. These data can be compared against the chemical characteristics specified by the drinking water standards to determine the treatment required before domestic or industrial use.

Reporting results in milligrams per liter in tabular form is not convenient for visualizing the chemical composition of a water. Therefore, results are often expressed in milliequivalents per liter, which permits graphical presentation and a quick check on the accuracy of the analyses for major ions. The sum of the milliequivalents per liter of the cations (positive radicals) must equal the sum of the anions (negative radicals). In a perfect evaluation they would be exactly the same, since a water in equilibrium is electrically balanced. Graphical presentation of a water analysis using milliequivalents is performed by plotting the milliequivalents per liter values to scale, for example, by letting 1 meq/l equal 1 inch. The bar graph shown in Figure 2–1 is based on the water data in Table 2–4. The top row consists of the major cations arranged in the order calcium, magnesium, sodium, and potassium. The under row is arranged in the sequence of bicarbonate, sulfate, and chloride.

Hypothetical combinations of the positive and negative ions can be developed from a bar graph as in Figure 2–1. These combinations are particularly helpful in considering lime water softening. In Figure 2–1, the carbonate hardness [$Ca(HCO_3)_2$ + $Mg(HCO_3)_2$] is 2.15 meq/l and the noncarbonate is 0.45 meq/l ($MgSO_4$).

TABLE 2–4

Chemical Analysis of a Surface Water

Parameter	(ml/l)	(meq/l)	Parameter	(ml/l)	(meq/l)
Alkalinity (as $CaCO_3$)	108		Magnesium	9.9	0.81
Arsenic	0		Nitrate	2.2	
Barium	0		pH (in pH units)	7.6	
Bicarbonate	131	2.15	Phosphorus (total inorganic)	0.5	
Cadmium	0	1.79	Potassium	3.9	0.10
Calcium	35.8		Selenium	0	
Chloride	7.1	0.20	Silver	0	
Chromium	0		Sodium	4.6	0.20
Copper	0.10		Sulfate	26.4	0.55
Fluoride	0.2		Total dissolved solids	220	
Iron plus manganese	0.13		Turbidity (in NTU[a])	5	
Iron	0.10		Trihalomethanes	0	
Lead	0		Units of color (in color units)	5	

[a]Nephelometric turbidity units.

Figure 2–1

Milliequivalents per liter bar graph for water analysis given in Table 2–4.

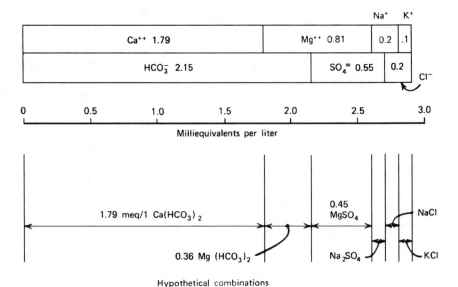

Hypothetical combinations

■ EXAMPLE 2–2

The results of a water analysis are calcium 29.0 mg/l, magnesium 16.4 mg/l, sodium 23.0 mg/l, potassium 17.5 mg/l, bicarbonate 171 mg/l (as HCO_3), sulfate 36.0 mg/l, and chloride 24.0 mg/l. Convert milligrams per liter concentrations to milliequivalents per liter, list hypothetical combinations, and calculate hardness as mg/l of $CaCO_3$ for this water.

Solution

Using Eq. 2–1,

Component	MG/L	Equivalent Weight	MEQ/L
Ca^{++}	29.0	20.0	1.45
Mg^{++}	16.4	12.2	1.34
Na^+	23.0	23.0	1.00
K^+	17.5	39.1	0.45
		Total Cations	4.24
HCO_3^-	171	61.0	2.81
$SO_4^=$	36.0	48.0	0.75
Cl^-	24.0	35.5	0.68
		Total Anions	4.24

Hypothetical combinations: $Ca(HCO_3)_2$, 1.45 meq/l; $Mg(HCO_3)_2$, 1.34; $NaHCO_3$, 0.02; Na_2SO_4, 0.75; NaCl, 0.23; and KCl, 0.45 meq/l.

Hardness $(Ca^{++} + Mg^{++}) = 2.79$ meq/l

Equivalent weight of $CaCO_3 = 50.0$ (Table 2–3)

By Eq. 2–2,

$$Hardness = 2.79 \times 50.0 = 140 \text{ mg/l}$$
$$= 1.40 \text{ mg/l } CaCO_3$$

■ ■ ■

2–3   HYDROGEN ION CONCENTRATION AND pH

Water (H_2O) dissociates to only a slight degree, yielding a concentration of hydrogen ions equal to 10^{-7} mole per liter. Because water yields one hydroxyl (basic) ion for each hydrogen (acid) ion, pure water is considered neutral.

$$H_2O \Longleftrightarrow H^+ + OH^- \quad (2\text{–}3)$$

The acidic nature of a water is related to the concentration of hydrogen ions in water solution by use of the symbol pH, where

$$pH = \log \frac{1}{[H^+]} \quad (2\text{–}4)$$

Since the logarithm of 1 over 10^{-7} is 7, the pH at neutrality is 7.

When an acid is added to water, the hydrogen ion concentration increases, resulting in a lower pH number. Conversely, when an alkaline substance is added, the OH^- ions unite with the free H^+, lowering the hydrogen ion concentration and causing a higher pH. The pH scale, ranging from 0 to 14, is acid from 0 to 7 and basic from 7 to 14.

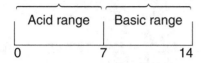

2–4 CHEMICAL EQUILIBRIA

Most chemical reactions are reversible to some degree, with the final state of equilibrium determining the concentration of reactants and products. For the general reaction expressed by Eq. 2–5, an increase in either A or B shifts the equilibrium to the right, while an increase in the concentration of C or D drives the reaction to the left. A chemical reaction in true equilibrium can be expressed by the mass-action relationship in Eq. 2–6.

$$aA + bB \rightleftharpoons cC + dD \qquad (2–5)$$

$$\frac{[C]^c[D]^d}{[A]^a[B]^b} = K \qquad (2–6)$$

where A and B = reactants
C and D = products
[ ] = molar concentrations
K = equilibrium constant

Homogeneous Equilibria

In homogeneous chemical equilibria, all reactants and products occur in the same physical state, for instance, all are in solution. Inorganic strong acids and bases approach 100 percent ionization in dilute solutions. In contrast, weak inorganic and organic acids and bases are poorly ionized. The degree of ionization of a weak compound is expressed by the mass-action relationship. For example, Eqs. 2–7 through 2–10 characterize the equilibria of carbonic acid.

$$H_2CO_3 \rightleftharpoons H^+ + HCO_3^- \qquad (2–7)$$

$$\frac{[H^+][HCO_3^-]}{[H_2CO_3]} = K_1 = 4.45 \times 10^{-7} \text{ at } 25°C \quad (2–8)$$

$$HCO_3^- \rightleftharpoons H^+ + CO_3^= \qquad (2–9)$$

$$\frac{[H^+][CO_3^=]}{[H_2CO_3]} = K_2 = 4.69 \times 10^{-11} \text{ at } 25°C \quad (2–10)$$

Section 2–7 discusses bicarbonate-carbonate equilibrium relative to the alkalinity of a water, and Section 2–9 discusses ionization of organic acids.

Heterogeneous Equilibria

These refer to equilibria existing between substances in two or more physical states. For example, at greater than pH 10, solid calcium carbonate in water reaches stability with the calcium and carbonate ions in solution.

$$CaCO_3 \rightleftharpoons Ca^{++} + CO_3^= \qquad (2–11)$$

This kind of equilibrium, between crystals of a substance in a solid state and its ions in solution, can be expressed mathematically as homogeneous equilibrium as follows:

$$\frac{[Ca^{++}][CO_3^=]}{[CaCO_3]} = K \qquad (2–12)$$

The concentration of the solid substance can be treated as a constant, K_S, in mass-action equilibrium. Hence, $[CaCO_3]$ can be assumed equal to KK_S such that Eq. 2–12 becomes

$$[Ca^{++}][CO_3^=] = KK_S = K_{SP}$$
$$= 5.0 \times 10^{-9} \text{ at } 25°C \quad (2–13)$$

where K_{SP} = solubility-product constant

The solution is unsaturated if the product of the ionic molar concentrations is less than the solubility-product constant. Conversely, the solution is supersaturated if the product of the molar concentrations is greater than K_{SP}. In this case, crystals form and precipitation progresses until the ionic concentrations are reduced to equal those of a saturated solution. Based on the K_{SP} given in Eq. 2–13, the theoretical solubility of $CaCO_3$ in pure water at 25°C is 0.000,071 moles per liter, or 7.1 mg/l.

Shifting of Chemical Equilibria

Chemical reactions in water and wastewater processing are used to shift chemical equilibria in order to remove substances from solution. The most common methods for completing reactions are formation of insoluble precipitates, weakly ionized compounds, and gases that can be stripped from solution.

Precipitation Reactions. Forming insoluble precipitates of undesirable substances by the addition of treatment chemicals is the most common method of shifting equilibrium. Examples are the use of coagulants such as aluminum sulfate to remove turbidity and the use of lime to precipitate the multivalent ions of hardness. After formation by chemical reactions, the precipitates are usually removed from suspension by gravity sedimentation followed by filtration of the water through granular media.

An illustration of a precipitation reaction is lime softening. Water containing a high concentration of calcium (that is, hard water) tends to deposit scale on the inside of pipes and on appliances during water use. In the process of precipitation softening, calcium ions are removed from solution by raising the pH of the water by addition of lime (CaO). This results in a shift of the equilibrium in Eq. 2–14 to the right. (The arrow pointing down on the right side of the equation indicates that the $CaCO_3$ is a solid precipitate.)

$$CaO + Ca(HCO_3)_2 = 2CaCO_3\downarrow + H_2O \quad \textbf{(2–14)}$$

A chemical equation, such as Eq. 2–14, gives the formulas for reactants and products and the number of moles of each. It is said to be balanced when the number of atoms of each element on the left side equals the number on the right side. By using a balanced equation, one can use molecular relationships to calculate the quantities of reactants and products. The process of using a balanced chemical equation for making calculations is called *stoichiometry*. In some reactions, equivalent weights can be used to determine amounts of reactants and products. (Refer to Examples 2–3 and 2–4.)

Oxidation and Reduction Reactions. Many chemical changes involve the addition of oxygen and/or the change of valence of one of the reacting substances. Oxidation is the addition of oxygen or removal of electrons, and reduction is the removal of oxygen or addition of electrons. A classic oxidation reaction is the rusting of iron by oxygen.

$$4Fe + 3O_2 = 2Fe_2O_3 \quad \textbf{(2–15)}$$

The iron is oxidized from Fe to Fe^{+++}, while the oxygen is reduced from O to $O^=$.

An oxidation–reduction reaction in water treatment is the removal of soluble ferrous iron (Fe^{2+}) from solution by oxidation using potassium permanganate. In this reaction, Eq. 2–16, the iron gains one positive charge while the manganese in the permanganate ion is reduced from a valence of 7^+ to a valence of 4^+ in manganese dioxide. (Note that an arrow is used instead of an equal sign in Eq. 2–16 since the equation is not balanced.)

$$3Fe^{2+} + Mn^{(7+)}O_4^- \longrightarrow 3Fe^{(3+)}O_x\downarrow + Mn^{(4+)}O_2\downarrow$$
$$\textbf{(2–16)}$$

The precipitates of iron oxides and manganese dioxide are removed from suspension by sedimentation and filtration of the water.

Acid–Base Reactions (Neutralization). Acid added to water dissociates into hydrogen ions and anion radicals, resulting in an acid solution. For example, sulfuric acid, H_2SO_4, forms H^+ and $SO_4^=$ ions in solution. A basic solution is formed by adding an alkali such as NaOH to water. If both an acid and a base are put in the same water, the H^+ ions combine with the OH^- ions to form water; if equivalent amounts are added, they neutralize each other, forming a salt solution, as shown in the following equation:

$$2H^+ + SO_4^= + 2Na^+ + 2OH^-$$
$$= 2H_2O + 2Na^+ + SO_4^= \quad \textbf{(2–17)}$$

Gas-Producing Reactions. Chemical processes involving formation of a gaseous product approach completion as the gas escapes from solution. An example is ammonia stripping to remove nitrogen from wastewater. Raising the pH above 10.8 by lime addition converts the ammonia nitrogen in solution from NH_4^+ ions to NH_3 gas.

$$NH_4^+ + OH^- \longrightarrow NH_3 + H_2O \qquad (2\text{--}18)$$

By the application of wastewater to a counter-current tower, where air enters at the bottom and exhausts from the top while the wastewater sprinkles down over the tower packing, ammonia gas is stripped out of solution and discharged with the air.

■ EXAMPLE 2–3

The chemical reaction for removal of calcium hardness by lime precipitation softening is given by Eq. 2–14. What dosage of lime with a purity of 78 percent CaO is required to combine with 70 mg/l of calcium?

Solution

1 mole of $Ca(HCO_3)_2$ (162 g) contains 40.1 g of calcium. Therefore, 70 mg/l of calcium is equivalent to

$$70 \text{ mg/l Ca}^{++} \frac{162 \text{ g/mole Ca(HCO}_3)_2}{40.1 \text{ mole Ca}^{++}}$$

$$= 283 \text{ mg/l Ca(HCO}_3)_2$$

56.1 g of CaO combines with 162 g of $Ca(HCO_3)_2$. Therefore, 283 mg/l $Ca(HCO_3)_2$ reacts with

$$\frac{56.1}{162} \times 283 = 98.0 \text{ mg/l CaO}$$

For a purity of 78 percent, the dosage of commercial lime required is

$$\frac{98.0 \text{ CaO}}{0.78} = 126 \text{ mg/l} = 1050 \text{ lb/mil gal}$$

■ ■ ■

■ EXAMPLE 2–4

How many pounds of pure sodium hydroxide are required to neutralize an industrial waste with an acidity equivalent to 50 lb of sulfuric acid per million gallons of wastewater?

Solution

Using Eq. 2–17,

2 moles NaOH neutralize 1 mole H_2SO_4
2×40.0 lb NaOH reacts with 1×98.1 lb H_2SO_4

$$\text{lb NaOH required} = \frac{80.0 \text{ lb NaOH}}{98.1 \text{ lb H}_2SO_4}$$

$$\times 50 \text{ lb H}_2SO_4$$

$$= 40.8 \text{ lb NaOH/mil gal}$$

■ ■ ■

Alternate Solution

In neutralization reactions one equivalent of base reacts with one of acid. Therefore, by using equivalent weights (Table 2–3) rather than molecular values, the calculation is

$$\text{lb NaOH required} = \frac{40.0 \text{ NaOH}}{49.0 \text{ H}_2SO_4} \times 50 \text{ lb H}_2SO_4$$

$$= 40.8 \text{ lb NaOH/mil gal}$$

■ ■ ■

■ EXAMPLE 2–5

Theoretically, how many milligrams per liter of reduced iron can be oxidized by 1 milligram per liter of potassium permanganate?

Solution

From Eq. 2–16, $3Fe^{++}$ ($3 \times 55.8 = 167$) is oxidized for each MnO_4^- reduced ($KMnO_4$ weight is 158).

$$\frac{158 \text{ KMnO}_4}{167 \text{ Fe}} = \frac{1 \text{ mg/l}}{X \text{ mg/l}} \qquad X = 1.06 \text{ mg/l}$$

■ ■ ■

2–5   CHEMICAL KINETICS

Process kinetics define the rate of a reaction. The kinetics discussed in this section are for irreversible reactions occurring in one phase with uniform distribution of the reactants throughout the liquid. In irreversible reactions, the combination of reactants leads to nearly complete conversion to products, which is true for most processes in water and wastewater treatment. Hence, this assumption is valid for the purpose of interpretation of process kinetics data. Of the reaction types discussed here, first-order kinetics most commonly define the reactions that occur in chemical and biological processing of water and wastewater.

Zero-order reactions proceed at a rate independent of the concentration of any reactant or product; therefore, the disappearance of reactants and appearance of products are linear. Consider the following conversion of a single reactant to a single product:

$$A(\text{reactant}) \longrightarrow P(\text{product}) \qquad (2\text{--}19)$$

Using C to represent the concentration of A at any time t, the disappearance of A with respect to time is

$$\frac{dC}{dt} = kC^0 = -k \qquad (2\text{--}20)$$

where $\dfrac{dC}{dt}$ = rate of change in the concentration of A with time

k = reaction-rate constant, per day

By integrating Eq. 2–20 and rearranging the terms,

$$C = C_0 - kt \qquad (2\text{--}21)$$

where C = concentration of A at any time t, milligrams per liter

C_0 = initial concentration of A, milligrams per liter

A plot of the zero-order reaction, represented by Eq. 2–21, is shown by the straight line in Figure 2–2. The negative slope, $-k$, means that the concentration of A decreases with time.

First-order reactions proceed at a rate directly proportional to the concentration of one reactant. If one considers a single reactant converting to a single product (A → P), the change in concentration C of A with respect to time is

$$\frac{dC}{dt} = -kC^1 = -kC \qquad (2\text{--}22)$$

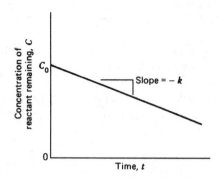

Figure 2–2

A plot of concentration of reactant remaining for a zero-order kinetics reaction, Eq. 2–21.

Integrating Eq. 2–22 and letting C equal C_0 when t equals 0 gives

$$\log_e \frac{C}{C_0} = -kt \quad \text{or} \quad C = C_0 e^{-kt} \qquad (2\text{--}23)$$

A plot of the first-order reaction, represented by Eq. 2–23, on arithmetic paper is a curve as shown in Figure 2–3a. The slope of a tangent at any point along the curve is equal to $-kC$. Thus, with passage of time from the start of the reaction, the rate decreases with decreasing concentration of A remaining. The first-order reaction can be linearized by graphing on semilogarithmic paper as shown in Figure 2–3b.

Second-order reactions proceed at a rate proportional to the second power of a single reactant being converted to a single product (2A → P). The reaction rate is

$$\frac{dC}{dt} = -kC^2 \qquad (2\text{--}24)$$

The integrated form for Eq. 2–24 is

$$\frac{1}{C} - \frac{1}{C_0} = kt \qquad (2\text{--}25)$$

A plot of the second-order reaction, represented by Eq. 2–25, on arithmetic paper is shown in Figure 2–4. The reciprocal of the concentration of reactant remaining versus time graphs as a straight line with a slope equal to k.

Figure 2–3

Arithmetic and semilogarithmic plots of concentration of reactant remaining for a first-order kinetics reaction, Eq. 2–23.

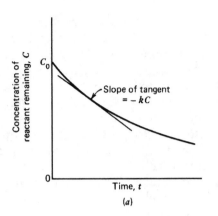

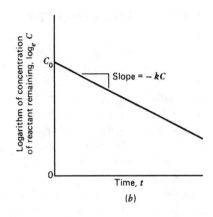

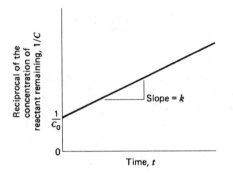

Figure 2–4

A plot of reciprocal of the concentration of reactant remaining for a second-order kinetics reaction, Eq. 2–25.

Effect of Temperature on Reaction Rate

The rate of chemical reactions increases with increasing temperature, provided the higher temperature does not affect a reactant or catalyst. The basic relationship between temperature and the reaction-rate constant for chemical reactions was formulated by Arrhenius in 1889. When his original equation is modified and simplified,[1] the common expression for temperature dependence of reaction rates becomes

$$k_2 = k_1 \Theta^{T_2 - T_1} \qquad (2\text{–}26)$$

where k = reaction-rate constant at respective temperature T, per day

Θ = temperature coefficient, dimensionless

T = temperature, °C

This equation allows adjustment of the value of a rate constant for a temperature change based on an assumed or experimentally determined temperature coefficient.

Temperature coefficients are relatively constant within the narrow range of temperatures normally associated with operation of water and wastewater processes. Prudent use of a Θ coefficient limits its application to a few degrees (approximately $\pm 5°C$) from the experimental temperature used to establish its value. A commonly used temperature coefficient for both biological and chemical reactions is 1.072, which doubles or halves the reaction rate for a 10°C temperature change. A value of 1.047 doubles or halves the k-rate for a 15°C temperature change.

■ EXAMPLE 2–6

A chemical reaction that follows first-order kinetics has a measured reaction rate of 50 per day at 20°C. Using a temperature coefficient of 1.072, calculate the k-rate at 18°C for the reactant to decrease in concentration from 100 mg/l to 10 mg/l.

Solution

From Eq. 2–26, the reaction rate at 18°C is

$$k_2 = 50 \, (1.072)^{18-20} = 43.5 \text{ per day}$$

Using Eq. 2–23,

$$\log_e \frac{10}{100} = -43.5t$$

$$t = 0.053 \text{ days} = 1.27 \text{ hr}$$

■ ■ ■

2–6 GAS SOLUBILITY

Most gases either are soluble in water to some degree or react chemically with water. An exception is methane gas (CH_4), commonly referred to as illuminating gas, which does not interact with water to a measurable extent. (Methane is produced in anaerobic decomposition of waste sludge and can be collected and burned for its heat value.) The two major atmospheric gases, nitrogen and oxygen, while not reacting chemically with water, dissolve to a limited degree. The solubility of each gas is directly proportional to the pressure it exerts on the water. At a given pressure, solubility of oxygen varies greatly with water temperature and to a lesser degree with salinity (chloride concentration). Dissolved oxygen saturation values, based on normal atmosphere consisting of 21 percent oxygen, for various temperatures and chloride concentrations are listed in Table 2–5. The table footnote explains how to correct oxygen solubility values for barometric pressure. The change of barometric pressure with altitude depends on the rate of decrease in density of the air. Near the earth's surface, it is sufficiently accurate for most engineering computations to say that pressure decreases at a rate of 25 mm (1.0 in.) of mercury per 1000 ft of increased elevation.

Gases may be formed in solution by the decomposition of organic substances in water. Ammonia biochemically released from nitrogenous matter appears in solution as the ammonium radical if the solution is acidic; however, if the water is basic, it remains as ammonia gas. In the laboratory test for ammonia nitrogen, the pH of the sample is increased by sodium hydroxide to extract the ammonia by distillation. Another gas released from septic wastewater is hydrogen sulfide (H_2S), identifiable by its rotten egg odor. The SH^- group is biochemically produced in solution, converted to H_2S under reduced conditions, and released from solution as a gas. In sewers this can lead to crown corrosion of the pipe by having the H_2S oxidized to sulfuric acid (H_2SO_4) in the condensation moisture hanging on the interior of the pipe.

Chlorine, the most common oxidizing agent used for disinfection of municipal water supplies in the United States, reacts with water as shown in Eq. 2–27. Liquid chlorine is shipped to treatment plants in pressurized steel cylinders. On pressure release, the liquid converts to chlorine gas and is applied to the water. In dilute solution at pH levels above 4, chlorine combines with water molecules to form hydrochloric acid plus hypochlorous acid, which, in turn, yields the hypochlorite ion. The degree of ionization is dependent on pH, with HOCl occurring below 6 and almost complete ionization taking place above pH 9. The pH of a water during disinfection is important, since HOCl is more effective than OCl^- in killing harmful organisms.

$$Cl_2 + H_2O \underset{pH<4}{\overset{pH>4}{\rightleftharpoons}} HCl + HOCl$$

$$\underset{pH<6}{\overset{pH>9}{\rightleftharpoons}} H^+ + OCl^- \quad (2\text{–}27)$$

Carbon dioxide, although only about 0.03 percent of atmospheric air, plays a major role in water chemistry, since it reacts readily with water, forming bicarbonate and carbonate radicals. CO_2 is absorbed from the air or can be produced by bacterial decomposition of organic matter in the water. Once in solution, it reacts to form carbonic acid.

$$CO_2 + H_2O \rightleftharpoons H_2CO_3$$

$$\underset{pH\,4.5}{\rightleftharpoons} H^+ + HCO_3^-$$

$$\underset{pH\,8.3}{\rightleftharpoons} H^+ + CO_3^= \quad (2\text{–}28)$$

When the pH of the water is greater than 4.5, carbonic acid ionizes to form bicarbonate, which, in turn, is transformed to the carbonate radical if the pH is above approximately 8.3. Carbon dioxide is very aggressive and leads to corrosion of water pipes; therefore, water supplies with low pH are frequently neutralized with a base to reduce pipe corrosion. On the other hand, alkaline waters with $CO_3^=$ ions are hard and form scale by precipitation of $CaCO_3$. Water treatment practice to lower the pH or soften such a water may be advantageous.

■ EXAMPLE 2–7

What is the dissolved oxygen concentration in water at 18°C, 800 mg/l chloride concentration, and barometric pressure of 660 mm (4000-ft altitude)?

TABLE 2–5

Saturation Values of Dissolved Oxygen in Water Exposed to Water-Saturated Air Containing 20.90 Percent Oxygen Under a Pressure of 760 mm of Mercury[a]

| TEMPERATURE IN °C | DISSOLVED OXYGEN (MG/L) | | | DIFFERENCE PER 100 MG CHLORIDE | TEMPERATURE IN °C | VAPOR PRESSURE (MM) |
| | CHLORIDE CONCENTRATION IN WATER (MG/L) | | | | | |
	0	5000	10,000			
0	14.6	13.8	13.0	0.017	0	5
1	14.2	13.4	12.6	0.016	1	5
2	13.8	13.1	12.3	0.015	2	5
3	13.5	12.7	12.0	0.015	3	6
4	13.1	12.4	11.7	0.014	4	6
5	12.8	12.1	11.4	0.014	5	7
6	12.5	11.8	11.1	0.014	6	7
7	12.2	11.5	10.9	0.013	7	8
8	11.9	11.2	10.6	0.013	8	8
9	11.6	11.0	10.4	0.012	9	9
10	11.3	10.7	10.1	0.012	10	9
11	11.1	10.5	9.9	0.011	11	10
12	10.8	10.3	9.7	0.011	12	11
13	10.6	10.1	9.5	0.011	13	11
14	10.4	9.9	9.3	0.010	14	12
15	10.2	9.7	9.1	0.010	15	13
16	10.0	9.5	9.0	0.010	16	14
17	9.7	9.3	8.8	0.010	17	15
18	9.5	9.1	8.6	0.009	18	16
19	9.4	8.9	8.5	0.009	19	17
20	9.2	8.7	8.3	0.009	20	18
21	9.0	8.6	8.1	0.009	21	19
22	8.8	8.4	8.0	0.008	22	20
23	8.7	8.3	7.9	0.008	23	21
24	8.5	8.1	7.7	0.008	24	22
25	8.4	8.0	7.6	0.008	25	24
26	8.2	7.8	7.4	0.008	26	25
27	8.1	7.7	7.3	0.008	27	27
28	7.9	7.5	7.1	0.008	28	28
29	7.8	7.4	7.0	0.008	29	30
30	7.6	7.3	6.9	0.008	30	32

[a]Saturation at barometric pressures other than 760 mm (29.92 in.), C_s' is related to the corresponding tabulated values, C_s, by

$$C_s' = C_s \frac{P - p}{760 - p}$$

where C_s' = solubility at barometric pressure P and given temperature, in milligrams per liter
C_s = saturation at given temperature from table, milligrams per liter
P = barometric pressure, millimeters
p = pressure of saturated water vapor at temperature of the water selected from table, millimeters

Solution

From Table 2–5, saturation at 18°C with 800 mg/l chloride concentration is

$$9.5 - 8 \times 0.009 = 9.4 \text{ mg/l}$$

Correcting for a barometric pressure of 660 mm by using the equation in the footnote to Table 2–5,

$$C_s' = 9.4 \frac{660 - 16}{760 - 16} = 8.1 \text{ mg/l}$$

■ ■ ■

2–7   ALKALINITY

The bicarbonate-carbonate character of a water can be analyzed by slowly adding a strong acid solution to a sample of water and reading resultant changes in pH. This process, called titration, is used to measure the alkalinity of a water. Acidity is measured by titrating with a strong basic solution. If two samples of pure water (pH 7) are titrated with sulfuric acid and sodium hydroxide solutions, respectively, the composite titration curve is as shown by Figure 2–5a. Very small initial additions of either titrant result in significant changes in pH, since addition or withdrawal of hydrogen ions is reflected immediately in changing pH readings. Curve b is a titration curve of a water containing a high initial concentration of carbonate ions prepared by adding sodium carbonate to distilled water. When acid is added, the majority of the hydrogen ions from acid combine with the carbonate ions to form bicarbonates (Eq. 2–28). The excess hydrogen ions lower the pH gradually, until at pH 8.3 all carbonate radicals have been converted to bicarbonates. Additional hydrogen ions reduce the bicarbonates to carbonic acid below pH 4.5. Stirring the sample at this time results in release of carbon dioxide gas formed from the original carbonates. Figure 2–5c is the reverse titration of curve b, where the addition of a strong base results in conversion of carbonic acid to bicarbonate and, in turn, to carbonate; this follows Eq. 2–28 from left to right. Substances in solution, such as the various ionic forms of carbon

dioxide that offer resistance to change in pH as acids or bases are added, are referred to as buffers. An understanding of buffering action in water and wastewater chemistry is essential, since many chemical reactions in water treatment and biological reactions in wastewater treatment are pH dependent and rely on pH control. Curve d is a typical titration curve of relatively hard well water having an initial pH of 7.8. Inflections of the curve near pH values 4.5 and 8.3 indicate that the primary buffer is the bicarbonate-carbonate system. Irregularities and deviations from the ideal bicarbonate buffer curve, Figure 2–5b, are frequently observed in actual titrations of water and wastewater samples; these may be caused by any number of buffering compounds frequently found in low concentrations, such as phosphates ($H_2PO_4^-$, $HPO_4^=$, $PO_4^\equiv$).

Alkalinity of a water is a measure of its capacity to neutralize acids, in other words, to absorb hydrogen ions without significant pH change. Alkalinity is measured by titrating a given sample with 0.02 normal (N) sulfuric acid. (Normality is a method of expressing the strength of a chemical solution. A 1.00 N solution contains 1 gram of available hydrogen ions, or its equivalent, per liter of solution. A 0.02 N H_2SO_4 solution contains 0.98 gram of pure sulfuric acid, which is 0.02 times the equivalent weight of sulfuric acid.) For highly alkaline samples, the first step is titrating to a pH of 8.3. The second phase, or first in the case of a water with an initial pH of less than 8.3, is titrating to an indicated pH of 4.5. Colorimetric indicators, chemicals that change color at specific pH values, can be used to determine the end points of these titrations if a pH meter is not used. Phenolphthalein turns from pink to colorless at pH 8.3, and mixed bromocresol green-methyl red indicator changes from bluish-gray at pH 4.8 to light pink at pH 4.6. Methyl orange indicator, a substitute for mixed bromocresol green-methyl red solution, changes from orange at pH 4.6 to pink at 4.0. That part of the total alkalinity above 8.3 is referred to as phenolphthalein alkalinity. Alkalinity is conventionally expressed in terms of milligrams $CaCO_3$ per liter and is calculated by Eq. 2–29. Note, if the water sample is 100 ml and normality of acid is 0.02 N, 1 ml of titrant represents 10 mg/l alkalinity.

Figure 2–5

Titration curves for pure water (strong acids and bases), carbonate (salt of a weak acid), carbonic acid (weak acid), and typical well water. (*a*) Pure water. (*b*) Carbonate in water. (*c*) Carbonic acid. (*d*) Well water.

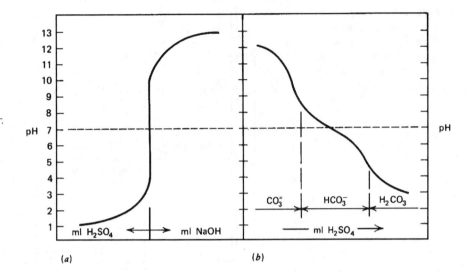

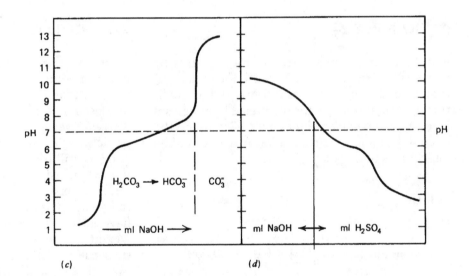

Alkalinity as mg/l $CaCO_3$

$$= \frac{\text{ml titrant} \times \text{normality of acid} \times 50{,}000}{\text{ml sample}}$$

$$(2\text{–}29)$$

Figure 2–6 is a graphical representation of the various forms of alkalinity found in water samples. This diagram can be related to Eq. 2–28. If the pH of a sample is less than 8.3, the alkalinity is all in the form of bicarbonate. Samples containing carbonate and bicarbonate alkalinity have a pH greater than 8.3, and titration to the phenolphthalein end point represents one half of the carbonate alkalinity. If the number of milliliters of phenolphthalein titrant equals the number of milliliters of titrant to the mixed bromocresol green-methyl red end point, all of the alkalinity is in the form of carbonate. Any alkalinity in excess of this amount is due to hydroxide (OH^-). If abbreviations of P for the measured phenolphthalein alkalinity and T for the total alkalinity are used, the following relationships define the bar graphs in Figure 2–6: (a) Hydroxide = $2P - T$ and carbonate = $2(T - P)$; (b) Carbonate = $2P = T$; (c) Carbonate = $2P$ and bicarbonate = $T - 2P$; and (d) Bicarbonate = T.

Figure 2–6

Graphical representation of various forms of alkalinity in water samples related to titration end points.

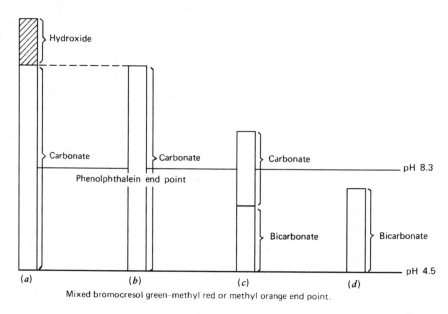

■ EXAMPLE 2–8

A 100-ml water sample is titrated for alkalinity by using 0.02 N sulfuric acid. To reach the phenolphthalein end point requires 3.0 ml, and an additional 12.0 ml is added for the mixed bromocresol green-methyl red color changes. Calculate the phenolphthalein and total alkalinities. Give the ionic forms of the alkalinity present.

Solution

Using Eq. 2–29,

$$\text{Phenolphthalein alkalinity} = \frac{3.0 \times 0.02 \times 50{,}000}{100}$$

$$= 30 \text{ mg/l (as } CaCO_3)$$

$$\text{Total alkalinity} = \frac{15.0 \times 0.02 \times 50{,}000}{100}$$

$$= 150 \text{ mg/l (as } CaCO_3)$$

Referring to Figure 2–6,

$$\text{Alkalinity as carbonate} (CO_3^=) = 2 \times 30$$

$$= 60 \text{ mg/l}$$

$$\text{Alkalinity as bicarbonate} (HCO_3^-) = 150 - 60$$

$$= 90 \text{ mg/l}$$

■ ■ ■

2–8   COLLOIDS AND COAGULATION

A large variety of turbidity-producing substances found in polluted waters do not settle out of solution, for example, color compounds, clay particles, microscopic organisms, and organic matter from decaying vegetation or municipal wastes. These particles, referred to as colloids, range in size from 1 to 500 millimicrons (nanometers) and are not visible when using an ordinary microscope. A colloidal dispersion formed by these particles is stable in quiescent water, since the individual particles have such a large surface area relative to their weight that gravity forces do not influence their suspension. Colloids are classified as hydrophobic (water hating) and hydrophilic (water loving).

Hydrophilic colloids are stable because of their attraction for water molecules, rather than because of the slight charge that they might possess. Typical examples are soap, soluble starch, synthetic detergents, and blood serum. Because of their affinity for water, they are not as easy to remove from suspension as hydrophobic colloids, and coagulant dosages of 10 to 20 times the amount used in conventional water treatment are frequently necessary to coagulate hydrophilic materials.

Hydrophobic particles, possessing no affinity for water, depend on electrical charge for their stability in suspension. The bulk of inorganic and organic matter in a turbid natural water is of this

Figure 2–7

Schematic actions of colloids and coagulation. (*a*) Forces acting on hydrophobic colloids in stable suspension. (*b*) Compression of the double-layer charge on colloids (destabilization) by addition of chemical coagulants. (*c*) Agglomeration resulting from coagulation with metal salt and polymer aid.

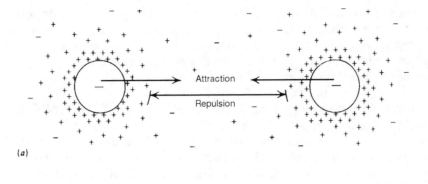

(*a*)

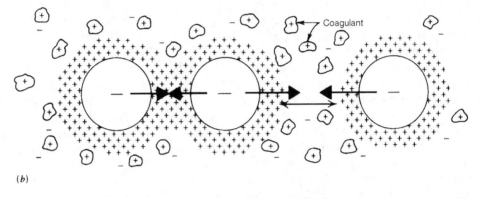

(*b*)

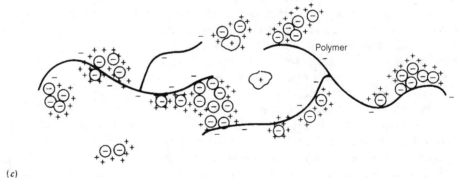

(*c*)

type. The forces acting on hydrophobic colloids are illustrated in Figure 2–7*a*. Individual particles are held apart by electrostatic repulsion forces developed by positive ions adsorbed onto their surfaces from solution. An analogy is the repulsive force that exists between the common poles of two bar magnets. The magnitude of the repulsive force developed by the charged double layer of ions attracted to a particle is referred to as the zeta potential.

A natural force of attraction exists between any two masses (Van der Waals force). Random motion of colloids (Brownian movement), caused by bombardment of water molecules, tends to enhance this physical force of attraction in pulling the particles together. However, a colloidal suspension remains dispersed indefinitely when the forces of repulsion exceed those of attraction and the particles are not allowed to contact.

The purpose of chemical coagulation in water treatment is to destabilize suspended contaminants such that the particles contract and agglomerate, forming flocs that drop out of solution by sedimentation. Destabilization of hydrophobic colloids is accomplished by the addition of chemical coagulants, for instance, salts of aluminum and iron. Highly charged hydrolyzed metal ions produced by these salts in solution reduce the repulsive forces between colloids by compressing the diffuse double layer surrounding individual particles (Figure 2–7b). With the forces of repulsion suppressed, gentle mixing results in particle contact, and the forces of attraction cause particles to stick to each other, producing progressive agglomeration. Coagulant aids may be used to enhance the process of flocculation. For example, organic polymers provide bridging between particles by attaching themselves to the absorbent surfaces of colloids, building larger flocculated masses (Figure 2–7c).

Removal of turbidity by coagulation depends on the type of colloids in suspension; the temperature, pH, and chemical composition of the water; the type and dosage of coagulants and aids; and the degree and time of mixing provided for chemical dispersion and floc formation. Although in chemistry the term *coagulation* means the destabilization of a colloidal dispersion by suppressing the double layer (Figure 2–7a) and *flocculation* refers to aggregation of the particles, engineers have not traditionally restricted the use of these terms to describe the chemical mechanisms only. More frequently, coagulation and flocculation are associated with the physical processes used in chemical treatment. Mixing, involving violent agitation, is used to dissolve and disperse coagulant chemicals throughout the water being treated. Flocculation is a slow mixing process, following chemical dispersion, during which the destabilized particles form into well-developed flocs of sufficient size to settle. The word *coagulation* is commonly used to describe the entire mixing and flocculation operation. Actual units used in chemical treatment may be constructed in series (mixing—flocculation—sedimentation) or may be housed in a single compartmented tank. An in-line system, as shown in Figure 7–4, normally provides about 1 min for flash mixing, 30 min flocculation, and 2 to 4 hr sedimentation, followed by filtration to remove nonsettleable matter. A flocculator-clarifier (Figure 7–10) mixes raw water and applied chemicals with previously flocculating solids in the center mixing compartment, thus contacting settleable solids with untreated water and fresh

chemicals. Solids settled in the peripheral zone automatically return to the mixing area; excess solid buildup is withdrawn from the bottom for disposal.

2–9 ORGANIC COMPOUNDS

All organic compounds contain carbon atoms connected to each other in chain or ring structures with other elements attached. Major components are carbon, hydrogen, and oxygen; minor elements include nitrogen, phosphorus, sulfur, and certain metals. Each carbon exhibits four connecting bonds. Organics are derived from nature, for example, plant fibers and animal tissues; produced by synthesis reactions producing rubber, plastics, and the like; and through fermentation processes, for instance, alcohols, acids, antibodies, and others. In contrast to inorganic compounds, organic substances are usually combustible, high in molecular weight, only sparingly soluble in water—reacting as molecules rather than ions—and a source of food for animal consumers and microbial decomposers.

The molecular formula for an inorganic compound is specific; however, in organic chemistry an empirical formula may represent more than one compound. For example, the formula C_2H_6O may be arranged to represent the following two substances:

The one on the left is ethyl alcohol; the one on the right is dimethyl ether. Compounds that have the same molecular formula but different structural formulas are called isomers. Because of isomerism and the frequency of ring structures, organic compounds are normally given as graphic formulas rather than as molecular ones. For ease of printing, formulas not containing rings are frequently written on a single line. For example, ethyl alcohol and dimethyl ether become

$$CH_3CH_2OH \quad \text{and} \quad CH_3{-}O{-}CH_3$$

Hydrocarbons

Saturated hydrocarbons (paraffins and alkanes) contain only the elements of carbon and hydrogen with single bonds between carbon atoms. Methane (CH_4), the simplest hydrocarbon, is a gas produced in the anaerobic decomposition of waste sludge. When mixed with approximately 90 percent air, it is highly explosive and can be used as a fuel for gas engines or sludge heaters. Short-chain paraffin hydrocarbons can be obtained from petroleum by fractional distillation. Propane is available in high-pressure cylinders for a heating fuel (Eq. 2–30).

$$CH_3CH_2CH_3 + 5O_2 \longrightarrow 3CO_2 + 4H_2O + \text{energy} \tag{2-30}$$

Names and formulas of one- to four-carbon saturated hydrocarbons along with the radicals of these parent compounds are given in Table 2–6.

Unsaturated hydrocarbons (olefins) are distinguished from paraffins by the presence of multiple bonds between some carbon atoms. For example, ethene (ethylene) and acetylene are, respectively,

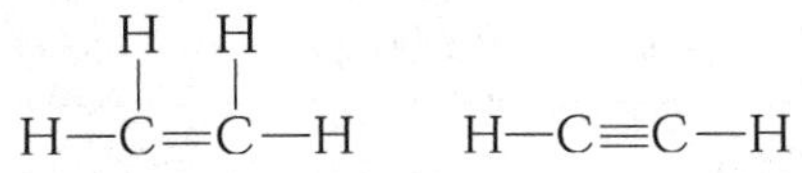

The multiple bonds between carbons displace hydrogen atoms, creating units containing fewer hydrogens than could be attached. Hence, the term *unsaturated* is used. Natural vegetable oils containing a large number of unsaturated bonds are liquids at room temperature. Popular vegetable shortenings available as solid fats are commercially produced from oils through the process of hydrogenation—the addition of hydrogen gas under controlled conditions. Reducing the number of unsaturated bonds increases the melting point, converting an oil to solid fat.

The parent compound of aromatic hydrocarbons is benzene. It is a six-carbon ring with double bonds between alternate atoms. Frequently, the carbon atoms are not shown in writing the graphic formula. Benzene is used in the manufacture of a wide variety of commercial products, including explosives, insecticides, certain plastics, solvents, and dyes.

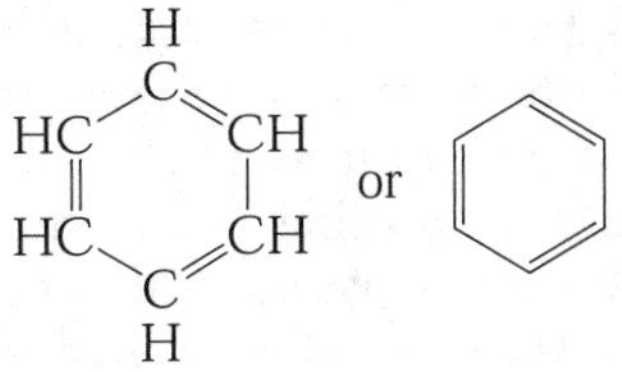

Benzene

TABLE 2–6

Saturated Hydrocarbons

Name	Formula	Radical	Formula
Methane	CH_4	Methyl	$CH_3{-}$
Ethane	CH_3CH_3	Ethyl	$CH_3CH_2{-}$
Propane	$CH_3CH_2CH_3$	n-Propyl	$CH_3CH_2CH_2{-}$
		Isopropyl	$\begin{array}{c} CH_3 \\ {>}CH{-} \\ CH_3 \end{array}$
n-Butane	$CH_3CH_2CH_2CH_3$	n-Butyl	$CH_3CH_2CH_2CH_2{-}$

TABLE 2–7

Primary Alcohols

COMMON NAME	IUPAC[a] NAME	FORMULA
Methyl alcohol	Methanol	CH_3OH
Ethyl alcohol	Ethanol	CH_3CH_2OH
n-Propyl alcohol	1-Propanol	$CH_3CH_2CH_2OH$
Isopropyl alcohol	2-Propanol	$\begin{matrix} CH_3 \\ \diagdown \\ CH_3 \diagup \end{matrix} CHOH$

[a]International Union of Pure and Applied Chemistry.

Alcohols

These are formed from hydrocarbons by replacing one or more hydrogen atoms by hydroxyl groups (—OH). Names and formulas of three primary alcohols are listed in Table 2–7. Methanol is manufactured synthetically by a catalytic process from carbon monoxide and hydrogen. It is used extensively in manufacturing organic compounds, such as formaldehyde, solvents, and fuel additives. Ethyl alcohol for beverage purposes is produced by fermentation of a variety of natural materials—corn, wheat, rice, and potatoes. Industrial ethanol is produced from fermentation of waste solutions containing sugars, for example, blackstrap molasses, a residue resulting from the purification of cane sugar. Propanol has two isomers, the more common being isopropyl alcohol, which is widely used by industry and sold as a medicinal rubbing alcohol. The three primary alcohols listed in Table 2–7 have boiling points less than 100°C, and they are miscible with water.

The derivative of benzene containing one hydroxyl group, known as phenol, has a molecular formula of C_6H_5OH and a graphic representation of

The formula and -ol name ending indicate the characteristics of an alcohol, but phenol, commonly known as carbolic acid, ionizes in water,

yielding hydrogen ions, and exhibits features of an acid. It occurs as a natural component in wastes from coal-gas petroleum industries and in a wide variety of industrial wastes where phenol is used as a raw material. Phenol is a strong toxin that makes these wastes particularly difficult to treat in biological systems. Phenols also impart undesirable taste to water at extremely low concentrations.

Aldehydes and Ketones

These are compounds containing the carbonyl group. Formaldehyde, in addition to being a preservative for biological specimens, is used in producing a variety of plastics and resins. Acetone (dimethyl ketone) is a good solvent of fats and is a common cleaning agent for laboratory glassware.

$$\underset{\text{Carbonyl group}}{-\overset{\overset{\displaystyle O}{\|}}{C}-} \qquad \underset{\text{Formaldehyde}}{H-\overset{\overset{\displaystyle O}{\|}}{C}-H} \qquad \underset{\text{Acetone}}{CH_3-\overset{\overset{\displaystyle O}{\|}}{C}-CH_3}$$

Carboxylic Acids

All organic acids contain the carboxyl group, written as

$$-\overset{\overset{\displaystyle O}{\|}}{C}-OH \quad \text{or} \quad -COOH$$

This is the highest state of oxidation that an organic radical can achieve. Further oxidation results in the formation of carbon dioxide and water.

$$\underset{\text{Hydrocarbon}}{CH_4} \xrightarrow{+O} \underset{\text{Alcohol}}{CH_3OH} \xrightarrow{-2H} \underset{\text{Aldehyde}}{H_2C=O} \xrightarrow{+O}$$

$$\underset{\text{Acid}}{HCOOH} \xrightarrow{+O} CO_2 + H_2O$$

Names, formulas, and ionization constants for the six simplest acids are given in Table 2–8. Acids through nine carbons are liquids, but those with longer chains are greasy solids, hence the common name fatty acid. Since organic acids are

TABLE 2–8

Carboxylic Acids

COMMON NAME	IUPAC[a] NAME	FORMULA	IONIZATION CONSTANT, K
Formic	Methanoic	HCOOH	0.000,214
Acetic	Ethanoic	CH_3COOH	0.000,018
Propionic	Propanoic	CH_3CH_2COOH	0.000,014
n-Butyric	Butanoic	$CH_3CH_2CH_2COOH$	0.000,015
Valeric	Pentanoic	C_4H_9COOH	0.000,016
Caproic	Hexanoic	$C_5H_{11}COOH$	Almost insoluble

[a]International Union of Pure and Applied Chemistry.

weak and ionize poorly, equilibrium constants are used to express their degrees of ionization. (Refer to Section 2–4.) The values of the ionization constants in Table 2–8 are for dilute solutions. For a typical organic acid, consider the dissociation of acetic acid.

$$CH_3{-}\overset{\overset{\displaystyle O}{\|}}{C}{-}OH \rightleftharpoons CH_3{-}\overset{\overset{\displaystyle O}{\|}}{C}{-}O^- + H^+ \quad \text{(2–31)}$$

The ionization (equilibrium) constant is equal to the molar concentration of acetate ion times that of hydrogen ion divided by the molar quantity of molecular acetic acid in solution, Eq. 2–32. The K-value for acetic acid is 0.000,018 at 25°C, which means that in moderately dilute solution only 18 out of every million acetic acid molecules in water dissociate, releasing hydrogen ions.

$$K = \frac{[CH_3{-}\overset{\overset{\displaystyle O}{\|}}{C}{-}O^-][H^+]}{[CH_3{-}\overset{\overset{\displaystyle O}{\|}}{C}{-}OH]} = 0.000,018 \quad \text{(2–32)}$$

Formic, acetic, and propionic acids have sharp penetrating odors, while butyric and valeric have extremely disagreeable odors associated with rancid fats and oils. Anaerobic decomposition of long-chain fatty acids results in production of two- and three-carbon acids, which are then converted to methane and carbon dioxide gas in digestion of waste sludge.

Basic compounds react with acids to produce salts. For example, NaOH reacts with acetic acid to produce sodium acetate. Soaps are salts of long-chain fatty acids.

$$CH_3C\overset{\displaystyle O}{\underset{\displaystyle O^-Na^+}{}} \qquad CH_3(CH_2)_{13}CH_2C\overset{\displaystyle O}{\underset{\displaystyle O^-Na^+}{}}$$

Sodium acetate   Sodium palmitate (a soap)

Other derivatives of carboxylic acids include esters, such as ethyl acetate, and amides, for example, urea.

$$CH_3{-}C\overset{\displaystyle O}{\underset{\displaystyle OCH_2CH_3}{}} \qquad C\overset{\displaystyle NH_2}{\underset{\displaystyle NH_2}{}}{=}O$$

Ethyl acetate   Urea (diamide)

Table 2–9 is a summary of functional groups of organic compounds. These data are useful in identifying an organic compound based on a formula or name. Amines, shown at the bottom of the table, may be regarded as derivatives of ammonia in which hydrogen atoms have been replaced by carbon chains or rings.

2–10 ORGANIC MATTER IN WASTEWATER

Biodegradable organic matter in municipal wastewater is classified into three major categories: carbohydrates, proteins, and fats. Carbohydrates

TABLE 2–9

Functional Groups of Organic Compounds

FUNCTIONAL GROUP	TYPE OF COMPOUND	PROPER NAME ENDING	EXAMPLE	COMMON NAME	IUPAC[a] NAME
$-\overset{\mid}{\underset{\mid}{C}}-\overset{\mid}{\underset{\mid}{C}}-$	Saturated hydrocarbon	–ane	CH_3CH_3	Ethane	Ethane
$-\overset{\mid}{C}=\overset{\mid}{C}-$	Olefin	–ene	$H_2C=CH_2$	Ethylene	Ethene
$-C\equiv C-$	Acetylene	–yne	$H-C\equiv C-H$	Acetylene	Ethyne
$-OH$	Alcohol	–ol	CH_3CH_3-OH	Ethyl alcohol	Ethanol
$-O-$	Ether	—	$CH_3CH_2-O-CH_2CH_3$	Ethyl ether	Ethoxyethane
$-\overset{O}{\overset{\|}{C}}-H$	Aldehyde	–al	$CH_3-\overset{O}{\overset{\|}{C}}-H$	Acetaldehyde	Ethanal
$-\overset{O}{\overset{\|}{C}}-$	Ketone	–one	$CH_3-\overset{O}{\overset{\|}{C}}-CH_3$	Acetone	Propanone
$-\overset{O}{\overset{\|}{C}}-OH$	Carboxylic acid	–oic acid	$CH_3-\overset{O}{\overset{\|}{C}}-OH$	Acetic acid	Ethanoic acid
$-N\overset{H}{\underset{H}{\diagdown}}$	Amine	—	CH_3-NH_2	Methylamine	Aminomethane

[a]International Union of Pure and Applied Chemistry.

consist of sugar units containing the elements of carbon, hydrogen, and oxygen. A single sugar ring is known as a monosaccharide; few of these occur naturally. Disaccharides are composed of two monosaccharide units. Sucrose, common table sugar, is glucose plus fructose, and the most prevalent sugar in milk is lactose, consisting of glucose plus galactose. Polysaccharides, long chains of sugar units, can be divided into two groups: readily degradable starches, abundant in potatoes, rice, corn, and other edible plants; and cellulose, found in wood, cotton, paper, and similar plant tissues. Cellulose compounds degrade biologically at a much slower rate than starches. (See figure on p. 27.)

Proteins in the simple form are long strings of amino acids containing carbon, hydrogen, oxygen, nitrogen, and phosphorus. They form an essential part of all living tissue and constitute a diet necessity for all higher animals. The following is a small section of a protein showing four amino acids connected together.

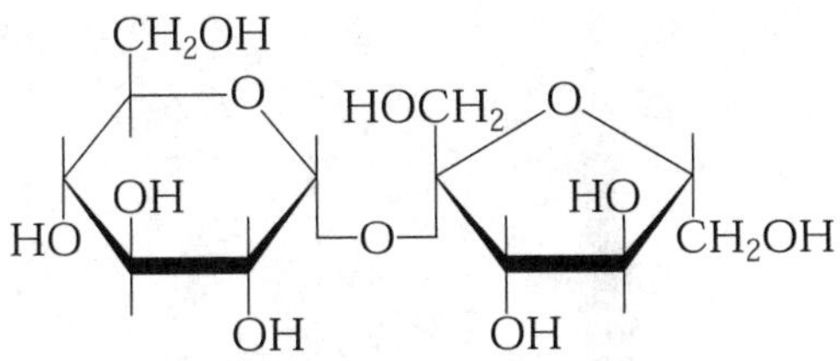

Sucrose

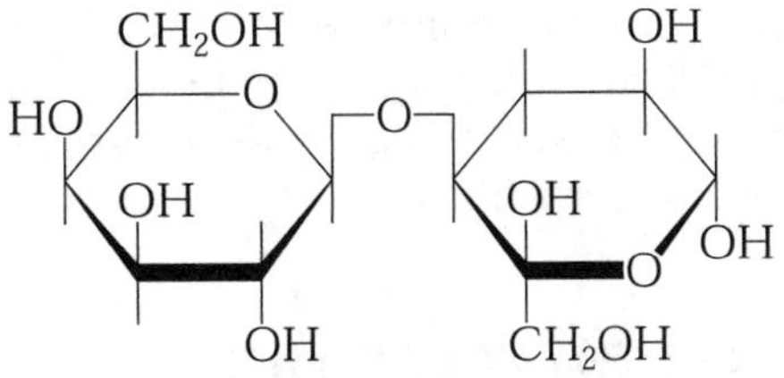

Lactose

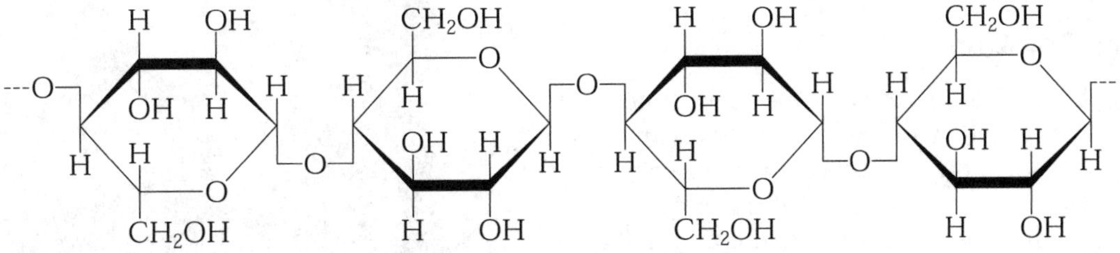

Section of a cellulose molecule

Fats refer to a variety of biochemical substances that have the common property of being soluble to varying degrees in organic solvents (ether, ethanol, acetone, and hexane) while being only sparingly soluble in water. Because of their limited solubility, degradation by microorganisms is at a very slow rate. A simple fat is a triglyceride composed of a glycerol unit with short- or long-chain fatty acids attached. The formula for glycerol oleobutyropalmitate, found in butter fat, is

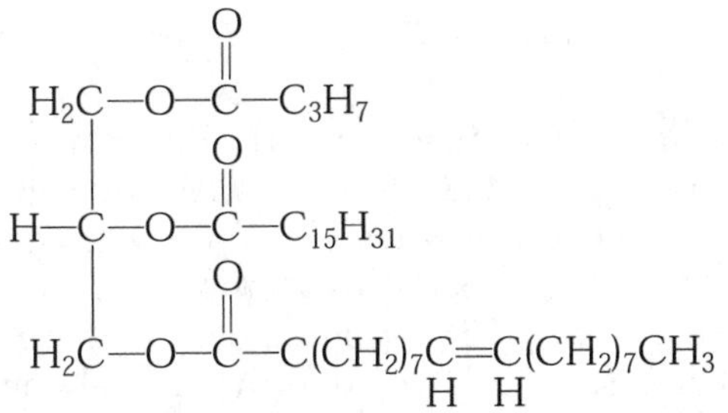

$$H_2C-O-\overset{\overset{O}{\|}}{C}-C_3H_7$$
$$H-C-O-\overset{\overset{O}{\|}}{C}-C_{15}H_{31}$$
$$H_2C-O-\overset{\overset{O}{\|}}{C}-C\underset{H}{(CH_2)_7}\overset{}{C}=\underset{H}{C}(CH_2)_7CH_3$$

The majority of carbohydrates, fats, and proteins in wastewater are in the form of large molecules that cannot penetrate the cell membrane of microorganisms. Bacteria, in order to metabolize high-molecular-weight substances, must be capable of breaking down the large molecules into diffusible fractions for assimilation into the cell. The first step in bacterial decomposition of organic compounds is hydrolysis of carbohydrates into soluble sugars, proteins into amino acids, and fats into short fatty acids. Further aerobic biodegradation results in the formation of carbon dioxide and water. By anaerobic digestion, decomposition in the absence of oxygen, the end products are organic acids, alcohols, and other liquid intermediates as well as gaseous entities of carbon dioxide, methane, and hydrogen sulfide.

Of the organic matter in wastewater, 60 to 80 percent is readily available for biodegradation. Several organic compounds, such as cellulose, long-chain saturated hydrocarbons, and other complex compounds, although available as a bacterial substrate, are considered nonbiodegradable because of the time and environmental limitations of biological wastewater treatment systems. Petroleum derivatives, detergents, pesticides, and other synthetic organic compounds are also resistant to biodegradation, and some are toxic and inhibit the activity of microorganisms in biological treatment processes.

Although some waste odors are inorganic compounds, for example, hydrogen sulfide gas, many are caused by volatile organic compounds, such as mercaptans (organics with SH groups) and butyric acid. Industries may produce a variety of medicinal odors in the processing of raw materials. Surface-water supplies plagued with blooms of blue-green algae have fishy or pigpen odors. Actually very little is known about the origin or characteristics of the organic compounds that produce disagreeable odors. Each case of a smelly wastewater treatment plant or disagreeable taste in a water supply must be investigated separately, keeping in mind that the cause may be anaerobic decomposition, industrial chemicals, or growths of obnoxious microorganisms.

2–11 LABORATORY CHEMICAL ANALYSES

Standard Solutions

Solutions of chemical reagents with known concentrations are commonly used in laboratory testing. A 1 molar (1 M) solution contains 1 gram molecular weight of chemical per liter of solution. For example, a molar solution of NaOH would contain 40.0 grams of the pure salt dissolved in a total solution volume of 1 liter. A 1 normal (1 N) solution contains 1 gram equivalent weight of reagent per liter of solution. A normal NaOH solution would contain the same amount of salt as the 1 molar solution, since the molecular weight and equivalent weight are identical (see Table 2–3). However, in the case of sulfuric acid, a 1 M solution would contain 98.1 g/l, whereas a 1 N solution would contain 49.0 g/l.

Solution concentrations can also be expressed in terms of milligrams per liter (mg/l), milliequivalents per liter (meq/l), grains per gallon (gpg), pounds per million gallons (lb/mil gal), and pounds per million cubic feet (lb/1,000,000 cu ft). The relationship between milliequivalents per liter and milligrams per liter is given in Eq. 2–1. One mg/l, being 1 part by weight per 1,000,000 parts by weight, is equivalent to 8.34 lb/mil gal, since the weight of 1 gal of water is 8.34 lb. Also, since 1 cu ft of water weighs 62.4 lb, 1 mg/l equals 62.4 lb/1,000,000 cu ft. By definition, 1 lb contains 7000 grains; hence, 1 mg/l equals 0.0584 gpg, or 1 gpg equals 17.1 mg/l. Equation 2–33 summarizes these concentration factors.

$$1.00 \text{ mg/l} = 8.34 \text{ lb/mil gal}$$
$$= \frac{62.4}{1,000,000 \text{ cu ft}} = 0.0584 \text{ gpg} \quad \textbf{(2–33)}$$

Hydrogen Ion Concentration

The term *pH*, defined by Eq. 2–4, is used to express the intensity of an acid or alkaline solution. Hydrogen ion concentration is measured

Figure 2–8

A bench-top pH/ISE (ion selective electrode) meter shown with a stand supporting an electrode to measure pH of the water sample. The meter can also measure ion concentrations of selected ions using ion selective electrodes. (Courtesy of Hach Company, Loveland, CO.)

using a meter that reads directly in pH units, as shown in Figure 2–8. The pH electrode in the stand is a platinum series electrode with a glass bulb membrane and a built-in temperature sensor. The bench-top pH/ISE (ion selective electrode) meter can also be calibrated for linear and nonlinear calibration standards for measuring ion concentrations using ion-specific electrodes, such as chloride, fluoride, and nitrate.

Alkalinity and Acidity

Section 2–7 discusses alkalinity and gives the common forms (Figure 2–6), equation for calculation (Eq. 2–29), and sample computation (Example 2–8). A typical laboratory apparatus for titration analysis is shown in Figure 2–9. The container

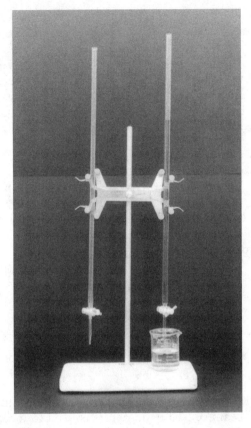

Figure 2–9

Titration apparatus including dispensing burettes, stand, and sample container.

Acidity is a measure of carbon dioxide and other acids in solution. Analysis is by titration, similar to that used in the determination of alkalinity. Strong inorganic-acid acidity exists below pH 4.5; carbon dioxide acidity (carbonic acid) is between pH 4.5 and 8.3 (Figure 2–5c). A measured volume of water sample is titrated with 0.02 N NaOH from the existing pH to a pH of 8.3, noting the amount of milliliters of titrant used below pH 3.7 and that neutralized between 3.7 and 8.3. Acidity, conventionally expressed in terms of milligrams $CaCO_3$ per liter, is calculated by Eq. 2–34. Methyl orange acidity is that below pH 4.5, phenolphthalein exists in the pH range of 3.7 to 8.3, and total acidity is the sum of these two values.

Acidity as mg/l $CaCO_3$

$$= \frac{\text{ml titrant} \times \text{normality of base} \times 50{,}000}{\text{ml sample}}$$

$$(2\text{–}34)$$

Hardness

Hardness is caused by multivalent metallic cations; those most abundant in natural waters are calcium and magnesium. Hard waters from both underground and surface supplies are most common in areas having extensive geological formations of limestone. Although satisfactory for human consumption, Ca^{++} and Mg^{++} precipitate soap, reducing its cleaning action, and cause scale ($CaCO_3$ and $Mg(OH)_2$) in water distribution mains and hot-water heaters. Even with the advent of synthetic detergents and methods for dealing with scaling problems, partial softening of extremely hard waters by municipal treatment plants is desirable. Waters with less than 50 mg/l hardness are considered soft, up to 150 mg/l moderately hard, and in excess of 300 mg/l very hard.

The most common testing method for hardness is the EDTA titrimetric method. Disodium ethylenediaminetetraacetate (Na_2EDTA) forms stable complex ions with Ca^{++}, Mg^{++}, and other divalent cations causing hardness, thus removing them from solution. If a small amount of dye, Eriochrome Black T, is added to the water containing hardness ions at pH 10, the solution becomes wine red; in the absence of hardness the color is

with a measured sample volume is stirred while a standardized titrant solution in a calibrated burette is dispensed into it. The end point of titration is determined either by a colorimetric indicator added to the sample or by use of a glass pH electrode and meter.

The primary need for alkalinity measurements is related to water processing, although this measurement is routinely included in any water analysis. Since lime in water softening and coagulants for turbidity removal react with alkalinity, it is essential that this parameter be monitored in both raw and treated water to ensure optimum dosages of treatment chemicals. Buffering actions for control of pH in biological systems is the carbon dioxide-bicarbonate system; therefore, alkalinity measurements are performed on aerating wastewaters and digesting sludges in evaluating environmental conditions.

blue. The test procedure uses a titration apparatus like that shown in Figure 2–9. The burette is filled with a standardized EDTA solution for dispensing into a measured water sample containing indicator dye. The EDTA added complexes hardness ions until all have been removed from solution and the water color changes from wine-red to blue, indicating end of titration, Eq. 2–35.

$$\underset{\text{Wine-red color}}{Ca^{++} + Mg^{++} + EDTA}$$

$$\xrightarrow{pH=10} \underset{\text{Blue color}}{Ca \cdot EDTA + Mg \cdot EDTA} \quad (2\text{–}35)$$

Hardness is conventionally expressed in the units of mg/l as $CaCO_3$ and is calculated from the laboratory data by the following:

Hardness as mg/l $CaCO_3$

$$= \frac{\text{ml titrant} \ \times \ CaCO_3 \ \text{equivalent of EDTA} \times 1000}{\text{ml sample}}$$

$$(2\text{–}36)$$

Iron and Manganese

These metals at very low concentrations are highly objectionable in water supplies for domestic or industrial use. Traces of iron and manganese can cause staining of bathroom fixtures, can impart a brownish color to laundered clothing, and can affect the taste of water. Groundwaters devoid of dissolved oxygen can contain appreciable amounts of ferrous iron (Fe^{++}) and manganous manganese (Mn^{++}), which are soluble (invisible) forms. When exposed to oxidation, they are transformed to the stable insoluble ions of ferric iron (Fe^{+++}) and manganic manganese (Mn^{++++}), giving water a rust color. Supplies drawn from anaerobic bottom water of reservoirs, or rivers that have contacted natural formations of iron- and manganese-bearing rock, can contain both reduced and oxidized forms, the latter being complexed frequently with organic matter.

The most popular methods of determining iron and manganese in water use colorimetric procedures. Color development as a technique in testing has the major advantage of being highly specific for the ion involved and generally requires minimum pretreatment of the water sample. A photoelectric colorimeter or spectrophotometer is used to measure the intensity of color developed in the treated sample, which can be related to the concentration of iron and manganese.

The phenanthroline method is preferred for measuring iron in water. The first step is to ensure that all iron in solution is in the ferrous state by treating the sample with hydrochloric acid and hydroxylamine as the reducing agent, as follows:

$$4Fe^{+++} + 2NH_2OH$$

$$\xrightarrow{pH=3.2} 4Fe^{++} + N_2O + H_2O + 4H^+ \quad (2\text{–}37)$$

Three molecules of 1,10-phenanthroline sequester each atom of ferrous iron to form an orange-red complex.

1,10-phenanthroline   Orange-red complex

A spectrophotometer set for a monochromatic light beam of 510 nanometers is used to measure the light absorption of the colored solution. The concentration of iron in the sample is determined from a percentage transmission versus iron concentration calibration curve prepared from a series of standard iron solutions previously tested.

Manganese is also determined by a colorimetric procedure—the persulfate method. Chloride interference is overcome by adding mercuric sulfate to form relatively insoluble $HgCl_2$. All manganese in the sample is then converted to permanganate by persulfate in the presence of silver as a catalyst.

$$Mn^{++} + S_2O_8^= \xrightarrow{Ag^+} MnO_4^- + SO_4^= \quad (2\text{–}38)$$

The concentration of permanganate ion produced and its resultant color represent the amount of manganese in the sample.

Colorimetric Measurements

The colorimeter and the spectrophotometer are used to measure color intensity in chemical testing of water parameters. The portable colorimeter

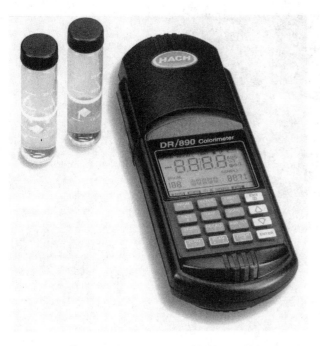

Figure 2–10

Portable colorimeter and sample cells to test parameters in water whose concentrations can be quantitatively determined by color intensity in the water sample after treatment with prescribed chemical reagents. (Courtesy of Hach Company, Loveland, CO.)

shown in Figure 2–10 with two sample cells has an internal light source, a chamber to insert a sample cell, an automatic wavelength selector, and a phototube to measure the intensity of the light passing through the sample. This particular model comes with preprogrammed procedures and automatic wavelength selection. Simplified instrument operation allows push-button program selection and step-by-step prompts that guide each testing procedure. The display shows the concentration of the measured parameter, absorbance, and percent of transmittance.

In addition to iron and manganese determinations, common water parameters determined by colorimeter testing include hardness, dissolved oxygen, and chlorine residual. The water sample is prepared by addition of reagents to develop a particular color with an intensity related to the concentration of the test substance. The sample cell is inserted into the colorimeter so the path of the monochromatic beam of light of the desired wavelength passes through the sample cell and is

measured by a phototube. Test results are displayed on the colorimeter screen and electronically stored.

The most accurate laboratory apparatus used in colorimetric measurements is the spectrophotometer. It operates similarly to a colorimeter, but the monochromatic light is developed by an optical system that allows the selection of any wavelength in the visible spectrum with great precision. Figure 2–11 pictures a spectrophotometer and schematic diagram of the optical system.

Trace Metals

The toxic trace metals specified in drinking water standards and several other common elements, such as calcium and sodium, are detected by atomic absorption spectrophotometry. The concentration of an element in solution is determined by measuring the quantity of light of a specific wavelength absorbed by atoms of the element released in a flame. As represented graphically in Figure 2–12, an atomic absorption spectrophotometer consists of an atomizer-burner to convert the element in solution to free atoms in an air-acetylene flame, a monochromator (prism and slit) to disperse and isolate the lightwaves emitted, and a photomultiplier to detect and amplify the light passing through the monochromator. The light source is a lamp with a cathode formed of the same element being determined, since each element has characteristic wavelengths that are readily absorbed. The light passing through the sample is separated in the monochromator into its component wavelength. The photomultiplier then receives only the isolated resonance wavelength and any absorption of its light by the sample atoms. After the proper lamp for the test element has been inserted, the intensity of the light is measured passing through the unrestricted flame. Then, the sample is introduced into the flame and the concentration of the element in the sample determined by the decrease in light intensity.

Color

Hues in water may result from natural minerals, such as iron and manganese, vegetable origins—humus material and tannins—or colored wastes

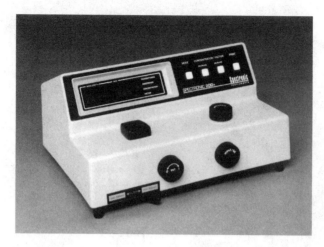

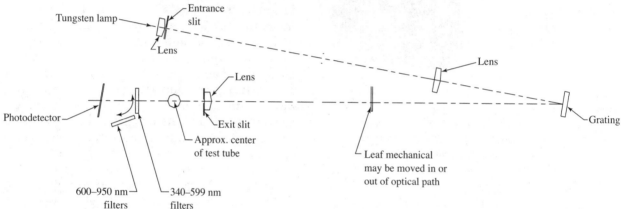

Figure 2–11

Spectrophotometer and optical system schematic. (Courtesy of Milton Roy Company, Rochester, NY, a subsidiary of Sundstrand Corporation.)

Figure 2–12

Simplified schematic diagram of an atomic absorption spectrophotometer.

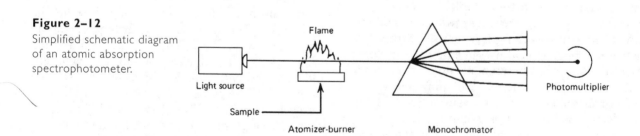

discharged from a variety of industries, including mining, refining, pulp and paper, chemicals, and food processing. The true color of water is considered to be only that attributable to substances in solution after removal of suspended materials by centrifuging or filtration. In domestic water, color is undesirable aesthetically and may dull clothes or stain fixtures. Stringent color limits are required for water use in many industries—beverage production, dairy and food processing, paper manufacturing, and textiles.

Standard color solutions are composed of potassium chloroplatinate (K_2PtCl_6) tinted with small amounts of cobalt chloride. The color produced by

1 mg/l of platinum in combination with $\frac{1}{2}$mg/l of metallic cobalt is taken as 1 standard color unit. The yellow-brownish hue produced by these metals in solution is similar to that found in natural waters. Comparison tubes containing standard platinum-cobalt solutions ranging from 0 to 70 color units are used for visual measurements; however, laboratories often employ a colorimeter for readings.

Turbidity

Insoluble particles of soil, organics, microorganisms, and other materials impede the passage of light through water by scattering and absorbing the rays. This interference of light passage through water is referred to as turbidity. In excess of 5 units, it is noticeable by visual observation. Turbidity in a typical clear lake water is about 25 units, and muddy water exceeds 100 units.

The earliest method for determination of turbidity used a Jackson candle turbidimeter in which a candle flame was viewed through a column of water contained in a calibrated glass tube. Units of turbidity using this apparatus are expressed as Jackson Turbidity Units (JTU). Since the lowest value that could be measured directly by this technique was 25 units, the Jackson candle turbidimeter was limited in application to turbid waters.

Measurement of turbidity in treated drinking water, commonly less than 1 unit, is measured using a precalibrated commercial turbidimeter (nephelometer). Units of turbidity using a nephelometer are expressed as Nephelometric Turbidity Units (NTU). The turbidimeter shown in Figure 2–13 can measure turbidities in the nonratio mode in excess of 40 NTU up to 10,000 NTU and in the ratio mode from less than 1.00 NTU down to 0. Light from a tungsten-filament lamp is focused and passed through the water sample. Transmitted and forward scatter detectors receive light passing through the sample. The back scatter detector measures light scattered back toward the light source. The 90° scatter detector receives light scattered by particles in the water at a right angle to the light beam. In the nonratio mode for moderate to high turbidity, the measurement is the light received by the 90° scatter detector. The back scatter detector is incorporated to permit measurement of very

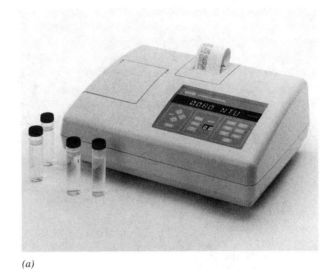
(a)

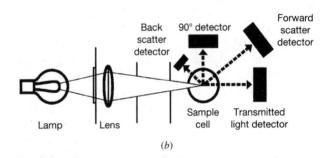
(b)

Figure 2–13
Turbidimeter (nephelometer) and light path diagram.
(Courtesy of Hach Company, Loveland, CO.)

high turbidity. In the ratio mode for low turbidity, forward scatter is negligible compared to transmitted light and the measurement is a ratio of 90° scattered light to transmitted light. This ratio mode provides calibration stability, linearity over a wide range, and negates the effect of color in the water sample.

Jar Tests

The effectiveness of chemical coagulation of water or wastewater can be experimentally evaluated in the laboratory by using a stirring device, as illustrated in Figure 2–14. The stirrer consists of six paddles capable of variable-speed operation between 5 and 300 rpm. In making tests, 1 liter or more of water is placed in each of the square

Figure 2–14
Stirring device used in jar tests for chemical coagulation with outlets for sample withdrawal from the chambers. (Courtesy of Phipps & Bird, Inc.)

TABLE 2–10

Results of Jar Test for Coagulation

Jar Number	Aluminum (ML/JAR)	Sulfate (MG/L)	Dosage (GPG)	Floc Formation
1	1	10	0.58	None
2	2	20	1.17	Smoky
3	3	30	1.85	Fair
4	4	40	2.34	Good
5	5	50	2.92	Good
6	6	60	3.50	Heavy

containers and is dosed with different amounts of coagulant. After rapid mixing to disperse the chemicals, the samples are stirred slowly for floc formation and then are allowed to settle under quiescent conditions. The containers are mixed at a speed of 60 to 80 rpm for 1 min after adding the coagulant solution and then are stirred at a speed of 30 rpm for 15 min. The stirrer can run sequentially through a programmed sequence of paddle speeds and time duration. Programmed information is retained, even with the power off, permitting day-to-day test reproducibility. After the stirrer is stopped, the nature and settling characteristics of the floc are observed and recorded in qualitative terms, as poor, fair, good, or excellent. A hazy sample indicates poor coagulation; properly coagulated water contains floc that is well formed, with the liquid clear between particles.

The lowest dosage that provides good turbidity removal during a jar test is considered the first trial dosage in plant operation. Ordinarily, a full-scale treatment plant gives better results than a jar test at the same dosage. For example, results from a jar test for coagulation of a turbid river water using alum are given in Table 2–10. The lowest recommended dosage in treating this water is 40 mg/l of coagulant. Since such other factors as temperature, alkalinity, and pH influence coagulation, jar tests can also be run to evaluate these parameters and to determine optimum dosage under differing conditions.

Fluoride

Excessive fluoride ions in drinking water cause dental fluorosis, or mottling of teeth. On the other hand, communities whose drinking water contains no fluoride have a high prevalence of dental caries. Optimum concentrations of fluoride provided in public water supplies, generally in the range of 0.8 to 1.2 mg/l, reduce dental caries to a minimum without causing noticeable dental fluorosis. Several fluoride compounds (Table 2–3) are used in treating municipal water; all of these dissociate readily, yielding the fluoride ion (F^-).

Electrode and colorimetric methods are currently most satisfactory for determining fluoride ion concentration. Both methods are susceptible to interfering substances, for example, chlorine in the colorimetric method and polyvalent cations, such as Al^{+++}, through which complex fluoride ions hinder electrode response. Fluoride can be separated from other constituents in water by distillation of hydrofluoric acid (HF) after acidifying the sample with sulfuric acid. Pretreatment separation is performed using a distillation apparatus. Samples not containing significant amounts of interfering ions can be tested directly.

The fluoride-ion-activity electrode is a specific ion sensor designed for use with a calomel reference electrode and a pH meter having a millivolt scale. The key element in the fluoride electrode is the laser-type doped single lanthanum fluoride crystal across which a potential is established by the presence of fluoride ions. The crystal contacts the sample solution at one face and an internal reference solution at the other. An appropriate calibration curve must be developed that relates meter reading in millivolts to concentration of fluoride. Fluoride activities dependent on total ionic strength of the sample and the electrode do not respond to fluoride that is bound or complexed. These difficulties are overcome by the addition of CDTA (cyclohexylene-diaminetetraacetic acid) solution, which is a buffer of high total ionic strength to swamp out variations in sample ionic strength.

The colorimetric method is based on the reaction between fluoride and a zirconium-dye lake. (The term *lake* refers to the color produced when zirconium ion is added to SPADNS dye.) Fluoride reacts with the reddish color-dye lake, dissociating a portion of it into a colorless complex anion $(ZrF_6^=)$ and the yellow-color dye. As the amount of fluoride is increased, the color becomes progressively lighter and of a different hue. This bleaching action is directly proportional to fluoride ion concentration. Either a spectrophotometer (Figure 2–11) or a colorimeter (Figure 2–10) may be used to measure sample absorbance for comparison against a standard curve.

Chlorine

The most common method for disinfecting water supplies in the United States is by chlorination. As a strong oxidizing agent, it is also used effectively in taste and odor control and removal of iron and manganese from water supplies.

Chlorine combines with water to form hypochlorous and hydrochloric acids, Eq. 2–27. Hypochlorous acid reacts with ammonia in water to form monochloramine (NH_2Cl), dichloramine $(NHCl_2)$, and trichloramine (NCl_3), depending on the relative amounts of acid and ammonia. Chlorine, hypochlorous acid, and hypochlorite ion are free chlorine residual, while the chloramines are designated as combined chlorine residual.

The N,N-diethyl-p-phenylenediamine (DPD) test differentiates and measures both free and combined chlorine residuals in samples of clear water. In the absence of iodide ion, free available chlorine reacts instantaneously with DPD indicator to produce a red color. Subsequent addition of a small amount of iodide ion acts catalytically with monochloramine to produce color. Addition of excess iodide causes a rapid response from dichloramine. The intensity of the color development, which is proportional to the amount of chlorine residual, can be measured either by titration with ferrous ammonium sulfate (FAS) or by use of a spectrophotometer at a wavelength of 515 nanometers or a photoelectric colorimeter with a filter having maximum transmission in the range of 490 to 530 nanometers.

The procedure for the colorimeter method is as follows: (1) The sample is diluted with chlorine-demand-free water when the available chlorine in the sample exceeds 4 mg/l. (2) For free chlorine residual, a phosphate buffer solution and DPD indicator solution are mixed with the water, and color development is read immediately and recorded as reading A. (3) For determining monochloramine, one small crystal of KI is added and the sample mixed. The intensity of color is read again and recorded as reading B. (4) For dichloramine, a few more crystals of KI are dissolved in the sample. After about 2 min, the final color intensity is measured and recorded as reading C. The color readings are interpreted as follows: A is free chlorine residual, $B - A$ is monochloramine (NH_2Cl), $C - B$ is dichloramine $(NHCl_2)$, $C - A$ is the combined residual, and C is the total residual.

The portable colorimeter illustrated in Figure 2–15 can measure free and total chlorine residuals in clear water by the DPD method. This colorimeter has a preprogrammed calibration and digital readout. A test kit with a color disk requiring visual comparison of color intensity in the treated sample does not have sufficient accuracy for measuring the chlorine residual in drinking water. This simpler apparatus is often used to monitor residuals in swimming pools.

Residual chlorine in wastewater samples cannot be reliably determined by the DPD technique because of interference from organic matter. The indirect procedure of the iodometric method is much more precise. A wastewater sample is prepared by adding a measured volume of standardized phenylarsine oxide, or thiosulfate, solution; excess potassium iodide; and acetate buffer solution to reduce the pH to between 3.5 and 4.2. Chlorine residual oxidizes an equivalent portion of the excess iodide ion to free iodine, which, in turn, is immediately converted back to iodide by a portion of the reducing agent. The amount of

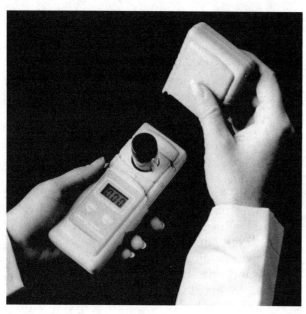

Figure 2–15
Battery-operated colorimeter to measure free and total chlorine residuals in samples of clear water by the DPD method. (Courtesy of Hach Company, Loveland, CO.)

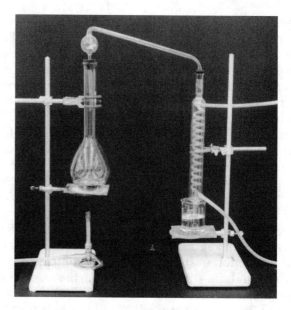

Figure 2–16
Distillation apparatus for ammonia and organic nitrogen tests.

phenylarsine oxide remaining is then quantitatively measured by titration with a standard iodine solution. In other words, this analysis reverses the end point by determining the unreacted standard phenylarsine oxide in the treated sample using standard iodine, rather than directly titrating the iodine released by the chlorine. Thus, contact between the full concentration of liberated iodine and the wastewater is avoided. The titration end point is indicated by using starch, which produces a blue color in the presence of free iodine.

Nitrogen

Common forms of nitrogen are organic, ammonia, nitrite, nitrate, and gaseous nitrogen. Nitrogenous organic matter, such as protein, is essential to living systems. Industrial wastes are often tested for nitrogen and phosphorus content to ensure that there are sufficient nutrients for biological treatment. Inorganic nitrogen, principally in the ammonia and nitrate forms, is used by green plants in photosynthesis. Since nitrogen in natural

waters is limited, pollution from nitrogen wastes can promote the growth of algae, causing green-colored water. Ammonia is also considered a serious water pollutant because of its toxic effect on fishes.

The test for ammonia nitrogen is a distillation process using the apparatus illustrated in Figure 2–16. The analysis procedure is based on shifting the equilibrium between the ammonium ion and free ammonia and the release of gaseous ammonia along with the steam that is produced when the water is boiled. After the sample is placed in a flask, a buffer solution is added to increase the pH and to shift the equilibrium ammonia to the right (Eq. 2–39).

$$NH_4^+ \underset{\text{acidic}}{\overset{\text{basic}}{\rightleftharpoons}} NH_3\uparrow + H^+ \qquad (2\text{–}39)$$

A condenser is then attached to the flask and the mixture is distilled, driving off steam and free ammonia. The steam containing ammonia is condensed and collected in a boric acid solution. The reaction with boric acid forms ammonium ions, Eq. 2–39 to the left, while producing borate ions from the acid. The amount of ammonia in the sample is determined by the quantity of boric acid consumption in the collecting beaker. This may be measured by back titration of the solution with a standard acid to determine the amount of borate ion produced.

Organic nitrogen in wastewater is determined by digestion of the organic matter, thus releasing ammonia, and then proceeding with the ammonia nitrogen test. Since the bulk of organic nitrogen is in the form of proteins and amino acids, it can be converted to ammonium ion by boiling in an acidified solution with chemical catalysts (Eq. 2–40).

$$\text{Organic nitrogen} \xrightarrow[\text{catalysts}]{\text{sulfuric acid}} \text{NH}_4^+ \qquad \textbf{(2–40)}$$

When acid is first added to the wastewater sample, the organics turn black. Complete destruction and conversion after 1 to 2 hr of boiling is indicated by a clearing of the liquid. On cooling, a sodium hydroxide reagent is added to raise the pH and to convert the ammonium ion to ammonia. The flask is attached to the connecting bulb of the condenser, and the test is completed by distillation for measurement of the ammonia nitrogen. If both ammonia and organic nitrogen determinations are desired, the sample may be tested first for ammonia and then can be digested and run for organic nitrogen. The sum of these two results is often referred to as total Kjeldahl nitrogen (TKN). If separation of organic and ammonia nitrogens is not needed, the total Kjeldahl nitrogen may be determined directly by digesting the entire sample and distilling off both the ammonia nitrogen that originally existed in the sample and that released from digesting the organic nitrogen.

Nitrite (NO_2^-) and nitrate (NO_3^-) in natural and treated waters are routinely determined by colorimetric techniques. Nitrite is determined through formation of a reddish-purple azo dye produced at pH 2.0 to 2.5 by coupling diazotized sulfanilic acid with NED dihydrochloride. If the sample is turbid, suspended solids are removed by filtering through a membrane with a pore diameter of 0.45 µm. The intensity of color development can be determined by a photometric measurement or visual comparison with color standards prepared in Nessler tubes.

Nitrate concentration is determined by the cadmium reduction method. NO_3^- is reduced almost quantitatively to NO_2^- in the presence of cadmium. After turbidity removal and pH adjustment, the reduction is performed by pouring the sample through a 30-cm column packed with commercially available Cd granules treated with CuSO_4 to form a copper coating. The nitrite produced by reduction, plus any nitrite in the original sample, is determined by diazotizing with sulfanilamide and coupling with N-(1-naphthyl)-ethylenediamine to form a highly colored azo dye. Color intensity is measured using either a spectrophotometer or a colorimeter.

The specific nitrogen tests conducted on a natural water, treated water, or wastewater depend on the objectives of the study. A drinking water test may require only analysis for nitrate; a stream pollution survey may be primarily interested in ammonium nitrogen. Total nitrogen in a water is equal to the sum of the organic, ammonia, nitrite, and nitrate nitrogen concentrations. All nitrogen test results are expressed in units of milligrams of N/l.

Phosphorus

The common compounds are orthophosphates (H_2PO_4^-, $\text{HPO}_4^=$, $\text{PO}_4^\equiv$); polyphosphates, such as $\text{Na}_3(\text{PO}_3)_6$, used in synthetic detergent formulations; and organic phosphorus. All polyphosphates gradually hydrolyze in water to the stable ortho form, while decaying organic matter decomposes biologically to release phosphate. Orthophosphates are, in turn, synthesized back into living animal or plant tissue.

The prime concern in biological waste treatment is ensuring sufficient phosphorus to support microbial growth. Although sanitary wastes normally have a surplus of phosphates, some industrial wastes may be nutrient deficient because of high carbohydrate or hydrocarbon chemical content. In phosphorus pollution control, the primary concern is overfertilization (eutrophication) of surface waters, resulting in nuisance growths of algae and aquatic weeds.

A popular procedure for determining orthophosphate is the stannous chloride colorimetric method. Ammonium molybdate and stannous chloride reagents combine with phosphate to produce a blue-colored colloidal suspension. A second technique for orthophosphate is the vanadomolybdophosphoric acid method. Vanadium is used in combination with ammonium molybdate to produce a yellow color with phosphate ion. Color development is quantitatively measured using a spectrophotometer.

Acid hydrolysis of a water sample at boiling-water temperature converts condensed phosphates (pyro, tri, poly, and higher-molecular-weight species, such as hexametaphosphate) to orthophosphate. This mild hydrolysis unavoidably releases some phosphate from organic compounds, but this is reduced to a minimum by selection of the acid strength and hydrolysis time and temperature. The condensed phosphate fraction is equal to the acid-hydrolyzable phosphorus determined by this test minus the direct orthophosphate measurement.

Digestion using a strong acid converts all of the phosphorus in a sample, including organic phosphorus, to orthophosphate. Therefore, in a total phosphorus test, the sample is boiled in either a concentrated perchloric acid or sulfuric acid–nitric acid solution to digest the organic matter, releasing bound phosphorus. The total phosphorus is measured by testing the digested sample for orthophosphate content.

The classification of phosphorus fractions in a sample includes the physical state of filterable (dissolved) and particulate as well as chemical types. Separating dissolved from particulate forms is accomplished by filtration through a 0.45-μm membrane filter. A complete phosphorus analysis consists of conducting ortho, acid-hydrolyzable, and total phosphate tests on measured portions of unfiltered and filtered samples. The particulate contents for each of the three phosphorus fractions are calculated by subtracting the filterable measurements from the respective test determinations on the whole sample.

The variety of techniques for pretreatment of samples, measurement of phosphorus concentrations, and expression of results can lead to confusion. The particular procedure used in analysis should be recorded with test results. The most common shortcoming in presenting phosphorus data in printed literature is failure to document collection technique, storage, and filtration or other pretreatment. The thirteenth and subsequent editions of *Standard Methods*[1] have determined total phosphorus by acid digestion, whereas earlier editions considered that portion available after acid hydrolysis as being the total amount, when in fact this includes only the ortho and poly forms. Another modification involved changing the expression of results. *Standard Methods* now calculates the results of all phosphate tests in terms of milligrams per liter as phosphorus; prior to 1971, milligrams per liter as phosphate were used. To convert milligrams per liter as phosphate to milligrams per liter as phosphorus, the value given should be divided by 95 (molecular weight of PO_4) and multiplied by 31 (atomic weight of phosphorus). In other words, 1.00 mg/l PO_4 equals 0.34 mg/l P, or 1.00 mg/l P equals 3.06 mg/l PO_4.

Dissolved Oxygen

Dissolved oxygen (DO) is a significant factor in water quality, pollution control, and several treatment processes. Biological decomposition of organic matter uses dissolved oxygen. Levels significantly below saturation values often occur in polluted surface waters. Since fishes and most aquatic life are stifled by a lack of oxygen, dissolved oxygen determination is a principal measurement in pollution surveys. The rate of air supply to aerobic treatment processes is monitored by dissolved oxygen testing to maintain aerobic conditions and to prevent waste of power by excessive aeration. DO tests are used in the determination of biochemical oxygen demand of a wastewater. Small samples of wastewater are mixed with dilution water and placed in BOD bottles for dissolved oxygen testing at various intervals of time. Oxygen is a significant factor in corrosion of piping systems. Removal of oxygen from boiler feed waters is common practice, and the DO test is the means of control.

The azide modification of the iodometric method is the most common chemical technique for measuring dissolved oxygen. The standard test uses a 300-ml BOD bottle for containing the water sample (Figure 2–17). The chemical reagents used in the test are manganese sulfate solution, alkali-iodide-azide reagent, concentrated sulfuric acid, starch indicator, and standardized sodium thiosulfate titrant. The first step is to add 1.0 ml of each of the first two reagents to the BOD bottle, restopper with care to exclude air bubbles, and mix by repeatedly inverting the bottle. If no oxygen is

Figure 2–17
Apparatus for chemical determination of dissolved oxygen by the azide modification of the iodometric method. The items from left to right are graduated cylinder, BOD bottle, beaker on a magnetic stirrer, titration apparatus with dispensing burette containing sodium thiosulfate titrant, measuring pipettes, and reagent bottles of magnesium sulfate solution, alkali-iodide-azide reagent, concentrated sulfuric acid, and starch solution.

present, the manganous ion (Mn^{++}) reacts only with the hydroxide ion to form a pure white precipitate of $Mn(OH)_2$, Eq. 2–41. If oxygen is present, some of the Mn^{++} is oxidized to a higher valence (Mn^{++++}) and precipitates as a brown-colored oxide (MnO_2), Eq. 2–42.

$$Mn^{++} + 2OH^- \longrightarrow Mn(OH)_2\downarrow \qquad (2\text{–}41)$$

$$Mn^{++} + 2OH^- + \tfrac{1}{2}O_2 \longrightarrow MnO_2\downarrow + H_2O \quad (2\text{–}42)$$

After the bottle has been shaken and sufficient time has been allowed for all the oxygen to react, the chemical precipitates are allowed to settle, leaving clear liquid in the upper portion. Then, 1.0 ml of concentrated sulfuric acid is added. The bottle is restoppered and mixed by inverting until the suspension is completely dissolved and the yellow color is uniform throughout the bottle. The reaction that takes place with the addition of acid is shown in Eq. 2–43; the manganic oxide is reduced to manganous manganese while an

equivalent amount of iodide ion is converted to free iodine. The quantity of I_2^o is equivalent to the dissolved oxygen in the original sample.

$$MnO_2 + 2I^- + 4H^+ \longrightarrow Mn^{++} + I_2^o + 2H_2O$$
$$(2\text{–}43)$$

A volume of 201 ml, corresponding to 200 ml of the original sample after correction for the loss of sample displaced by reagent additions, is poured from the BOD bottle into a container for titration with 0.0250 N thiosulfate solution. Thiosulfate in the titrant is oxidized to tetrathionate while the free iodine is converted back to iodide ion, Eq. 2–44.

$$2S_2O_3^= + I_2^o \xrightarrow[\text{indicator}]{\text{starch}} S_4O_6^= + 2I^- \qquad (2\text{–}44)$$

Since it is impossible to titrate accurately the yellow-colored iodine solution to a colorless liquid, an end point indicator is needed. Soluble starch in the presence of free iodine produces a blue color. Therefore, after titration to a pale straw color, a few drops of starch solution are added and titration is continued to the first disappearance of blue color. If 0.0250 N sodium thiosulfate is used to measure the dissolved oxygen in a volume equal to 200 ml of original sample, 1.0 ml of titrant is equivalent to 1.0 mg/l DO.

Membrane electrodes are available for measurement of dissolved oxygen without chemical treatment of a sample. A dissolved oxygen probe is composed of two solid metal electrodes in contact with a salt solution that is separated from the water sample by a selective membrane (Figure 2–18). The recessed end of the probe containing the metal electrodes is filled with potassium chloride solution and is covered with a polyethylene or Teflon™ membrane held in place by a rubber O-ring. The probe also has a sensor for measuring temperature. The unit inserted in the bottle in Figure 2–18 is designed specifically for measuring dissolved oxygen in nondestructive BOD testing; the same bottle can be measured for oxygen depletion at various time intervals, restoppering between readings. The field probe, as shown on the right, is submersible and can be lowered into water for DO and temperature measurements in lakes and streams. The same unit attached to a

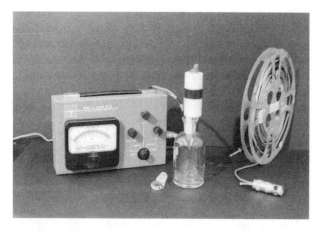

Figure 2–18
Dissolved oxygen meter with the laboratory probe inserted in a BOD bottle. At the right is a field probe with the extension wire wound on a reel.

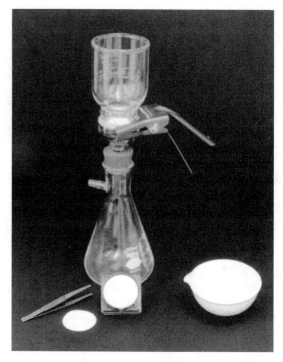

Figure 2–19
Apparatus for solids determinations, including a suction flask with attached funnel, filter pads, and a drying dish.

long rod can record dissolved oxygen in aeration tanks. Meters used with the probes have both temperature and dissolved oxygen scales and can be operated by line power in the laboratory or battery operated in the field. Membrane electrodes are calibrated by reading against air saturated with moisture or a water sample of known DO concentration determined by the iodometric method.

Solids

Suspended and dissolved solids, both organic and inert, are common tests of polluted waters. In drinking water, the maximum recommended total dissolved solids concentration is 500 mg/l. Suspended and volatile solids are common parameters used in defining a municipal or industrial wastewater. Operational efficiency of various treatment units is defined by solids removal, for example, suspended solids removal in a settling tank and volatile solids reduction in sludge digestion.

Total solids, total residue on evaporation, is the term applied to material left in a dish after evaporation of a sample of water or wastewater and subsequent drying in an oven. A measured volume of sample is placed in an evaporating dish, usually porcelain, as shown in Figure 2–19. Water is evaporated from the dish on a steam bath or in a drying oven at approximately 2°C below boiling to prevent splattering. The evaporated sample is dried for a least 1 hr in an oven at 103°C to 105°C and

cooled in a desiccator to a constant weight. The milligrams of total residue are equal to the difference between the cooled weight of the dish and the original weight of the empty dish. Concentration of total residue is calculated using Eq. 2–45.

mg/l total solids

$$= \frac{\text{mg of dried residue} \times 1000 \text{ ml/l}}{\text{ml of sample}} \quad (2\text{–}45)$$

The terms *suspended solids* and *dissolved solids* refer to matter that is retained and passed through a standard glass-fiber filter, respectively. A measured portion of sample is drawn through a glass-fiber filter, retained in a funnel, by applying a vacuum to the suction flask pictured in Figure 2–19. After filtration, the disk is removed from the funnel, dried, and weighed to determine the increase as a result of the residue retained. Calculation of total suspended solids uses the same type of equation as Eq. 2–45. Dissolved matter (nonfilterable residue) is not determined directly but is calculated by subtracting suspended solids concentration from the total solids concentration.

Volatile solids are determined by igniting the residue on evaporation, or the filtered solids, at 500°C ± 50°C in an electric muffle furnace. Dried solids are burned for 15 to 20 min. Loss of weight on ignition is reported as milligrams per liter of volatile solids, and residue after burning is referred to as fixed solids, or ash. The evaporating dish used in volatile solids analysis must be pretreated by burning in the muffle furnace to determine an accurate initial (empty) weight. Glass-fiber filters do not tolerate pretreatment by burning; therefore, several oven-dried filters (from the batch being used in testing) can be ignited to determine the average weight loss in burning. This average value, which is very small compared to the loss of volatile suspended solids from the sample, is added to the burned weight of the filter and fixed solids. In this way, the volatile loss from the glass filter by ignition is compensated for in calculating volatile suspended solids.

Volatile solids in a wastewater are often interpreted as being a measure of the biodegradable organic matter. However, this is not precisely true, since combustion of many pure organic compounds leaves an ash and many inorganic salts volatilize during ignition.

Following are typical data and calculations in the analysis of a wastewater sludge to determine total solids and total volatile solids:

Weight of empty dish on analytical balance		64.532 g
Weight of dish on simple balance	64.5 g	
Weight of dish plus sludge sample	144.7	
Weight of sample	80.2	
Weight of dish plus dry sludge solids		68.950
Weight of sludge solids		4.418

Percentage of total solids $\dfrac{4.42}{80.2} \times 100 = 5.51$ percent

Weight of dish plus ignited solids	65.735
Weight of volatile solids	3.215

Percentage of volatile solids $\dfrac{3.22}{80.2} \times 100 = 4.02$ percent

Percentage of total solids that are volatile $\dfrac{3.22}{4.42} \times 100 = 72.8$ percent

Chemical Oxygen Demand

Chemical oxygen demand (COD) is widely used to characterize the organic strength of wastewaters and pollution of natural waters. The test measures the amount of oxygen required for chemical oxidation of organic matter in the sample to carbon dioxide and water. The apparatus used in the dichromate reflux method consists of a flask fitted with a condenser and a hot plate. The test procedure is to add a known quantity of standard potassium dichromate solution, sulfuric acid reagent containing silver sulfate, and a measured volume of sample to the flask. This mixture is refluxed (vaporized and condensed) for 2 hr. Most types of organic matter are destroyed in this boiling mixture of chromic and sulfuric acid, Eq. 2–46.

$$\text{Organics} + Cr_2O_7^= + H^+ \xrightarrow[Ag^-]{\text{heat}} CO_2 + H_2O + 2Cr^{+++} \quad \textbf{(2–46)}$$

After the mixture has been cooled and diluted with distilled water, and the condenser has been washed down, the dichromate remaining in the specimen is titrated with standard ferrous ammonium sulfate (FAS) using ferroin indicator. Ferrous ion reacts with dichromate ion as in Eq. 2–47,

$$6Fe^{++} + Cr_2O_7^= + 14H^+ \longrightarrow 6Fe^{+++} + 2Cr^{+++} + 7H_2O \quad \textbf{(2–47)}$$

with an end point color change from blue-green to reddish-brown. A blank sample of distilled water is carried through the same COD testing procedure as the wastewater sample. The purpose of running a blank is to compensate for any error that may result because of the presence of extraneous organic matter in the reagents. COD is calculated using Eq. 2–48; the difference between amounts of titrant used for the blank and the sample is divided by volume of sample and multiplied by normality of the titrant. The 8000 multiplier is to express the results in units of milligrams per liter of oxygen, since 1 liter contains 1000 ml and the equivalent weight of oxygen is 8.

$$\frac{\text{COD}}{\text{mg/l}} = \frac{(\text{ml blank} - \text{ml sample titrant}) \times (\text{molarity of FAS})8000}{\text{ml sample}} \quad \textbf{(2–48)}$$

Total Organic Carbon

The total organic carbon (TOC) is the carbon bound in a variety of organic compounds in water and wastewater. The fractions of total carbon are defined as: inorganic carbon—the carbonate, bicarbonate, and dissolved carbon dioxide (CO_2); TOC—all carbon atoms covalently bonded in organic molecules; dissolved organic carbon (DOC)—the fraction of TOC that passes through a 0.45-µm-pore-diameter filter; and suspended organic carbon—the fraction of TOC retained by a 0.45-µm filter. The inorganic carbon interference in testing for TOC can be eliminated by acidifying the sample to pH 2 or less to convert inorganic carbon species to CO_2, and the subsequent purging with purified gas or vacuum degassing to remove the CO_2.

All of the methods for TOC analysis use specific laboratory equipment to convert TOC to CO_2 and detect the CO_2 released from the oxidation of the organic carbon in the sample. The high-temperature combustion method[1] is suitable for samples with higher concentrations of TOC that would require dilution for the persulfate method. In the persulfate-ultraviolet and heated-persulfate oxidation methods,[1] organic carbon is oxidized to CO_2 by persulfate in the presence of ultraviolet light or heat. The CO_2 produced may be purged from the sample, dried, and transferred with a carrier gas to a nondispersive infrared analyzer, coulometrically titrated, or separated from the liquid stream by a membrane that allows the specific passage of CO_2 to high-purity water where a change in conductivity is measured and related to the CO_2 passing the membrane. The persulfate and/or ultraviolet oxidation methods can quantify TOC as low as 10 µg/l.

Sources of water supply for an increasing number of municipalities are augmented by indirect reuse of treated wastewater. The augmentation of surface water may result from inadvertent upstream wastewater discharges or planned augmentation of groundwater by infiltration or injection of reclaimed water (wastewater given tertiary and advanced treatment). Water-quality issues arise from residual organic compounds resistant to treatment processes and natural degradation in the environment. The organic compounds in TOC include substances contributed by domestic and industrial usage, disinfection by-products from chlorination, and microbial products formed during biological wastewater treatment. Some of these compounds have been detected in measurable, albeit low, concentrations in reclaimed water, raising concerns about health risks of indirect reuse of treated wastewater to augment water-supply sources. The absence of reliable techniques for detecting these compounds in reclaimed water creates significant uncertainty regarding health risks to the water's consumers.

The use of TOC as a surrogate or composite parameter for quality of reclaimed water and drinking water derived from augmented sources is being considered to establish a measure of safety even when individual contaminants cannot be identified.[2] Some would argue that a parameter as indiscriminating as TOC provides negligible value for indicating health risks. Nevertheless, removing TOC or DOC from a drinking water supply reduces the concentration of potentially hazardous though unidentified organic compounds. Some states, where groundwater and surface water sources are either already augmented with reclaimed water or augmentation is being considered because of diminishing freshwater, are currently establishing a risk management strategy by setting maximum TOC concentrations in reclaimed water used to augment groundwater in drinking water aquifers.

Trace Organic Chemicals

Minute quantities of the many different organic substances that appear in surface waters have been detected in drinking water. Naturally occurring trace organic chemicals include humic and fulvic acids derived from decaying vegetation. Upon chlorination of water containing these compounds, trihalomethanes such as chloroform (trichloromethane, $CHCl_3$) are formed. Trihalomethanes are suspected carcinogens and limited by drinking water standards.

Other organic compounds of health significance are synthetic chemicals from industrial wastewaters and pesticides. The analytical technique for detection of these trace organics is gas chromatography. The methods of sample preparation are complex and involve initial separation of the

Figure 2–20

Schematic diagram of a gas chromatograph.

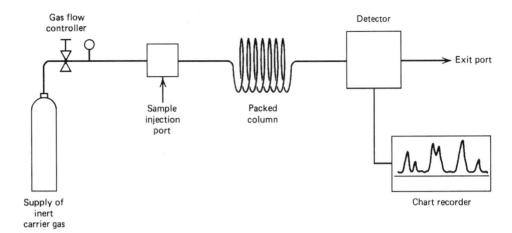

trace organics from water solution. Chlorinated-hydrocarbon pesticides are extracted by a solvent and concentrated by evaporation prior to injection into a gas chromatograph. Organohalides (trihalomethanes) are purged by bubbling an inert gas through the water sample, followed by trapping in a short column containing a suitable sorbant. The compounds are desorbed thermally from the trap and backflushed into a gas chromatographic column.

The main components of a gas chromatograph are a supply of carrier gas with flow control, an injection port for introducing the sample, a temperature-controlled column containing a specific packing, a detector to identify compounds that elute from the column, and a recorder to graph the signal from the detector (Figure 2–20). The carrier gas (nitrogen, argon-methane, helium, or hydrogen) flows continuously through the column at constant temperature and flow rate. A small sample for analysis is injected with a microsyringe through the septum into the sample port, where flash vaporization converts the volatile components into gases. Conveyed by the carrier gas, the gaseous organic compounds travel through the column at different rates, depending on differences in partition coefficients between the moving carrier gas phase and the stationary phase. The stationary phase is a liquid that has been coated on an inert granular solid, called the column packing, held in a glass or metal tube. Components of the sample that are relatively soluble in the liquid phase move through the column slower than relatively insoluble constituents. As each component

of the sample exits the column and passes through the detector, a quantitatively proportional change in electrical signal is measured on a strip-chart recorder. The retention time in the column permits identification of a particular substance, and the peak height or peak area graphed by the recorder is proportional to its quantity. Methods of detection include thermal conductivity, flame ionization, electron capture, and mass spectrometry. The stationary-phase material and concentration, column length and diameter, operating temperature, carrier-gas flow, and detector type are the controlled variables that allow analysis for different trace organics.

Oil and Grease

A variety of organic substances including hydrocarbons, fats, oils, waxes, and high-molecular-weight fatty acids are collectively referred to as oil and grease. Their importance in municipal and industrial wastes is related to their difficulty in handling and treatment. Because of low solubility, grease separates from water and adheres to the interior of pipes and tank walls, reduces biological treatability of a wastewater, and produces greasy sludge solids difficult to process.

The EPA method for monitoring wastewater discharges under the National Pollutant Discharge Elimination System for oil and grease uses n-hexane as the extraction solvent and gravimetry to determine the quantity of material extracted. Since the extractable material may adhere to sampling equipment, samples are collected as 1-liter grab samples and tested without compositing. If analysis is to be delayed more than a few hours, the sample pH is acidified to less than 2 with HCl or H_2SO_4 (generally, 5 ml is sufficient for a 1-liter sample). After acidification, the sample is transferred to a separatory funnel, which is then rinsed with 30 ml of n-hexane and added to the separatory funnel. The funnel is shaken vigorously and allowed to stand quiescent to allow a layer to separate. The aqueous layer along with a small amount of the organic layer is drained to the original sample container. The solvent layer is drained through a funnel containing a filter paper and 10 g Na_2SO_4. The aqueous layers and any remaining emulsion or solids are recombined in the separatory funnel. Extraction is done twice more with 30 ml n-hexane each time by first rinsing the sample container with each n-hexane portion. The extracts are combined in a tarred distilling flask including a final rinse of the filter and Na_2SO_4 with an additional 10 to 20 ml of n-hexane. The flask is fitted with a distillation adapter. The n-hexane solvent is distilled from the flask in a water bath at 85°C. When visible solvent condensation stops, the flask is removed from the water bath. Air is drawn through the flask by vacuum and the flask cooled in a desiccator. The gain in weight of the tarred distilling flask is the extract. By dividing this weight in milligrams by the volume of the original sample in liters, the concentration of oil and grease is expressed in milligrams per liter.

Organic Acids

Measurement of organic acid concentration in anaerobically digesting sludge is used to monitor the digestion process. The acids, principally acetic, propionic, and butyric, are separated from water solution by column chromatography. A small clarified sample is drawn into a Gooch or fritted-glass crucible packed with granular silicic acid to adsorb the organic acids. Chloroformbutanol reagent is then added to the column and drawn through to elute selectively the organic acids from the column packing. After the extract has been purged with nitrogen or carbon dioxide free air, it is titrated with a standard base to measure acid content. Total organic acids are expressed in milligrams per liter as acetic acid.

Gas Analysis

Anaerobic decomposition of sludges produces methane, carbon dioxide, and traces of hydrogen sulfide. Digester operation can be monitored by observing the relative amounts of carbon dioxide and methane in the head gases; the ratio is approximately one-third and two-thirds, respectively. A common laboratory apparatus consists of a water-jacketed burette, catalytic combustion chambers, and a series of glass units containing selective gas absorption reagents for carbon dioxide, oxygen, and carbon monoxide. The catalytic burners are for combustion of methane. A gas sample is drawn into the burette and the initial volume recorded. The confined gas can then be passed through a manifold into any one of the absorbing units or through the combustion chambers and drawn back to the burette again. After several passages through one of the absorbing units, the quantity of gas remaining is measured, and the percentage of that gas constituent in the sample can be calculated.

REFERENCES

1. *Standard Methods for the Examination of Water and Wastewater*, 20th Ed., 1998. Published jointly by American Public Health Association, Washington, DC 20005, American Water Works Association, and Water Environment Federation.
2. National Research Council, 1998, *Issues in Potable Reuse: The Viability of Augmenting Drinking Water Supplies with Reclaimed Water*. National Academy Press, Washington, DC 20418.

PROBLEMS

2–1 Using atomic weights given in Table 2–1, calculate the molecular and equivalent weights of (a) alum ($Al_2(SO_4)_3 \cdot 14.3H_2O$), (b) lime, (c) ferrous sulfate ($FeSO_4 \cdot 7H_2O$), (d) fluorosilicic acid, and (e) soda ash. [*Answers* (a) 600 and 100; others are given in Table 2–3]

2–2 What ions are formed when the following compounds dissolve in water: (a) sodium nitrate, (b) sulfuric acid, (c) calcium hypochlorite, and (d) sodium carbonate. [*Answers* (a) through

(c) ions and radicals are listed in Tables 2–1 and 2–2, (d) Na^+ and $CO_3^=$ or HCO_3^- depending on pH, Figure 2.6]

2–3 All of the fluoridation chemicals listed in Table 2–3 yield F^- ions in solution. If 1.0 mg of fluorosilicic acid is added to water, what is the increase in fluoride ion concentration? (*Answer* refer to Table 7–2)

2–4 If a water contains 40 mg/l of Ca^{++} and 10 mg/l of Mg^{++}, what is the hardness expressed in milligrams per liter as $CaCO_3$? (*Answer* 141 mg/l)

2–5 If a water contains 175 mg/l of calcium hardness and 40 mg/l of magnesium hardness, what are the concentrations of Ca^{++} and Mg^{++} ions?

2–6 The alkalinity of a water consists of 12 mg/l of $CO_3^=$ and 100 mg/l of HCO_3^-. Calculate the alkalinity in milligrams per liter as $CaCO_3$. (*Answer* 102 mg/l)

2–7 If a water contains 6.0 mg/l of $CO_3^=$ and 45 mg/l of HCO_3^-, what is the alkalinity as $CaCO_3$?

2–8 Draw a milliequivalents-per-liter bar graph and list the hypothetical combinations of compounds for the following water analysis.

$Ca^{++} = 94$ mg/l	$HCO_3^- = 317$ mg/l
$Mg^{++} = 24$ mg/l	$SO_4^= = 67$ mg/l
$Na^+ = 14$ mg/l	$Cl^- = 24$ mg/l

(*Answer* The sums of the cations and anions both equal 7.28 meq/l)

2–9 Draw a milliequivalents-per-liter bar graph for the following water analysis:

$Ca^{++} = 60$ mg/l	$HCO_3^- = 115$ mg/l as $CaCO_3$
$Mg^{++} = 10$ mg/l	$SO_4^= = 96$ mg/l
$Na^+ = 7$ mg/l	$Cl^- = 11$ mg/l
$K^+ = 20$ mg/l	

2–10 A brackish groundwater in an arid region has the following chemical characteristics:

$Ca^{++} = 108$ mg/l	$HCO_3^- = 146$ mg/l
$Mg^{++} = 44$ mg/l	$SO_4^= = 110$ mg/l
$Na^+ = 138$ mg/l	$Cl^- = 366$ mg/l

Draw the milliequivalents-per-liter bar graph. Calculate the carbonate hardness (associated with the bicarbonate ion), noncarbonated hardness, total hardness, sodium ion concentration, and alkalinity.

2–11 Draw a milliequivalents-per-liter graph and list the hypothetical combinations of chemicals in solution for the following:

Calcium hardness	= 150 mg/l
Magnesium hardness	= 65 mg/l
Sodium ion	= 8 mg/l
Potassium ion	= 4 mg/l
Alkalinity	= 190 mg/l
Sulfate ion	= 29 mg/l
Chloride ion	= 10 mg/l
pH	= 7.7

2–12 A sulfuric acid solution is added to scale-forming water to convert calcium carbonate to calcium bicarbonate. Write the chemical equation for this reaction, and calculate the amount of sulfuric acid in milligrams per liter to neutralize 20 mg/l of calcium carbonate. (*Answer* 9.8 mg/l)

2–13 Calculate the pH of a solution containing 10 mg/l of sulfuric acid. (*Answer* 4.0)

2–14 Calculate the pH of a solution containing (a) 3.0 mg/l of sulfuric acid and (b) 1.0 mg/l of sulfuric acid. [H^+] in Eq. 2–4 is the molar concentration, and the molecular weight of sulfuric acid in grams is given in Table 2–3.

2–15 Carbon dioxide gas is added to water containing excess dissolved lime to form calcium carbonate precipitate. Write a balanced equation of CO_2 reacting with $Ca(OH)_2$. Calculate the milligrams per liter of carbon dioxide required to react with 35 mg/l of calcium hydroxide and the milligrams per liter of precipitate formed. (*Answer* Eq. 7–19, 20.8 mg/l, 47.2 mg/l)

2–16 If additional carbon dioxide gas is added to the water in Problem 2–15, the calcium carbonate is converted to calcium bicarbonate. Write a balanced equation for this reaction and calculate the milligrams per liter of carbon dioxide required to convert 47.2 mg/l of $CaCO_3$ to $Ca(HCO_3)_2$.

2–17 Metal-plating wastes containing sodium cyanide (NaCN) are chemically treated with chlorine and sodium hydroxide to oxidize the cyanide (CN^-) to nitrogen gas according to the following reaction:

$$2NaCN + 5Cl_2 + 12NaOH$$
$$= N_2\uparrow + Na_2CO_3 + 10NaCl + 6H_2O$$

How many pounds of chlorine are needed to destroy 1 pound of cyanide?

2–18 In softening of water, lime slurry $Ca(OH)_2$ is added to precipitate the calcium ion, associated with the bicarbonate radical, as $CaCO_3$. Write a balanced equation for this reaction. Calculate the amount of lime as calcium oxide necessary to react with 100 mg/l of calcium hardness. (*Answer* Eq. 7–20, 56 mg/l CaO)

2–19 Equation 2–16 is an oxidation–reduction reaction in which Fe^{++} is oxidized by losing 3 electrons while the Mn in MnO_4 gains 3 electrons. Write a balanced equation for oxidation of soluble manganese Mn^{++} by the reduction of Mn in MnO_4 to precipitate MnO_2. Theoretically, 1.0 mg/l of potassium permanganate can oxidize how many milligrams per liter of manganous manganese? (Refer to Eq. 7–29 and discussion in Section 7–18.)

2–20 After desalination of seawater, lime and carbon dioxide are added to the distillate to form $Ca(HCO_3)_2$ in solution to make a stable, noncorrosive water. Write a balanced equation with CaO, water, and carbon dioxide combining to form $Ca(HCO_3)_2$. If the quantity of calcium ion in the treated water is to be 45 mg/l, what concentrations of carbon dioxide and lime as CaO are to be added?

2–21 In zero-order kinetics, what parameter determines the rate of disappearance of reactants? In first-order kinetics, what is the parameter?

2–22 If the rate of a chemical reaction doubles with a 10°C temperature increase, calculate the increase in the rate resulting from a 5°C increase. (*Answer* 1.42)

2–23 In biological aeration, the rate of biological activity is often established by a water temperature of 15°C. If the rate doubles for every 15°C increase, calculate the increase in rate at 25°C and the decrease in rate at 5°C.

2–24 What is the saturation value of dissolved oxygen in pure water at 15°C at sea level? At an altitude of 2000 ft? (*Answer* 10.2 mg/l, 9.5 mg/l)

2–25 What is the saturation value of dissolved oxygen in a water containing 1000 mg/l of chloride ion with a temperature of 22°C at sea level? At an altitude of 4000 ft?

2–26 The rate of oxygen transfer from air bubbles into solution as dissolved oxygen (DO) is directly related to the DO deficit of the water

$(C_s - C_t)$, where C_s is the DO at saturation and C_t is the DO existing in the water. (Refer to Chapter 11, Aeration and Oxygen Transfer.) At the standard conditions of 20°C, $C_t = 0$, and 760 mm of pressure, what is the value of the DO deficit? If the actual conditions for oxygen transfer are 25°C and $C_t = 1.0$ mg/l at an elevation of 5000 ft, what is the value of the DO deficit? (*Answer* 9.2 mg/l, 7.0 mg/l)

2–27 Refer to the titration curves in Figure 2–5a for pure water and 2–5c for water containing carbonic acid. Why does the curve in 2–5c have a sloping section in the central portion while the curve in 2–5a is vertical in this same pH range?

2–28 A 100-ml water sample with an initial pH of 7.5 is titrated to pH 4.5 using 16.5 ml of 0.02 N sulfuric acid. Calculate the alkalinity. What is the ionic form of the alkalinity present in the water?

2–29 In Eq. 2–29, why is the value of the constant 50,000 to calculate alkalinity as $CaCO_3$?

2–30 What are colloids? Why are colloidal particles resistant to settling by gravity? How is intensity of a colloidal suspension measured and what are the units of expression? (Refer to the appropriate subsection in Section 2–11.)

2–31 The common chemicals for destabilization of a colloidal suspension in treatment of surface water are alum and polymer. How do these chemicals react with colloids so that they are removed by granular-media filtration?

2–32 How does the structural arrangement of carbon elements in organic compounds differ from inorganic compounds? Give examples of specific organic chemicals for the different structures.

2–33 Name the following compounds: CH_3CH_2COOH, $CH_3—NH_2$, and $CH_3CH_2—O—CH_2CH_3$. Write formulas for the following chemicals: propane, acetic acid, and sodium acetate.

2–34 Describe the three major categories of biodegradable organic compounds.

2–35 Explain why not all of the organic matter in wastewater is converted to carbon dioxide and water in biological treatment.

2–36 (a) Why is 8.34 the conversion factor between milligrams per liter and pounds per million gallons? Convert the following units: (b) 50 mg/l to pounds per million gallons and pounds per million cubic feet, (c) 100 lb/mil gal to milligrams per liter, and (d) 5.0 gpg to milligrams per liter. (*Answers* (a) refer to Section 2–11, Standard Solutions, (b) 417 lb/mil gal, 3120 lb/mil cu ft, (c) 12.0 mg/l, (d) 86 mg/l)

2–37 A dosage of 30 mg/l of alum is applied in treating a water supply of 5.0 mgd. How many pounds are applied per day?

2–38 A chlorine dosage of 0.7 mg/l is applied to groundwater pumped from a well at 400 gpm. How many pounds of chlorine are added per hour of operation?

2–39 If 48.0 lb of pure sodium silicofluoride that contains 61 percent fluoride ion is dosed into 3.5 mil gal of water, what is the concentration of fluoride ion added?

2–40 What chemicals cause hardness in water, and how is hardness determined in laboratory testing?

2–41 What laboratory apparatus is used to measure color development in colorimetric techniques? Name the common tests performed by colorimetric procedures.

2–42 What is the purpose of jar tests? How are the results of the jar test listed in Table 2–10 interpreted to determine the alum dosage for a full-scale treatment plant?

2–43 Why is the colorimetric method by the DPD technique not satisfactory for measuring residual chlorine in wastewater? What method should be used?

2–44 Describe the test procedure for ammonia nitrogen. How is this test modified to include organic nitrogen? What is the definition of Kjeldahl nitrogen?

2–45 List the sequence of reagents used in determining dissolved oxygen by the azide modification of the iodometric method.

2–46 What are suspended solids in wastewater as defined by the laboratory test procedure?

2–47 Define the term *total organic carbon* (TOC). Why is TOC considered a potential risk to human health in the indirect reuse of reclaimed water to augment water-supply sources?

2–48 How are trace organic chemicals detected in drinking water?

2–49 How is the concentration of organic acids in anaerobically digesting sludge expressed?

Biology

An understanding of the key biological organisms—bacteria, viruses, algae, protozoa, and crustaceans—is essential in sanitary technology. Bacteria and protozoa are the major groups of microorganisms in the "living" system that is used in secondary treatment of wastewaters. A mixed culture of these microorganisms also performs the reaction in the biochemical oxygen demand test to determine wastewater strength. The aquatic food chain, involving organisms in natural waters, can be distressed by water pollution. Several waterborne diseases of humans are caused by bacteria, viruses, and protozoa. Indicator organisms, particularly coliforms, are used to evaluate the sanitary quality of water for drinking and recreation.

3–1 BACTERIA AND FUNGI

Bacteria (*sing.* bacterium) are simple, colorless, one-celled plants that use soluble food and are capable of self-reproduction without sunlight. As decomposers, they fill an indispensable ecological role of decaying organic matter in nature and in stabilizing organic wastes in treatment plants. Bacteria range in size from approximately 0.5 to 5 μm and therefore are only visible through a microscope. Figures 3–1*a* and *b* are photomicrographs of bacteria magnified 400 times. Individual cells may be spheres, rods, or spiral-shaped, and may appear singly or in pairs, packets, or chains.

Bacterial reproduction is by binary fission, that is, a cell divides into two new cells, each of which matures and again divides. Fission occurs every 15 to 30 min in ideal surroundings of abundant food, oxygen, and other nutrients. As a means of survival, some species form spores with tough coatings that are resistant to heat, lack of moisture, and loss of food supply.

A binomial system is used to name bacteria and all other biological organisms. The first word is the genus; the second is the species name. In activated sludge growing in domestic wastewater, a wide variety of bacteria are found, the majority of which appear to be of the genera *Alcaligenes*, *Flavobacterium*, *Bacillus*, and *Pseudomonas*. Identification of particular types in biological floc

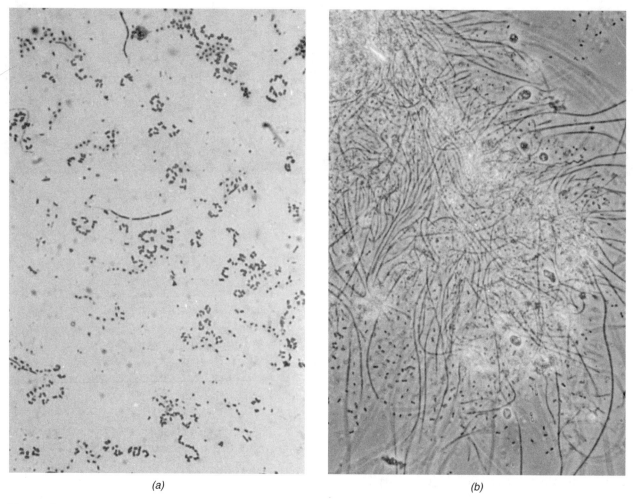

(a)   (b)

Figure 3–1

Photomicrographs of bacterial growths in wastewater. (*a*) Dispersed growth of bacteria (400×). (*b*) Strands of *Sphaerotilus* with swarming cells (400×).

is performed only in research studies, since it is extremely difficult and of limited value in treatment operations to isolate them. One of the most common problems in aerobic biological treatment is poor settleability of the activated-sludge floc because of filamentous growths that produce more buoyant floc (Figure 3–2*a*). Often this is caused by the bacterium *Sphaerotilus natans* (Figure 3–1*b*), the cells of which grow protected in a long sheath; on maturity, individual motile cells swarm out of the protective tube seeking new sites for growth. Perhaps the most frequently referred to bacterium in sanitary work is *Escherichia coli*, a common coliform used as an

indicator of the bacteriological quality of water. *E. coli* cells under microscopic examination at 1000 magnification appear as individual short rods.

Bacteria are classified into two major groups as heterotrophic or autotrophic, depending on their source of nutrients. Heterotrophs, sometimes referred to as saprophytes, use organic matter as both an energy and a carbon source for synthesis. These bacteria are further subdivided into three groups depending on their action toward free oxygen. Aerobes require free dissolved oxygen in decomposing organic matter to gain energy for growth and multiplication, Eq. 3–1. Anaerobes oxidize organics in the complete absence of dissolved

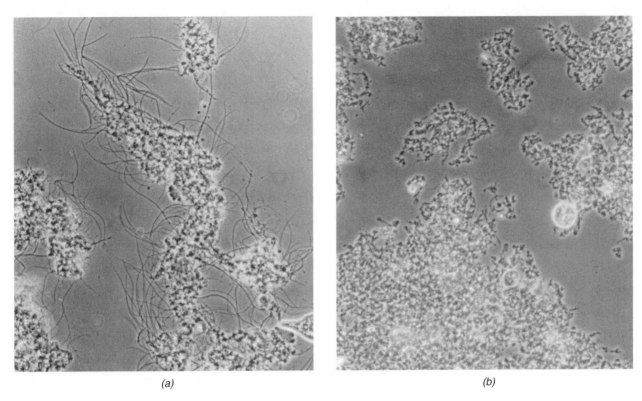

(a)   *(b)*

Figure 3–2
Photomicrographs of biological floc in treatment of wastewater by the activated sludge process. (*a*) Filamentous floc (100×). (*b*) Normal floc (100×).

oxygen by using oxygen bound in other compounds, such as nitrate and sulfate. Facultative bacteria compose a group that uses free dissolved oxygen when available but that can also live in its absence by gaining energy from anaerobic reaction. In wastewater treatment, aerobic microorganisms are found in activated sludge and trickling filters, but anaerobes predominate in sludge digestion. Facultative bacteria are active in both aerobic and anaerobic treatment units.

Aerobic:

$$\text{Organics} + \text{oxygen} \rightarrow \text{CO}_2 + \text{H}_2\text{O} + \text{energy}$$
(3–1)

Anaerobic:

$$\text{Organics} + \text{NO}_3^- \rightarrow \text{CO}_2 + \text{N}_2 + \text{energy}$$
(3–2)

$$\text{Organics} + \text{SO}_4^= \rightarrow \text{CO}_2 + \text{H}_2\text{S} + \text{energy}$$
(3–3)

$$\text{Organics} \rightarrow \text{organic acids} + \text{CO}_2$$
$$+ \text{H}_2\text{O} + \text{energy} \quad \text{(3–4)}$$
$$\rightarrow \text{CH}_4 + \text{CO}_2 + \text{energy} \quad \text{(3–5)}$$

The primary reason heterotrophic bacteria decompose organics is to gain energy for synthesis of new cells and for respiration and motility. A small fraction of the energy is lost in the form of heat, Eq. 3–6.

Synthesis: Energy + organics → new cell growth
$$\rightarrow \text{respiration and motility}$$
$$\rightarrow \text{lost heat} \quad \text{(3–6)}$$

The amount of energy biologically available from a given quantity of matter depends on the oxygen source used in metabolism. The greatest amount is available when dissolved oxygen is used in oxidation, Eq. 3–1, and the least energy yield is derived from strict anaerobic metabolism, Eq. 3–4. Microorganisms growing in wastewater seek the greatest energy yield in order to have maximum synthesis. For example, consider the reactions that would occur in a sample of fresh aerated wastewater placed in a closed container. Aerobic and facultative bacteria immediately start to decompose waste organics, depleting the dissolved oxygen. Although the strict aerobic organisms could not continue to function, facultative bacteria operating anaerobically can use the bound oxygen in nitrate, releasing nitrogen gas, Eq. 3–2. The next most accessible oxygen is available in sulfate, Eq. 3–3, by conversion to hydrogen sulfide (rotten egg odor). Simultaneously, other facultative and anaerobic bacteria partially decompose material to organic acids, alcohols, and other compounds, Eq. 3–4, producing the least amount of energy. If methane-forming bacteria are present, the digestion process is completed by converting the organic acids to gaseous end products of methane and carbon dioxide, Eq. 3–5.

Autotrophic bacteria oxidize inorganic compounds for energy and use carbon dioxide as a carbon source. Nitrifying, sulfur, and iron bacteria are of greatest significance. Nitrifying bacteria oxidize ammonium nitrogen to nitrate in a two-step reaction, as follows:

$$NH_3 + oxygen \xrightarrow{\text{Nitrosomonas}} NO_2^- + energy \quad (3\text{–}7)$$

$$NO_2^- + oxygen \xrightarrow{\text{Nitrobacter}} NO_3^- + energy \quad (3\text{–}8)$$

Nitrification can occur in biological secondary treatment under the conditions of low organic loading and warm temperature. Although providing a more stable effluent, nitrification is often avoided to reduce oxygen consumption in treatment and to prevent floating sludge on the final clarifier. The latter is caused when sludge solids are buoyed up by nitrogen gas bubbles that are formed as a result of nitrate reduction, Eq. 3–2.

A common sulfur bacterium performs the reaction given in Eq. 3–9, which leads to crown corrosion in sewers.

$$H_2S + oxygen \rightarrow H_2SO_4 + energy \quad (3\text{–}9)$$

Wastewater flowing through sewers often turns septic and releases hydrogen sulfide gas, Eq. 3–3. This occurs most frequently in sanitary sewers constructed on flat grades in warm climates. The hydrogen sulfide is absorbed in the condensation moisture on the side walls and crown of the pipe. Here sulfur bacteria, able to tolerate pH levels of less than 1.0, oxidize the weak acid H_2S to strong sulfuric acid using oxygen from air in the sewer. The sulfuric acid reacts with concrete, reducing its structural strength. If the concrete is sufficiently weakened, corrosion can lead to collapse under heavy overburden loads. Using corrosion-resistant pipe material, such as vitrified clay or PVC plastic, is the best protection from corrosion in sanitary sewers. In large collectors, where size and economics dictate concrete pipe, crown corrosion can be reduced by ventilation to expel the hydrogen sulfide and to reduce the moisture of condensation, or by chemical treatment of the wastewater as it flows through the sewer to control generation of hydrogen sulfide. Protection of the concrete pipe interior by coatings and linings should be considered.

Iron bacteria are autotrophs that oxidize soluble inorganic ferrous iron to insoluble ferric, Eq. 3–10.

$$Fe^{++} \text{ (ferrous)} + oxygen \rightarrow$$
$$Fe^{+++} \text{ (ferric)} + energy \quad (3\text{–}10)$$

The filamentous bacteria *Leptothrix* and *Crenothrix* deposit oxidized iron, $Fe(OH)_3$, in their sheath, forming yellow or reddish-colored slimes. Iron bacteria thrive in water pipes where dissolved iron is available as an energy source and bicarbonates are available as a carbon source. With age, the growths die and decompose, releasing foul tastes and odors. There is no easy or inexpensive way of controlling bacteria in water distribution systems. The most certain procedure is to eliminate the ferrous iron from solution by water treatment and by controlling internal pipe corrosion. An alternative to removing iron from the water is a continuous ongoing maintenance program of treating and flushing water mains. A section of main to be treated is isolated, and the water is then dosed with an excessive concentration of chlorine or other chemical to kill the bacteria. After several hours of contact, the pipe section is flushed and returned to service.

Fungi (*sing.* fungus) refer to microscopic non-photosynthetic plants, including yeasts and molds. Yeasts are used for industrial fermentations in baking, distilling, and brewing. Under anaerobic conditions, yeast metabolizes sugar, producing alcohol with minimum synthesis of new yeast cells. Under aerobic conditions, alcohol is not produced and the yield of new cells is much greater. Therefore, if the objective is to grow fodder yeast on waste sugar or molasses, aerobic fermentation is used.

Molds are filamentous fungi that resemble higher plants in structure with branched, threadlike growths. They are nonphotosynthetic, multicellular, heterotrophic, and aerobic and grow best in acid solutions high in sugar content. Growths are frequently observed on the exterior of decaying fruits. Because of their filamentous nature, molds in activated-sludge systems can lead to a poor-settling floc, preventing gravity separation from the wastewater effluent in the final clarifier. Undesirable fungi growths are most frequently created by low pH conditions often associated with treatment of industrial wastes high in sugar content. Addition of an alkali to increase the pH, and sometimes ammonium nitrogen in high-carbohydrate wastewater, is necessary to suppress molds and to enhance the growth of bacteria.

The primary waterborne pathogenic bacteria include *Salmonella* spp., *Vibrio cholerae,* and *Shigella* spp.[1] The forms of salmonellosis in humans are gastroenteritis, enteric fever, and septicemia. As gastroenteritis, the infection is characterized by diarrhea, fever, and abdominal pain that is usually self-limiting and lasts a few days. Waterborne outbreaks commonly result from inadequate disinfection or contamination of the distribution system. As enteric fever, the infection is caused by *S. typhi* (typhoid) or *S. paratyphi* (paratyphoid) and has much more serious symptoms that may involve fatal liver, spleen, respiratory, and neurological damage. *Salmonella* septicemia is characterized by chills, high remittent fever, anorexia,

and bacteremia. Cholera, as a disease resulting from poor sanitation, may be waterborne or spread by contaminated food. A waterborne cholera outbreak is often associated with untreated or inadequate disinfection of drinking water. Waterborne outbreaks of salmonellosis from *Vibrio cholerae* and *Salmonella typhi* are unlikely in the United States because of control of source water quality, water treatment, and protection of treated water in the distribution system. *Shigella* species cause acute gastroenteritis and dysentery, characterized by diarrhea, fever, nausea, vomiting, and cramps. Waterborne outbreaks of shigellosis have usually been the result of fecal contamination of private and noncommunity water supplies. Outbreaks have also been associated with recreational exposure to fecal-contaminated swimming pools and natural surface waters.

3–2 PROTOZOA AND MULTICELLULAR ANIMALS

Protozoa (or protozoans, *sing.* protozoan) are single-celled aquatic animals that multiply by binary fission. (See Figure 3–3a and b.) They have complex digestive systems and use solid organic matter as food. Protozoa are aerobic organisms found in activated sludge, trickling filters, and oxidation ponds treating wastewater, as well as in natural waters. By ingesting bacteria and algae, they provide a vital link in the aquatic food chain.

Flagellated protozoa are the smallest type, ranging in size from 10 to 50 μm. Long, hairlike strands (flagella) provide motility by a whiplike action. While many ingest solid food, some flagellated species take in soluble organics. Amoebas move and take in food through the action of a mobile protoplasm. Although not as common as other forms of protozoa, amoebas are often found in the slime coating on trickling filter media and aeration basin walls. Free-swimming protozoa (Figure 3–4a and b) have cilia, small hairlike processes, used for propulsion and gathering in organic matter. These are easily observed in a wet preparation under a microscope because of their rapid movement and relatively large size, 50 to 300 μm. Stalked forms (Figure 3–4c) attach by a stem to suspended solids and use cilia to propel their heads about and bring in food.

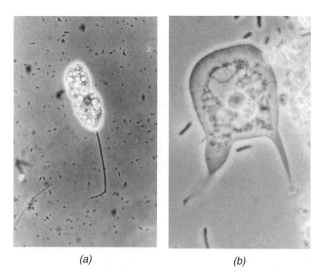

(a)   (b)

Figure 3–3

Photomicrographs of protozoa found in wastewater aeration basins. (a) Flagellated protozoan (1000×). (b) Amoeba, mobile protoplasm (1000×).

Giardia lamblia and *Cryptosporidium parvum*, which occur worldwide, are parasitic protozoa that complete their life cycles by infecting many species of mammals, including humans.[1] The *Giardia* cyst is round to oval with dimensions ranging from 8–18 µm long by 5–15 µm wide. Once inside the host, the cyst releases a trophozoite that actively feeds, grows, and reproduces, causing gastrointestinal distress ranging from no symptoms to illness requiring hospitalization, depending on the immune competency and health status of the host. *Giardia* trophozoites also produce resistant cysts that pass out in the host's feces, which can infect another host by the fecal-oral route. The *Cryptosporidium* thick-walled oocyst, which is the transmissible stage of cryptosporidiosis, is spherical with a diameter of 4–6 µm. When ingested, the oocyst opens in the small intestine, releasing sporozoites that attach to and invade the epithelial cells of the intestinal tract and develop into trophozoites. The most common symptom of cryptosporidiosis is profuse and watery diarrhea. The illness in immunodeficient persons may be life-threatening. Oocysts are released in feces and can transmit infection. Both *Giardia* cysts and *Cryptosporidium* oocysts can be transmitted by close person-to-person contact, for example, among family members, and by poor hygiene in institutions such as day-care centers and nursing homes. Drinking water is the largest potential common source of transmission.

Entamoeba histolytica, a protozoan parasite with spheroid 10–20-µm cysts, causes amoebic dysentery by the fecal-oral route. Transmission is by direct person-to-person contact or exposure to contaminated food, water, or fomites, with water-borne transmission relatively uncommon. The prevalence of infection is worldwide. With symptomatic infection, recurrent diarrhea of varying degrees is common; however, most infections with *E. histolytica* are asymptomatic.

Rotifers are simple, multicelled, aerobic animals that metabolize solid food. The rotifer illustrated in Figure 3–5 uses two circular rows of head cilia for catching food. Its head and foot are telescopic, and it moves with a leechlike action, attaching the foot to a surface. An entirely different type of rotifer with neither appendages nor telescopic head or foot is pictured with algae in Figure 3–8. *Keratella* has a hard protective shell, an anus, and, on the front, six spines along with cilia. Rotifers are found in natural waters, stabilization ponds, and extended aeration basins under low organic loading.

Microcrustaceans are multicellular animals, typically 2 mm in size, easily visible with the naked eye. They have branched swimming feet or a shell-like covering with a variety of appendages. In the aquatic food chain they serve as herbivores, ingesting algae and, in turn, being eaten by fishes. Most protozoa and higher animals can be identified by microscopic examination using a recognition key as an aid.

3–3   VIRUSES

Viruses are obligate, intracellular parasites that replicate only in living hosts' cells. Composed largely of nucleic acid and protein, they lack the metabolic systems for self-reproduction. Their small size requires viewing under an electron microscope; most viruses of interest in sanitary technology are 20 to 100 nanometers (millimicrons) in size, which is about one-fiftieth the size of bacteria. As illustrated in Figure 3–6, virus particles are generally helical, polyhedral, or combination T-even. Viruses that infect only bacteria are called bacteriophages, or simply phages.

Figure 3–4

Photomicrographs of protozoa
found in wastewater aeration basins.
(*a, b*) Free-swimming protozoa
(400×). *Euplotes* (left) and
Blepharisma (right). (c) Stalked
protozoa, *Vorticella* (400×).

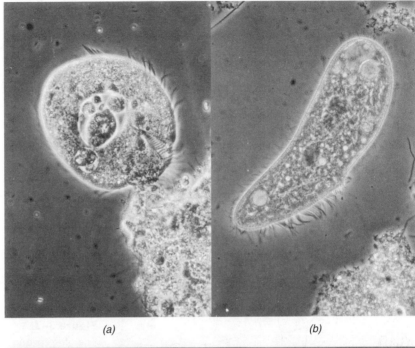

(a)   (b)

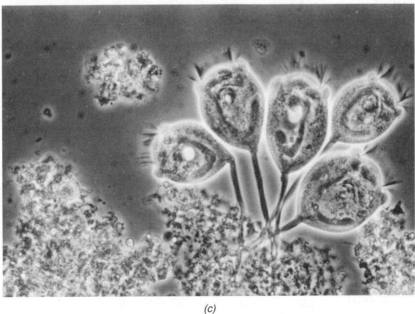

(c)

The life-reproductive cycle of a virus of a T-even phage infecting a bacterium consists of six stages, as diagramed in Figure 3–7. In the dormant stage, viruses have no interaction with their surroundings. Adsorption of a virus onto the cell wall takes place by means of a complementary association between the attachment site of the virus particle and a receptor site on the cell. Following attachment, the bacteriophage injects genetic material into the bacterium; the protein coat does not enter the cell. In contrast, animal viruses break the host cell's membrane and drift in to uncoat the genetic material. Once inside, the virus takes over the cell's metabolism to synthesize new virus proteins

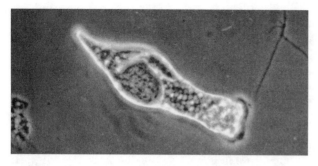

Figure 3–5

Photomicrograph of a rotifer. *Rotaria*, found in aerobic wastewater treatment (200×).

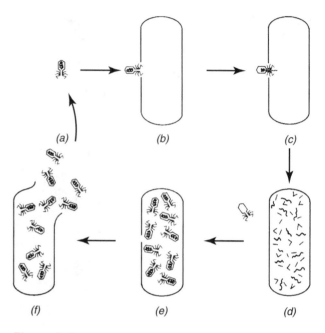

Figure 3–7

The life-reproductive cycle of a virus as illustrated by a T-even bacteriophage infecting a bacterial cell: (*a*) dormancy, (*b*) adsorption, (*c*) penetration, (*d*) replication of new proteins and nucleic acids, (*e*) maturation, and (*f*) release.

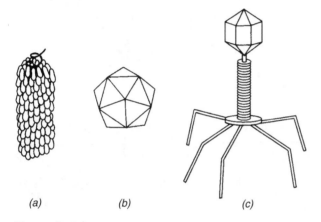

Figure 3–6

Typical structural shapes of viruses. (*a*) Helical. (*b*) Polyhedral. (*c*) Combination T-even.

and nucleic acids. The subsequent maturation process assembles the protein coat and nucleic acids to form complete virus particles, and finally the bacterium lyses (breaks open) to release the new viruses. For phages infecting bacteria, the whole cycle takes about 30 to 40 min.

Viruses of concern in water pollution are those found in the intestinal tracts of humans. Since viruses do not normally reside in humans, they are excreted with the feces from infected persons, who are mostly infants and children. The frequent means of transmission is from person to person by the fecal-oral route. Nevertheless, viruses are present in wastewater treatment plant effluents discharged to surface waters.

The viruses found in relatively large numbers in human feces are: adenoviruses associated with respiratory and eye infections; caliciviruses, which cause diarrhea and gastroenteritis; enteroviruses, which include coxsackie, echoviruses, and poliovirus and cause a wide range of illnesses including meningitis, myocarditis, and paralytic disease; and the hepatitis A virus (HAV), which causes infectious hepatitis, an acute inflammation of the liver. Caliciviruses, enteroviruses, and the hepatitis A virus have all been documented in waterborne-disease outbreaks.

3–4 ALGAE

Algae (*sing.* alga) are microscopic photosynthetic plants of the simplest forms, having neither roots, stems, nor leaves. They range in size from tiny single cells, giving water a green color, to branched forms of visible length that often appear as attached green slime. The term *diatom* is sometimes used in referring to single-celled algae encased in intricately etched silica shells. Figure 3–8 pictures algae filtered from water of a eutrophic lake, magnified 100 times. *Anacystis, Anabaena,* and *Aphanizomenon* are blue-green algae associated with polluted water. Long strands of the latter clumped together appear to the naked eye as short grass clippings when suspended in water. *Oocystis* and *Pediastrum* are green algae. There are hundreds of algal species in a wide variety of cell structures in various shades of green and, less commonly, brown and red. An alga is identified by microscopic observation of its essential characteristics. *Standard Methods*[1] contains colored illustrations of common algae associated with taste and odor, filter clogging, polluted water, clean water, and others.

The process of photosynthesis is illustrated by the equation

$$CO_2 + PO_4 + NH_3 \xrightarrow{\text{enery from sunlight}} \text{new cell growth} + O_2 \quad \textbf{(3–11)}$$

Figure 3–8

Photomicrographs of algae filtered from water of a eutrophic lake (100×).

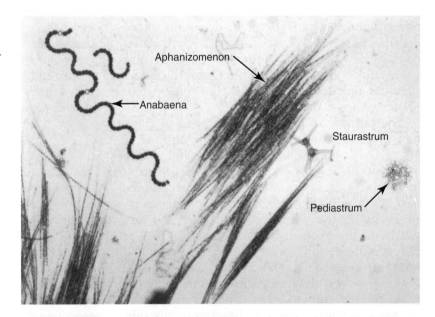

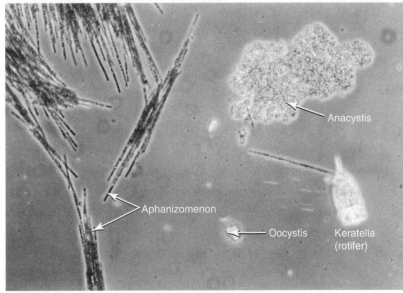

Algae are autotrophic, using carbon dioxide or bicarbonates as a carbon source and inorganic nutrients of phosphate and nitrogen as ammonia or nitrate. In addition, certain trace nutrients are required, such as magnesium, boron, cobalt, and calcium. Certain species of blue-green algae are able to fix gaseous nitrogen if inorganic nitrogen salts are not available. The products of photosynthesis are new plant growth and oxygen. Energy for photosynthesis is derived from light. Pigments, the most common being chlorophyll, which is green, biochemically convert sunlight energy into useful energy for plant growth and reproduction. In the absence of sunlight, plants use some of the previously formed photosynthate, which when combined with oxygen provides the necessary energy for continued survival. The rate of this survival reaction is significantly slower than photosynthesis.

The purpose of photosynthesis is to produce new plant life, thereby increasing the number of algae. Given a suitable environment and proper nutrients, algae grow and multiply in abundance. In natural waters the growth of algae may be limited by turbidity blocking sunlight, low temperatures during the winter, or depletion of a key nutrient. Clear, cold, mountain lakes tend to support few algae, while warm-water lakes enriched with nitrogen and phosphorus from land runoff, or wastewaters, exhibit heavy algal growths, creating green-colored, turbid water during the plant growing season. Wastewater stabilization ponds support luxurious blooms of algae to the point where the suspension becomes self-shading, that is, surplus nitrogen, phosphorus, and carbon nutrients cannot be synthesized because turbidity caused by the algae limits penetration of sunlight.

3–5   WATERBORNE DISEASES

Many infectious enteric (intestinal) diseases of humans are transmitted through fecal wastes. Pathogens (disease-producing agents in the feces of infected persons) include all major categories: viruses, bacteria, protozoa, and helminths (parasitic worms). The resultant diseases are most prevalent in those areas of the world where sanitary disposal of human feces and treatment of drinking water are not adequately practiced.

Transmission is by the feces of an infected person getting into the mouth of another person. This travel from anus to mouth, referred to as the fecal-oral route, may be direct from person to person by contaminated fingers or indirectly through food or water. Some may be transmitted by feces of diseased animals, insect vectors, inhalation of dust or aerosol droplets, and a few (notably hookworm) can penetrate through the skin. Transmissibility varies with each species of pathogen, and human carriers exist for all enteric diseases. Effective disease control is achieved by instituting a comprehensive environmental health program that incorporates personal and household hygiene, control of fly species and other insects, monitoring of food processing, proper waste disposal, protection of water sources and drinking water treatment, immunization of the people when possible, and treatment of diseased persons.

Pathogens Excreted in Human Feces

The kinds and concentrations of pathogens in domestic wastewater from a community depend on the health of the population, namely, the pathogens excreted in human feces.

Viruses are obligate, intracellular parasites that replicate only in living hosts' cells (Section 3–3). Human feces contain over 140 serotypes of viruses. The groups most likely to be transmitted by water are listed in Table 3–1 along with their associated diseases and transmissibility. Persons infected by ingesting these viruses do not always become sick; however, disease is possible in persons infected with any of the enteric viruses. Enteroviruses cause several diseases involving the central nervous system and, more rarely, the skin and heart. The hepatitis A virus is particularly virulent. The symptoms of the disease are loss of appetite, fatigue, nausea, and pain. The most characteristic feature is a yellow color that appears in the white of the eyes and skin, hence the common name yellow jaundice. Waterborne outbreaks of infectious hepatitis are documented, but in the United States none have occurred in municipal water systems.

Bacteria are microscopic single-celled plants capable of self-reproduction (Section 3–1). The feces of healthy persons contain 1 to 1000 million per gram of each of the following groups of

TABLE 3–1

Typical Pathogens Excreted in Human Feces

Pathogen Group and Name	Associated Diseases	Category for Transmissibility[a]
Virus		
Adenoviruses	Respiratory, eye infections	I
Caliciviruses	Diarrhea	I
Enteroviruses		
Polioviruses	Aseptic meningitis, poliomyelitis	I
Echoviruses	Aseptic meningitis, diarrhea, respiratory infections	I
Coxsackie viruses	Aseptic meningitis, herpangina, myocarditis	I
Hepatitis A virus	Infectious hepatitis	I
Other viruses	Gastroenteritis, diarrhea	I
Bacterium		
Salmonella typhi	Typhoid fever	II
Salmonella paratyphi	Paratyphoid fever	II
Other salmonellae	Gastroenteritis	II
Shigella species	Bacillary dysentery	II
Vibrio cholerae	Cholera	II
Other vibrios	Diarrhea	II
Yersinia enterocolitica	Gastroenteritis	II
Protozoan		
Giardia lamblia	Diarrhea	I
Cryptosporidium species	Diarrhea	I
Entamoeba histolytica	Amoebic dysentery	I
Helminth		
Ancylostoma duodenale (hookworm)	Hookworm	III
Ascaris lumbricoides (roundworm)	Ascariasis	III
Hymenolepis nana (dwarf tapeworm)	Hymenolepiasis	I
Necator americanus (hookworm)	Hookworm	III
Strongyloides stercoralis (threadworm)	Strongyloidiasis	III
Trichuris trichiura (whipworm)	Trichuriasis	III

[a] I = Nonlatent, low infective dose.
II = Nonlatent, medium to high infective dose, moderately persistent.
III = Latent, persistent.

bacteria: enterobacteria, enterococci, lactobacilli, clostridia, bacteroides, bifidobacteria, and eubacteria. *E. coli,* the common fecal coliform, is in the enterobacteria group. For many bacterial infections of the intestines, the major symptom is diarrhea. The most serious waterborne bacterial diseases are typhoid fever caused by *Salmonella typhi,* cholera (*Vibrio cholera*), and bacillary dysentery (*Shigella dysenteriae*). Typhoid is an acute infectious disease characterized by a continued high fever and infection of the spleen, gastrointestinal tract, and blood. Cholera symptoms include diarrhea, vomiting, and dehydration. Dysentery causes diarrhea, bloody stools, and sometimes fever. All of these diseases are debilitating and can cause death if not treated. Transmission is by direct contact, food, milk, shellfish, and water. Although these diseases are still prevalent in underdeveloped countries and have been responsible for millions of deaths historically, they have been virtually eliminated in the United States by environmental control through pasteurization of milk and chlorination of water supplies. Although rare, waterborne outbreaks of intestinal diseases in the United States have been attributed to dysentery from *Shigella* and gastroenteritis from salmonellae. Gastroenteritis is inflammation of the lining membrane of the stomach and intestines.

Protozoa infecting humans are intestinal parasites that replicate in the host and exist in two forms. Trophozoites live attached to the intestinal wall, where they actively feed and reproduce. At some time during the life of a trophozoite, it releases and floats through the intestines while making a morphologic transformation into a cyst, or oocyst, for protection against the harsh environment outside the host. The cyst form is infectious for other persons by the fecal-oral route of transmission. The cysts are usually 5 to 15 μm in size, making them significantly larger than intestinal bacteria. The common diseases are diarrhea and dysentery caused by the three pathogenic protozoa listed in Table 3–1. *Entamoeba histolytica* causes amoebic dysentery, which is severely debilitating to the human host. While common in tropical climates, amoebic dysentery is considered nontransmittable in temperate climates. In contrast, giardiasis and cryptosporidiosis are worldwide diseases and waterborne outbreaks have occurred in community systems in the United States.

The disease caused by *Giardia lamblia* is characterized by diarrhea usually lasting one week or more and may be accompanied by abdominal cramps, bloating, flatulence, fatigue, and weight loss. Depending on the immune competency and health status of a sick person, the parasite may produce a continuum of deviations from normal diarrhea disease and require hospitalization. The incubation period in the host before symptoms appear is usually one to two weeks.

A unique feature of giardiasis is transmission to humans through beavers, which serve as amplifying hosts. In mountainous regions, beavers have been infected by upstream contamination with human excreta containing *G. lamblia*. After being infected, they return millions of cysts to the water for every one ingested, amplifying the number of Giardia cysts in clear mountain streams. The inactive cyst stage in the protozoal life cycle can exist for long periods of time in water, with longer survival times at cooler temperatures. After cysts are ingested, trophozoites come out of the cysts and infect the host's intestines. Many of the initial outbreaks of waterborne giardiasis in the United States occurred in downstream resorts and towns in mountainous regions where the water supplies were unfiltered or filtered without adequate chemical coagulation prior to filtration.

The disease caused by *Cryptosporidium* species is profuse and watery diarrhea with variable severity for which no effective remedial treatment is known. Many infected persons have mild, nonspecific symptoms, while in others the diarrhea is prolonged and accompanied by weight loss. In persons with immunodeficiency syndrome, serious diarrhea can cause life-threatening dehydration.

Cryptosporidium oocysts have been found in surface waters contaminated by runoff washing cattle or sheep feces into the water. Thus, infected livestock in a watershed may contribute to the transmission of cryptosporidiosis. As with *Giardia* cysts, oocysts are very resistant and can remain viable for extended periods of time of several months in cool surface waters.

Outbreaks of waterborne cryptosporidiosis have occurred in public water systems in the United States, with documented cases in communities with filtered surface-water supplies. In one case, a substantial percentage of the more than ten thousand people served by the water system became ill with fever, diarrhea, and intestinal cramps. *Cryptosporidium* oocysts were identified in the treated water and in stools of sick persons. No other bacterial, viral, or parasitic pathogens were implicated in the outbreak. All samples of treated water in the system tested negative for coliforms.

The likely source of contamination of the raw water supply was a wastewater overflow from a sewer into the river upstream from the water plant intake. Passage of oocysts through chemical coagulation and filtration was attributed to the mechanical flocculators, which were out of service, and inefficient filtration because of damaged equipment and improper operation. In a second case, several thousand persons out of a population of one million served by a treated surface-water supply were reported to have contracted gastrointestinal illness. *Cryptosporidium* oocysts were identified in the raw and treated waters. Failure of adequate chemical coagulation and filtration was primarily attributed to inadequate operation. Recommendations were to optimize treatment by improving chemical coagulation, shortening filter runs, decreasing filtered-water turbidity, and testing waters for *Cryptosporidium* oocysts. (Refer to Section 7–15 for a comprehensive discussion on disinfection of surface-water supplies.)

Many cases of giardiasis and cryptosporidiosis reported to disease control agencies are person-to-person transmission in day-care centers. Lack of hygiene and frequent mouthing of objects among diapered children can lead to rapid spread in day-care centers and homes. These diseases are also found to cause traveler's diarrhea, often where the sick persons ate contaminated food or drank untreated surface waters in underdeveloped countries. These kinds of cases should be distinguished from waterborne outbreaks in community water systems.

Helminths are parasitic intestinal worms that (except for *Strongyloides*) do not multiply in the human host. Therefore, the worm burden in an infected person is directly related to the number of infective eggs ingested. The worm burden is also related to the severity of the infected person's disease symptoms. Eggs are excreted in the host's feces. Of the helminths listed in Table 3–1, most can be transmitted by ingestion of contaminated water or food after a latent period of several days. Hookworms live in the soil and after molting are infectious to humans by penetrating the skin. With a heavy worm infection, the symptoms can be anemia, digestive disorder, abdominal pain, and debility. Helminth eggs are commonly 40 to 60 µm in length and denser than water.

Factors Affecting Transmission of Diseases

The transmission of waterborne diseases is influenced by latency, persistence, and infective dose of the pathogens. Latency is the period of time between excretion of a pathogen and its becoming infective to a new host. No excreted viruses, bacteria, and protozoa have a latent period. Among the helminths, only a few have eggs or larvae passed in feces that are immediately infectious to humans. The majority of helminths require a distinct latent period either for eggs to develop to the infectious stage or to pass through an intermediate to complete their life cycles. For example, *Ascaris lumbricoides* has a latency of 10 days.

Infective dose is the number of organisms that must be ingested to result in disease. Usually, the minimum infective dose for viruses and protozoans is low and less than that for bacteria, but a single helminth egg or larva can infect. Median infective dose is that dose required to infect half of those persons exposed.

Persistence is measured by the length of time that a pathogen remains viable in the environment outside a human host. The transmission of persistent microorganisms can follow a long route, for example, through a wastewater treatment system, and still infect persons located remotely from the original host. In general, persistence increases, starting with bacteria (which have the least persistence) to viruses, to protozoal cysts, to helminth eggs, which have persistence measured in months.

The transmission characteristics of pathogens are categorized based on latency, infective dose, and persistence, as shown in the right column of Table 3–1. Category I contains infections that have a low median infective dose (less than 100) and are infective immediately upon excretion. These infections are transmitted person to person where personal and domestic hygiene are poor. Therefore, control of these diseases requires improvements in personal cleanliness and environmental sanitation, including food preparation, water supply, and wastewater disposal.

Category II contains all bacterial diseases having a medium to high-medium infective dose (greater than 10,000) and are less likely to be transmitted by person-to-person contact than

Category I infections. In addition to the control measures given for Category I, wastewater collection, treatment, and reuse are of greater importance, particularly if personal hygiene and living standards are high enough to reduce person-to-person transmission.

Category III contains soil-transmitted helminths that are both latent and persistent. Their transmission is less related to personal cleanliness because these helminth eggs are not immediately infective to human beings. Most relevant is the cleanliness of vegetables grown in fields exposed to human excreta by reuse of wastewater for irrigation and sludge for fertilization. Effective wastewater treatment is necessary to remove helminth eggs, and sludge stabilization is necessary to inactivate the removed eggs.

Human carriers exist for all enteric diseases. Thus, in communities where a disease is endemic, a proportion of the healthy persons excrete pathogens in feces. In some infections, the carrier condition may cease along with symptoms of the illness; in others, it may persist for months, years, or a lifetime. The carrier condition exists for most bacterial and viral infections, including the dreaded diseases of cholera and infectious hepatitis. Human carriers without symptoms of disease are primarily responsible for continued transmission of the intestinal protozoa *Giardia lamblia* and *Cryptosporidium* species. In light helminthic infections, the human host may have only minor symptoms of illness while passing eggs in feces for more than a year.

Safety surveys of sanitation workers have shown that the incidence of waterborne disease is no greater in this group than in the population as a whole. Although wastewater must be considered potentially pathogenic, the reduced incidence of waterborne disease in the United States has diminished the possibility of treatment plant employees becoming infected. Most safety manuals stress that the best defense against infection is the practice of good personal hygiene and prompt medical care for any injury that breaks the skin. The latter appears to be the most critical problem, since infected wounds are far more common than enteric diseases. The most common source of *Clostridium tetani* is human feces; therefore, it is recommended that sanitation workers receive artificial immunity by tetanus toxoid injections.

3–6 TESTING FOR ENTERIC VIRUSES

Viruses of particular importance in drinking water and reclaimed water are those that infect the gastrointestinal tract of humans and are excreted with feces of infected persons. Although viruses are transmitted most frequently from person to person by the fecal-oral route, they may also be present in surface waters and groundwaters contaminated by domestic wastewater. The viruses known to be excreted in relatively large numbers with feces include polioviruses, coxsackieviruses, echoviruses, and other enteroviruses, adenoviruses, reoviruses, rotaviruses, the hepatitis A (infectious hepatitis) virus, and the Norwalk-type agents that can cause acute infectious nonbacterial gastroenteritis.[2] Since each group consists of a number of different serological types, more than 100 different enteric viruses are recognized.

Testing for viruses requires extraction, concentration, and identification. The process of extraction is by pumping a large volume of water through a cartridge or a large disk filter. For drinking water and other relatively unpolluted water, the virus levels are likely to be so low that a sample of several hundred liters up to 1000 liters must be processed to increase the probability of virus detection. The different techniques used to concentrate the eluate from the cartridge or disk filter are: adsorption followed by elution from microporous filters, aluminum hydroxide adsorption-precipitation, or polyethylene glycol hydroextraction-dialysis.[2] During these processes of extraction and concentration, only a portion of the viruses present in the original sample are captured. The efficiency of separation varies widely depending on water quality. Therefore, to determine the precision of separation, the procedure must be conducted on samples to which known suspensions of one or more test virus types have been added to a water sample to establish recovery efficiency.

Assay and identification of viruses in sample concentrates rely on the fact that viruses are obligate, intracellular parasites that multiply in, and thereby destroy, their host cells. The two major host cell systems for human enteric viruses are mammalian cell cultures of primate origin and whole animals. Most of the known enteric viruses can be detected by using two or more cell culture systems and perhaps suckling mice. No single universal host system exists for all enteric viruses. Nevertheless, a continuous line developed from African green monkey cells is a sensitive host for several enteric viruses. In cell cultures, monolayers of cells are grown attached to glass surfaces in covered culture containers. A measured portion of virus concentrate is diluted and spread on the surface of the cell tissue. Following incubation, the concentration of viruses in the applied portion of sample is determined by microscopically counting the plaques and correcting for dilution. (A plaque is a clear area in the cell monolayer produced by viral destruction of the cells.) Plaque assay cultures are examined periodically for appearance of plaques over a 14-day period. Virus concentration in the water sample is expressed as the number of plaque-forming units (PFU) per liter. Further examination is required to identify the virus type creating a plaque. Preliminary identification can be made from microscopic examination of the visible effects of the virus on the infected culture cells. Precise identification involves recovering viruses from an individual plaque and inoculating them into different cell cultures and assay in mice.

Testing for enteric viruses is beyond the capability of the majority of water and wastewater microbiology laboratories. Laboratories planning to concentrate viruses from water and wastewater samples should do so with the clear understanding that the available methodology has important limitations.[2] The efficiency of virus extraction and concentration methods vary widely depending on water quality. Sample processing requires special equipment, and virus assay and identification procedures require cell cultures and related virology laboratory facilities. Virus assay and identification should be done only by a trained virologist working in a specially equipped virology laboratory facility.

Testing for enteric viruses is prudent or essential for water quality investigations in special circumstances such as research studies, wastewater reclamation, or disease outbreaks. Investigations that include testing for enteric viruses by the Sanitation Districts of Los Angeles County are presented in Section 13–3 (Pathogen Removal). An extensive research project in the mid-1970s investigated enteric virus removal by tertiary wastewater treatment systems on a pilot-plant scale.

Currently, eight tertiary plants process wastewater by direct filtration and chlorination with an extended contact time to remove viruses (Figure 13–10). The effluent standards with daily monitoring are a 7-day median coliform concentration not to exceed 2.2 total coliforms per 100 ml and a turbidity limit not to exceed 2 NTU. To ensure removal of viruses, a long-term virus monitoring program was established to periodically test 24-hour composite effluent samples. From 1979 to 1999, 1045 samples, each with a reclaimed-water volume of at least 1000 liters from the eight tertiary plants, were tested for enteric viruses. The results yielded only one positive sample—*Coxsackie B*.

3–7 TESTING FOR *GIARDIA* AND *CRYPTOSPORIDIUM*

These pathogenic intestinal protozoa are common in waterborne outbreaks causing diarrhea or gastroenteritis. Person-to-person contact is also considered to be a common mode of transmission. Although these parasites occur in domestic and feral animals, direct animal-to-person transmission appears to be limited. Nevertheless, serious outbreaks of waterborne disease have occurred where drinking waters were not given adequate treatment after the surface-water source was contaminated by animal feces. During the 1970s in the United States, waterborne outbreaks due to *Giardia lamblia* were reported with increasing frequency, especially in community water systems using unfiltered surface-water sources. By the mid-1980s, waterborne outbreaks due to *Cryptosporidium parvum* were occurring as a result of inadequate filtration and disinfection or accidental contamination of water in the distribution system by domestic wastewater. Cryptosporidiosis is a severe, life-threatening illness in patients with AIDS. While a self-limiting diarrheal illness in immunocompetent persons, it is chronic and more severe in immunodeficient persons. No safe and effective form of specific treatment has been identified to date.

The test for *Giardia* cysts consists of filtration of the water being tested (raw water source, at the treatment plant before disinfection or from the distribution system), extraction from the filter material, extract concentration, and microscopic examination by immunofluorescence detection. A large volume of water, usually 100 gal (308 l), is pumped through a polyproylene-wound filter with a 1-µm nominal porosity. After cutting the filter fibers from the supporting core, the retained particulates and cysts are eluted and the extract concentrated by centrifugation. The cysts are separated to some extent from particulate debris by flotation on a Percoll-sucrose solution. A portion of the water/Percoll-sucrose interface is placed on a membrane filter to form a monolayer, stained with an indirect fluorescent antibody, and examined using epifluorescent microscopy.[2] Because slide examination requires subjective judgment, extraneous organisms may be misidentified as *Giardia* cysts. As a result, greater confidence should be given to counts of cysts in which appropriate internal structures have been identified rather than counts including empty organisms.

The test for *Cryptosporidium* oocysts, which is similar to the test for *Giardia* cysts, includes filtration of a large volume of water for extraction, removal of particulates from the filter, concentration by centrifugation, separation of oocysts from debris, and staining with a fluorescent antibody for microscopic detection. Oocysts are confirmed by size, shape, and internal morphological characteristics.

Laboratories conducting tests for *Cryptosporidium* oocysts must be audited and approved for quality assurance. Case histories reveal that boil-water advisories had been unnecessarily issued in two major cities resulting in false public health concerns, anxiety among the citizens, and embarrassment for public officials. The suspected contamination was based on "observed" oocysts in water samples collected in the distribution system, although no cases of human disease were apparent. In one case, the laboratory providing the monitoring data had significant quality assurance and quality control problems, making all the test data suspect. In part, the misidentification of oocysts resulted from algae in both raw and finished water that mimicked oocysts in microscopy. In the second case, a formal audit revealed a poor-quality laboratory situation where the method of analysis was inappropriate, resulting in false results.

Method 1622 and method 1623 were developed by the Environmental Protection Agency (EPA) for simultaneous detection and enumeration of *Giardia* cysts and *Cryptosporidium* oocysts by revised laboratory techniques that use immuno-magnetic separation (IMS) and immunofluorescence assay (FA) on a recommended water sample volume of only 10 l. Method 1622 has been evaluated for detection of *Cryptosporidium* oocysts in stream waters in anticipation of a proposed modification of the surface-water treatment rule by establishing treatment requirements based on the average concentration of *Cryptosporidium* oocysts in surface-water sources. Replicate 10-l stream water samples, both spiked and unspiked, were tested. (A spiked sample has 100–250 oocysts added to the 10-l stream water to increase the number of oocysts for detection. An unspiked sample is the natural stream water.) Oocyst recoveries from spiked samples were always positive but highly variable in the number of oocysts recovered (from lows of less than 5 percent to highs of greater than 20 percent). Few unspiked stream water samples were positive for oocysts. Recovery efficiency and sensitivity of the test method are strongly influenced by the characteristics of the water, particularly interference from turbidity. Consequently, the probability of accurate detection and enumeration of oocysts in natural surface waters is significantly reduced.

3–8 COLIFORM BACTERIA AS INDICATOR ORGANISMS

Testing a water for pathogenic bacteria might at first glance be considered a feasible method for determining microbiological quality. Upon closer examination, testing for pathogens to determine microbiological quality of drinking water has major shortcomings. Laboratory tests for disease-producing bacteria, viruses, and protozoa are difficult to perform and generally are not quantitatively reproducible. Furthermore, the demonstrated absence of one pathogen does not exclude the possible presence of a different pathogen. Most utilities have neither qualified personnel nor laboratories equipped to monitor for pathogens either routinely or even occasionally. Some large utilities with surface-water sources are testing for

Cryptosoridium oocysts and *Giardia* cysts at a frequency of 1 to 4 samples per month. Since pathogens are most likely to enter a water distribution system because of a treatment breakthrough on an intermittent rather than continuous basis, conducting limited monitoring provides little value as a means of protecting public health. In the case of groundwater (not under the influence of surface water), the most likely pathogens are selected species of enteric viruses for which tests are impractical and, for some viruses, impossible to perform. The available methods for detection and identification of human pathogens do not produce credible data for making public health decisions. For these reasons, the microbiological quality of water is based on testing for nonpathogenic indicator organisms, principally the coliform group.

Coliform bacteria, as typified by *Escherichia coli*, reside in the intestinal tract of humans and are excreted in large numbers in feces of humans and warm-blooded animals, averaging about 50 million coliforms per gram. Untreated domestic wastewater generally contains more than 3 million coliforms per 100 ml. Pathogenic bacteria, viruses, and protozoa causing enteric diseases in humans originate from the same source, namely, fecal discharges of diseased persons. Consequently, water contaminated by fecal pollution is identified as being potentially dangerous by the presence of coliform bacteria.

Some genera of the coliform group of bacteria found in water and soil are not of fecal origin but grow and reproduce on organic matter outside the intestines of humans and animals. These coliforms indicate neither fecal contamination nor the possible presence of pathogens. The term *total coliforms* in laboratory testing refers to all coliform bacteria from feces, soil, or other origin. The term *fecal coliforms* refers to coliform bacteria originating from human or warm-blooded animal feces. Fecal coliforms residing in the intestinal tract of humans are generally viewed as being nonpathogenic.

Escherichia coli O157:H7, which is pathogenic to humans, is an antibiotic-resistant mutant strain found in feces of infected cattle. The primary health concerns have been with contaminated ground beef, raw milk, and person-to-person spread. The only waterborne outbreak of *E. coli* O157:H7 occurred in 1987 in a community with a population of 2000. Antibiotic-resistant infections resulted in bloody and nonbloody diarrhea, causing four deaths. The water supply was contaminated with wastewater during repair of water main breaks.

The reliability of coliform bacteria in indicating the presence of pathogens in water depends on the persistence of the pathogens relative to coliforms. For pathogenic bacteria, the die-off rate is greater than it is for coliforms outside the intestinal tract of humans. Thus, exposure in the water environment reduces the number of pathogenic bacteria relative to coliform bacteria. Viruses also reduce in number in the water environment but not necessarily faster than coliforms. Protozoal cysts and helminth eggs are much more persistent than coliform bacteria. For example, the threshold chlorine residual effective in killing bacteria may not inactivate resistant species of enteric viruses and is not effective in inactivating protozoal cysts or harming helminth eggs. In contrast, filtration through natural soils and sand aquifers for a sufficient distance, or granular media in a treatment plant after chemical coagulation, can entrap cysts and eggs because of their relatively large size while allowing viruses and bacteria to be carried through suspended in the water.

Coliforms are a reliable indicator of the safety of the processed surface waters for human consumption provided the treatment includes chemical coagulation and filtration to remove cysts, eggs, and suspended matter for effective chlorination of the clear water to inactivate viruses and kill bacteria. In a similar manner, coliforms can be used as an indicator of water quality for reuse of reclaimed water, provided the biologically treated wastewater is chemically coagulated and filtered to physically remove the protozoal cysts and helminth eggs, followed by long-term chlorination to inactivate viruses and kill bacteria.

Extension of coliform criteria to water quality for purposes other than drinking is poorly defined. Since these bacteria can originate from warm-blooded animals, soil, and cold-blooded animals in addition to feces of humans, presence of coliforms in surface waters indicates any one or a combination of three sources: wastes of humans, farm animals, or soil erosion. Although a special test can be run to separate fecal coliforms from soil types, there is no way of distinguishing between the human bacteria and those of animals. The significance of coliform testing in pollution surveys then depends on a knowledge of the watershed and the most probable source of the observed coliforms.

3–9 TESTS FOR THE COLIFORM GROUP

The coliform group consists of several genera of bacteria in the family Enterobacteriaceae, which includes *Escherichia coli*. The historical definition is this: all aerobic and facultative anaerobic, non-spore-forming, Gram-stain negative rods that ferment lactose with gas production within 48 hr of incubation at 35°C. Laboratory methods for detection of coliforms are the multiple-tube fermentation technique, presence-absence technique, fecal coliform procedure, and membrane filter technique. The following are brief overviews of these methods. For detailed information, refer to *Standard Methods.*[1]

Extensive laboratory apparatus is needed to conduct bacteriological tests, including a hot-air sterilizing oven, an autoclave, incubators, sample bottles, graduated pipettes and pipette containers, a wire inoculating loop, culture media, and preparation utensils. The hot-air oven is used for sterilizing glassware, empty sample bottles, pipettes, and culture dishes. Heating to a temperature of 160 to 180°C for a period of $1\frac{1}{2}$ hr is adequate to kill microbial cells and spores. An autoclave is used to sterilize culture media and dilution water under steam pressure. Recommended operation is for 15 min at 15 psi steam pressure, which corresponds to a temperature of 121.6°C. Sample bottles with a volume of about 120 ml are usually soft glass with screw-top closures suitable to withstand sterilization temperatures. Pipettes used to transfer samples from one container to another are hot-air treated to prevent sample contamination. A small, thin, wire loop of Chromel or platinum is used to aseptically transfer small quantities of culture. The wire is sterilized by heating it red-hot in the flame of a bunsen burner. Culture media are special formulations of organic and inorganic nutrients to support the growth of microorganisms. Dehydrated culture

media (powders) are available commercially. Preparation involves placing a measured amount in distilled water and heating to dissolve without scorching.

Multiple-Tube Fermentation Technique

This technique tests for *total coliforms*, including coliform bacteria from feces, soil, or other origins, and for *fecal coliforms*, which are coliform bacteria from human or warm-blooded animal feces. It is the most reliable technique for enumerating total coliforms and thermotolerant fecal coliforms in wastewater and surface waters. Common usage is to monitor treated wastewater for compliance with a coliform standard and to survey flowing and impounded waters for compliance with a water-quality standard.

Figure 3–9 diagrams the laboratory procedures for determining presence and enumeration of total coliforms and thermotolerant fecal coliforms. Fermentation tubes are glass tubular containers designed to accept either a screw cap or a slip-on stainless steel closure. Inside the fermentation tubes are inverted vials (small tubes) placed upside down in the culture medium to collect gas generated by the bacterial growth.

The culture medium for the total coliform test is lauryl tryptose broth containing tryptose, lactose (milk sugar), and inorganic nutrients. If the water sample added to the lauryl tryptose broth produces gas at 35°C (95°F human body temperature), the test is positive and the sample is presumed to contain coliforms. (Occasionally, noncoliform aerobic, spore-forming bacteria, or a group of bacteria growing together, can produce gas in lauryl tryptose broth.) Fecal coliforms from human and warm-blooded animal feces grow in EC medium, producing lactic acid and gas after 22 to 26 hr of incubation at the elevated temperature of 44.5°C ± 0.2°C (112.1°F). Nonfecal coliforms do not grow at this high temperature. EC medium is an enrichment broth containing tryptose, lactose, bile salts mixture, and inorganic nutrients. A sterile wire loop is used to transfer a tiny drop of growth from a positive lauryl tryptose tube to a fermentation tube of EC medium. Growth with gas production, observed by a bubble in the inverted vial, is a positive test showing the presence of fecal coliforms. If no gas

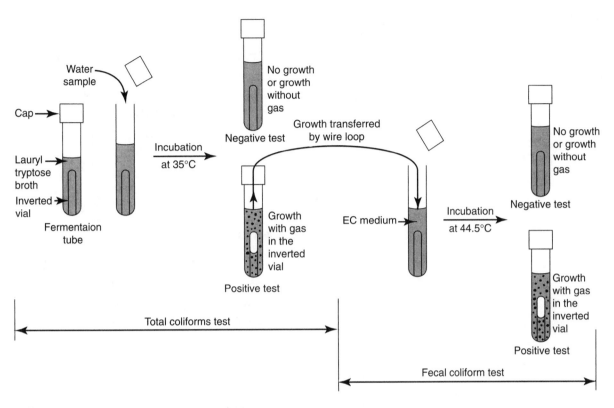

Figure 3–9
Diagram of the multiple-tube fermentation technique for determining and enumerating total coliforms and thermotolerant fecal coliforms present in a treated wastewater or surface-water sample.

appears, the test is negative and no fecal coliform bacteria were present in the positive test for total coliforms.

Bacterial density in a water sample can be determined by serial dilutions in multiple tubes using the fermentaion technique. The result is reported as the most probable number (MPN) based on tabulated probability tables and expressed as MPN Index/100 ml. The common procedure for MPN determination involves the use of sterile pipettes calibrated in 0.1-ml increments, sterile screw-top dilution bottles containing 99 ml of water, and a rack containing six sets of five lauryl tryptose broth fermentation tubes. Figure 3–10 is a sketch showing test preparation. A sterile pipette is used to transfer 1.0-ml portions of the sample into each of five fermentation tubes, followed by dispensing of 0.1 ml to a second set of five. For the next higher dilution, the third, only 0.01 ml of sample water is required, which it is impossible to pipette

accurately. Therefore, 1.0 ml of sample is placed in a dilution bottle containing 99 ml of sterile water and mixed. Now, 1.0-ml portions containing 0.01 ml of the surface-water sample can be pipetted into the third set of five tubes. The fourth set receives 0.1 ml from this same dilution bottle. The process is then carried one more step by transferring 1.0 ml from the first dilution bottle into 99 ml of water in the second for another 100-fold dilution. Portions from this dilution bottle are pipetted into the fifth and sixth tube sets. After incubation for 48 hr at 35°C, the tubes are examined for gas production, and the number of positive reactions for each of the serial dilutions is recorded.

The MPN index and confidence limits for various combinations of positive and negative multiple-tube fermentation results can be determined from Table 3–2. The first three columns in the table refer to the number of positive tubes out of

Figure 3–10

Procedure for preparation of the multiple-tube fermentation technique for determining the most probable number (MPN) of coliform bacteria in a water sample.

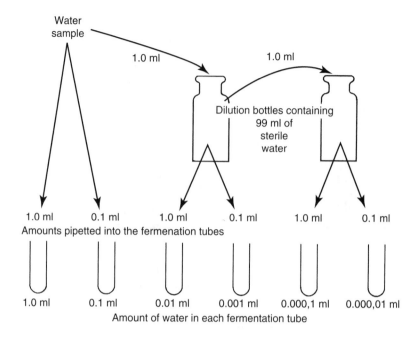

five containing 10-ml, 1-ml, and 0.1-ml sample portions. If smaller amounts of sample were used in the tubes, which is usually the case, the tabulated answers must be adjusted. If 1.0, 0.1, and 0.01 ml are used, the resultant MPN is 10 times the value given in the table; if the sample quantities are 0.1, 0.01, and 0.001 ml, 100 times the tabulated value is recorded; and so on for other combinations. When more than three dilutions are employed in a decimal series of dilutions, the results from only three of these are used in computing the MPN. The three dilutions selected are the highest dilution giving positive results in all five portions tested (no lower dilution giving any negative results) and the two next succeeding higher dilutions.

Serial Dilution in Multiple Tubes

Tests for bacterial density by inoculating fermentation tubes with only one dilution can be used for waters with low bacterial counts, for example, wastewater after tertiary treatment. The MPN Index and confidence limits for various combinations of positive and negative results for 10 fermentation tubes using 10-ml portions are listed in Table 3–3. The MPN Index/100 ml for this procedure is limited to the range of 1.1 to 23.0.

■ EXAMPLE 3–1

The following are results from multiple-tube fermentation analyses of a polluted river-water sample for total and fecal coliforms. The serial dilutions were set up as is illustrated in Figure 3–10. Determine the MPN for both tests.

| | | NUMBER OF POSITIVE REACTIONS OUT OF FIVE TUBES | |
SERIAL DILUTION	SAMPLE PORTION (ML)	LAURYL TRYPTOSE BROTH	EC MEDIUM
0	1.0	5	5
1	0.1	5	4
2	0.01	5	1
3	0.001	1	1
4	0.000,1	0	0
5	0.000,01	0	0

Solution

Total Coliform MPN. The three dilutions selected for MPN determination show 5, 1, and 0 positive tubes. The index value from Table 3–2, without

TABLE 3–2

MPN Index and 95 Percent Confidence Limits for Coliform Counts by the Multiple-Tube Fermentation Technique for Various Combinations of Positive and Negative Results When Five 10-ml, Five 1-ml, and Five 0.1-ml Portions Are Used

No. of Tubes Giving Positive Reaction out of			MPN Index per 100 ml	95 Percent Confidence Limits		No. of Tubes Giving Positive Reaction out of			MPN Index per 100 ml	95 Percent Confidence Limits	
5 of 10 ml Each	5 of 1 ml Each	5 of 0.1 ml Each		Lower	Upper	5 of 10 ml Each	5 of 1 ml Each	5 of 0.1 ml Each		Lower	Upper
0	0	0	<2			4	2	1	26	12	65
0	0	1	2	1.0	10	4	3	0	27	12	67
0	1	0	2	1.0	10	4	3	1	33	15	77
0	2	0	4	1.0	13	4	4	0	34	16	80
1	0	0	2	1.0	11	5	0	0	23	9.0	86
1	0	1	4	1.0	15	5	0	1	30	10	110
1	1	0	4	1.0	15	5	0	2	40	20	140
1	1	1	6	2.0	18	5	1	0	30	10	120
1	2	0	6	2.0	18	5	1	1	50	20	150
						5	1	2	60	30	180
2	0	0	4	1.0	17						
2	0	1	7	2.0	20	5	2	0	50	20	170
2	1	0	7	2.0	21	5	2	1	70	30	210
2	1	1	9	3.0	24	5	2	2	90	40	250
2	2	0	9	3.0	25	5	3	0	80	30	250
2	3	0	12	5.0	29	5	3	1	110	40	300
						5	3	2	140	60	360
3	0	0	8	3.0	24						
3	0	1	11	4.0	29	5	3	3	170	80	410
3	1	0	11	4.0	29	5	4	0	130	50	390
3	1	1	14	6.0	35	5	4	1	170	70	480
3	2	0	14	6.0	35	5	4	2	220	100	580
3	2	1	17	7.0	40	5	4	3	280	120	690
						5	4	4	350	160	820
4	0	0	13	5.0	38	5	5	0	240	100	940
4	0	1	17	7.0	45	5	5	1	300	100	1300
4	1	0	17	7.0	46	5	5	2	500	200	2000
4	1	1	21	9.0	55	5	5	3	900	300	2900
4	1	2	26	12	63	5	5	4	1600	600	5300
4	2	0	22	9.0	56	5	5	5	≤2400		

American Public Health Association. *20th Edition of Standard Methods for the Examination of Water and Wastewater*, APHA, 1998: Section 9221. Reprinted with permission from the American Public Health Association.

regard to the size of the sample portion, is 30. However, since the sample portions corresponding to the 5, 1, and 0 serial dilutions are 0.01, 0.001, and 0.000,1 ml, the MPN value must be multiplied by 1000. Therefore, the MPN index for the total coliforms test series is 30,000, with 95 percent confidence limits of 10,000 and 120,000. (MPN is the most probable number in 100 ml of

TABLE 3–3

MPN Index and 95 Percent Confidence Limits for Various Combinations of Positive and Negative Results in Ten Fermentation Tubes When 10-ml Portions Are Used

No. of Tubes Out of Ten Giving Positive Reaction	MPN Index per 100 ml	95 Percent Confidence Limits	
		Lower	Upper
0	<1.1	0	3.0
1	1.1	0.03	5.9
2	2.2	0.26	8.1
3	3.6	0.69	10.6
4	5.1	1.3	13.4
5	6.9	2.1	16.8
6	9.2	3.1	21.1
7	12.0	4.3	27.1
8	16.1	5.9	36.8
9	23.0	8.1	59.5
10	>23.0	13.5	Infinite

American Public Health Association. *20th Edition of Standard Methods for the Examination of Water and Wastewater*, APHA, 1998: Section 9221. Reprinted with permission from the American Public Health Association.

water, and the confidence limits mean that 95 percent of all MPN analyses run on the same sample should, statistically speaking, fall within this range of limiting values.)

Fecal Coliform MPN. Selected dilutions are 5, 4, and 1, with a correction factor of 10. MPN equals 1700, with confidence limits of 700 to 4800.

■ ■ ■

Presence-Absence Technique

The presence-absence (P-A) test can determine if coliform bacteria or *Escherichia coli* are present in a water sample without indicating the number of coliforms in a positive result. This test is intended for use in routine monitoring of drinking water immediately after treatment and in the distribution system pipe network. This test is based on the concept that the MCL for coliform bacteria is zero.

The drinking water to be tested is collected in a sterile bottle containing 15–30 mg of sodium thiosulfate, which is sufficient to neutralize and stop

the disinfecting action of 10 mg/l of chlorine. The test procedure starts by vigorously shaking the sample bottle to suspend the bacteria and the particulate matter. The lid of the P-A culture bottle is aseptically removed to add 100 ml of the water sample and to add the contents of one packet of Colilert® reagent. The reagent contains bacterial nutrients and two special compounds (ONPG, O-nitrophenyl-β-d-galactopyranoside, and MUG, 4-methylumbelliferyl-β-d-glucuronide) that indicate the growth and presence of coliform bacteria and *E. coli*. (Colilert® is based on IDEXX's patented Defined Substrate Technology®.)

The culture bottle is incubated for 24 hr at 35°C. Growth of coliform bacteria in the Colilert® uses the enzyme β-galactosidase to metabolize ONPG and change the color of the sample from clear to yellow. If a yellow color does not appear, the test is negative and no coliforms were present in the water sample. *E. coli* uses the enzyme β-galactosidase to metabolize MUG and create fluorescence under 365-nanometer ultraviolet light. (Fluorescence is the emission of radiation as visible light resulting from the absorption of radiation from another source.) If the yellow-color broth glows fluorescent with a bluish color, the test is positive for *E. coli*. Absence of a blue color is a negative test.

Colilert® can also be used for a serial-dilution MPN analysis using a Quanti-Tray™ for enumeration of total coliform and *E. coli*. After preparing the 100-ml water sample with Colilert® reagent in a P-A culture bottle, aseptically remove the lid and pour the sample into a Quanti-Tray, filling the 10 wells, and seal the tray. Incubate for 24 hr at 35°C. Count the number of wells with yellow color (for total coliform) and the number of wells that fluoresce (for *E. coli*). For 10 fermentation wells, each with 10-ml portions, the MPN Index and confidence limits for various combinations of positive and negative results are listed in Table 3–3.

■ EXAMPLE 3–2

The effluent from a tertiary wastewater treatment plant was tested for both total coliforms and fecal coliforms. Ten fermentation tubes with lauryl tryptose broth were inoculated with 10-ml portions of a 100-ml sample and incubated at 35°C.

Four of the 10 tubes showed growth with gas production. Growths from these 4 positive tubes were transferred to 4 fermentation tubes containing EC medium and incubated at 44.5°C. After 22 hr, 2 of the 4 tubes showed gas production. What are the MPN Indexes for total coliforms and fecal coliforms? What are the 95 percent confidence limits for the fecal-coliform index?

Solution

From Table 3–3 for 4 of 10 positive reactions, the count for total coliforms is 5.1 MPN Index/100 ml.

For 2 out of 10 positive reactions, the count for fecal coliforms is 2.2 MPN Index/100 ml in the confidence limits of 0.26 and 8.1.

■ ■ ■

Sampling and Testing for Different Waters

Coliform tests are performed as soon as possible after collection. If laboratory processing cannot be started within an hour, the samples are stored in a refrigerator or ice chest at a temperature less than 10°C immediately after collection. The recommended maximum time in cold storage is 6 hr. The exception is for community drinking-water samples that must be sent by mail or bus to a central laboratory. In this case, the sample can be placed in an ice pack for up to 30 hr prior to testing.

The typical sampling bottle for drinking water is a brown-glass, wide-mouth bottle with a screw cap and a volume of 120 ml to collect a 100-ml sample for testing. The bottle is sterilized in an autoclave with a few drops of dechlorinating chemical inside to neutralize up to 15 mg/l of residual chlorine.

The best water faucet in a distribution system for collecting a sample is a clean, single, cold-water inside faucet that is supplied water directly from the service pipe connected to the community main. Avoid the following: faucets where the water passes through a household softener, filter, tank, or cistern; leaking faucets; swivel faucets and faucets with screens; and outside hydrants and sill cocks. If a cold- and hot-water mixing faucet must be used, remove the screen if possible and run only the hot water for 2 min, then, turn off the hot water and turn on the cold water for flushing and sampling. Neither washing a faucet with bleach nor heating with a flame are recommended.

With the proper faucet selected for collecting the sample, open the faucet fully and let the cold water run to waste for 2 to 3 min or for a time sufficient to allow clearing of the service pipe. Then, partially close the faucet to reduce the flow and collect the sample. Keep the sample bottle closed until it is to be filled, remove the cap, fill the bottle (without rinsing) to the base of the bottle neck, leaving an air space at the top, and replace the cap immediately. During collection, be careful not to touch, or allow any object to contact, the inside of the cap, inside of the bottle, or threaded neck of the bottle. Do not set the cap down. The person collecting the sample should wash his or her hands before opening the bottle. The strict maximum contaminant level standards for coliform contamination of drinking water do not allow any tolerance for unintentional sample contamination by the person collecting the sample.

The presence-absence technique is the coliform test recommended by the Environmental Protection Agency (EPA) for monitoring the microbiological quality of drinking water. If the test is positive for the coliform group, the yellow color in the P-A broth is tested by fluorescence to determine if the coliform growth is E. coli.

Wastewater effluent after biological treatment and chlorination is normally limited to a microbiological standard of a fecal coliform count of 200 per 100 ml. The preferred test is the multiple-tube fermentation technique, with the first phase for total coliforms and the second phase for fecal coliforms. An alternate is the membrane filter technique.

Wastewater effluent after tertiary treatment with chemical coagulation and granular-media filtration is commonly limited to a median coliform count of 2.2 per 100 ml, with a maximum in any sample not to exceed 23 per 100 ml. The usual testing procedure is serial dilution in multiple-tube fermentation using 10 tubes, each inoculated with 10 ml from a 100-ml sample.

Surveys of surface waters in lakes and rivers enumerate total coliforms and/or fecal coliforms to monitor water quality. For example, numerical coliform criteria for body-contact recreation are often established at the upper limit of 200 fecal coliforms per 100 ml and 1000 total coliforms per 100 ml. The common testing procedures are the multiple-tube fermentation technique and the membrane filter technique.

3–10 BIOCHEMICAL OXYGEN DEMAND

Biochemical oxygen demand (BOD) is the most commonly used parameter to define the strength of a municipal or organic industrial wastewater. Its widest application is in measuring waste loadings to treatment plants and in evaluating the efficiency of such treatment systems. In addition, the BOD test is used to determine the relative oxygen requirements of treated effluents and polluted waters. However, it is of limited value in measuring the actual oxygen demand of surface waters, and extrapolation of test results to actual stream oxygen demands is highly questionable, since the laboratory environment cannot reproduce the physical, chemical, and biological stream conditions.

BOD is by definition the quantity of oxygen utilized by a mixed population of microorganisms in the aerobic oxidation (of the organic matter in a sample of wastewater) at a temperature of $20°C \pm 1°C$ in an air incubator or water bath. A plastic cup or foil cap is placed over the flared mouth of the BOD bottle during incubation to reduce evaporation of the water seal. Measured amounts of a wastewater, diluted with prepared water, are placed in 300-ml BOD bottles (Figure 3–11). The dilution water, containing phosphate buffer (pH 7.2), magnesium sulfate, calcium chloride, and ferric chloride, is saturated with dissolved oxygen. Seed microorganisms are supplied to oxidize the waste organics if sufficient microorganisms are not already present in the wastewater sample. The general biological reaction that takes place is Eq. 3–12. The wastewater supplies the organic matter (biological food), and the dilution water furnishes the dissolved oxygen. The primary reaction is metabolism of the organic matter and uptake of dissolved oxygen by bacteria, releasing carbon dioxide and producing a substantial increase in bacterial population. The secondary reaction results from the oxygen used by the protozoa-consuming bacteria, a predator-prey

reaction. Depletion of dissolved oxygen in the test bottle is directly related to the amounts of degradable organic matter. The BOD of a wastewater where microorganisms are already present in the sample, requiring no outside seed, is calculated using Eq. 3–13. (The standard test has an incubation period of 5 days at 20°C.)

$$\text{Organic matter} \xrightarrow[\text{bacteria}]{\text{dissolved oxygen}} CO_2 + \text{bacterial cells} \xrightarrow[\text{protozoa}]{\text{dissolved oxygen}} CO_2 + \text{protozoal cells} \quad \text{(3–12)}$$

$$BOD = \frac{D_1 - D_2}{P} \quad \text{(3–13)}$$

where BOD = biochemical oxygen demand, milligrams per liter

D_1 = initial DO of the diluted wastewater sample about 15 min after preparation, milligrams per liter

D_2 = final DO of the diluted wastewater sample after incubation for 5 days, milligrams per liter

P = decimal fraction of the wastewater sample used

$= \dfrac{\text{milliliters of wastewater sample}}{\text{milliliter volume of the BOD bottle}}$

The biochemical oxygen demand of a wastewater is in reality not a single point value but time dependent. The curve in Figure 3–12 shows BOD exerted, dissolved oxygen depleted, as the biological reactions progress with time. Carbonaceous oxygen demand, Eq. 3–12, progresses at a decreasing rate with time, since the rate of biological activity decreases as the available food supply diminishes. The shape of the hypothetical curve is best expressed mathematically by first-order kinetics in the form of Eq. 3–14, where t is time in days and k is a rate constant. Based on this equation for a k equal to 0.1 per day, a common value for domestic waste, 68 percent of the ultimate carbonaceous BOD is exerted after 5 days.

$$\begin{array}{c}\text{BOD at any} \\ \text{time, } t\end{array} = \begin{array}{c}\text{ultimate} \\ \text{BOD}\end{array}(1 - 10^{-kt}) \quad \text{(3–14)}$$

Figure 3–11

Essential constituents in the biochemical oxygen demand (BOD) test are a measured amount of the wastewater sample being analyzed, prepared dilution water containing dissolved oxygen, and seed microorganisms if these are not present in the wastewater.

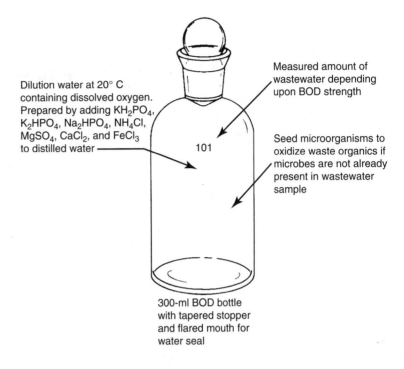

Dilution water at 20° C containing dissolved oxygen. Prepared by adding KH_2PO_4, K_2HPO_4, Na_2HPO_4, NH_4Cl, $MgSO_4$, $CaCl_2$, and $FeCl_3$ to distilled water

Measured amount of wastewater depending upon BOD strength

Seed microorganisms to oxidize waste organics if microbes are not already present in wastewater sample

101

300-ml BOD bottle with tapered stopper and flared mouth for water seal

Figure 3–12

Hypothetical biochemical oxygen demand reaction showing the carbonaceous and nitrification demand curves.

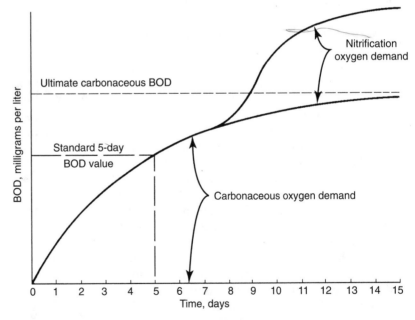

Nitrification oxygen demand

Ultimate carbonaceous BOD

Standard 5-day BOD value

Carbonaceous oxygen demand

BOD, milligrams per liter

Time, days

Nitrifying bacteria can exert an oxygen demand in the BOD test as in Eqs. 3–7 and 3–8. Fortunately, the growth of nitrifying bacteria lags behind that of the microorganisms performing the carbonaceous reaction. Nitrification generally does not occur until several days after the standard 5-day incubation period for BOD tests on untreated wastewaters. Treatment plant effluents and stream waters may show early nitrification where the sample has a relatively high population of nitrifying bacteria. The standard method recommended for preventing nitrification is the addition of 10 mg/l of 2-chloro-6 (trichloromethyl) pyridine to the dilution water to stop nitrate formation.

The degree of reproducibility of the BOD test cannot be defined precisely because of variations that occur in bacterial decomposition of various organic substances. Laboratory technicians can periodically evaluate their procedures by conducting BOD analyses on a standard solution of glucose and glutamic acid containing 150 mg/l of each component. The mean BOD value should fall in the range of 200 ± 30 mg/l, and the *k*-rate in the bounds of 0.16 to 0.19. This represents a deviation of about ±15 percent from the average value. Tests on real wastewaters normally show observations varying from 10 to 20 percent on either side of the mean. The larger variations occur in testing industrial wastewaters that require seeding or contain substances that inhibit biological activity and in treatment plant effluent samples that are affected by nitrification.

Contaminated dilution water and dirty incubation bottles can distort BOD test results. *Standard Methods*[1] emphasizes proper cleaning of the bottles and analysis of dilution water blanks during BOD testing as a check on the quality of unseeded dilution water. The DO uptake for unseeded dilution water with phosphate buffer and inorganic nutrient solutions added should not be more than 0.2 mg/l and preferably not more than 0.1 mg/l after 5 days incubation at 20°C. If the quality of the dilution water is questionable, it can be checked for acceptability before using it in BOD analyses. The procedure is to add sufficient seed material to the dilution water being tested to produce a DO uptake of 0.05 to 0.1 mg/l in 5 days. The actual oxygen utilization should not exceed 0.1 to 0.2 mg/l.

No uniform relationship exists between the COD and BOD of wastewaters, except that the COD value must be greater than the BOD. This is because chemical oxidation decomposes nonbiodegradable organic matter, and the standard BOD test measures only the oxygen used in metabolizing the organic matter for 5 days. The correlation of COD to BOD for a particular wastewater can be determined by statistical comparison of several laboratory analyses. Unfortunately, such a relationship may be invalidated by the simple day-to-day variations in quality of a municipal wastewater. Occasionally, out of the necessity for converting oxygen demand data, even if the results are of questionable accuracy, the COD of a soluble wastewater is assumed to be numerically equal to the ultimate BOD.

BOD of Municipal Wastewater

Following is the description of a typical BOD test on a municipal wastewater. The first step is to determine the portion of sample to be placed in each BOD bottle. The information in either Table 3–4 or Eq. 3–13 may be used to select appropriate amounts. For example, for a wastewater with an estimated BOD of 350 mg/l, Table 3–4 indicates sample volumes in the range of 2 to 5 ml per 300-ml bottle would be appropriate; or substituting into Eq. 3–13 a BOD equal to 350 mg/l, a bottle volume of 300 ml, and a desired dissolved oxygen decrease of 5 mg/l, the calculated wastewater portion is 4.3 ml. For a valid BOD test, at least 2 mg/l of dissolved oxygen should be consumed, but the final DO should not be less than 1 mg/l. Since the initial dissolved oxygen in dilution water is about 8 mg/l, an average amount of 5 mg/l is available for biological uptake between the minimum desired 2 mg/l and the maximum of 8 mg/l.

TABLE 3–4

Suggested Wastewater Portions and Dilutions in Preparing BOD Tests

By Direct Measurement of Wastewater into a 300-ml BOD Bottle		By Mixing Wastewater into Dilution Water [Wastewater Volume / Total Volume of Mixture]	
Wastewater (ml)	Range of BOD (mg/l)	Percentage of Mixture	Range of BOD (mg/l)
0.20	3000 to 10,500	0.10	2000 to 7000
0.50	1200 to 4200	0.20	1000 to 3500
1.0	600 to 2100	0.50	400 to 1400
2.0	300 to 1050	1.0	200 to 700
5.0	120 to 420	2.0	100 to 350
10.0	60 to 210	5.0	40 to 140
20.0	30 to 105	10.0	20 to 70
50.0	12 to 42	20.0	10 to 35
100	6 to 21	50.0	4 to 14

TABLE 3–5

Typical BOD Data from the Analysis of a Municipal Wastewater to Determine the 5-Day BOD Value and Estimate the *k*-Rate of the Biological Reaction[a]

Bottle Number	Wastewater Portion (ml)	Initial DO (mg/l)	Incubation Period (days)	Final DO (mg/l)	DO Drop (mg/l)	Calculated BOD (mg/l)
		BOD Tests to Determine Average 5-Day Value				
1	2.0	8.3	0			
2	4.0	8.4	0			
3	6.0	8.4	0			
		Average = 8.4				
4	2.0	8.4	5.0	5.9	2.5	375
5	2.0	8.4	5.0	6.0	2.4	360
6	4.0	8.4	5.0	3.8	4.6	345
7	4.0	8.4	5.0	3.5	4.9	365
8	6.0	8.4	5.0	0	(Invalid test)	
9	6.0	8.4	5.0	0	(Invalid test)	
						Average = 360
		BOD Analyses to Permit Calculation of k-Rate				
10	4.0	8.4	0.5	7.2	1.2	90
11	4.0	8.4	0.5	7.4	1.0	75
12	4.0	8.4	1.0	6.2	2.2	165
13	4.0	8.4	1.0	5.9	2.5	190
14	4.0	8.4	2.0	5.2	3.2	240
15	4.0	8.4	2.0	5.2	3.2	240
16	4.0	8.4	3.0	4.4	4.0	300
17	4.0	8.4	3.0	4.6	3.8	285

[a]Results are plotted in Figure 3–13. All BOD bottles had a volume of 300 ml.

The distilled water for dilution is stabilized at 20°C by placing it in the BOD incubator. It is then removed and prepared by aeration and addition of the four buffer and nutrient solutions (Figure 3–11). Outside seed microorganisms are not required for municipal wastewater, since they are already present in the sample. To ensure that some of the bottles prepared provide valid test data, three different dilutions of two or three bottles each are set up, plus three bottles for initial dissolved oxygen measurement and three blanks of dilution water only. Measured portions of the wastewater are pipetted directly into empty BOD bottles, which are then filled with prepared dilution water by siphoning through a hose to avoid entraining air. The sample preparation and dissolved oxygen measurements for this illustration are summarized in Table 3–5. Wastewater volumes of 2.0 ml, 4.0 ml, and 6.0 ml were used. The first three bottles were titrated immediately for dissolved oxygen, using the azide modification of the iodometric method. The other nine test bottles and three blank bottles of dilution water were incubated for 5 days at 20°C and then were titrated for remaining dissolved oxygen. The BOD value for each bottle was calculated using Eq. 3–13. The average BOD for four of the six tests was 360 mg/l; two bottles were considered invalid, since the dissolved oxygen was depleted prior to the end of the 5-day incubation period. The three blanks showed less than 0.2 mg/l of DO uptake.

Figure 3–13

BOD-time curve and graphical determination of the k-rate for the wastewater data given in Table 3–5.

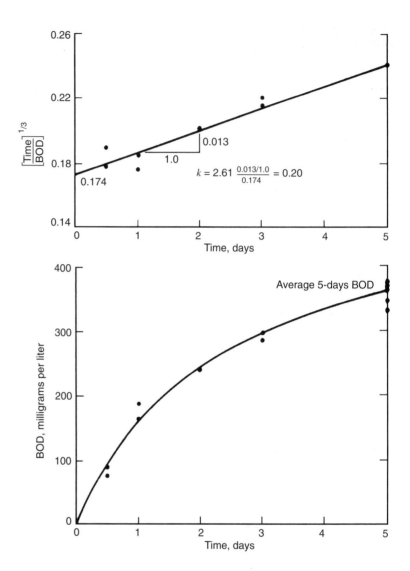

Determination of BOD k-Rate

The rate constant k in Eq. 3–14 can be computed from BOD values measured at various times. Bottles numbered 10 to 17 of the analyses in Table 3–5 were analyzed after incubation periods ranging from 0.5 to 3.0 days. These data along with the average 5-day value are plotted in Figure 3–13. The shape of the BOD-time curve is typical of a municipal wastewater derived primarily from domestic sources. The upper portion of Figure 3–13 illustrates a graphical method for estimating the k-rate. Values of the cube root of time in days over BOD in milligrams per liter are calculated from the laboratory data and are plotted as ordinates against the corresponding times. The best-fit line drawn

through these points is used to calculate the k-rate by the following relationship:

$$k = 2.61 \frac{B}{A} \qquad (3-15)$$

where k = rate constant, per day
A = intercept of the line on the ordinate axis
B = slope of the line

BOD Analysis of Industrial Wastewaters

Most organic wastes from food-processing industries and other sources that are susceptible to biological decomposition can be tested for BOD. However, particular care must be taken in properly neutralizing the wastewaters, seeding the test bottles or dilution water with microorganisms from aged wastewater or waste-polluted stream water, and using sufficient dilution so that the effect of any toxicity is reduced and the maximum BOD value is obtained. In addition to careful collection and compositing of samples according to wastewater flow variation, special data should be recorded for industrial wastes. The industry's production—namely, the type and quantity of product manufactured—and specific operational conditions existing during the sampling period should be recorded for correlation with the quantity and strength of the wastewater produced. Several composite samples over an extended period of time are often required for industries with variable production schedules to key the amount of the wastes produced to the quantity of product manufactured.

Initial pretreatment is to neutralize, if necessary, the sample to pH 7.0 with sulfuric acid or sodium hydroxide to remove caustic alkalinity or acidity. The pH of the dilution water should not be changed by addition of the wastewater in preparing a BOD test bottle of lowest dilution. Samples containing residual chlorine must be dechlorinated prior to setup. Often, the residual dissipates if samples are allowed to stand for 1 or 2 hr, but higher chlorine residuals should be destroyed by adding sodium sulfite solution. Industrial wastewaters containing other toxic substances require special study and pretreatment. In extreme cases where a technique for neutralizing the toxic chemical cannot be developed, BOD testing is abandoned and replaced with COD analyses.

Unknown substances that inhibit biological growth are often detected by carefully conducted BOD tests. Table 3–6 and Figure 3–14 summarize the results from analysis on a food-processing wastewater containing an interfering compound. The source of the toxicity was a bactericide used in disinfecting production pipelines and tanks. The first evidence of the potential problem is revealed by comparing the rather wide range of BOD values observed in test bottles at the same dilution—at 0.67 percent, the numbers range from 220 mg/l to 485 mg/l. More important, the tests show increasing BOD values with increasing dilutions. The concentration of the toxic chemical, and consequently the inhibition of biological activity, is greater in lower dilutions (3.0 ml

TABLE 3–6

BOD Data from the Analysis of a Food-Processing Wastewater Containing a Bactericidal Agent Causing an Increase in BOD Value with Increasing Dilution[a]

Wastewater Portion (ml)	Percentage Dilution of Wastewater	Measured 5-day BOD (mg/l)
3.0	1.00	240
3.0	1.00	205
3.0	1.00	250
3.0	1.00	145
		Average = 210
2.0	0.67	365
2.0	0.67	485
2.0	0.67	315
2.0	0.67	220
		Average = 350
1.0	0.33	520
1.0	0.33	440
1.0	0.33	565
1.0	0.33	520
		Average = 510

[a]Figure 3–14 is a graphical presentation of the results.

wastewater in a 300-ml bottle) than at higher dilutions (1.0 ml/300 ml). The reported BOD should be the highest value obtained in valid tests, this being the average of the highest dilutions providing a minimum uptake of 2.0 mg/l dissolved oxygen.

Very few industrial wastewaters have sufficient biological populations to perform BOD testing without providing an acclimated seed. The ideal seed is a mixed culture of bacteria and protozoa adapted to decomposing the specific industrial wastewater organics, with a low number of nitrifying bacteria. Microorganisms for food-processing wastes and similar organics can be obtained from aged untreated domestic wastewater. The seed material is the supernatant liquid from a sample of domestic wastewater that has been allowed to age and settle in an open container for about 24 hr at room temperature. Biota for industrial wastewaters with biodegradable organic compounds not abundant in municipal wastewater should be obtained from a source having microorganisms

Figure 3–14

Plot of BOD values, given in Table 3–6, showing increasing BOD for the same industrial sample set up at increasing dilutions. The sample was a food-processing wastewater containing a biologically inhibiting substance.

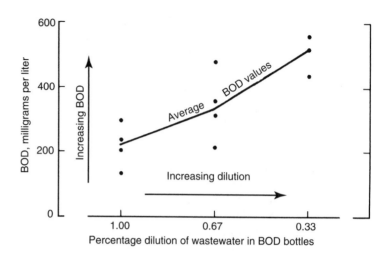

acclimated to these wastewaters. If the industrial discharge is being treated by a biological system, activated sludge from an aeration basin, slime growth from a trickling filter, or water from a stabilization pond contains acclimated microorganisms. When the wastewater is disposed of in a watercourse, the stream water several miles below the point of discharge may be a good source for seed. An adapted seed culture can be developed by using a small laboratory draw-and-fill activated-sludge unit on a mixed feed of industrial and domestic wastewater if a natural source of microbial seed is not readily available or proves to be unsatisfactory.

The amount of seed material added to the dilution water, or each individual BOD bottle, must be sufficient to provide substantial biological activity without adding too much organic matter. When aged domestic or polluted water is used, the rule of thumb is to add an amount of seed wastewater such that 5 to 10 percent of the total BOD exerted by the sample results from oxygen demand of the seed alone. For example, assume that the water being used for seed has an estimated BOD of 150 mg/l. Based on Table 3–4, a 10-ml volume per bottle would be used to run a BOD test on this wastewater; however, for seeding purposes, only 5 to 10 percent of this amount is desired. Hence, 0.5 to 1.0 ml should be added in setting up each industrial wastewater test bottle. If the culture is to be placed in the dilution water, rather than directly into each BOD bottle, the percentage-of-mixture data from Table 3–4 is helpful.

To conduct a BOD test on a 150-mg/l waste, a 3.5 percent mixture is recommended. Therefore, seeded dilution water should be between 0.17 and 0.35 percent seed wastewater, prepared by adding 1.7 and 3.5 ml to each liter of dilution water.

Oxygen demand of the seed is compensated for in computing BOD of a seeded industrial wastewater test by using Eq. 3–16. The terms D_1 and D_2 are the initial and final dissolved oxygen concentrations in the seeded wastewater bottles containing an industrial wastewater fraction of P. The terms B_1 and B_2 are initial and final dissolved oxygen values from a separate BOD test on the seed material, and f is the ratio of seed volume used in the industrial wastewater test to the amount used in the test on the seed. Hence, $(B_1 - B_2)f$ is the oxygen demand of the seed.

$$\text{BOD} = \frac{(D_1 - D_2) - (B_1 - B_2)f}{P} \qquad (3\text{–}16)$$

where D_1 = DO of diluted seeded wastewater sample about 15 min after preparation

D_2 = DO of wastewater sample after incubation

B_1 = DO of diluted seed sample about 15 min after preparation

B_2 = DO of seed sample after incubation

f = ratio of seed volume in seeded wastewater test to seed volume in BOD test on seed

$= \dfrac{\text{percentage or milliliters of seed in } D_1}{\text{percentage or milliliters of seed in } B_1}$

P = decimal fraction of wastewater sample used

$$= \frac{\text{volume of wastewater}}{\text{volume of dilution water plus wastewater}}$$

A time lag, resulting from insufficient seed, unacclimated microorganisms, or the presence of inhibiting substances, often occurs in the oxygen-demand reaction on an industrial wastewater. To determine BOD accurately, existence of a time lag must be identified during the first few days of incubation. This can be accomplished by either preparing extra bottles for dissolved oxygen testing after $\frac{1}{2}$, 1, 2, and 3 days or, if a dissolved oxygen probe is available, monitoring the rate of oxygen depletion in two or three bottles each day during incubation. Figure 3–15 illustrates dissolved oxygen testing on a chemical-manufacturing wastewater that exhibits a 2-day lag period. The most probable BOD can be approximated in a test of this nature by reestablishing time zero at the end of the lag period and by measuring the 5-day value based on this new origin.

Data throughout the incubation period may also reveal other irregularities influencing selection of a 5-day value. Figure 3–16a is an effluent from an extended aeration plant. Because of the high population of nitrifying bacteria, nitrification started on the third day of incubation and exerted approximately one-half of the total oxygen demand by the fifth day of the test. The diphasic curve in Figure 3–16b is a result of rapid oxygen uptake during the first 2 days caused by a large amount of discharge from a pudding factory into the municipal sewer. Since soluble pudding waste is more easily decomposed by bacteria, it is metabolized first at a very high rate $(k_1 = 0.65)$ while the remaining domestic waste organics continue their oxygen demand at a slower rate throughout the duration of the test.

Industrial wastes frequently have high strengths that make it difficult, or even impossible, to pipette accurately the small quantity desired for a single test bottle. In this case, the wastewater can be diluted by serial dilution to volumes that can be accurately measured. For example, only 0.5 ml of a 3000-mg/l BOD waste is required per BOD bottle. However, if 100 ml of the 3000-mg/l waste is diluted to 1000 ml with distilled water, 5.0 ml of the mixture can be pipetted accurately into each bottle. Wastewaters high in suspended solids may be difficult to mix with water; one alternative is to homogenize the sample in a blender to aid dispersion in the dilution water.

The purpose of this discussion is to present some of the problems and pitfalls in BOD testing of industrial wastewaters. Although difficult to perform, these tests are common in evaluating manufacturing wastes to assess sewer use fees or treatment plant loadings. Improper seeding and unrecognized influence of inhibiting substances cause erroneous results that lead to confusion and consternation. Although industrial wastewaters

Figure 3–15

Reaction-rate curve from a seeded BOD test on a chemical-manufacturing wastewater exhibiting a time lag in the biological reaction.

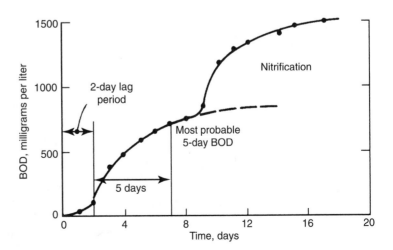

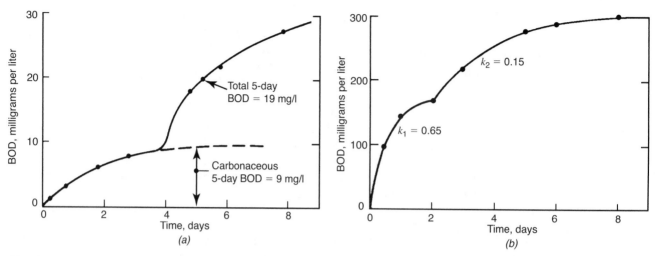

Figure 3–16

BOD analyses showing early nitrification and a diphasic carbonaceous stage. (a) BOD curve of treated wastewater from an extended aeration plant showing nitrification starting about the fourth day of incubation. (b) BOD curve of raw municipal wastewater exhibiting a diphasic carbonaceous stage resulting from a large amount of soluble pudding-manufacturing waste combined with the domestic wastewater.

occasionally defy biological testing, more often they can be tested with reasonable accuracy by following very careful analytical procedures.

■ EXAMPLE 3–3

Data from an unseeded domestic wastewater BOD test are 5.0 ml of wastewater in a 300-ml bottle, initial DO of 7.8 mg/l, and 5-day DO equal to 4.3 mg/l. Compute (a) the BOD and (b) the ultimate BOD, assuming a k-rate of 0.10 per day.

Solution

(a) From Eq. 3–13,

$$\text{BOD} = \frac{7.8 - 4.3}{5.0/300} = 210 \text{ mg/l}$$

(b) Using Eq. 3–14,

$$\text{Ultimate BOD} = \frac{\text{BOD}}{1 - 10^{-kt}} = \frac{210}{1 - 10^{-0.1 \times 5.0}}$$

$$= \frac{210}{1 - 0.32} = 310 \text{ mg/l}$$

■ ■ ■

■ EXAMPLE 3–4

A seeded BOD test is to be conducted on meat-processing wastewater with an estimated strength of 800 mg/l. The seed is supernatant from aged, settled, domestic wastewater with a BOD of about 150 mg/l. (a) What sample portions should be used for setting up the middle dilutions of the wastewater and seed tests? (b) Calculate the BOD value for the industrial wastewater if the initial DO in both seed and sample bottles is 8.5 mg/l and the 5-day DOs are 4.5 mg/l and 3.5 mg/l for the seed test bottle and seeded wastewater sample, respectively.

Solution

(a) Based on data in Table 3–4,

Volume required for BOD test on 150-mg/l seed = 10.0 ml pipetted directly into a 300-ml bottle

Volume of meat-processing wastewater required, assuming 800 mg/l = 2.0 ml, pipetted into a BOD bottle

Amount of aged wastewater for seeding wastewater sample = 0.10 × 10.0 = 1.0 ml pipetted into the BOD bottle, or the addition of 3.33 ml/l of dilution water for a mixture of 0.33 percent

(b) Substituting into Eq. 3–16,

$$\text{BOD} = \frac{(8.5 - 3.5) - (8.5 - 4.5)1.0/10.0}{2.0/300}$$

$$= 690 \text{ mg/l}$$

■ ■ ■

3–11 BIOLOGICAL TREATMENT SYSTEMS

Biological processing is the most efficient way of removing organic matter from municipal wastewaters. These living systems rely on mixed microbial cultures to decompose and remove colloidal and dissolved organic substances from solution. The treatment chamber holding the microorganisms provides a controlled environment; for example, activated sludge is supplied with sufficient oxygen to maintain an aerobic condition. Wastewater contains the biological food, growth nutrients, and inoculum of microorganisms. Persons who are not familiar with wastewater operations often ask where the "special" biological cultures are obtained. The answer is that the wide variety of bacteria and protozoa present in domestic wastes seeds the treatment units. Then, by careful control of wastewater flows, recirculation of settled microorganisms, oxygen supply, and other factors, the desirable biological cultures are generated and retained to process the pollutants. The slime layer on the surface of the media in a trickling filter is developed by spreading wastewater over the bed. Within a few weeks, the filter is operational, removing organic matter from the liquid trickling through the bed. Activated sludge in a mechanical, or diffused-air, system is started by turning on the aerators and feeding the wastewater. Initially, a high rate of recirculation from the bottom of the final clarifier is necessary to retain sufficient biological culture. However, within a short period of time a settleable biological floc matures that efficiently flocculates the waste organics. An anaerobic digester is the most difficult treatment unit to start up, since the methane-forming bacteria, essential to digestion, are not abundant in raw wastewater. Furthermore, these anaerobes grow very slowly and require optimum environmental conditions. Startup of an anaerobic digester can be hastened considerably by filling the tank with wastewater and seeding with a substantial quantity of digesting sludge from a nearby treatment plant. Raw sludge is then fed at a reduced initial rate, and lime is supplied as necessary to hold pH. Even under these conditions, several months may be required to get the process fully operational.

Enzymes are organic catalysts that perform biochemical reactions at temperatures and chemical conditions compatible with biological life. Chemical decomposition of potato or meat requires boiling in a strong acid solution, as performed in the COD test. However, these same foods can be readily digested by microorganisms, or in the stomach of an animal, at a much reduced temperature without strong mineral acids through the action of enzymes. Most enzymes cannot be isolated from living organisms without impairing their functioning capability. Although a discussion of enzymes is beyond the scope of this book, technicians in the field of sanitary science must be aware that enzyme additives sold to enhance biological treatment processes are ineffective. The label on the container generally uses highly scientific terms to convince the purchaser of product worthiness, for example, "enzymes for wastewater" (or "for anaerobic digestion, stabilization ponds, septic tanks," etc.), "minimum of 10 billion colonies per gram," "excellent diastic, proteolytic, amylolytic, and lipolytic activity," "a special formulation of enzymes, aerobic and anaerobic bacteria," and the like. In reality, domestic wastewater contains an abundant supply of all these enzymes, and to pour in more at excessive cost can be figuratively described as pouring money down the drain.

Factors Affecting Growth

The most important factors affecting biological growth are temperature, availability of nutrients, oxygen supply, pH, presence of toxins, and, in the case of photosynthetic plants, sunlight. Bacteria are classified according to their optimum temperature range for growth. Mesophilic bacteria grow in a temperature range of 10 to 40°C, with an optimum of 37°C. Aeration tanks and trickling filters generally operate in the lower half of this range, with wastewater temperatures of 20 to 25°C in warm climates and 8 to 10°C during the winter in northern regions. If cold well water serves as a water supply, wastewater temperatures can be lower than 20°C during the summer, and winter operation during extremely cold weather may result in ice formation on the surface of final clarifiers and freezing of stabilization ponds. Anaerobic digestion tanks are normally heated to near the optimum level of 35°C (98°F).

The rate of biological activity generally doubles or halves for every 10 to 15°C temperature rise or decrease within the range of 5 to 35°C. The mathematical relationship for the change in the reaction-rate constant with temperature (adopted from chemical kinetics, Section 2–5, Eq. 2–26) is expressed as

$$k = k_{20}\Theta^{T-20} \qquad (3\text{--}17)$$

where k = reaction-rate constant at temperature T, per day
k_{20} = reaction-rate constant at 20°C, per day
Θ = temperature coefficient, dimensionless
T = temperature of biological reaction, degrees Celsius

A hot, arid climate changes the biological and chemical kinetics in wastewater treatment processes from those in continental and temperate climates. In a desert region, wastewater temperatures are significantly higher because of hot air temperature and warm drinking water.[3] If the source of drinking water is desert wells or desalinated seawater, the year-round wastewater temperature can be 25 to 30°C and warmer in the summer. Warm wastewater has substantial influences on treatment processes by increasing the rate of bacterial activity and rate of chemical reactions compared with the rates applied in process design in North America and Europe. The adverse effects of warm wastewater are: greater release of hydrogen sulfide from anaerobic wastewater, increasing corrosion and emission of odors; in sedimentation of raw wastewater and activated sludge, reduction in the density of settled sludge due to fermentation of organic matter in the sludge blanket; in biological aeration processes, increase in the rate of nitrification of ammonia, resulting in greater oxygen demand; and difficulty

in dewatering unstabilized waste sludge for disposal. The benefits of warm wastewater are: higher allowable design loadings for wastewater aeration processes because of the increased rate of biological activity; reduced retention time for wastewater effluent disinfection because of the increased rate of the oxidative reaction of chlorine; and, in stabilization ponds, improved biological treatment and more rapid die-off of pathogenic microorganisms.

The value of Θ is 1.072 if the rate of biological activity doubles or halves with a 10°C temperature change and is 1.047 if it doubles or halves with a 15°C change. The rate of a biological reaction does not remain constant over the entire 5 to 35°C range; in fact, considerable variation may occur. The general reaction-rate–temperature curve sketched in Figure 3–17 shows the reaction rate increasing with rising temperature; the slope of the curve becomes steeper with increasing temperature. Therefore, for this reaction-rate–temperature curve, the value of Θ becomes greater with higher reaction temperatures.

Above 40°C, mesophilic activity drops off sharply and thermophilic growth starts. Thermophilic bacteria have a range of approximately 45 to 75°C, with an optimum near 55°C. This higher temperature range is rarely used in waste treatment because it is difficult to maintain that high an operating temperature and because thermophilic bacteria are more sensitive to small temperature changes. Of particular importance is the dip between mesophilic and thermophilic ranges; operation in this region should be avoided. A technician seeking increased activity in an anaerobic digester may gradually increase the temperature of the digesting sludge and thus realize improved gas production and efficiency. However, if the sludge is heated in excess of the mesophilic optimum, the rate of biological activity will decrease sharply, adversely affecting operation.

Municipal wastewaters commonly contain sufficient concentrations of carbon, nitrogen, phosphorus, and trace nutrients to support the growth of a microbial culture. Theoretically, a BOD-to-nitrogen-to-phosphorus ratio of 100/5/1 is adequate for aerobic treatment, with small variations depending on the type of system and mode of operation. Average domestic wastewater exhibits a surplus of nitrogen and phosphorus with a

Figure 3–17

Effect of temperature on the rate of biological activity and relative positioning of the mesophilic and thermophilic zones.

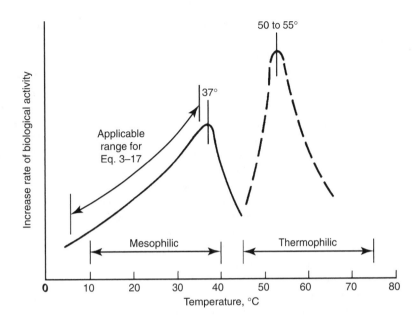

BOD/N/P ratio of about 100/17/3. If a municipal wastewater contains a large volume of nutrient-deficient industrial waste, supplemental nitrogen is generally supplied by the addition of anhydrous ammonia (NH_3) or ammonium nitrate (NH_4NO_3), and phosphate as phosphoric acid (H_3PO_4).

Diffused and mechanical aeration basins must supply sufficient air to maintain dissolved oxygen for the biota to use in metabolizing the waste organics. The rate of microbial activity is independent of the dissolved oxygen concentration above a minimum critical value, below which the rate is reduced by the limitation of oxygen required for respiration. The exact minimum depends on the type of activated sludge process and the characteristics of the wastewater being treated. The most common design criterion for critical dissolved oxygen is 2.0 mg/l, but in actual operation, values as low as 0.5 mg/l have proved satisfactory. Anaerobic systems must, of course, operate in the complete absence of dissolved oxygen; consequently, digesters are sealed with floating or fixed covers to exclude air.

Hydrogen ion concentration has a direct influence on biological treatment systems, which operate best in a neutral environment. The general range of operation of aeration systems is between pH 6.5 and 8.5. Above this range microbial activity is inhibited, and below pH 6.5 fungi are favored over bacteria in the competition for metabolizing

the waste organics. Normally, the bicarbonate buffer capacity of a wastewater is sufficient to prevent acidity and reduced pH, and carbon dioxide production by the microorganisms tends to control the alkalinity of high-pH wastewaters. Where industrial discharges force the pH of a municipal wastewater outside the optimum range, addition of a chemical may be required for neutralization. In that case, it is more desirable to have the industry pretreat its waste by equalization and neutralization prior to disposal in the sewer rather than contend with the problem of pH control at the city's disposal plant.

Anaerobic digestion has a small pH tolerance range of 6.7 to 7.4, with optimum operation at pH 7.0 to 7.1. Domestic waste sludge permits operation in this narrow range except during startup or periods of organic overloads. Limited success in digester pH control has been achieved by careful addition of lime with the raw sludge feed. Unfortunately, the buildup of acidity and reduction of pH may be a symptom of other digestion problems, for example, accumulation of toxic heavy metals, that the addition of lime cannot cure.

Biological treatment systems are inhibited by toxic substances. Industrial wastes from metal-finishing industries often contain toxic ions, such as nickel and chromium; chemical manufacturing produces a wide variety of organic compounds that can adversely affect microorganisms. Since little can be done to remove or neutralize toxic compounds in municipal treatment, pretreatment by industries prior to discharging wastes to the city sewer is required.

Population Dynamics

In biological processing of wastes, the naturally occurring biota are a variety of bacteria growing in mutual association with other microscopic plants and animals. Three of the major factors in population dynamics are competition for the same food, predator-prey relationship, and symbiotic association. When organic matter is fed to a mixed population of microorganisms, competition arises for this food, and the primary feeders that are most competitive become dominant. Under normal operating conditions, bacteria are the primary feeders in both aerobic and anaerobic operations. Protozoa consuming bacteria is the common predator-prey relationship in activated-sludge and trickling filters. In stabilization ponds, protozoa and rotifers graze on both algae and bacteria. Symbiosis is the living together of organisms for mutual benefit such that the association produces more vigorous growth of both species. An excellent example of this is the relationship between bacteria and algae in a stabilization pond.

In an activated-sludge process, waste organics serve as food for the bacteria and the small population of fungi that might be present. Some of the bacteria die and lyse, releasing their contents, which are resynthesized by other bacteria. The

secondary feeders (protozoa) consume several thousand bacteria for a single reproduction. The benefit of this predator-prey action is twofold: (1) removal of bacteria stimulates further bacterial growth, accelerating metabolism of the organic matter; and (2) the settling characteristics of the biological floc are improved by reducing the number of free bacteria in solution. The effluent from the process consists of nonsettleable organic matter and dissolved inorganic salts (Figure 3–18).

Control of the microbial populations is essential for efficient aerobic treatment. If wastewater were simply aerated, the liquid detention times would be intolerably long, requiring a time period of about 5 days at $20°C$ for 70 percent reduction. However, extraction of organic matter is possible within a few hours of aeration provided a large number of microorganisms are mixed with the wastewater. In practice, this is achieved by settling the microorganisms out of solution in a final clarifier and returning them to the aeration tank to metabolize additional waste organics (Figure 3–18). Good settling characteristics occur when an activated sludge is held in the endogenous (starvation) phase. Furthermore, a large population of underfed biota remove BOD very rapidly from solution. Excess microorganisms are wasted from the process to maintain proper balance between food supply and biological mass in the aeration tank. This balance is referred to as the food-to-microorganism ratio (F/M), which is normally expressed in units of pounds of BOD applied per day per pound of MLSS in the aeration basin (MLSS is the mixed liquor suspended solids).

Figure 3–18

Generalized biological population dynamics in the activated-sludge wastewater treatment process. (a) Predator-prey relationship between protozoa and bacteria in the activated-sludge process. (b) Sedimentation and recirculation maintains the desired food-to-microorganism (F/M) ratio in the aeration basin.

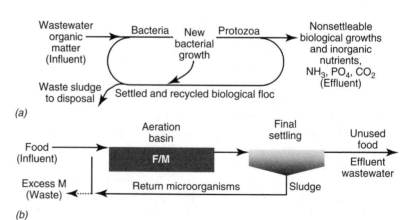

Figure 3–19

Schematic diagram of the mutually beneficial association (symbiotic relationship) between bacteria and algae in a stabilization pond.

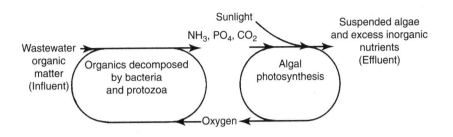

Operation at a high F/M ratio results in incomplete metabolism of the organic matter, poor settling characteristics of the biological floc, and, consequently, poor BOD removal efficiency. At a low F/M ratio, the mass of microorganisms are in a near starvation condition that results in a high degree of organic matter removal, good settleability of the activated sludge, and efficient BOD removal.

The relationship between bacteria and algae in a small pond is illustrated in Figure 3–19. Bacteria decompose the organic matter, yielding inorganic nitrogen, phosphates, and carbon dioxide. Algae use these compounds, along with energy from sunlight, in photosynthesis, releasing oxygen into solution. This oxygen is, in turn, taken up by the bacteria, thus closing the cycle. The effluent from a stabilization pond contains suspended algae and excess bacterial decomposition end products. In the summer, BOD reduction in lagoons is very high, commonly in excess of 95 percent. However, during cold-temperature operation, microbial activity is reduced, and BOD removal relies to a considerable extent on dilution of the inflowing raw wastewater into the large volume of impounded water. Liquid detention in a stabilization pond is rarely less than 90 days and generally is considerably greater. Following a cold winter, particularly if the pond was covered with ice and snow, odorous conditions can be anticipated during the spring thaw. With increased temperature, the organic matter accumulated during the winter is rapidly decomposed by the bacteria, using dissolved oxygen at a faster rate than can be absorbed from the air or supplied by the algae. After a few days to several weeks, depending on climatic conditions and waste load on the lagoon, the algae become reestablished and again supply oxygen to the bacterial cycle. Once this symbiotic relationship is again operational, aerobic conditions are firmly established and odorous emissions cease.

3–12 BIOLOGICAL KINETICS

The characteristic growth pattern for a single bacterial species in a batch culture is sketched in Figure 3–20. This growth with time occurs when a sterile liquid substrate is inoculated with a small number of bacteria. After a short lag period, the bacteria reproduce exponentially by binary fission, rapidly increasing the number of viable cells and biomass in the medium. The presence of excess substrate promotes the maximum rate of growth possible, limited only by the ability of the bacteria to reproduce. In this *exponential growth phase*, the increase in both the number of viable cells and the accumulation of biomass is represented by Eq. 3–18, in which μ is a proportionality constant.

$$r_g = \mu X \qquad (3\text{–}18)$$

where r_g = biomass growth rate, milligrams per liter per day

μ = specific growth rate (rate of growth per unit of biomass), per day

X = concentration of biomass, milligrams per liter

The *declining growth phase* is the result of diminishing substrate limiting bacterial growth. The rate of reproduction decreases and some cells die so that the total biomass exceeds the mass of viable cells. When the substrate is depleted at the end of declining growth, the number of viable bacteria and biomass remain relatively constant, resulting in a *stationary phase*. Rate of growth in

Figure 3–20

Characteristic growth curves for biomass and number of viable bacteria in pure, batch culture.

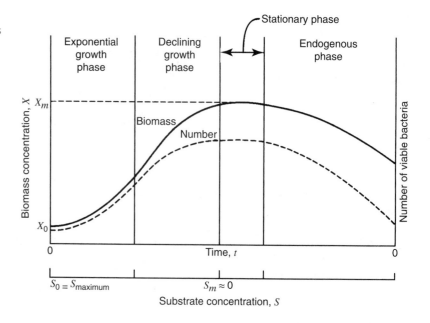

the declining growth phase is described by the Monod equation:

$$\mu = \mu_m\left(\frac{S}{K_s + S}\right) \quad\quad (3\text{–}19)$$

where μ_m = maximum specific growth rate, per day

S = concentration of growth-limiting substrate, milligrams per liter

K_s = saturation constant (equal to the limiting substrate concentration at half the maximum growth rate), milligrams per liter

This mathematical relationship between the concentration of the growth-limiting substrate and the specific growth rate of biomass is a hyperbolic function as plotted in Figure 3–21. The constant K_s is equal to the concentration of substrate S when the specific growth rate μ equals half of the growth rate μ_m. If the specific growth rate relationship in Eq. 3–19 is substituted into Eq. 3–18, the rate of biomass growth in a substrate-limiting solution is

$$r_g = \frac{\mu_m X S}{K_s + S} \quad\quad (3\text{–}20)$$

The *endogenous growth phase* (Figure 3–20) is a period of decreasing metabolism with a resulting decrease in both the biomass and the number of viable bacteria. The bacteria remaining compete for the small amount of substrate still in solution. Aged cells die and lyse, releasing nutrients back into solution. The action of lysis decreases both the number of cells and total biomass. The rate of

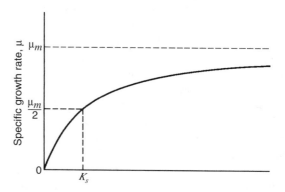

Figure 3–21

Curve of specific growth rate versus concentration of the growth-limiting substrate for a pure bacterial culture during the declining growth phase, defined mathematically by the Monod relationship in Eq. 3–19.

biomass decrease is proportional to the biomass present, so that

$$r_d = -k_d X \qquad \text{(3–21)}$$

where r_d = biomass decay rate, milligrams per liter per day
k_d = biomass decay coefficient, per day
X = concentration of biomass, milligrams per liter

The net growth rate of biomass during the endogenous growth phase is the sum of Eqs. 3–20 and 3–21

$$r_g' = \frac{\mu_m X S}{K_s + S} - k_d X \qquad \text{(3–22)}$$

where r_g' = net biomass growth rate, milligrams per liter per day

The Monod relationship (Eq. 3–19) modified for net specific growth rate is

$$\mu' = \mu_m \left(\frac{S}{K_s + S} \right) - k_d \qquad \text{(3–23)}$$

where μ' = net maximum specific growth rate, per day

Growth yield is the incremental increase in biomass resulting from metabolism of an incremental amount of substrate. In a batch culture (Figure 3–21), the maximum yield is the biomass increase during the exponential and declining growth phase $(X_m - X_0)$ relative to the substrate used during this same period $(S_0 - S_m)$. The substrate concentration S_m approaches zero, therefore,

$$Y = \frac{X_m - X_0}{S_0} = \frac{r_g}{r_{su}} \qquad \text{(3–24)}$$

where Y = growth yield, milligrams per liter of biomass increase per milligram per liter of substrate metabolized
r_{su} = rate of substrate utilization, milligrams per liter per day

If the growth yield is incorporated into Eq. 3–22, the net growth rate in the endogenous phase is

$$r_g' = Y r_{su} - k_d X \qquad \text{(3–25)}$$

The following equation is for the observed growth yield that accounts for the effect of the endogenous decay in biomass.

$$Y_{obs} = \frac{r_g'}{r_{su}} \qquad \text{(3–26)}$$

where Y_{obs} = observed growth yield in a laboratory test

REFERENCES

1. *Standard Methods for the Examination of Water and Wastewater*, 20th Ed., 1998. Published jointly by American Public Health Association, Washington, DC 20005, American Water Works Association, and Water Environment Federation.
2. *Waterborne Pathogens*, AWWA M48, 1st Ed., 1999. American Water Works Association, Denver, CO 80235.
3. Hammer, M. J. *Wastewater Treatment in Dry Climates (Desert or Desert with Some Rain)*, 2001. Water Environment Federation, Alexandria, VA 22314-1994.

PROBLEMS

3–1 Define the term *bacteria*. What are heterotrophic bacteria? In what preferential sequence do facultative bacteria use available sources of oxygen? In this sequence, when are obnoxious odors released? In wastewater aeration, how do heterotrophic bacteria stabilize organic wastes? (Refer to Section 3–1, Section 11–6.)

3–2 State why some bacteria convert ammonia to nitrate (Eqs. 3–7 and 3–8) and others reduce nitrate to nitrogen gas (Eq. 3–2).

3–3 Define the term *autotrophic bacteria*. What are the bacterial reactions in nitrification? (Refer to Sections 3–1 and 13–7.) In sewer corrosion? (Refer to Sections 3–1 and 10–4.) In discoloration of drinking water? (Refer to Sections 3–1 and 7–18.)

3–4 Name the three primary waterborne pathogenic bacteria. What are the common reasons for waterborne outbreaks of salmonellosis? In what categories of water systems are outbreaks of shigellosis most likely to occur?

3–5 Define the term *protozoa*. What is the principal role of protozoa in aerobic biological wastewater treatment? (Refer to Sections 3–2 and 11–6.)

3–6 What are the symptoms of the diseases produced by the two pathogenic protozoa that are found worldwide? What is the common mode of transmission and in what environments is transmission most likely to occur?

3–7 Define the terms *virus* and *bacteriophage*. What size are viruses relative to bacteria?

3–8 Compare the life-reproductive cycles of viruses and heterotrophic bacteria.

3–9 What are algae? Define the process of photosynthesis.

3–10 Define the meaning of the term *pathogen*, and give the names of pathogen groups. What determines the kinds and concentrations of pathogens in wastewater?

3–11 Define the meaning of the fecal-oral route in the transmission of diseases.

3–12 Compare the latency, persistence, and infective dose of *Ascaris* and *Salmonella*.

3–13 Historically in the United States, the prevalent infectious diseases were typhoid, cholera, and dysentery. How have these diseases been virtually eliminated? Currently, the prevalent infectious diseases are giardiasis and cryptosporidiosis, causing diarrhea that can be life-threatening for persons with immunodeficiency syndrome. What actions are being taken to reduce the probability of waterborne transmission of these diseases? (Refer to Sections 3–5 and 7–15.)

3–14 Discuss the significance of human carriers in transmission of enteric diseases. What major waterborne diseases in the United States are spread by carriers? How is the spread of two of these diseases amplified by animals?

3–15 Outline the three phases in testing for enteric viruses in relatively unpolluted water. Since extraction and concentration capture only a portion of the viruses in a water sample, how is the precision of separation determined? When is testing for enteric viruses recommended?

3–16 List some of the problems associated with detection and identification of viruses. In cell-culture testing, how are viruses in a sample concentrate detected and enumerated? Why is the viral count expressed in terms of PFU per liter?

3–17 In one statement, what is the general process in testing for *Giardia* cysts and *Cryptosporidium* oocysts? In method 1622, the water sample is only 10 l for testing natural stream water for *Cryptosporidium* oocysts. Using this method to test stream samples at a variety of locations, why was the accuracy for detection and enumeration of oocysts low?

3–18 Why must laboratories conducting tests for *Cryptosporidium* oocysts be audited and approved for quality assurance?

3–19 Why are coliform bacteria used as indicators of quality of drinking water? Under what circumstances is the reliability of coliform bacteria to indicate the presence of pathogens questioned?

3–20 Coliform bacteria in surface waters can originate from feces of humans, wastes of farm animals, or soil erosion. Can the coliforms from these three different sources be distinguished from one another?

3–21 Why is a positive test for fecal coliforms in a public water supply considered more serious than a positive test for total coliforms?

3–22 What are the significant differences between *Escherichia coli* and *Escherichia coli* O157:H7?

3–23 Why is lactose (milk sugar), an ingredient in all culture media, used to test for the coliform group?

3–24 A multiple-tube fermentation analysis of a river water yielded the following results. What are the MPN and confidence limits?

Serial Dilution	Sample Portion (ml)	Positive Tubes
0	1.0	5 of 5
1	0.1	4 of 5
2	0.01	2 of 5
3	0.001	1 of 5
4	0.000,1	0 of 5

(*Answers* 2200, 1000–5800)

3–25 Multiple-tube fermentation analyses of a river water yielded the following results.

SERIAL DILUTION	SAMPLE PORTION (ML)	NUMBER OF POSITIVE TUBES OUT OF FIVE	
		LAURYL TRYPTOSE	EC
0	1.0	5	5
1	0.1	5	2
2	0.01	2	2
3	0.001	0	0
4	0.000,1	0	0

What are the MPN and confidence limits for total coliforms and thermotolerant fecal coliforms?

3–26 Filtered chlorinated effluent from a tertiary wastewater treatment plant was tested for total coliforms by inoculating 10 tubes containing lauryl tryptose broth with 10-ml portions of a 100-ml sample. After incubation, 4 of the 10 tubes were clear and 6 showed growth, with 3 of the 6 containing bubbles in the inverted vials. What are the MPN Index/100 ml and the 95 percent confidence limits?

3–27 Fecal coliform standards for reuse of reclaimed water are often written as <2.2/100 ml and <23/100 ml. Based on the bacterial density test procedure of inoculating 10-ml portions of a sample in 10 tubes, what is the maximum number of tubes that can give a positive reaction to meet the <2.2 MPN Index per 100 ml? For the <23 MPN Index per 100 ml?

3–28 In the presence-absence test for coliform bacteria, what indicates a positive test for total coliforms? For fecal coliforms?

3–29 Describe the procedure for collecting a drinking-water sample for coliform testing to reduce the probability of unintentional contamination.

3–30 Calculate the 5-day BOD of a domestic wastewater based on the following data: volume of wastewater added to 300-ml bottle = 6.0 ml, initial DO = 8.1 mg/l,   5-day DO = 4.2 mg/l. What is the ultimate BOD assuming a k-rate of 0.1 per day? (*Answer* 195 mg/l and 290 mg/l)

3–31 A seeded BOD test is to be conducted on a poultry-processing waste with an estimated 5-day value of 600 mg/l. The seed taken from an existing preaeration tank at the industrial site has an estimated BOD of 200 mg/l. (a) What sample portions should be used for setting up the middle dilutions of the wastewater and seed tests? (b) Compute the BOD value for the poultry waste if the initial DO in both the seed and sample bottles is 8.2 mg/l and the 5-day values are 3.5 mg/l and 4.0 mg/l for the seed test bottle and seeded wastewater sample, respectively. Use the volumetric additions from part (a) and assume that the volume of seed used in the waste BOD bottle is 10 percent of that used in the seed test. [*Answers* (a) 2.0 ml poultry waste plus 0.7 ml of seed, 7.0 ml of seed sample (b) 560 mg/l]

3–32 Determine the 5-day BOD and k-rate for a raw domestic wastewater based on the following test data. Draw the BOD-time curve and graph for determination of the k-rate. The 300-ml test bottles were filled with 2.00 percent wastewater mixed with dilution water. The initial dissolved oxygen (DO) in all bottles was 8.3 mg/l based on three bottles titrated after preparation. The time-residual DO data for the incubated bottles were as follows: t = 1.1 days, DO = 6.7 mg/l; 2.2 days, 6.0 mg/l; 4.3 days, 4.7 mg/l; 6.3 days, 3.8 mg/l; and 14.4 days, 2.4 mg/l.

3–33 Determine the carbonaceous 5-day BOD and k-rate for an unchlorinated effluent after treatment by activated-sludge aeration. Draw the BOD-time curve and graph for determination of the k-rate. Nine identical test bottles were prepared, with 60 ml of wastewater added to each bottle followed by 240 ml of dilution water. Three bottles were titrated after setup to determine the initial DO of 8.1 mg/l. The remaining 6 bottles were incubated and tested for time-residual DO data with the following results: 1.3 days, DO = 7.2 mg/l; 3.3 days, 6.4 mg/l; 4.3 days, 6.2 mg/l; 5.3 days, 5.8 mg/l; 8.3 days, 4.5 mg/l; and 11.3 days, 3.5 mg/l.

3–34 A BOD analysis was performed on a municipal wastewater containing food-processing wastewater discharged to the sewer system. The purpose was to draw a BOD-time curve, determine the 5-day BOD, and make a graphical determination of the k-rate. The wastewater was mixed with dilution water for a 1.5 percent mixture, which is equal to 4.5 ml of wastewater in each 300-ml bottle. The initial DO was 8.1 mg/l based on titration of 3 bottles. The results from titrating the incubated bottles are as follows: 0.6 days, DO = 6.0 mg/l; 0.9 days,

5.6 mg/l; 1.7 days, 4.2 mg/l; 2.8 days, 3.1 mg/l; 3.6 days, 2.3 mg/l; 3.9 days, 2.1 mg/l; and 5.0 days, 1.7 mg/l.

3–35 The BOD data given here are a series of tests on a 24-hr composite effluent from a tertiary plant treating domestic wastewater with granular-media filtration after biological aeration. The samples were collected before chlorination so seeding of the bottles was not necessary.

Time, days	1	2	3	5	6	9	13	17
BOD, mg/l	4	6.3	12.5	13.6	14.3	18.2	20.2	23.8

Graph BOD versus time and locate the 5-day BOD value. Calculate the values of (time/BOD)$^{1/3}$, and graph these values versus time to determine k-rate.

3–36 The BOD data listed here are from a series of tests on a 24-hr composite wastewater from a factory producing canned fruits.

Time, days	2	3	5	6.1	9	12
BOD, mg/l	50	150	252	280	345	390

Graph BOD versus time and locate the approximate 5-day BOD value. (The graph has a lag period; review Figure 3–15.) Calculate the values of [(time – lag period)/BOD]$^{1/3}$, and graph these values versus time to determine k-rate.

3–37 Determine the BOD of a wastewater from candy manufacturing based on the following data. The seed was settled, aged, domestic wastewater.

WASTE-WATER PORTION (ML)	SEED PORTION (ML)	INITIAL DO (MG/L)	5-DAY DO (MG/L)
BOTTLES TESTED FOR INITIAL DISSOLVED OXYGEN			
5.0	1.5	8.8	
6.0	1.5	8.7	
7.0	1.5	9.0	

(Continued)

TESTS FOR DETERMINING THE **BOD** OF THE SEED		
15	0	5.4
15	0	5.5
15	0	5.7
TEST FOR DETERMINING THE **BOD** OF THE WASTEWATER		
8.0	1.5	0.5
8.0	1.5	0.8
8.0	1.5	0.4
6.0	1.5	3.1
6.0	1.5	2.9
6.0	1.5	2.9
4.0	1.5	4.6
4.0	1.5	4.5
4.0	1.5	4.8

3–38 A seeded BOD analysis was conducted on high-strength, food-processing wastewater. Twenty-milliliter portions were used in setting up the 300-ml bottles of aged, settled, wastewater seed. The seeded sample BOD bottles contained 1.0 ml of industrial waste and 2.0 ml of seed material. The results for this series of test bottles are listed in the following table. Calculate the BOD values. (B_1 = 8.2 mg/l, D_1 = 8.2 mg/l, f = 0.10, P = 1/300.) Plot a BOD-time curve; what is the 5-day value? Determine the k-rate. (*Answers* 5-day BOD = 900 mg/l and k = 0.07 per day)

TIME (DAYS)	SEED TESTS B_2 (MG/L)	SAMPLE TESTS D_2 (MG/L)	TIME (DAYS)	SEED TESTS B_2 (MG/L)	SAMPLE TESTS D_2 (MG/L)
0	8.2	8.2	7.9	4.5	4.1
1.0	7.3	7.2	9.8	4.1	3.8
1.9	6.5	6.4	11.8	3.9	3.5
2.9	5.9	5.8	14.0	3.4	2.9
3.8	5.5	5.3	15.6	3.6	2.7
4.7	5.1	5.0	19.0	3.8	2.1
6.0	4.8	4.6			

3–39 A seeded BOD analysis was conducted on a meat-processing wastewater. Dissolved oxygen data for the seed test and seeded wastewater follow. The BOD bottles for the seed (aged, settled, domestic wastewater) were set up by adding 15 ml per 300-ml bottle. The test bottles for the meat-processing wastewater

49.727

were set up by adding 2.0 ml of wastewater and 1.5 ml of seed.

	DO Measurements in Bottles	
Time (days)	Seed (mg/l)	Seeded Waste-water (mg/l)
0	7.9	8.1
1.0	6.5	5.5
2.2	5.4	4.2
3.0	4.9	3.7
4.0	4.0	2.5
5.0	3.8	2.1
6.0	3.8	2.1
7.0	3.6	2.0

Based on the averages of three tests, the initial DO in the seed bottles was 7.9 mg/l and the initial DO in the wastewater bottles was 8.1 mg/l. Sketch BOD-time curves for both the seed wastewater and meat-processing wastewater. Determine the 5-day BOD values and k-rates using the graphical procedure.

3–40 Solids analyses were conducted on the domestic wastewater that was tested for BOD in Problem 3–32. The tests were done in triplicate. Based on the following test data, calculate the concentrations of total solids, total volatile solids, suspended solids, and volatile suspended solids.

Total Solids Test Data			
Dish number	1	2	3
Weight of empty dish, g	52.842	53.252	53.073
Weight of dish + sample, g	110.0	104.0	102.8
Weight of dish + dry solids, g	52.902	53.308	53.126
Weight of dish + ignited solids, g	52.859	53.287	53.092

Suspended Solids Test Data			
Test number	1	2	3
Weight of filter, g	0.1160	0.1165	0.1152
Sample volume, g	48	72	72
Weight of filter + dry solids, g	0.1270	0.1353	0.1332
Weight of filter + ignited solids, g	0.1170	0.1208	0.1198

(*Answers* TS = 1070 mg/l (1050, 1100, 1070), TVS = estimated 720 mg/l (750, 410, 690) 69 percent, SS = 250 mg/l (229, 261, 250), VSS = 198 mg/l (208, 201, 186) 79 percent)

3–41 Solids analyses were conducted on the meat-processing wastewater that was tested for BOD in Problem 3–39. The tests were conducted in triplicate. Based on the test data that follow, determine the concentrations of total solids, total volatile solids, suspended solids, and volatile suspended solids.

Total Solids Test Data			
Dish number	1	2	3
Weight of empty dish, g	51.494	51.999	50.326
Weight of dish + sample, g	109.5	103.4	118.4
Weight of dish + dry solids, g	51.587	52.081	50.437
Weight of dish + ignited solids, g	51.541	52.042	50.383

Suspended Solids Test Data			
Test number	1	2	3
Weight of filter, g	0.1154	0.1170	0.1170
Sample volume, ml	25	25	30
Weight of filter + dry solids, g	0.1241	0.1259	0.1278
Weight of filter + ignited solids, g	0.1160	0.1173	0.1178

3–42 Calculate the percentage of total solids and percentage of volatile solids in the sample and percentage of total solids that are volatile from the following laboratory analysis of a wastewater sludge. (Refer to Solids in Section 2–11.)

Weight of empty dish on analytical balance, g	50.160
Weight of dish on simple balance, g	50.2
Weight of dish plus sludge sample, g	90.8
Weight of dish plus dry sludge solids, g	52.783
Weight of dish plus ignited solids, g	50.969

3–43 What is the source of the bacteria and protozoa for biological treatment of wastewater? Must "special" microbial cultures be purchased to inoculate biological processes?

3–44 List the major factors influencing the rate of biological growth.

3–45 How does temperature affect biological activity in the mesophilic range? For a Θ value of 1.072, how many degrees of temperature increase is required to increase the rate of biological activity 100 percent (double the rate)? If the temperature is increased by 5°C, what is the increase in activity?

3–46 At a BOD loading of 640 g/m³ · d, a biological aeration process operating at 15°C yields an excellent effluent. If this process were operating at 25°C, what would be the allowable BOD loading for the same quality effluent? (*Answer* 1000 g/m³ · d)

3–47 The rate of BOD reduction in aeration of a synthetic wastewater decreased by 20 percent when the temperature of the laboratory fermentation tank was lowered from 20°C to 16°C. Using Eq. 3–17, calculate the temperature coefficient.

3–48 In the activated-sludge aeration process, what microorganisms are the primary feeders? The secondary feeders?

3–49 Why must an activated-sludge process be operated in the starvation (low-F/M) stage?

3–50 Why do stabilization ponds support dense populations of algae?

3–51 Describe the conditions of biological growth defined mathematically by the Monod equation.

3–52 A series of fermentation tubes containing varying concentrations of glucose in a nutrient broth were inoculated with a pure bacterial culture. The concentrations of cells in the broth media were determined after 16 hr of incubation at 37°C. Rates of growth and initial glucose concentrations follow. Plot the growth rate expressed as cell divisions per hour versus initial glucose concentration. Estimate the maximum growth rate and the saturation constant (glucose concentration at one-half the maximum growth rate). Write an equation in the form of Eq. 3–19, and draw the curve for this equation on the graph with the plotted data. Does the growth of this pure culture appear to be a hyperbolic function as defined by the Monod relationship?

GLUCOSE (MOLES $\times 10^{-4}$)	CELL DIVISIONS (PER HR)	GLUCOSE (MOLES $\times 10^{-4}$)	CELL DIVISIONS (PER HR)
0.1	0.23	0.8	0.94
0.1	0.28	1.6	1.06
0.2	0.32	3.2	1.15
0.4	0.71		

Hydraulics and Hydrology

The following sections present the basic principles of hydraulics applicable to water distribution and wastewater collection systems. Low-flow characteristics of streams and stratification of lakes provide background for the study of water pollution. The basic concepts of groundwater hydrology are essential to understanding water well construction and operation.

4–1 WATER PRESSURE

Mass per unit volume is referred to as the density of a fluid. The density of water at a temperature of 60°F (15°C) and a pressure of 1 atmosphere is 1.94 slugs/cu ft (999 kg/m^3). The force exerted by gravity on 1.0 cu ft (1.0 m^3) of water is 62.4 lb (9.80 kN). This is equal to the density multiplied by the acceleration of gravity, which is 32.2 ft/sec^2 (9.81 m/s^2).

Pressure is the force exerted per unit area. Figure 4–1a shows that the pressure exerted on the bottom of a 1.0-cu ft container filled with water is equal to 0.433 psi. In engineering hydraulics, water pressure is frequently expressed in terms of feet of head as well as psi. The relationship between these units is visually shown in Figure 4–1a, where a height of 1.0 ft of water head exerts a pressure of 0.433 psi. Water pressure increases with depth below the surface linearly such that the pressure in psi is equal to 0.433 times the depth in feet. The sketch in Figure 4–1b shows the pressure acting only horizontally for ease of illustration. In fact, water pressure is exerted equally in all directions. The computation given under the diagram states that a head of 2.31 ft produces a water pressure of 1.0 psi. Thus, 1.0 ft of head is equivalent to 0.433 psi, or, alternately, a pressure of 1.0 psi is equivalent to 2.31 ft of head.

The relationships for water pressure and water head in SI metric units are illustrated in Figure 4–2. The pressure exerted on the bottom of a 1.0-m^3 container filled with water at 15°C equals 9.80 kPa. Therefore, 1.0 m of water head exerts a pressure of 9.80 kPa, or 1.0 kPa of pressure is equal to 0.102 m of water head.

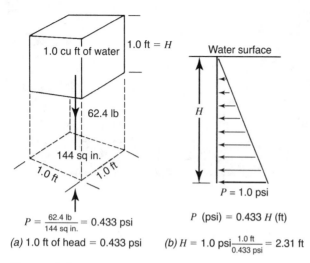

Figure 4–1

Basic relationships between water pressure in psi (pounds per square inch) and feet of water head.

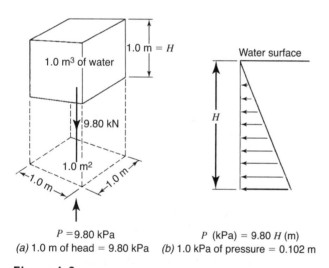

Figure 4–2

Basic relationships between water pressure in kilopascals and meters of water head.

A simple piezometer, illustrated in Figure 4–3a, consists of a small tube rising from a container of water under pressure. The height of the water in the piezometer tube denotes the pressure of the confined water. Water piezometers are rarely practical for pressure measurements, since values generally exceed the height of water columns that are convenient to read. A mercury column can be

used to measure relatively high pressure values within a limited head range, since mercury is approximately 13.6 times heavier than water; that is, a 1.0-ft head of mercury is equivalent to a pressure of 5.9 psi. Figure 4–3b is a sketch of a mercury manometer. The pressure reading is equal to the difference of the fluid pressures in the two legs of the U-shaped tube. The use of manometers is generally restricted to indoor applications where the unit is in a fixed location.

Water pressure is commonly measured by a Bourdon gauge (Figure 4–3c). A hollow metal tube of elliptical cross section is bent in the form of a circle with a pointer attached to the end through a suitable linkage. As pressure inside the tube increases, the elliptical cross section tends to become circular, and the free end of the Bourdon tube moves outward. The dial of the instrument can be calibrated to read gauge pressure in psi. Bourdon tubes and manometers actually measure pressure relative to the atmospheric pressure. [Atmospheric or barometric pressure is caused by the weight of the thick mass of air above the earth's surface. Under standard atmospheric conditions, the barometric pressure at sea level is 14.7 psi (101 kPa).] If the pressure measured is greater than atmospheric, this value is sometimes called gauge pressure. If the pressure measured is less than atmospheric, it is referred to as a vacuum. Absolute pressure is the term used for a pressure reading that includes atmospheric pressure, that is, the pressure relative to absolute zero.

4–2 PRESSURE-VELOCITY-HEAD RELATIONSHIPS

The association between quantity of water flow, average velocity, and cross-sectional area of flow is given by the equation

$$Q = VA \qquad (4\text{–}1)$$

where Q = quantity, cubic feet per second (cubic meters per second)

V = velocity, feet per second (meters per second)

A = cross-sectional area of flow, square feet (square meters)

Figure 4–3

Water pressure measuring
devices. (*a*) Piezometer.
(*b*) Manometer.
(*c*) Pressure gauge.

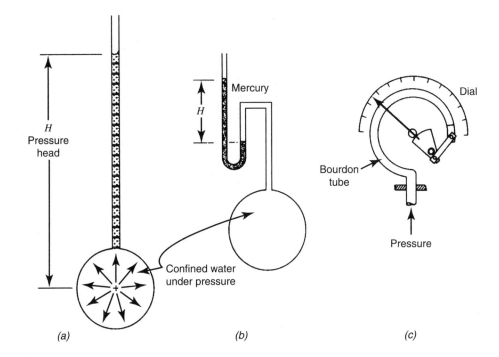

(a)   (b)   (c)

This formula is known as the continuity equation. For an incompressible fluid such as water, if the cross-sectional area decreases (the water main becomes smaller), the velocity of flow must increase; conversely, if the area increases, the velocity decreases (Figure 4–4).

The total energy at any point in a hydraulics system is equal to the sum of the elevation head, pressure head, and velocity head.

$$\frac{\text{Total}}{\text{energy}} = \frac{\text{Elevation}}{\text{head}} + \frac{\text{Pressure}}{\text{head}} + \frac{\text{Velocity}}{\text{head}} \quad \textbf{(4–2)}$$

$$E = Z + \frac{P}{w} + \frac{V^2}{2g}$$

where E = total energy head
  Z = elevation above datum
  P = pressure
  V = velocity of flow
  w = unit weight of liquid
  g = acceleration of gravity

The vertical line at point 1 on the left-hand side of Figure 4–5 graphically illustrates the total energy equation. The pressure head is equal to the height to which the water would rise in a piezometer tube

inserted in the pipe. The elevation head is equal to the vertical distance Z from an assumed datum plane to the pipe. The sum of these two heads is the hydraulic head. The velocity head is equal to the kinetic energy in the water flow and when added to the hydraulic head yields total energy. If ft-lb-sec units are used in Eq. 4–2, each of the separate head calculations will be in feet.

If the energy is compared in English units at two different points in a piping system (Figure 4–5), Eq. 4–3 results. The term h_L represents the energy losses that occur in any real system. The major loss of energy is due to friction between the moving water and pipe wall; however, energy losses also occur from flow disturbance caused by valves, bends in the pipeline, and changes in diameter. The

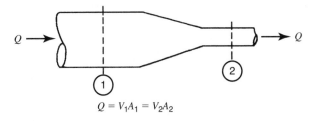

$$Q = V_1A_1 = V_2A_2$$

Figure 4–4

Flow equation for incompressible liquids.

Figure 4–5

Energy equation parameters as related to water flow in a pipe.

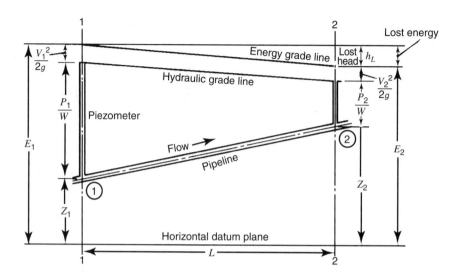

imaginary line connecting all points of total energy is called the energy gradient or grade line. This line must always slope in the direction of flow, showing a decrease in energy, unless external energy is added to the system, for example, by a pump. The hydraulic gradient is defined as the line connecting the level of elevation plus pressure energies, this being defined by the water surfaces in imaginary piezometer tubes inserted in the piping.

$$Z_1 + \frac{P_1}{62.4} + \frac{V_1^2}{64.4} = Z_2 + \frac{P_2}{62.4} + \frac{V_2^2}{64.4} + h_L$$

(4–3)

where Z = elevation, feet
$\quad P$ = pressure, pounds per square foot
$\quad V$ = velocity, feet per second
$\quad h_L$ = head loss, feet
$\quad 62.4$ = unit weight of water, pounds per cubic foot
$\quad 64.4$ = 2g, feet per second squared

The expression for Eq. 4–3 in SI metric units is

$$Z_1 + \frac{P_1}{9.80} + \frac{V_1^2}{19.6} = Z_2 + \frac{P_2}{9.80} + \frac{V_2^2}{19.6} + h_L$$
(SI units) (4–4)

where Z = elevation, meters
$\quad P$ = pressure, kilopascals (kilonewtons per square meter)

$\quad V$ = velocity, meters per second
$\quad h_L$ = head loss, meters
$\quad 9.80$ = specific weight of water, kilonewtons per cubic meter
$\quad 19.6$ = 2g, meters per second squared

Head loss as a result of pipe friction can be computed by using the Darcy Weisbach equation:

$$h_L = f\frac{LV^2}{D2g}$$

(4–5)

where h_L = head loss, feet
$\quad f$ = friction factor, Figure 4–6
$\quad L$ = length of pipe, feet
$\quad V$ = velocity of flow, feet per second
$\quad D$ = diameter of pipe, feet

The friction factor f is related to the relative roughness of the pipe material and the fluid flow characteristics. For turbulent water flow, the f value for Eq. 4–5 can be taken from Figure 4–6 based on pipe diameter and material roughness.

Valves, fittings, and other appurtenances disturb the flow of water, causing losses of head in addition to the friction loss in the pipe. These units in a typical distribution system are at infrequent intervals; consequently, losses due to appurtenances are relatively insignificant in comparison with pipe friction losses. In the case of pumping stations and treatment plant piping, the minor losses in valves and fittings are significant and

Figure 4–6

Relative roughness of pipe materials and friction factors for turbulent pipe flow.

(From *Flow of Fluids*, Courtesy of Crane Co. All rights reserved.)

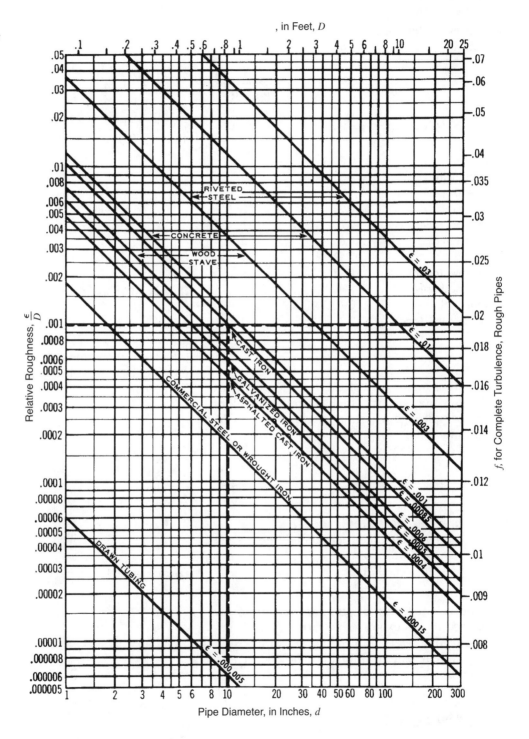

TABLE 4-1

Approximate Minor Head Losses in Fittings and Valves

FITTING OR VALVE	LOSS COEFFICIENT k	EQUIVALENT LENGTH (DIAMETERS OF PIPE)
Tee (run)	0.60	20
Tee (branch)	1.80	60
90° bend		
Short radius	0.90	32
Medium radius	0.75	27
Long radius	0.60	20
45° bend	0.42	15
Gate valve (open)	0.48	17
Swing check valve		
(open)	3.7	135
Butterfly valve (open)	1.2	40

constitute a major part of the total losses. Unit head losses may be expressed as being equivalent to the loss through a certain length of pipe or by the formula

$$h_L = \frac{kV^2}{2g} \qquad (4\text{--}6)$$

where h_L = head loss, feet (meters)
      V = velocity, feet per second (meters per second)
      k = loss coefficient

The equivalent length of pipe that results in the same head loss is expressed in terms of the number of pipe diameters. Table 4–1 gives values for k and equivalent lengths of pipe for various fittings and valves.

■ EXAMPLE 4-1

Calculate the head loss in a 24-in.-diameter, 5000-ft-long, smooth-walled concrete ($\epsilon = 0.001$) pipeline carrying a water flow of 10 cu ft/sec.

Solution

Using Eq. 4–1,

$$V = \frac{Q}{A} = \frac{10 \text{ cu ft/sec}}{\pi \text{ sq ft}} = 3.2 \text{ ft/sec}$$

From Figure 4–6, entering with $d = 24$ in. and $\epsilon = 0.001$, $f = 0.017$. Then substituting into Eq. 4–5,

$$h_L = 0.017 \frac{5000}{2} \times \frac{(3.2)^2}{2 \times 32.2} = 6.8 \text{ ft}$$

■ ■ ■

■ EXAMPLE 4-2

A pump discharge line consists of 200 ft of 12-in. new cast-iron pipe, three 90° medium-radius bends, two gate valves, and one swing check valve. Compute the head loss through the line at a velocity of 3.0 ft/sec.

Solution

If equivalent lengths for the fittings and valves as given in Table 4–1 are used, the total equivalent pipe length is

$$200 + 3 \times 27 + 2 \times 17 + 1 \times 135 = 450 \text{ ft}$$

From Figure 4–6, $f = 0.019$; then using Eq. 4–5,

$$h_L = 0.019 \frac{450}{1.0} \times \frac{(3.0)^2}{2 \times 32.2} = 1.2 \text{ ft}$$

■ ■ ■

■ EXAMPLE 4-3

Calculate the head loss in the pipeline illustrated in Figure 4–5 based on the following: $Z_1 = 4.5$ m, $P_1 = 280$ kPa, $V_1 = 1.2$ m/s, $Z_2 = 9.3$ m, $P_2 = 200$ kPa, $V_2 = 1.2$ m/s.

Solution

Substituting into Eq. 4–4,

$$4.5 + \frac{280}{9.80} + \frac{(1.2)^2}{19.6} = 9.3 + \frac{200}{9.80} + \frac{(1.2)^2}{19.6} + h_L$$

$$4.5 + 28.6 + 0.07 = 9.3 + 20.4 + 0.07 + h_L$$

$$h_L = 3.4 \text{ m} = 33 \text{ kPa}$$

■ ■ ■

4–3 FLOW IN PIPES UNDER PRESSURE

The Darcy Weisbach equation for computing head loss is cumbersome and not widely used in water-works design and evaluation. A trial-and-error solution is required to determine pipe size for a given flow and head loss, since the friction factor is based on the relative roughness, which involves the pipe diameter. Because of this practical short-coming, exponential equations are commonly used for flow calculations.

The most common pipe flow formula used in the design and evaluation of a water distribution system is the Hazen Williams, Eq. 4–7.

$$Q = 0.281 \, CD^{2.63} \, S^{0.54} \qquad \textbf{(4–7)}$$

This equation relates the quantity of turbulent water flow through a circular pipe flowing full with diameter of the pipe, slope of the hydraulic gradient, and coefficient of friction depending on the roughness of the pipe. Coefficient values for different pipe materials are given in Table 4–2.

where Q = quantity of flow, gallons per minute
$\quad\quad C$ = coefficient, Table 4–2
$\quad\quad D$ = diameter of pipe, inches
$\quad\quad S$ = hydraulic gradient, feet per foot

The nomograph shown in Figure 4–7 in English units solves the equation for a coefficient equal to 100, representing 15- to 20-year-old, ductile-iron

TABLE 4–2

Values of Coefficient C for the Hazen Williams Formula, Equations 4–7, 4–8, and 4–9

Pipe Material	C
Asbestos cement	140
Ductile iron	
Cement lined	130 to 150
New, unlined	130
5-year-old, unlined	120
20-year-old, unlined	100
Concrete	130
Copper	130 to 140
Plastic	140 to 150
New welded steel	120
New riveted steel	110

pipe. Given any two of the parameters (discharge, diameter of pipe, loss of head, or velocity), the remaining two can be determined from the intersections along a straight line drawn across the nomograph. For example, the flow of 500 gpm in an 8-in.-diameter pipe has a velocity of 3.2 ft/sec with a head loss of 8.5 ft/1000 ft (0.0085 ft/ft).

Head losses in pipes with coefficient values other than 100 can be determined by using the correction factors in Table 4–3. For example, if the head loss at $C = 100$ is 8.5 ft/1000 ft, the head loss at $C = 130$ would be equal to $0.62 \times 8.5 = 5.3$ ft/1000 ft.

The Hazen Williams equation can also be expressed to calculate head loss directly by substituting $S = h_L/L$ and rearranging

$$h_L = 0.002083 \times L \left(\frac{100}{C}\right)^{1.85} \frac{Q^{1.85}}{D^{4.8655}} \qquad \textbf{(4–8)}$$

where $\quad h_L$ = head loss, feet
$\quad\quad L$ = length of pipe, feet
$\quad\quad C$ = coefficient, Table 4–2
$\quad\quad Q$ = quantity of flow, gallons per minute
$\quad\quad D$ = diameter of pipe, inches

Figure 4–7
Nomograph in English units for the Hazen Williams formula based on a $C = 100$.

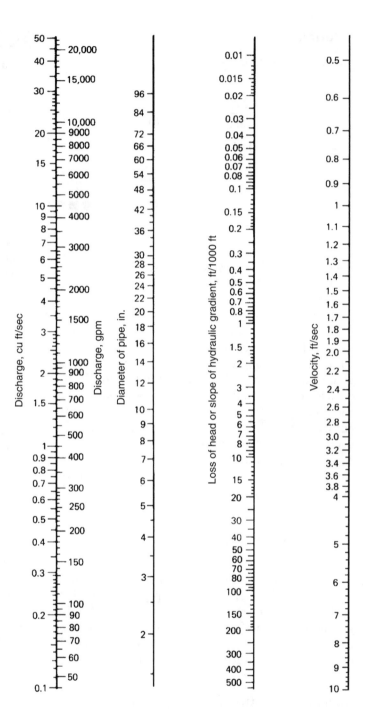

The Hazen Williams formula, Eq. 4–8, expressed in SI metric units is

$$Q = 0.278 \, CD^{2.63}S^{0.54} \text{ (SI units)} \qquad \textbf{(4–9)}$$

where Q = quantity of flow, cubic meters per second

C = coefficient, Table 4–2

D = diameter of pipe, meters

S = hydraulic gradient, meters per meter

To calculate head loss directly,

$$h_L = 0.002131 \times L\left(\frac{100}{C}\right)^{1.85} \frac{Q^{1.85}}{D^{4.8655}} \text{ (SI units)}$$

$$\textbf{(4–10)}$$

TABLE 4–3

Correction Factors to Determine Head Losses from Figure 4–7 at Values of C Other than $C = 100$

	CORRECTED $h_L = K \times h_L$ AT $C = 100$		
C	K	C	K
80	1.51	120	0.71
100	1.00	130	0.62
110	0.84	140	0.54

where h_L = head loss, meters
L = length of pipe, meters
Q = quantity of flow, cubic meters per second
C = coefficient, Table 4–2
D = diameter of pipe, meters

The nomograph shown in Figure 4–8 in SI units solves the Hazen Williams equation for a coefficient equal to 100, representing 15- to 20-year-old, ductile-iron pipe. In the nomograph, given any two parameters (discharge, diameter of pipe, loss of head, or velocity), the remaining two can be determined from the intersections along a straight line drawn across the nomograph. For example, the flow of 30 l/s in a 200-mm-diameter pipe has a head loss of 0.0080 m/m and velocity of 0.95 m/s.

■ EXAMPLE 4–4

Groundwater from a well field is pumped through a 14-in.-diameter transmission main for 10,500 ft to the treatment plant. Calculate the head loss for a flow rate of 1400 gpm assuming $C = 100$ and assuming $C = 140$.

Solution

Using Eq. 4–8 with $C = 100$,

$$h_L = 0.002083 \times 10,500 \left(\frac{100}{100}\right)^{1.85} \times \frac{(1400)^{1.85}}{(14)^{4.8655}}$$
$$= 38.3 \text{ ft}$$

Using the nomograph in Figure 4–7 for $C = 100$, draw a straight line through a discharge of 1400 gpm

and diameter of 14 in. The intercept on the loss-of-head line reads 3.6 ft/1000 ft.

$$h_L = 3.6 \times 10.5 = 38 \text{ ft}$$

Using Eq. 4–8 with $C = 140$,

$$h_L = 0.002083 \times 10,500 \left(\frac{100}{140}\right)^{1.85} \times \frac{(1400)^{1.85}}{(14)^{4.8655}}$$
$$= 20.6 \text{ ft}$$

Using a loss of head value of 3.6 from the nomograph and the correction factor for C from Table 4–3,

$$h_L = 3.6 \times 10.5 \times 0.54 = 20 \text{ ft}$$

■ ■ ■

■ EXAMPLE 4–5

If a 200-mm water main ($C = 100$) is carrying a flow of 30 l/s, what is the velocity of flow and head loss?

Solution

(a) Using Figure 4–8, a straight line extended through a discharge of 30 l/s and a diameter of 200 mm intersects the head loss at 0.008 m/m.

(b) $Q = 30 \, l/s = 0.030 \text{ m}^3/s$ and $D = 200 \text{ mm}$
  $= 0.20 \text{ m}$

The cross-sectional area of flow $= \pi(0.20/2)^2 = 0.0314 \text{ m}^2$. Using Eq. 4–1,

$$0.030 \text{ m}^3/s = V \times 0.0314 \text{ m}^2$$
$$V = 0.030/0.0314 = 0.955 \text{ m/s}$$

Substituting into Eq. 4–9,

$$0.030 \text{ m}^3/s = 0.278 \times 100(0.20 \text{ m})^{2.63}S^{0.54}$$
$$S^{0.54} = \frac{0.030}{0.278 \times 100(0.20)^{2.63}} = \frac{0.030}{0.403} = 0.0744$$
$$S = 0.0744^{1/0.54} = 0.0744^{1.85} = 0.00817 \text{ m/m}$$
$$= 8.17 \text{ m}/1000 \text{ m}$$

Figure 4–8

Nomograph in SI units for the Hazen Williams formula based on a $C = 100$.

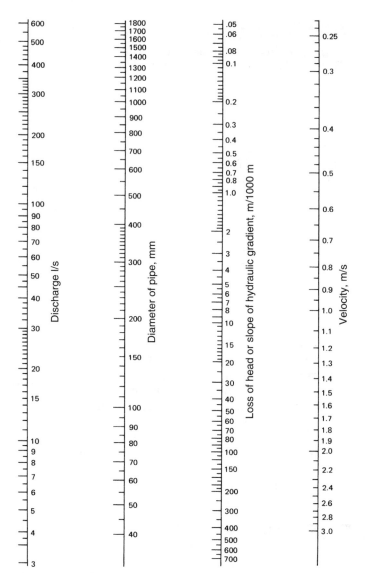

(c) Substituting into Eq. 4–10,

$$\frac{h_L}{L} = 0.002131\left(\frac{100}{100}\right)^{1.85}\frac{(0.030)^{1.85}}{(0.20)^{4.8655}}$$

$$= 0.00817 \text{ m/m}$$

■■■

■ EXAMPLE 4–6

An extremely simplified water supply system consisting of a reservoir with lift pumps, elevated storage, piping, and load center (withdrawal point)

is shown in Figure 4–9. (a) Based on the following data, sketch the hydraulic gradient for the system:

$Z_A = 0$ ft, $P_A = 80$ psi, $Z_B = 30$ ft, $P_B = 30$ psi, $Z_C = 40$ ft, $P_C = 100$ ft (water level in tank)

(b) For these conditions, compute the flow available at point B from both supply pumps and elevated storage. Use $C = 100$ and pipe sizes as shown in the diagram.

Solution

(a) Hydraulic head at $A=0$ ft $+ 80$ psi $\times 2.31$ ft/psi $= 185$ ft; at $B = 30 + 30 \times 2.31 = 99$ ft; at $C = 40 + 100 = 140$ ft.

Figure 4–9
Simplified water system for Example 4–6.

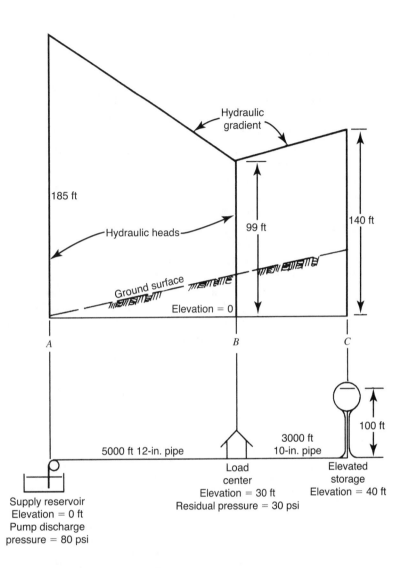

The hydraulic gradient is shown as straight lines connecting these hydraulic heads drawn vertically.

(b) h_L between A and $B = 185 - 99 = 86$ ft

$$h_L \text{ per } 1000 \text{ ft} = 86 \div 5 = 17.2 \text{ ft}$$

Using Figure 4–7, align $h_L = 17.2$ ft/1000 ft and 12-in. diameter, read $Q = 2160$ gpm.

$$h_L \text{ per } 1000 \text{ ft } B \text{ to } C = \frac{140 - 99}{3} = 13.7 \text{ ft}$$

For $h_L = 13.7$ ft/1000 ft and 10 in., $Q = 1180$ gpm. Hence, total Q available at $B = 2160 + 1180 = 3340$ gpm.

■ ■ ■

4–4   CENTRIFUGAL PUMP CHARACTERISTICS

Pumps are used for a variety of functions in water and wastewater systems. Low-lift pumps are used to elevate water from a source, or wastewater from a sewer, to the treatment plant; high-service pumps are used to discharge water under pressure to a distribution system or wastewater through a force main; booster pumps are used to increase pressure in a water distribution system; recirculation and transfer pumps are used to move water being processed within a treatment plant; well pumps are used to lift water from shallow or deep wells for water supply; still other types are used for chemical feeding, sampling, and fire fighting.

Centrifugal pumps are used commonly for low and high service to lift and transport water, reciprocating positive-displacement and progressing cavity pumps are used to move sludges, vertical turbine pumps are used for well pumping, and pneumatic ejectors are used for small wastewater lift stations. Air-lift, peristaltic, rotary displacement, and turbine pumps are used in special applications. The discussion in this section, however, is restricted to the characteristics of centrifugal pumps; specific types are illustrated and discussed in relation to their applications in other sections of the book.

Centrifugal pumps are popular because of their simplicity, compactness, low cost, and ability to operate under a wide variety of conditions. The essential parts are a rotating member with vanes, the impeller, and a surrounding case (Figure 4–10). The impeller, driven at a high speed, throws water into the volute, which channels it through the nozzle to the discharge piping.

This action depends partly on centrifugal force, hence the particular name given to the pump. The function of the pump passages is to develop water pressure by efficient conversion of kinetic energy.

A closed impeller is generally used in pumping water for higher efficiency, while an open unit is used for wastewater containing solids. The casing may be in the form of a volute (spiral) or may be equipped with diffuser vanes. Several other variations in design involve casing and impeller modifications to provide a wide range of pumps with special operating features.

Pump Head-Discharge Curve

The head developed by a particular pump at various rates of discharge at a constant impeller speed is established by pump tests conducted by the manufacturer. The head given is the discharge pressure with the inlet static water level at the elevation of the pump centerline and excluding losses in suction and discharge piping. Consider the test arrangement illustrated schematically in Figure 4–11, where the discharge is controlled by a throttling valve, the discharge pressure measured by a gauge, and the rate of discharge recorded by a flow meter. Simultaneously, the power input is measured and efficiency determined. With the

Figure 4–10

Cross-sectional diagrams showing the features of a centrifugal pump.

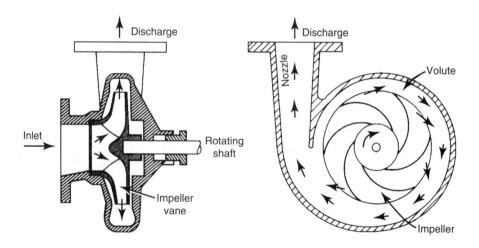

Figure 4–11

Schematic of a pump head-discharge test.

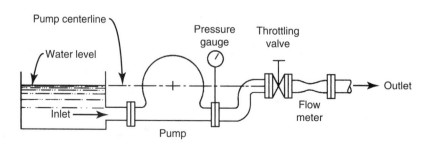

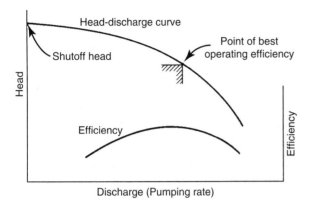

Figure 4–12
Characteristic curves for a centrifugal pump operating at a constant speed.

Pump Characteristics

Normally, a pump can be furnished with impellers of different diameters, within a specified range, for each size casing. Since the discharge of a pump changes with impeller diameter and operating speed, manufacturers publish data showing the characteristic curves for impellers of several diameters operating at two or more speeds. For a given impeller diameter operated at different speeds, the discharge is directly proportional to the speed (Eq. 4–11), the head is proportional to the square of the speed (Eq. 4–12), and the power input varies with the cube of the speed (Eq. 4–13).

$$\frac{Q_1}{Q_2} = \frac{N_1}{N_2} \tag{4–11}$$

$$\frac{H_1}{H_2} = \frac{N_1^{\,2}}{N_2^{\,2}} \tag{4–12}$$

$$\frac{P_{i_1}}{P_{i_2}} = \frac{N_1^{\,3}}{N_2^{\,3}} \tag{4–13}$$

where Q = discharge, gallons per minute (liters per second)
  H = head, feet (meters)
  P_i = power input, horsepower (kilowatts)

For a pump operating at the same speed, a change in impeller diameter affects discharge, head, and power input approximately as follows:

$$\frac{Q_1}{Q_2} = \frac{D_1}{D_2} \tag{4–14}$$

$$\frac{H_1}{H_2} = \frac{D_1^{\,2}}{D_2^{\,2}} \tag{4–15}$$

$$\frac{P_{i_1}}{P_{i_2}} = \frac{D_1^{\,3}}{D_2^{\,3}} \tag{4–16}$$

where D = impeller diameter, inches (centimeters).

valve in the discharge pipe closed, the rotating impeller simply churns in the water, causing the pressure at the outlet of the pump to rise to a value referred to as the shutoff head. As the valve is gradually opened, allowing increasing flow of water, the pump head decreases, as drawn in Figure 4–12. The pump efficiency rises with increasing rate of discharge to an optimum value and then decreases. The flow rate at peak efficiency is determined by the pump design and the rotational speed of the impeller.

A centrifugal pump is designed to operate near the point of best operating efficiency. Radial loads on the bearings are at a minimum when the impeller is operating at a balanced load at this point. As the pump discharge increases beyond optimum operation, the radial loads increase and cavitation is a potential problem. Cavitation occurs when vapor bubbles form at the pump inlet and are carried into a zone of higher pressure where they collapse abruptly. The force of the surrounding water rushing in to fill the voids creates a hammering action and high localized stresses that can pit the pump impeller. When the rate of pump discharge decreases toward the shutoff head, recirculation of water within the casing can cause vibration and hydraulic losses in the pump. For these reasons, operating a pump at a rate of discharge within a range between 60 and 120 percent of the best efficiency point (bep) is good practice.

Power and Efficiency

Power input is the motor power applied to a pump. The power output is the work done per unit of time lifting the water to a higher elevation. The efficiency of a pump is the ratio of power output to measured power input.

$$E_p = \frac{P_o}{P_i} \qquad (4\text{-}17)$$

where E_p = pump efficiency, dimensionless
  P_i = motor power input (brake horsepower), horsepower (kilowatts)
  P_o = power output, horsepower [kilowatts (kilonewton · meters per second)]

Centrifugal pump efficiency is usually in the range of 0.60 to 0.85 (60 to 85 percent).

The equation relating discharge to motor power input is

$$P_i = \frac{wQH}{(550 \times 60)E_p} = \frac{QH}{3960E_p} \qquad (4\text{-}18)$$

where P_i = motor power input, horsepower
  Q = discharge, gallons per minute
  H = head, feet
  w = unit weight of water = 8.34 lb/gal
  E_p = pump efficiency, dimensionless
  550 = foot pounds/second per horsepower
  60 = seconds per minute
  3960 = 550 × 60/8.34

In SI metric units, Eq. 4–18 is expressed as

$$P_i = \frac{wQH}{E_p} = \frac{0.0098\,QH}{E_p} \quad \text{(SI units)} \qquad (4\text{-}19)$$

where P_i = motor power input, kilowatts
  Q = discharge, liters per second
  H = head, meters
  w = unit weight of water = 0.0098 kN/l

■ EXAMPLE 4–7

The characteristics of a centrifugal pump operating at two different speeds are listed in the following chart. Graph these curves and connect the best efficiency points (bep) with a dashed line.

Calculate head-discharge values for an operating speed of 1450 revolutions per minute (rpm) and plot the curve. Finally, sketch the pump operating envelope between 60 and 120 percent of the best efficiency points.

SPEED = 1750 RPM			SPEED = 1150 RPM		
DISCHARGE (GPM)	HEAD (FT)	EFFI-CIENCY (%)	DISCHARGE (GPM)	HEAD (FT)	EFFI-CIENCY (%)
0	220	—	0	96	—
1500	216	63	1000	93	65
2500	203	81	1500	89	77
3000	192	85	2000	82	83
3300	182	86[a]	2200	77	84[a]
3500	176	85	2500	70	83
4500	120	72	3000	49	71

[a]Best efficiency point.

Solution

The characteristic curves for 1750 rpm and 1150 rpm are drawn in Figure 4–13, with efficiency values listed at the plotting points. The best efficiency points are connected with a dashed line.

The head-discharge values for an operating speed of 1450 rpm are calculated using Eqs. 4–11 and 4–12. For example, using the data of 1500 gpm and 216 ft at 1750 rpm, one plotting point is calculated as

$$Q_2 = Q_1\frac{N_2}{N_1} = 1500\,\frac{1450}{1750} = 1240 \text{ gpm}$$

$$H_2 = H_1\left(\frac{N_2}{N_1}\right)^2 = 216\left(\frac{1450}{1750}\right)^2 = 148 \text{ ft}$$

All of the values calculated from the data at 1750 rpm are 1240 gpm at 148 ft, 2040 at 140, 2450 at 132, 2730 at 125 (bep), 2860 at 121, and 3730 gpm at 82.

The boundaries of the operating envelope at 1750 rpm are calculated as 60 and 120 percent of the discharge at the best operating point of 3300 gpm.

$$0.60 \times 3300 = 2000 \text{ gpm}$$
$$1.20 \times 3300 = 4000 \text{ gpm}$$

Figure 4–13

Characteristic pump curves for Example 4–7 showing the pump operating envelope.

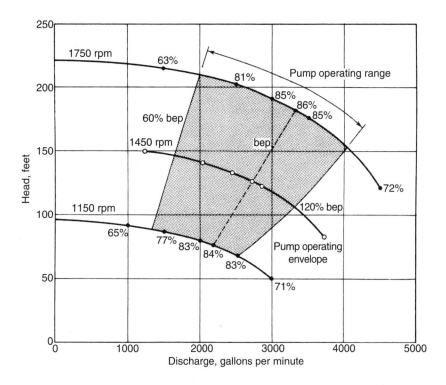

At 1150 rpm for a bep of 2200 gpm, the limits are 1300 gpm and 2600 gpm. The pump operating envelope is shown shaded in Figure 4–13.

■ ■ ■

4–5 SYSTEM CHARACTERISTICS

When a centrifugal pump lifts water from a reservoir into a piping system, the resistance to flow at various rates of discharge is described by a system head curve. The two components of discharge resistance are the static head, which is the elevation difference between the water levels in the suction reservoir and the discharge tank or point of discharge, and friction head loss, which increases with pumping rate.

Consider the simplified water system illustrated in Figure 4–14. Water is being pumped through a pipe to discharge either into elevated storage (outlet 1) or at the load center (outlet 2) or both. Hydraulic grade line 1 is for the entire pump flow Q discharging into elevated storage at outlet 1. Hydraulic grade line 2 is for the entire pump flow discharging from outlet 2. Under the condition of grade line 2, water is being pumped

also flowing out of elevated storage and discharging from outlet 2. The system head-discharge curves for these hydraulic grade lines are shown in Figure 4–15. Curve 1 is for the pump flow entering elevated storage, and curve 2 is for the pump flow discharging at outlet 2, the load center. The head-discharge operating point for this piping and elevated storage system can lie anywhere between these two curves, depending on the relative amounts of water discharging at outlets 1 and 2. For a real water distribution network, the system head curve is actually a curved band enveloping a series of individual curves for different conditions of pumping, storage, and withdrawal.

Constant-Speed Pumps

Continuity of flow rate and water pressure must exist at the common boundary between a pump and piping system. Hence, a constant-speed pump operates at the head-discharge point defined by the intersection of the pump head-discharge curve and the system head-discharge curve. Figure 4–16 graphs the system curves 1 and 2 from Figure 4–15 with a constant-speed pump

Figure 4–14

Schematic of a simplified water-supply system consisting of a pipeline, a lift pump, elevated storage, and a withdrawal outlet at the load center. The hydraulic grade lines are 1 for discharge at outlet 1 only and 2 for discharge at outlet 2 only.

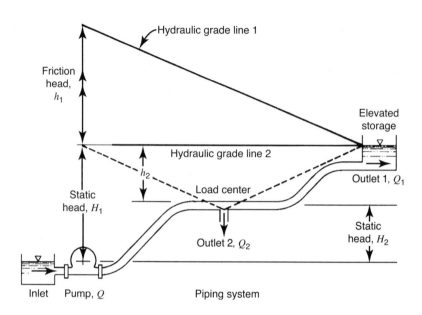

head-discharge curve. When the entire discharge flows into elevated storage, pump operation is at the point labeled A. If free discharge is allowed only from the piping system at outlet 2, the pump operates at point B on the head-discharge curves. The quantity of flow, Q_2, is greater than Q_1, since the operating head (static head plus friction head loss) at B under condition 2 is less than the operating head at A under condition 1. Starting from the condition of hydraulic grade line 2, consider throttling the discharge at outlet 2.

The result is the raising of both the hydraulic grade line and the system head-discharge curve so that the new operating point along the pump head-discharge curve is at some point C between A and B. A constant-speed centrifugal pump always operates at some point along its pump head-discharge curve, and the system

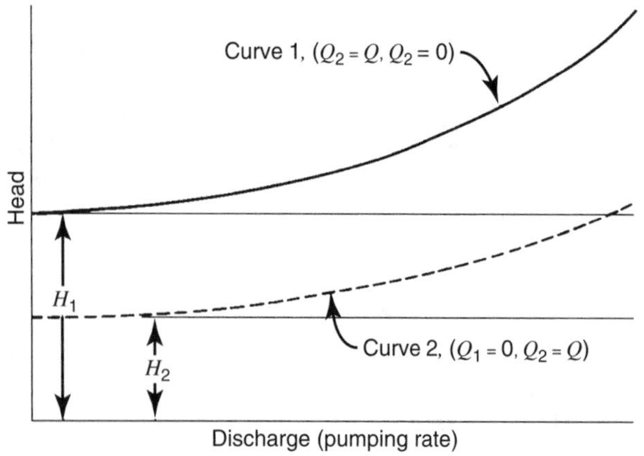

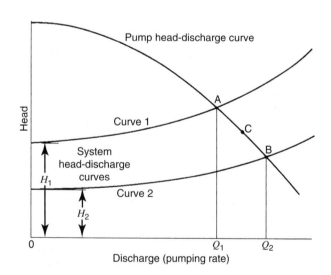

Figure 4–16

The head-discharge curve for a constant-speed pump drawn with system head-discharge curves to illustrate pump operating points. The related hydraulic scheme is given in Figure 4–14, and the system head-discharge curves are redrawn from Figure 4–15.

Figure 4–15

System head-discharge curves for the simplified water system illustrated in Figure 4–14. Curve 1 for discharge at outlet 1 only, and curve 2 is for discharge at outlet 2 only.

head-discharge curve must intersect at this operating point.

Pumping stations with two constant-speed pumps of the same capacity may be used in small water systems. For example, the water can be pumped directly into elevated storage that, in turn, supplies the distribution network. The pump operates intermittently, controlled by the fluctuation of water level in the tank, and the elevated storage maintains pressure in the distribution system independent of pump operation. In larger systems, the installation of at least three pumps is desirable to cover the extremes of water demand and to provide a standby pump in case one unit is out of service. Since a larger system has continuous water demand, the pumps can discharge directly into the distribution piping, and elevated storage tanks are connected to the pipe network.

Parallel operation of constant-speed pumps is illustrated graphically in Figure 4–17. Pumps 1 and 2 are identical units, pump 3 has a greater capacity and higher shutoff head, and head-discharge curves include friction losses in suction and discharge piping. The pumps are operated individually or in combination to meet the water demand by discharging into a common header and outlet pipe. When two or more pumps operate simultaneously, the combined head-discharge curve is determined by adding the rates of discharge at the same head of the individual curves. The point of intersection of the combined head-discharge curve and the system curve gives the combined rate of discharge of the pumps and the operating head of the pumps.

Variable-Speed Pumps

The best way to maintain a constant pump discharge pressure over a wide range of flow rates is by varying the rotational speed of the pump impeller. Although the discharge of a pump running at constant speed can be controlled by the use of a throttling valve installed in the pump outlet, restricting discharge causes the impeller to recirculate water in the casing, reducing efficiency and possibly damaging the pump bearings and impeller. Speed control of a centrifugal pump is accomplished by using an electric motor designed for stepless-speed drive.

The head-discharge curves for a pump operating at two impeller speeds are given in Figure 4–18a, with the lines of equal efficiency plotted between the two curves. Actually, the pump can operate on an infinite number of curves between the minimum and maximum speeds. The demand-head curve, which is the required head boost by the pump versus flow demand of the system, turns downward from the theoretical constant-control pressure, as sketched in Figure 4–18b. Pump speed increases when the pump discharge pressure reduces as a result of increasing demand and decreases with increasing discharge pressure. The transducer of a variable-speed drive transfers the signal from the pressure switches to the drive motor. To prevent cycling between pressure sensors, the decrease pressure switch is usually adjusted to be actuated when the pressure is slightly higher than the pressure that actuates the increase pressure switch. This transducer span results in an outlet pressure at maximum pump discharge (maximum demand) approximately 10 percent less than the outlet pressure at zero

Figure 4–17
Head-discharge curves for multiple-pump operation in parallel. (a) Parallel installation arrangement. Pump numbers 1 and 2 are the same size; number 3 is larger. (b) Head-discharge curves for the pumps operating individually and in various combinations, with the system head-discharge curve shown as a dotted line.

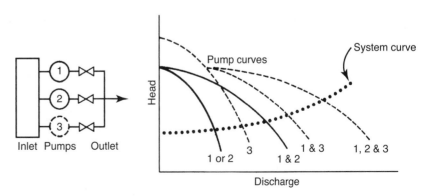

(a)   (b)

Figure 4–18

Characteristic curves for a variable-speed pump. (*a*) Head-discharge curves with efficiency values for a pump operating at two impeller speeds. (*b*) System demand-head curve for a constant pressure discharge corrected for transducer operation. (*c*) Superimposed pump head-discharge and demand curves. (*d*) Curves of speed and efficiency versus demand.

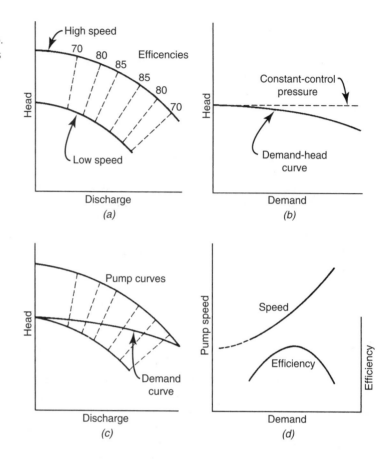

discharge. Transducer span and downward curvature of the demand-head curve can be reduced to zero with a more complex pump control system that includes error detection and a correction loop; however, this feature is usually unnecessary in water–pumping applications. The pump head-discharge and demand-head curves are graphed together in Figure 4–18*c*. Values from the intersections of the demand-head curve with the pump speed and efficiency lines can be used to plot speed and efficiency versus demand (Figure 4–18*d*).

A variable-speed drive must be prevented from operating a pump at extremely low speeds. When the demand is less than the minimum required discharge, the pump is protected from damage by recirculating water through the pump. The recommended minimum discharge rate is generally 25 to 35 percent of the pumping rate at the best operating efficiency. Two control techniques are shown schematically in Figure 4–19. The first system uses a flow meter to operate a modulating valve in a by-pass to maintain a nearly constant flow

Figure 4–19

Arrangements for protecting variable-speed pumps at low discharge. (*a*) Flow-meter modulating valve system. (*b*) Back-pressure regulating valve arrangement.

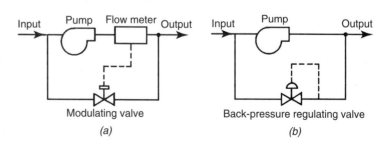

through the pump when the demand is less than the minimum recommended rate of discharge. A second system employing a back-pressure regulating valve in the by-pass can be used if the discharge curve turns downward. The regulating valve is adjusted to maintain a pump discharge pressure approximately 10 percent above the desired output head at a pump discharge equal to the minimum rate of discharge. Therefore, at high output pressure corresponding to lowering demand, the valve in the by-pass begins to open, allowing recirculation flow.

Variable-speed pumps can be operated in parallel in multiple-pump installations. Based on the choice of design, the pumps may function by load sharing or staggered operation. In load sharing, all pumps run at the same speed and discharge at equal rates. In staggered operation, one or more of the pumps runs at optimum efficiency (constant speed) while the speed of only one pump is varied to meet changing demand. The power input for alternative systems should be calculated to determine the design for most economical operation. Variable-speed pumps can also be used in combination with constant-speed pumps. Often, the variable-speed unit meets the low-flow demands in staggered operation. The flow head-discharge curves and demand-head curves can be added for multiple-pump operation in the manner described for multiple constant-speed pump installations (Figure 4–17).

■ EXAMPLE 4–8

Draw system head-discharge curves for the two operating conditions in the simplified water system diagramed in Figure 4–20. The highest system head-discharge curve (upper boundary) occurs when pump discharge enters the elevated storage tank with no system withdrawal. The lowest anticipated system head-discharge curve occurs when the pressure at the load center is 45 psi and flow is entering the system from both pump discharge and elevated storage. On the same head-discharge diagram, draw the characteristic pump curve from Figure 4–13 at an operating speed of 1750 rpm. Label the range of pump operation.

Solution

Under the condition of no system withdrawal, the static head transmitted to the pumps by elevated storage is 150 ft. The friction head losses in the 8000 ft of 16-in. pipe is determined for different pump discharges. The total pump head is the static head plus the friction head loss. For example, at 2000 gpm the head loss in 8000 ft of 16-in. pipe equals 8000 ft × 0.0037 ft/ft, or 30 ft, and the total pump head is 150 ft plus 30 ft, for 180 ft. Therefore, one plotting point on the upper system head-discharge curve is 2000 gpm and 180 ft. Other calculated points are shown along the curve plotted in Figure 4–21.

For discharge at the load center, the static head is 45 psi or 104 ft, and the friction head loss is determined for 5000 ft of 16-in. pipe. For example, at 2000 gpm the total head is 104 ft plus 5000 × 0.0037 ft/ft, equaling 123 ft. This is one of the plotting points on the lower system head-discharge curve.

The pump head-discharge curve is the same as the pump characteristic curve at 1750 rpm from Figure 4–13. During operation along the upper

Figure 4–20
Simplified water system for Example 4–8.

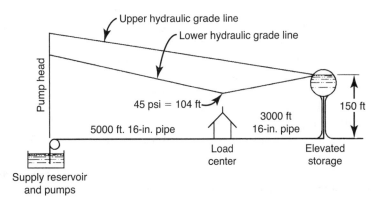

Figure 4–21

Solution for Example 4–8. System head-discharge curves for simplified water system diagramed in Figure 4–20, and characteristic pump curve at 1750 rpm from Figure 4–13.

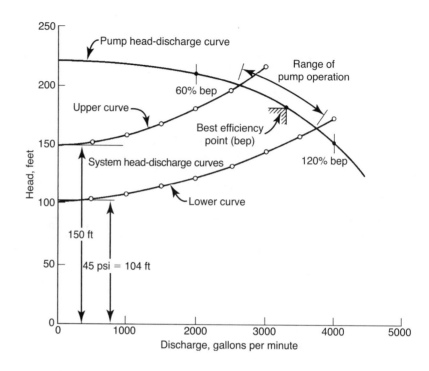

system head-discharge curve, the pump head is 200 ft and the discharge is 2620 gpm. The intersection of the lower system head-discharge curve and the pump curve is at 165 ft and 3770 gpm. The range of pump operation straddles the best efficiency point and is within the normal pump operating range of 60 to 120 percent of the bep.

■ ■ ■

4–6 EQUIVALENT PIPES

An equivalent pipe is an imaginary conduit that replaces a section of a real system such that the head losses in the two systems are identical for the quantity of flow. For example, pipes of differing diameters connected in series can be replaced by an equivalent pipe of one diameter as follows: Assume a quantity of flow and determine the head loss in each section of the line for this flow; then, using the sum of the sectional head losses and the assumed flow, enter the nomograph to find the equivalent pipe diameter. For parallel pipe systems, a head loss is assumed, and the quantity of flow through each of the pipes is calculated for that head loss. Then the sum of the flows and the assumed head loss are used to determine the equivalent pipe

size. Example 4–9 illustrates equivalent pipe calculations for lines in series and parallel.

The method of equivalent pipes cannot be applied to complex systems, since crossovers result in pipes being included in more than one loop and because normally a number of withdrawal points are throughout the system. Flows in pipes, head losses, and water pressures in distribution systems are determined by computer analysis.

■ EXAMPLE 4–9

Determine an equivalent pipeline 2000 ft in length to replace the pipe system illustrated in Figure 4–22.

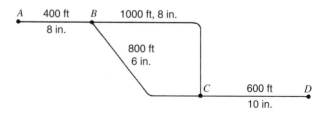

Figure 4–22

Pipe system for Example 4–9.

Solution

First replace the parallel pipes between B and C with one equivalent line 1000 ft in length. Assume a head loss of 10 ft between B and C. Based on the nomograph, the flow in 1000 ft of 8-in. pipe at a head loss of 10 ft/1000 ft = 550 gpm, and the flow in the 6-in. pipe at a loss of 10 ft/800 ft (12.5 ft/1000 ft) = 290 gpm. The equivalent pipe B to C, 1000 ft in length, is then that size pipe that has a 10 ft/1000 ft head loss at a discharge of 550 + 290 = 840 gpm. From the nomograph, diameter = 9.4 in.

Next consider the three pipes in series: 400 ft of 8 in., 1000 ft of 9.4 in., and 600 ft of 10 in. At an assumed flow of 500 gpm, the head losses in the pipes are 0.4 × 8.3 = 3.3 ft, 3.8 ft, and 0.6 × 2.7 = 1.6 ft, respectively, for a total of 8.7-ft head loss in 2000 ft. The equivalent pipe A to D is then that diameter that exhibits a loss of 4.4 ft/1000 ft at a flow = 500 gpm. The answer is 9.2 in.

■ ■ ■

4–7 COMPUTER ANALYSIS OF PIPE NETWORKS

Several mathematical programs have been developed for computer analyses of water distribution systems using desktop computers. The common choice is a steady-state simulation model that represents the system for any prescribed set of flow and pressure conditions. An unsteady-state model simulates the behavior of a system during a series of successive steps over time. After an appropriate computer program has been selected, the data on the water system are entered and the model is calibrated based on field measurements. The computer printout sheets usually organize the results in a columnar format, listing flows in pipes, pressures at pipe junctions, hydraulic gradients, pump discharge pressures, and all of the input data. The physical components describing a water distribution system are the pipe network, storage reservoirs, pumps, valves, and other appurtenances.

A computer program stores the pipe network as a mathematically defined map by incorporating data on pipe lengths and diameters, roughness values (Hazen Williams C values), pipe junctions (nodes) including elevations and connecting pipes

by number, check valves, and pressure regulators. Pipe roughness factors are generally assumed based on the age of the pipe and then adjusted during calibration of the model. In some cases, actual C values for larger mains are determined by field testing. Some computer programs allow the exclusion of selected pipes when running an analysis. With this option, the model can incorporate anticipated main extensions that can then be either included to evaluate proposed expansions or deleted for analysis of the existing network. To minimize the size of the network model, small-diameter pipes can be either ignored or combined with adjacent lines by substituting hydraulically equivalent pipes. A carefully selected skeletal pipe network generally simplifies the model with no loss of accuracy in analysis. Storage reservoirs are defined by location in the pipe network and their operating water level.

Pumping stations are described by elevation and station head-discharge curves, which are defined mathematically in the normal operating range of the pumps. The pump head-discharge curves from manufacturers cannot be incorporated directly into the program, since they do not account for suction water level and head losses in station piping, column head losses in wells, wear of impellers, or other factors. Pump characteristics for modeling are best determined by field measurements at the pump station or well under a range of discharge pressures. Simultaneous measurements of water level in the suction reservoir or well casing are necessary to adjust the pumping curves for different conditions.

With the physical system modeled, the next step is to incorporate water use data by assigning water withdrawals at the nodes (junctions in the pipe network) for a known condition of water consumption. One approach is to assign water withdrawals for commercial and industrial users to nearby nodes and distribute nodal domestic consumptions based on residential population density and type of dwellings. If the sum of the withdrawals does not equal the average daily water production, the unaccounted-for water may be distributed among the nodes. An alternate approach is to start by assigning the average daily water withdrawal to large commercial and industrial users and assign the remaining consumption to nodes on the basis of land area or length of piping. Obviously, actual water consumption records for residential, commercial, and industrial areas from meter readings provide the best data on water use.

The mathematical model of the system in the computer program must be calibrated to be sure that it represents the real system as closely as possible. Normally, the procedure involves recording of flows, pressures, and operational conditions for selected test days. These data are then entered for computer analysis and the results checked against field measurements. Preliminary calibration is done by measuring pressures at various locations throughout the pipe network when the distribution of withdrawals is typical of the average consumption, which is usually 8 A.M. to 4 P.M. on weekdays. In addition to static pressure recordings at hydrants, essential data include elevations of water in reservoirs, discharge pressures and rates of flow from pumping stations and wells, and flow measurements of meters in the distribution system and at major water customers. If simulation is not satisfactory because of a localized deviation, the possibility of a closed valve or other irregularity in the distribution system should be investigated. A model is generally brought into calibration by changing C values in the piping and, if necessary, adjusting the distribution of water withdrawals. If the model is calibrated to reproduce the pressure observed during average water consumption, accurate prediction is not assured under extreme conditions, such as fire demand or maximum daily use. A major problem is that compensating errors can be introduced into the model during preliminary calibration. For example, higher estimates than actual water use offset the selection of incorrectly high C values, yet the model will appear to be calibrated.

The predictive capability of a model necessitates collection and analysis of independent sets of data for both calibration and verification to eliminate compensating errors. Verification can be accomplished by measurements of peak flows during periods of high seasonal water consumption and by conducting fire flow tests at important locations in the distribution system. Accuracy of calibration is checked by comparing the computer output values of hydraulic parameters with field measurements during verification studies. Predicted pressures should be within ±5 psi. Accuracy in reading a hydrant pressure gauge is

approximately 2 psi, and a 10-ft error in elevation of a test hydrant results in a 4-psi error. Since pressure readings are subject to the influence of inaccurate elevation data, an alternate method to determine the degree of calibration is to compare flows predicted by field test and computer simulation at a common arbitrary residual pressure, such as the 20-psi value used in calculating available fire flows. A third criterion is the ratio of observed to predicted head loss, which also indicates whether to modify the estimated C values or water withdrawals to achieve calibration.

A reliable computer model of a distribution system has several advantages. The hydraulics of the system can be evaluated for optimum energy efficiency. For periods of both low and high water demand, pumps and wells can be analyzed for best operation. Emergency situations, such as a major fire demand or main break, can be studied, and the effect of a potential major industrial customer can be analyzed. In the planning of new facilities, a calibrated model can be used to test alternative schemes and establish priority for improvements. Finally, the model is essential for design in determining the size of new mains, location of distribution storage, and selection of pumps.

The process of writing and calibrating a computer model requires both complete understanding of the water distribution system being modeled and expertise in network analysis. Although writing the mathematical model can be accomplished in a short period of time, calibration and verification require collection of field data intermittently throughout at least one entire year. The activity involving model calibration and optimization of operations provides an educational opportunity for system employees, whereas the program analyses evaluating improvements are often of particular interest in management of the system.

4–8 GRAVITY FLOW IN CIRCULAR PIPES

Sanitary and storm sewers are designed to flow as open channels, not under pressure, although storm sewers may occasionally be overloaded when water rises in the manholes, surcharging the sewer. The wasterwater flows downstream in the pipe, moved by the force of gravity. Velocity of flow depends on steepness of the pipe slope and frictional resistance. The Manning formula, Eq. 4–20, is used for uniform, steady, open channel flow. The coefficient of roughness, n, depends on the condition of the pipe surface, alignment of pipe sections, and method of jointing. The common sewer pipe materials of vitrified clay and smooth concrete have n values in the range of 0.011 to 0.015. The lower value is applicable to clear water and smooth joints, whereas the greater roughness is for wastewater and poor joint construction. Corrugated steel pipe used for culverts has considerably greater roughness, in the range of 0.021 to 0.026. The common value of n adopted for sewer design is 0.013. Even though certain pipe materials, such as plastic, in a new condition may exhibit a lower n, after pipes are placed in use they accumulate grease and other solids that modify the interior and disturb wastewater flow.

$$Q = \frac{1.49}{n} AR^{2/3}S^{1/2} \qquad \textbf{(4–20)}$$

where Q = quantity of flow, cubic feet per second
  n = coefficient of roughness depending on material
  A = cross-sectional area of flow, square feet
  R = hydraulic radius, feet (cross-sectional area divided by the wetted perimeter)
  S = slope of the hydraulic gradient, feet per foot

The Manning formula, Eq. 4–21, expressed in SI metric units is

$$Q = \frac{1.00}{n} AR^{2/3}S^{1/2} \text{ (SI units)} \qquad \textbf{(4–21)}$$

where Q = quantity of flow, cubic meters per second
  n = coefficient of roughness depending on material
  A = cross-sectional area of flow, square meters
  R = hydraulic radius, meters
  S = slope of hydraulic gradient, meters per meter

The nomograph shown in Figure 4–23 solves the Manning equation (Eq. 4–20) for circular pipes flowing full based on a coefficient of roughness of 0.013. Given any two of the parameters (flow, pipe diameter, slope of pipe, or velocity), the remaining two can be determined from the intersections along a straight line drawn across the nomograph. For example, an 8-in.-diameter pipe set on a slope of 0.02 ft/ft (2.0 percent) conveys a quantity of flow of 760 gpm (1.7 cu ft/sec) flowing full with an average velocity of 4.9 ft/sec. The quantity of flows at n values other than 0.013 can be calculated by multiplying the nomograph value by 0.013 and dividing the resultant by the desired n factor.

Hydraulic problems of circular pipes flowing partly full are solved by using Figure 4–24. To plot these curves, ratios of hydraulic elements were calculated at various depths of flow in a circular pipe using the Manning equation. The symbols q, v, and a relate to the partial-flow condition; Q, V, and A are at full flow. Consider flow in a pipe at a depth of 30 percent of the pipe diameter, represented by the horizontal line labeled 0.3. Where this line intersects the curved lines for quantity of flow, area of flow, and velocity, project downward to the horizontal scale and read, respectively, the values 0.2, 0.25, and 0.78; these values mean that partial flow at this depth conveys a quantity of flow equal to 20 percent of the flowing-full quantity, the cross-sectional area of flow is 25 percent of the total open area of the pipe, and average velocity of flow is 78 percent of the flowing-full velocity. This diagram shows that a pipe flowing at one-half depth conveys one-half the flowing-full quantity of flow at a velocity equal to the flowing-full velocity. Also, the greatest quantity of flow occurs when the pipe is flowing at about 0.93 of the depth and maximum velocity is at about 0.8 depth. The reason for this is that as the section approaches full flow, the additional frictional resistance caused by the crown of the pipe has a greater effect than the added cross-sectional area. For practical application, Figure 4–24 yields only approximate results, since the theoretical curves are based on steady, uniform flow, which does not precisely characterize the nature of actual flow in sewers. Examples 4–10 through 4–13 illustrate the use of the Manning equation, nomographs, and partial-flow diagram.

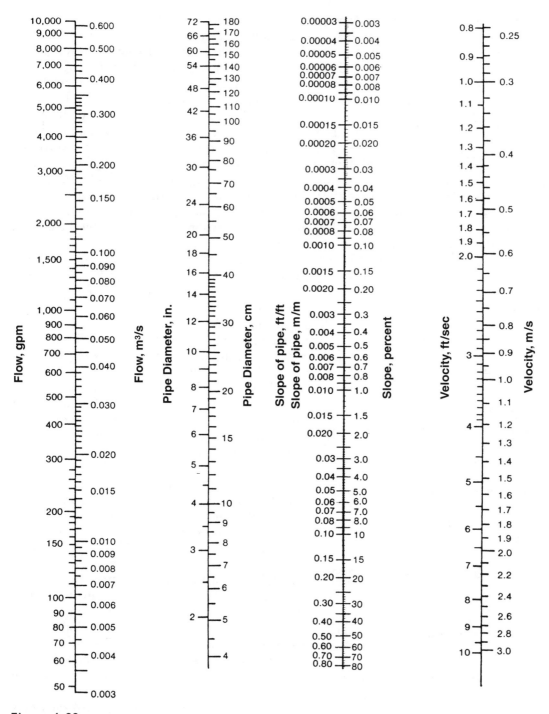

Figure 4–23

Nomograph for Manning formula in English and SI metric units for circular pipes flowing full based on *n* = 0.013. (Courtesy of U.S. Pipe & Foundry Co.)

Figure 4–24

Relative quantity, velocity, and cross-sectional area of flow in a circular pipe for any depth of flow.

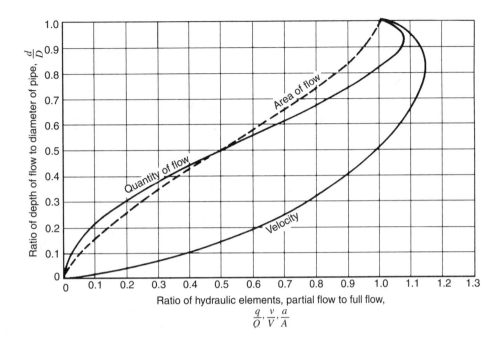

■ EXAMPLE 4–10

If a 10-in. sewer is placed on a slope of 0.010, what is the flowing-full quantity and velocity for (a) $n = 0.013$ and (b) $n = 0.015$?

Solution

(a) Project across Figure 4–23 through slope = 0.010 and diameter = 10 in. to read Q = 990 gpm and $V = 4.0$ ft/sec.

(b) From Eq. 4–20, Q and V are both inversely proportional to n; therefore,

$$Q \text{ (at } n = 0.015) = \frac{0.013 \times 990}{0.015} = 860 \text{ gpm}$$

$$V \text{ (at } n = 0.015) = \frac{0.013 \times 4.0}{0.015} = 3.5 \text{ ft/sec}$$

■ ■ ■

■ EXAMPLE 4–11

The measured depth of flow in a 48-in. storm sewer on a grade of 0.00015 ft/ft is 30 in. What is the calculated quantity and velocity of flow?

Solution

From Figure 4–23, the flowing-full Q and V are 7700 gpm and 1.4 ft/sec, respectively. The depth of flow to diameter of pipe ratio is

$$\frac{d}{D} = \frac{30}{48} = 0.62$$

Entering Figure 4–24 with a horizontal line at this ratio, read q/Q and v/V by projecting downward at the intersections with the flow and velocity curves. Read $q/Q = 0.72$ and $v/V = 1.08$. Then, flow at a depth of 30 in. = 0.72 × 7700 = 5500 gpm and $v = 1.08 \times 1.4 = 1.5$ ft/sec.

■ ■ ■

■ EXAMPLE 4–12

An 18-in. sewer pipe, $n = 0.013$, is placed on a slope of 0.0025. At what depth of flow does the velocity of flow equal 2.0 ft/sec?

Solution

Velocity flowing full = 3.0 ft/sec (Figure 4–23). Calculate $v/V = 2.0/3.0 = 0.67$.

Enter Figure 4–24 vertically at 0.67, intersect the velocity line, and project horizontally to read $d/D = 0.23$.

Depth at a velocity of 2.0 ft/sec = $0.23 \times 18 = 4.1$ in.

■ ■ ■

■ EXAMPLE 4–13

What is the flowing-full quantity and velocity of flow for a 450-mm-diameter sewer, $n = 0.013$, on a slope of 0.00412 m/m? Determine the quantity of flow for a depth of flow equal to 300 mm.

Solution

Using Figure 4–23 for $D = 45$ cm and $S = 0.00412$ m/m,

$$Q_{full} = 0.18 \text{ m}^3/\text{s} = 180 \text{ l/s}$$
$$V_{full} = 1.14 \text{ m/s}$$

Alternate solution for Q_{full} and V_{full} using Eq. 4–21:

Cross-sectional area of pipe

$$= \pi\left(\frac{0.450}{2}\right)^2 = 0.159 \text{ m}^2$$

Hydraulic radius flowing full

$$= \frac{\pi d^2/4}{\pi d} = \frac{d}{4} = \frac{0.45}{4} = 0.112 \text{ m}$$
$$Q_{full} = \frac{1.00}{0.013}(0.159)(0.112)^{2/3}(0.00412)^{1/2}$$
$$= 0.182 \text{ m}^3/\text{s} = 180 \text{ l/s}$$
$$V_{full} = \frac{Q}{A} = \frac{0.182}{0.159} = 1.14 \text{ m/s}$$

Enter Figure 4–24 with

$$\frac{d}{D} = \frac{300 \text{ mm}}{450 \text{ mm}} = 0.667$$

and read $q/Q = 0.78$. Hence, Q at d of 300 mm is $0.78 \times 180 = 140$ l/s.

■ ■ ■

4–9 FLOW MEASUREMENT IN PIPES

The common positive-displacement water meter has a measuring chamber of known volume containing a disk that goes through a cyclic motion as water passes through (Figure 4–25). The rotation resulting from the filling and emptying of the chamber is transmitted to a recording register. The advantages of this meter are simplicity of construction, high sensitivity and accuracy, small loss of head, and low maintenance costs. Also, the accuracy of registration is not materially affected by position. This oscillating-disk meter is commonly used for small-customer services, such as individual households and apartments.

Several meter manufacturing firms have developed remote registration systems. The simplest is a remotely mounted digital register that is placed so that the person reading the register can do so from outside the house (Figure 4–25). A more sophisticated system consists of an outdoor receptacle read by plugging in a small portable recording device. The recorder punch card, or magnetic tape, is later used in data processing equipment at the central office for billing and recording in the memory system. Automatic remote meter reading and billing by computer processing reduces costs for reading of meters and preparation of water bills.

The measuring device in a current (velocity) meter is a bladed wheel that rotates at a speed in proportion to the quantity of water that passes through the blades (Figure 4–26). A recording register is geared to the turbine wheel. The shortcoming of a current meter is poor accuracy at low flow rates when the water is not moving at sufficient velocity to rotate the blades. Consequently, current meters are of limited value in water systems and are used only in restricted applications, for instance, on sprinkler carts that draw water from hydrants or on water towers. Current meters are unsuitable for customer services because of their inability to register low flow rates.

A general-service compound meter consists of a current meter, a positive-displacement meter, and an automatic valve arrangement that directs the water to the current meter during high rates of flow and to the displacement meter at low rates (Figure 4–27). The advantages are high accuracy at

Figure 4–25
Household water meter with an oscillating disk that transmits a metering signal to the register by a magnetic drive. The meter can be connected to a remote register mounted outside of the house.
(Courtesy of Kent Meters, Inc.)

Figure 4–26
Turbine-type water meter used for measuring continuous high rates of flow.

all flow rates and a relatively wide operating range. The switch from low- to high-flow operation is controlled by head loss through the disk meter; when it exceeds a certain amount, a valve automatically opens and the current meter in the main line begins to measure the higher flow. Compound meters are used on large services where widely varying flow rates are typical, such as motels, office buildings, factories, and commercial properties.

Proportional meters are sometimes referred to as "compound meters." The name is derived from the fact that only a portion of the total flow through the meter passes through a by-pass measuring chamber, which is the actual measuring device. The by-pass meter, either positive-displacement or current type, is adjusted to register the total flow through the meter. Different methods of by-passing the water are utilized, such as an orifice in the main throat or a diverging tube. The proportional meter is capable of measuring large volumes of flow at comparatively low head loss and unrestricted flow, as well as accuracy of measurement.

Differential pressure meters—including venturi tubes, orifices, and nozzles—placed in a

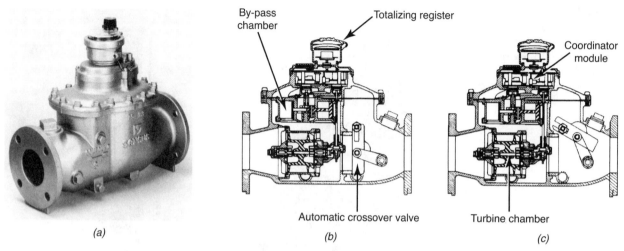

By-pass chamber

Totalizing register

Coordinator module

Automatic crossover valve

Turbine chamber

(a) *(b)* *(c)*

Figure 4–27

Compound water meter designed for services requiring accurate measurements of both high and low flows. (*a*) At low flows, the automatic crossover valve is closed, directing the water through the by-pass chamber with a positive-displacement oscillating-disk meter. (*b*) At high flows the valve is open, so the water passes through the turbine chamber with an axial-flow turbine meter. A coordinator module transfers the two separate chamber drives to a single totalizing register.

pipe section are the primary flow-measuring devices in water systems. The meter shown in Figure 4–28 consists of a venturi tube with a converging portion, throat, and diverging portion and differential pressure gauge. As water flows through the tube, velocity is increased in the constricted portion and temporarily lowers the static pressure, in accordance with the energy, Eq. 4–2. The pressure difference between inlet and throat is measured and correlated to the rate of flow. A flow recorder with a moving chart and ink pen plots the variation of flow with respect to time, and a digital totalizer registers total flow. Venturi meters are shaped to maintain streamline flow for minimum head loss. The equation for calculating flow through a venturi meter is

$$Q = CA_2[2g(H_1 - H_2)]^{1/2} \qquad \textbf{(4–22)}$$

where Q = quantity of flow, cubic feet per second (cubic meters per second)

C = discharge coefficient, commonly in the range of 0.98 to 1.02

A_2 = cross-sectional area of the throat, square feet (square meters)

g = acceleration of gravity, 32.2 ft/sec^2 (9.81 m/s^2)

$H_1 - H_2$ = difference of pressure heads between the inlet and throat, feet (meters)

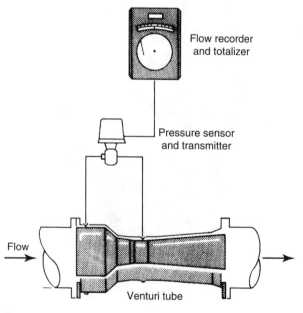

Flow recorder and totalizer

Pressure sensor and transmitter

Flow

Venturi tube

Figure 4–28

Venturi (differential pressure) meter inserted in a water pipeline, with sensor and recorder.

4–10 FLOW MEASUREMENT IN OPEN CHANNELS

Wastewater contains suspended and floating solids that prohibit the use of enclosed meters. Furthermore, wastewater is commonly conveyed by open channel flow rather than in pressure conduits. Therefore, the Parshall flume is the most common device used to measure wastewater flows. A typical flume (Figure 4–29) consists of a converging and dropping open channel section. Flow moving freely through the unit can be calculated by measuring the upstream water level. A stilling well is normally provided to hold a float, bubble tube, or other depth-measuring device, which is connected to a transmitter and flow recorder similar to that shown in Figure 4–28. The advantages of an open channel flume are low head loss and self-cleansing capacity.

For a Parshall flume under free-flowing (unsubmerged) conditions, the formula for calculating discharge for a flume with a throat width between 1 and 8 ft is

$$Q = 4BH^{1.522B^{0.026}} \qquad (4\text{–}23)$$

where Q = quantity of flow, cubic feet per second
B = throat width, feet
H = upper head, feet

Palmer-Bowlus flumes, described in Section 10–3, are commercially available to set in half-sections of sewer pipes to measure flow from portions of a collection system or individual industrial wastewater discharges. Parshall flumes cannot be fitted into a half-section of sewer pipe to measure flow.

■ EXAMPLE 4–14

Depth of flow measured in the upstream channel of a Parshall flume with a throat width of 4.00 ft is 18.0 in. Calculate the quantity of flow.

Solution

Using Eq. 4–23 gives

$$Q = 4 \times 4.00 \times 1.50^{1.522 \times 4.00^{0.026}} = 30.0 \text{ cu ft/sec}$$

■ ■ ■

4–11 AMOUNT OF STORM RUNOFF

The rational method for calculating quantity of runoff for storm sewer design is defined by the relationship

$$Q = CIA \qquad (4\text{–}24)$$

where Q = maximum rate of runoff, cubic feet per second
C = coefficient of runoff based on type and character of surface, Table 4–4
I = average rainfall intensity, for the period of maximum rainfall of a given frequency of occurrence having a duration equal to the time required for the entire drainage area to contribute flow, inches per hour
A = drainage area, acres

Equation 4–24 in SI metric units is

$$Q = 0.278 \, CIA \text{ (SI units)} \qquad (4\text{–}25)$$

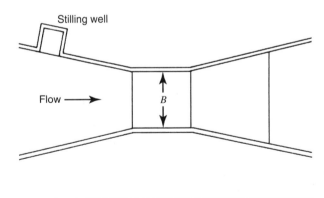

Stilling well

Flow →

B

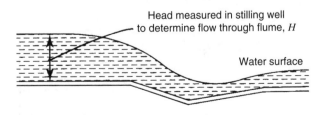

Head measured in stilling well to determine flow through flume, H

Water surface

Figure 4–29
Parshall flume for measuring flow in an open channel.

TABLE 4–4

Coefficients of Runoff for the Rational Method, Eqs. 4–24 and 4–25, for Various Areas and Types of Surfaces

Description	Coefficient
Business areas depending on density	0.70 to 0.95
Apartment-dwelling areas	0.50 to 0.70
Single-family areas	0.30 to 0.50
Parks, cemeteries, playgrounds	0.10 to 0.25
Paved streets	0.80 to 0.90
Watertight roofs	0.70 to 0.95
Lawns, depending on surface slope and character of subsoil	0.10 to 0.25

where Q = maximum rate of runoff, cubic meters per second

C = coefficient of runoff, Table 4-4

I = rainfall intensity, millimeters per hour

A = drainage area, square kilometers

Rainfall may be intercepted by vegetation, retained in surface depressions and evaporate, infiltrate into the soil, or drain away over the surface. The coefficient of runoff is that fraction of rainfall that contributes to surface runoff from a particular drainage area. The coefficients given in Table 4–4 show that the majority of rain falling on paved and built-up areas runs off, while open spaces with grassed surfaces retain the bulk of rain water. The size of the drainage area is determined by field survey or measurement from a map.

The most complex parameter in the rational formula is the rainfall intensity. The U.S. Weather Bureau maintains recording gauges that automatically chart rainfall rates with respect to time. These data can be compiled and organized statistically into intensity-duration curves like those shown in Figure 4–30. The following illustrates how to read this diagram: For storms with a duration of 30 min, the maximum average rainfall anticipated once every 5 years is 2.9 in./hr and the 25-year frequency is 3.9 in./hr.

Although curves are a popular method of presenting rainfall information, precipitation formulas have been developed for various parts of the United States. In design, 5-year storm frequency is used for residential areas, 10-year frequency for business sections, and 15-year frequency for high-value districts where flooding would result in considerable property damage.

The duration of rainfall used to enter Figure 4–30 depends on the time of concentration of the watershed. It is the time required for the maximum runoff rate to develop during a continuous

Figure 4–30

Typical rainfall intensity-duration curves used in the rational method for calculating quantity of storm water runoff.

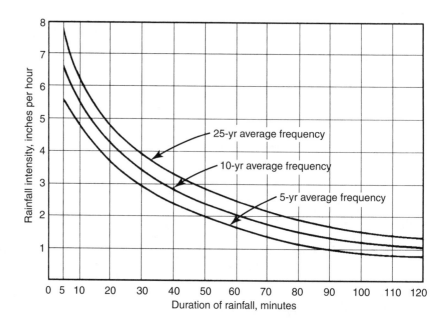

uniform rain, or, in other words, the duration of rainfall required for the entire watershed to be contributing runoff. If a portion of the area being considered drains into an inlet and through a storm sewer, the time of concentration equals the inlet time plus the time of flow through the pipe. Inlet times generally range from 5 to 20 min based on characteristics of the drainage area, including amount of lawn area, slope of street gutters, and spacing of street inlets. Example 4–15 illustrates an application of the rational formula.

■ EXAMPLE 4–15

Compute the diameter of the outfall sewer required to drain storm water from the watershed described in Figure 4–31, which gives the lengths of lines, drainage areas, and inlet times. Assume the following: a rainfall coefficient of 0.30 for the entire area, the 5-year frequency curve from Figure 4–30, and a flowing-full velocity of 2.0 ft/sec in the sewers.

Solution

Flow time in sewer from Manhole 1 to Manhole 2

$$= \frac{400 \text{ ft}}{2.0 \text{ ft/sec} \times 60 \text{ sec/min}} = 3.3 \text{ min}$$

Flow time from Manhole 2 to Manhole 3

$$= \frac{600}{2.0 \times 60} = 5.0 \text{ min}$$

Times of concentration from remote points of the three separate areas to Manhole 3 are 5.0 + 3.3 + 5.0 = 13.3 min for Area 1; 5.0 + 3.3 = 8.3 min for Area 2; and 8.0 min (inlet time only) for Area 3. Entering Figure 4–30 with the maximum time of concentration (duration of rainfall) for the watershed of 13.3 min, the rainfall intensity I is 4.4 in./hr for a 5-year frequency.
The sum of the CA values

$$= 0.3 \times 3.0 + 0.3 \times 6.0 + 0.3 \times 4.5 = 4.1$$

Substituting into Eq. 4–24,

$$Q = 4.4 \times 4.1 = 18 \text{ cu ft/sec (acre-in./hr)}$$

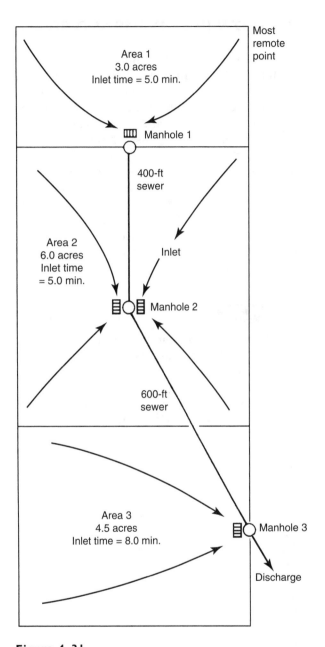

Figure 4–31
Watershed for Example 4–15 illustrating storm-runoff calculations using the rational method.

For $Q = 18$ cu ft/sec (8080 gpm) and $V = 2.0$ ft/sec, from Figure 4–23 the required sewer diameter is 42 in. set on a slope of 0.0004 ft/ft.

■ ■ ■

4–12 FLOW IN STREAMS AND RIVERS

Permanent gauging stations have been located on streams and rivers throughout the United States by several governmental agencies, including the Corps of Engineers, Bureau of Reclamation, Soil Conservation Service, and Geological Survey. A typical recording gauge installation is a stilling well, connected to the river by an intake pipe, with a float-operated water-level recorder housed in a shelter above the well. The river stage is continuously scribed with a pen on a roll of paper driven by a clock mechanism. In some instances, a control weir is constructed across a stream to provide sensitivity at low flows, but more frequently a

rating curve is developed for the natural river channel. Data for a stage-discharge curve are gathered by using current meter measurements for various river discharges. Water levels during the times of survey are plotted against the measured flows. Rating curves must be reexamined periodically to account for shifts that occur as a result of channel filling or scouring. Continuous stage readings recorded on the chart paper are translated into average daily flow by using the rating curve; these are tabulated and published, usually under the title of surface-water records, and are distributed to libraries and interested governmental agencies.

Streamflows are of concern in water pollution, since treated wastewater effluents are often

TABLE 4–5

Streamflow Records Listing the Lowest Mean Discharge for 7 Consecutive Days for Each Year from 1961 to 1982[a]

Year	Lowest Mean Flow in Cubic Feet per Second for 7 Consecutive Days
1961	19.6
1962	28.6
1963	18.1
1964	34.3
1965	29.3
1966	35.7
1967	35.0
1968	27.0
1969	35.0
1970	36.9
1971	90.3
1972	50.6
1973	35.3
1974	59.4
1975	26.3
1976	30.1
1977	29.4
1978	29.7
1979	30.4
1980	49.6
1981	36.6
1982	59.1

[a]The average annual discharge for this period was 178 cu ft/sec.

TABLE 4–6

Streamflow Data from Table 4–5 Organized for Statistical Evaluation as Shown in Figure 4–32

Minimum Flows in Order of Severity (cu ft/sec)	Serial Number, M	$\dfrac{M}{N+1}$
90.3	1	$\dfrac{1}{23} = 0.0435$
59.4	2	$\dfrac{2}{23} = 0.087$
59.1	3	0.130
50.6	4	0.174
49.6	5	0.217
36.9	6	0.261
36.6	7	0.305
35.7	8	0.347
35.3	9	0.390
35.0	10	0.435
35.0	11	0.477
34.3	12	0.520
30.4	13	0.565
30.1	14	0.605
29.7	15	0.650
29.4	16	0.695
29.3	17	0.740
28.6	18	0.782
27.0	19	0.825
26.3	20	0.870
19.6	21	0.912
18.1	22 = n	0.955

disposed of by dilution in rivers and streams; of greatest interest are low flows when the least dilution capacity is provided. The stream classification system of water-quality standards establishes maximum allowable concentration of pollutants; for example, ammonia nitrogen in warm-water streams is set at a maximum allowable level of 3.5 mg/l. These standards must be keyed to a specific quantity of flow to determine the amount of pollutant that can be discharged to the watercourse without exceeding the specified concentration. The 1-in-10-year, 7-consecutive-day low flow is the value that is commonly adopted. Of course, if the stream has intermittent flow, the critical condition is during the dry period.

The following procedure is used to develop a frequency curve of 7-consecutive-day low flows for determining the 1-in-10-year value. All of the daily discharge records for gauging stations on the stream being evaluated are assembled. After the data for each year are scanned, the 7 consecutive days of lowest flows are identified and averaged arithmetically. These values are listed by year, as shown in Table 4–5. The following statistical

method, illustrated in Table 4–6 and Figure 4–32, is used to calculate and plot the frequency curve from the yearly low-flow values.

1. Arrange the minimum annual flows from historical records in order of severity, that is, from highest to lowest flow.
2. Assign a serial number m to each of the n values: 1, 2, 3, . . . , n.
3. Compute the probability plotting position for each serial value as m divided by $n + 1$.
4. Plot flow on the vertical logarithmic scale and the corresponding probability value along the horizontal axis.
5. Draw the best-fit line through the plotted data.

The frequency curve in Figure 4–32 is read by entering the diagram from either the top or bottom, the return period or probability, and reading the corresponding low flow on the vertical scale. For a return period of 10 years, or probability of 90 percent, the minimum flow is 22.6 cu ft/sec. In other words, during one 7-day period every 10 years, the lowest average flow for those 7 days

Figure 4–32

Plot of yearly low flows from Table 4–6 on logarithmic probability paper to determine the 1-in-10-year, 7-consecutive-day low flow.

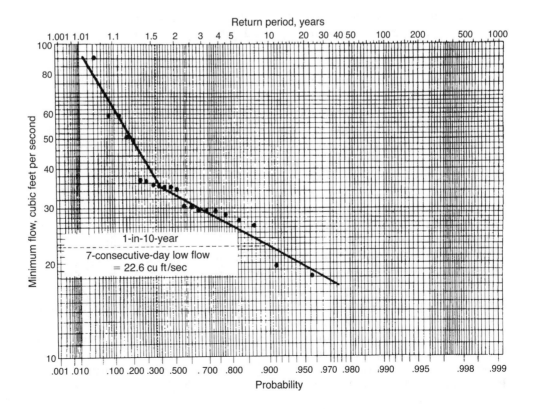

is expected to be 22.6 cu ft/sec, no other 7-day period having a lower flow. Ninety percent of the flows during 7-day intervals are expected to be greater than this value. Logarithmic probability paper is used because the hydrologic extremes, floods and droughts, are skewed and do not follow a normal symmetrical distribution. The break in the line drawn in Figure 4–32 indicates that drought flows did not occur during the 5 years when the low flows were greater than 40 cu ft/sec. The importance of understanding the concept of low flows for wastewater dilution is illustrated by the fact that the average flow in this stream throughout the period of record was 178 cu ft/sec. If the waste assimilative capacity were to be based on average flow rather than the calculated low flow, excessive pollution of the stream would occur a considerable portion of the time. On the other hand, using the lowest daily flow, in this case 18.1 cu ft/sec, leads to extreme conservation in the opinion of many authorities.

4–13 HYDROLOGY OF LAKES AND RESERVOIRS

The parameters used to define the physical characteristics of a lake are surface area, mean depth, volume, retention time (volume divided by influent flow), color and turbidity of the water, currents, surface waves, thermodynamic relations, and stratification. All of these influence the chemistry and biology of a lake or reservoir and, hence, water quality. Thermal stratification is the most important phenomenon with regard to water supply and eutrophication. Lakes in the temperate zone, or in higher altitudes in subtropical regions, have two circulations each year, in spring and autumn. Thermal stratification is inverse in winter and direct in summer. Lakes of the warmer latitudes in which the water temperature never falls below 4°C at any depth have one circulation each year in winter and directly stratify during the summer. For example, a lake may stratify from May through September and may circulate continuously from October to April.

The seasons in a lake are defined graphically in Figure 4–33. In winter, the densest water sinks to the bottom and ice near 0°C covers the surface; the maximum density of fresh water is 4°C.

Shading caused by ice and snow cover inhibits photosynthesis and, if the lake is rich in organic matter, the dissolved oxygen near the bottom gradually decreases. In spring, after melting of the ice, the surface waters warm to 4°C and begin to sink, while the less dense bottom waters rise. These convection currents, aided by wind, mix the lake thoroughly for several weeks while the water temperature gradually increases. This is called the spring overturn, or spring circulation. With the approach of summer, the surface waters warm more rapidly and brisk spring winds subside, such that a lighter surface layer is formed. As the summer season progresses, resistance to mixing between the top and bottom layers of different density becomes greater and thermal stratification is established. The epilimnion (the warm surface layer) is continuously mixed by wind and density currents and supports the growth of algae. The hypolimnion (the cooler bottom layer) is dark and stagnant. Although the bulk of fish food is found in the epilimnion, many species find the cooler bottom water a more suitable environment. In nutrient-rich bodies of water, the hypolimnion increases in carbon dioxide content and may become devoid of dissolved oxygen after many weeks of stratification. The thermocline is the thin zone of rapid temperature drop between the water layers. The approach of autumn with shorter, cooler days causes the lake to lose heat faster than it is absorbed. When the surface waters are cooled to a higher density than the water in the hypolimnion, vertical currents lead to autumnal circulation. This mixing is supported by wind action until finally the densest water stays on the bottom and the surface freezes.

Thermal stratification in reservoirs and lakes has a direct influence on the quality of the water supply. In the summer, water drawn from near the surface is warm and may contain algae that cause filter clogging and taste and odor problems. Stagnant, cooler hypolimnion water may be devoid of dissolved oxygen and high in carbon dioxide and may contain the products of anaerobic conditions, such as hydrogen sulfide, odorous organic compounds, or reduced iron. Usually, the region just below the thermocline provides the most satisfactory water quality during stratification. During winter stagnation, water closer to the surface is likely to be more desirable, since the quality

Figure 4–33
Thermal stratification and
circulation of a dimictic lake in
the northern United States.
(*a*) Thermal stratification during
late summer. (*b*) Spring and
autumn overturns.
(*c*) Temperature profiles of a
lake showing stratification and
mixing.

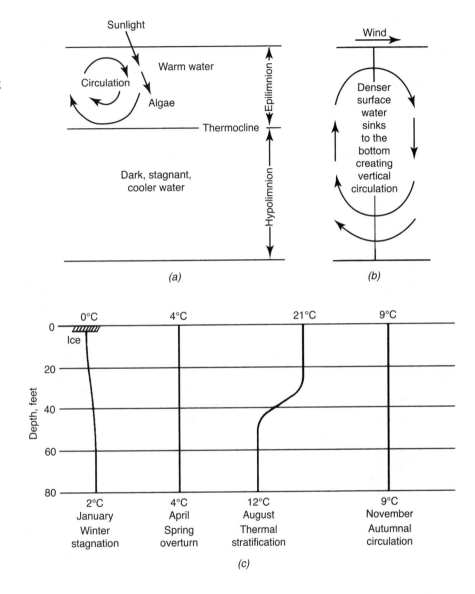

adjacent to the bottom may be poor because of contact with decaying organic matter. The vertical variation that may occur in a body of water illustrates the importance of having a water intake tower with ports at various depths so that the water supply can be drawn from the most advantageous level in the water profile. Spring and autumn circulation mixes the water, spreading any undesirable matter throughout the entire profile. Treatment for taste and odor control may have to be intensified, particularly in the autumn when decaying algae and anaerobic bottom waters are mixed.

4–14 GROUNDWATER HYDROLOGY

Groundwater originates as infiltration from precipitation, streamflow, lakes, and reservoirs. As the water percolates vertically down through the near-surface soils, thin layers of water referred to as soil moisture are left behind, coating soil grains. Eventually, the water enters the zone of saturation where porous soil or fissured rock is filled with water. The surface of the saturated zone is called the water table, and its depth is described by the level of free water in an observation well extending into the saturated zone. A water table may fluctuate up and down with the seasonal supply and demand of groundwater. Horizontal movement is dictated by the hydraulic gradient, flowing downgrade with little vertical mixing.

The porosity of a soil or fissured rock is an expression of the void space defined as

$$n = \frac{V_v}{V} \qquad (4\text{-}26)$$

where n = porosity
V_v = volume of voids
V = total volume

Typical values of porosity are 0.2 to 0.4 for sands and gravels, depending on grain size, size distribution, and degree of compaction; 0.1 to 0.2 for sandstone; and 0.01 to 0.1 for shale and limestone, depending on texture and the size of the fissures. When groundwater drains from an aquifer as a result of lowering the water table, some water is retained in the voids. The quantity draining out is the specific yield or effective porosity; the amount retained is the specific retention. The specific yield for alluvial sand and gravel deposits ranges from 90 to 95 percent. For instance, if a coarse sand has a porosity of 0.40 and specific yield of 90 percent, the volume draining from 1.0 cu ft of aquifer is calculated as follows: $0.40 \times 0.90 = 0.36$ cu ft of water.

Aquifers are defined as permeable geologic strata that convey groundwater. They act as reservoirs, discharging by gravity flow or well extraction and recharging by infiltration. Common aquifers are valley fills composed of sand, gravel, and silt adjacent to streams; granular deposits along coastal plains; alluvium and loess of high plains; water-borne deposits from glaciation; terrains of volcanic origin; fractured limestone or dolomite rock; and poorly cemented sandstones. As illustrated in Figure 4–34, aquifers may be either unconfined or confined. The upper boundary of an unconfined aquifer is the water table, which is free to move up and down, changing the saturation zone. A well in an unconfined aquifer is referred to as a water table well. A perched water table is a special case of an unconfined aquifer where groundwater lies on a relatively impermeable stratum of small areal extent above the main body of groundwater. Confined aquifers, also known as artesian or pressure aquifers, exist where groundwater is confined by relatively impermeable strata. The groundwater under pressure in a confined aquifer rises in a well to the piezometric level.

Permeability is the ability of a porous medium to transmit water. The coefficient of permeability, K, is defined by Darcy's law stating that the velocity of flow is directly proportional to the hydraulic gradient.

$$v = Ki \qquad (4\text{-}27)$$

where v = velocity of flow, feet per second (millimeters per second)
K = coefficient of permeability, feet per second (millimeters per second)
i = hydraulic gradient, feet per foot (meters per meter)

Figure 4–34

A profile of unconfined and confined aquifers illustrating water table and artesian wells.

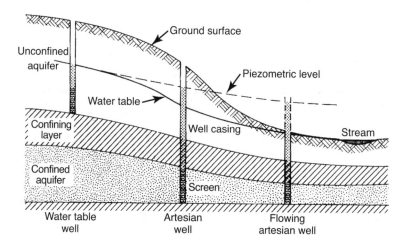

Values for K cover a wide range from less than 10^{-5} ft/sec (0.003 mm/s) for fine-grain deposits to over 1 ft/sec (300 mm/s) for coarse gravels.

A typical water well is constructed by drilling a deep, small-diameter borehole into the ground, which is held open by inserting a casing. The well is screened in the aquifer to allow groundwater to enter the borehole. Pump impellers and a suction pipe are hung down into the casing to lift the water out of the well. The movement of groundwater near a pumping well depends on the characteristics of the aquifer and construction of the well. When pumping is started, the water table is drawn down, forming a cone of depression. Water flows radially toward the well with steadily increasing velocity. Based on Darcy's law, the hydraulic gradient must also increase as groundwater approaches the well; therefore, the drawdown curve has a continuously steeper slope near the well. The size and shape of the cone of depression for a particular well are related to the withdrawal rate, duration of pumping, slope of the water table, sources of recharge, and aquifer characteristics.

Steady radial flow to a well penetrating an ideal unconfined aquifer is drawn in Figure 4–35. Ideal steady-state flow is defined by uniform withdrawal, a stable drawdown curve, uniform horizontal and laminar groundwater flow, a velocity of flow proportional to the tangent of the hydraulic gradient, and a homogeneous aquifer. Assuming these conditions, the well discharge is related to the coefficient of permeability, depth of aquifer, and shape of the drawdown curve as follows:

$$Q = \pi K \frac{h_0^2 - h_w^2}{\log_e(r_0/r_w)} \qquad (4\text{--}28)$$

where Q = well discharge, cubic feet per second (liters per second)

K = coefficient of permeability, feet per second (millimeters per second)

h_0 = saturated thickness of aquifer before pumping, feet (meters)

r_0 = radius of the cone of depression, feet (meters)

h_w = depth of water in well while pumping, feet (meters)

r_w = radius of well, feet (meters)

If this equation is used, prediction of well discharge Q based on a known permeability K, or determination of K from a measured Q, is reasonably accurate for natural unconfined aquifers that approach ideal conditions. Application of Eq. 4–28 requires estimating the radius of influence, r_0, or installation of observation wells in the drawdown area to determine the radius of influence.

Figure 4–35

Steady radial flow to a well penetrating an ideal unconfined aquifer, Eq. 4–30.

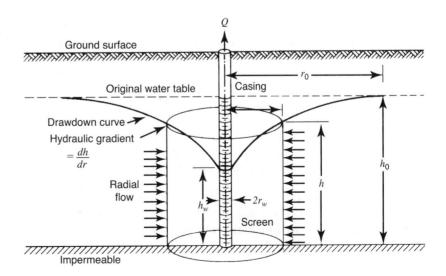

Figure 4–36
Steady radial flow to a well completely penetrating an ideal confined aquifer, Eq. 4–29.

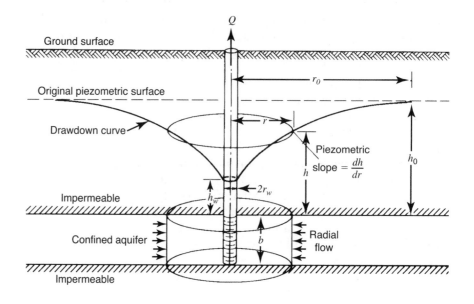

Figure 4–37
Flow of water toward a well during a pumping test in an unconfined coarse-grained aquifer. The dimensions shown apply to Eq. 4–30.

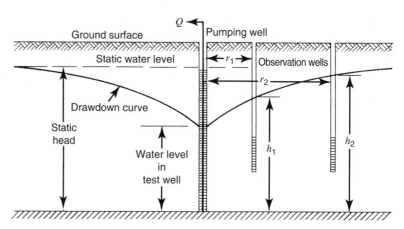

Steady radial flow to a well penetrating a confined aquifer is shown in Figure 4–36. For ideal conditions, the equation for well discharge is

$$Q = 2\pi K b \, \frac{h_0 - h_w}{\log_e(r_0/r_w)} \qquad (4\text{–}29)$$

where b = thickness of aquifer in feet (meters). Values of r_0 and h_0 may be assumed or measured from observation well data.

The permeability of an aquifer surrounding a well can be determined in the field by conducting a pumping test. Observation wells are needed to record drawdown at various distances from the test well, as illustrated in Figure 4–37. After establishing steady-state conditions under continuous well discharge, the permeability of an unconfined aquifer is calculated as

$$K = \frac{Q}{\pi(h_2^2 - h_1^2)} \log_e\left(\frac{r_2}{r_1}\right) \qquad (4\text{–}30)$$

For a confined aquifer (Figure 4–36), the permeability is calculated from observation well data as

$$K = \frac{Q}{2\pi b(h_2 - h_1)} \log_e\left(\frac{r_2}{r_1}\right) \qquad (4\text{–}31)$$

■ EXAMPLE 4–16

A well with a diameter of 2.0 ft is constructed in a confined aquifer as illustrated in Figure 4–36. The sand aquifer has a uniform thickness of 50 ft overlain by an impermeable layer with a depth of 115 ft. A pumping test was conducted to determine the coefficient of permeability of the aquifer. The initial piezometric surface was 49.0 ft below the ground surface datum of the test well and observation wells. After water was pumped at a rate of 0.46 cu ft/sec for several days, water levels in the wells stabilized with the following drawdowns: 21.0 ft in the test well, 12.1 ft in the observation well 30 ft from the test well, and 7.9 ft in a second observation well at a distance of 100 ft. From these test data calculate the permeability of the aquifer. Then, using this K value, estimate the well discharge with the drawdown in the well lowered to the top of the confined aquifer.

Solution

Assume a datum at the top of the aquifer as in Figure 4–36.

$$h_0 = 115 - 49.0 = 66.0 \text{ ft}$$

$$h_w = 66.0 - 21.0 = 45.0 \text{ ft}$$

$$h_1 = 66.0 - 12.1 = 53.9 \text{ ft}$$

$$h_2 = 66.0 - 7.9 = 58.1 \text{ ft}$$

Substituting into Eq. 4–31 gives

$$K = \frac{0.46 \text{ cu ft/sec}}{2\pi(50 \text{ ft})(58.1 \text{ ft} - 53.9 \text{ ft})}$$

$$\times \log_e\left(\frac{100 \text{ ft}}{30 \text{ ft}}\right) = 0.00042 \text{ ft/sec}$$

With the drawdown in the well at the top of the sand aquifer, $h_0 = 66.0$ ft and $h_w = 0.0$ ft. The radius of the well is 1.0 ft, and the distance to the edge of the cone of depression is assumed to be 700 ft. The estimated well discharge at maximum drawdown is calculated using Eq. 4–29 as follows:

$$Q = 2\pi(0.00042 \text{ ft/sec})(50 \text{ ft})\frac{66.0 \text{ ft} - 0.0 \text{ ft}}{\log_e(700 \text{ ft}/1.0 \text{ ft})}$$

$$= 1.3 \text{ cu ft/sec}$$

■ ■ ■

PROBLEMS

4–1 (a) What is the hydraulic head in feet equivalent to 45 psi? (b) What is the static pressure equivalent to 100 ft of head? [Answers (a) 104 ft, (b) 43 psi]

4–2 If the pressure in a water main is 50 psi, what is the remaining pressure at a faucet in a building 30 ft above the main, assuming a head loss of 20 psi in the service connection?

4–3 (a) What is the hydraulic head in meters equivalent to 320 kPa? (b) What is the static pressure equivalent to 40 m of head?

4–4 Calculate the velocity of flow in an 8-in.-diameter pipe when the quantity of flow is 400 gpm. Check your answer using Figure 4–7. (Answer 2.56 ft/sec)

4–5 Calculate the velocity of flow in a 200-mm-diameter pipe when the quantity of flow is 40 l/s. Check your answer using the nomograph in Figure 4–8. (Answer 1.27 m/s)

4–6 Compute the total energy in a pipeline with an elevation head of 100 ft, water pressure of 50 psi, and velocity of flow equal to 2.0 ft/sec; use Eq. 4–3.

4–7 Compute the total energy in a pipeline with an elevation head of 9.0 m, water pressure of 410 kPa, and velocity of flow equal to 1.2 m/s; use Eq. 4–4). (Answer 51 m)

4–8 Calculate the head loss in 2000 ft of 14-in.-diameter pipe for a flow rate of 1000 gpm. The friction factor f for the pipe material is 0.025. (Answer 2.9 ft)

4–9 Calculate the head loss in 1000 ft of 8-in.-diameter cast-iron pipe at a flow rate of 500 gpm. Use the Darcy Weisbach equation, and select the appropriate f value from Figure 4–6.

4–10 Using the nomograph in Figure 4–7 for a C of 100, determine the head loss and velocity of flow in a 10-in. ductile-iron pipe carrying 800 gpm. What are the values of head loss and velocity for a $C = 140$?

4–11 Using the nomograph in Figure 4–8 for a C of 100, determine the head loss and velocity of flow in a 150-mm pipe with a discharge of 22 l/s. Check your answers using Eq. 4–10 and Eq. 4–1. (*Answers* 0.019 m/m, 1.24 m/s)

4–12 What size pipe should be used to supply 1300 gpm so that the velocity does not exceed 3.0 ft/sec assuming a $C = 100$? (*Answer* 14 in.)

4–13 A water main 8 in. in diameter, $C = 100$, and 800 ft in length is laid 40 ft uphill to residential apartments. When the discharge is 750 gpm as required for fire protection, the minimum pressure at the end fire hydrant must be 20 psi. Calculate the pressure at the inlet of the main to satisfy these conditions. Calculate h_L by Eq. 4–8 and check using Figure 4–7.

4–14 A flow test was conducted on an old existing pipeline to determine the coefficient C for the Hazen Williams formula. The pipeline was a straight, horizontal section with a diameter of 8 in. The pressure loss was measured as 14.3 psi in 2000 ft of pipe at a water flow of 500 gpm. Calculate the C value of the pipe.

4–15 Review Example 4–6. What is the calculated discharge at point B in Figure 4–9 if water is flowing neither to nor from the elevated storage tank?

4–16 A simplified water system is illustrated in Figure 4–38. Calculate the rate of discharge from elevated storage, and draw the hydraulic gradient for the system based on a pump inflow of 2000 gpm at A and outflows of 2500 gpm at B and 1500 gpm at D. Assume a value of $C = 100$ for all pipes.

4–17 Draw the hydraulic gradient for the system shown in Figure 4–39. The pump inflow at A is 100 l/s at a discharge pressure of 550 kPa, and

Figure 4–38
Illustration for Problem 4–16.

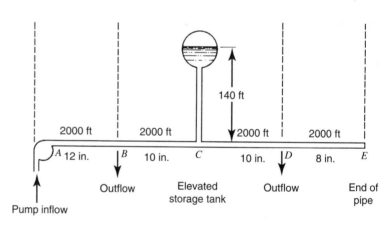

Figure 4–39
Illustration for Problem 4–17.

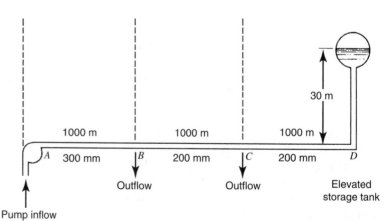

the outflow at B is 40 l/s. Determine the outflow at point C in liters per second. Assume a value of $C = 100$ for all pipes.

4–18 Refer to the simplified water system illustrated in Figure 4–9. For the same physical layout, calculate the pump discharge pressure required at A to provide fire flow and maximum daily discharge of 5000 gpm at 20 psi at point B. Assume $C = 100$. (*Answer* 300 ft)

4–19 (a) For the conditions shown in Figure 4–9 and as calculated in Example 4–6, calculate the power input (Eq. 4–18) for the pumps supplying water from the supply reservoir. Assume a pump efficiency of 70 percent. (b) What is the horsepower input required for the hydraulic conditions in Problem 4–14? Assume $E_p = 0.70$. (c) What is the power input required for the fire-flow conditions in Problem 4–17? Assume $E_p = 0.70$.

4–20 Assume for the system illustrated in Figure 4–9 that two primary pumps deliver water from the supply reservoir. (a) Plot pump curves for pumps 1 and 2 from the head-discharge data that follows. (b) Plot the head-discharge curve for pumps 1 and 2 operating in parallel. The combined head curve can be determined by adding the individual pump head curves horizontally. (Refer to the discussion of parallel operation of constant-speed pumps associated with Figure 4–17.) (c) Plot the estimated system head-discharge curve from the data that follows. (d) Locate the point on the pump head-discharge curve when the demand for water is 1500 gpm (mean summer usage), 2160 gpm (maximum daily usage), and 3600 gpm (maximum daily usage plus fire demand).

PUMP CURVE 1		PUMP CURVE 2		SYSTEM CURVE	
DISCHARGE (GPM)	HEAD (FT)	DISCHARGE (GPM)	HEAD (FT)	DISCHARGE (GPM)	HEAD (FT)
900	300	1500	350	0	100
1200	200	1900	300	1700	150
1500	100	2300	200	2200	200
		2600	150	2900	300

4–21 At night, water is pumped from a treatment plant reservoir through distribution piping to an elevated storage tank. Using the energy equation, calculate the pump discharge pressure required to supply a flow of 1000 gpm to the tank. Assume no other withdrawals from the system. The water surface in the supply reservoir is at 10-ft elevation; the lift-pump elevation is 20 ft; the water level in the elevated tank is 140 ft. The pipe network may be considered equivalent to 5000 ft of 10-in.-diameter cast-iron pipe, $C = 100$. (*Answer* 74 psi)

4–22 The discharge, head, and efficiency data of a centrifugal pump follow. The best efficiency point (bep) is at 2500 gpm. The impeller diameter is 15 in., and the operating speed is 2500 rpm. Plot the characteristic curves of head discharge and efficiency as diagramed in Figure 4–12, and locate the bep and the recommended pump operating range. Calculate the power input at the bep.

DISCHARGE (GPM)	HEAD (FT)	EFFICIENCY (%)	DISCHARGE (GPM)	HEAD (FT)	EFFICIENCY (%)
0	105	—	2000	83	85
500	100	41	2500	72	88
1000	96	63	3000	58	84
1500	91	78	3500	42	65

4–23 Draw a head-discharge curve for the pump described in Problem 4–22 operating at 1700 rpm. Locate points along the curve at 60 percent bep, and 120 percent bep, and sketch the pump operating envelope as shown in Figure 4–13. (*Answer* bep: 1700 gpm, 33 ft)

4–24 Draw a head-discharge curve for the pump described in Problem 4–22 operating at 2500 rpm with a 12-in.-diameter impeller. (*Answer* bep: 2000 gpm, 46 ft)

4–25 Water is pumped into an elevated storage tank through a 1000-m, horizontal pipeline with a $C = 100$, as shown in Figure 4–40. Draw system head-discharge curves at the low and high water levels in the tank for the range of inflows from 0 to 75 l/s.

4–26 Draw system head-discharge curves for the simplified water system diagramed in Figure 4–9. The lowest anticipated system head-discharge curve is when the hydraulic gradient at the load center is 99 ft, as drawn in Figure 4–9. The highest anticipated head-discharge curve is when the discharge at B is zero and the hydraulic head at C is 140 ft.

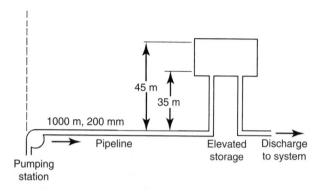

Figure 4–40

Illustration for Problem 4–25.

4–27 Draw hydraulic gradients for the system in Figure 4–41. The ground levels at points *A*, *B*, and *C* are at the same elevation. The lift pumps at *A* provide flow at a discharge of 400 kPa, the water level elevation in the elevated storage tank at *C* is 35 m, and the equivalent pipelines to the load center at point *B* are as given in the illustration. Use the nomograph in Figure 4–8. (a) Draw the hydraulic gradient and calculate the total discharge at *B* from the lift pumps and outflow from elevated storage when the water pressure at *B* equals 275 kPa. (b) Draw the hydraulic gradient and calculate the discharge into elevated storage with no discharge at the load center. Determine the equivalent pipe for the combined pipelines between *A* and *C*. Use this pipe diameter in the nomograph to determine quantity of flow. (*Answers* (a) 86 l/s, (b) 170 mm, 20 l/s)

4–28 Using the data from Problem 4–27 and its solution, plot the lower head-discharge curve

[part (a)] and the upper head-discharge curve [part (b)]. Three identical constant-speed pumps are provided with one as a standby. The operating range of each pump is from 44 m and 18 l/s to 36 m and 36 l/s with the best efficiency point (bep) at 40 m and 30 l/s. With two pumps in operation, the range is from 44 m and 36 l/s to 36 m and 72 l/s with the bep at 40 m and 60 l/s. On the graph of head-discharge curves, plot pump curves for one and two pumps in operation. Draw a small circle around the operating points of the pumps on the head-discharge curves.

4–29 How many parallel 4-in.-diameter pipes are needed to be equivalent to one 8-in.-diameter pipe? (*Answer* 6)

4–30 Is a 150-mm pipe equivalent in flow capacity to two 100-mm pipes in parallel?

4–31 A 6-in. pipe with a *C* = 100 is laid in parallel with an 8-in. pipe with a *C* = 140. What is the diameter of an equivalent pipe with a *C* value of 120?

4–32 Find an equivalent, 1200-ft single pipe to replace the parallel pipes shown in Figure 4–42.

4–33 Determine the diameter of an equivalent, 4100-ft pipe between points *A* and *B* in Figure 4–43.

4–34 Determine the diameter of an equivalent pipe with a *C* = 100 for a 1000-m, 200-mm pipe with a *C* = 100 in series with a 1000-m, 150-mm pipe with a *C* = 140. (*Answer* 183 mm at *C* = 100)

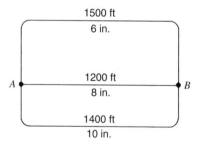

Figure 4–42

Pipes in parallel for Problem 4–32.

Figure 4–43

Pipes in series for Problem 4–33.

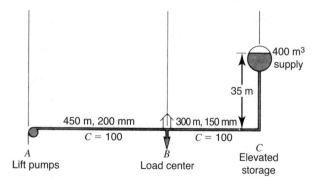

Figure 4–41

Illustration for Problems 4–27 and 4–28.

4–35 Find the equivalent single pipe 1000 m in length to replace three parallel pipes as illustrated in Figure 4–42 if the pipes between *A* and *B* are as follows: upper pipe 1800 m in length and 150 mm in diameter, center pipe 1000 m and 200 mm, and bottom pipe 1400 m and 250 mm.

4–36 Outline the data on a water distribution system that are necessary to program the system for computer analysis. How is a program calibrated and verified?

4–37 What is the flowing-full quantity and velocity of flow for a 12-in. sewer on a 0.0060 ft/ft slope using (a) *n* = 0.013 and (b) *n* = 0.011? [*Answers* (a) 1210 gpm, 3.5 ft/sec; (b) 1430 gpm, 4.1 ft/sec]

4–38 What is the flowing-full quantity and velocity of flow for a 50-cm.-diameter sewer on a slope of 0.003 m/m using (a) *n* = 0.013 and (b) *n* = 0.011? (c) For a roughness of 0.013, compute the quantity of flow for a depth of flow equal to 30 cm.

4–39 What is the minimum slope required for a 24-in.-diameter sewer to maintain an average velocity of flow equal to 3.0 ft/sec when the quantity of flow is 20 percent of the flowing-full capacity? (*Answer* 0.0028 ft/ft)

4–40 What size sewer pipe do you recommend to convey 5.4 cu ft/sec based on the following limitations: maximum allowable velocity of 10 ft/sec and a maximum allowable pipe slope of 3.0 ft/100 ft? (*Answer* 12 in.)

4–41 What size sewer pipe do you recommend to convey 15 cu ft/sec based on the following limitations: minimum allowable flowing-full velocity of 2.0 ft/sec, maximum allowable velocity of 15 ft/sec, and a maximum pipe slope of 3.0 ft/100 ft, which is specified to reduce the quantity of trench cut requiring rock excavation?

4–42 Compute the design population that can be served by a 200-mm sanitary sewer laid on a slope of 0.40 percent. Assume that the design flow per person is 1500 l/d. (*Answer* 1200 persons)

4–43 A 33-in. sewer pipe (*n* = 0.013) is placed on a slope of 0.40 ft/100 ft. (a) At what depth of flow is the velocity equal to 2.0 ft/sec? (b) If the depth of flow is 18 in., what is the discharge?

4–44 What slope is required to produce a velocity of flow equal to 1.0 m/s when the depth of flow is

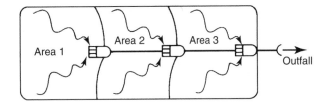

Figure 4–44
Drainage diagram for Problem 4–50.

90 mm in a 450-mm-diameter vitrified clay sewer?

4–45 A 60-in.-diameter storm sewer conveys a flow of 55 cu ft/sec at a depth of flow equal to 40 in. What is the unused capacity of the pipe in cubic feet per second? (*Answer* 14 cu ft/sec)

4–46 Explain why a compound meter is suitable for customer services while a current meter is not.

4–47 Referring to the energy equation, Eq. 4–2, explain the operating principle of a venturi meter (Figure 4–28).

4–48 A venturi water meter with a throat of 4.0 in. registers a pressure head difference of 120 in. of water between the inlet and throat. Calculate the quantity of flow through the meter using a discharge coefficient of 0.98. (*Answer* 2.2 cu ft/sec)

4–49 A Parshall flume to measure flow with a throat width of 1.50 ft has an upper wastewater head of 0.82 ft. Calculate the quantity of flow through the flume. (*Answer* 4.4 cu ft/sec)

4–50 Given the drainage area in Figure 4–44, calculate the discharge at the outfall using the rational method. Use the 5-year rainfall intensity-duration curve in Figure 4–30. Other data are: for Area 1, *C* = 0.50, area = 1.3 acres, and inlet time = 7 min; for Area 2, *C* = 0.40, area = 2.5 acres, and inlet time = 5 min; for Area 3, *C* = 0.70, area = 3.9 acres, and inlet time = 5 min; sewer lines in Areas 2 and 3 are each 500 ft in length; and the average velocity of flow in the sewers may be assumed to be 3.0 ft/sec. (*Answer* 20 cu ft/sec)

4–51 Determine the size of outfall sewer below MH1 needed to serve the 12-acre drainage area in Figure 4–45. Each area has an inlet time of 10 min. The coefficient of runoff for the two housing areas is 0.45, and the park is 0.15. The distance between manholes is 600 ft, and all

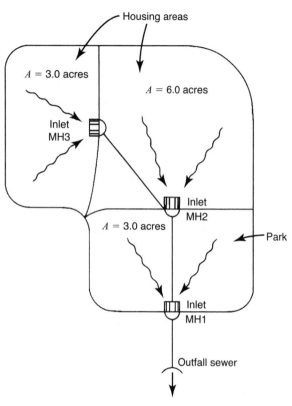

Figure 4–45
Drainage diagram for Problem 4–51.

pipes are set on a slope of 0.0020. The rainfall intensity-duration relationship is $i = 131$ divided by $t + 19$, where i = inches per hour and t = minutes.

4–52 Determine the 1-in-10-year 7-consecutive-day low flow given the following flow data. (Two-cycle logarithmic probability paper is provided in the appendix.) (*Answer* 85 cu ft/sec)

YEAR	CU FT/SEC	YEAR	CU FT/SEC
1965	120	1973	170
1966	162	1974	82
1967	142	1975	74
1968	137	1976	110
1969	254	1977	184
1970	367	1978	121
1971	145	1979	208
1972	153	1980	145
		1981	198

4–53 Determine the 1-in-10-year, 7-consecutive-day low flow given the following flow data.

YEAR	CU FT/SEC	YEAR	CU FT/SEC	YEAR	CU FT/SEC
1953	62.7	1963	89.6	1973	77.4
1954	15.1	1964	149.0	1974	88.4
1955	54.1	1965	121.0	1975	36.1
1956	27.4	1966	108.0	1976	23.0
1957	48.4	1967	81.4	1977	123.0
1958	52.9	1968	98.0	1978	136.0
1959	39.0	1969	191.0	1979	159.0
1960	42.4	1970	188.0	1980	186.0
1961	123.0	1971	271.0	1981	135.0
1962	145.0	1972	191.0	1982	117.0

4–54 Explain why lakes thermally stratify and why the impounded water in a dimictic lake circulates twice a year.

4–55 A pumping test was conducted in an unconfined aquifer using a test well penetrating to the underlying impervious stratum at a depth of 60 ft. Two observation wells were located at distances of 60 ft and 360 ft from the main well. Before the test was started, the static water level in all three wells was at a depth of 15.0 ft below the ground surface datum. Upon reaching equilibrium conditions after pumping the test well at a rate of 550 gpm for 2 days, the water level drawdowns were measured as 9.1 ft and 2.4 ft in the observation wells at distances of 60 ft and 360 ft, respectively. Calculate the field coefficient of permeability. (*Answer* $K = 0.0013$ ft/sec)

4–56 A well is installed in a 50-ft-thick sandstone aquifer confined under an impervious overburden with a depth of 100 ft. The diameter of the screen and gravel pack in the aquifer is 2.0 ft. Based on test well data, the piezometric surface is at a depth of 30 ft below the ground surface, the radius of the cone of depression is 1000 ft, and the permeability of the sandstone is 2.5×10^{-4} ft/sec. Calculate the estimated pumping rate that will lower the water level in the well to the top of the sandstone aquifer.

4–57 A well with a diameter of 0.6 m is constructed in a confined aquifer as illustrated in Figure 4–36. The sand aquifer has a uniform thickness of 15 m overlain by a surface clay layer with a depth of 35 m. A pumping test is conducted to determine the permeability of the aquifer. The initial piezometric surface was 15.0 m below the ground surface datum in the test well and observation wells. After pumping at a rate of 13.0 l/s for several days, water levels in the wells stabilized with the following drawdowns: 6.4 m in the test well, 3.7 m in the observation well, 10 m from the test well, and 2.4 m in the second test well at a distance of 30 m. Calculate the coefficient of permeability for the aquifer. Then, using this K value and an assumed distance of 200 m to the edge of the cone of depression, estimate the well discharge with the drawdown in the well lowered to the surface of the confined aquifer. (*Answer* $K = 0.12$ mm/s, $Q = 35$ l/s)

Water Quality

The many ways in which water promotes the economic and general welfare of society are known as beneficial uses. The major ones are public and industrial water supplies, agricultural irrigation, fish propagation, and wildlife, recreation, and aesthetics. Many applications are restricted within narrow ranges of water quality, particularly public water supply. Unregulated wastewater disposal conflicts with use of the water as a municipal source. Therefore, control of quality is required to ensure that the best employment of water is not prevented by indiscriminate use of watercourses for disposition of wastes.

The Public Health Service first established drinking water standards in 1914 to protect the health of the traveling public and assist in the enforcement of Interstate Quarantine Regulations. They have been changed and updated several times, with the last revision in 1962. Although the standards were enforceable only on interstate carriers, they were widely adopted and provided criteria to measure the quality of municipal water supplies. The Safe Drinking Water Act of 1974 gave the U.S. Environmental Protection Agency (EPA) the power to set maximum limits on the level of contaminants permitted in drinking water and to enforce those standards if the states fail to do so. The bill also directed the EPA to develop rules for the operation and maintenance of drinking water systems.

The Water Pollution Control Act of 1948 was the earliest attempt to control pollution of surface waters. Amendments in 1956 and 1965 established limited federal interest in enforcement; nevertheless, water quality standards and pollution control were still the responsibility of the states. The Clean Water Act of 1972 gave the U.S. Environmental Protection Agency the dominant role in directing and defining water pollution control programs for all the states. The initial national goals were to eliminate the discharge of pollutants into navigable waters and to achieve a water quality to protect fish, shellfish, and wildlife and to provide recreation in and on the water wherever attainable. The precepts of this environmental law further state that the right to pollute does not exist, a permit for discharge is required, violation of the conditions of a permit is subject to prosecution, and any discharge controls higher than the federal minimum technology requirements must be based on water quality of the receiving watercourse.

5–1   SAFE DRINKING WATER ACT

The Safe Drinking Water Act, initially enacted in 1974, authorizes the U.S. Environmental Protection Agency (EPA) to establish comprehensive national regulations to ensure the drinking water quality of public water systems.[1] The following are the three categories of public water systems. A community water system serves at least 25 people at their primary residences (or at least 15 residences that are primary residences). Examples are a municipality, mobile home park, and homeowner subdivision. A nontransient noncommunity water system regularly serves at least 25 of the same people for at least 6 months per year but not at their primary residences. Examples are schools, commercial facilities, or manufacturing plants that have their own water systems. A transient noncommunity water system serves 25 or more people for at least 60 days per year but not the same people or not on a regular basis. Examples are highway rest areas, recreation areas, gas stations, and motels that have their own water available for employees and the public.

The Safe Drinking Water Act has been regularly amended, resulting in regulations to ensure against previously unknown or unrecognized pathogens and toxic substances. The Total Coliform Rule requires a maximum concentration of zero for total coliform, fecal coliform, and *Escherichia coli*, which is much more stringent than the original coliform rule (Section 5–2). The Lead Contamination Control Act contains requirements for special surveillance of water in distribution systems and a ban on lead solders, flux, and pipe in public systems (Sections 5–3 and 7–19). The Surface Water Treatment Rule and the Enhanced Surface Water Treatment Rule require filtration followed by chemical disinfection to ensure removal of *Giardia* and *Cryptosporidium* protozoa and enteric viruses from surface-water supplies and groundwater under the influence of surface water (Section 7–15). Every 5 years, the

EPA is required to make determinations on whether to regulate at least five contaminants from the Drinking Water Candidate List[1] of unregulated contaminants.

Regulatory Basis for MCLGs

A maximum contaminant level goal (MCLG) is a nonenforceable, health-based goal set at a level with an adequate margin of safety to ensure no adverse effect on human health. To establish a MCLG, the EPA conducts a risk assessment that includes qualitative determination of adverse effects, quantitative effects at different doses, route and amount of human exposure, and description of risk assessment results and underlying assumptions. A MCLG is established based on toxicology data, including human epidemiology or chemical studies and animal exposure studies. Since epidemiology data often do not exist, a contaminant's potential risk is estimated by the response of laboratory animals based on the assumption that effects in animals may occur in humans, which is the subject of ongoing controversy.[1] The two major problems in assessing risk based on animal studies are extrapolation from observed effects at relatively high exposure levels used in laboratory animal studies to the low levels of exposure in humans and extrapolation of the estimated risk from laboratory animals to humans. For a contaminant that is a known human carcinogen (cancer-causing agent), or if evidence of carcinogenicity is strong, the MCLG is set at zero, which means no safe threshold exists. Zero means drinking water should not contain carcinogens.

Regulatory Basis for MCLs

Maximum contaminant level (MCL) is an enforceable standard (for noncarcinogenic contaminants) set at a numerical value with an adequate margin of safety to ensure no adverse effect on human health. The lowest practical level is selected to minimize the amount of toxic substance contributed by water when other sources like food, beverages, smoking, and air pollution are known to represent exposure to humans. The EPA determines the no-effect level (known as the reference dose, RfD) for chronic or lifetime exposure without significant risk to humans, including sensitive subgroups such as infants, children, pregnant women, the elderly, and immunodeficient persons. Human and/or animal toxicology data are used to identify the highest no-observed-adverse-effect level (NOAEL) or the lowest-observed-adverse-effect level (LOAEL). The RfD, measured in milligrams per kilogram of body weight per day, is calculated as

$$RfD = \frac{NOAEL \text{ or } LOAEL}{\text{uncertainty factor}} \qquad (5-1)$$

The uncertainty factor, in the range of 10 to 1000, is to account for differences in response to toxicity within the human population and between humans and animals.

Using the RfD, the drinking water equivalent level (DWEL) is calculated as

$$DWEL \text{ (mg/l)} = \frac{RfD \times \text{body weight (kg)}}{\text{drinking water consumption (l/d)}}$$

$$(5-2)$$

The DWEL represents a lifetime exposure with no adverse health effects assuming only exposure from drinking water. For regulatory purposes, a body weight of 70 kg and drinking water consumption rate of 2 l/d are assumed for adults or, if based on effects on infants, a body weight of 10 kg and consumption of 1 l/d are applied to Eq. 5–2.

The MCLG for a noncarcinogenic contaminant takes into account contributions from other sources of exposure such as air and food. If quantitative data are available on each source of exposure and if such exposures are between 20 percent and 80 percent, the sum of the exposures is used. If the exposure from drinking water is 80 percent or greater, a value of 80 percent is used. If the exposure is 20 percent or less, a value of 20 percent is used. Once the relative source contribution from drinking water is determined, the MCLG is calculated as

$$MCLG \text{ (mg/l)} = DWEL \times \text{(fraction}$$
$$\text{of drinking water contribution)} \qquad (5-3)$$

Treatment Technique

A treatment technique in lieu of a MCL may be specified by the EPA if monitoring for a contaminant is not economically or technologically feasible. A good example is the Surface Water Treatment Rule to control *Giardia* and *Cryptosporidium*, for which monitoring is not feasible for public health protection. The treatment technique requires effective coagulation and filtration reducing turbidity to less than a specified level and effective disinfection determined by the chemical concentration and time of contact. Copper and lead are treatment techniques rather than MCLs because reducing contaminants' concentrations requires modification or improvement of water processing.

Multiple-Barrier Concept

Although not written as a separate rule, the concept of multiple barriers is incorporated in the Safe Drinking Water Act. The use of several barriers in a drinking water system is effective in preventing pathogens and other contaminants from reaching water consumers. The first barrier is source protection. For a surface-water supply, watershed use should be controlled and wastewater contamination from human and animal sources minimized. For a groundwater supply, a wellhead protection program should be adopted to prevent surface contamination from entering the aquifer, and proper design and construction are needed to prevent surface water from entering the borehole of the well. The most important barrier for a surface supply is water treatment that commonly consists of chemical coagulation, sedimentation, and filtration. Treatment facilities for well water vary depending on the quality of the groundwater. The third barrier is disinfection of treated water by chlorination in a contact tank and an adequate protective chlorine residual maintained in the distribution system. The final barrier is properly maintained mains and storage reservoirs in the distribution system, operated at a high enough water pressure to prevent outside water from entering the system. Enforcement of a cross-connection-control program is essential to prevent backflow of contaminated water through service connections from industries, hospitals, and other institutions handling hazardous substances.

5–2 MICROBIOLOGICAL QUALITY OF DRINKING WATER

Drinking water must be free of all pathogenic microorganisms. The viruses, bacteria, protozoa, and helminths most likely to be transmitted by water are listed in Table 3–1 (Section 3–4, Waterborne Diseases). Testing water for this broad diversity of pathogens is not feasible because of the difficulty of performing laboratory analyses and their poor quantitative reproducibility. Therefore, the microbial quality of drinking water is controlled by specified treatment techniques and monitoring for the presence of coliform bacteria.

Treatment techniques are prescribed physical and/or chemical processes required by the EPA to ensure removal of microorganisms (or contaminants) that are health risks (Section 7–15, Disinfection). The disinfection of surface waters is defined by treatment techniques for removal of protozoal cysts by chemical coagulation and granular-media filtration, such as *Giardia* and *Cryptosporidium* resistant to chlorine residual. Inactivation of any remaining cysts and enteric viruses is by chemical treatment, commonly chlorination. Effective coagulation and filtration of the treated water is determined by a turbidity equal to or less than 0.3 NTU in at least 95 percent of the measurements. Effective chemical disinfection of the water prior to entering the distribution system is determined by the $C \cdot t$ product, which is the disinfectant concentration multiplied by the time of contact.

The treatment technique for groundwater (Section 7–15, Disinfection) assesses a well site for natural disinfection by travel time of filtration through surface soils and the aquifer. If a groundwater is susceptible to contamination from fecal wastewaters, chemical disinfection for inactivation of enteric viruses is based on the $C \cdot t$ product determined by site conditions.

Water in the distribution system is monitored for microbiological quality by testing for coliform bacteria as indicator organisms (Section 3–8). The primary objective is to detect microbial contamination of the treated water in the pipe network resulting from backflow through customer service connections and other system outlets (Section 6–9). The number of water samples tested for coliforms in a small public water system serving under 1000 persons is one per month. In larger systems, the minimum number required increases with size of population at the rate of approximately one additional test per 1000 residents. For example, for a population of 28,001 to 37,000, the minimum number of samples per month is 40; for 83,001 to 96,000, 96; and for 450,000 to 600,000, 232. Sample collection sites are located throughout the distribution system based on population densities and system zones, with at least 16 locations for communities with populations greater than 11,000.

The Total Coliform Rule specifies maximum contaminant level goals (MCLGs) of zero for total coliforms, fecal coliforms, and *Escherichia coli*. The Surface Water Treatment Rule specifies MCLGs of zero for *Giardia lamblia*, *Cryptosporidium* species, enteric viruses, and *Legionella*. Nevertheless, allowance is made for inadvertent sample contamination by a dirty faucet or coliforms from the hands of the person collecting the sample. For a system that collects fewer than 40 samples per month, no more than 1 sample per month may test positive for total coliforms. For a system that collects more than 40 samples per month, no more than 5.0 percent may test positive. If a routine sample is positive, 3 or 4 repeat samples must be collected within 24 hr. If one of these is positive, an additional set of repeat samples is required. If one of these is again positive, another repeat set is required and the system is evaluated to determine the source of coliforms. Positive repeat samples are included in determining if the coliform maximum contaminant level (MCL) has been violated. Furthermore, if any routine or repeat sample tests positive for total coliforms, the positive culture medium must be tested for fecal coliforms. Since the MCL for fecal coliforms is zero, their presence in any sample constitutes a violation of the MCL for total coliforms. Water laboratories using the presence-absence technique test for both total coliforms and the fecal coliform *E. coli*. (Tests for the coliform group are discussed in Section 3–9.)

5–3 CHEMICAL QUALITY OF DRINKING WATER

The chemicals listed in Table 5–1 are currently regulated in drinking water by the EPA. The chemicals in this list and/or the associated MCLs may change depending on future risk assessments. Deletion may result from prohibiting use of a chemical resulting in significantly reduced contamination of water, or future studies showing a lack of convincing evidence of risk to human health. Addition of new chemicals will occur, since approximately 10 more inorganic and 40 more organic chemicals are listed for regulation. Also, disinfection by-products and radionuclides are being reevaluated. Drinking water standards are likely to remain flexible, with continuous adjustments to accommodate the changing chemical usage in the environment. (A reader interested in a specific contaminant is advised to contact the EPA or state regulatory agency to confirm the most recent regulatory status and MCL.)

The MCL is an enforceable standard for protection of human health. MCLs are set as close to MCLGs as feasible based on best control management, treatment technology, and other means, while taking cost into consideration. Monitoring requirements are very specific, with a prescribed schedule of routine sampling and repeat sampling to confirm the results if a MCL is exceeded. The frequency of sampling varies depending on the chemical, population served by a public water system, and whether the source of supply is surface or groundwater. Testing is performed in an approved laboratory by specified methods. The concentration of a regulated chemical in excess of the MCL for human health constitutes grounds for rejection of the water supply.

TABLE 5–1

Chemical Drinking Water Standards Established by the Environmental Protection Agency for Protection of Health, Maximum Contaminant Levels in Milligrams per Liter

Inorganic Chemicals			
Antimony	0.006		
Arsenic	0.01	Mercury	0.002
Barium	2.	Nickel	0.1
Beryllium	0.004		
Cadmium	0.005	Nitrate (as N)	10.
Chromium (total)	0.1	Nitrite (as N)	1.
Copper	TT[a]	Nitrate + Nitrite (as N)	10.
Cyanide	0.2		
Fluoride[b]	4.0	Selenium	0.05
Lead	TT[a]	Thallium	0.002
Asbestos	7 million fibers/liter (longer than 10 μm)		

Volatile Organic Chemicals			
Benzene	0.005	Ethylbenzene	0.7
Carbon tetrachloride	0.005	Monochlorobenzene	0.1
Chlorobenzene	0.1	Styrene	0.1
Dichloromethane	0.005	Tetrachloroethylene	0.005
p-Dichlorobenzene	0.075	Toluene	1.
o-Dichlorobenzene	0.6	1,2,4-Trichlorobenzene	0.07
1,2-Dichloroethane	0.005	1,1,1-Trichloroethane	0.2
1,1-Dichloroethylene	0.007	1,1,2-Trichloroethane	0.005
cis-1,2-Dichloroethylene	0.07	Trichloroethylene	0.005
trans-1,2-Dichloroethylene	0.1	Vinyl chloride	0.002
1,2-Dichloropropane	0.005	Xylenes (total)	10.

Continued

TABLE 5–1 Continued

SYNTHETIC ORGANIC CHEMICALS

Acrylamide	TT[a]	Ethylene dibromide	0.00005
Alachlor	0.002	Glyphosate	0.7
Aldicarb	0.003	Heptachlor	0.0004
Aldicarb sulfone	0.002	Heptachlor epoxide	0.0002
Aldicarb sulfoxide	0.004	Hexachlorobenzene	0.001
Atrazine	0.003	Hexachlorocyclopentadiene	0.05
Carbofuran	0.04	Lindane	0.0002
Chlordane	0.002	Methoxychlor	0.04
Dalapon	0.2	Oxlamyl (Vydate)	0.2
Di(2-ethylhexyl)adipate	0.4	PAHs (benzo[a]pyrene)	0.0002
Dibromochloropropane	0.0002	Pentachlorophenol	0.001
Diethylhexyl phthalate	0.006	Picloram	0.5
Dinoseb	0.007	Polychlorinated byphenyls	0.0005
Diquat	0.02	Simazine	0.004
Endothall	0.1	Toxaphene	0.003
Endrin	0.002	2,3,7,8-TCDD (Dioxin)	0.00000003 (3×10^{-8})
Epichlorohydrin	TT[a]	2,4-D	0.07
2,4,5-TP (Silvex)	0.05		

DISINFECTION BY-PRODUCTS

Total trihalomethanes	0.080	Five haloacetic acids	0.060

RADIONUCLIDES

Radium 226 + Radium 228	5 pCi/l	Beta particle and photon radioactivity	4 mrem/yr
Gross alpha particle activity	15 pCi/l	Uranium	30 µg/l

[a]Treatment technique (TT) requires modification or improvement of water processing to reduce the contaminant concentration.
[b]Many states require public notification at least annually if fluoride is in excess of 2.0 mg/l to warn consumers of potential dental fluorosis.

The regulation of specific MCLs depends on the kind of water system. All of the standards are applicable to community systems and nontransient noncommunity systems that supply water to the same people for a long period of time, for example, schools and factories. Transient noncommunity systems that serve different people for a short period of time—for example, campgrounds, parks, and highway rest stops—are required to meet the MCLs only of those contaminants with health effects caused by short-term exposure, such as nitrate and coliforms.

A treatment technique rather than a MCL is specified for selected chemicals. Acrylamide and epichlorohydrin are used during water treatment in flocculants to decrease turbidity. The treatment technique requirements limit the concentration of these chemicals in polymers and their application. The sources of lead and copper in tap water are service connections of copper pipe with lead-solder joints, lead pipe, and lead goosenecks used to connect galvanized pipe to water mains. If the concentrations of these metals exceed the action levels, the treatment technique is to reduce the corrosivity of the water (Section 7–19) or replace the service lines with plastic pipe.

Inorganic Chemicals

The sources of trace metals are associated with both human activities, such as mining and manufacturing, and natural processes of chemical weathering and soil leaching. Corrosion in distribution piping and customers' plumbing can also add trace metals to tap water.

Antimony, arsenic, barium, beryllium, cadmium, chromium, mercury, nickel, selenium, and thallium are toxic metals affecting the internal organs of the human body. *Antimony* is a trace metal used as a constituent of alloys. It is rare in natural waters; however, ingestion effects the blood, decreasing longevity. *Arsenic* is widely distributed in waters at low concentrations, with isolated instances of higher concentrations in well waters. It is also found in trace amounts in food. *Barium*, one of the alkaline earth metals, occurs naturally in low concentrations in most surface waters and in many treated waters. *Beryllium* is not likely to be found in natural waters in greater than trace amounts because beryllium oxides and hydroxides are relatively insoluble. Soluble beryllium sulfate is transported in the bloodstream to bone, where it is found to induce bone cancer in animals. *Cadmium* can be introduced into surface waters in amounts significant to human health by improper disposal of industrial wastewaters. Nevertheless, the major sources are food, cigarette smoke, and air pollution; hence, the MCL is set so that less than 10 percent of the total intake is expected to be from water consumption. The health effects of cadmium can be either acute, resulting from overexposure at a high concentration, or chronic, caused by the accumulation in the liver and renal cortex. *Chromium* held in rocks is essentially the insoluble forms of trivalent chromium; thus, the content in natural waters is extremely low. Acute poisoning can result from high exposures to hexavalent chromium from industrial wastes; trivalent is relatively innocuous. *Mercury* is a scarce element in nature, and it has been banned for most applications with environmental exposure, for example, mercurial fungicides. The biological magnification of mercury in freshwater food fish is the most significant hazard to human health. The mercury enters the food chain through the transformation of inorganic mercury to organic methylmercury by microorganisms present in the sediments of lakes and rivers. Thus, toxicity via the oral route is related mainly to methylmercury compounds rather than to inorganic mercury salts or metallic mercury. Symptoms of methylmercury poisoning include mental disturbance and impairment of speech, hearing, vision, and movement. *Nickel* salts

cause gastrointestinal irritation without inherent toxicity; however, large oral doses have produced toxicity in laboratory animals. *Selenium* is a trace metal naturally occurring in soils derived from some sedimentary rocks. Surface streams and groundwater in seleniferous regions contain variable concentrations. Effects on human health have not been clearly established. A low-selenium diet is beneficial, whereas very high doses can produce undesirable physical effects, such as loss of hair and fingernails.

Lead exposure occurs through air, soil, dust, paint, food, and drinking water. Lead toxicity affects the red blood cells, nervous system, and kidneys, with young children, infants, and fetuses being most vulnerable. Depending on local conditions, the contribution of lead from drinking water can be either a minor or major exposure for children. Lead is not a natural contaminant in either surface waters or groundwaters, and the MCL of 0.005 mg/l in source waters is rarely exceeded. It is a corrosion by-product from high-lead solder joints in copper piping, old lead-pipe goosenecks connecting the service lines to the water main, and old brass fixtures. Lead pipe and brass fixtures with high lead content are not installed today, and lead-free solder (less than 0.2 percent lead) has replaced the old tin solder of 50 percent lead and 50 percent tin in copper water piping. Since dissolution of lead requires an extended contact time, lead is most likely to be present in tap water after being in the service connection piping and plumbing for a 6- to 8-hr period. Therefore, the first-flush sample concentration is the highest expected, and it should not exceed 0.015 mg/l to be below the action level.

Copper is commonly found in drinking water. Trace amounts below 20 µg/l can derive from weathering rock, but the principal sources in house water supplies are from corrosion of copper service pipes and brass plumbing fixtures. As an essential element in human nutrition, copper intake is safe and adequate at 1.5 to 3 mg/day. Copper in a large oral dose causes gastrointestinal distress with nausea and vomiting within 60 min of ingestion; nevertheless, it causes no apparent chronic health effects. As an indicator of corrosivity, copper in first-flush samples of tap water from service connections of copper piping should not exceed 1.3 mg/l.

Routine monitoring for lead and copper in first-flush tap samples is required in service connections with copper pipe and lead solder or connections with lead pipe or lead goosenecks. The number of sampling points and the frequency of monitoring depend on size of the water system, results of previous monitoring, and, if implemented, effectiveness of treatment for corrosion control. Noncompliance is 10 percent or more of the first-flush samples exceeding either the 0.015 mg/l lead or 1.3 mg/l copper action levels. The treatment technique for noncompliance of either lead or copper is to perform corrosion studies to evaluate alternative methods of treatment to reduce corrosivity and implement the optimal treatment. If the lead action level is exceeded, a public education program on the health risks of lead and recommendations for reducing lead intake, such as flushing the service connection before using the water for drinking or food preparation, is required. Furthermore, lead service connection replacement is required for water systems that continue to exceed the action level after implementing corrosion control.

Fluoride is found in groundwaters as a result of dissolution from geologic formations. Surface waters generally contain much smaller concentrations of fluoride. Absence or low concentration of fluoride in drinking water causes the formation of tooth enamel less resistant to decay, resulting in a high incidence of dental caries in children's teeth. Excessive concentration of fluoride in drinking water causes fluorosis, also referred to as mottling. This dental disease in mildest form results in opaque whitish areas on the posterior teeth. With greater severity, the fluorosis is widespread and the color of the teeth is yellow-brown. The MCL of 4.0 mg/l in the drinking water standards is established to prevent unsightly fluorosis. The optimum fluoride concentration in drinking water protects teeth from decay without causing noticeable fluorosis. Cities with water supplies deficient in natural fluoride are successfully providing supplemental fluoridation to optimum levels to reduce the incidence of dental caries in children. Since water consumption is influenced by climate, the recommended optimum concentrations listed in Table 5–2 are based on the annual average of the maximum air temperatures obtained for a minimum record of 5 years.

TABLE 5–2

Recommended Optimum Concentrations of Fluoride in Drinking Water Based on the Annual Average of the Maximum Daily Air Temperatures

Temperature Range (°F)	Recommended Optimum (mg/l)
53.7 and below	1.2
53.8 to 58.3	1.1
58.4 to 63.8	1.0
63.9 to 70.6	0.9
70.7 to 79.2	0.8
79.3 to 90.5	0.7

Nitrate is the common form of inorganic nitrogen found dissolved in water. In agricultural regions, groundwater can have significant concentrations of nitrate from unused fertilizer leaching into the underlying aquifers. For good crop yields, the inorganic nitrogen content in the soil moisture surrounding the root zone is often much greater than the MCL allowed by drinking water standards. Porous soil profiles permit rainfall and irrigation water to transport this high-nitrate pore water to the groundwater table without measurable denitrification loss. Surface waters can be polluted by nitrogen from both discharge of municipal wastewater and drainage from agricultural lands. The health hazard of ingesting excessive nitrate in water is infant methemoglobinemia. In the intestines of an infant, particularly one under 3 to 6 months of age, nitrate can be reduced to nitrite that is absorbed into the blood, oxidizing the iron of hemoglobin. This interferes with oxygen transfer in the blood, resulting in cyanosis, which gives the baby a blue color. Incidents of infant methemoglobinemia, however, are extremely rare, since most mothers in regions of known high-nitrate drinking water use either bottled water or a liquid formula requiring little or no dilution. Methemoglobinemia is readily diagnosed by a medical doctor and rapidly reversed by injecting methylene blue into the infant's blood. Healthy adults are able to consume large quantities of nitrate in drinking water without adverse effects. The principal sources of nitrate in the adult diet are vegetables and saliva.

Occupational exposure to asbestos dust can lead to pulmonary fibrosis of the lungs, bronchogenic carcinoma, and other respiratory diseases. The fact that exposure to air polluted with asbestos fibers leads to these diseases does not necessarily indicate that drinking water contaminated with an equally large number of fibers may lead to these or other diseases. Nevertheless, the hypothesis is tenable to the degree that it cannot be ruled out of consideration without evidence to the contrary.

Organic Chemicals

Organic chemicals in drinking water are found in trace amounts often so low in concentration that predicting any potential effect on human health is difficult. Reliable data on the toxicity to humans of most organic compounds are not readily obtained, since the information available is usually from uncontrolled accidental or occupational exposures. Therefore, long-term animal studies are used to evaluate chronic exposure and carcinogenic risk of organic compounds to humans. In feeding studies of laboratory animals, the dosage of toxic substance is assumed to be physiologically equivalent to humans on the basis of body weight. After the acceptable daily intake value for a no-observed-adverse-effect level (NOAEL) in animals is determined, a safety factor is applied to reduce allowable human intake to account for the uncertainties involved in extrapolating from animals to humans. The safety factors used in establishing drinking water standards are a factor of 10 when chronic human exposure data are available and are supported by chronic oral toxicity data in animals, a factor of 100 when good chronic oral toxicity data are available in some animal species but not in humans, and a factor of 1000 with limited chronic animal toxicity data. The hazard of ingesting a chemical assessed as a confirmed or a suspected carcinogen is evaluated in terms of dose-related risk. The estimate of risk resulting from animal bioassays is made by first converting the laboratory animal dose to the physiologically equivalent human dose on the basis of relative surface areas of test animals and the human body. Then, a mathematical model is used to relate dose to effect, and the dose rate associated with cancer risk is calculated on the basis of daily ingestion over a human lifetime of 70 years.

Volatile organic chemicals (VOCs) are produced in large quantities for use in industrial, commercial, agricultural, and household activities. The adverse health effects of VOCs include cancer and chronic effects on the liver, kidney, and nervous system. Volatility reduces their concentrations in surface waters. Groundwater contamination is more common, since VOCs have little affinity for soils and are diminished only by dispersion and diffusion, which is often limited. Those most frequently detected in contaminated groundwaters are trichloroethylene, a degreasing solvent in metal industries and common ingredient in household cleaning products; tetrachloroethylene, a dry-cleaning solvent and chemical intermediate in producing other compounds; carbon tetrachloride, used in the manufacture of fluorocarbons for refrigerants and solvents; 1,1,1-trichloroethane, a metal cleaner; 1, 2-dichloroethane, an intermediate in manufacture of vinyl chloride monomers; vinyl chloride, used in the manufacture of plastics; and polyvinyl chloride, used in the manufacture of plastics and polyvinyl chloride resins.

The *synthetic organic chemicals* (SOCs) of greatest concern in contamination of drinking water sources are pesticides. Other SOCs are used in manufacturing consumer products, particularly plastic materials; and a few have had direct applications. Insecticides and herbicides may be present in surface waters receiving runoff from either agricultural or urban areas where these chemicals are applied. Groundwaters can be contaminated by pesticide manufacturing wastewaters, spillage, or infiltration of rainfall and irrigation water. Alachlor, aldicarb, atrazine, carbofuran, ethylene dibromide, and dibromochloropropane have been detected in drinking waters. Most pesticides can be absorbed into the human body through the lungs, skin, and gastrointestinal tract. From acute exposure, the symptoms in humans are dizziness, blurred vision, nausea, and abdominal pain. Chronic exposure of laboratory animals indicates possible neurologic and kidney effects and, for some pesticides, cancer.

Disinfection By-Products

Trihalomethanes (THMs) are organohalogen derivatives of methane in which three of the four hydrogen atoms have been replaced by three atoms of chlorine, bromine, or iodine. Detectable concentrations of chloroform (trichloromethane), bromodichloromethane, dibromochloromethane, bromoform (tribromomethane), and dichloroiodomethane have been found in drinking waters. Chloroform is the trihalomethane most commonly found in drinking water and is usually present in the highest concentration. In some cases, bromated trichloromethanes dominate as a result of naturally occurring bromide in the water. THMs are produced during treatment of surface waters as a result of the chemical interaction of chlorine applied for disinfection and organic substances naturally present in the raw water. The natural organic substances include humic and fulvic acids produced from decaying vegetation. The general reaction producing trihalomethanes is

$$\text{Chlorine} + (\text{bromide ion} + \text{iodide ion})$$
$$+ \text{ organic substances}$$
$$= \text{ trihalomethanes and other}$$
$$\text{halogenated compounds} \quad \textbf{(5–4)}$$

In addition to total trihalomethanes, which are the major disinfection by-products, five haloacetic acids (HAA5) are the next most significant fraction. Other by-products that have been detected by laboratory testing of chlorinated waters are not present in significant concentrations.

Trihalomethanes and haloacetic acids are not formed instantaneously but continue to increase in concentration for an extended period of time following chlorination. Thus, their concentrations can increase in chlorinated water held in the distribution system. The MCLs for total THMs at 80 µg/l and HAA5 at 60 µg/l have been established because of carcinogenicity in laboratory animals. These numerical limits are determined by calculating a 12-month running average value rather than by a single test. Several water samples are collected for a 3-month period from the distribution system, with 25 percent from the extremities of the pipe network and 75 percent based on population distribution. The values for each period are the arithmetrical averages of the total THMs and HAA5 concentrations of all samples tested. The maximum contaminant levels are the averages of the most recent quarterly values and the average concentrations recorded during the previous three quarters.

The Disinfectants/Disinfection By-products Rule promulgated by the EPA also establishes MCLs of 0.01 mg/l for bromate and 1.0 mg/l for chlorite. Maximum residual disinfection levels are set at 4.0 mg/l for chlorine plus chloramines and 0.8 mg/l for chlorine dioxide.

Radionuclides

Radioactive elements decay by emitting alpha, beta, or gamma radiations caused by transformation of the nuclei to lower energy states. An alpha particle is the helium nucleus (2 protons + 2 neutrons); for example, radon-222 decays to polonium-218, emitting helium-4. A beta particle is an electron emitted from the nucleus as a result of neutron decay; for example, radium-228 decays to actinium-228, emitting β^-. In these processes, the helium nucleus emitted as an alpha particle or the electron ejected as a beta particle changes the parent atom into a different element. A gamma ray is a form of electromagnetic radiation; other forms are light, infrared and ultraviolet radiations, and X rays. Gamma decay involves only energy loss and does not create a different element. Alpha, beta, and gamma radiations have different energies and masses, thus producing different effects on matter. Each is capable of knocking an electron from its orbit around the nucleus and away from the atom, a process referred to as ionization. Radiation is detected by ionization, and highly reactive ions taken into the human body can lead to deleterious health effects. The potential health risks from radiation are developmental and teratogenic, genetic, and somatic.

The ability to penetrate matter varies among nuclear radiations. Most alpha particles are stopped by one thickness of paper; most gamma rays pass through the human body, as do X rays. Since alpha particles are stopped by short penetrations, more energy is deposited, doing more damage per unit volume of matter receiving radiation.

The measurement unit of pCi/l is 10^{-12} curie per liter, with a curie being the activity of 1 gram of radium. The rem (radiation equivalent man) is a unit of radiation dose equivalence, which is numerically equal to the absorbed dose in rad multiplied by a quality factor to describe the actual damage to tissue from the ionizing radiation. A mrem is 1/1000 of a rem. The rad is the unit of dose or radiation absorbed. One rad deposits 100 ergs of energy in 1 gram of matter.

Radioactivity in drinking water can be from natural or artificial radionuclides. Background radiation resulting from cosmic ray and terrestrial sources is about 100 mrem/yr. Radium-226 is found in groundwater from geologic formations, and radioactivity from radium is widespread in surface waters because of fallout from testing of nuclear weapons. The majority of human exposure is unavoidable background radiation from these sources. Only a small portion results from drinking water containing small releases from nuclear power plants, hospitals, and industrial users of radioactive materials. The recommended allowable dose from radioisotopes in drinking water supplies is very low, amounting to 0.244 mrem/yr, which is less than 1 percent of the normal background. Although the dose to bone would be considerably higher because strontium and radium are bone seekers, even this dose constitutes less than 10 percent of the total average background. Nevertheless, because of the many uncertainties of the environmental effects, the EPA has adopted the MCLs listed in Table 5–1.

Secondary Standards

Secondary maximum contaminant levels (SMCLs) for aesthetics listed in Table 5–3 are recommended for characteristics that render the water less desirable for use. They are not related to health risks and are nonenforceable by the EPA.

Excessive color, foaming, or odor cause customers to question the safety of a drinking water and result in complaints from users. Chloride, sulfate, and total dissolved solids have taste and laxative properties, and highly mineralized water affects the quality of coffee and tea. Both sodium sulfate and magnesium sulfate are well-known laxatives with the common names of Glauber salt and Epsom salt, respectively. The laxative effect is commonly noted by travelers or new consumers drinking waters high in sulfates; however, most persons become acclimated in a relatively short time.

Aluminum salts, rarely found in natural waters, increase fecal excretion of fluoride, resulting in decreased absorption, and can cause constipation. Aluminum hydroxide is contained in

TABLE 5–3

Recommended Secondary Maximum Contaminant Levels for Aesthetics of Drinking Water

Aluminum	0.05 to 0.2 mg/l
Chloride	250 mg/l
Color	15 color units
Copper	1.0 mg/l
Corrosivity	noncorrosive
Fluoride	2 mg/l
Foaming agents	0.5 mg/l
Iron	0.3 mg/l
Manganese	0.05 mg/l
Odor	3 threshold odor numbers
pH	6.5 to 8.5
Silver	0.1 mg/l
Sulfate	250 mg/l
Total dissolved solids	500 mg/l
Zinc	5 mg/l

some antacid medications to treat ulcers and stomach upset caused by stomach acid. A noncorrosive water with a neutral or slightly alkaline pH is desirable to reduce corrosion of pipe, which can contribute trace metals to the water. Iron and manganese are objectionable because of the brownish-colored staining of laundry and porcelain and the bittersweet taste contributed by iron. Silver in trace amounts well below the health-effect level has been found in some natural waters. At high doses, the effect is cosmetic, causing blue-gray discoloration of the skin, eyes, and mucous membranes. Zinc is an essential element in the human diet, but excessive amounts act as a gastrointestinal irritant.

5–4 CLEAN WATER ACT

The Clean Water Act (Federal Water Pollution Control Act amendment of 1972) authorizes the U.S. Environmental Protection Agency (EPA) to direct and define natural water pollution control programs.[2] The objectives are to maintain the chemical, physical, and biological quality of surface waters, seawater, and groundwater by placing ecological considerations and protection of human health ahead of economic concerns. In 1972, many of the nation's waters were polluted; thus, the initial goals were to reduce discharge of pollutants and achieve an interim water quality to protect fish, shellfish, and wildlife and achieve fishable and swimmable waters wherever attainable. Congressional policy was to recognize and preserve the states' primary responsibility to meet these goals.

In the original Clean Water Act, the policies were to:

- prohibit the discharge of toxic pollutants in toxic amounts;
- provide financial assistance for construction of publicly owned treatment plants with state and local participation (which was phased out in 1990);
- implement planning and design for areawide treatment plant construction;
- sponsor development of new technology through research and demonstration projects at federal expense; and
- research, study, and establish water quality standards for surface waters to ensure that the goals of the act are met.

All surface waters should be capable of supporting aquatic life and be aesthetically pleasing. Additionally, if needed as a public supply, the water must be treatable by conventional processes to yield a potable water meeting the drinking water standards. Many impoundments and rivers are also maintained at a quality suitable for swimming, water skiing, and boating. Surface waters throughout the nation are classified according to intended uses that dictate the specific physical, chemical, and biological quality standards, thus ensuring the most beneficial uses will not be deterred by pollution. Criteria defining quality are dissolved oxygen, solids, coliform bacteria, toxic substances, pH, temperature, and other parameters as necessary.

The most important amendment to the act was the National Pollutant Discharge Elimination System (NPDES) permit program. Technology-based effluent limits backed by surface-water-quality standards were defined for treatment plants discharging pollutants through a pipe or conveyance. For compliance, the owner of the treatment plant must monitor and record discharge data and report any violations. For enforcement, a willful or negligent violator, or one making a false statement or representation regarding a discharge, can be fined and is also subject to possible imprisonment.

The NPDES program for wastewater treatment plants has significantly changed and expanded since initiation. Specific areas are effluent standards, water-quality-based permitting, control of toxic substances, industrial wastewater pretreatment program, new performance standards, inspection and monitoring provisions, and seawater discharge criteria. In addition, several other programs have been incorporated into the NPDES permit system, including the control of combined sewer overflows, use and disposal of waste sludge by regulating management practices and acceptable levels of toxic substances in sludge, and watershed protection strategy to integrate the NPDES to support states' basin management. Additional proposed rules to expand the scope of the NPDES permit system include more comprehensive reports of operational data, expanded monitoring of effluent parameters, effluent monitoring for biological toxicity, and waste sludge quality standards.[3]

5–5 NATIONAL POLLUTANT DISCHARGE ELIMINATION SYSTEM (NPDES)

Figure 5–1 shows how various water-quality standards relate to water use and treatment for managing an integrated water and wastewater system. Surface-water standards establish the desirable quality in flowing and impounded waters. Effluent standards specify the quality to be achieved in treatment of municipal and industrial wastewaters. Standards for pretreatment of industrial wastewaters discharged to municipal sewers ensure that hazardous and toxic wastes do not interfere with treatment of the domestic wastewater or pass through the plant to receiving waters. Drinking water standards are established for protection of public health. The locations designated by a circled letter M are monitoring points for testing quality of surface waters, wastewaters, and drinking water.

Figure 5–1

This illustration diagrams the various water quality standards used to manage an integrated water and wastewater system.

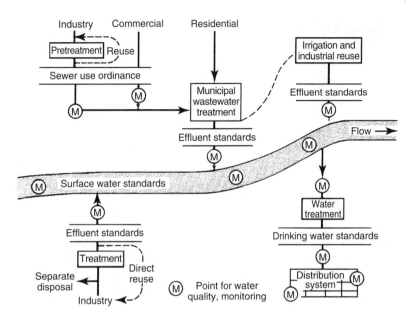

Surface waters are monitored for biological, chemical, and physical quality to ensure that degradation from pollution does not exceed established standards. The most common sampling locations are in the mixing zone downstream from the discharge of a wastewater effluent and immediately upstream from the intake of a water treatment plant.

Permit Program

NPDES permits are issued to individual wastewater treatment plants to control pollutants and the sources of such pollutants. Conventional pollutants contained in domestic, commercial, and industrial wastewaters are BOD, suspended solids, fecal coliforms, pH, and oil and grease. Toxic pollutants include synthetic organic chemicals and heavy metals. Nonconventional pollutants are any additional substances not in conventional or toxic categories, such as nitrogen and phosphorus.

Permits for discharging wastewaters contain the following: standard conditions common to all permits, site-specific effluent limits, compliance monitoring and reporting requirements, and site-specific conditions necessary for discharge control. Standard conditions describe the legal aspects of the permit and its revocability. They also describe the permittee's duties and obligations; for example, the permittee is required to report changed

conditions, to allow compliance inspections, and to reapply for a permit renewal. Finally, the permittee is informed regarding penalties that may be assessed for permit violations. Effluent limits imposed are based on technology-based and water-quality-based standards. The management of wastewater sludge is also controlled by conditions in a NPDES permit. The EPA is responsible for implementing and enforcing the NPDES program; however, individual states with primacy also perform these functions with oversight by the EPA.

The NPDES is a self-monitoring program whereby the owner of a treatment plant is required to be responsible for compliance monitoring and reporting the performance of wastewater and sludge processing, including any violation of permit conditions. Effluent limitations and monitoring provisions specify the procedures to be used for sampling, flow measurement, laboratory testing, and recordkeeping. Operation and maintenance require reporting of treatment system failure, bypassing of untreated wastewater, and process upset. Additional reporting requirements are any change in discharge, noncompliance notification, and compliance schedules. Enforcement for noncompliance is by administrative actions ranging from an informal communication to an administrative order to a monetary penalty. For serious violations, judicial action can be taken as either a civil or a criminal action.

Effluent Standards

Technology-based standards require secondary (biological) treatment for all municipal wastewater with the following limitations specified for conventional pollutants. For both BOD and suspended solids concentrations, the arithmetic mean of the values for 24-hr composite samples collected in a period of 30 consecutive days must not exceed 30 mg/l, and the arithmetic mean of values in any period of 7 consecutive days must not exceed 45 mg/l. Furthermore, the removals of BOD and suspended solids must be at least 85 percent. For concentrations of oil and grease, the arithmetic mean must not exceed 10 mg/l for any period of 30 days or 20 mg/l for any period of 7 days. Effluent pH must remain within the limits of 6.0 to 9.0. Beyond these basic requirements, more restrictive discharge quality standards for treated wastewater may be specified to meet specific limits for a particular treatment plant site (site-specific limits).

Water-quality-based standards apply to the waters receiving wastewater discharges whether surface water, groundwater, or seawater. Imposing technology-based effluent standards is no guarantee that a treated wastewater does not pollute the receiving water. Dilution in surface waters may not be adequate to prevent degradation, and evaporation of water from a discharge to a dry streambed can result in increased contaminant concentrations in the infiltration reaching groundwater. Therefore, water-quality-based limits take precedence over technology-based limits. Monitoring of the receiving water is essential to ensure compliance with the water-quality-based standards. The owner of the treatment plant is responsible for compliance monitoring, recordkeeping, and reporting as specified in the discharge permit.

An integrated strategy to water-quality-based pollution control to protect aquatic life uses two approaches: control of specific pollutants and testing of effluent toxicity by biological assays. Control of specific pollutants is done by establishing numerical limits on toxic metals and organic chemicals and on nonconventional pollutants, such as ammonia and chlorine, that adversely affect aquatic life. The most likely source of toxic chemicals in a community wastewater are industrial discharges to the sewer system; therefore, these discharges should be tested first to select the specific chemicals to be controlled. Others may be pollutants discharged by tank trucks unloading at the treatment plant. Since the water-quality-based chemical limits are monitored by testing the wastewater and surface water after mixing, the water-quality-based standards take into account dilution. In general, numerical limits of specific pollutants in the wastewater effluent are established if dilution in the mixing zone is inadequate.

The chemical contaminants that pass through wastewater treatment plants are very difficult, if not impossible, to remove by processes that can be effectively applied in drinking water treatment. The approach to protection of surface waters and groundwater as sources of supply for drinking water is to control specific chemicals. Therefore, the water-quality-based standards for wastewater discharges may be the same as drinking water standards. In other words, the monitored chemical quality of surface waters and groundwater must meet all of the maximum contaminant levels established for drinking water.

A commonly limited nontoxic, nonconventional pollutant is phosphorus, which is the key to controlling the rate of eutrophication of lakes, reservoirs, and estuaries. Nitrogen from wastewater discharges also contributes to fertilization of impounded waters. Nevertheless, significant nitrogen pollution comes from uncontrolled sources, such as runoff from agricultural land and the fact that blue-green algae in water can fix atmospheric nitrogen. The adverse effects of eutrophication are rapid plant growth resulting in blooms of algae, increasing water turbidity, and unsightly floating masses that are windblown to the shore. Decaying algae also settle to the bottom of the lake or reservoir, reducing the dissolved oxygen in the hypolimnion needed by preferred food fishes. In addition to algae, shorelines and shallow bays can become weed-choked with prolific growths of rooted aquatic plants.

The effluent standard for fecal coliforms is most limiting if the receiving water is a source for drinking water, has beneficial use as water contact recreation, or if the wastewater is discharged to a low-flow stream through a residential area. By lowering the coliform limit, the stronger disinfection processes required in treatment decrease the numbers of pathogenic microorganisms in the effluent. In some cases, secondary treatment with extended chlorine contact for disinfection may be adequate. In more restrictive cases, tertiary treatment by chemical coagulation, filtration, and chlorination is necessary to achieve the desired removal of coliforms and pathogens. Tertiary treatment also clarifies the effluent by reducing BOD and suspended solids.

Total Maximum Daily Load (TMDL) Rule

A TMDL is a calculation of the maximum amount of a single pollutant that a surface water can receive and still meet water-quality standards. Pollutant sources include both point sources (wastewater treatment plants) and nonpoint sources, such as agricultural land drainage, mining, and urban storm runoff. To compensate for seasonal variations in surface-water flows and in chemical characteristics, a calculated TMDL should include an appropriate margin of safety. A TMDL program for management of a watershed involves issuing revised NPDES permits to control discharges from treatment plants. Water-quality-based effluent limits for a pollutant are likely to be expressed in maximum allowable pounds discharged per day rather than milligrams per liter in the wastewater effluent.

Under the Clean Water Act, the states are required to identify surface waters that do not meet specified water-quality standards or do not meet the designated beneficial uses. In the 1990s, a significant percentage of the nation's waters were found to be impaired based on quality standards or beneficial uses. For the most part, these waters were not suitable for fishing, swimming, or drinking water sources. The major causes were excess sediments, plant nutrients (phosphorus and nitrogen), and human pathogens. Other causes were low dissolved oxygen, toxic metals, habitat alterations, pH, suspended solids, and pesticides. As a result, the EPA has proposed application of the TMDL Rule to restore such waters to meet the quality standards and beneficial uses.

Implementation of the TMDL Rule to improve water quality in impaired waters poses significant demands on the clean water program. Only one-tenth of impairments identified in the 1990s were caused by point sources alone. The other nine-tenths were caused either by nonpoint sources or by a combination of point and nonpoint sources. These facts limit the success of the NPDES permit system for point sources only. Quantifying nonpoint sources is challenging because they are diffuse, transient, and highly variable among different runoff events. Among the states, consistent and uniform identification of impaired waters needs clear guidelines from the EPA. Clean water standards are needed for the major contaminants of sediment, nutrients, and pathogens, and better definitions are needed for ecological impairments, such as habitat alteration. Better water quality models to mathematically trace the movements of pollutants through surface waters are also needed. Existing models for dry weather flow are too elementary, and many of the complex models have input requirements for which sufficient data are usually unavailable. Environmental conditions and allocation of all available loads among various point and nonpoint sources need to be defined for calculation of a TMDL. In summary, a watershed management program requires a reliable method to define impaired waters, appropriate water-quality standards and realistic beneficial uses, mathematical models that include estimated nonpoint inputs, and feasible guidelines for setting allocations among contaminant contributors.

Industrial Pretreatment Program

Wastewaters discharged by industries can contain toxic or otherwise harmful substances at concentrations not common in domestic wastewater. These pollutants often pose serious hazards to the sewer system and treatment plant because collection and treatment systems are not designed for these wastes. They can interfere with the operation of biological treatment, and many are difficult to remove and so pass through to contaminate the receiving water. Also, sludge contaminated with toxic substances adversely affects biological stabilization and increases environmental risks in disposal. These undesirable effects from the discharge of industrial wastewaters into municipal sewers can be prevented. Using proven treatment control technologies and manufacturing practices that promote recycling, industries can remove or eliminate pollutants before discharge of wastewaters. This practice is known as pretreatment.

EPA pretreatment regulations require municipal treatment plants to establish an approved local program for incorporation in the NPDES permit program. The objectives are to prevent discharge of pollutants to the treatment plant that adversely affect process operations, interfere with the disposal or reuse of wastewater sludge, or pass through the treatment processes. A treatment plant with control authority develops the approved pretreatment program, evaluates compliance of industrial dischargers, and initiates enforcement action against industries in noncompliance. Oversight of the program is by the municipality, the state, or the EPA. The control authority issues permits for discharge of industrial wastewaters to the municipal sewer and prepares a municipal sewer ordinance. The permits contain numerical limits for controlled pollutants and requirements for sampling, flow measurement, laboratory testing, and reporting.

Wastewater hauling to the treatment plant in tank trucks is also permitted and required to comply with the pretreatment standards. Each hauler should be required to complete a permit application submitted to the control authority.

Effluent Biological Toxicity Testing

For selected municipal treatment plants, whole effluent biological testing is required under their NPDES permit. This bioassay to determine toxicity is performed by exposing selected aquatic organisms to wastewater effluent in a controlled laboratory environment. In the static short-term test [defined by EPA as the Whole Effluent Toxicity (WET) test], effluent and uncontaminated water are placed in laboratory containers and test organisms are added to both containers and then monitored for toxic effects in a controlled laboratory environment. The objective is to estimate the safe, or no-observed-adverse-effect level, that will permit normal propagation of aquatic life in the receiving water.

The chambers are borosilicate glass or disposable polystyrene ranging in volume from 250 ml to 1000 ml, depending on the size of the test organism. During the test period, the dissolved oxygen concentration should be near saturation, with the temperature held at 25°C, and the pH

checked periodically. The effluent is a 24-hr composite filtered through a sieve to remove suspended solids. During effluent preparation, aeration should be limited to prevent loss of volatile organic compounds. The common warm-water test organisms are fathead minnow (*Pimephales promelas*), daphnid (*Ceriodaphnia dubia*), and a green alga (*Selenastrum capricornutum*). Some states have developed culturing and testing methods for indigenous species.

The test result is expressed in terms of mortality after a 24-hr exposure. A 25 percent reduction in survival is defined as the threshold of biological significance, indicating probable impairment of the receiving water.

Impairment in the receiving water and effluent toxicity are related to dilution of the effluent within the receiving water. Therefore, a more definitive toxicity test is to set up a series of laboratory containers at decreasing effluent dilution, such as, 100 percent, 50, 25, 12.5, and 6.25 percent (a geometric series). This range is likely to encompass the effluent flow relative to the recorded 1-in-10-year 7-consecutive-day low flow, which is usually the flow in a river or stream established for the maximum allowable concentration of pollutants set by surface-water standards. The dilution water is from the receiving stream above the outfall pipe or from the edge of the mixing zone. It is filtered through a plankton net to remove indigenous organisms that may attack or be confused with the test organisms. Alternatively, the dilution water may be a moderately hard mineral water. These tests may be conducted for either acute 24-hr toxicity or chronic toxicity for monitoring up to 7 days. In chronic testing, food is supplied to the organism cultures, for example, flake fish food or brine shrimp for fathead minnows. Also, chronic tests may be static-renewal tests where test organisms are transferred every 24 hr to a fresh solution of the same concentration of wastewater. Based on the results, the no-observed-adverse-effect concentration is determined for effluent in the diluted test samples. This can be compared against the ratio of wastewater diluted with the 1-in-10-year 7-consecutive-day low flow.

If toxicity is indicated by biological testing, a chemical evaluation of the effluent is required. This can be very costly unless toxic substances likely to be present can be reduced to a reasonable number. A release inventory of toxic pollutants from industries contributing wastewater to the sewer system is the best way to identify possible toxic substances and their sources. A proper pretreatment program requires sampling, flow measurement, laboratory testing, and reporting. After toxic pollutants have been quantitatively identified, remedial measures to reduce or eliminate effluent toxicity can be taken by additional industrial pretreatment and improved municipal wastewater treatment.

Groundwater Quality

Water-quality-based standards apply to groundwater as well as surface waters. The NPDES watershed protection strategy is to promote protection of surface water, groundwater, and natural habitat. Protection of groundwater is particularly important because approximately 50 percent of the U.S. population depends on groundwater as a source, with 30 percent delivered by community systems and 20 percent from domestic wells. Since groundwater is a high-quality economical source, future demand is expected to increase for domestic use.

The degree of treatment of wastewater discharged to areas where groundwater is potable and susceptible to contamination requires reclamation by tertiary or advanced wastewater treatment. Infiltration reaching the aquifer must meet drinking water standards because natural purification dispersion and/or dilution cannot be considered in groundwater. Requiring this high degree of treatment promotes effluent reuse for agricultural and urban irrigation as an alternative to discharge.

Prevention is the key to management of groundwater quality. After contamination, remedial actions are usually ineffective, since cleanup of most aquifers is not technically or economically feasible. Natural purification requires decades, and if the pollutants do not decay or are not flushed out of the aquifer by groundwater flow, natural cleansing may never occur. Knowledge of potential sources and comprehensive understanding of the hydrogeology of an area are both essential to preventing contamination.

MANAGEMENT OF GROUNDWATER QUALITY	
POINT SOURCES	NONPOINT SOURCES
PREVENTION (Groundwater Protection)	
Siting, design, construction, and operation.	Site preparation and control of chemical application.
MONITORING (Early Warning)	
Testing of groundwater from wells installed around the site.	Testing of groundwater from observation, irrigation, and water-supply wells.
ABATEMENT (Elimination of Source)	
Reconstruction of site or modifying or abandoning operation.	Changing method of application or restricting or banning chemical usage.
(Examples: Wastewater ponds, landfills, refuse piles, buried storage tanks, and deep injection wells.)	(Examples: Application of pesticides or fertilizers on agricultural land and land application of wastewater.)

Figure 5–2

Generalized schemes for controlling groundwater contamination from point and nonpoint sources of pollution.

The generalized scheme of prevention, monitoring, and abatement for point and nonpoint (diffuse) pollutant sources is outlined in Figure 5–2. Effective prevention from point sources, such as wastewater ponds, landfills, refuse piles, buried storage tanks, and deep injection wells, is based on site selection, controlled design, proper construction, and careful operation. Strategically located monitoring wells around the site for testing of groundwater are recommended for early warning. With the onset of contamination, abatement action must be instituted to stop further damage by eliminating the cause through reconstruction or modifying operations. Nonpoint sources from the application of chemicals on the land surface can generally be controlled only by regulatory actions. Agricultural use of pesticides and fertilizers and spreading of highway deicing salts are examples of potential pollutant sources. Prevention is through controlling the rate and method of application. Monitoring can be conducted by testing groundwater samples taken from wells in the area, for example, observation wells for recording groundwater levels, irrigation wells, and private and public groundwater supplies. Abatement is often difficult because private and regional economies may rely in part on the use of chemicals. Reducing applications or changing operational methods may be acceptable, or, in extreme cases, the chemicals may be banned from use.

5–6 SEAWATER QUALITY

The major beneficial uses of seawater to be protected are water-contact and noncontact recreation, commercial and sport fishing, marine habitat, shellfish harvesting (mussels, clams, and oysters), and industrial water supply.

Discharge of wastewater to seawater requires construction of an ocean outfall, as illustrated in Figure 5–3, composed of a pipeline to relatively deep water with a diffuser at the end with a series of ports spaced to provide initial dilution. By this method of discharge, sufficient initial dilution can be provided to minimize the concentration of substances not removed in biological treatment and effluent chlorination. The location of a discharge is determined by an assessment of the oceanographic characteristics and current patterns to ensure pathogens are not present in shellfish harvesting areas or water-contact recreational areas. Also, the location should provide maximum protection of the marine environment.

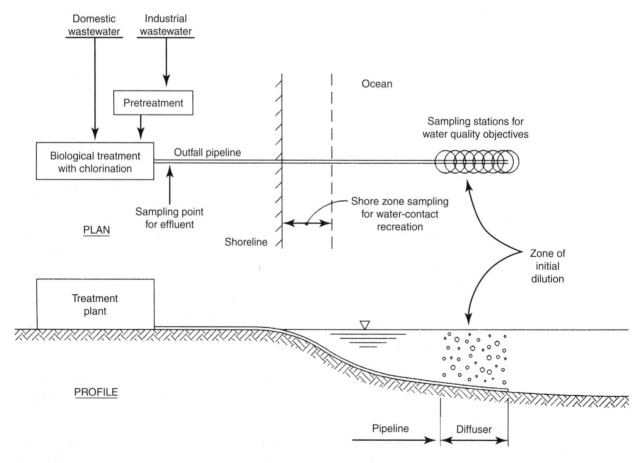

Figure 5–3
Schematic plan and profile diagrams of marine discharge.

Water Quality in the Zone of Initial Dilution

Compliance with bacterial, physical, and chemical water-quality standards is determined from samples collected at boat stations representative of the zone of initial dilution (Figure 5–3). For water-contact standards, coliform sampling is generally conducted from the shoreline into the ocean for a distance of 1000 ft or the 30-ft depth contour, whichever is further. A typical standard for coliforms is an average of less than 1000 total coliforms per 100 ml in any 30-day period and no single sample exceeding 10,000 per 100 ml; for fecal coliforms, the standard is not to exceed a geometric mean of 200 per 100 ml with no more than 10 percent exceeding 400 per 100 ml in any 60-day period (California standard). At all areas where shellfish are being harvested for human consumption, the median total coliform count is not to exceed 70 per 100 ml, with no more than 10 percent of the tests exceeding 230 per 100 ml.

Wastewater discharges must be essentially free of floatable materials, settled materials, toxic substances, turbidity, and color that can interfere with the indigenous marine life and a healthy and diverse marine community. The physical characteristics in the zone of initial dilution for compliance are no visible floating particulates, grease, or oil, and no aesthetically undesirable discoloration of the ocean surface. Natural light penetration shall not be significantly reduced outside the initial dilution zone, and deposition of solids shall not degrade biological benthic communities.

The chemical changes limit the decrease of dissolved oxygen to no more than 10 percent, lowering the pH more than 0.2 units, increasing the dissolved sulfide concentration, increasing the concentration of degrading substances in sediments, or increasing nutrients causing objectionable aquatic plant growths. Change in the biological characteristics of the water should not degrade vertebrate, invertebrate, and plant species; alter the natural taste, odor, and color of fish or shellfish for human consumption; or increase the bioaccumulation of toxic substances in fish or shellfish harmful to human health. Radioactivity cannot degrade marine life.

Effluent Quality Limits for Wastewater Discharge

Effluent quality requirements for wastewater discharges are specified as limitations on major wastewater constituents, on toxic substances for protection of marine aquatic life, and on noncarcinogens and carcinogens for protection of human health. These quality requirements apply to community and industrial wastewaters and other discharges, for example, cooling water from power plants. The limitations of major wastewater constituents specified in Table 5–4 by the state of California are grease and oil, suspended solids, settled solids, turbidity, pH, and acute toxicity. The limitations of specific toxic substances, mostly heavy metals and pesticides, are listed separately, and BOD is included for domestic wastewater. Fish bioassays for acute toxicity are conducted with the threespine stickleback, *Gasterosteus aculeatus*. The test species for chronic toxicity bioassays preferably include a fish (silversides), an invertebrate (shrimp, oyster), and an aquatic plant (red algae, giant kelp). Depending on the species, the duration of the toxicity tests are from 48 hr to 7 days.

The state of California lists effluent quality limitations for chemicals that are noncarcinogens and carcinogens for protection of human health. The allowable limits in the effluent from a particular plant can differ from the listed concentrations based on calculated initial dilution. The dilution factor is determined using an approved mathematical model. The characteristics of the outfall for inputs to the model include the length of the diffuser, number and spacing of ports, port diameter and angle from the horizontal, average depth of ports under mean sea level, and rate of wastewater discharge. If the dilution factor is sufficiently high, the limitations of selected chemicals in the wastewater discharge can be increased to utilize the greater dilution by initial mixing. The effluent limitations for several of the chemicals listed for protection of human health have been established at the minimum detection level for a specified testing technique. For this reason, the allowable limits of noncarcinogens may be increased to a practical quantification level equal to 10 times the detection level and carcinogens increased to 5 times the detection level.

TABLE 5–4

Effluent Quality Limits of Major Wastewater Constituents for Ocean Discharge to Protect Marine Aquatic Life

Parameter	Monthly (30-day average)	Weekly (7-day average)	Maximum (at any time)
Grease and oil, mg/l	25	40	75
Suspended solids, mg/l	60 with a minimum removal of 75%		
Settleable solids, ml/l	1.0	1.5	3.0
Turbidity, NTU	75	100	225
pH	within limits of 6.0 to 9.0 at all times		
Acute toxicity, TUa[a]	1.5	2.0	2.5

$$^a\text{TUa} = \frac{100}{96\text{-hour LC }50}$$

where TUa = toxicity units acute
LC = lethal concentration 50 percent

Source: Water Quality Control Plan, Ocean Waters of California, State Water Resources Control Board, Sacramento, California, 1990.

REFERENCES

1. *Water Quality & Treatment*, 5th Ed., 1999. American Water Works Association, Chapter: Drinking Water Quality Standards, Regulations, and Goals. Published by McGraw-Hill, Inc.
2. *The Clean Water Act*, 25th Anniversary Ed., 1997. Water Environment Federation, Alexandria, VA 22314–1994.
3. Environmental Protection Agency, National Pollutant Discharge Elimination System Permit Application Requirements for Publicly Owned Treatment Works and Other Treatment Works Treating Domestic Sewage; Proposed Rules, *Federal Register*, Vol. 60, No. 234 (Wednesday, December 6, 1995), pp. 62546–62659.

PROBLEMS

5–1 What is the definition of a community water system? What is the difference between a nontransient noncommunity water system and a transient noncommunity system? The regulation of specific MCLs depends on the kind of water system (Section 5–3). What kinds of contaminants are transient noncommunity water systems required to comply with?

5–2 What is the meaning of MCLG? Often a MCLG is established on toxicology data from animal exposure studies. What are the two major problems in assessing risk based on animal studies?

5–3 What is the meaning of MCL? How is the drinking water equivalent level (DWEL) for a noncarcinogenic contaminant determined?

5–4 Why is the control of *Giardia* and *Cryptosporidium* based on a specified treatment technique rather than a MCL?

5–5 What are the four barriers in a drinking water system to prevent pathogens and other contaminants from reaching water consumers?

5–6 Briefly describe the treatment technique for drinking water to eliminate pathogens from a surface-water source.

5–7 What does the Total Coliform Rule specify? If a routine coliform sample withdrawn from a water distribution system is positive, what is the required repeat sampling?

5–8 Why is the MCL for cadmium set so that less than 10 percent of the total intake is expected to be from water consumption?

5–9 How is monitoring conducted for lead and copper in drinking water? What are the action levels, and what are the treatment techniques if these action levels are exceeded?

5–10 What is the optimum concentration of fluoride in drinking water in a location where the average maximum daily air temperature is 60°F? What is the health benefit of drinking water containing the optimum concentration of fluoride?

5–11 What is the health risk of nitrate in drinking water?

5–12 What are the most frequently detected VOCs in contaminated groundwater? What pesticide SOCs have been detected in groundwater?

5–13 Write the chemical formulas for the five trihalomethane compounds that have been found in drinking water. What is the source of the THMs?

5–14 Why are iron and manganese included in secondary standards for aesthetics?

5–15 When the EPA was authorized to direct and define natural water pollution control programs, what were the objectives of the Clean Water Act?

5–16 What does the acronym NPDES refer to? Since inception, list several aspects of this permit program.

5–17 Viewing Figure 5–1, list the regulatory standards to control water quality from an industrial wastewater discharged to a municipal sewer to drinking water in the distribution system of a community downstream.

5–18 How is the NPDES program monitored?

5–19 What are the technology-based standards for secondary (biological) treatment for all municipal wastewater treatment plants?

5–20 When are water-quality-based standards necessary for a wastewater discharge?

5–21 Define the term *eutrophication*. What pollutant is commonly limited in wastewater effluent to control the rate of eutrophication?

5–22 How is a TMDL program expected to improve water quality in an impaired surface water? What are the major causes of surface-water impairment? What appears to be the major challenge to establishing a TMDL program for management of a watershed?

5–23 What are the primary objectives of the industrial pretreatment program?

5–24 How is the Whole Effluent Toxicity (WET) test performed?

5–25 What is the usual procedure following a negative effluent biological toxicity test that indicates probable impairment of the receiving water.

5–26 Why is prevention considered the only feasible approach to groundwater-quality management?

5–27 Based on the treatment limits given for ocean discharge in Table 5–4, recommend the minimum treatment processes to obtain the required removal of suspended solids. What steps should be taken if toxicity limits are exceeded?

CHAPTER **6**

Water Distribution Systems

The objectives of a municipal water system are to provide safe, potable water for domestic use; an adequate quantity of water at sufficient pressure for fire protection; and industrial water for manufacturing. A typical waterworks consists of a source, treatment, pumping, and distribution system. Sources for municipal supplies are deep wells, shallow wells, rivers, lakes, and reservoirs. About two-thirds of the water for public supplies comes from surface-water sources. Large cities generally use major rivers or lakes to meet their high demand, whereas the majority of towns use well water if available. Often groundwater is of adequate quality to preclude treatment other than chlorination and fluoridation. Wells can then be located at several points within the municipality, and water can be pumped directly into the distribution system. However, where processing is needed, the well pumps, or low-lift pumps from the surface-water intake, convey the raw water to the treatment plant site. A large reservoir of treated water (clearwell storage) provides reserves for the high demand

periods and the equalizing of pumping rates. The high-lift pumps deliver treated water under high pressure through transmission mains to distribution piping and storage.

The distribution system consists of a gridiron pattern of water mains to deliver water for domestic, commercial, industrial, and fire fighting purposes. Elevated storage tanks, or ground-level reservoirs with booster pumps, reserve water for peak periods of consumption and fire demand. A short lateral line connects each fire hydrant to a distribution main. Shutoff valves are located at strategic points throughout the piping system to provide control of any section or service outlet, including hydrants. These valves are used to isolate for required maintenance and to ensure that main breaks affect only a small section. A service connection to a residence includes a corporation stop tapped into the water main, a service line to a shutoff valve at the curb, and the owner's line into the dwelling, which incorporates a water meter and a pressure regulator or relief valve if necessary.

6–1 WATER QUANTITY AND PRESSURE REQUIREMENTS

The amount of water required by a municipality depends on population; climate (for example, lawn sprinkling and public landscape irrigation); commercial and industrial water users; conservation (for example, high-efficiency plumbing fixtures); water reuse for landscape irrigation; and economic conditions. In a typical water system, metered water sales are estimated to be about 70 percent of the total water supplied. The remainder is accounted for by inaccurate meters, at 10 percent, underground leakage from pipes, at 15 percent, and the balance from miscellaneous losses. State standards for unaccounted water range from 10 percent to 20 percent.

Municipal water use in the United States averages 600 gpd (2270 l/d) per metered service including residential, commercial, and industrial customers. For residential customers, water consumption in eastern and southern areas is 210 gpd (790 l/d) and in central states 280 gpd (1060 l/d); western regions use 460 gpd (1740 l/d) per household service. Only a small amount of water is sprinkled on lawns where the rainfall exceeds 40 in./yr (1000 mm/y); in semiarid climates lawns and gardens are maintained by irrigation.

Typical average-day municipal per capita demands, based on location, climate, and mix of residential versus commercial and industrial connections, vary between 100 and 200 gpcd (380 and 760 l/person · d) for all demands included. Residential demands alone are 80 and 150 gpcd (300 and 570 l/person · d). In some municipalities, unusual mixes of connections and large industrial demands can cause significant variations. As discussed in Section 8–5, high-efficiency plumbing fixtures can reduce indoor water consumption by 40 to 50 percent, to a demand of about 40 gpcd (150 l/person · d).

Residential water use varies seasonally, daily, and hourly. Typical daily winter consumption is about 80 percent of the annual daily average; summer use is 30 percent greater. Variations from these commonly quoted values for a particular community may be significantly greater depending on seasonal weather changes. Maximum daily demand can be considered to be 180 percent of the average daily, with values ranging from about 150 to more than 300 percent. Maximum hourly figures have been observed to range from about 2.5 to more than 5 times the average flow in extreme cases; a mean for the maximum hourly rate is 300 percent.

Public water demand is unique for every municipality. Projecting future demand, or establishing a conservation program, must be based on accumulated local records, including average-day usage, seasonal highs and lows, maximum hourly withdrawals, seasonal highs and lows for indoor and outdoor demands, and separate demands for residential, commercial, and industrial users.

Water flows used in waterworks design depend on the magnitude and variations in municipal water consumption and the reserve needed for fire fighting. The quantities of water required for fire demand, as detailed in Section 6–2, are of significant magnitude and frequently govern the design of distribution piping, pumping, and storage facilities. Water intakes, wells, treatment plants, pumping, and transmission lines are sized for peak demand, normally maximum daily use where hourly variations are handled by storage. Standby units in the source-treatment-pumping system may be installed for emergency use, for convenience of maintenance, or to serve as capacity for future expansion. The required design flow of maximum daily consumption plus fire flow frequently determines the size of distribution mains and results in additional pumping capacity and a need for storage reserves in addition to that required to equalize pumping rates. If the maximum hourly consumption exceeds the maximum daily plus fire fighting demand, it may be the controlling criterion in sizing some units.

The recommended water pressure in a distribution system is 65 to 75 psi (450 to 520 kPa), which is considered adequate to compensate for local fluctuations in consumption. This level of pressure can provide for ordinary consumption in buildings up to ten stories in height, as well as sufficient supply for automatic sprinkler systems for fire protection in buildings of four or five stories. For a residential service connection, the minimum pressure in the water distribution main should be 40 psi (280 kPa). Pressures in excess of 100 psi (690 kPa) are undesirable, and the maximum allowable pressure is 150 psi (1030 kPa). At excessive levels, leaks occur in domestic plumbing, requiring pressure reducers in service connections, and undue stress is placed on mains in the ground. Pipe and fittings used in ordinary water distribution systems are designed for a maximum working pressure of 150 psi. Nevertheless, high pressure in water mains results in more frequent pipeline breaks and greater leakage losses.

6–2 MUNICIPAL FIRE PROTECTION REQUIREMENTS*

The Insurance Services Office (ISO), a not-for-profit association of insurance companies, collects survey data used to establish rates for fire protection policies for commercial and residential properties. Grading of fire defenses includes evaluation of the water supply, fire department, and receiving and handling fire alarms. Reliability and adequacy of the following major water-supply components are considered in a municipal survey: water-supply source, treatment plant capacity, pumping capacity, power supply, water-supply mains, distribution mains, spacing of valves, and location of fire hydrants. All of these are essential facilities for fighting fires in a municipality. In addition, ISO recommends providing internal fire control in appropriate buildings, for example, by installation of automatic fire sprinkler systems.

The *Guide for Determination of Needed Fire Flow*[1] of the ISO defines procedures to determine the needed fire flows from the public water-supply system when all suppression systems inside the buildings fail or do not exist and fire occurs. The objective of the evaluations is to provide sufficient manual fire flows to minimize loss and avoid a total loss.

Needed Fire Flow

Needed fire flow (NFF) is the rate of water flow required for fire fighting to confine a major fire to the buildings within a block or other group complex with minimal loss. The ISO NFF is based on buildings without sprinklers or on failure of the

---

automatic sprinklers. Determination of this flow considers construction, occupancy, exposure, and communication of each building in a building complex. The construction factor C_i, based on the kind of construction and square footage of the building, is calculated by Eq. 6–1.

$$C_i = 18F\,(A_i)^{0.5} \qquad (6\text{–}1)$$

where C_i = construction factor, gallons per minute
 F = coefficient related to the class of construction
 F = 1.5 for class 1, wood-frame
 F = 1.0 for class 2, ordinary
 F = 0.8 for class 3, noncombustible
 F = 0.8 for class 4, masonry, noncombustible
 F = 0.6 for class 5, modified fire resistive
 F = 0.6 for class 6, fire resistive
 A_i = effective area, square feet. Effective area is the area of the largest floor in the building plus the following percentage of the other floors: (1) for buildings of construction class 1 through 4, 50 percent of all other floors and (2) for buildings of construction classes 5 or 6, if all vertical openings in the building have 1.5 hr or more protection, 25 percent of the area not exceeding the two largest floors. The doors shall be automatic or self-closing and labeled as class B fire doors (1.0 hr or more protection). In other buildings, 50 percent of the area not exceeding other floors.

$$C_i = 3.7F(A_i)^{0.5} \qquad (\text{SI units}) \qquad (6\text{–}2)$$

where C_i = construction factor, liters per second
 A_i = effective area, square meters

Regardless of the calculated value, the C_i shall not exceed the following: 8000 gpm (500 l/s) for construction classes 1 and 2; 6000 gpm (380 l/s) for construction classes 3, 4, 5, and 6; and 6000 gpm (380 l/s) for a one-story building of any class of construction. The minimum value of C_i is 500 gpm (32 l/s). ISO rounds the calculation value of C_i to the nearest 250 gpm (16 l/s).

The occupancy factors O_i for combustibility classes are listed in Table 6–1. They represent the influence of the occupancy on NFF. The following

TABLE 6–1

Occupancy Factors O_i for Combustibility Classes

COMBUSTIBILITY CLASS		O_i
C–1	Noncombustible	0.75
C–2	Limited combustible	0.85
C–3	Combustible	1.00
C–4	Free burning	1.15
C–5	Rapid burning	1.25

Source: Guide for Determination of Needed Fire Flow.[1] Copyright, Insurance Services Office, Inc., 2004.

are examples in each classification of 1 through 5 for nonmanufacturing:

Classification 1: Merchandise or materials that do not constitute an active fuel, for example, stored clay, glass, marble, stone, or metal products.

Classification 2: Merchandise or materials with limited combustibility, for example, banks, churches, shops, hospitals, motels, and offices. For manufacturing occupancies with over 20 percent of total floor area used for storage of combustibility class, occupancy shall not be less than C–3.

Classification 3: Merchandise or materials of moderate combustibility, for example, food markets, most wholesale and retail sales and service, and repair or service shops.

Classification 4: Merchandise or materials that burn freely, for example, furniture stock and wood products, freight depots and terminals, and theaters.

Classification 5: Merchandise or materials that burn with a great intensity, spontaneously ignite, and give off flammable or explosive vapors, for example, explosives, chemical or paint sales and storage, plastic products, and upholstering shops.

In multiple-occupancy buildings with different classifications, the occupancy classification applicable to the building shall be determined according to the total floor area occupied by each as follows:

C–1 Noncombustible class shall apply only where 95 percent or more of the total floor area is occupied by C–1 occupants with no C–5 occupancies.

C–2 Limited combustibility class shall apply to buildings that are classified as construction class 5 or 6 (F coefficients listed for Eq. 6–1) and where 80 percent or more of the floor area is occupied by C–1 and C–2.

C–4 Free-burning class shall apply to any building containing C–4 occupants, where the combined area occupied by C–4 and C–5 occupants is limited to 25 percent of the building area, provided the C–5 occupancies, if any, are less than 15 percent.

C–5 Rapid-burning class shall apply to any building where 15 percent or more is occupied by C–5 occupancies.

C–3 Combustible class shall apply to any building not provided for above.

Exposure X_i and communication P_i factors represent the influence of exposed and communicating buildings on the NFF. A value for $(X + P)_i$ is developed for each side of the building as follows:

$$(X + P)_i = 1.0 + (X_i + P_i) \qquad (6\text{–}3)$$

with a maximum value of 1.60

where X_i = exposure factor, Table 6–2
P_i = communication factor, Table 6–3
n = number of sides of building

The factor for X_i depends on the class of construction, length-height value (length of wall in feet times height in stories) of the exposed building, and distance between facing walls of the building being evaluated and the exposed building. Factors for exposure are given in Table 6–2.

The value for the exposure factor (X_i) to an adjacent building considers only the side of the building with the highest numerical factor. The following conditions rule out an exposure factor from an adjacent building if the adjacent building is rated as sprinklered, occupied, construction class 5 or 6, or construction class 3 or 4 with C–1 or C–2 combustibility class contents.

The value for the communication factor (P_i) to an adjacent building considers only the one communication with the highest factor. The following conditions rule out any communication factor if the adjacent building is rated as sprinklered, occupied, construction class 5 or 6, or construction

---

TABLE 6–2

Factors for Exposure X_i

			EXPOSURE FACTOR X_i			
			CONSTRUCTION OF FACING WALL OF EXPOSED BUILDING			
			CONSTRUCTION CLASSES			
			1, 3	2, 4, 5, 6		
CONSTRUCTION OF FACING WALL OF BUILDING BEING EVALUATED	DISTANCE TO THE EXPOSED BUILDING (FT)	LENGTH-HEIGHT[a] OF FACING WALL OF EXPOSED BUILDING (FT · STORIES)		UNPROTECTED OPENINGS	SEMIPROTECTED OPENINGS (WIRE GLASS OR OUTSIDE OPEN SPRINKLERS)	BLANK WALL
Frame, metal, or masonry with openings	0–10	1–100	0.22	0.21	0.16	0
		101–200	0.23	0.22	0.17	0
		201–300	0.24	0.23	0.18	0
		301–400	0.25	0.24	0.19	0
		Over 400	0.25	0.25	0.20	0
	11–30	1–100	0.17	0.15	0.11	0
		101–200	0.18	0.16	0.12	0
		201–300	0.19	0.18	0.14	0
		301–400	0.20	0.19	0.15	0
		Over 400	0.20	0.19	0.15	0
	31–60	1–100	0.12	0.10	0.07	0
		101–200	0.13	0.11	0.08	0
		201–300	0.14	0.13	0.10	0
		301–400	0.15	0.14	0.11	0
		Over 400	0.15	0.15	0.12	0
	61–100	1–100	0.08	0.06	0.04	0
		101–200	0.08	0.07	0.05	0
		201–300	0.09	0.08	0.06	0
		301–400	0.10	0.09	0.07	0
		Over 400	0.10	0.10	0.08	0

Blank masonry wall	When facing wall of the exposed building is higher than building being evaluated, use the above information, except use only the length-height of facing wall of the exposed building above the height of the facing wall of the building being evaluated. Buildings five stories or over in height, consider as five stories. When the height of the facing wall of the exposed building is the same or lower than the height of the facing wall of the subject building, $X_i = 0$.

Source: *Guide for Determination of Needed Fire Flow.*[1] Copyright, Insurance Services Office, Inc., 2004.
[a]The length-height factor is the length of the wall of the exposed building in feet times its height in stories.

TABLE 6–3

Factors for Communication P_i

Protection of Passageway Openings	Fire-Resistive, Noncombustible, or Slow-Burning Communications				Communication With Combustible Construction					
	Open	Enclosed			Open			Enclosed		
	Any Length	10 ft or less	11 to 20 ft	21 to 50 ft[a]	10 ft or less	11 to 20 ft	21 to 50 ft[a]	10 ft or less	11 to 20 ft	21 to 50 ft[a]
Unprotected	0	0.30	0.30	0.20	0.30	0.20	0.10	0.30	0.20	0.30
Single class A fire door at one end of passageway	0	0.20	0.10	0	0.20	0.15	0	0.30	0.20	0.10
Single class B fire door at one end of passageway	0	0.30	0.20	0.10	0.25	0.20	0.10	0.35	0.25	0.15
Single class A fire door at each end or double class A fire doors at one end of passageway	0	0	0	0	0	0	0	0	0	0
Single class B fire door at each end or double class B fire doors at one end of passageway	0	0.10	0.05	0	0	0	0	0.15	0.10	0

Source: Guide for Determination of Needed Fire Flow.[1] Copyright, Insurance Services Office, Inc., 2004.
[a]For communication length over 50 ft, $P_i = 0$.

class 3 or 4 with C–1 or C–2 combustibility class contents. The factor for P_i depends on the protection of passageway openings, length of the communication, and combustibility of the construction. Factors for communication are given in Table 6–3. When more than one kind of communication exists in any one side wall, only the largest value for P_i is applied for that side. When no communication exists on a side, P_i is zero.

NFF is calculated as follows:

$$\text{NFF} = C_i \times O_i \times (X + P)_i \qquad \textbf{(6–4)}$$

where NFF = needed fire flow, gallons per minute (liters per second)

C_i = construction factor, gallons per minute (liters per second)
O_i = occupancy factor, Table 6–1
$(X + P)_i$ = exposure and communication factors, Eq. 6–3

An additional 500 gpm (32 l/s) of flow is added to the NFF when a wood-shingle roof on a building, or on exposed buildings, can contribute to spreading of a fire. The NFF shall neither exceed 12,000 gpm (760 l/s) nor be less than 500 gpm (32 l/s). The calculated NFF is rounded to the nearest 250 gpm if less than 2500 gpm and to the nearest 500 gpm if greater than 2500 gpm.

TABLE 6–4

Needed Fire Flows for Single-Family and Two-Family Residential Areas Not Exceeding Two Stories in Height

Distance Between Dwelling Units (ft)[a]	Needed Fire Flow (gpm)[b]
Over 100	500
31 to 100	750
11 to 30	1000
10 or less	1500

Source: Guide for Determination of Needed Fire Flow.[1] Copyright, Insurance Services Office, Inc., 2004.
[a] 1.0 ft = 0.305 m
[b] 1.0 gpm = 0.0631 l/s

NFFs for single- and two-family dwellings not exceeding two stories in height are listed in Table 6–4. The NFF can be up to a maximum of 3500 gpm (220 l/s) for other habitable buildings not listed in Table 6–4. For a building with an automatic sprinkler system, the NFF is the flow needed for the sprinkler system converted to 20 psi residual pressure with a minimum of 500 gpm (32 l/s).

Evaluation of Automatic Sprinkler Systems

The following are the minimum conditions to classify a building as being sprinklered. The building must have assured maintenance with a responsible caretaker who visits the premises not less than weekly. The usable unsprinklered area cannot exceed: (a) 25 percent of the total area in buildings with occupancy combustibility class C–1, (b) 20 percent with occupancy class C–2 or C–3, (c) 10,000 sq ft or 15 percent of the total area with occupancy class C–4, and (d) 5000 sq ft or 10 percent with occupancy C–5. (Occupancy factors for combustibility are listed in Table 6–1.)

The sprinkler system installation has to have evidence of flushing and hydrostatic tests and a full-flow main drain test within the last 48 months. The fire-pump installation requires evidence and results of a fire-pump test conducted within the last 48 months.

Practical Limits of Fire Flow

Withdrawal of a large quantity of water from a water system is not the preferred method of fire suppression. For many buildings, automatic sprinkler systems are more effective in protection of life and property than relying solely on the distribution system to provide fire protection. The minimum fire flow for buildings without sprinklers is 500 gpm (32 l/s) at a residual pressure of 20 psi (140 kPa). This represents the flow provided by two standard hose streams and the minimum amount to suppress a fire safely and effectively. Above this minimum, the calculated NFF based on ISO equations is a realistic assessment of the required flow for fire suppression.

The maximum fire flow that most municipalities are likely to be able to reliably provide for fire fighting is 3500 gpm (220 l/s). NFFs under the Guide for Determination of Needed Fire Flow[1] are for areas with buildings without automatic sprinklers. Since the value and effectiveness of sprinklers are proven, a municipal government may extend up to 100 percent credit for an area in which all of the buildings are equipped with sprinkler systems. If a 100 percent credit exists, no fire flow is required except for the water flow required for the sprinklers plus a hose stream allowance of 500 gpm (32 l/s). The flow requirement for sprinkler systems is generally in the range of 150 to 1600 gpm (10 to 100 l/s) depending on the combustibility class, the kind of sprinkler system, and other considerations. In determining public protection classification for the purpose of setting fire insurance rates for a municipality, the ISO procedure does not consider any major structure having a required fire flow greater than 3500 gpm (220 l/s). The municipal government may have an ordinance requiring isolated properties, which have NFFs greater than other properties in the area, to provide private protection. This may be accomplished by reducing the required fire flow by installing automatic sprinklers or on-site water storage and pumps.

Duration

The required duration for fire flow is 2 hr for up to 2500 gpm and 3 hr for fire flows of 3000 gpm and 3500 gpm. The major components of a water system on which the reliability of fire flow depends, including pumping capacity, supply mains, treatment plant, and power source, must have the ability to deliver the maximum daily water use rate for several days plus fire flow for the specified duration at any time during this interval. The period may be 5, 3, or 2 days depending on the system component under consideration and the anticipated out-of-service time required for maintenance and repair work.

Pressure

The pressure in a distribution system must be high enough to permit pumpers of the fire department to obtain adequate flows from hydrants. A positive water pressure is needed during withdrawal of fire flow from hydrants to overcome friction losses in the hydrant and section hose. The ISO specifies a minimum residual pressure of 20 psi (140 kPa) during fire flow in analyzing the adequacy of a water system.

Water-Supply Capacity

In evaluating a system, pumps should be credited at their effective capacities when discharging at normal operating pressures. The pumping capacity, in conjunction with storage, should be sufficient to maintain the maximum daily use rate plus maximum required fire flow with the single most important pump out of service. The fire flow must be sustainable at the required pressure of 20 psi (140 kPa) for the required duration.

Storage is frequently used to equalize pumping rates into the distribution system as well as provide water for fire fighting. Since the volume of stored water fluctuates, only the normal minimum daily amount maintained is considered available for fire fighting. In determining the fire flow from storage, it is necessary to calculate the rate of delivery during a specified period. Even though the amount available in storage may be great, the flow to a hydrant cannot exceed the carrying capacity of the mains, and the residual pressure at the point of use cannot be less than 20 psi (140 kPa).

Although a gravity system, that is, delivering water without the use of pumps, is desirable from a fire protection standpoint because of reliability, well-designed and properly safeguarded pumping systems can be developed to such a high degree that no distinction is made between the reliability of gravity-fed and pump-fed systems. Where electrical power is used, the supply should be so arranged that a failure in any power line or repair of a transformer, or other power device, does not prevent delivery of required fire flow. Electric power should be provided to all pumping stations and treatment facilities by two separate lines from different directions.

Reliability in water-system operations is essential to meet fire demand. Pumping stations, treatment plants, and operations control centers are constructed for protection from fire, flooding, and other disasters or accidents. Audible alarms and other notification systems by radio or automatic dialing telephones are installed to warn water personnel of problems. To take remedial actions, suitable equipment and transportation for emergency crews should be on call to assist the fire department to ensure water service in case of a system malfunction or failure, such as a broken main.

Distribution System

Proper layout of supply mains, arteries, and secondary distribution feeders is essential for delivering required fire flows in all built-up parts of the municipality with usage at the maximum daily rate.[2] Consideration must be given to the greatest effect that a break, joint separation, or other main failure could have on the supply of water to a system. With the most serious failure, no deficiency is considered if the remaining mains from the source of supply and storage can provide the fire flow for the specified duration during a period of 3 days with usage at the maximum daily rate.

Supply mains, arteries, and secondary feeders should extend throughout the system, properly spaced—about every 3000 ft (910 m)—and looped for mutual support and reliability of service. The gridiron pattern of small distribution mains supplying residential districts should consist of mains at least 6 in. (150 mm) in diameter. Where long lengths are necessary, exceeding about 600 ft (180 m), 8-in. (200-mm) or larger intersecting mains should be used. In new construction, 8-in. or larger pipes are used where dead ends and poor gridiron are likely to exist for a considerable time during development or because of the layout of streets and topography. Hydrants for fire protection should never be located on the dead end of 6-in. or smaller mains. In commercial districts, the minimum-size main should be 8 in. with intersecting lines in each street, with 12-in. (300-mm) or larger mains used on principal streets and for all long lines that are not connected to other mains at intervals close enough for mutual support.

A distribution system is equipped with a sufficient number of valves located so that a pipeline break does not affect more than $\frac{1}{4}$ mile of arterial mains, 500 ft (150 m) of mains in high-value districts, or 800 ft (240 m) of mains in other districts.

Fire hydrants are installed at spacings and in locations convenient for the fire department. The lineal distance between hydrants along streets in residential districts is normally 600 ft (180 m), with a maximum of 800 ft (240 m), and in high-value districts it is normally 300 ft (91 m), with a maximum of 500 ft (150 m). Common locations are at intersections, at the middle of long blocks, and near the end of dead-end streets. For evaluation of NFF at a particular location, ISO considers each hydrant within 1000 ft (300 m) of the location measured in a way the hose can be laid by fire apparatus. Credit up to 1000 gpm (63 l/s) is allowed for each hydrant within 300 ft, 670 gpm (42 l/s) for hydrants within 301 to 600 ft, and 250 gpm (16 l/s) for hydrants 601 to 1000 ft from the location.

Hydrants should have at least two outlets; one must be a pumper outlet and the other at least a $2\frac{1}{2}$-in. nominal-size hose outlet. ISO allows the maximum credit of 1000 gpm from a hydrant with at least one pumper outlet, 750 gpm from two or more hose outlets and no pumper outlet, and 500 gpm from a hydrant with only one hose outlet.

■ EXAMPLE 6–1

A dairy processing building of ordinary construction is 100 ft by 150 ft and one 12-ft story high. An adjacent building for food and beverage sales is masonry construction, 100 ft by 100 ft, and three stories high. The 100-ft walls of the two buildings face each other 35 ft apart, and both walls have doorways of combustible construction (open communication). Calculate the NFF for each building, including the exposure of the other building.

Solution

The dairy processing building has a coefficient F of 1.0 for ordinary construction. Using Eq. 6–1 for a 100-ft by 150-ft, single-story building, the coefficient factor is

$$C_i = 18 \times 1.0(100 \text{ ft} \times 150 \text{ ft})^{0.5} = 2200 \text{ gpm}$$

Dairy processing is a manufacturing classification 3 for an occupancy factor O_i of 1.00.

The facing wall of the exposed masonry building is class 4 construction with unprotected openings, and the distance to this exposed wall is 35 ft. The length-height of the facing wall of the exposed building is 100 ft by 3 stories for 300 ft · stories. From Table 6–2, the factor for exposure X_i is 0.13. From Table 6–3, the communication factor P_i for open combustible construction at a distance of 35 ft is 0.10. Using Eq. 6–3,

$$(X + P)_i = 1.0 + (0.13 + 0.10) = 1.23$$

Using Eq. 6–4, the needed fire flow is

$$\text{NFF} = 2200 \text{ gpm} \times 1.00 \times 1.23 = 2700 \text{ gpm}$$
$$= 2500 \text{ gpm (rounded to nearest 500 gpm)}$$

The required duration for fire flow is 2 hr.

The food and beverage sales building has a coefficient F of 0.8 for class 4 masonry, noncombustible construction. Using Eq. 6–1 for a 100-ft by 100-ft, three-story building of class 4 construction, the coefficient factor is

$$C_i = 18 \times 0.8(10,000 \text{ ft}^2 + 2 \times 5000 \text{ ft}^2)^{0.5}$$
$$= 2040 \text{ gpm}$$

The food and beverage sales are a nonmanufacturing classification 3 for an occupancy factor O_i of 1.00.

The facing wall of the exposed dairy building of ordinary construction is class 2 with unprotected openings, and the distance to this wall is 35 ft. The length-height of the facing wall of the exposed building is 100 ft by 1 story for 100 ft · stories. From Table 6–2, the factor for exposure X_i is 0.10. From Table 6–3, the communication factor P_i for open combustible construction at a distance of 35 ft is 0.10. Using Eq. 6–3,

$$(X + P)_i = 1.0 + (0.10 + 0.10) = 1.20$$

Using Eq. 6–4, the needed fire flow is

$$NFF = 2040 \times 1.00 \times 1.20 = 2450 \text{ gpm}$$
$$= 2500 \text{ gpm (rounded to the nearest}$$
$$250 \text{ gpm)}$$

The required duration for fire flow is 2 hr.

■■■

■ EXAMPLE 6–2

A wood-frame building on the first floor is a restaurant with floor dimensions of 30 ft by 75 ft (2250 sq ft). On one end of the second floor is a cabinet-making shop 30 ft by 30 ft (900 sq ft). The building has two sides with exposures to adjacent buildings. Opposite the long side of the restaurant at a distance of 11 ft is building A with two-story masonry walls and semiprotected openings. The length-height measurement is $120 \times 2 = 240$ ft · stories. The second exposure is building B with a width of 28 ft at a distance of 12 ft from the end of the restaurant. The building is constructed of frame walls with two stories for a length-height measurement of $28 \times 2 = 56$ ft · stories. Calculate the NFF for the restaurant-cabinet shop building.

Solution

Wood-frame construction has F equal to 1.5. The effective area A_i is the ground floor plus 50 percent of the second floor, equal to $2250 + 450 = 2700$ sq ft.

$$C_i = 18 \times 1.5(2700)^{0.5} = 1400 \text{ gpm}$$

The cabinet-making shop occupies over 25 percent of the total floor area of the building.

$$\text{Percentage} = [900/(2250 + 900)]/100 = 28.6$$

From Table 6–1, the $O_i = 1.15$.

The exposure from Table 6–2 for building A at a distance of 11 ft and length-height of 240 ft · stories is 0.14. The exposure from Table 6–2 for building B at a distance of 12 ft and length-height of 240 ft · stories is 0.17. The building with the largest factor is B.

$$X_i = 0.17$$

No communication exists so $P_i = 0$.
Using Eq. 6–3, $(X + P)_i = 1.0 + (0.17 + 0) = 1.17$.

Using Eq. 6–4, the needed fire flow is

$$NFF = 1400 \times 1.15 \times 1.17 = 1889 \text{ gpm}$$
$$= 2000 \text{ gpm (rounded to nearest}$$
$$500 \text{ gpm)}$$

■■■

6–3   WELL CONSTRUCTION

Water in the voids of underground sand or gravel beds can be tapped for municipal supply by using a drilled well, as shown in Figure 6–1. The main components of a well are a solid casing, sealed in the upper soil profile; a screen set into the aquifer, allowing water to enter the casing while precluding

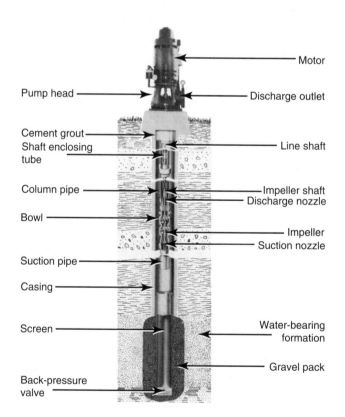

Figure 6–1

Gravel-packed water well in a sand aquifer equipped with a two-stage vertical turbine pump.

(Courtesy of Layne, Inc, formerly known as Layne-Western Company, Inc.)

sand and gravel; and a turbine pump suspended in the casing on a column pipe. Groundwater flows toward the well, enters the casing through screen openings, and is lifted for discharge by the pump.

Well drilling is commonly performed by the hydraulic rotary method. In rotary drilling, the borehole is advanced using a rotating bit attached to the lower end of a string of drill pipe held by a rotary table of the drilling machine under a mast used to lift the lengths of pipe. The cuttings are removed from the borehole by continuous circulation of a drilling fluid, which is a viscous, dense mud formed by mixing bentonite clay in water. The mud flows through the drill pipe and nozzles in the bit and in the opposite direction in the borehole outside the pipe. After returning to the ground surface, it flows through a pit to settle out cuttings prior to recycling. The drilling fluid also holds the side of the hole open by hydraulic pressure and caking of mud on the wall. After installation of the well screen, this clogging of the water-bearing formation can be corrected by hydraulic cleaning and flushing to remove the residue of mud.

The casing is constructed of iron pipe, corrosion-resistant tubular steel, or polyvinyl chloride. It may be installed as a single wall using pipe lengths of the same diameter or as a double wall formed by telescoping cylindrical sections halfway through one another. After drilling is completed, the casing is set in position and sealed in place by forcing cement grout between the casing and the drilled hole. Temporary casing, penetrating near surface strata, may be withdrawn as the grout is placed around the inserted well casing.

The screen length is commonly 70 to 80 percent of the aquifer thickness. Two common screens for municipal wells are spiral-wound and shutter; both can be sized for optimum hydraulic efficiency for groundwater to enter the well while precluding the passage of sand. The screen openings and grain size of the aquifer media dictate the grading of the gravel-pack material. The gravel must be carefully placed, by mechanical or hydraulic means. Merely dumping the material around the casing and screen results in separation of the grain sizes and in bridging, creating poor hydraulic characteristics and possible pumping of sand.

Well development is performed by vigorous hydraulic action through pumping, surging, or jetting. Mechanical surging is done by operating a plunger up and down in the casing like a piston in a cylinder. A heavy baler can be used to produce the surging action, but it is less effective. Air surging is accomplished by lowering an air pipe into the screen section and purging with compressed air. In high-velocity jetting, a tool is slowly rotated inside the screen, shooting water out through the openings. Development is intended to correct any damage or clogging of the water-bearing formation that may have occurred as a side effect of drilling, to increase the porosity and permeability of the natural formations surrounding the screen, and to stabilize the aquifer so that the water is free of sand. The net effect is reduction of water drawdown in the well and higher-quality water.

The common well pump is a multistage vertical-turbine pump, illustrated in Figure 6–1. The four major components are a motor drive that is mounted at ground level or is submerged below the pump, a discharge pipe column, pump bowls with enclosed impellers, and suction pipe. The pumping units are bottom-suction centrifugal impellers, mounted on a vertical shaft, that carry one or more stages; submerged below the water level is the well casing.

6–4 SURFACE-WATER INTAKES

An intake structure is required to withdraw water from a river, lake, or reservoir. Typical intakes are towers and shoreline structures. Their primary functions are to supply the highest-quality water from the source and to protect piping and pumps from damage or clogging as a result of wave action, ice formation, flooding, or floating and submerged debris.

Fish protection is important in the design and construction of intake structures. If endangered fish species are identified as being present, the National Oceanic & Atmospheric Administration fisheries regulations should be reviewed. The most widely accepted and successful method to repel fish from an intake is to use a physical barrier with an effective screen area for a low entrance velocity. Inclined or vertical flat-plate intake screens are mounted on a framework submerged in flowing waters. Plate screens can be cleaned of collected debris by a high-pressure

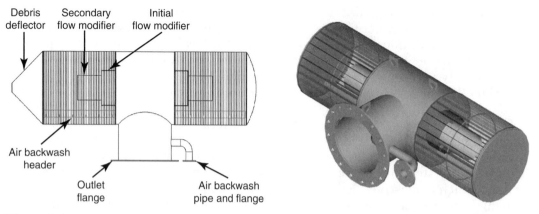

Debris deflector • Secondary flow modifier • Initial flow modifier • Air backwash header • Outlet flange • Air backwash pipe and flange

Figure 6–2

This cylindrical intake has a Vee-Wire® screen designed with a high open area for inflow of water while repelling against entrance of fish. The duel-pipe flow modifier inside the screen cage produces an even flow across the entire screen surface. When debris accumulates on the exterior of the cylinder, air bursts scour the screen from inside to outside for effective cleaning.

(Courtesy of Johnson Screens, Inc., A Weatherford Company.)

air-burst system or by a mechanical system. Figure 6–2 illustrates a cylindrical vee-screen that is installed with its long axis parallel to the water flow to reduce collection of debris. The screen is held in place by a supporting pipe connected to the water-receiving structure. Because the circular screens are cleaned with an air-burst system, any accumulated debris removed from the screen remains in the source water.

Towers are for lakes and reservoirs with fluctuating water levels or variations of water quality with depth. Ports at several depths permit selection of the most desirable water quality any season of the year. For example, in a eutrophic lake during the summer the surface layer is warm and supports abundant growths of algae, while bottom waters can be devoid of dissolved oxygen, causing foul tastes and odors. On the other hand, during the winter, water immediately under an ice cover may be of the highest quality. Submerged ports also have the advantage of being free from ice and floating debris. Underflows of sediment-laden water have been observed during spring runoff in reservoirs, making bottom waters undesirable; hence, a tower with upper entry ports is advantageous. The location, height, and selection of port levels must be related to characteristics of the water body. Therefore, a limnological survey should precede design to define the degree and depth of stratification, water currents, sediment deposits, undesirable eutrophic conditions, and other factors.

Shore intakes located adjacent to a river must be sited with consideration for water currents that might threaten safety of the structure, location of navigation channels, ice floes, formation of sandbars, and potential flooding. In addition, water-quality considerations and distances from the pumping station and treatment plant are important.

Figure 6–3 illustrates the type of screening equipment generally employed in shoreline intakes. A coarse screen of vertical steel bars, having openings of 1 to 3 in. placed in a near vertical position, is used to exclude large objects. It is equipped with a trash rack rake to remove accumulated debris. Leaves, twigs, and other material passing through the bar rack are removed by a finer screen with openings sized according to site conditions. A traveling screen consists of wire mesh trays that retain solids as the water passes through. A drive chain and sprockets raise the trays into a head enclosure, where the debris is removed by means of water sprays. Heat may be required in winter to prevent ice formation. Travel is intermittent and is controlled by the amount of accumulated material.

Low-lift pumping stations transporting raw water from the source to treatment are located as close as is practical to the intake. In the case of a tower or submerged intake, the station is situated

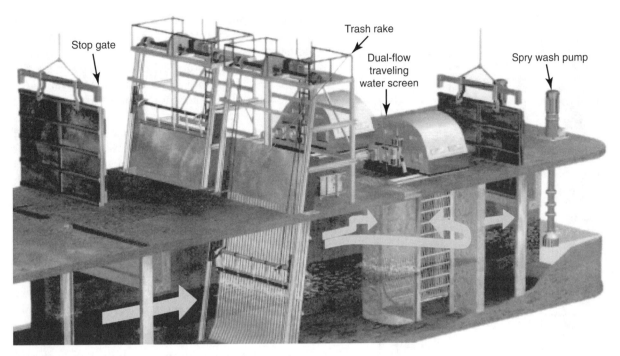

Figure 6–3
Shore water intake screens. (a) Coarse trash bar rack with power rake for cleaning and trash hopper. (b) Dual-flow traveling water screen with automatic cleaning mechanism to remove debris that passes through the coarse bar rack. (c) Stop gates are installed both ahead of the bar rack and after the traveling screen.
(Courtesy of US Filter, Envirex Products.)

on the near shore of the lake or reservoir. Pumps for a shoreline intake are frequently housed in the same structure as the screens. A typical pumping station consists of a suction well located behind the intake screens, or at the discharge end of the withdrawal pipe, pumping equipment, and associated motors, valves, piping, and control systems. With manual controls, an operator must be available to start and stop pumping equipment, to open and close valves, and to monitor other controls. Highly automated pumping stations can be supervised from a central control panel located in the station, or they can be supervised remotely in the treatment plant. Standby pumps and duplicate equipment are provided as necessary to ensure uninterrupted water supply during maintenance and repair and to meet emergency water demands.

The selection of low-lift pumps depends on station capacity, the height to which the water must be lifted, the number of pumps required, the anticipated method of operation, and costs. The centrifugal pump common in low-lift pumping at water intakes is a vertical-turbine pump (Figure 6–4). Liquid flow is parallel to the axis of the pump and is designed to operate efficiently under low-head discharge. The vertical turbine is a centrifugal pump with one or more impellers located in stages. As the water leaves the discharge of the first stage, it enters the suction of the second stage and continues in this manner through subsequent stages. Each stage increases the rate of flow and pressure. No priming is needed for turbine pumps because they are installed with the impellers submerged.

6–5 PIPING NETWORKS

A municipal water distribution system includes a network of mains with storage reservoirs, booster pumping stations (if needed), fire hydrants, and service lines. Arterial mains, or feeders, are pipelines of larger size that are connected to the transmission lines that supply the water for distribution. All loop, forming a complete gridiron system that services fire hydrants and domestic and commercial consumers.

The gridiron system, illustrated in Figure 6–5a, is the best arrangement for distributing water. All of the arterials and secondary mains are looped and interconnected, eliminating dead ends and permitting water circulation such that a heavy discharge from one main allows drawing water from other pipes. When piping repairs are necessary, the area removed from service can be reduced to one block if valves are properly located. Shutoff valves for sectionalizing purposes should be spaced at about 1200-ft (370-m) intervals and at all branches from arterial mains. At grid intersections no more than one branch, preferably none, should be without a valve. Feeder mains are usually placed in every second or third street in one direction and every fourth to eighth street in the other. Sizes are selected to furnish the flow for domestic, commercial, and industrial demands, plus fire flow. Distribution piping in intermediate streets should not be less than 6 in. (150 mm) in diameter.

The dead-end system, shown in Figure 6–5b, is avoided in new construction and can often be corrected in existing systems by proper looping. Trunk lines placed in the main streets supply submains, which are extended at right angles to serve individual streets without interconnections. Consequently, if a pipe break occurs, a substantial portion of the community may be without water. Under some conditions, the water in dead-end lines develops tastes and odors from stagnation. To prevent this, dead ends may require frequent flushing where houses are widely separated.

The ideal gridiron system, with duplicating transmission lines, arterial mains, and fully looped distribution piping, may not be feasible in all cases. For example, in communities located on steep hillsides, elevations may vary so greatly that two or more separate systems are needed. Pressures in a system with large elevation differences should preferably be in the range of 50 to 100 psi (340 to 690 kPa) under average flow conditions. Since a pressure differential of 50 psi (340 kPa) is equivalent to an elevation difference of 115 ft (35 m), the desirable elevation change between the high and low points should be limited to this value. Frequently, separate mains serve different elevation zones with either no direct connections

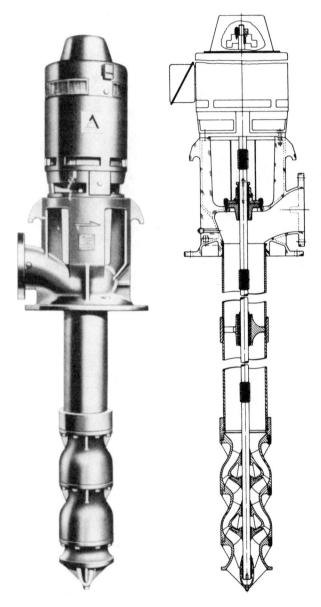

Figure 6–4

Vertical-turbine pump.

(Courtesy of Allis-Chalmers Corp.)

major water demand areas in a city should be served by a feeder loop; where possible, the arterial mains should be laid in duplicate. Two moderately sized lines a few blocks apart are preferred to a single large main. Parallel feeder mains are cross-connected at intervals of about 1 mile, with valving to permit isolation of sections in case of a main break. Distribution lines tie to each arterial

Figure 6–5
Pipe networks.
(a) Gridiron distribution system consisting of an arterial pipe network with a superimposed system of distribution mains.
(b) Undesirable dead-end distribution system.

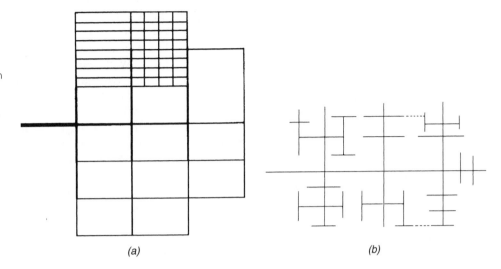

(a)   (b)

or pressure control valves inserted in the joining lines. Ideally, piping in each zone should be a gridiron system with multiple feeder mains; however, for economic reasons this may not always be feasible. The alternative arrangement is a single arterial main with branches at right angles and subbranches between them such that a grid of interconnected pipes surrounds the central feeder.

Service Connections

A typical service installation consists of a pipe from the distribution main to a turnoff valve located near the property line (Figure 6–6a). The pipe is generally attached to the main by means of a corporation stop that can be inserted by using a special tapping machine while the main is in service under pressure. Occasionally, outlets are provided in the main at the time of installation. Access to the curb stop is through a service box extending from the valve to the ground surface. A number of materials are used for service connections, the most popular being plastic, copper, and cast iron. Although copper pipe has been viewed as the standard, plastic materials, which are just as durable and less expensive, are preferred because of potential copper contamination of tap water if it is corrosive. Cast iron is used for larger services, generally in a 2-in. size or larger. The water meter, associated shutoff valve, and pressure regulator, if needed, are normally in the basement of the dwelling (Figure 6–6b). In some municipalities, the meter is placed outside in a

box. However, this has the disadvantage of exposing the meter and piping to possible freezing and burial under snow cover. Modern water meters (Figure 4–25) are equipped with remote readers extended to the outside of dwellings so that water utility personnel do not have to enter buildings to record water usage.

The maximum instantaneous residential consumption rate has increased in recent years because of the larger number of fixture units. The estimated rate for a typical house having two bathrooms, full laundry, kitchen, and one or two hose bibbs is 15 gpm. With the possible exception of lawn sprinkling, a pressure of 15 psi is normally adequate for the operation of any fixture. If one allows 10 psi (23 ft) to overcome static lift from the basement to upper floors, the generally accepted minimum standard of a service is 25 psi at 15 gpm (170 kPa at 0.95 l/s) on the customer side of the meter. For a typical residential service consisting of 40 ft of $\frac{3}{4}$-in. copper pipe and a $\frac{5}{8}$-in. disk meter, the friction loss for simultaneous fixture use (6 to 12 gpm) is in the range of 5 to 19 psi. Therefore, if 25 psi is to be available to the customer, the pressure in the main must be approximately 40 psi (280 kPa).

6–6 KINDS OF PIPE

Pipes used for distributing water under pressure include ductile iron, plastic, concrete, and steel. Small-diameter pipes for house connections are

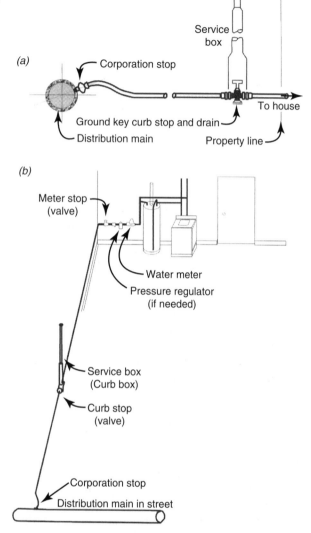

Figure 6–6

Typical residential service connection. (*a*) Service line and meter box installation. (*b*) Complete residential water service.

handling, and installation; a smooth, noncorrosive interior surface for minimum resistance to water flow; an exterior unaffected by aggressive soils and groundwater; and a pipe material that can be provided with tight joints and is easy to tap for making connections.

Ductile iron is noted for long life, toughness, imperviousness, and ease of tapping as well as the ability to withstand internal pressure and external loads. Ductile iron is produced by introducing a carefully controlled amount of magnesium into a molten iron of low sulfur and phosphorus content and treating with a silicon base alloy. This kind of pipe is stronger, tougher, and more elastic than gray cast iron. Iron pipes are available in sizes ranging from 2- to 48-in. in diameter (50 to 1200 mm), with several pressure classes in each size. The selection of a particular pressure class depends on the design internal pressure, including allowance for water hammer; the external load due to trench backfill and superimposed wheel loads; the allowance for corrosion and manufacturing tolerance; and the design factor of safety. The procedure for pressure design of water pipes has been simplified by the use of charts and tables that are published by pipe associations and manufacturers.

Although ductile iron is resistant to corrosion, aggressive waters may cause pitting of the exterior or tuberculation on the interior. Outside protection is usually provided by coating the pipe with bitumastic tar applied by spray nozzles. In corrosive soils, loosely wrapped polyethylene sleeves are used to encase the pipeline to isolate the pipe from the soil. A section of sleeve is placed over each section of pipe before it is lowered into the trench. While the connection is being made, the sleeve is pulled over the joint to overlap the adjoining section and then taped in position. The interior of the pipe is covered by a thin coating of cement mortar about $\frac{1}{8}$-in. thick. This lining, placed while the pipe is rotated at a high speed to compact the mortar, adheres closely to the iron surface such that cutting and tapping the pipe will not cause separation.

Pipe lengths, normally 18 ft, may be joined together by several types of joints; the most common are illustrated in Figure 6–7. One of the earliest was the bell and spigot connection, in which the plain end of one pipe was inserted into a flared

usually either copper or plastic. For use in transmission and distribution systems, pipe materials must have the following characteristics: adequate tensile and bending strength to withstand external loads that result from trench backfill and earth movement caused by freezing, thawing, or unstable soil conditions; high bursting strength to withstand internal water pressures; ability to resist impact loads encountered in transportation,

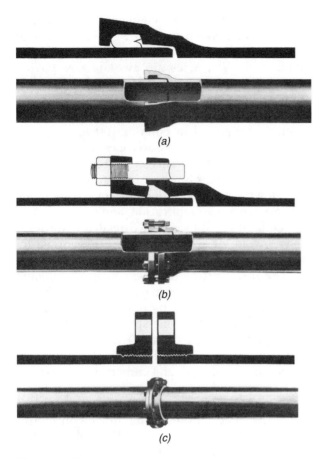

Figure 6–7
Common kinds of ductile-iron pipe joints.
(*a*) Compression-type (slip joint). (*b*) Mechanical joint.
(*c*) Flanged joint.
(Courtesy of Clow Water Systems Co.)

installation of pipelines on a radial curve. Typical allowable deflections are 5° for 3- to 12-in.-diameter pipe, 4° for 14- to 16-in., and 3° for 18- to 36-in. The mechanical joint utilizes the principle of the stuffing box to provide a fluid-tight, flexible connection for distribution piping; however, it is not as popular as the compression union. Flanged joints are made by threading plain-end pipe and screwing on flanges that are then faced and drilled to permit bolting together. Flanged connections are normally used for interior piping in water plants. In addition to these three common joints, particular types are available for special installations. For example, underwater pipelines or other applications requiring a very flexible connection use boltless, or bolted, flexible ball-and-socket joints.

The boltless restrained pipe joint for ductile-iron pipe in distribution system piping is illustrated in Figure 6–8. It can be used in place of the bolted mechanical joint in Figure 6–7*b*. The retainer casing is placed on the plain end of the pipe and the lock ring is welded into position at the factory. For field assembly, the plain end is pushed into the bell end, securing the push-on joint seal. Finally, the retainer is inserted over the

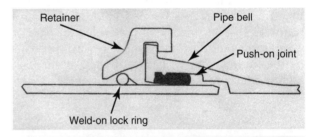

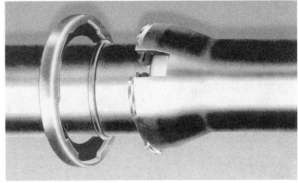

Figure 6–8
Boltless Super-Lock® push-on restrained joint for ductile-iron pipe 6 to 30 in. in diameter.
(Courtesy of Clow Water Systems Co.)

end of another and then was sealed with the caulking material, such as lead. This development led to the compression-type joint, also referred to as a push-on or slip joint, which is the most popular type used in water distribution piping today. The beveled spigot end pushes into the bell with a specially designed recess to accept a rubber ring gasket that provides a tight joint to seal the internal pressure. The pipes are held in place against separation and displacement by external means, including confinement of the surrounding soil, concrete thrust blocks, special joints with built-in restraint features, or special restraining gaskets. The chief advantages of the compression joint are ease of installation, water tightness, and flexibility. The joint permits deflection, making possible the bell end and locked behind the lugs on the bell. The retainer lock and roll pin are inserted as per directions. Any necessary deflection of the joint is made after the joint is completely assembled.

Plastic, or more correctly thermoplastic, pipe is not subject to corrosion or deterioration by electrolysis, chemicals, or biological activity and is exceptionally smooth, minimizing friction losses in water flow. The pipe is manufactured by an extrusion process, and fittings are formed in injection molds. The product is slowly cooled and further shaped through sizing devices to ensure precise dimensions. PVC (polyvinyl chloride) is the plastic preferred for water distribution piping because of its strength and resistance to internal pressure. Yet PVC has a significantly lower modulus of tensile elasticity than iron and is, therefore, less resistant to pressure surges. Commonly manufactured sizes are 4–12 in. (100–300 mm), with internal pressure classes of 100, 150, and 200 psi. PVC pipe is rated at a standard temperature of 73.4°F, which is satisfactory for most water distribution systems. At warmer temperatures, the pressure resistance decreases until the critical point is reached near 150°F. For joining PVC water mains, the compression joint uses a rubber seal that fits into a recess formed in the belled end (Figure 6–9). The spigot end is beveled for ease of installation, and the joint is made by simply pushing the lubricated spigot into the bell, compressing the rubber gasket for a pressure-tight fit.

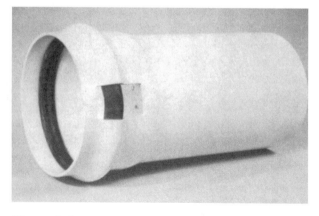

Figure 6–9
PVC bell-and-spigot compression-type joint employing a rubber gasket seal.
(Courtesy of Certain-teed Products Corporation.)

Plastic pipe for service connections and household plumbing systems includes ABS (polymers of acrylonitrile, butadiene, and styrene), PE (polyethylene), and PVC. Sizes are available over wide ranges: ABS, $\frac{1}{2}$ to 12 in.; PE, $\frac{1}{2}$ to 6 in.; and PVC, $\frac{1}{2}$ to 16 in. ABS and PVC, being semirigid products, are normally produced in 20- to 39-ft lengths. Flexible PE pipe is supplied in coils of 100 to 500 ft in length. PE is often used in service connections, and ABS is primarily for drainage, waste and vent fittings, and piping for interior applications.

Joining plastic pipe is done by several methods that vary with size, application, and material. ABS may be connected by screw-threaded couplings, solvent weld, or slip couplings. PE is joined by using insert fittings, flaring, or compression jointing. The insert-fitting method consists of placing an internal support tube inside the pipe ends and holding the connection by outside clamping. PE pipe can be flared by using special tools with the application of heat to soften the material. This allows the joining of plastic service lines directly into conventional curb and corporation stops. Yet another method uses internal metal stiffener sleeves and O-rings to form a watertight compression joint. The simplest and most commonly applied joining methods for PVC pipe are the solvent-weld system and the belled-end coupling. PVC can be chemically bonded through the use of solvents that dissolve the plastic. The solvated plastic surfaces mix when pressed together and leave a monolithic joint on evaporation of the solvent. Belled-end PVC joints may employ a solvent weld or rubber gasket. In the former, the plain pipe end is cemented into a straight bell formed on the opposite end of a second pipe length.

Three types of reinforced concrete pipes are used for pressure conduits: steel cylinder, prestressed with steel cylinder, and noncylinder reinforced that is not prestressed. Concrete pipe has the advantages of durability, watertightness, and low maintenance costs. It is particularly applicable in larger sizes that can be manufactured at or near the construction site by using local labor and materials, insofar as they are available. Nonprestressed steel-cylinder pipe is a welded steel pipe surrounded by a cage of steel reinforcement covered inside and out with concrete. The concrete protects the steel from corrosion, provides a smooth interior surface, and contributes high compressive strength to resist stresses from external loading. It is used for internal pressures ranging from 40 to 260 psi. For higher pressures, 50 to 350 psi, steel-cylinder pipe is prestressed by winding wire directly on a core consisting of either a steel cylinder lined with concrete or a pipe of concrete imbedded with a steel cylinder. The exterior of the prestressed core is then finished with a coating of mortar. Reinforced concrete pipe using steel bars or wire fabric, without a steel cylinder, is applicable for low-pressure water transmission requirements, not exceeding 45 psi. Reinforced concrete pipe comes in a variety of diameters ranging from 16 to 144 in. Laying lengths vary with size, the common manufactured lengths being 8, 12, or 16 ft. The usual concrete pipe joint is a modified bell and spigot with steel rings and a sealing element including a rubber seal. A gasket is placed in the outside groove on the steel-faced spigot ring and then is forced into the steel bell ring of another pipe. The rubber gasket provides a watertight seal by filling the groove in a tightly compressed condition. The narrow space between the bell and pipe surfaces is then sealed with a flexible, adhesive plastic gasket, and the exposed steel joint faces are painted.

Steel pipe, used in transmission lines, exhibits the characteristics of high strength, ability to yield without breaking, and resistance to shock, but careful protection against corrosion is absolutely necessary. A common outside protective coating includes paint primer, coal tar enamel, and wrapping. The particular materials applied depend on the corrosive environment of the pipe and placement, aboveground or underground. Interior linings may be either coal tar enamel or cement mortar. The two types of steel pipe fabricated are electrically welded and mill type. Pipe ends are manufactured in a variety of ways for field connection: plain or beveled ends for field welding, flanges for bolting, bumped or lap ends for riveting, and bell and spigot ends with rubber gasket.

6–7 DISTRIBUTION PUMPING AND STORAGE

High-lift pumps move processed water from a basin at the treatment plant into the distribution system. Different sets of pumps may be needed to pump against unequal pressures to different service areas. If so, some pumps connect directly for main service in lower areas, while booster units are used to reach high elevations in the system. The most common types of pumps for servicing high areas are vertical turbine and horizontal split-case centrifugal pumps because of good efficiency and capability to deliver water against high discharge heads. The model illustrated in Figure 6–10 is a double-suction design such that water is drawn into both sides of the double volute case, ensuring both radial and axial balance for minimum pressure on bearings. The impeller discharges water to the ever-increasing spiral casing, gradually reducing the velocity head while producing an increase in pressure head. This type of pump operates at a range of capacities from design flow to shutoff without excessive loss of discharge pressure or efficiency. Centrifugal pump hydraulics and system characteristics are discussed in Sections 4–4 and 4–5.

Distribution storage may be provided by elevated tanks, standpipes, underground basins, or

Figure 6–10
Horizontal split-case double-suction centrifugal pump applicable for high-service pumping in waterworks. (Courtesy of Fairbanks Morse Pump Corporation.)

covered reservoirs. Elevated steel tanks are manufactured in a variety of ellipsoidal and spherical shapes and in capacities ranging from 50,000 to 3.0 mil gal (Figure 6–11). The advantage of elevated storage is the pressure derived from holding water higher than surrounding terrain. Steel tanks are normally protected from corrosion by painting the exterior and installing cathodic protection equipment to safeguard interior surfaces. Where gravity water pressure either is not necessary or is provided by booster pumping, ground-level standpipes or reservoirs are used. Steel standpipes (Figure 6–12) are usually available in sizes up to 5 mil gal. In general, the term *standpipe* applies where the height of a tank exceeds its diameter;

Figure 6–12

Ornamented steel standpipe for ground-level water storage, designed and built in any capacity.

(Courtesy of Chicago Bridge and Iron Co.)

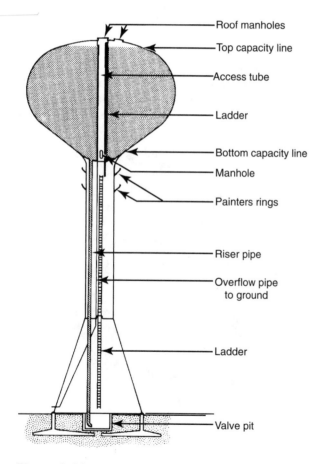

Figure 6–11

Elevated spheroid-shaped water storage tank available in capacities from 0.50 to 3.0 mil gal and specified pedestal height.

(Courtesy of Chicago Bridge and Iron Co.)

one that has a diameter greater than its height is referred to as a *reservoir*.

Concrete reservoirs may be constructed above- or belowground. Recent designs in circular, prestressed concrete reservoirs have increased watertightness and reduced maintenance costs.

Clear-well storage at a treatment plant is commonly located under the filter beds, which provide a roof structure. Storage capacity at a treatment plant allows for differences in water production rates and high-lift pump discharge to the distribution system. Additional clear-well volume may be supplied to act as distribution storage.

The choice between elevated and ground storage in water distribution depends on topography, size of community, reliability of water supply, and economics. Ground storage facilities on hills high enough to provide adequate pressures are preferred; however, seldom are hilltop locations in a suitable position. Therefore, elevated tanks, or ground-level reservoirs with booster pumping to provide required water pressure, are needed. The economy and desirability of ground storage with booster pumping, as compared with elevated storage, must be determined in each individual area. In general, elevated tanks are most economical and are recommended for small water systems. Reservoirs and booster pumping facilities are often less expensive in large systems where adequate supervision can be provided. Reliable instrumentation and automatic controls are available for remote operation of automatic booster stations.

The principal function of distribution storage is to permit continuous treatment and uniform pumping rates of water into the distribution system while storing it in advance of actual need at one or more locations. The major advantages of storage are that the demands on source, treatment, transmission, and distribution are more nearly equal, reducing needed sizes and capacities; that the system flow pressures are stabilized throughout the service area; and that reserve supplies are available for contingencies, such as fire fighting and power outages. In determining the amount of storage needed, both the volume used to meet variations in demand and the amount related to emergency reserves must be considered. The storage to equalize supply and demand is determined from hourly variations in consumption on the day of maximum water usage.

The location of distribution storage as well as capacity and elevation are closely associated with water demands and their variation throughout the day in different parts of the system. Normally, it is more advantageous to provide several smaller storage units in different locations than an equivalent capacity at a central site. Smaller distribution pipes are required to serve decentralized storage, and more uniform water pressures can be established throughout the system. In normal operating service, some stored water should be used each day to ensure recirculation, and on peak days the drawdown should not consume fire reserves.

Storage within the distribution system for fire flow is provided by elevated tanks or ground-level reservoirs with high-service pumps. Reserve capacity for fire fighting is computed from required fire flow and duration. Properly sized elevated water tanks provide this dedicated volume

for fire storage at system pressure. Domestic supplies are fed to the pipe network from the top 10 to 15 ft of water in an elevated tank. As the water level drops in a tank beyond normal demand, an automatic pressure sensor turns on additional high-service pumps so that 70 to 75 percent of water is held in reserve as dedicated fire storage. Ground-storage of water to be delivered by pumps must have the following characteristics to sustain adequate fire flow: excess pumping capacity to deliver the peak demand for domestic uses as well as fire demand; standby power for these pumps maintained at all times; and an oversized distribution pipe network to handle peak domestic delivery plus fire flow. Example 6–3 illustrates common techniques for calculating the amount of storage required for a community. Capacity to balance supply and demand, at a constant pumping rate, is usually in the range of 15 to 20 percent of the daily consumption.

■ EXAMPLE 6–3

Calculate the distribution storage needed for both equalizing demand and for fire reserve based on the following information. Hourly demands on the day of maximum water consumption are given in Table 6–5. Listed are hourly consumption expressed in gallons per minute, gallons consumed each hour of the day, and cumulative consumption starting at 12 midnight. Fire flow requirements are 6000 gpm for a duration of 6 hr for the high-value district, with 2000 gpm from storage.

Solution

Figure 6–13 is a diagram of the consumption rate-time data given in Table 6–5. When the consumption rate is less than the 1860-gpm pumping rate, the reservoir is filling. When it is greater than 1860 gpm, the reservoir is emptying. The area under the emptying, or filling, curve is the storage volume needed to equalize demand for the average 24-hr pumping rate of 1860 gpm.

Since the areas on Figure 6–13 are difficult to measure, a mass diagram is commonly used to determine equalizing storage. Figure 6–14 is a plot of the cumulative flow, column 4 in Table 6–5, versus time. A straight line connecting the origin

TABLE 6–5

Peak Water Consumption Data on Day of Maximum Water Usage for Example 6–3

Time	Hourly Consumption (GPM)	Hourly Consumption (GAL)	Cumulative Consumption (GAL)
12 P.M.	0	0	0
1 A.M.	866	52,000	52,000
2	866	52,000	104,000
3	600	36,000	140,000
4	634	38,000	178,000
5	1000	60,000	238,000
6	1330	80,000	318,000
7	1830	110,000	428,000
8	2570	154,000	582,000
9	2500	150,000	732,000
10	2140	128,000	860,000
11	2080	125,000	985,000
12	2170	130,000	1,115,000
1 P.M.	2130	128,000	1,243,000
2	2170	130,000	1,373,000
3	2330	140,000	1,513,000
4	2300	138,000	1,651,000
5	2740	164,000	1,815,000
6	3070	184,000	1,999,000
7	3330	200,000	2,199,000
8	2670	160,000	2,359,000
9	2000	120,000	2,479,000
10	1330	80,000	2,559,000
11	1170	70,000	2,629,000
12	933	56,000	2,685,000
	Average = 1860		

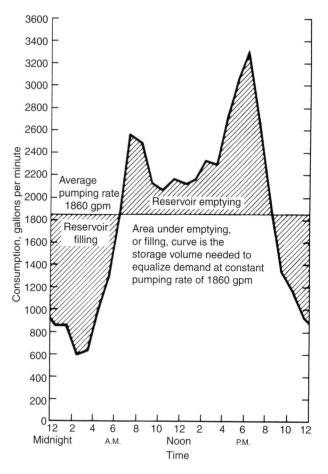

Figure 6–13

Plot of hourly water consumption rates, from Table 6–5 for Example 6–3, to determine the storage needed to equalize demand at constant pumping rate.

and final point of this mass curve is the cumulative pumpage necessary to meet the consumptive demand; the slope of this line is the constant 24-hour pumping rate. To find the required storage capacity, construct lines that are parallel to the cumulative pumping rate tangent to the mass curve at the high and low points. The vertical distance between these two parallels is the required tank capacity, in this case, 500,000 gal.

However, sometimes it is not expedient to pump water into the distribution system at a constant rate throughout the day and night. For example, a small community may limit treatment plant operations to daylight hours or operation of pumps to off-peak periods when power rates are low. Assume in this instance that the lowest power rates may be obtained during the 8-hr period between 12 midnight and 8 A.M. The graphical procedure for determining storage requirements using this 8-hr pumping period is illustrated in Figure 6–14 using dashed lines. An accumulated pumping line is drawn from the origin, 0 flow and time of 12 P.M., to the end of the pumping period at 8 A.M. on the maximum cumulative consumption for the day. The storage required is then equal to the vertical distance at 8 A.M. between the accumulated demand line and the maximum daily

Figure 6–14

Mass diagram of water consumption, from Table 6–5 for Example 6–3, to determine storage needed to equalize demand at constant pumping rates.

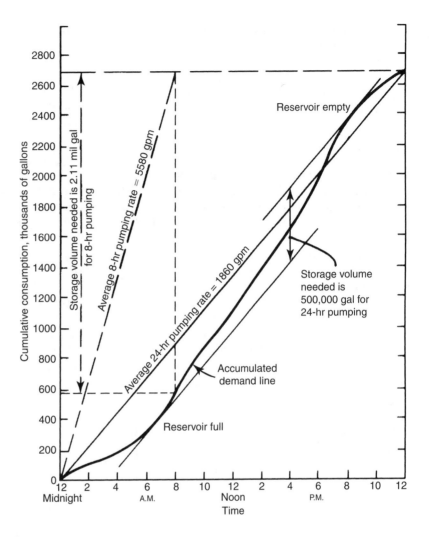

pumpage. (If a starting time for pump operation other than 12 P.M. is to be considered, the data in Table 6–5 must be shifted to the selected time, cumulative consumption values must be recomputed, and Figure 6–14 will have to be redrawn with an origin at the new time.)

The storage required to provide the entire fire reserve is equal to flow rate times duration:

$$2000 \frac{\text{gal}}{\text{min}} \times 60 \frac{\text{min}}{\text{hr}} \times 6 \text{ hr} = 720,000 \text{ gal}$$

The total storage capacity required for equalizing demand for a continuous 24-hr pumping rate plus fire protection is equal to

$$0.50 + 0.72 = 1.22 \text{ mil gal}$$

The total, considering an 8-hr pumping period plus fire reserve, equals

$$2.11 + 0.72 = 2.83 \text{ mil gal}$$

■ ■ ■

■ EXAMPLE 6–4

Consider a water-supply system serving a city with the following demand characteristics: average daily demand, 4.0 mgd (2780 gpm); maximum day, 6.0 mgd (4170 gpm); peak hour, 9.0 mgd (6250 gpm); and required fire flow, 7.2 mgd (5000 gpm), resulting in a maximum 5-hr rate of 13.2 mgd (9170 gpm) maximum daily demand plus fire flow.

Assume that the minimum pressure to be maintained in the main district is 50 psi (115 ft) except during fire flow and that the piping system is equivalent to a 24-in.-diameter main with a $C = 100$. Consider the system without storage and with storage beyond the load center.

Solution

Effect of No Storage. At each given demand rate the pumping station discharge head must be sufficient to overcome system losses and to maintain a hydraulic gradient with a minimum of 115 ft of head at the load center. Thus, at the average daily demand, the pumping head required is 115 ft plus the head loss in 29,000 ft of 24-in. pipe at 4.0 mgd (using Figure 4–7 for head loss):

$$115 + (0.9 \times 29) = 140 \text{ ft}$$

At maximum daily rate (6.0 mgd):

$$115 + (1.9 \times 29) = 170 \text{ ft}$$

At peak hourly demand (9.0 mgd):

$$115 + (4.0 \times 29) = 230 \text{ ft}$$

And, at maximum daily rate plus fire flow (13.2 mgd):

$$115 + (8.2 \times 29) = 350 \text{ ft}$$

These results are plotted in Figure 6–15.

Storage Beyond Load Center. In the arrangement shown in Figure 6–16, 1.0 mil gal of storage is provided 10,000 ft beyond the load center, 39,000 ft from the pumping station at an elevation of 120 ft. When no water is being taken from storage, the pumping head must be sufficient to pump against the head at the tank and to overcome losses between the pumping station and the load center. When part of the demand is being supplied from storage, however, the pumping head need only be sufficient to discharge against the head at the load center and to overcome losses in the pipeline between the pumping station and load center.

Figure 6–15

Hydraulic gradients with no storage for Example 6–4.

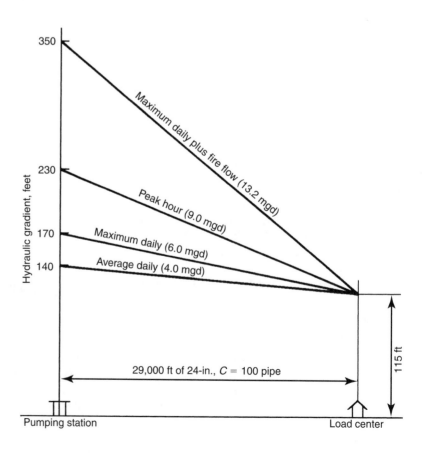

Figure 6–16

Hydraulic gradients with elevated storage beyond load center for Example 6–4.

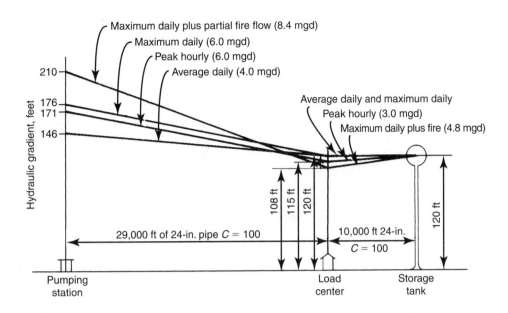

At the average daily demand, the required pumping rate is 4.0 mgd with no water taken from storage. The hydraulic gradient at the load center is, thus, identical to that at the tank, namely, 120 ft. The pumping head required is equal to the hydraulic gradient at the load center plus the head loss in 29,000 ft of pipe,

$$120 + (0.9 \times 29) = 146 \text{ ft}$$

During the maximum day, the required pumping rate is 6.0 mgd with no water taken from storage. The pumping head required equal to the hydraulic gradient at the load center plus the head loss in 29,000 ft at 6.0 mgd is

$$120 + (1.9 \times 29) = 176 \text{ ft}$$

At the peak hourly demand, consider supplying 3.0 mgd from storage and the remaining 6.0 mgd from pumping. The hydraulic gradient at the load center is that at the tank minus the head loss in 10,000 ft of pipe between the tank and load center at the storage discharge rate of 3.0 mgd:

$$120 - (0.5 \times 10) = 115 \text{ ft}$$

The pumping head is then

$$115 + (1.9 \times 29) = 171 \text{ ft}$$

Maximum Daily Plus Fire Flow. If the 1.0 mil gal in storage is supplied at a uniform rate for the required fire duration of 5 hr, the flow is 3330 gpm (4.8 mgd) with a resulting hydraulic head at the load center of

$$120 - (1.2 \times 10) = 108 \text{ ft}$$

The pumping rate needed is 13.2 − 4.8 = 8.4 mgd, and the pumping head required is

$$108 + (3.5 \times 29) = 210 \text{ ft}$$

In this analysis the 1.0-mil-gal storage supplied about 35 percent of the fire flow, with the pumping station providing the balance. This is realistic, since about half of the storage is to equalize supply and demand. Considering only 500,000 gal is available for fire demand, the head at the load center is 116 ft and the pumping head increases to 275 ft.

A comparison of Figures 6–15 and 6–16 illustrates the benefits of distribution storage. If no storage is provided, significantly greater pumping heads are required to furnish peak demands. The pumping rate at maximum daily plus fire flow more than doubles the power needed. The same is true at peak hourly pumpage: without storage, 9.0 mgd at 230 ft compared to 6.0 mgd at 171 ft. During average and maximum daily demands, the pumping heads are approximately the same.

■ ■ ■

6–8   VALVES

Valves are installed throughout water systems in treatment plants, pumping stations, and pipe networks, as well as at storage reservoirs. Their purpose is to control the magnitude or direction of water flow. To regulate flow, all valves have a movable part that extends into the pipeline for opening or closing the interior passage. The four basic kinds of valves are slide, rotary, globe, and swing; others less common are sphere, diaphragm, sleeve, and vertical-lift disk. Valves are also commonly classified by operating purpose (for example, shutoff and altitude) and function (by-pass and flow control) without regard to the kind of device used. Since the primary purpose of this discussion is to describe the use of valves in water systems, the method of presentation is by function with illustrations showing a commonly used kind of valve. Other kinds may often be used for the same purpose.

The means of operating the movable element of a valve are by screw, gears, or water pressure. Screw stems are common in gate, globe, and needle valves and can be opened or closed by a manually operated handwheel or by a powered operator. In some designs, the screw stem rises as the shutoff element closes, and in others the element rides up the screw inside the body of the valve as the stem turns, the latter nonrising stem being more common. In large valves, where water pressure prevents use of a screw, a gear train can be employed to allow the shutoff element to be moved slowly by applying a minimum of torque to the stem. The gear system may be operated manually or by electric, hydraulic, or pneumatic power operators. The kinds of valves that may be equipped with a geared operator include butterfly, gate, globe, and ball. Water pressure can open or close some kinds of valves by direct pressure on the movable element. The simplest is a hinged swing gate that opens under water pressure and closes under the influence of gravity or back pressure. Another example is the automatic globe valve with a body design that controls movement of the shutoff element by system water pressure.

Shutoff Valves

Valves to stop the flow of water through a pipeline are the most abundant valves in a water system. Pipe networks are sectionalized by installation of shutoff valves so that any area affected by a main break or pipe repair can be isolated with a minimum reduction in service and fire protection. Depending on the district within the city and the size of the water mains, valve spacings range from 500 to 1200 ft (150 to 370 m). Ideally, a minimum of three of the four pipes connected at a junction are valved and arterial mains have a shutoff every 1200 ft. The pipe connecting a fire hydrant to a distribution main contains a valve to facilitate hydrant repair. In treatment plants and pumping stations, shutoff valves are installed in inlet, outlet, and by-pass lines so that valves and pumps can be removed for maintenance and repair. Gate valves are the shutoffs; however, rotary butterfly valves may be installed in large-diameter pipes.

A gate valve has a solid sliding gate that moves at right angles to the direction of water flow by a screw-operated stem. When installed in a pipeline, the gate is drawn up into the housing of the case by a nonrising stem. The gate is lowered to fit snugly against the sides and bottom to block water flow. The modern valve shown in Figure 6–17 has the gate encapsulated with rubber and the inside of the valve body coated with epoxy. Guides fit into slots on both sides of the gate to keep it in alignment so the rubber sealing surfaces are compressed when the gate is closed to prevent leakage. Older gate valves, still commonly found in distribution pipe networks, are double-disk, parallel-seat valves constructed of cast iron and subject to incrustation and corrosion. Normally, underground valves in mains are provided with a valve box extending to the street surface. The valve is opened, or closed, by using an extension rod to reach down into the enclosure and turn the nut located on top of the valve stem. Larger gate valves may be placed in vaults or manholes to facilitate operation and maintenance. They may be equipped with small by-pass valves to reduce pressure differentials on opening and closing and frequently have spur or beveled gearing to reduce the force required for operation.

A butterfly valve (Figure 6–18) has a movable disk that rotates on a spindle or axle set in the shell. The circular disk rotates in only one direction, from full closed to full open, and seats against a ring in the casing. The main disadvantages of a butterfly valve result from the disk always being in the flow stream, restricting the use of pipe-cleaning tools. On the other hand, the advantages of this valve are tight shutoff, low head loss, small space requirement, and throttling capabilities. The latter is one of the most popular applications of butterfly valves, for example, use in a rate of flow controller to regulate the rate of discharge from gravity filters in a water treatment plant. Recently, and more often in larger sizes, rubber-seated butterfly valves are being used in distribution systems. Because pressure differences across a butterfly valve disk tend to close the valve, a mechanical operator must be employed to overcome the torque in opening the valve and to resist this force during closing to prevent slamming. For treatment plant applications, the operator is often a hydraulic cylinder and piston rod assembly used to hold the valve disk in any intermediate position, as well as in opening and closing action. Operators for large valves may also be motor driven.

Check Valves

A check valve is a semiautomatic device that permits water flow in only one direction. It opens under the influence of pressure and closes automatically when flow ceases. Usual installations are in the discharge pipes of centrifugal pumps to prevent backflow when the pump is not operating and in conjunction with altitude valves in connections between storage reservoirs and the distribution network.

The two basic configurations of check valves in water-system applications are lift check and swing

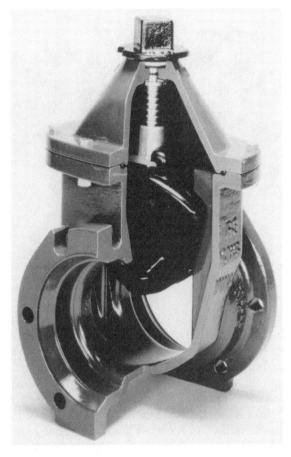

Figure 6–17
Resilient-seated gate valve with a rubber encapsulated gate guided for alignment so the rubber sealing surfaces compress when the gate is closed.
(Courtesy of U.S. Pipe & Foundry Co.)

Figure 6–18
Views of a butterfly valve in open and closed positions.

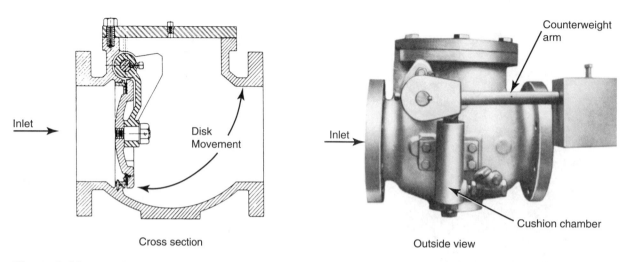

Figure 6–19

Counterbalanced, swing check valve with a cushion chamber to prevent valve slam.
(Courtesy of GA Industries Inc., Golden-Anderson Valve Div.)

check. A lift check has a flat disk that moves vertically within the valve body, opening with vertical flow and closing by gravity or spring action when flow stops. Of the swing checks, the most common is the kind illustrated in Figure 6–19 with a closed disk that rests at right angles to the direction of water flow and where the disk can be lifted to provide full flow. Closure may be by gravity or assisted by a counterweight arm attached to the disk. For tight closing, the swing check is manufactured to fit tightly over a rubber or metal seat. The function of the cushion chamber and counterweight is to prevent valve slam.

Small Pressure-Reducing and Pilot Valves

The function of these valves is to reduce high inlet pressure to a predetermined lower outlet pressure. The applications of small pressure-reducing valves, manufactured in sizes from 0.5 to 2 in., are to protect house plumbing from excessive main pressures and as pilot valves to control the action of large automatic valves that are operated by water pressure. By control of a variable spring, flow through the valve is regulated, causing a head loss resulting in a pressure differential between the inlet and outlet. For the unit depicted in Figure 6–20, vertical movement of the stem, predetermined by the position of the diaphragm attached on top, controls the size of the opening through the valve. Outlet

pressure, transmitted through the opening connecting the valve chamber to the space under the diaphragm, tends to close the valve by moving the valve disk upward. The force of the spring above the diaphragm pushes downward in an attempt to open the valve. If the pressure on the outlet side decreases below the desired level, the change is transmitted to the diaphragm and the spring forces the valve disk downward to a more open position, thus allowing greater flow and higher water pressure through the valve. Conversely, if the outlet pressure increases, the diaphragm pulls the stem upward, partially closing the valve and reducing outlet pressure. By turning the handwheel, the force of the spring against the diaphragm changes to allow adjustment of the outlet pressure.

Automatic Control Valves

An automatic valve uses water pressure in the system to operate the movable element to close or open the valve. In large valves, an external pilot valve or a spring-loaded diaphragm assembly directs water pressure to the proper side to set the movable element in the valve body. Sensing devices for signalling valve operation may be hydraulic, electric, pneumatic, or solenoid. Of the many applications for automatic valves, the common units in a water system are pressure-reducing, altitude, controlled check, surge relief, and shutoff valves.

Figure 6–20

Pressure-reducing valve available in sizes from 0.5 to 2.0 in. for use in service connections (Figure 6–6) and as a pilot valve to control the action of automatic valves (Figures 6–22 and 6–23). (Courtesy of GA Industries Inc., Golden-Anderson Valve Div.)

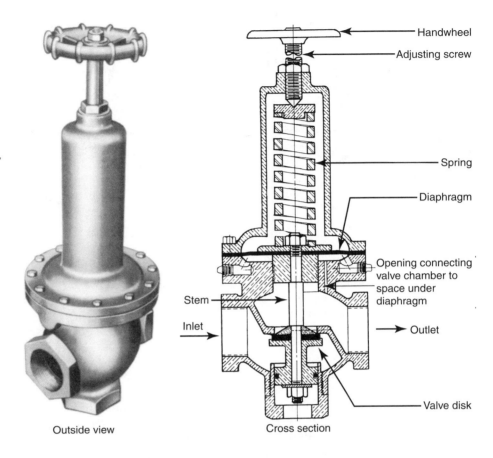

Outside view   Cross section

The operating principle of one type of automatic globe valve is shown schematically in Figure 6–21. The movable element in the upper chamber of the valve is shaped like a piston, with the top surface area greater than the bottom area. When the chamber above the piston is exhausted to the atmosphere, the piston is held open by inlet water pressure in the body of the valve (Figure 6–21a). Changing the position of the three-way valve on top closes the exhaust and allows inlet water to enter the chamber above the piston. The same magnitude of pressure exerted against both the top and bottom of the piston creates a greater downward force, since the top area of the piston is greater than the bottom area. Thus, when inlet water is allowed to enter the upper chamber, the valve automatically closes (Figure 6–21b). Air moving in or out of the piston chamber passes through a small opening to slow the movement of the piston and prevent water hammer. The three-way valve on top is usually controlled by a hydraulic or solenoid pilot valve.

Pressure-Reducing Valves

Pressure-reducing valves are automatic valves operated by pilots to maintain a preset outlet pressure against a higher inlet pressure. In water distribution systems, the common application is installation in a main connecting separate pipe networks located on two different elevations. As the water flows through the valve from the higher zone to the lower zone, the water pressure is reduced to prevent excessive pressure in the pipe network at the lower elevation.

A pressure-reducing valve is illustrated in Figure 6–22. The inlet pressure of the automatic valve is transmitted to the top of the valve piston, which tends to close the valve, while outlet pressure is transmitted to the top of the piston through the pilot valve. The automatic valve's inlet pressure under the diaphragm of the pilot valve presses upward, trying to close the pilot. This force is opposed by the spring pushing downward on the diaphragm, trying to open the valve. By turning

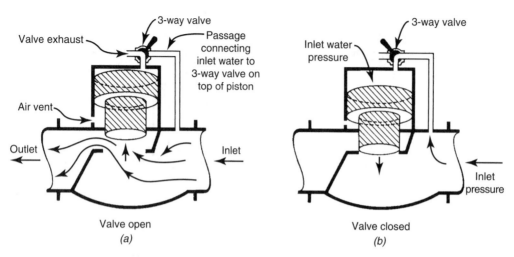

3-way valve
Valve exhaust
Passage connecting inlet water to 3-way valve on top of piston
Air vent
Outlet
Inlet

Valve open
(a)

3-way valve
Inlet water pressure
Inlet pressure

Valve closed
(b)

Figure 6–21

Diagrams illustrating the operation of an automatic globe valve. (*a*) The valve is held open by inlet water pressure when the chamber above the piston is vented to the atmosphere. (*b*) The valve is held closed by inlet pressure exerting a greater force on the top of the piston.

(Courtesy of GA Industries Inc., Golden-Anderson Valve Div.)

the handwheel, the force of the spring is adjusted, presetting the reduced outlet pressure of the automatic valve to the desired value. If the pressure at the inlet of the automatic valve is relatively low, the pressure under the diaphragm of the pilot is lowered, thus opening the pilot valve, which, in turn, relieves the pressure above the valve piston and allows increased flow through the automatic valve. Conversely, a high inlet pressure entering the automatic valve reduces flow through the pilot valve, which increases the pressure on top of the valve piston, reducing flow through the automatic valve. Therefore, the outlet pressure is held constant even though the inlet pressure varies.

Pressure-reducing valves can be equipped with two pilot valves. By this means, the automatic valve can function as both a pressure-reducing and a sustaining valve. It maintains a preset outlet pressure, provided the inlet pressure is above a predetermined value, and closes when the inlet pressure drops below a preset sustaining pressure to protect water services on the inlet side of the valve. Other control arrangements can be used to operate an automatic valve as a pressure-reducing and relief valve or as a pressure-reducing and check valve. A relief pilot opens the automatic valve to equalize inlet and outlet pressures if the inlet pressure drops below a preset value. In the pressure-reducing and check valve, backflow is prevented by closing the

valve if the inlet pressure drops below the outlet pressure. The type of inlet pressure control selected for a pressure-reducing valve depends on the operation desired for a given installation.

Altitude Valves

Altitude valves are used to automatically control the flow into and out of an elevated storage tank or standpipe to maintain desired water-level elevations. These valves are usually placed in a valve pit adjacent to the tank riser. An altitude valve designed as a double-acting sequence valve (Figure 6–23) is installed in the pipe connecting the tank to the system. The valve automatically closes when the tank is full to prevent overflow. When the pressure on the distribution side of the valve is less than on the tank side, it opens, allowing water to enter (or leave) the tank.

The piping arrangement for a double-acting valve is shown schematically in Figure 6–23*a*. The altitude valve is installed in the inlet-outlet pipe between two isolating gate valves. The pilot valve is connected through two small pipes to the tank riser on one side and to the main from the distribution system on the other. An outside view of a double-acting altitude valve is pictured in Figure 6–23*b*. Speed of valve closing is controlled by adjusting the needle valve in the pipe between

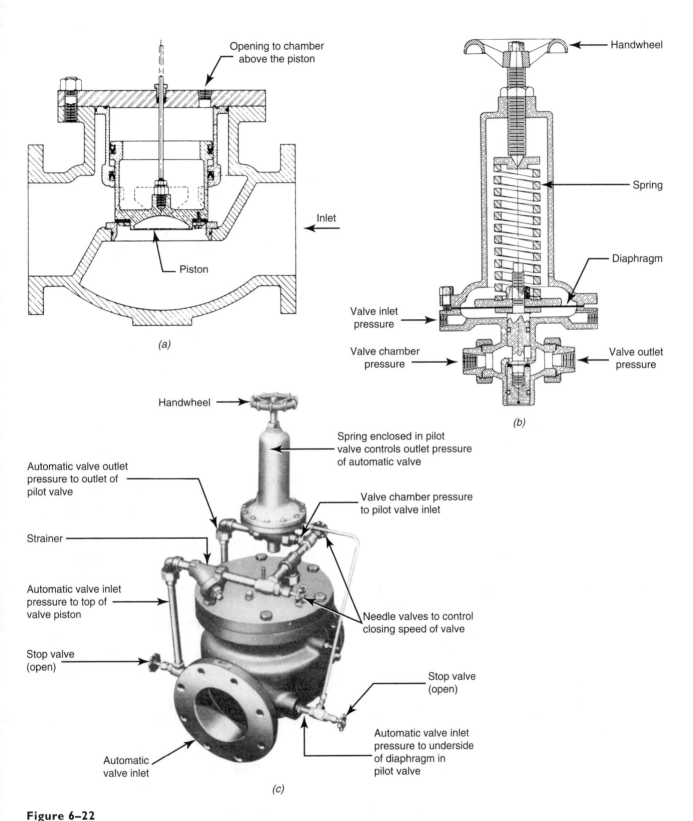

Figure 6–22

Pressure-reducing valve. (*a*) Cross section of the automatic valve. (*b*) Cross section of the pilot valve. (*c*) Outside view of pressure-reducing valve with pilot valve and control piping.

(Courtesy of GA Industries Inc., Golden-Anderson Valve Div.)

Figure 6–23

Double-acting altitude valve to automatically control water flow into and out of an elevated storage tank. (*a*) Piping arrangement. (*b*) Outside view of the valve. (*c*) Pilot valve operation when water is flowing out of the tank. (*d*) Pilot valve when the tank is full and automatic valve is closed.

(Courtesy of GA Industries Inc., Golden-Anderson Valve Div.)

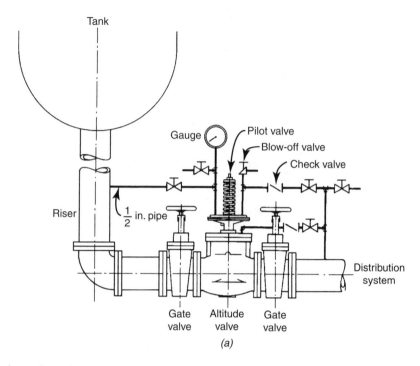

the valve body and the base of the pilot. The pilot valve on top is a diaphragm-operated, spring-loaded, three-way valve that transmits water pressure to the top of the piston in the main valve. By turning the nut threaded onto the stem, the force of the spring on the diaphragm can be adjusted to preset the automatically maintained water level in the storage tank. Figure 6–23c illustrates operation of the pilot valve, allowing stored water to flow out of the tank. Water drains out of the chamber above the pilot diaphragm through an open check valve in the small-diameter pipe to the low-pressure system side of the valve. This permits the stem of the pilot to be forced upward when the piston is pushed up by high-pressure water entering the valve from the tank. Therefore, water flows out of the tank through the open valve into the system. During filling of the tank, the direction of flow is reversed, passing from the system into the tank. The pilot does not close the automatic valve during this reverse flow, since the check valve in the pipe to the pilot is closed as a result of the higher pressure from the system side, preventing the force of the spring from lowering the diaphragm in the pilot. Thus, in these pilot positions, water is free to flow in and out of the tank and the tank is said to be "floating" on the system. If water continues to enter the tank, eventually the water level rises to the top and the

water pressure plus force from the spring lowers the diaphragm, closing the valve between the chamber under the diaphragm and the top of the valve piston. Water can now flow through a by-pass from the valve body to the top of the piston from the system side, as shown in Figure 6–23d, and the automatic valve closes. (The by-pass pipe is the one attached to the valve body, as pictured in Figure 6–23b.) A globe altitude valve is either fully opened or completely closed and, therefore, does not act as a throttling valve.

A single-acting altitude valve is installed in the inlet pipe to control only water flow into a storage tank. Outflow is through another pipe with a check valve to prevent system water from entering. Figure 6–24 illustrates typical arrangements for installing a single-acting sequence valve.

Altitude valves come in a variety of other designs. For example, a differential altitude valve closes when a tank is full and remains closed until the tank level drops to a predetermined water level before it opens to refill the tank. This operation ensures renewal of the stored water after each filling. An altitude valve may also be a combination valve composed of a swing check built into a butterfly control vane assembly. Outflow is through the check valve. When the water level in the tank drops below a preset elevation, the butterfly vane in the connecting pipe to the system opens.

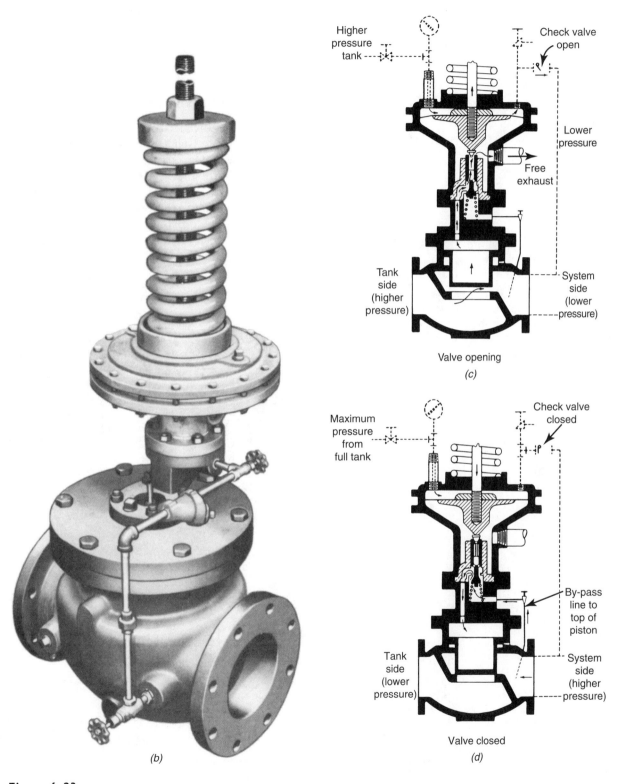

Higher pressure tank

Check valve open

Lower pressure

Free exhaust

Tank side (higher pressure)

System side (lower pressure)

Valve opening

(c)

Maximum pressure from full tank

Check valve closed

By-pass line to top of piston

Tank side (lower pressure)

System side (higher pressure)

Valve closed

(d)

(b)

Figure 6–23
(continued)

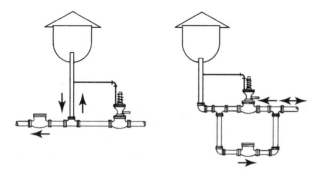

Figure 6–24

Typical arrangements for installing a single-acting altitude valve to automatically control water level in an elevated storage tank.

(Courtesy of GA Industries Inc., Golden-Anderson Valve Div.)

Figure 6–25

Solenoid pilot valve. (*a*) Solenoid angle valve in open position. (*b*) Solenoid angle valve in closed position. (*c*) Several kinds of electrical controls that may be used to operate a solenoid valve.

(Courtesy of GA Industries Inc., Golden-Anderson Valve Div.)

Refilling of the tank occurs when the butterfly control vane is open and the system pressure exceeds tank pressure. When the tank is full, the butterfly vane closes; thus, this kind of valve provides a double-acting operating sequence.

Solenoid Pilot Valves

The force required to change the position of a three-way valve that controls the operation of an automatic valve is very small and can be produced by a solenoid. A solenoid is a coil of wire wound in a helix so that when electric current flows through the wire a magnetic field is established within the helix. The force created by this magnetic field is sufficient to open or close a pilot valve. Figure 6–25

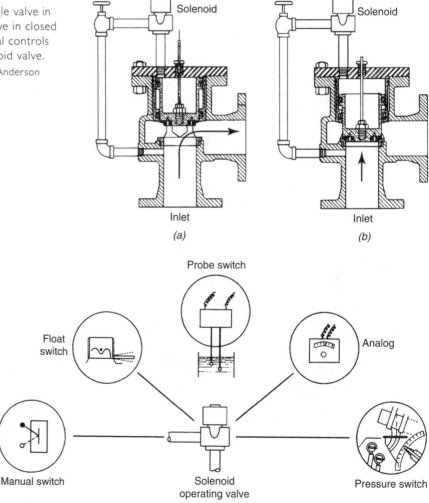

illustrates the operation of a solenoid pilot mounted on top of an angle valve. When the solenoid pilot is open, the chamber above the valve piston is vented to the atmosphere; simultaneously, a valve in the pipe from the inlet side of the angle valve is closed. This allows the piston in the angle valve to be held open by inlet water pressure. When the solenoid pilot is closed, the vent is sealed and pressure from the inlet water pushing on the larger top area of the piston forces it downward, closing the angle valve.

Several kinds of electrical controls may be used to energize a solenoid pilot valve. All of these allow operation of a valve from a remote location, which is an essential feature for telemetry and automation of a water distribution system. The simplest electrical control is a manual on-off switch. Automatic control can be achieved by using sensing devices to transmit a signal to the solenoid valve. Discrete level sensors provide only on or off, open or close control. Included in this category are storage-tank-level controls such as float, mercury, and on-off pressure switches.

Analog level sensors continuously measure the water level over a predetermined range. Examples of analog sensors are pressure-measuring diaphragms, long probes, and sonic signals that reflect back from the water surface. Of these, pressure-measuring devices using diaphragm deformation to measure pressure on a load cell,

or some other principle of measurement, are particularly advantageous since they can be used with both altitude valves and pressure-reducing valves. Analog level sensors can also be used to control the operation of pumps.

Air-Release Valves

Air can enter a pipe network from a pump drawing air into the suction pipe, through leaking joints, and by entrained or dissolved gases being released from the water. Air pockets increase the resistance to the flow of water by accumulating in the high points of distribution piping, in valve domes and fittings, and in discharge lines from pumps. Air-release valves are installed at these locations to discharge the trapped air. The common air-release valve shown in Figure 6–26 contains a ball that floats at the top of a cylinder, sealing a small opening. When air accumulates in the valve chamber, the ball drops away from the outlet orifice, allowing the air to escape. With return of the water level, the ball reseals the outlet.

6–9   BACKFLOW PREVENTERS

The water in a distribution system must be protected against contamination from backflow through customer service lines and other system

Figure 6–26
Small-orifice air-release valve and common locations of installation in a water distribution system (a) at the high points in pipelines and (b) on the domes of valves.
(Courtesy of GA Industries Inc., Golden-Anderson Valve Div.)

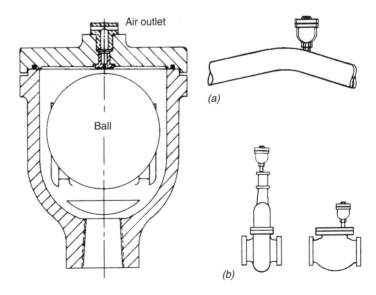

outlets. A cross-connection refers to an actual or potential connection between a potable water supply and an industrial or residential source of contamination. By practicing cross-connection control through enforcement of a plumbing code and inspection of backflow preventers in service connections, a municipal utility or private water purveyor can ensure against distribution contamination under foreseeable circumstances. Of greatest concern is backflow of toxic chemicals and wastewaters that may contain pathogens. The risk of back siphonage is reduced by maintaining adequate pressure in the supply mains to prevent reversal of flow. In undersized water distribution networks, low pressures can result from undersized piping, inadequate pumping capacity, or excessive peak water consumption.

Back siphonage is backflow resulting from negative or reduced pressure in the supply piping. Back siphonage can also result from a break in a pipe, repair of a water main at an elevation lower than the service point, and reduced pressure from the suction side of booster pumps. In contrast, back pressure causes reversal of flow when the pressure in a customer's service connection exceeds the pressure in the distribution main supplying the water. An example of backflow resulting from back pressure is chemically treated water from a malfunctioning boiler system being forced back through the feed line into the potable water supply.

The simplest method of preventing backflow is to provide an air space between the free-flowing discharge end of a supply pipe and an unpressurized receiving vessel. To have an acceptable air gap, the end of the discharge pipe has to be at least twice the diameter of the pipe above the highest rim of the receiving vessel, but in no case should this distance be less than 1 in. (Figure 6–27a). Although air gaps are common in small installations, the loss of water pressure as a result of physical separation in the supply line to a large facility requires repressurizing the system. Installation of a storage tank and pumps is often more costly than installing a mechanical backflow preventer that transmits supply pressure. In many countries, household water supplies are isolated from the public water-supply system by air-gap separation, using a storage tank in the attic or on the roof of the dwelling. Flow into the tank is controlled by a float valve in the inlet pipe, and an emergency overflow pipe ensures that failure of the inlet valve does not result in flow over the tank rim (Figure 6–27b). The disadvantage of this arrangement is that a direct pressure connection is necessary for supplying adequate pressure to operate some appliances, such as an automatic washing machine, and for lawn sprinkling.

The four kinds of mechanical backflow preventers are the atmospheric vacuum breaker, pressure vacuum breaker, double check valve, and reduced-pressure-principle device. The one selected depends on the type of installation and the hazard involved if backflow occurs. For direct water connections subject to back pressure, only the reduced-pressure-principle backflow preventer is considered adequate as an alternate to an air-gap separation. Still, air gaps are recommended in industrial applications where toxic chemicals are being mixed with the water, for example, vats containing acids and

Figure 6–27

Air-gap separation to prevent backflow into the supply pipe. (a) Minimum recommended air gap is twice the diameter of the supply pipe. (b) Storage tank with air-gap separation and overflow pipe to prevent overfilling the tank if supply valve fails to close tightly.

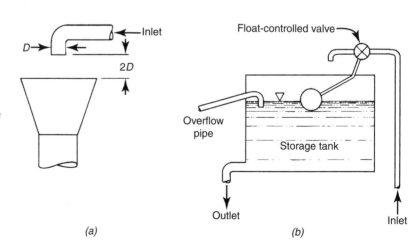

solutions for metal plating. For applications not subject to back pressure, vacuum breakers can be installed. The atmospheric vacuum breaker is used in flush valve toilets, and either an atmospheric or a pressure vacuum breaker may be connected to a lawn sprinkler system.

Vacuum and Pressure Vacuum Breakers

An atmospheric vacuum breaker has an inside moving element that prevents water from discharging through the top of the breaker during flow and drops down to provide a vent opening when flow stops. As diagramed in Figure 6–28a, the check-float element is open, raised by the pressure of water flowing up through the valve. When flow stops, the element drops onto the valve seat by force of gravity. This prevents back siphonage by allowing air to enter and break the siphon in the elevated pipe loop. A typical installation for lawn sprinklers is shown in Figure 6–28b. Another application of an atmospheric-vacuum breaker is

installation on the overhead pipe of a water filling station where water can be added to mobile chemical tanks to mix fertilizer, herbicide, or other chemical solutions. Since the hose attached to the end of the fill pipe can be submerged, back siphonage of the solution is possible unless the system is designed to prevent backflow. A vacuum breaker on the top of the riser pipe will break the siphon in the overhead pipe when the water supply is shut off, thus preventing backflow. An atmospheric breaker should not remain under pressure for long periods of time and cannot have a shutoff valve installed on the discharge side of the breaker.

A pressure vacuum breaker contains an assembly consisting of a spring-loaded check valve and a spring-loaded air valve, as shown in Figure 6–29. The check valve prevents backflow, and the air valve opens to admit air when the pressure within the body of the breaker approaches atmospheric pressure. If the check valve does not close tight because of interference from foreign matter, air is drawn in through the automatically operating vent

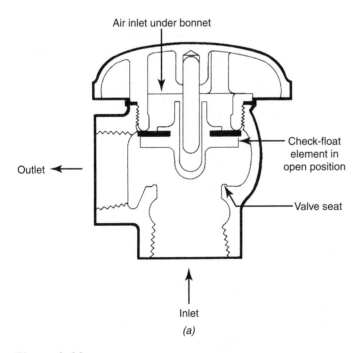

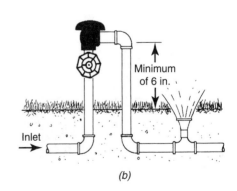

Figure 6–28

Atmospheric vacuum breaker. (a) The check-float element seals the air inlet with water flow through the unit. When flow stops, the element drops to cover the inlet and allows air to enter to break the siphon. (b) Typical installation on a sprinkler system.

(Courtesy of FEBCO® Backflow Prevention.)

Figure 6–29

Cross section of pressure vacuum breaker showing the spring-loaded check valve over the inlet and the spring-loaded air valve over the air inlet port under the canopy.

(Courtesy of FEBCO® Backflow Prevention.)

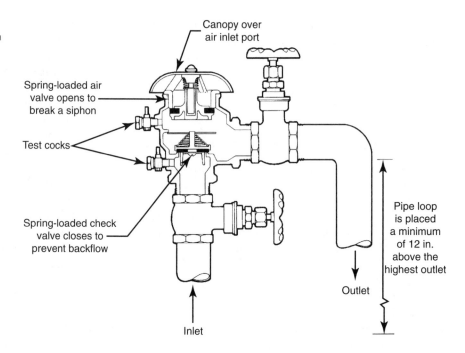

valve to preclude backflow. The breaker body is fitted with two test cocks for checking valve tightness against reverse flow and a shutoff valve on each side of the breaker. Pressure vacuum breakers are installed where low inlet water connections are not subject to back pressure; however, they may be attached to direct water connections subject to backflow provided the cross-connection is with water containing only nontoxic substances.

Double Check Valve Assembly

A check valve backflow preventer is composed of two single, independently acting check valves. An installation also has two tightly closing shutoff valves located on each side of the four test cocks for checking tightness of the valves against backflow. In Figure 6–30, both check valves are closed. As with pressure vacuum breakers, double check

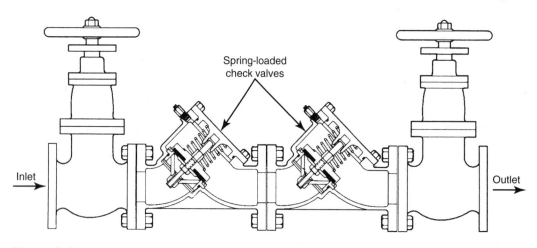

Figure 6–30

A double check valve backflow preventer consists of two spring-loaded check valves that operate independently.

(Courtesy of FEBCO® Backflow Prevention.)

valves are used to protect against backflow of non-toxic substances in direct water connections subject to backflow and to protect against backflow of nontoxic and toxic substances in water connections not subject to back pressure.

Reduced-Pressure-Principle Backflow Preventer

This backflow device consists of two independent, spring-loaded check valves with an automatically operating pressure-differential relief valve located between the two checks. The two isolating gate valves are for testing the operation of the three interior valves and to facilitate removal of the unit for maintenance. The four modes of operation described here are illustrated by diagrams in Figure 6–31.

1. **Normal Flow.** Both check valves are open, allowing flow through the preventer. Head loss through the first check valve reduces the pressure in the zone between the two valves by 5 to 11 psi below the supply pressure. Controlled by differential pressure, the relief valve is held closed by the greater supply pressure.
2. **Static Pressure.** With no flow through the unit, the check valves are held closed by force of

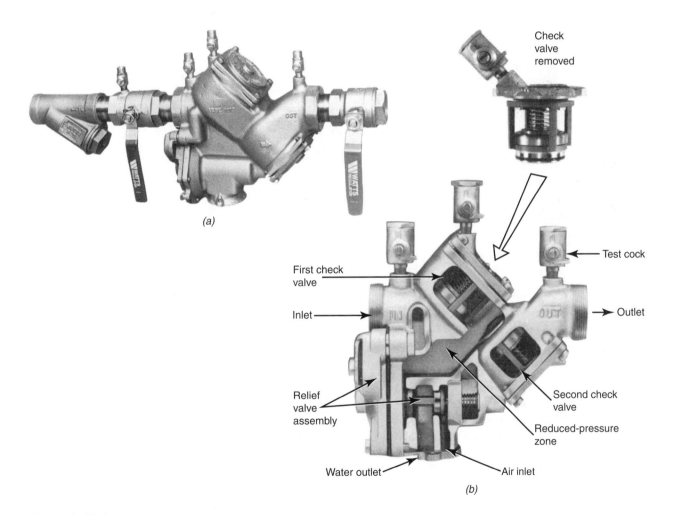

(a)

Check valve removed

First check valve

Inlet

Relief valve assembly

Test cock

Outlet

Second check valve

Reduced-pressure zone

Water outlet

Air inlet

(b)

Figure 6–31

Reduced-pressure-principle backflow preventer. (a) Outside view. (b) Cutaway section showing the three interior valves. (c) Illustrations of the four modes of operation.

(Courtesy of Watts Regulator Co.)

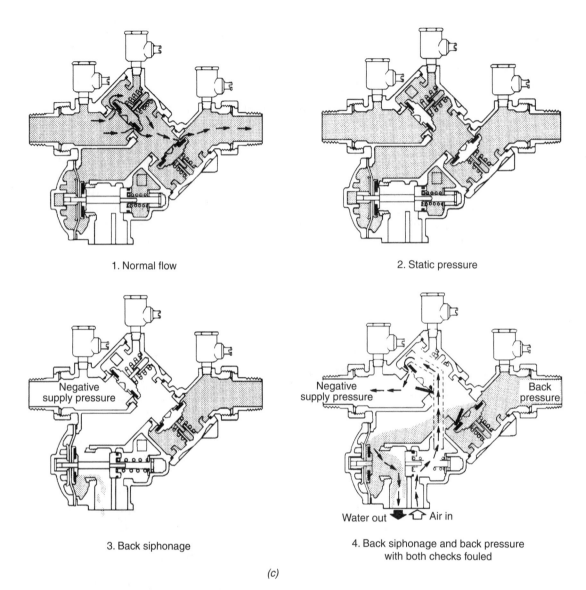

1. Normal flow

2. Static pressure

3. Back siphonage

4. Back siphonage and back pressure
with both checks fouled

(c)

Figure 6–31

(continued)

the valve springs. The relief valve is held closed by the pressure difference of 5 to 11 psi between the supply and the intermediate (reduced-pressure) zone.

3. **Back Siphonage.** With a negative supply pressure, both the check valves are closed. As a result of pressure equalization (loss of pressure differential) between the supply and the intermediate zone, the relief valve opens, allowing discharge of the water trapped in the intermediate zone. Thus, if any water seeps through the

second check valve from the discharge side, it drains out through the relief valve.

4. **Back Siphonage and Back Pressure with Both Checks Fouled.** As shown in diagram 4, the negative supply pressure causes the relief valve to open. Now, assume the worst condition of both check valves held partly open with pieces of pipe scale or other material. The condition permits a negative pressure in the intermediate zone and allows potentially contaminated water to enter this zone. Since the relief valve is double

seated, backflow entering through the second check valve is drained out, while the other half of the relief valve allows air to be drawn in by the negative supply pressure to enter the intermediate zone above the first check valve.

Reduced-pressure-principle backflow preventers are comparable in safety to air-gap separation. In addition, they have the advantage of transmitting water pressure to the user's piping system. This allows operation of water fixtures with main pressure, including an automatic sprinkler system for fire protection. Common preventer installations are in service connections to industries, chemical plants, hospitals, mortuaries, and irrigation systems.

All mechanical backflow preventers must be inspected periodically to ensure proper operation of the interior valves. The procedure involves attaching a differential pressure gauge to selected test cocks and closing one of the isolating shutoff valves to apply backflow water pressure to the discharge side of an internal valve. Then, by release of the water pressure from the inlet side of the valve, the tightness of the closure is checked by the ability of the valve to hold backflow pressure without leaking.

6–10 FIRE HYDRANTS

Hydrants provide access to underground water mains for the purposes of extinguishing fires, washing down streets, and flushing out water mains. Figure 6–32 shows the principal parts of a hydrant. The cast-iron barrel is fitted with outlets on top, and a shutoff valve at the base is operated by a long valve stem that terminates above the barrel. A typical unit has two $2\frac{1}{2}$-in.-diameter hose nozzles and one $4\frac{1}{2}$-in. pumper outlet for a suction line. The barrel and

Figure 6–32

Cutaway section of a fire hydrant and typical installation adjacent to a street.

(Reprinted by permission of Mueller Co., A Grinnell Company.)

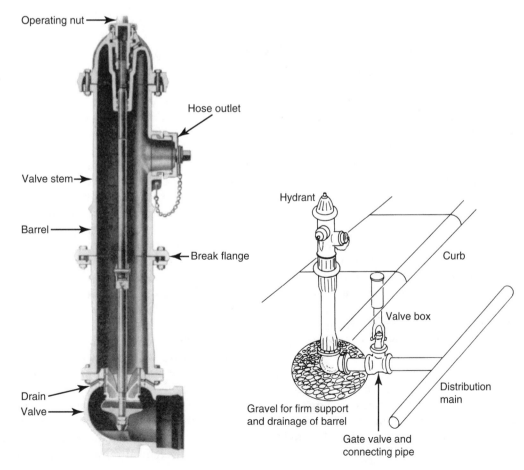

valve stem are designed such that accidental breaking of the barrel at ground level will not unseat the valve, thus preventing water loss. Hydrants are installed along streets behind the curb line a sufficient distance, usually 2 ft, to avoid damage from overhanging vehicles. The pipe connecting a hydrant to a distribution main is normally not less than 6 in. (150 mm) in diameter and includes a gate valve to allow isolation of the hydrant for maintenance purposes. A firm gravel or broken-rock footing is necessary to prevent settling and to permit drainage of water from the barrel after hydrant use. In cold climates the water remaining in a hydrant can freeze and break the barrel. Where groundwater stands at levels above the hydrant drain, generally the drain is plugged at the time of installation and, for service in cold climates, is pumped out after use.

6–11 DESIGN LAYOUTS OF DISTRIBUTION SYSTEMS

The arrangement of a water system is dictated by the source of water supply, the topography of the distribution area, and variations in water consumption. If water is supplied at only one point in a pipe network, elevated storage or ground-level storage with booster pumping is required in remote areas to maintain residual pressures. Also, the arterial mains between supply and storage must be large enough to transmit water without excessive head loss. On the other hand, if water enters a distribution network at several points from individual wells, the storage capacity can be reduced and the pipe sizes can be smaller because the water supply does not have to travel as far before consumption.

Topography can be the major factor in system design. Consider the two extremes, one where the water is supplied at a high elevation and flows by gravity through the pipe network and the other where the supply enters at a low elevation and must be pumped up into the pipe network and storage tanks. Distribution systems may have independent pipe networks at different elevations for pressure control; these may be completely separated or connected by pipes with pressure-reducing valves or through a common reservoir that acts as storage on the lower network and pumping suction for the upper system.

The pattern for water consumption is directly a function of industrial, commercial, and residential demands. For municipalities, land-use planning and zoning are commonly employed to control variations of water consumption in the distribution network. Climate has a definite effect, since lawn watering creates a major demand in residential areas in semiarid regions.

Each water distribution system is unique and is influenced by local conditions. Since pumping capacity, sizes of network pipes, and volume of storage are all interrelated, an increase in one of these can compensate for deficiencies in the others. Although founded on economic considerations, the options for expansion or modification of a system are governed to a considerable extent by the arrangement of existing facilities. Furthermore, the costs of construction and operation vary with time. For example, as a result of increasing energy costs, larger pipes to reduce pumping pressures and bigger storage reservoirs to equalize pumping rates are desirable options. In all cases, however, the primary engineering objectives are to provide a stable hydraulic gradient pattern for maintaining adequate pressure throughout the service area and sufficient pumping and storage capacities to meet fire demands and emergencies, such as main breaks and power failures.

The following simplified distribution systems illustrate the basic principles of design. In Figure 6-33a, an arterial pipe network for a small system is shown with a high-service pumping station and an elevated storage tank. The ground-level storage at the pumping station receives water after treatment or directly from wells. By providing larger mains between the pumping station and the elevated tank, an adequate quantity of stored water can be maintained to supply peak demands in the area of the system around the elevated tank. The hydraulic gradient over the system, produced by the high-service pumps, is supported by the water level in the elevated storage tank. In Figure 6–33b, wells distributed throughout the pipe network pump water directly into the system at several locations, allowing the installation of smaller-diameter pipes. The primary functions of the storage tank are to stabilize pressures and equalize pumping rates. Wells are operated as needed to provide water, and standby power is installed at a sufficient number of locations to meet demand in

Figure 6–33

Simplified layouts of distribution systems. (a) The high-service pumping and elevated storage tank support the hydraulic gradient over the distribution system. (b) Wells located throughout the pipe network support the hydraulic gradient.

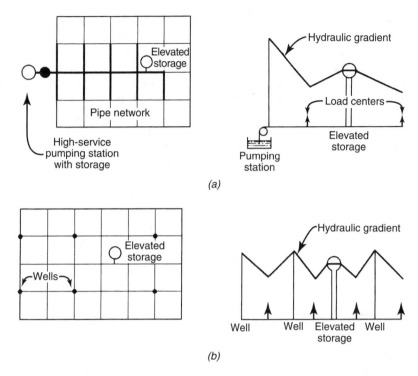

emergencies. Some wells are used during only part of the year, for example, during a dry summer season when consumption reaches a maximum. The hydraulic gradient is supported by the operating wells and the level in the elevated storage tank.

6–12 EVALUATION OF DISTRIBUTION SYSTEMS

Quantity

The supply source plus storage facilities should be capable of yielding enough water to meet both the current daily demands and the anticipated consumption 10 years hence. To quantitatively evaluate water usage, a record of average daily, peak daily, and peak hourly rates of consumption for the past 10 years should be recorded. Projected future needs can be estimated from these values and other factors relating to community growth.

The minimum amount of water available from a supply source should always be sufficient to ensure uninterrupted service. Consideration must be given to the probability of a succession of drought years equal to the previous worst drought

experience and the possibility of lowering groundwater levels. For surface supplies, the tributary watershed should be able to yield the estimated maximum daily demand 10 years in the future. As a general rule, the storage capacity of an impounding reservoir should be equal to at least 30 days of maximum daily demand 5 years into the future. Ideally, for well supplies there should be no mining of water; that is, neither the static groundwater level nor the specific capacity of the wells (gpm/ft of drawdown) should decrease appreciably as demand increases. Preferably, these values should be constant over a period of 5 years except for minor variations that correct themselves within 1 week.

Intake Capacity

A surface-water intake must be large enough to deliver sufficient water to meet municipal use and treatment plant needs (for example, filter backwashing) during any day of peak demand. With respect to fire flow requirements, if no storage is available, the intake capacity must be large enough to meet fire demand, maximum hourly flow, and in-plant processing needs simultaneously. On the other hand, where the quantity of distribution storage is sufficient to meet all fire flow requirements, the intake capacity involves comparing total storage of the system to the maximum amount of water needed, for both present and future demands. Intake facilities should be sized to meet maximum needs projected for at least 5 years into the future.

A water intake system must be reliable; it must be located, protected, or duplicated such that no interruption of service to customers or to fire protection occurs by reason of floods, ice, or other weather conditions or for reasons of breakdown, equipment repair, or power failure. In other words, intake facilities should be so reliable that no conceivable interruption in any part of the facility could cause curtailment of water-supply service to a community.

Pumping Capacity

In a typical surface-water supply system, low-lift pumps draw water from the source and transport it to the treatment plant. After processing, high-lift pumps deliver the water from clear-well storage to the distribution system. In the case of a groundwater supply, the well pumps deliver the raw water to treatment. However, if processing is not needed, the wells may discharge water directly to the transmission mains. In large communities, or in areas with widely varying elevations in topography, booster pumping stations may be needed to increase pressure in the distribution network and to extend the system for greater distances from the main pumping station.

With due consideration for the amount of water storage available, a pumping system must have sufficient capacity to provide the amount of water at pressures and flow rates needed to meet both daily and hourly peak demand with required fire flow. Pumping facilities must also be able to meet demands, taking into account common system failures and maintenance requirements. For example, the specified maximum pumping capacity should be obtainable with the largest pump out of service. In many instances, system storage, either elevated or ground level with booster pumping, is a component of the pumping system for meeting peak hourly demands.

Pumping facilities should be sufficiently reliable through duplication of units, standby equipment, and alternative sources of power such that no interruptions of service occur for any reason. In the event of power outage, standby power sources must be capable of sufficiently quick response to prevent exhaustion of water available in distribution storage during peak hourly demands, including required fire flow. Where unattended automatic booster stations exist, the control system should report back to a central station both the condition of operation and any departure from normal.

Piping Network

Arterial and secondary feeder mains should be designed to supply water service for 40 or more years after installation. Actual useful service life of mains under normal conditions is 50 to 100 years. Submains should be at least 6 in. in diameter in residential districts, and the minimum size in important districts should be 8 in. in diameter with 12-in. intersecting mains. Distribution lines are laid out in a gridiron pattern, avoiding dead ends by proper looping. An adequate number of valves should be inserted to permit shutoff in case of a break so that no more than one block is out of service. The distribution of hydrants is based on Insurance Services Office standards.

REFERENCE

1. *Guide for Determination of Needed Fire Flow*, 2004. Insurance Services Office, Inc., 545 Washington Boulevard, Jersey City, NJ 07310-1686, 2004 (www.iso.com).
2. *Distribution System Requirements for Fire Protection (M31)*, 3rd Ed., 1998. American Water Works Association, 6666 West Quincy Avenue, Denver, CO 80235-3098 (www.awwa.org).

PROBLEMS

6–1 What is the average municipal water use in the United States? In what region is water use greatest for residential customers, and in that region what is a major influence on excess usage?

6–2 Based on demand of 100 gpcd, estimate the maximum daily demand and the mean maximum hourly rate.

6–3 What is the range of water pressure recommended in a distribution system? What are the minimum and maximum pressures for residential service connections?

6–4 Define needed fire flow (NFF). Based on Eq. 6–4, list the factors taken into consideration in determining NFF.

6–5 Recalculate the NFFs for the buildings in Example 6–1 with the following modifications: (a) The two buildings are 65 ft apart. (b) The food and beverage sales building is a one-story, wood-frame structure, and the walls of the two buildings are 10 ft apart. [*Answers* (a) 2500 gpm, 2 hr, 2000 gpm, 2 hr, (b) 3500 gpm, 3 hr, 4000 gpm, 3 hr]

6–6 Calculate the C_i and duration required for a 210-ft by 450-ft five-story residential condominium. The structure is of fire-resistive construction with all interior vertical openings protected by self-closing class B fire doors. A second identical building is located next to this building with the 450-ft walls containing unprotected openings facing each other 90 ft apart, but without a communication passageway. Calculate the NFF including exposure of the second building. If two hydrants are within 120 ft of the building, does this reduce the NFF? Are automatic sprinkler systems in the buildings justified?

6–7 A two-story restaurant, 115 ft by 105 ft, is of masonry (noncombustible) construction. A one-story office building, 75 ft by 55 ft, of masonry (noncombustible) construction is adjacent to the restaurant. The 115-ft wall of the restaurant with an entrance of combustible construction (open communication) faces the 55-ft blank brick wall of the office building at a distance of 48 ft. Two fire hydrants, each with one pumper and two hose connections, are located 200 ft and 500 ft from the restaurant and 350 ft and 450 ft from the office building. Calculate the NFF for each building, determine the duration required, and evaluate the proximity of the fire hydrants.

6–8 A one-story supermarket of ordinary construction is 130 m by 30 m with a 3.5-m story height. An adjacent two-story office building of ordinary construction with open vertical stairways is 30 m by 20 m. The 30-m supermarket wall facing the office building is a blank wall. The 20-m office building wall facing the supermarket has windows and doors of combustible construction (unprotected openings). The adjacent walls are 6.0 m apart. Calculate the NFF for each building including exposure of the other building.

6–9 What is the NFF for a residential area of single- and two-story family houses of ordinary construction with asphalt roofing? The minimum distance between dwellings is 20 ft. What is the required spacing of fire hydrants in this residential area?

6–10 What is the maximum fire flow that most municipalities are likely to be able to reliably provide for fire fighting? For major structures with needed fire flows greater than this practical limit, how can adequate fire suppression be provided?

6–11 In addition to water-supply capacity, what other major components of a water system are evaluated for reliability in meeting fire demand?

6–12 A well is drilled by the hydraulic rotary method into an unconsolidated sand aquifer. How is the borehole held open during drilling? How is the well developed before being placed in service? After the well is placed in service, what prevents sand grains from the aquifer passing through the screen openings with the pumped water?

6–13 How does the cylindrical intake in Figure 6–2 repel fish when installed in a flowing surface water?

6–14 What kinds of centrifugal pumps are commonly installed in low-lift pumping stations? How are pumps in shore intake protected from debris in the river water?

6–15 Why is a gridiron pipe network the best arrangement for distributing water? What minimum-size mains are used in residential districts? In commercial districts? (Refer to Section 6–5 and Section 6–2, subsection Distribution System.)

6–16 List the major components of a house water service connection.

6–17 Where are the ductile-iron pipe joints illustrated in Figure 6–7 used? In Figure 6–8? How is PVC distribution piping joined? What kinds of plastic pipe are used for service connections and household plumbing? Describe the applications for the three types of reinforced concrete pipe.

6–18 Discharge pressure and flow measurements were recorded under a variety of pumping conditions at a city's high-lift station in order to gather data to draw pump head-discharge curves. The following is a summary of the results:

NUMBER OF PUMPS OPERATING	DISCHARGE HEAD (FT)	PUMPING RATE (MGD)
3	210	3.0
3	175	4.6
3	125	6.0
4	200	5.0
4	150	6.8
5	220	6.0
5	200	7.2
5	150	9.0

(a) Plot these data as three pump head-discharge curves on graph paper, as discussed in Section 4–4. (b) Assume these curves are for the pumping station in Figure 6–16. Plot a system head-discharge curve for the conditions shown in Figure 6–16 on the same graph, as discussed in Section 4–5. (c) At maximum daily flow (6.0 mgd), how many pumps are operating and what is the discharge pressure? (d) Assume the station is operating with four pumps at a discharge head of 175 ft and 6.0 mgd. What happens to the discharge flow and pressure conditions if a fifth pump is turned on? If one of the four pumps is turned off?

6–19 A new vertical-turbine well pump is being installed to discharge groundwater directly into the pipe network of a small community. Operating head-discharge data based on specifications from the pump manufacturer are 210 gpm discharge at a pump head of 170 ft measured from ground level, 300 gpm at 165 ft, 410 gpm at 150 ft, and 470 gpm at 130 ft. The water pressure at ground level into the pipe network varies from 71 psi to 48 psi, with an average pressure of 64 psi. Plot the head-discharge curve and locate the pressure range and average value. What is the pump discharge against the average pressure of 64 psi?

6–20 Discharge pressure and flow measurements were recorded at a city's high-lift station under a variety of pumping conditions at different

system demands and varying water levels in the elevated storage tank. Based on these measurements, the system head-discharge curves were defined as follows:

		SYSTEM HEADS AT DIFFERENT WATER LEVELS IN ELEVATED STORAGE		
CONSUMPTION	SYSTEM DISCHARGE (l/s)	LOW (m)	MID-HEIGHT (m)	FULL (m)
(Static pressure)	0	30	35	40
Winter average	140	37	42	47
Annual average	180	40	45	50
Summer average	225	45	50	55
Maximum day	270	50	55	60
Maximum day and fire flow	350	60	65	70

The pumping station has a total of six constant-speed pumps. Numbers 1, 2, and 3 are identical with a best efficiency point (bep) at 100 l/s and 45 m; 60 percent bep is 60 l/s and 60 m; and 120 percent bep is 120 l/s and 30 m. Numbers 4, 5, and 6 are identical with a bep at 150 l/s and 60 m; 60 percent bep is 90 l/s and 80 m; and 120 percent bep is 180 l/s and 35 m. Draw the pump head-discharge curves for the following combinations of pumps operating in parallel: pump 1 alone; pumps 1 and 2; pumps 1, 2, and 3; pump 4 alone; pumps 1 and 4; pumps 4 and 5; and pumps 4, 5, and 6. Draw the system head-discharge curves on the same graph.

What pumps would normally be operated during days when the consumption is at (a) winter average, (b) annual average, (c) summer average, and (d) the maximum day? Assume that the volume of the storage tank is adequate to equalize the pumping rate on the maximum day with reserve storage for fire demand. Is the pumping capacity of the station adequate?

6–21 From the results of Example 6–3, the storage volume needed to equalize the 24-hr pumping rate on the day of maximum water usage is 500,000 gal. What percentage of the total consumption for that day is this required storage capacity? How does this value compare to the rule-of-thumb percentages given in Section 6–7 for balancing pumping?

6–22 The water usage on the day of maximum consumption is as follows:

Time	Hourly Rate (gpm)	Time	Hourly Rate (gpm)
12 P.M.	—	1 P.M.	6600
1 A.M.	2200	2	6400
2	2100	3	6300
3	1800	4	6400
4	1400	5	6400
5	1300	6	6700
6	1200	7	7400
7	2000	8	9200
8	3500	9	8400
9	5000	10	5000
10	6000	11	3200
11	6400	12	2800
12	7000		

Plot a consumption-time curve like that shown in Figure 6–13. Calculate hourly cumulative consumption values and plot a mass diagram as illustrated in Figure 6–14. (a) What is the constant 24-hr pumping rate and required storage capacity to equalize demand over the 24-hr period? Calculate the required storage capacity as a percentage of the total cumulative consumption. (b) What is the constant 12-hr pumping rate and required storage capacity to equalize demand during a pumping period between 6 A.M. and 6 P.M.? Calculate the required storage capacity as a percentage of the total cumulative consumption.

6–23 What are the principal functions of distribution storage? In evaluating water storage in a distribution system, what amount of the storage capacity is considered to be available for fire fighting?

6–24 The water system for a town with a population of 900 has four wells that have yields of 400 gpm, 400 gpm, 600 gpm, and 800 gpm, and an elevated storage tank with a capacity of 100,000 gal. The annual average water consumption is 120,000 gpd, and the maximum daily usage is 280,000 gal. The required fire flow for the commercial district is 1500 gpm with a duration of 2 hr. (a) Assuming this town has no large industrial water users, do the average and maximum daily consumptions appear to be reasonable? (b) Assume the piping in the distribution network is adequate to convey water from the wells and elevated storage to the commercial district under conditions of peak demand without excess pressure losses. Are the well pumps in combination with distribution storage able to meet the specified fire demand with no insurance deficiency?

6–25 What is the design feature of a fire hydrant that prevents the discharge of water if the barrel is struck and broken by a motor vehicle? Why is a hydrant inspected after use in a cold climate?

6–26 Why is a cushion chamber with a counterweight arm installed on a check valve?

6–27 Refer to the pressure-reducing valve in Figure 6–20. Explain the hydraulic action of the valve that results in reduction of water pressure.

6–28 Refer to the pressure-reducing valve in Figure 6–22. Assume that this automatic valve is installed in a pipeline and is adjusted to reduce pressure from 90 psi to 70 psi. In what direction should the handwheel be turned to decrease the outlet pressure to 60 psi? Explain the resulting change in operation of the valve.

6–29 The pilot valve on top of the double-acting altitude valve illustrated in Figure 6–23 is a three-way valve. With reference to both Figure 6–21 and Figure 6–23, explain how the pilot valve automatically opens and closes the globe valve to maintain the water level in the elevated storage tank below the preset maximum elevation.

6–30 The operation of high-service pumps can be controlled by the water level in an elevated storage tank in the distribution system. As the water level in the tank lowers below predetermined elevations, additional high-service pumps can be turned on; conversely, as the water level rises above preset elevations, selected pumps can be turned off. How can this kind of control be achieved?

6–31 What kind of backflow preventers are usually installed on residential lawn sprinkler systems? On large irrigation systems?

6–32 A large house is being converted into a mortuary. What special provisions should be considered in modifying the current residential water service connection?

6–33 Explain why the backflow preventer illustrated in Figure 6–31 is referred to as a reduced-pressure-principle backflow preventer. For a large building such as a hospital, why is installation of a reduced-pressure-principle backflow preventer preferred to air-gap separation?

6–34 A homeowner has requested installation of a pressure vacuum breaker (Figure 6–29) in the basement to prevent back siphonage from his lawn sprinker system. His request was refused. Why? What kind of backflow preventer is required for installation in the basement at an elevation below the sprinker system?

6–35 In Figure 6–33*a*, describe the benefits of locating elevated storage near the load centers away from the pumping station. In Figure 6–33*b*, how is the hydraulic gradient supported during high water consumption?

Water Processing

The objective of municipal water treatment is to provide a potable supply—one that is chemically and microbiologically safe for human consumption. For domestic uses, treated water must be aesthetically acceptable—free from apparent turbidity, color, odor, and objectionable taste. Quality requirements for industrial uses are frequently more stringent than for domestic supplies. Thus, additional treatment may be required by the industry. For an example, boiler feed water must be demineralized to prevent scale deposits.

Common water sources for municipal supplies are deep wells, shallow wells, rivers, natural lakes, and reservoirs.

Pollution and eutrophication are major concerns in surface-water supplies. Water quality depends on agricultural practices in the watershed, location of municipal and industrial outfall sewers, river development such as dams, season of the year, and climatic conditions. Periods of high rainfall flush silt and organic matter from cultivated fields and forest land, and drought flows may result in higher concentrations of wastewater pollutants from sewer discharges. River temperature may vary significantly between summer and winter. The quality of water in a lake or reservoir depends considerably on the season of the year. Municipal water-quality control actually starts with management of the river basin to protect the source of water supply. Highly polluted waters are both difficult and costly to treat. Although some communities are able to locate groundwater supplies, or alternate less polluted surface sources within feasible pumping distance, the majority of the nation's population draws from nearby surface supplies. The challenge in waterworks operation is to process these waters to a safe, potable product acceptable for domestic use.

7–1 SURFACE-WATER TREATMENT

The primary process in surface-water treatment is chemical clarification by coagulation, sedimentation, and filtration, as illustrated in Figure 7–1. Lake and reservoir water has a more uniform year-round quality and requires a lesser degree of treatment than river water. Natural purification results

Figure 7–1

Schematic patterns of traditional surface-water treatment systems.

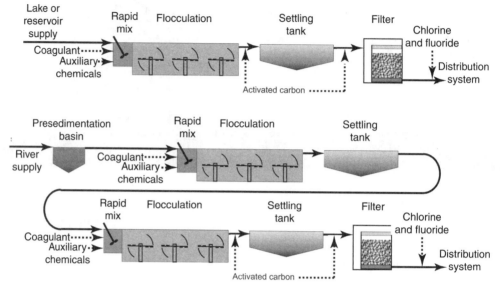

in reduction of turbidity, coliform bacteria, and color, and elimination of day-to-day variations. On the other hand, growths of algae cause increased turbidity and may produce difficult-to-remove tastes and odors during the summer and fall. The specific chemicals applied in coagulation for turbidity removal depend on the character of the water and economic considerations. The most popular coagulant is alum (aluminum sulfate). As a flocculation aid, the common auxiliary chemical is a synthetic polymer. Activated carbon is applied to remove taste- and odor-producing compounds. Chlorine and fluoride are posttreatment chemicals. Prechlorination may be used for disinfection of the raw water only if it does not result in formation of disinfection by-products.

River supplies normally require the most extensive treatment facilities with greatest operational flexibility to handle the day-to-day variations in raw water quality. The preliminary step is often presedimentation to reduce silt and settleable organic matter prior to chemical treatment. As illustrated in Figure 7–1, many river-water treatment plants have two stages of chemical coagulation and sedimentation to provide greater depth and flexibility of treatment. The units may be operated in series or by split treatment, with softening in one stage and coagulation in the other. As many as a dozen different chemicals may be used under varying operating conditions to provide a satisfactory finished water.

7–2 MIXING AND FLOCCULATION

In water treatment, rapid mixing provides dispersion of chemicals so that dissolution occurs in 10 to 30 sec. Most common is mechanical mixing, as illustrated in Figure 7–2, using a vertical shaft impeller in a tank with stator baffles to reduce vortexing about the impeller shaft. Vortexing hinders mixing and reduces impeller power usage. Other methods of rapid mixing are a suspended propeller, a static mixing element inserted into a pipe for in-line mixing, and hydraulic mixing, such as injection of chemicals into the inlet of a centrifugal pump.

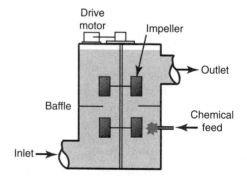

Figure 7–2

Rapid mixing chamber with impellers and baffles for rapid dispersion of chemicals into raw water prior to flocculation.

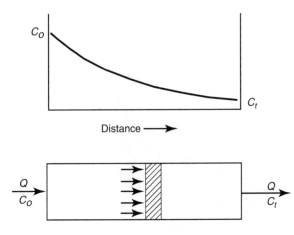

Figure 7–3
Ideal plug flow. Symbols are defined in Eq. 7–1.

In an ideal plug-flow system, the water flows through a long chamber at a uniform rate without intermixing. The concentration of reactant decreases along the direction of flow, remaining within the imaginary plug of water moving through the basin (Figure 7–3). For steady-state conditions, the relationship between detention time and concentration, applying first-order kinetics, is given in Eq. 7–1. In practice, ideal plug flow is very difficult to achieve because of short-circuiting and intermixing caused by frictional resistance along the walls, density currents, and turbulent flow. Flocculation basins in chemical water processing are designed as plug-flow units using baffles to reduce short-circuiting.

$$t = \frac{V}{Q} = \frac{L}{v} = \frac{1}{k}\left(\log_e\frac{C_0}{C_t}\right) \qquad (7\text{–}1)$$

where t = detention time
V = volume of basin
Q = quantity of flow
L = length of rectangular basin
v = horizontal velocity of flow
k = rate constant for first-order kinetics
C_0 = influent reactant concentration
C_t = effluent reactant concentration

An in-line treatment plant consisting of rapid mix, flocculation, and sedimentation is illustrated in Figure 7–4. Raw water withdrawn from the reservoir visible at the top of the picture is pumped through two rapid mixers for chemical

addition and into two flocculation tanks. When this picture was taken, only the left half of the plant was in service. The flocculation tank arrangement is horizontal-shaft wooden paddles in a series of compartments separated by baffles to direct water flow through the paddle flocculators (Figure 7–4b). Figure 7–4c shows a paddle flocculator. During flocculation, the chemically treated water is given gentle, slow mixing to build large settleable floc. The rotational speed of each stage of flocculators can be controlled by a variable-speed drive for optimum slow mixing. The water then passes through a slotted baffle wall into the sedimentation tank, where it slowly flows to overflow channels at the discharge end (Figure 7–4d). Since this plant is located in a warm climate, the tanks are outside and the settled sludge is scraped to the discharge end by a blade supported by a bridge that spans the tank and runs on rails on the side walls.

The *Recommended Standards for Water Works*[1] recommends that rapid mixing for dispersion of chemicals throughout the water be performed with mechanical mixing devices with a detention time not more than 30 sec. These recommendations are made for flocculation basins:

1. Inlet and outlet design shall prevent short-circuiting and destruction of floc.
2. Minimum flow-through velocity shall not be less than 0.5 nor greater than 1.5 ft/min (2.5 to 7.5 mm/s) with a detention time for floc formation of at least 30 min.
3. Agitators shall be driven by variable-speed drives with the peripheral speed of paddles ranging from 0.5 to 3.0 ft/sec (0.15 to 0.91 m/s). Flocculation and sedimentation basins shall be as close together as possible. The velocity of flocculated water through conduits to settling basins shall not be less than 0.5 nor greater than 1.5 ft/sec (0.15 to 0.45 m/s). Allowances must be made to reduce turbulence at bends and changes in direction.

■ EXAMPLE 7–1

Based on laboratory studies, the rate constant for a chemical coagulation reaction was found to be first-order kinetics with a k equal to 75 per day.

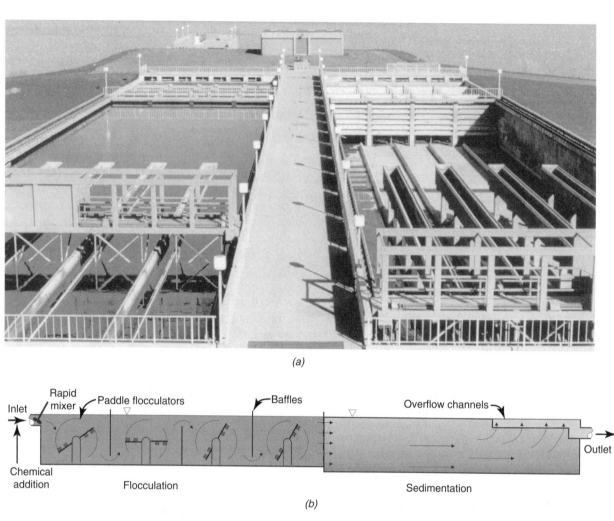

(a)

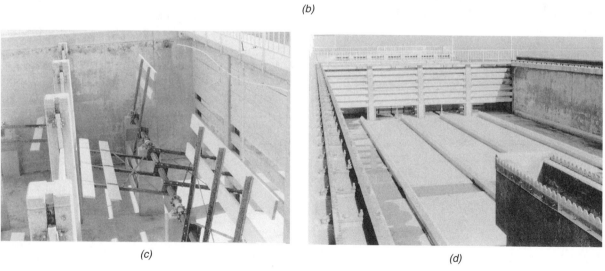

(b)

(c)   *(d)*

Figure 7–4

In-line rapid mixing, flocculation, and sedimentation in water treatment.

Calculate the detention time required in a plug-flow reactor for an 80 percent reduction, $C_0 = 200$ mg/l and $C_t = 40$ mg/l.

Solution

For plug flow, substituting into Eq. 7–1,

$$t = \frac{1440}{75}\left(\log_e\frac{200}{40}\right) = 19.2\ (2.3 \log 5)$$

$$= 19.2 \times 2.3 \times 0.7 = 31 \text{ min}$$

■ ■ ■

7–3   SEDIMENTATION

Sedimentation, or clarification, is the removal of particulate matter, chemical floc, and precipitates from suspension through gravity settling. The common criteria for sizing settling basins are detention time; overflow rate; weir loading; and, with rectangular tanks, horizontal velocity. Detention time, expressed in hours, is calculated by dividing the basin volume by average daily flow, Eq. 7–2.

$$t = \frac{V \times 24}{Q} \qquad (7\text{–}2)$$

where t = detention time, hours
V = basin volume, million gallons (cubic meters)
Q = average daily flow, million gallons per day (cubic meters per day)
24 = number of hours per day

The overflow rate (surface loading) is equal to the average daily flow divided by total surface area of the settling basin, expressed in units of gallons per day per square foot, Eq. 7–3.

$$V_0 = \frac{Q}{A} \qquad (7\text{–}3)$$

where V_0 = overflow rate (surface loading), gallons per day per square foot (cubic meters per square meter per day)

Q = average daily flow, gallons per day (cubic meters per day)
A = total surface of basin, square feet (square meters)

Most settling basins in water treatment are essentially upflow clarifiers where the water rises vertically for discharge through effluent channels; hence, the ideal basin shown in Figure 7–5 can be used for explanatory purposes. Water flows horizontally through the basin and then rises vertically, overflowing the weir of a discharge channel at the tank surface. Flocculated particles settle downward, in a direction opposite to the upflow of water, and are removed from the bottom by a continuous mechanical sludge removal apparatus. The particles with a settling velocity v greater than the overflow rate Q/A are removed, while lighter flocs, with settling velocities less than the overflow rate, are carried out in the basin effluent.

Weir loading is computed by dividing the average daily quantity of flow by the total effluent weir length and expressing the results in gallons per day per foot (cubic meters per meter per day).

Sedimentation basins, either circular or rectangular, are designed for slow uniform water movement with a minimum of short-circuiting. In a circular tank, as diagramed in Figure 7–6a, the influent enters through a vertical riser pipe in the

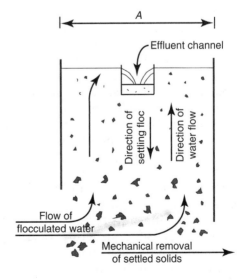

Figure 7–5
Ideal sedimentation tank. Symbols are defined in Eq. 7–3.

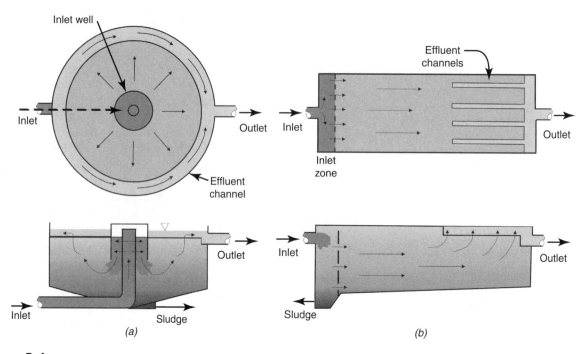

Figure 7–6

Plan and cross-sectional diagrams of sedimentation tanks used in water treatment. *(a)* In a circular tank, water enters behind a central inlet well and flows radially to an effluent channel around the perimeter of the tank. *(b)* In a rectangular tank, water enters an inlet zone for distribution, flows horizontally through the tank, and rises to overflow into multiple effluent channels.

center with outlet ports discharging behind a circular inlet well. This baffle dissipates the horizontal velocity by directing the flow downward. Radial flow from the center is collected into the effluent channel attached to the outside wall by overflowing a V-notch weir mounted along the edge of the channel. The settled solids that accumulate on the bottom are gently pushed by scraper blades into a central hopper for discharge. The sludge scraper can be driven by a central turntable supported by a bridge or pier. In a warm climate without ice and snow, it can be driven around by a wheel running on the peripheral wall of the tank.

In water treatment, sedimentation basins are more commonly rectangular, as diagramed in Figure 7–6b. The influent is spread across the end of the basin by a baffle structure to dissipate velocity and uniformly distribute the flow. The water passes horizontally to effluent overflow channels with V-notch weirs mounted along the edges. The effluent channels are commonly long "finger" channels extending into the basin to increase weir length and, thus, reduce the vertical velocity as the water rises up to the overflow channels. Sludge is removed by flat scrapers attached to continuous chains that are supported and driven by sprocket wheels mounted on the inside walls of the tank. In a warm climate, the scrapers can be hung in the water from an overhead bridge that spans the width of the tank. The wheels of the bridge roll on steel rails mounted on the side walls (Figure 7–4).

For sedimentation basins following chemical flocculation, the *Standards for Water Works* recommend a detention time of not less than 4 hr, which may be reduced 2 hr by approval when equivalent effective settling is demonstrated. The maximum recommended horizontal velocity through the sedimentation basin is 0.5 ft/min (2.5 mm/s), and the maximum weir loading is 20,000 gpd/ft of weir length (250 m^3/m · d). The overflow rate is generally in the range of 500 to 800 gpd/sq ft (20 to 33 m^3/m^2 · d).

Presedimentation basins may be installed to settle out heavy solids from muddy river water prior to chemical flocculation and sedimentation. Generally, circular sedimentation tanks with hopper bottoms and scraper arms for heavy sludge are used. The recommended *Standard* is a detention time of not less than 3 hr.

Residence-Time Distribution

The hydraulic character of an actual tank is defined by the residence-time distribution of individual particles of the liquid flowing through the tank. Since routes through the tank differ in travel times, the effluent age distribution of a nonideal reactor may extend from less than to greater than the theoretical detention time as calculated in Eq. 7–2.

Evaluation of an actual process tank (flocculation, sedimentation, or chlorination) is performed by introducing a tracer to the influent during steady-state flow and measuring the concentration in the output over an extended period of time. The general shapes of residence-time distributions for a dispersed plug-flow reactor are illustrated in Figure 7–7. If the tracer input is suddenly initiated and then held at a constant application rate, the tracer output curve of concentration in the effluent slowly rises and approaches the influent tracer concentration, C_0, at a time interval greater than the mean residence time, t_R. Applying a pulse input of dye solution results in an effluent tracer concentration curve as shown in Figure 7–7b. The centroid of the curve is located to the left of time t_{50} as a result of flow dispersion with backmixing and short-circuiting of liquid in the tank because of stagnant pockets.

The mean residence time, calculated from experimental data of a pulse tracer input, equals the summation of time-concentration values divided by the summation of the concentrations.

$$t_{50} = \frac{\sum t_i C_i}{\sum C_i} \qquad (7\text{-}4)$$

where t_{50} = mean residence time (time to centroid)
$\sum$ = summation
t_i = time from injection of dye to collection of effluent sample
C_i = concentration of dye in effluent sample collected at t_i

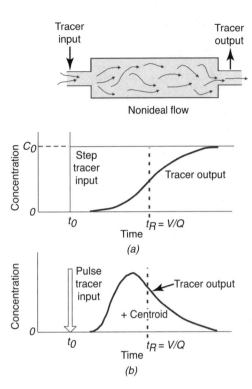

Figure 7–7
Output time-distribution curves for flow through a nonideal tank in response to (*a*) a continuous tracer input, and (*b*) a pulse tracer input.

■ EXAMPLE 7–2

Each half of the in-line treatment plant illustrated in Figure 7–4 has the following sized units: rapid mixing chamber with a volume of 855 cu ft—flocculation tank 140 ft wide, 58 ft long, and 14.5 ft in liquid depth; and a sedimentation tank 140 ft wide, 280 ft long, and 17.0 ft in depth. The length of the effluent weir along the four channels and the end of the tank is 1260 ft. Calculate the major parameters used in sizing these units based on a design flow of 40.0 mgd for each half of the plant. Compare the calculated values to the *Standards for Water Works*.

Solution

Flow = 40.0 mgd = 2780 gpm
    = 5,348,000 cu ft/day
    = 3710 cu ft/min

Detention time in rapid mixer:

$$t = \frac{855 \text{ cu ft}}{3710 \text{ cu ft/min}} \times 60 \frac{\text{sec}}{\text{min}}$$

$$= 14 \text{ sec } (<30 \text{ sec OK})$$

Detention time and horizontal velocity in flocculation:

$$t = \frac{140 \text{ ft} \times 58 \text{ ft} \times 14.5 \text{ ft}}{3710 \text{ cu ft/min}}$$

$$= 32 \text{ min } (>30 \text{ min OK})$$

$$v = \frac{Q}{A} = \frac{3710 \text{ cu ft/min}}{140 \text{ ft} \times 14.5 \text{ ft}}$$

$$= 1.8 \text{ ft/min } (>1.5 \text{ ft/min})$$

Sedimentation time and horizontal velocity:

$$t = \frac{140 \text{ ft} \times 280 \text{ ft} \times 17.0 \text{ ft}}{3710 \text{ cu ft/min}} = 180 \text{ min}$$

$$= 3.0 \text{ hr } (<4.0 \text{ hr})$$

$$v = \frac{Q}{A} = \frac{3710 \text{ cu ft/min}}{140 \text{ ft} \times 17.0 \text{ ft}}$$

$$= 1.6 \text{ ft/min } (>0.5 \text{ ft/min})$$

$$\text{Weir loading} = \frac{40,000,000 \text{ gal/day}}{1260 \text{ ft}}$$

$$= 32,000 \text{ gpd/ft } (<20,000 \text{ gpd/ft})$$

$$\text{Overflow rate } V_0 = \frac{40,000,000 \text{ gal/day}}{140 \text{ ft} \times 280 \text{ ft}}$$

$$= 1020 \text{ gpd/sq ft}$$

■ ■ ■

■ EXAMPLE 7–3

Dispersed plug flow through a baffled flocculation tank (Figure 7–4b) was analyzed to determine the residence-time distribution curve (Figure 7–8) by injecting a pulse of dye tracer into the influent and

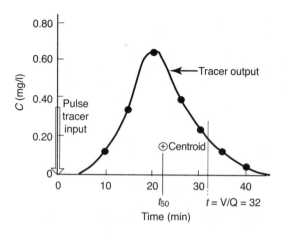

Figure 7–8
Residence-time distribution curve for Example 7–3.

measuring the tracer concentrations in the effluent at time intervals of 5 min after injection, as listed below. Sketch the tracer distribution curve by plotting C versus t and locate the centroid based on the following data:

t_i (min)	5	10	15	20	25	30	35	40	45
C_i (mg/l)	0.00	0.10	0.35	0.65	0.40	0.22	0.11	0.05	0.00

Solution

From these tracer data, calculate the t_{50} using Eq. 7–4.

t_i (min)	C_i (mg/l)	$t_i C_i$ (min · mg/l)
5	0.00	0.0
10	0.10	1.0
15	0.35	5.2
20	0.65	13.0
25	0.40	10.0
30	0.22	6.6
35	0.11	3.8
40	0.05	2.0
45	0.00	0.0
	1.88	41.6

Centroid, $t_{50} = 41.6/1.88 = 22$ min

The calculated detention time from Example 7–2 is 32 min.

■ ■ ■

7–4   DIRECT FILTRATION

The process of direct filtration does not include sedimentation prior to filtration. The impurities removed from the water are collected and stored in the filter. Although rapid mixing of chemicals is necessary, the flocculation stage is either eliminated or reduced to a mixing time of less than 30 min. Contact flocculation of the chemically coagulated particles in the water takes place in the granular media. Successful advances in direct filtration are attributed to the development of coarse-to-fine multimedia filters with greater capacity for "in-depth" filtration, improved backwashing systems using mechanical or air agitation to aid cleaning of the media, and the availability of better polymer coagulants.

Surface waters with low turbidity and low color both less than 25 units are most suitable for processing by direct filtration. Nevertheless, water with low color and higher turbidity or low turbidity with higher color may be satisfactory. Operational problems are expected when color exceeds 40 units or turbidity is greater than 15 NTU (Nephelometric Turbidity Unit) on a continuous basis. The feasibility of filtration without prior flocculation and sedimentation relies on a comprehensive review of water-quality data. The incidence of high turbidities caused by runoff from storms and blooms of algae must be evaluated. Often, pilot testing is valuable in determining the efficiency of direct filtration compared to conventional treatment, design of filter media, and selection of chemical conditioning. Filtration rates in direct filtration are usually in the range of 5 to 15 gpm/sq ft (3.4 to 10.2 l/m$^2 \cdot$ s), but tend to be nearer 5 gpm/sq ft. The major advantage of direct filtration is the reduced capital costs for flocculation and sedimentation. Also with direct filtration, chemical costs may be significantly reduced compared to traditional treatment.

7–5   BALLASTED FLOCCULATION

Ballast is defined as a heavy substance that gives weight; in this case, microsand is used to enhance flocculation and sedimentation. The components of the ACTIFLO® process are shown in Figure 7–9. A chemical coagulant, commonly alum, is rapidly mixed with the raw water for approximately 2 min in the coagulation tank. In the injection tank, polymer as a flocculation aid and microsand (60–210 µm) as ballast are added to initiate floc formation for a hydraulic detention time of approximately 2 min. In the maturation tank, moderate mixing provides ideal conditions for the

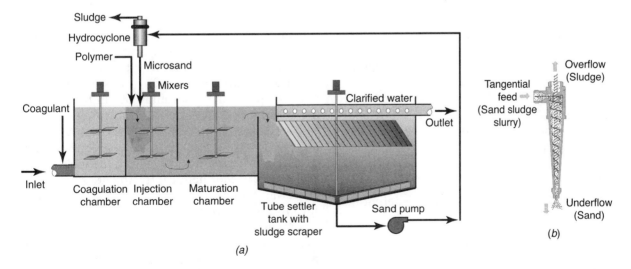

Figure 7–9
ACTIFLO® microsand ballasted clarification process. (*a*) Schematic flow diagram. (*b*) Hydrocyclone for recycling microsand. (Courtesy of Krüger.)

formation of polymer bridging between colloids to develop stronger and larger floc growth. Typical hydraulic detention time is approximately 6 min.

The hydrocyclone draws the sand-slurry from the bottom of the settling tank and recycles cleaned sand to the injection tank and sludge to waste. In the settling tank, laminar upflow through the Lamella settling tubes provides rapid and effective removal of the microsand flocs. The settling time has a detention of approximately 10 min. Clarified water exits the settling system through collection troughs for subsequent granular-media filtration. Sand-slurry from the bottom of the settling tank is cleaned in the hydrocyclone. The sludge is sent to waste and the cleaned sand may be reused. Some health departments limit the recycle of sand due to the potential recycle of *Cryptosporidium* and *Giardia* on the sand.

Ballasted flocculation is appropriate to treat surface waters with fluctuating quality, such as low to high turbidity in a short period of time, excessive algae from blooms, and naturally occurring organic matter from land drainage. Source waters with relatively high total organic carbon and dissolved organic carbon contents benefit from the additional floc density from ballasted flocculation. The ACTIFLO® process can be tested on the local source water in a trailer-mounted pilot unit. The trailer contains all process equipment, chemical feed systems, controls, and laboratory facilities. A typical test period takes 2–3 weeks.

■ EXAMPLE 7–4

Compare the total detention times suggested for the ACTIFLO® process with detention times for rapid mixing and flocculation recommended by the *Standards for Water Works* for a traditional system and flocculation-clarifiers for turbidity removal of surface waters.

Solution

For the ACTIFLO® process approximate detention times are

Rapid mixing in the coagulation tank = 2 min
Initial floc formation in injection tank = 2 min
Flocculation in maturation tank = 6 min
Laminar upflow in settling tank = 10 min
Total = 20 min

For traditional processing specified by the *Standards for Water Works,* detention times are

Rapid mixed tank chamber = 10 to 20 sec
Flocculation basin = 30 min
Sedimentation basin = 4 hr or 2 hr, if
reduced
Total = 270 min or
150 min, if
reduced

For flocculation-clarifiers based on total tank volume specified by the *Standards for Water Works,* the detention time is

Detention time = 2 to 4 hr = 120 to 240 min

■ ■ ■

7–6 FLOCCULATOR-CLARIFIERS

Flocculator-clarifiers, also referred to as solids contact units or upflow tanks, combine the processes of mixing, flocculation, and sedimentation in a single compartmented tank. One such unit, shown in Figure 7–10, introduces coagulants or softening chemicals in the influent pipe and mixes the water under a central cone-shaped skirt, where a high-floc concentration is maintained. Flow passing under the hood is directed through the sludge blanket at the bottom of the tank to promote growth of larger conglomerated clusters where the heavier particles have settled. Overflow rises upward in the peripheral settling zone to radial weir troughs or circular inboard troughs suspended at the surface. These units are particularly advantageous in lime softening of groundwater, since the precipitated solids help seed the floc, growing larger crystals of precipitate to provide a thicker waste sludge. Recently, flocculator-clarifiers have been receiving wider application in the chemical treatment of industrial wastewaters and surface-water supplies. The major advantages promoting their use are reduced space requirements and less costly installation. However, the unitized nature of construction generally results in a sacrifice of operating flexibility.

The *Standards for Water Works* state that flocculator-clarifiers (solids contact units) are most acceptable for clarification in combination

Figure 7–10
A flocculator-clarifier provides rapid mixing, flocculation, and sedimentation in a single compartmented tank. (a) Photograph of a flocculator-clarifier in a water softening plant (Metropolitan Utilities District, Omaha, NE).
(b) Cross-sectional view of unit pictured below.
(Courtesy of Walker Process Equipment, Division of McNish Corporation.)

(a)

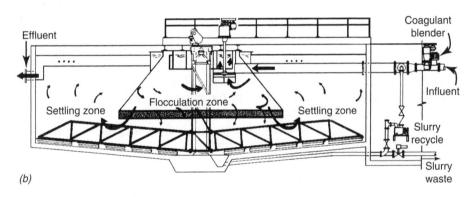

(b)

with softening for the treatment of waters with relatively uniform characteristics and flow rates. They recommend the following in sizing units: mixing and flocculation time of not less than 30 min based on the total volume of mixing and flocculation zones; minimum detention time of 2 to 4 hr for turbidity removal in treatment of surface waters and 1 to 2 hr for precipitation in lime-soda ash softening of groundwaters, with the calculated detention time based on the entire volume of the flocculator-clarifier; weir loadings not exceeding 10 gpm/ft (2.1 l/m · s) for turbidity removal units and 20 gpm/ft (4.1 l/m · s) for softening units; and upflow rates not exceeding 1.0 gpm/sq ft (0.68 l/m² · s) and 1.75 gpm/sq ft (1.19 l/m² · s), respectively. The volume of sludge removed from these units should not exceed 5 percent of the

water treated for turbidity removal or 3 percent for softening.

7–7   FILTRATION

The granular-media gravity filter is the most common type used in water treatment to remove nonsettleable floc remaining after chemical coagulation and sedimentation. A typical filter bed (Figure 7–11) is placed in a concrete box with a depth of about 9 ft. The granular media, about 2 ft deep, are supported by a graded gravel layer over underdrains. During filtration, water passes downward through the filter bed by a combination of water pressure from above and suction from the bottom. Filters are cleaned by backwashing

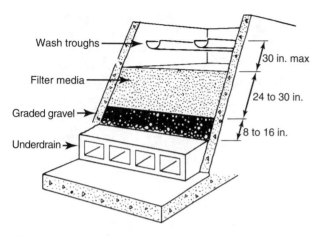

Figure 7–11

Cross section of a granular-media gravity filter.

(reversing the flow) upward through the bed. Wash troughs suspended above the filter surface collect the backwash water and carry it out of the filter box.

The traditional flow scheme for municipal water processing incorporates flocculation with a chemical coagulant and sedimentation prior to filtration (Figure 7–1). Instead of separate in-line tanks, flocculator-clarifiers can be used for mixing and sedimentation, particularly in treatment of groundwaters. Filtration rates following flocculation and sedimentation are in the range of 2 to 10 gpm/sq ft (1.4 to 6.8 l/m$^2 \cdot$ s), with 5 gpm/sq ft (3.4 l/m$^2 \cdot$ s) normally the maximum design value.

Flow Control

Traditional gravity systems regulate the rate of filtration by controlling the rate of discharge from the filter underdrain. Influent entering each filter

box is controlled automatically, or manually, to equal outflow so that a constant water level is maintained above the filter bed.

The following description of filter operation follows the valve numbering in Figure 7–12. Initially, valves 1 and 4 are opened, and 2, 3, and 5 are closed for filtration. Overflow from the settling basin applied to the filter passes through the bed and underdrain system to the clear well underneath. The depth of water above the filter surface is between 3 and 4 ft. The underdrain pipe is trapped in the clear well to provide a liquid connection to the water being filtered, thus preventing backflow of air into the underdrain. The maximum head available for filtration is equal to the difference between the elevation of the water surface above the filter and the level in the clear well; this is commonly 9 to 12 ft. When the filter is clean, flow through the bed must be regulated to prevent an excessive filtration rate. A rate-of-flow controller, consisting of a valve controlled by a venturi meter, throttles flow in the discharge pipe, restraining the rate of filtration. As the filter collects impurities, the resistance to flow increases and the flow controller valve opens wider to maintain a preset rate.

A filter is cleaned by backwashing when the measured head loss through the media is approximately 8 ft. Valves 1 and 4 are closed (3 remaining closed) and 2 is opened. This valving arrangement drains away the water from above the filter bed down to the elevation of the outlet-channel wall. If air agitation is used, compressed air is introduced through the underdrain to loosen accumulated impurities from the grains of media by rubbing them against each other in the turbulence caused by the air mixing. If rotary washer units are

Figure 7–12

Diagrammatic sketch showing operation of a gravity filter with a rate-of-flow controller regulating the rate of filtration.

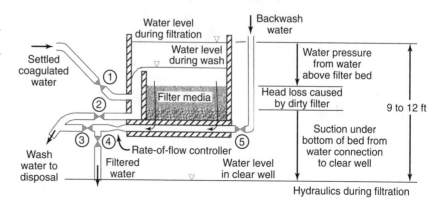

employed, they are placed in operation prior to starting the water backwash. By opening valve 5, clean water flows into the filter underdrain and passes up through the bed hydraulically, expanding the stationary bed depth about 50 percent. Dirty wash water is collected by troughs and conveyed to disposal or for treatment to remove the impurities before recycling to the plant influent for reuse (Figure 7–43 in Section 7-30). The first few minutes of filtered water at the beginning of the next run may be wasted to flush the wash water remaining in the bed out through the drain. This is accomplished by opening valve 3 when valve 1 is opened to start filtration (valves 2, 4, and 5 are shut). Opening valve 4 while closing 3 permits filtration to proceed again.

Filter Media

The action that takes place in a granular-media filter is extremely complex, consisting of straining, flocculation, and sedimentation. Gravity filters do not function properly unless the applied water has been chemically treated and, if necessary, settled to remove the large floc. Coagulant carryover is essential in removing microscopic particulate matter that would otherwise pass through the bed. If an excessive quantity of large floc overflows the settling basin, a heavy mat forms on the filter surface by straining action and clogs the bed. However, the impurities in improperly coagulated water may penetrate too far into the bed and can be flushed through before being trapped, causing a turbid effluent.

Optimum filtration occurs when nonsettleable coagulated floc is held in the pores of the bed and produces "in-depth" filtration. The ideal filter medium possesses the following characteristics: coarse enough for large pore openings to retain large quantities of floc yet sufficiently fine to prevent passage of suspended solids, adequate depth to allow relatively long filter runs, and graded to permit effective cleaning during backwash. Dual-media beds of coarser anthracite overlaying the sand filter provide an upper layer of increased porosity to reduce surface plugging.

Filtration can be stopped because of a low rate of filtration or passage of excess turbidity through the bed. Under average operating conditions, granular-media filters are backwashed about once in 24 hr at a rate of about 15 gpm/ft^2 (10 l/m$^2 \cdot$ s) for a period of 5 to 10 min. Initial filtered water may be wasted for 3 to 5 min. A bed is out of operation for 10 to 15 min to complete the cleaning process. The amount of water used in backwashing varies from 2 to 4 percent of the filtered water. During backwashing, the bed of filter media is expanded hydraulically about 50 percent, and the released impurities are conveyed by upflow of wash water to the wash troughs.

Problems in backwashing of dual-media filters can result if the cleaning action is limited to wash-water fluidization. Nonuniform expansion and poor scouring can result in mud balls dropping through the coarser coal media and lodging on top of the sand layer. The common methods of cleaning dual-media filters use either air scouring or air-and-water scouring prior to water backwash. For separate air scouring, the water level is lowered below the wash-water troughs to near the surface of the media and only air is introduced to mix and scour the media. Wash water is then used to fluidize the bed and purge contaminants. For air-and-water scouring, the wash cycle also begins by draining off the water above the filter. After backwash flow has started, at about one-quarter of the fluidizing rate, air is supplied and the simultaneous flow of air and water scours the bed as the wash-water level rises in the filter box. When the water surface approaches the bottom of the wash troughs, air injection is terminated and the backwash rate increases to the desired fluidization velocity to carry the impurities out of the expanded bed.

The granular media are thoroughly mixed in the agitated, turbulent flow of an expanded bed during backwashing. When the upward flow of wash water is stopped, the suspended grains settle down to form a stratified bed with the finest grains of each medium on top. In a mixed-media bed, the medium of lowest density settles on top, that is, on the anthracite layer above the sand bed.

Granular media are specified by effective size and uniformity coefficient. The effective size is the 10-percentile diameter, which means that 10 percent of the filter grains by weight are smaller than this diameter. The uniformity coefficient is the ratio of the 60-percentile diameter to the 10 percentile diameter. The common range in effective

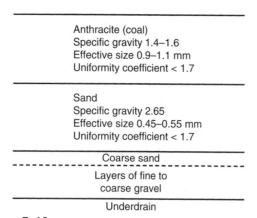

Figure 7–13

Cross section of the media in a coal-sand dual-media filter and supporting gravel layer showing typical grain sizes, specific gravities, effective sizes, and uniformity coefficients.

size for single-medium sand filters is 0.45 to 0.55 mm, and the specific gravity is 2.65. The uniformity coefficient is less than 1.7, with a lower uniformity coefficient beneficial in reducing backwash flow rate. The depth of the sand bed is 24 to 30 in.

The profile of a coal-sand dual-media filter is shown in Figure 7–13. This filter uses a relatively coarse anthracite medium with an effective size between 0.9 to 1.1 mm and specific gravity of 1.4 to 1.6 over a finer sand layer with the characteristics previously described. The uniformity coefficient of the anthracite coal medium is less than 1.7, with the lowest practical uniformity coefficient available commercially of 1.5. The upper layer of coarser anthracite has voids about 20 percent larger than the sand, and thus a coarse-to-fine grading of media is provided in the direction of flow. After backwashing, the bed stratifies, with the heavier sand on the bottom and the lighter, coarser coal medium on top. Larger floc particles are adsorbed and trapped in the surface coal layer, while finer material is held in the sand filter; therefore, the bed filters in greater depth, preventing premature surface plugging.

Filter Underdrains

The function of an underdrain is to support the filter media, collect the filtered water, and distribute backwash water and air for scouring. A coarse sand and several graded gravel layers separate the filter media from the underdrain, with the gradation of the lowest gravel layer dependent upon the size of openings in the underdrain (Figure 7–13). Their function is to prevent loss of the media during filtration and to uniformly distribute the backwash water. Several different kinds of filter underdrain systems are listed in Table 7–1. Some systems allow separate air scour before water backwash, some do not allow air scour, and others allow concurrent air-and-water scour.

The filter bottom in Figure 7–14 is a plastic block with a dual-parallel lateral design for uniform distribution of air scour, or concurrent air-and-water scour, and water backwash. The graded gravel layer placed between the block and the filter sand is 8 to 12 in. in thickness, with particle sizes ranging from $\frac{3}{4}$ in. to $\frac{1}{8}$ in. Air, concurrent air and water, and water enter from the lower lateral through control orifices to the upper compensating lateral and proceed through dispersion holes in the top of the block. By this arrangement, the

TABLE 7–1

Common Filter Underdrain Systems

Underdrain System	Characteristics
Pipe laterals with orifices	Deep gravel layer Medium head loss No air scour
Pipe laterals with nozzles	Shallow gravel layer High head loss Air scour
Vitrified tile block	Shallow gravel layer Medium head loss No air scour
Plastic dual-lateral block	Shallow gravel layer Low head loss Air scour or concurrent air-and-water scour
Plastic nozzles	Shallow or no gravel layer High head loss Air scour or concurrent air-and-water scour

Figure 7–14

Dual-media gravity filter, (a) Cutaway view of a concrete filter box with an air/water underdrain, gravel supporting layer, sand-anthracite filter media, and wash-water troughs.
(b) High-density polyethylene underdrain block with dual-parallel design for either air scour followed by water backwashing or concurrent air-and-water scouring and water backwashing.
(Courtesy of the F. B. Leopold Company, Inc.)

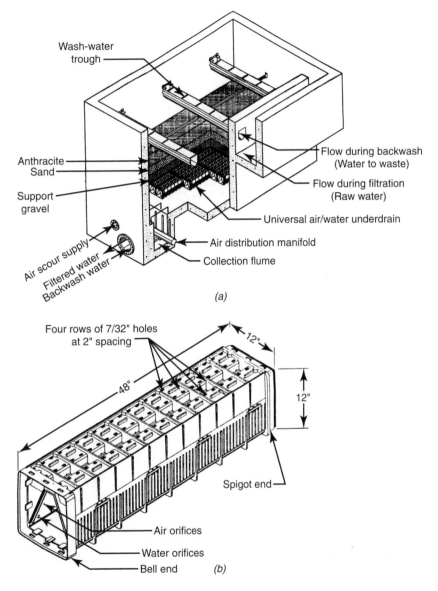

(a)

(b)

pressure of air or wash water is equalized to produce uniform upward flow through the holes along the top of the block. The layer of coarser gravel disperses the wash-water flow, and the finer overlying gravel serves as a barrier to prevent intermixing of gravel and filter sand. Air scour or concurrent air-and-water scour causes aggressive scrubbing to loosen impurities from the grains of sand and anthracite media. After scouring, backwash hydraulically fluidizes the filter media, increasing the bed depth by about 50 percent. This upward flow of wash water conveys impurities

out of the bed and allows the media mixture to reclassify, with the anthracite above the sand. After cleaning, backwash is terminated, permitting the media to stratify and settle out of suspension.

Filters are commonly cleaned every 24 hr, and the water backwash rate is about 15 gpm/ft^2 (10 l/m^2 · s) for a period of 5 to 10 min. A filter is out of operation for an additional 10 to 15 min for the complete cleaning process, including scour and settling. Initial filtered water may be wasted for 3 to 5 min to flush out any impurities remaining in

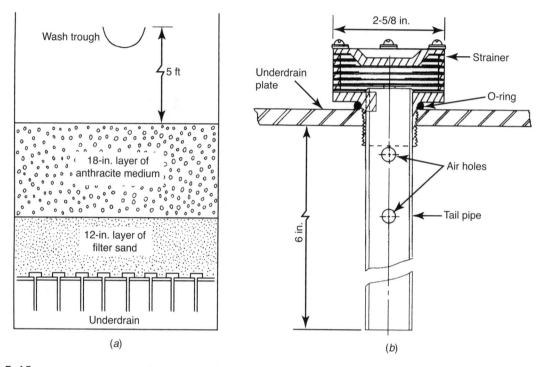

Figure 7–15

Underdrain system for air scouring and water backwashing of a granular-media filter (a) Cross section of the filter. (b) Detail of the air-water nozzle.
(Courtesy of General Filter Co., Ames. IA.)

the filter. The amount of water used in cleaning varies from 2 to 4 percent of the filtered water.

The underdrain system in Figure 7–15 consists of plastic nozzles placed through holes in a steel or concrete underdrain. Each nozzle is constructed of wedge-shaped segments assembled with stainless steel bolts to form media retaining slots that are tapered larger in the direction of water flow during filtration. The recommended media for this filter nozzle are 12 in. of sand and 18 in. of anthracite. After the water in the filter box is lowered, air and water are simultaneously introduced into the filter bottom. The nozzle tail pipes have air holes sized to maintain a 2- to 3-in. air blanket under the plate to ensure equal nozzle pressure and uniform air distribution to the bottom of the bed. Concurrent air-and-water backwash is continued for 7 to 10 min to clean the media. The dirty wash water is collected by baffled wash troughs and conveyed to disposal. After the air supply is turned off, water backwash is continued for 1 to 3 min to purge trapped air

from the underdrain and expanded media prior to restratification.

Filter Control

The control console for each filter unit has a head-loss gauge, flow meter, and rate controller. A filter run is normally terminated when head loss through the filter reaches a prescribed value between 6 and 9 ft or when effluent turbidity exceeds the amount considered acceptable. As suspended matter collects in the pores of the filter media and head loss increases through the bed, the lower portion of the filter is under suction from the water column connected through the piping to the clear well (Figure 7–12). This partial vacuum permits release of dissolved gases that tend to collect in the filter pores and, as a result, cause air binding and a reduced rate of filtration. In addition to shortening filter runs, the gases accumulated in a bed may tend to discharge violently during the start of the backwash cycle, potentially causing disruption of the gravel base.

The rate controller prevents rapid changes in filtration rates and controls the velocity of flow through a clean bed. Continuous-flow photoelectric turbidimeters are used for monitoring filter effluent quality. By detecting turbidity in the filtered water, the operator can terminate a filter run if the quality deteriorates prior to reaching the maximum head loss.

Pressure Filters

Pressure filters have media and underdrains contained in a steel tank. The media are similar to those used in gravity filters, and filtration rates range from 2 to 4 gpm/sq ft. Water is pumped through the bed under pressure, and the unit is backwashed by reversing the flow, flushing out impurities. Pressure filters are not generally employed in large treatment works because of size limitations; however, they are popular in small municipal water plants that process groundwater for softening and iron removal. Their most extensive application has been treating water for industrial purposes.

■ EXAMPLE 7–5

The filter unit illustrated in Figure 7–14 is 15 ft by 30 ft. After filtering 2.5 mil gal in a 24-hr period, the filter is backwashed at a rate of 15 gpm/sq ft for 12 min. Compute the average filtration rate and the quantity and percentage of treated water used in backwashing.

Solution

$$\text{Filtration rate} = \frac{2{,}500{,}000 \text{ gal/day}}{15 \text{ ft} \times 30 \text{ ft} \times 1440 \text{ min/day}}$$

$$= 3.9 \frac{\text{gpm}}{\text{sq ft}}$$

Quantity of wash water

$$= \frac{15 \text{ gal}}{\text{min} \times \text{sq ft}} \times 12 \text{ min} \times 15 \text{ ft} \times 30 \text{ ft}$$

$$= 81{,}000 \text{ gal}$$

$$\frac{\text{Wash water}}{\text{Treated water}} = \frac{81{,}000 \text{ gal}}{2{,}500{,}000 \text{ gal}} \times 100$$

$$= 3.2 \text{ percent}$$

■ ■ ■

7–8 CHEMICAL COAGULATION

Surface waters generally contain suspended and colloidal solids from land erosion, decaying vegetation, microorganisms, and color-producing compounds. Coarser materials, such as sand and silt, can be eliminated to a considerable extent by plain sedimentation, but finer particles must be chemically coagulated to produce larger floc that is removable in subsequent settling and filtration. Destabilization of colloidal suspensions is discussed in Section 2–8, and typical flow schemes for surface-water treatment plants are diagramed in Figure 7–1.

Coagulation and flocculation are sensitive to many variables, for instance, the nature of the turbidity-producing substances, type and dosage of coagulant, pH of the water, and the like. Of the many variables that can be controlled in plant operation, pH adjustment appears to be most important. Generally, the types of coagulants and aids available are defined by the plant process scheme; of course, dosages of these substances can be regulated to meet changes in raw-water quality. Also, mechanical mixing can be adjusted by varying the speed of the flocculator paddles.

Jar tests are widely used to determine optimum chemical dosages for treatment. This laboratory test (Section 2–11) attempts to simulate the full-scale coagulation-flocculation process and can be conducted for a wide range of conditions. The interpretation of test results involves visual and chemical testing of the clarified water. Ordinarily, a treatment plant gives better results than the jar test for the same chemical dosages. Of course, to confirm optimum chemical treatment, the water should be tested at various processing stages, including the finished effluent. One common monitoring technique is analysis of the filtered water for turbidity; another is to record the length of time between backwashing of filters.

Coagulants

Commonly used metal coagulants in water treatment are (1) those based on aluminum, such as aluminum sulfate, sodium aluminate, potash alum, and ammonia alum; and (2) those based on iron, such as ferric sulfate, ferrous sulfate, chlorinated ferrous sulfate, and ferric chloride. The description that follows gives some of the relevant properties and chemical reactions of these substances in the coagulation of water. The latter are presented in the form of hypothetical equations with the understanding that they do not represent exactly what happens in water. Studies have shown that the hydrolysis of iron and aluminum salts is far more complicated than these formulas would indicate; however, they are useful in approximating the reaction products and quantitative relationships.

Aluminum sulfate, $Al_2(SO_4)_3 \cdot 14.3H_2O$, is by far the most widely used coagulant; the commercial product is commonly known as alum, filter alum, or alumina sulfate. Filter alum is a grayish-white crystallized solid containing approximately 17 percent water-soluble Al_2O_3 and is available in lump, ground, or powdered forms as well as concentrated solution. Ground alum is commonly measured by a gravimetric-type feeder into a solution tank from which it is transmitted to the point of application by pumping. Amber-colored liquid aluminum sulfate contains about 8 percent available Al_2O_3.

The hydrolysis of aluminum ion in solution is complex and is not fully defined. In pure water at low pH, the bulk of aluminum appears as Al^{+++}, while in alkaline solution complex species such as $Al(OH)_4^-$ have been shown to exist. In the hypothetical coagulation equations, aluminum floc is written as $Al(OH)_3$. This is the predominant form found in a dilute solution near neutral pH in the absence of complexing anions other than hydroxide. The reaction between aluminum and natural alkalinity is given in Eq. 7–5. If lime or soda ash is added to the water with the coagulant, the theoretical reactions are as shown in Eqs. 7–6 and 7–7.

$$Al_2(SO_4)_3 \cdot 14.3H_2O + 3Ca(HCO_3)_2$$
$$= 2Al(OH)_3\downarrow + 3CaSO_4 + 14.3H_2O + 6CO_2$$
$$(7–5)$$

$$Al_2(SO_4)_3 \cdot 14.3H_2O + 3Ca(OH)_2$$
$$= 2Al(OH)_3\downarrow + 3CaSO_4 + 14.3H_2O \qquad (7–6)$$

$$Al_2(SO_4)_3 \cdot 14.3H_2O + 3Na_2CO_3 + 3H_2O$$
$$= 2Al(OH)_3\downarrow + 3Na_2SO_4 + 3CO_2 + 14.3H_2O$$
$$(7–7)$$

Based on these reactions, 1.0 mg/l of alum with a molecular weight of 600 reacts with 0.50 mg/l natural alkalinity, expressed as $CaCO_3$; 0.39 mg/l of 95 percent hydrated lime as $Ca(OH)_2$, or 0.33 mg/l 85 percent quicklime as CaO; or 0.53 mg/l soda ash as Na_2CO_3. When lime or soda ash is reacted with the aluminum sulfate, the natural alkalinity of the water is unchanged. Sulfate ions added with the alum remain in the finished water. In the case of natural alkalinity and soda ash, carbon dioxide is produced. The dosages of alum used in water treatment are in the range of 5 to 50 mg/l, with the higher concentrations needed to clarify turbid surface waters. Alum coagulation is generally effective within the pH limits of 5.5 to 8.0.

Sodium aluminate, $NaAlO_2$, may be used as a coagulant in special cases. The commercial grade has a purity of approximately 88 percent and may be purchased either as a solid or as a solution. Because of its high cost, sodium aluminate is generally used to aid a coagulation reaction instead of being the primary coagulant. It has been found effective for secondary coagulation of highly colored surface waters and as a coagulant in the lime-soda ash softening process to improve settleability of the precipitate.

Ferrous sulfate, $FeSO_4 \cdot 7H_2O$ (also commonly known as copperas), is a greenish-white crystalline solid that is obtained as a by-product of other chemical processes, principally from the pickling of steel. Although available in liquid form from processed spent pickle liquor, the common commercial preparations are granular. Ferrous iron added to water precipitates in the oxidized form of ferric hydroxide; therefore, the addition of lime or chlorine is generally needed to provide effective coagulation. Ferrous sulfate and lime coagulation, shown in Eq. 7–8, is effective for clarification of turbid water and other reactions conducted at high pH values, for example, in lime softening.

$$2FeSO_4 \cdot 7H_2O + 2Ca(OH)_2 + \tfrac{1}{2}O_2$$

$$= 2Fe(OH)_3\downarrow + 2CaSO_4 + 13H_2O \qquad (7\text{–}8)$$

Chlorinated copperas is prepared by adding chlorine to oxidize the ferrous sulfate. The advantage of this method, relative to lime addition, is that coagulation can be obtained over a wide range of pH values, 4.8 to 11.0. Theoretically, according to Eq. 7–9, each milligram per liter of ferrous sulfate requires 0.13 mg/l of chlorine, although additional chlorine is generally added to ensure complete reaction.

$$3FeSO_4 \cdot 7H_2O + 1.5Cl_2$$

$$= Fe_2(SO_4)_3 + FeCl_3 + 21H_2O \qquad (7\text{–}9)$$

Ferric sulfate, $Fe_2(SO_4)_3$, is available as a commercial coagulant in the form of a reddish-brown granular material that is readily soluble in water. It reacts with the natural alkalinity of water according to Eq. 7–10, or with added alkaline materials, such as lime or soda ash, Eq. 7–11.

$$Fe_2(SO_4)_3 + 3Ca(HCO_3)_2$$

$$= 2Fe(OH)_3\downarrow + 3CaSO_4 + 6CO_2 \qquad (7\text{–}10)$$

$$Fe_2(SO_4)_3 + 3Ca(OH)_2$$

$$= 2Fe(OH)_3\downarrow + 3CaSO_4 \qquad (7\text{–}11)$$

In general, ferric coagulants are effective over a wide pH range. Ferric sulfate is particularly successful when used for color removal at low pH values; at high pH, it may be used for iron and manganese removal and as a coagulant in precipitation softening.

Ferric chloride, $FeCl_3 \cdot 6H_2O$, is used primarily in the coagulation of wastewater and industrial wastes, and it finds only limited use in water treatment. Normally, it is produced by chlorinating scrap iron and is available commercially in solid and liquid forms. Being highly corrosive, the liquid must be stored and handled in corrosion-resistant tanks and feeders. The reactions of ferric chloride with natural and added alkalinity are similar to those of ferric sulfate.

Coagulant Aids

Difficulties with coagulation often occur because of slow-settling precipitates or fragile flocs that are easily fragmented under hydraulic forces in basins and filters. Coagulant aids benefit flocculation by improving settling and toughness of flocs. The most widely used materials are polymers; others are adsorbent-weighting agents and oxidants.

Synthetic polymers are long-chain, high-molecular-weight, organic chemicals commercially available under a wide variety of trade names. Polymers are classified according to the type of charge on the polymer chain. Those possessing negative charges are called anionic, those positively charged are called cationic, and those carrying no electric charge are nonionic. Anionic or nonionic polymers are often used with metal coagulants to provide bridging between colloids to develop larger and tougher floc growth (Figure 2–7c). The dosage required as a flocculent aid is generally on the order of 0.1 to 1.0 mg/l. In the coagulation of some waters, polymers can promote satisfactory flocculation at significantly reduced alum dosages. The potential advantages of polymer substitution are in reducing the quantity of waste sludge produced in alum coagulation and in changing the character of the sludge such that it can be more easily dewatered.

Cationic polymers have been used successfully in some waters as primary coagulants for clarification. Although the unit cost of cationic polymers is about 10 to 15 times higher than the cost of alum, the reduced dosages required may nearly offset the increased cost. Furthermore, unlike the gelatinous and voluminous aluminum hydroxide sludges, polymer sludges are relatively dense and easier to dewater for subsequent handling and disposal. Sometimes cationic and nonionic polymers may be used together to provide an adequate floc, the former being the primary coagulant and the latter a coagulant aid. Although significant strides have been made in the application of polymers in water treatment, their main application is still as an aid rather than as a primary coagulant. Many waters cannot be treated by using polymers alone but require aluminum or iron salts. Jar tests and actual plant operation must be used to determine the effectiveness of particular proprietary polymers in flocculation of a given water.

Acids and alkalies can be added to adjust the pH for optimum coagulation. The common acid used to lower the pH is sulfuric acid. Increasing pH is done by the addition of lime, soda ash, or sodium hydroxide. The latter is purchased and fed as a concentrated solution. The characteristics of lime and soda ash are discussed in Section 7–17.

Chemical Feeders

A chemical feeder is a mechanical device for measuring a quantity of chemical and applying it to the water at a preset rate. Liquid feeders apply chemicals in solutions or suspensions. Dry feeders apply chemicals in granular or powdered forms. Some chemicals, such as ferric chloride, polyphosphates, and sodium silicate, must be fed in solution form, whereas others, such as ferrous sulfate and alum, can be fed dry. If a chemical does not dissolve readily, it can be applied dry or in suspension provided that the solution is continuously stirred.

Gravimetric dry feeders are extremely accurate, available in large sizes to deliver high feed rates, and readily adaptable to recording and to automatic control. Diaphragm and peristaltic metering pumps are the two general types used for delivery of chemical solutions or suspensions to treatment units. Applying lime requires a special type of unit that is referred to as a slaker. (*Slake* means to cause to heat and crumble by treatment with water.) The chemical reaction for slaking is powdered lime combining with water to form lime slurry [$Ca(OH)_2$].

Figure 7–16 shows chemical feeders in a water treatment plant. Each is supplied from a hopper-bottom, chemical storage tank that extends through the ceiling to the floor above. Mechanical vibrators attached to the hoppers keep the chemicals flowing freely out into the feeders. After the dry chemicals are measured into continuously stirred dissolving chambers, the solutions either flow by gravity into an open flume or are transferred by pumps into a pressure main.

■ EXAMPLE 7–6

A dose of 50 mg/l of alum is used in coagulating a turbid surface water. (a) How much natural alkalinity is consumed? (b) What changes take place in

Figure 7–16

Photograph of gravimetric feeders in a water treatment plant. The unit on the right is a lime slaker; the one in the center is feeding alum.

the ionic character of the water? (c) How many milligrams per liter of aluminum hydroxide are produced?

Solution

(a) From the text, 1.0 mg/l of alum reacts with 0.5 mg/l of natural alkalinity; therefore,

$$50 \text{ mg/l (alum)} \frac{0.50 \text{ mg/l (alk)}}{1.0 \text{ mg/l (alum)}}$$
$$= 25 \text{ mg/l of alkalinity as } CaCO_3$$

(b) 50 mg/l of alum is equivalent to 0.50 meq/l; therefore, based on Eq. 7–5, 0.50 meq/l of sulfate ion is added to the water. The aluminum ions precipitate out of solution, the calcium content is not affected, and 0.50 meq/l of bicarbonate is converted to carbon dioxide.

(c) From Eq. 7–5, 1 mole of alum (600 g) reacts to produce 2 moles of aluminum hydroxide (2 × 78 = 156 g). Therefore,

$$\frac{156}{600} = \frac{\text{mg/l of } Al(OH)_3 \text{ produced}}{50 \text{ mg/l alum dose}}$$

or $Al(OH)_3$ = 13 mg/l.

■ ■ ■

■ EXAMPLE 7–7

A surface water is coagulated by adding 30 mg/l of ferrous sulfate and an equivalent dosage of lime. How many pounds of coagulant are used per million gallons of water processed? How many pounds of lime are required at a purity of 80 percent CaO?

Solution

Consumption of ferrous sulfate equals:

$$30 \text{ mg/l} \times 8.34 \frac{\text{lb/mil gal}}{\text{mg/l}} = 250 \text{ lb/mil gal}$$

One equivalent weight of ferrous sulfate (139) reacts with one equivalent of 80 percent CaO [28/0.08 = 35]. Therefore,

$$\text{Lime dosage} = 250 \frac{\text{lb}}{\text{mil gal}} \times \frac{35}{139} = 63 \text{ lb/mil gal}$$

■ ■ ■

■ EXAMPLE 7–8

The dosage of alum in the alum-lime coagulation of a water is 50 mg/l. It is desired to react only 10 mg/l (as $CaCO_3$) of the natural alkalinity with the alum. Based on the theoretical Eqs. 7–5 and 7–6, what dosage of lime is required, in addition to 10 mg/l of natural alkalinity, to react with the alum dosage?

Solution

Using equivalent weights, the alum that reacts with 10 mg/l of natural alkalinity is

$$10 \times (100/50) = 20 \text{ mg/l}$$

The amount of alum remaining to react with lime is 50 − 20 = 30 mg/l.

The lime dosage required to react with 30 mg/l of alum is

$$30 \times (28/100) = 8.4 \text{ mg/l as CaO}$$

■ ■ ■

7–9   TASTE AND ODOR CONTROL

One of the objectives in water treatment is to produce a palatable water that is aesthetically pleasing. The repeated presence of objectionable taste, odor, or color in a water supply may cause the public to question its safety for consumption. Taste may be affected by inorganic salts or metal ions, a variety of organic chemicals found in nature or resulting from industrial wastes, or products of biological growths. Algae are the most frequent cause of taste and odor problems in surface supplies. Their metabolic activities impart odorous compounds that produce fishy, grassy, musty, or foul odors. Actinomycetes are moldlike bacteria that create an earthy odor.

Problems related to the palatability of water are generally unique in each system, and they must be individually studied to determine the best approach for prevention and cure. In groundwater treatment, aeration is frequently effective, since the odor compounds are often dissolved gases that can be stripped from solution. However, aeration is rarely effective in processing surface waters where the odor-producing substances are generally nonvolatile.

First consideration should be given to preventative measures in surface-water supplies. If the interference can be traced to an industrial waste discharge, the source may be removed. In a eutrophic reservoir or lake, regular herbicide applications to the impounded water are effective in suppressing algal blooms that affect taste and cause odor and filter clogging.

Carbon adsorption is usually the most effective way to reduce the level of taste and odor in the treatment of water. Activated carbon can be prepared from hardwood charcoal, lignite, nut shells, or other carbonaceous materials by controlled combustion to develop adsorptive characteristics. Each particle is honeycombed with thousands of molecular-sized pores that adsorb odorous substances. Activated carbon is available in powdered and granular forms. Granular carbon adsorption beds have not been used extensively in municipal water processing because of economic considerations; however, they are used by food and beverage industries in purification of product water. Carbon in municipal treatment is a finely ground, insoluble black powder that can be applied either through dry feed machines or as a slurry. It can be introduced in any stage of processing before filtration where adequate mixing is available to disperse the carbon and where the contact time is 15 min or more before sedimentation or filtration. The optimum point of application is generally determined by trial and error and previous experience. Carbon reacts with chlorine; therefore, these two chemicals should not be applied simultaneously or in sequence without an appropriate time interval.

Oxidation of taste and odor compounds can be done by chlorination, chlorine dioxide, potassium permanganate, or ozone. Heavy chlorination may not be desirable because of formation of trihalomethanes and haloacetic acids (Table 5–1 in Section 5–3). Chlorine dioxide, produced by combining solutions of sodium chlorite and chlorine, has approximately the same oxidative power as chlorine without forming trihalomethanes. Although it oxidizes some aromatic organic compounds, it is ineffective for many other odorous substances. Potassium permanganate is a strong chemical effective in oxidizing a variety of taste- and odor-producing compounds. Since one of the reactions in water produces manganese dioxide (Eq. 7–28), it must be applied in processing before filtration. The use of ozone is limited in the United States.

7–10 SYNTHETIC ORGANIC CHEMICALS

Synthetic organic chemicals (SOCs) include a large number of manufactured chemicals used in industry, agriculture, and household applications. Some are potentially toxic substances associated with cancer risk and damage to vital organs such as the liver, heart, and kidneys if ingested at low doses for an extended period of time. Others impair the nervous system. The maximum contaminant levels for organic and volatile organic chemicals in the drinking water standards are very low, ranging from 0.1 mg/l down to 0.0002 mg/l (Table 5–1 in Section 5–3).

Surface waters may be polluted by runoff from agricultural lands, discharge of industrial wastewaters, and spillage of chemicals. Conventional water treatment by coagulation-sedimentation-filtration provides limited removal,

since dissolved SOCs rarely adsorb to metal hydroxides formed by coagulant chemicals and to polymers. Changing coagulants or polymers, adjustment of pH, and application of powdered activated carbon (PAC) are alternatives that may improve treatment.

PAC is a fine powder applied in a water slurry at any location in the treatment process ahead of filtration. At the point of application, the mixing must be adequate to ensure dispersion and the contact time long enough for adsorption. Although PAC is an effective adsorbent of organic compounds that cause taste and odor at a dosage up to 5 mg/l with a contact time of 10 to 15 min, this application is not successful in adsorbing SOCs. Poor adsorption is attributed to the pore structure of the PAC, short contact time, and interference by adsorption of other organic compounds.

More effective removal of SOCs can be achieved by filtration through a bed of granular activated carbon (GAC) because of the close contact with the water for an extended period of time for adsorption. GAC made from coal has the best physical properties of density, particle size, abrasion resistance, and ash content, which are essential for filter backwashing and reactivation of the GAC by heating in a furnace. Prior to design of a filter system, pilot plant studies are conducted for selection of the best carbon, determination of required contact time, effects of quality variations in the water being treated, and effectiveness of carbon reactivation.

Groundwaters may be contaminated by leaching of agricultural pesticides or by seepage from improper disposal of industrial volatile organic chemicals (VOCs). While VOCs are stable in groundwaters, they are rarely found in surface waters. The two processes for removal from groundwater are stripping by aeration and GAC adsorption. The only efficient method of aeration is air stripping in a countercurrent packed tower, where the water trickles downward through the packing while air is forced up through the voids of the packing. However, because of low allowable maximum contaminant levels, aeration may not be able to reduce the concentrations to drinking water standards. Furthermore, in cold climates, aeration is not feasible because of poor removal at low temperature and the possibility of ice formation on the tower packing. The more costly process of adsorption by a GAC filter may be used instead of air stripping or be applied as a second stage following partial removal by aeration.

7–11 FLUORIDATION

During the past three decades, hundreds of studies have been undertaken to correlate the concentration of waterborne fluoride and the incidence of dental caries. Consuming water with natural fluoride led to the following three distinct relationships upon examination of children's teeth: An excessive fluoride level increases the occurrence and severity of dental fluorosis (mottling of teeth) while decreasing the incidence of decaying, missing, or filled teeth; based on climate, the optimum occurs at a concentration of 0.6 to 1.2 mg/l, resulting in maximum reduction in caries with no aesthetically significant mottling; below optimum some benefit occurs, but decay reduction is not as great and decreasing fluoride levels relate to increasing incidence of caries. Controlled fluoridation in water treatment to bring the natural content up to optimum levels achieves the same beneficial results. Recommended limits given in Table 5–2 are based on air temperature, since this influences the amount of water ingested by people. Recent studies have indicated that fluoride also benefits older people in reducing the prevalence of osteoporosis and hardening of the arteries. Without doubt, fluoride in drinking water prevents dental caries, and controlled fluoridation

is an acceptable public health measure. Over half the nation's population consumes water containing near optimum fluoride content, with the bulk of these residing in communities that deliberately add a chemical compound to provide fluoride ion.

The three most commonly used fluoride compounds in water treatment are sodium fluoride, sodium silicofluoride, and fluorosilicic acid, also known as hydrofluorosilicic, hexafluorosilicic, or silicofluoric acid. Table 7–2 lists some of the characteristics of these compounds. Sodium fluoride, although one of the most expensive for the amount of available fluoride, is one of the most widely used chemicals. The crystalline type is preferred when manual handling is involved, since the absence of fine powder results in a minimum of dust. Fluorosilicic acid is a colorless, transparent, fuming, corrosive liquid having a pungent odor and an irritating action on the skin. The major source of acid is a by-product from phosphate fertilizer manufacture. Sodium silicofluoride, the salt of fluorosilicic acid, is the most widely used compound primarily because of its low cost. The white, odorless, crystalline powder is commercially available in different gradations for optimum application by various feeders.

No one specific type of fluoridation system is applicable to all water treatment plants. Selection is based on size and type of water facility, chemical availability, cost, and type of operating personnel available. For small utilities, some type of solution feed is almost always selected and batches are manually prepared. A simple system consists of a solution tank placed on a platform scale, for

TABLE 7–2

Characteristics of the Three Commonly Used Fluoride Compounds

	SODIUM FLUORIDE	SODIUM SILICOFLUORIDE	FLUOROSILICIC ACID
Formula	NaF	Na_2SiF_6	H_2SiF_6
Fluoride ion, percent	45	61	79
Molecular weight	42	188	144
Commercial purity, percent	90 to 98	98 to 99	22 to 30
Commercial form	Powder or crystal	Powder or fine crystal	Liquid
Dosage in pounds per million gallons required for 1.0 mg/l at indicated purity	18.8 (98%)	14.0 (98.5%)	35.2 (30%)

the convenience of weighing during preparation and feed, and a solution metering pump with appropriate piping from the tank to the water main for application. If fluorosilicic acid is used, it may be diluted with water in the feed tank or applied at full strength directly from the shipping drum. When sodium fluoride is used, the feed solution may be prepared to a desired strength or as a saturated solution in a dissolving tank. Sodium fluoride has a maximum solubility of 4.0 percent (18,000 mg F/l), which is essentially independent of water temperature. Saturators are commercially available to prepare a saturated solution by allowing water to trickle through a bed containing a large excess of sodium fluoride. Water used for dissolution should be softened whenever the hardness exceeds 75 mg/l. While NaF is quite soluble, calcium and magnesium fluorides form precipitates that can scale and clog feeders and lines.

Fluoride solution feed must be paced to water flow. If a pump has a fixed delivery rate, the feeder can be tied electrically to turn on and off with pump operation, and the fluoride application can be adjusted to the rate of water discharge. For small water plants, a water meter contactor can be used to pace the feeder. Essentially, a contactor is a switch that is geared to the water meter movement so it makes contact at specific volumetric discharges. A pulse from the meter contactor can be used to energize the feeder in proportion to the meter's response to flow.

Large waterworks use gravimetric dry feeders to apply sodium silicofluoride or solution feeders to apply fluorosilicic acid directly. (See Figure 7–20.) Automatic control systems use flow meters and recorders to adjust feed rate. In small waterworks, positive-displacement electronic diaphragm pumps are used to meter fluorosilicic acid in proportion to the rate of water flow.

Fluoride must be injected at a point where all the water being treated passes. If no single common point exists, separate fluoride feeding installations are required for each water facility. In a well-pump system, application can be in the discharge line of each pump or in a common line leading to a storage reservoir. Fluoride applied in a water plant can be introduced in a channel or main coming from the filters or directly to the clear well. Whenever possible it should be added after filtration to avoid losses that may occur as a result of reactions with other chemicals. Of particular concern are coagulation with heavy alum doses and lime-soda softening. Fluoride injection points should be as far away as possible from any chemical that contains calcium so as to minimize loss by precipitation.

Surveillance of water fluoridation involves testing both the raw and treated water for fluoride. The concentration in treated water should be the amount recommended by drinking water standards. Records of the weight of chemical applied and the volume of water treated should be kept to confirm that the correct amount of fluoride is being added. The amount added to the water should equal the quantity calculated from the increase in concentration and quantity of water treated.

■ EXAMPLE 7–9

A liquid feeder applies a 4.0 percent saturated sodium fluoride solution to a water supply, increasing the fluoride concentration from the natural level of 0.4 mg F/l to 1.0 mg F/l. The commercial NaF powder contains 45 percent F by weight. (a) How many pounds of NaF are required per million gallons treated? (b) What is the dosage of 4.0 percent NaF solution per million gallons?

Solution

$$\text{(a)} \quad \frac{\text{NaF required}}{\text{Million gallons}} = \frac{(1.0 \text{ mg F/l} - 0.4 \text{ mg F/l})}{0.45 \text{ mg F/mg NaF}}$$

$$\times 8.34 \frac{\text{lb/mil gal}}{\text{mg/l}} = 11.11 \text{ lb}$$

A 4.0% NaF solution contains 40,000 mg NaF/l, and $0.45 \times 40,000 = 18,000$ mg F/l

$$\text{(b)} \quad \frac{\text{Solution dosage}}{\text{Million gallons}}$$

$$= \frac{1,000,000 \text{ gal} (1.0 \text{ mg/l} - 0.4 \text{ mg/l})}{18,000 \text{ mg/l}} = 33.3 \text{ gal}$$

■ ■ ■

■ EXAMPLE 7–10

A dry feeder recorded a weight loss of 10 kg of sodium silicofluoride in the treatment of 7000 m^3 of water. From Table 7–2, commercial powder is 98 percent pure Na_2SiF_6 and 61 percent of the pure compound is F. Calculate the concentration of fluoride ion added to the treated water.

Solution

Fluoride concentration

$$= \frac{10 \text{ kg} \times 1000 \text{ g/kg} \times 0.98 \times 0.61}{7000 \text{ m}^3} = 0.85 \text{ mg/l}$$

■ ■ ■

7–12 CHLORINATION

Chlorine is the common chemical used for disinfection of water. One alternative to chlorine is chlorine dioxide, although it is usually applied for taste and odor control. Another alternative is ozone, but because of high cost it is rarely applied solely for disinfection. Other benefits attributed to ozonation are reduction of disinfection by-products, taste and odor control, and improved coagulation.

Chlorine Chemistry

Chlorine is a heavier-than-air, greenish-yellow-colored, toxic gas. One volume of liquid chlorine confined in a container under pressure yields about 450 volumes of gas. It is a strong oxidizing agent, reacting with most elements and compounds. Moist chlorine is extremely corrosive; consequently, conduits and feeder parts in contact with chlorine are either special alloys or nonmetal. The vapor is a respiratory irritant that can cause serious injury if exposure to a high concentration occurs.

Hypochlorites are salts of hypochlorous acid (HOCl). Calcium hypochlorite [$Ca(OCl)_2$] is the predominant dry form used in the United States. High-test calcium hypochlorites, available commercially in granular, powdered, or tablet forms, readily dissolve in water and contain about 70 percent available chlorine. Sodium hypochlorite (NaOCl) is commercially available in liquid form

at concentrations between 5 and 15 percent available chlorine. Most water treatment plants use liquid chlorine, since it is less expensive than hypochlorites. The latter are used in swimming pools, small waterworks, and emergencies.

Chlorine combines with water to form hypochlorous acid, which, in turn, can ionize to the hypochlorite ion. Below pH 7, the bulk of the HOCl remains un-ionized, while above pH 8 the majority is in the form of OCl$^-$, Eq. 7–12.

$$Cl_2 + H_2O \longrightarrow HCl + HOCl \underset{pH<7}{\overset{pH>8}{\rightleftharpoons}} H^+ + OCl^-$$

$$(7\text{--}12)$$

Hypochlorites added to water yield the hypochlorite ion directly, Eq. 7–13.

$$Ca(OCl)_2 + H_2O = Ca^{++} + 2OCl^- + H_2O \quad (7\text{--}13)$$

Chlorine existing in water as hypochlorous acid and hypochlorite ion is defined as free available chlorine.

Chlorine readily reacts with ammonia in water to form chloramines, as follows:

$$HOCl + NH_3$$
$$= H_2O + NH_2Cl \quad \text{(monochloramine)}$$
$$(7\text{--}14)$$

$$HOCl + NH_2Cl$$
$$= H_2O + NHCl_2 \quad \text{(dichloramine)}$$
$$(7\text{--}15)$$

$$HOCl + NHCl_2 = H_2O + NCl_3 \quad \text{(trichloramine)}$$
$$(7\text{--}16)$$

The reaction products formed depend on pH, temperature, time, and initial chlorine-to-ammonia ratio. Monochloramine and dichloramine are formed in the pH range of 4.5 to 8.5. Above pH 8.5 monochloramine generally exists alone, but below pH 4.4 trichloramine is produced. Chlorine existing in chemical combination with ammonia nitrogen or organic nitrogen compounds is defined as combined available chlorine. (The laboratory test procedure for available chlorine is given in Section 2–11.)

Free available residual chlorine is that residual chlorine existing in water as hypochlorous acid or

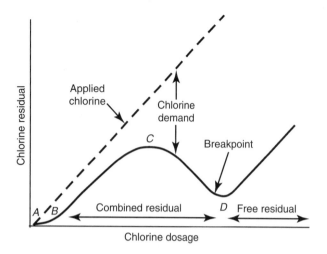

Figure 7–17

Typical breakpoint chlorination curve.

hypochlorite ion. *Combined available residual chlorine* is that residual existing in chemical combination with ammonia (chloramines) or organic nitrogen compounds. *Chlorine demand* is the difference between the amount added to a water and the quantity of free and combined available chlorine remaining at the end of a specified contact period.

When chlorine is added to water containing reducing agents and ammonia, residuals develop that yield a curve similar to that in Figure 7–17. Chlorine reacts first with the reducing agents present and develops no measurable residual, as shown by the portion of the curve extending from *A* to *B*. The chlorine dosage at *B* is the amount required to meet the demand exerted by the reducing agents (those common to water and wastewater include nitrites, ferrous ions, and hydrogen sulfide).

The addition of chlorine in excess of that required up to point *B* results in the formation of chloramines. Monochloramines and dichloramines are usually considered together because there is little control over which will be formed. The quantities of each are determined primarily by pH. Chloramines thus established show available chlorine residual and are effective as disinfectants. When all the ammonia has been reacted with, a free available chlorine residual begins to develop (point *C* on the curve). As the free

available chlorine residual increases, the previously produced chloramines are oxidized. This results in the creation of oxidized nitrogen compounds, such as nitrous oxide, nitrogen, and nitrogen trichloride, which in turn reduce the chlorine residual, as seen on the curve between *C* and *D*.

Once most of the chloramines are oxidized, additional chlorine applied to the water creates an equal residual, as indicated by the rising curve at point *D*. Point *D* is generally referred to as the *breakpoint*; beyond it, all added residual is free available chlorine. Some resistant chloramines can still be present beyond *D*, but their relative importance is small.

Disinfection

The most common application of chlorination is disinfection of drinking water to destroy microorganisms that cause diseases in humans. The disinfecting action of chlorine results from a chemical reaction between HOCl and the microbial cell structure, inactivating required life processes. The rate of disinfection depends on the concentration and form of available chlorine residual, time of contact, pH, temperature, and other factors. Hypochlorous acid is more effective than hypochlorite ion; therefore, the power of free chlorine residual decreases with increasing pH. The disinfecting action of combined available chlorine is significantly less than that of free chlorine residual.

Minimum chlorine residuals and contact times for virus inactivation and protozoal cyst destruction are considerably greater than for bacteria. Therefore, treatment of surface waters includes coagulation and filtration to physically remove protozoal cysts and helminth eggs and to reduce the density of viruses, followed by chlorination to inactivate the remaining viruses and bacteria. Establishing a free residual for disinfection and carrying a residual in the water entering the distribution system has proven to be satisfactory for protection. This requires breakpoint chlorination if the surface water contains dissolved ammonia (Figure 7–17).

Sometimes a combined chlorine residual, rather than a free residual, is established in the treated water entering the distribution system to maintain a protective residual and to control bacterial growths in the distribution piping network.

Compared to free residual chlorination, the advantages are the chloramines are less reactive and a residual can be maintained for a longer period of time without rechlorination. For instance, a combined residual can be applied to a treated water before it is pumped through a long pipeline to a municipal distribution system. If insufficient natural ammonia is present in the water, gaseous anhydrous ammonia is applied by feeding equipment similar to that used for chlorine.

Oxidation

Hydrogen sulfide present in groundwater can be rapidly converted to the sulfate ion using chlorine.

$$H_2S + 4Cl_2 + 4H_2O = H_2SO_4 + 8HCl \qquad (7\text{–}17)$$

Breakpoint chlorination in the treatment of surface waters may be used to destroy objectionable tastes and odors and to eliminate bacteria, minimizing biological growths on filters and aftergrowths in the distribution system. Breakpoint chlorination of polluted surface waters, however, can result in formation of trihalomethanes and haloacetic acids.

Distribution System Chlorination

A new main after installation should be pressure tested, flushed to remove all dirt and foreign matter, and disinfected by one of the following methods prior to being placed in service. The continuous-feed method involves supplying water to the new main with a chlorine concentration of at least 50 mg/l. Either a solution-feed chlorinator or a hypochlorite feeder injects chlorine into the water that fills the main. The chlorinated water should remain in the pipe for a minimum of 24 hr while all valves and hydrants along the main are operated to ensure their disinfection. At the end of 24 hr, no less than 25 mg/l chlorine residual should be remaining. In the slug method, a continuous flow of water is fed to the main with a chlorine concentration of at least 300 mg/l. The rate of flow is set so that the column of chlorinated water contacts the interior surfaces of the main for a period of at least 3 hr. As the slug passes other connections, the valves are operated to ensure

disinfection of appurtenances. This method is used principally for large-diameter mains where the continuous-feed technique is impractical. The tablet method, although commonly used for small-diameter mains, is least satisfactory, since it precludes preliminary flushing. Calcium hypochlorite tablets are placed in each section of the pipe, hydrants, and other appurtenances during construction. The new main is then slowly filled with water to dissolve the tablets without flushing them to one end of the pipe. The final solution, with a residual of at least 50 mg/l, should remain in contact for a minimum of 24 hr. Following disinfection by any of these methods, the chlorinated water should be flushed to waste by using potable water. Microbiological tests should then be conducted before placing the main in service.

A broken main is isolated by closing the nearest valves. The first step in repair involves flushing the broken section to remove contamination while pumping the discharge out of the trench. Minimum disinfection includes swabbing the new pipe sections and fittings with a 5 percent hypochlorite solution before installation and flushing the main from both directions before returning the system to service. Where conditions permit, the repaired section should be isolated and disinfected by the procedures prescribed for new mains.

Tanks and reservoirs should be disinfected before being placed into service or following inspection and cleaning. One method is to add chlorine directly to the filling water, using either hypochlorite or a portable solution-feed chlorinator. Standard procedure is a 50 mg/l residual maintained for a minimum of 6 hr. An alternative method, where convenient, is to spray the walls and other surfaces with a solution containing about 500 mg/l of available chlorine.

After the construction or repair of wells, disinfection is performed by using a 50- to 100-mg/l free chlorine residual for 12 to 24 hr. A new well should be operated until the water is practically free of turbidity. A quantity of chlorine solution equal to at least twice the volume of water in the well is added through a pipe or hose. The pump cylinder and drop pipe are then washed with a strong chlorine solution as the assembly is lowered into the casing. After a minimum contact of 12 hr, the chlorinated water is pumped to waste.

Equipment and Feeders

Liquid chlorine is shipped in pressurized steel cylinders. The most popular sizes are 100 or 150 lb, but 1-ton containers may be used in large installations. Storage areas should be cool, well ventilated, and protected from corrosive vapors and continuous dampness. Gas is withdrawn from a container through a valve connection at the top. Liquid chlorine in the cylinder vaporizes to allow continuous gas withdrawal. The design and operation of all facilities should minimize hazards associated with connecting, emptying, and disconnecting containers. The characteristic odor of chlorine provides warning of leaks. Since it reacts with ammonia to form dense white fumes, a leak can be readily detected by holding a cloth swab saturated with strong ammonia water near the suspected area. Calcium hypochlorite is relatively stable under normal conditions; however, reactions may occur with organic substances. Preferably, it should be stored in a location segregated from other chemicals and materials.

The most essential unit in a chlorine gas feeder is the variable orifice inserted in the feed line to control rate of flow out of the cylinder.

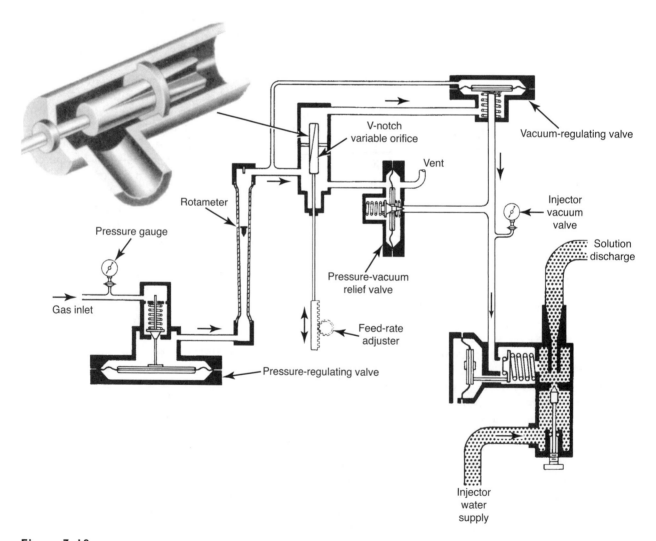

Figure 7–18
Flow diagram of a typical solution-feed chlorinator and a detail of the V-notch variable orifice. (Courtesy of Wallace & Tiernan Inc.)

Its operation is somewhat analogous to a water faucet with a constant supply pressure behind it. The amount of water discharged can be governed by opening the faucet, and if supply pressure does not change, a constant rate of flow is maintained for any given setting. The orifice shown in detail in Figure 7–18 consists of a grooved plug sliding in a fitted ring. The feed rate is adjusted by varying the V-shaped opening. However, since chlorine cylinder pressure varies with temperature, the discharge through such a throttling valve does not remain constant without frequent adjustments of the valve setting. Furthermore, the condition on the outlet side may vary as a result of pressure changes at the point of application. To ensure that these variable conditions do not affect control, a pressure-regulating valve is inserted between the cylinder and the orifice, with a vacuum-compensating valve on the discharge side. A safety pressure-relief valve is held closed by vacuum. If vacuum is lost and chlorine under pressure passes the inlet valve, the relief valve opens and the chlorine gas is vented outside the building. The visible flow meter (rotameter), pressure gauges, and feeder-rate adjusting knob are located on the front panel of the chlorinator cabinet.

Direct feed of chlorine gas into a pipe or channel has certain limitations; for example, a considerable safety problem would exist if gas piping were to be extended throughout the treatment plant. To overcome the difficulties of conveyance and direct introduction of chlorine gas, an injector is employed to permit solution feed. Water flowing through the injector creates a vacuum that draws in chlorine gas from the feeder and mixes it with the water supply. This concentrated solution is relatively stable and can be safely piped to various points in the treatment plant for discharge into an open channel, closed pipeline, or suction end of a pump.

Chlorine feeders can be controlled manually or automatically based on flow or chlorine residual or both. Manual adjustment establishes a continuous feed rate. This type of regulator is satisfactory only where chlorine demand and flow are reasonably constant and where an operator is available to make adjustments as necessary. Automatic proportional control equipment adjusts the feed rate to provide a constant preestablished dosage for all rates of flow. This is accomplished by metering the main flow and using a transmitter to signal the chlorine feeder. Automatic residual control uses an analyzer downstream from the point of application to pace the chlorinator. Combined automatic flow and residual control (Figure 7–19) maintains a preset chlorine residual in the water independent of

Figure 7–19

Automatic flow and residual control of a chlorine feeder.

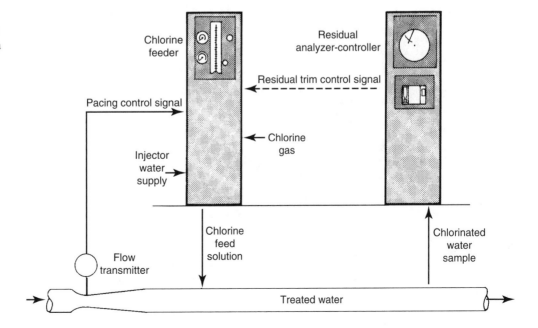

demand and flow variations. The feeder is responsive to signals from both the flow meter transmitter and the chlorine residual analyzer. The system is most effective when flow pacing is the primary chlorine feed regulator, and residual monitoring is used to trim the dosage.

Positive-displacement diaphragm pumps are used for metering hypochlorite solutions. (These pumps are also used for metering fluorosilicic acid in fluoridation.) Figure 7–20 illustrates an electric diaphragm pump showing front and rear views and a diagram labeling the major components. Pumps of this kind are available with capacities from 0 to 25 gal/hr, maximum injection pressures of 20 to 300 psi, powered to be either 120 volts or 240 volts AC or 12 volts DC, and fitted with heads made of different chemical-resistant materials. Ball valves with O-ring seats in the suction inlet and discharge outlet of the head provide tight seals when closed for positive displacement without backflow. The four-function valve with a return line to the chemical container is primarily for safety. An internal check valve prevents backflow into the head of the pump. If the discharge tubing becomes plugged, flow from the pump returns to the chemical supply container, and by pressing a button on the side of the valve, the

pump can be primed by discharging air in the head through the pressure-relief line, thus filling the head with the chemical being fed. The rate of chemical feed is controlled by the speed of rotation of the disk in the head and stroke length of the shaft pressing on the diaphragm. For the unit pictured in Figure 7–20, the feed rate is manually controlled. Other models are designed for flow-proportional metering, where chemical injection is paced by flow variations to maintain constant dosage, or by instrument-response control, where chemical injection is paced in response to a milliampere electric input signal to the pump.

Chlorine Dioxide

Chlorine dioxide (ClO_2) is applied for disinfection or taste and odor control. It is produced in the water treatment plant by mixing solutions of sodium chlorite and chlorine in controlled proportions, as shown in Eq. 7–18.

$$NaClO_2 + Cl_2 \longrightarrow 2ClO_2 + 2NaCl \qquad (7\text{–}18)$$

The use of chlorine dioxide in the United States is limited. The greatest disadvantage is the potential

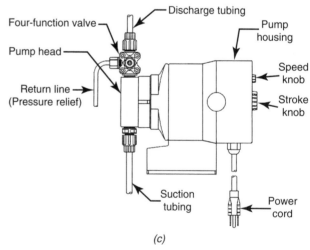

(a)   (b)   (c)

Figure 7–20

Positive-displacement electronic diaphragm pump for metering liquid chemicals, such as sodium hypochlorite and fluorosilicic acid. (a) Front view showing the discharge end assembly. (b) Rear view showing the rate-of-speed knob at the top and length-of-stroke knob in the center. (c) Component diagram. (Courtesy of LMI, Milton Roy.)

formation of chlorate and chlorite residuals, which are toxic chemicals. The recommended limit for these residuals in drinking water is 1.0 mg/l. Second, the high cost of sodium chlorite makes chlorine dioxide more expensive than chlorination. The advantages of chlorine dioxide are strong disinfecting action over a wide pH range, formation of a residual maintained in the treated water entering the distribution system, and in the treatment of surface waters, no reaction with ammonia nitrogen to form chloramines nor with humic acids to form trihalomethanes.

■ EXAMPLE 7–11

Chlorine usage in the treatment of 5.0 mgd of water is 17.0 lb/day. The residual after 10 min contact is 0.20 mg/l. Compute the dosage in milligrams per liter and chlorine demand of the water.

Solution

$$\text{Dosage} = \frac{17.0 \text{ lb/day}}{5.0 \text{ mil gal/day}} \times \frac{\text{mg/l}}{8.34 \text{ lb/mil gal}}$$

$$= 0.41 \text{ mg/l}$$

Chlorine demand $= 0.41 - 0.20 = 0.21$ mg/l

■ ■ ■

■ EXAMPLE 7–12

A new water main is disinfected using a 50-mg/l chlorine dosage by applying a 2.0 percent hypochlorite solution. (a) How many kilograms of dry hypochlorite powder, containing 70 percent available chlorine, must be dissolved in 100 l of water to make a 2.0 percent (20,000 mg/l) solution? (b) At what rate should this solution be applied to the water entering the main to provide a concentration of 50 mg/l? (c) If 34,000 l of water is used to fill the main at a dosage of 50 mg/l, how many liters of hypochlorite solution are used?

Solution

(a) Kilograms of hypochlorite powder for 2.0 percent solution:

$$= \frac{100 \text{ l} \times 1.0 \text{ kg/l} \times 0.02}{0.70}$$

$$= 2.86 \text{ kg/100 l}$$

(b) Feed rate for 50 mg/l:

$$= \frac{50 \text{ mg/l}}{20,000 \text{ mg/l}}$$

$$= \frac{1 \text{ volume of 2.0\% solution}}{400 \text{ volumes of water}}$$

(c) Solution usage $= 34,000 \text{ l} \times \dfrac{50 \text{ mg/l}}{20,000 \text{ mg/l}} = 85 \text{ l}$

■ ■ ■

7–13 DISINFECTION BY-PRODUCTS

Chlorination of water containing humic organic substances from natural sources, usually surface waters containing decaying vegetation from land runoff, produces toxic chlorinated by-products. The principal kinds are volatile hydrophobic compounds, referred to as trihalomethanes (THMs)—with chloroform ($CHCl_3$) and bromodichloromethane ($CHBrCl_2$) being the most common chemical forms—and haloacetic acids (HAA5). In addition, a variety of nonvolatile hydrophilic compounds are produced, including chlorinated and unchlorinated aromatic and aliphatic compounds. THMs and HAA5 are suspected carcinogens and have maximum contaminant levels of 0.080 mg/l and 0.060 mg/l, respectively, in drinking water (Section 5–3).

Control of Disinfection By-Products

The practice of breakpoint chlorination as an initial step in processing surface waters for disinfection and control of taste and odor contributes to the formation of THMs and HAA5. Also, sustained high free chlorine residual increases the concentrations of by-products with time. The following modifications to existing methods of water treatment are recommended as ways of reducing the level of by-products in treated water: application of chlorine after chemical coagulation and sedimentation; improvement of the clarification process where raw water is high in precursors—for example, greater than 2.0 mg/l of total organic carbon; use of alternative disinfectants, such as chlorine dioxide; and application of powdered activated carbon to adsorb by-products and precursor substances.

Applying chlorine to later stages in the treatment of surface waters is the easiest method for reducing formation. Chlorine can be added after coagulation and sedimentation or after filtration, if prior control of microbial populations is not necessary. The benefits of delayed chlorination are a reduction in the required dosage and removal of humic substances before applying chlorine. If necessary, powdered activated carbon can be applied during early treatment stages to adsorb humic substances, and improved coagulation can be implemented to provide better removal of organic matter prior to filtration.

Alternative disinfectants that reduce by-product formation are chloramines, chlorine dioxide, and ozone. The use of chloramines, which are weak disinfectants for bacteria and much less effective in inactivating viruses, increases the risk of pathogens in treated drinking water. Despite this, they can be used as a secondary disinfectant to establish a combined residual in the water entering the distribution system. Chlorine dioxide has had limited application as a water disinfectant, although it is used for taste and odor control. The advantage is production of a persistent residual without forming chloramines. A disadvantage is the possible health effects of products formed when chlorine dioxide reacts with organic substances. Ozone is an effective disinfectant that forms no known products that could be detrimental to human health. The disadvantages of ozone oxidation are high cost and lack of residual. If used as a primary disinfectant, chlorine can be added as a secondary disinfectant to provide a protective residual.

By-products are difficult to remove from water. Aeration can reduce the total level, but practical problems with aeration are purification of the air to prevent water contamination and enclosure to reduce cooling. The effectiveness of adsorbing by-products by powdered activated carbon and granular activated carbon is variable. Bromine THMs are more readily adsorbed than chloroform, which is the most common THM. The preferred approach to control by-products is prevention rather than removal after formation.

Disinfection/Disinfection By-Products Rule

The conflicting requirements of providing effective disinfection and of reducing the adverse health effects of disinfection by-products led to the Stage 1 Disinfection/Disinfection By-Products (D/DBP) rule. Under this EPA rule, public water systems are required to limit trihalomethanes (THMs) and five haloacetic acids (HAA5) and to maintain a residual chlorine in the distribution system. The rule also dictates the maximum contaminant levels for THMs and HAA5 (given in Table 5–1) and maximum allowable residual disinfection concentrations in the distribution system for chlorine of 4.0 mg/l, chloramines of 4.0 mg/l, and chlorine dioxide of 0.8 mg/l. In the future, revision and expansion of this rule is expected as Stage 2 Disinfection/Disinfection By-Products rule.

Water treatment plants providing chemical coagulation and filtration under the Stage 1 D/DP must also achieve prescribed reductions of total organic carbon (TOC), depending on source water quality. This contaminant limit is a treatment technique requiring modification or improvement of water processing to reduce TOC concentration in the treated water, which is likely to reduce the precursors in disinfection by-product formation. Although this rule has no exceptions, alternative compliance criteria exist. These criteria are[2]

- The system's source water TOC is less than 2.0 mg/L.
- The system's treated water TOC is less than 2.0 mg/L.
- The system's source water TOC is less than 4.0 mg/l, the source water alkalinity is greater than 60 mg/l (as $CaCO_3$), and the system's DBP (disinfection by-products) levels for TTHM (total trihalomethanes) and HAA5 (5 haloacetic acids) are less than 40 μg/L and 30 μg/L, respectively.

- The system is using only chlorine as its disinfectant and the DBP (disinfection by-products) levels for TTHM and HAA5 are less than 40 µg/L and 30 µg/L, respectively.
- The system's source water specific ultraviolet absorbance (SUVA) prior to any treatment is less than 2.0 L/mg-m.
- The system's treated SUVA is less than 2.0 L/mg-m.

Total organic carbon is the covalently bonded carbon in a wide variety of organic compounds in runoff containing decaying vegetation and contributed by domestic and industrial water usage, disinfection by-products from chlorination, and microbiological products from wastewater treatment. The use of TOC as a surrogate (composite parameter) for quality of drinking water is considered to establish a measure of safety even when individual contaminants cannot be identified. Removing TOC from a drinking water supply reduces the concentration of potentially hazardous, although unidentified, compounds.

7–14 OZONE

As a strong oxidizing gas, ozone (O_3) is both an effective disinfectant and an oxidant of taste and odor compounds. Its reaction is rapid in inactivating microorganisms—oxidizing iron, manganese, sulfide, and nitrite—and slower in oxidizing organic substances, such as pesticides, volatile organic chemicals, and other organic compounds. Whereas chlorine reacts with water to produce disinfecting species, ozone decomposes in water to produce oxygen and free hydroxyl radicals. The rapid oxidation reactions occur in short contact times relative to oxidation by chlorine. Since ozone does not produce a disinfecting residual, chlorine must be added to treated drinking water to establish a protective residual and control bacterial growths in the distribution piping network. Because of the rapid decay of stored ozone gas, it must be generated on the treatment plant site.

An ozonation system consists of (1) air preparation or oxygen feed, (2) electrical power supply, (3) ozone generation, (4) ozone contacting, and (5) ozone contactor exhaust gas destruction. Ambient air is dried to prevent fouling of the ozone production tubes and to reduce corrosion. A common system uses desiccant dryers in conjunction with compression and refrigerant dryers. By varying the voltage or frequency of the electrical supply to the ozone generator, the rate of ozone production is controlled. A generator for water treatment usually employs a corona discharge cell consisting of two electrodes separated by a discharge gap and a dielectric material across which a high voltage potential is maintained. As the dried air or oxygen flows between the electrodes, ozone is produced. Using ambient air, the ozone concentration is 1 to 3.5 percent by weight, which is an adequate concentration to dissolve enough ozone to attain the concentration-contact time necessary in water treatment. Using pure oxygen, the concentration of ozone is approximately doubled. As illustrated in Figure 7–21, the typical ozone contractor has two compartments in series with porous diffusers for introducing ozonated air at the bottom of water columns. The water flows downward in each compartment opposite to the rising fine bubbles of ozonated air. By using a covered tank, the partial pressure of ozone on the water surface is increased and gases are collected for disposal. Ozone in the exhaust gases must be destroyed, or removed by recycling, prior to venting because the concentration of ozone in the exhaust gases exceeds air quality standards. Destroying the ozone in exhausted gases is commonly the option selected because it is usually less expensive than recirculating the exhaust gases through the air preparation system.

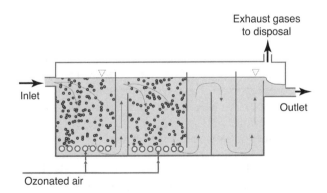

Figure 7–21

Ozone contractor with two compartments in series, with ozonated air entering at the bottom and rising through the downward flow of water, followed by three compartments for additional contact time.

Ozone is rarely applied solely for disinfection because of the high cost relative to chlorine and the absence of a disinfecting residual. Its application is usually for a combination of disinfection plus additional beneficial reasons such as taste, odor, or color control; oxidation of humic organic substances that react with chlorine to form trihalomethanes and haloacetic acids; and destabilization of colloids for improved flocculation at reduced coagulant dosage. The only by-products identified with ozonation in drinking water treatment are trace levels of aldehydes. The increased interest in applying ozone is due to the absence of health-related by-products and the potential for destruction of toxic trace organic chemicals.

7–15 DISINFECTION OF POTABLE WATER

The EPA Interim Enhanced Surface Water Treatment Rule[3] has established maximum goals of zero levels for the contaminants *Giardia lamblia*, *Cryptosporidium* species, enteric viruses, and *Legionella* for public water supplies. Since routine tests cannot be used to determine the presence of these microorganisms, treatment techniques have been established to ensure their removal and inactivation during water processing. Furthermore, because *G. lamblia* cysts, *Cryptosporidium* oocysts, and viruses represent the most persistent pathogens, treatment for their removal ensures the absence of other pathogens. Although some water supplies may not contain significant numbers of these pathogens, demonstrating their absence by water-quality monitoring is not feasible and cannot be used in lieu of applying the specified treatment techniques. The three categories of water supplies are (1) surface water open to the atmosphere and subject to surface runoff, (2) groundwater under the direct influence of surface water (that is, containing algae, insects, or other macroorganisms, or experiencing significant and relatively rapid shifts in water characteristics), and (3) groundwater.

Disinfection of surface waters requires coagulation and granular-media filtration followed by chlorination. Pathogen groups associated with polluted waters are bacteria, viruses, and protozoans (Section 3–5). Although chlorination is effective in killing enteric bacteria and inactivation of viruses, these pathogens may be protected in suspended and colloidal solids if the water has not been filtered for turbidity removal. Cysts of protozoans are resistant to chlorine disinfection, and their removal requires effective coagulation followed by filtration. Waterborne outbreaks of giardiasis and cryptosporidiosis have occurred in communities with surface-water supplies improperly coagulated prior to filtration or unfiltered, even though chlorine was applied for disinfection. In mountainous regions, infected beavers serve as amplifying hosts for *G. lamblia*, contributing cysts to clear mountain streams that may be used as water sources. Outbreaks of cryptosporidiosis have occurred in chlorinated water supplies where filtration was inadequate. Surface waters can be contaminated with *Cryptosporidium* oocysts from cattle and sheep feces as well as domestic wastewater. Waterborne outbreaks of protozoal diseases have emphasized the importance of proper chemical coagulation for effective turbidity removal in filtration of surface waters.

Groundwater not under the direct influence of surface water from properly constructed wells can be disinfected by chlorination, if necessary. Natural disinfection occurs by filtration through surface soils and the subsurface aquifer by straining and trapping of microorganisms and the natural attenuation of pathogens in the adverse underground environment. *Cryptosporidium* oocysts and *Giardia* cysts are filtered out because of their large size. In contrast, enteric viruses are of concern because of their submicroscopic size. Nevertheless, few data exist to substantiate a significant presence of viruses in groundwater not under the direct influence of polluted surface water. Where waterborne outbreaks have occurred in communities supplied by groundwater, most have been caused by backflow of wastewater into the distribution piping or by well contamination due to poor construction and proximity of a source of fecal pollution. A wellhead protection program that delineates a safe infiltration area is important to manage potential sources of groundwater contamination.

Concept of the C · t Product

Inactivation of a specific species of microorganism is a function of concentration of the chemical disinfectant and time of contact; both parameters are of equal importance. Other functions are the kind of disinfectant, temperature, pH, viability of the microorganisms, and turbidity. The $C \cdot t$ product is expressed in units of milligrams per liter times minutes [(mg/l) · min].

Each disinfectant (chlorine, chlorine dioxide, chloramine, or ozone) has distinctive characteristics that result in different $C \cdot t$ values for the same microorganism and same conditions. All disinfectants are influenced by water temperature such that a 10°C increase results in a two- to threefold increase in the inactivation rate of the microorganisms. Of the chlorine disinfectants, free chlorine is influenced by pH because of the dissociation of HOCl to OCl⁻. Hypochlorous acid, the stronger disinfectant form, is present in a proportion of 85 percent or more below pH 7; and the hypochlorite ion, the weaker form, is present in a proportion of 85 percent or more above pH 8.5. The inactivation rates of chloramines, chlorine dioxide, and ozone are relatively independent of pH in a range of 6 to 9.

The biological order of resistance to chemical disinfection from least to greatest follows: bacteria, viruses, protozoal cysts, helminth eggs. Each group has a variety of species encompassing a wide diversity of sizes, life cycles, and other biological characteristics, including resistance to chemical disinfectants. Even within the same species, resistance can vary between those cultured in the laboratory and those found naturally in the environment. Protection by organic matter affects the rate of inactivation, which increases the $C \cdot t$. Consequently, turbidity removal by coagulation and filtration is essential for optimum disinfection action.

Table 7–3 lists values of $C \cdot t$ for 99.9 percent (3.0 log) inactivation of *G. lamblia* cysts by free chlorine at various temperatures and pH values. These data are based on animal infectivity studies.[4] Isolates of *G. lamblia* from infected humans were maintained by passage in Mongolian gerbils. Clean cysts from these animals were exposed to several free chlorine residuals at different temperature and pH conditions for specified $C \cdot t$ values. At the end of the specified contact time, the cyst suspension was concentrated and washed. Several gerbils were fed these cysts and, after 7 days, examined to determine the number that were infected with giardiasis. After extensive trials, the animal infectivity data were

statistically analyzed to correlate $C \cdot t$ values to specific rates of inactivation for different temperature and pH conditions. For example, the $C \cdot t$ values in Table 7–3 inactivated 99.9 percent (3.0 log) of the cysts, leaving 0.1 percent viable. (Each log unit reduction is equivalent to a 90 percent reduction.)

Estimated $C \cdot t$ values for different inactivation rates can be calculated assuming first-order kinetics. For example, $C \cdot t$ values for 99 percent and 90 percent would be, respectively, three-quarters and one-half of the value for 99.9 percent. For temperature correction, the inactivation rate can be assumed to double (a twofold increase) for a 10°C temperature increase. For example, the $C \cdot t$ values for 15°C and 25°C would be, respectively, one-half and one-quarter of the $C \cdot t$ for 5°C. For intermediate temperatures, use Eq. 2–26 to calculate the change in inactivation rate. Extrapolated values are subject to error and should be considered only as estimated values.

Table 7–4 lists values of $C \cdot t$ for 0.5-log and 1.0-log inactivation of *G. lamblia* cysts at various temperatures and pH values for free chlorine, preformed chloramine, chlorine dioxide, and ozone. The data for free chlorine are based on animal infectivity studies. The data for the other three chemicals are based on disinfection studies in in vitro excystation (laboratory testing of cysts in disinfected water suspension followed by examination for viability) of *G. muris*, which has a response to inactivation similar to *G. lamblia*. The experimental $C \cdot t$ values for chlorine dioxide and ozone were multiplied by a safety factor of 1.5 and 2, respectively, to compensate for laboratory in vitro excystation testing rather than animal infectivity studies. Chloramine data were not increased by a safety factor, since chloramination (combining of free chlorine and ammonia in water) in actual practice is more effective than applying preformed chloramine. Ozone, with its extremely low $C \cdot t$ values, is the strongest disinfectant. Chlorine dioxide is next, followed by free chlorine. Even though free chlorine is a slightly weaker chemical, it is preferred because of its lower cost relative to ozone and chlorine dioxide. Chloramine is such a very weak disinfectant that it is considered ineffective against *G. lamblia* cysts.

Table 7–5 lists values of $C \cdot t$ for 2.0-log and 3.0-log inactivation of viruses at various temperatures and pH values. The free chlorine data are for a

TABLE 7–3

$C \cdot t$ Values for 99.9 Percent (3.0-log) Inactivation of *Giardia lamblia* Cysts by Free Chlorine at Various Temperatures and pH Values

FREE RESIDUAL CHLORINE (mg/l)	pH	WATER TEMPERATURE			
		0.5°C [(mg/l) · min]	5°C [(mg/l) · min]	10°C [(mg/l) · min]	20°C [(mg/l) · min]
≤ 0.4	6.5	163	117	88	44
	7.0	195	139	104	52
	7.5	237	166	125	62
	8.0	277	198	149	74
1.0	6.5	176	125	94	47
	7.0	210	149	112	56
	7.5	253	179	134	67
	8.0	304	216	162	81
2.0	6.5	197	138	104	52
	7.0	236	165	124	62
	7.5	286	200	150	75
	8.0	346	243	182	91
3.0	6.5	217	151	113	57
	7.0	261	182	137	68
	7.5	316	221	166	83
	8.0	382	268	201	101

Source: Adapted from *Guidance Manual for Compliance with the Filtration and Disinfection Requirements for Public Water Systems Using Surface Water Sources*. U.S. Environmental Protection Agency.

residual of 1.0 mg/l. The free chlorine $C \cdot t$ values are based on inactivation of Hepatitis A virus in laboratory testing and an applied safety factor of 3. Chlorine dioxide inactivation data using the Hepatitis A virus were multipled by a safety factor of 2. The inactivation by preformed chloramine was tested using rotavirus with no safety factor applied. Laboratory testing of ozone was performed on poliovirus with the $C \cdot t$ values calculated by applying a safety factor of 3. The effectiveness of these four chemicals in the inactivation of enteric viruses from strongest to weakest follows: ozone, free chlorine, chlorine dioxide, and chloramine.

Determination of Actual $C \cdot t$ in Water Treatment

The *Guidance Manual*[4] published by the EPA describes the procedure for determining $C \cdot t$ in water treatment. The $C \cdot t$ for a system is the summation of the calculated $C \cdot t$ values for tanks, reservoirs, and piping transporting chlorinated water before it arrives at the first customer. The C is the free chlorine residual (or other disinfectant residual) measured at the end of each chlorination segment in milligrams per liter, and the t is the calculated contact time of the segment in minutes. For example, if chlorine were added at a pumping station as the water entered a pipeline and again as the water discharged into a contact tank, the overall $C \cdot t$ would be the chlorine residual measured in the discharge of the pipeline multiplied by the contact time in transit plus the residual at the discharge of the contact tank multiplied by the t_{10} time. The contact time in the pipeline is calculated by dividing the length of the pipeline by the velocity of flow. With water entering a tank at a constant rate, the contact time is the t_{10} time, equal to the length of time for 10 percent of the water entering the tank to

TABLE 7–4

$C \cdot t$ Values for 0.5-log and 1.0-log Inactivation of *Giardia lamblia* Cysts

	pH	Log Inactivation	Water Temperature				
			0.5°C [(mg/l) · min]	5°C [(mg/l) · min]	10°C [(mg/l) · min]	15°C [(mg/l) · min]	20°C [(mg/l) · min]
Free chlorine[a]	6	0.5	25	18	13	9	7
	6	1.0	49	35	26	18	13
	7	0.5	35	25	19	13	9
	7	1.0	70	50	37	25	18
	8	0.5	51	36	27	18	14
	8	1.0	101	72	54	36	27
Preformed chloramine	6–9	0.5	640	370	310	250	190
	6–9	1.0	1300	740	620	500	370
Chloride dioxide	6–9	0.5	10	4.3	4.0	3.2	2.5
	6–9	1.0	21	8.7	7.7	6.3	5.0
Ozone	6–9	0.5	0.48	0.32	0.23	0.16	0.12
	6–9	1.0	0.97	0.63	0.48	0.32	0.24

[a]Free chlorine values are based on a residual of 1.0 mg/l.
Source: Adapted from *Guidance Manual for Compliance with the Filtration and Disinfection Requirements for Public Water Systems Using Surface Water Sources*. U.S. Environment Protection Agency.

TABLE 7–5

$C \cdot t$ Values for Inactivation of Viruses at Various Temperatures and pH 6–9

	Log Inactivation	Water Temperature				
		0.5°C [(mg/l) · min]	5°C [(mg/l) · min]	10°C [(mg/l) · min]	15°C [(mg/l) · min]	20°C [(mg/l) · min]
Free chlorine	2.0	6	4	3	2	1
	3.0	9	6	4	3	2
	4.0	12	8	6	4	3
Preformed chloramine	2.0	1200	860	640	430	320
	3.0	2100	1400	1100	710	530
Chlorine dioxide	2.0	8.4	5.6	4.2	2.8	2.1
	3.0	25.6	17.1	12.8	8.6	6.4
Ozone	2.0	0.9	0.6	0.5	0.3	0.2
	3.0	1.4	0.9	0.8	0.5	0.4

Source: Adapted from *Guidance Manual for Compliance with the Filtration and Disinfection Requirements for Public Water Systems Using Surface Water Sources*. U.S. Environmental Protection Agency.

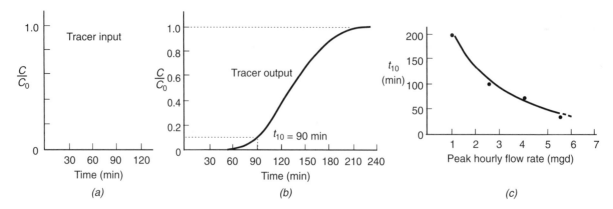

Figure 7–22

Tracer study diagrams to determine t_{10} times for calculating $C \cdot t$ values at peak hourly flow rates by the step-dose method. *(a)* Tracer input at a constant concentration. *(b)* Normalized residual tracer output. *(c)* The t_{10} times for four tracer analyses at different peak hourly rates of flow.

discharge from the tank. In tanks and reservoirs, the t_{10} contact time must be determined by tracer studies.

The hydraulic character of a tank or reservoir is defined by the residence time distribution of individual particles of water in the discharge. Because of short-circuiting and back mixing, the particle travel times through a tank vary from less than to greater than the theoretical detention time of volume divided by rate of inflow. Figure 7–7*a* in Section 7–3 illustrates the residence time-distribution curve from a step tracer input, which is the preferred tracer test method, as illustrated in Figure 7–22. The common tracer ions are chloride and fluoride, since they are nontoxic and approved for potable water use. A solution of the trace is applied at a constant concentration into the water entering the reservoir. The starting time is noted as time zero and the constant concentration of tracer as C_0. Residual tracer concentrations are measured at regular time intervals, the measurements normalized by dividing by the applied concentration C_0, and plotted as shown in Figure 7–22*b*. A horizontal line is drawn from C/C_0 of 0.10 to the tracer output curve and then vertically down to locate t_{10}. At this time, 10 percent of the tracer has passed through the reservoir. A comprehensive tracer study is conducted for at least four different flow rates from expected minimum to maximum flows through the plant during periods of peak flow. Figure 7–22*c* is a plot of t_{10} times

versus flow rate that can be used to determine t for daily $C \cdot t$ calculations. The contact time t used in computing the t_{10} is the peak hourly flow for the day.

Surface-Water Disinfection

The EPA Enhanced Surface Water Treatment Rule[3] requires at least 99.9 percent (3.0 log) removal and/or disinfection of *G. lamblia* cysts, 99 percent (2.0 log) removal of *Cryptosporidium* species oocysts, and at least 99.99 percent (4.0 log) removal and/or inactivation of enteric viruses. Surface waters are defined as all open waters receiving runoff and groundwaters subject to contamination by *G. lamblia* and *Cryptosporidium* by direct influence of nearby surface waters. Groundwaters under the direct influence of surface water are defined by the presence of macroorganisms, algae, or large-diameter pathogens (such as *G. lamblia*) or significant changes in turbidity, temperature, conductivity, or pH that correlate to nearby surface-water conditions.

The rule requires chemical coagulation and granular-media filtration followed by disinfection in water treatment. For traditional and direct filtration systems, turbidity must be continuously monitored in the filtered water from each individual filter in order to identify poor performance so that corrective action can be taken. Turbidity of the combined filter effluent taken every 15 min must be less than or equal to 0.3 NTU for at least 95 percent of measurements taken each month. The combined effluent must not exceed 1 NTU. If these filtered water turbidity criteria are met, the requirement of 2 log *Cryptosporidium* removal is assumed to be achieved.

The free plus combined chlorine residual in the water entering the distribution system cannot be less than 0.2 mg/l for more than 4 hr, and the chlorine residual in the distribution pipe network must be detectable in at least 95 percent of the samples tested each month. Where the residual is undetectable, a measurement of heterotrophic bacteria by plate count of equal or less than 500 per ml may be considered equivalent to a detectable chlorine residual.

Filtration of all surface waters is the major objective of the EPA rules, and only in an unusual circumstance would a community water system provide unfiltered surface water. Nevertheless, the use of unfiltered water is permitted under very stringent conditions. The disinfection requirement is the same: 99.9 percent (3.0 log) inactivation of *Giardia*, 99 percent (2.0 log) inactivation of *Cryptosporidium*, and 99.99 percent (4.0 log) inactivation of viruses. The source water cannot exceed a fecal coliform concentration greater than 20 per 100 ml, or total coliform concentration of 100 per 100 ml, in at least 90 percent of the samples tested. The turbidity cannot exceed 5 NTU except in the case of an unexpected event but not more than two events in the past 12 months. A comprehensive watershed control program is required to minimize pathogen contamination, a water-quality monitoring schedule must be instituted, and yearly on-site inspection is required. No waterborne outbreaks of disease can occur.

Traditional treatment by coagulation, sedimentation, filtration, and chlorine disinfection in a storage reservoir is diagramed in Figure 7–23*a*.

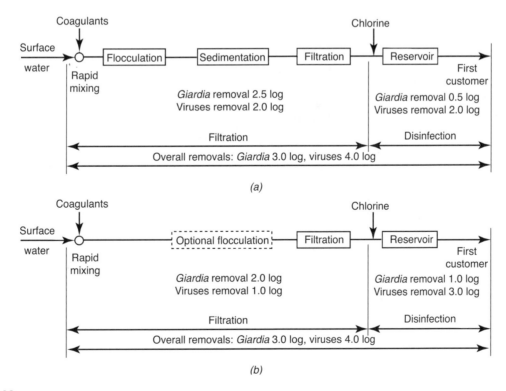

Figure 7–23
Surface-water treatment schemes listing the expected removals of *Giardia* and viruses in filtration and minimum inactivation of *Giardia* and viruses in disinfection. (*a*) Conventional water treatment. (*b*) Direct filtration treatment.

The expected minimum removals in filtration are 2.5 log of *G. lamblia* cysts and 2.0 log of enteric viruses, and the required inactivations by disinfection are 0.5 log of *G. lamblia* and 2.0 log of enteric viruses. Direct filtration by coagulation, filtration, and chlorine disinfection in a storage reservoir is diagramed in Figure 7–23*b*. The expected minimum removals in filtration are 2.0 log of *G. lamblia* cysts and 1.0 log of enteric viruses, which are less than by conventional treatment. Therefore, the required inactivations by disinfection are 1.0 log of *G. lamblia* and 3.0 log of enteric viruses. In actual practice, chlorine may also be applied in earlier stages of treatment to increase the overall $C \cdot t$ for disinfection provided this prechlorination does not produce an excessive concentration of trihalomethanes or haloacetic acids.

■ EXAMPLE 7–13

A new baffled reservoir at a surface-water treatment plant was constructed for storage of filtered water prior to distribution and for disinfection by free chlorine residual. A tracer study by the step-dose method was performed at the design flow of 2.5 mgd to determine t_{10}. The volume of the water in the full reservoir at the time of testing was 0.25 mil gal, which is equivalent to a theoretical detention time of 144 min. The tracer was fluoride ion by applying fluorosilicic acid with a metering pump at a dosage of 2.0 mg/l. The concentration of fluoride ion in the discharge from the reservoir was measured every 6 min from time zero (when the tracer was started at the inlet) for the first hour and then at intervals of 12 min or more. The tracer test data are listed in Table 7–6. Calculate C/C_0 values, plot C/C_0 versus t, and determine t_{10}.

Solution

Calculated values for C/C_0 are given in Table 7–6, and the plot of C/C_0 versus t is in Figure 7–24. The t_{10} determined graphically from the C/C_0 equal to 0.1 is 56 min.

TABLE 7–6

Tracer Test Data for Example 7–13 to Determine t_{10} as Shown in Figure 7–24

MEASUREMENT TIME, T (min)	FLUORIDE TRACER, C (mg/l)	CALCULATED C/C_0[a] (dimensionless)
6	0	0
12	0	0
18	0	0
24	0	0
30	0	0
36	0	0
42	0	0
48	0.09	0.045
54	0.22	0.11
60	0.47	0.24
72	0.74	0.37
84	0.84	0.42
96	1.24	0.62
120	1.32	0.66
144	1.73	0.86
168	1.72	0.86
192	1.77	0.88
216	1.86	0.93
240	1.90	0.95

[a] C_0 = 2.0 mg/l.

■ EXAMPLE 7–14

A traditional surface-water plant with coagulant addition, flocculation, sedimentation, and filtration produces a filtered water with a turbidity less than 0.3 NTU, pH 7, and temperature of 10°C at design flow of 2.5 mgd. After filtration, the water is chlorinated in a baffled reservoir with the hydraulic character as shown in Figure 7–24. (a) What is the required disinfection of the filtered water using free chlorine? (b) If the plant were direct filtration without sedimentation, what would be the required disinfection of the filtered water?

Solution

(a) For conventional treatment, *Giardia* removal is 2.5 log and virus removal is 2.0 log by filtration

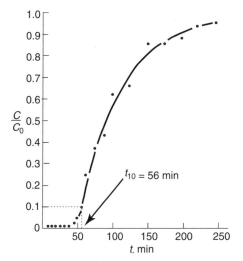

Figure 7–24
The plot of C/C_0 versus t for Example 7–13 to determine t_{10}.

leaving 0.5 log inactivation of *Giardia* and 2.0 log inactivation of virus for disinfection (Figure 7–23*a*).

The $C \cdot t$ required for 0.5 log inactivation of *Giardia* from Table 7–4 is 19 (mg/l) · min based on a free residual of 1.0 mg/l. (The $C \cdot t$ at a lower chlorine residual would be slightly lower.) The $C \cdot t$ required for 2.0 log inactivation of viruses is 3 (mg/l) · min, which is considerably less than for *Giardia*.

From Figure 7–24 at design flow of 2.5 mgd, the t_{10} is equal to 56 min. Therefore, the required free chlorine residual in the discharge for the baffled reservoir is

$$C = \frac{19 \ (\text{mg/l}) \cdot \text{min}}{56 \ \text{min}} = 0.34 \ \text{mg/l}$$

This chlorine residual also satisfies the requirement of a minimum residual of 0.2 mg/l for water entering the distribution system. Also, the drinking water will be acceptable to the taste of most consumers. Free chlorine residual greater than 0.5 mg/l is objectionable to most consumers.

(b) For direct filtration treatment, *Giardia* is a 2.0 log removal and virus is a 1.0 log removal, leaving 1.0 log of *Giardia* and 3.0 log of virus inactivation by disinfection (Figure 7–23*b*).

The $C \cdot t$ required for 1.0 log inactivation of *Giardia* is 37 (mg/l) · min based on a free residual of 1.0 mg/l. The required $C \cdot t$ for 3.0 log inactivation of viruses is 4. From Figure 7–24, the t_{10} is 56 min.

The required free chlorine residual in the discharge from the reservoir is

$$C = \frac{37 \ (\text{mg/l}) \cdot \text{min}}{56 \ \text{min}} = 0.66 \ \text{mg/l}$$

This residual is more than adequate for the minimum 0.2 mg/l required but will most likely be objectionable to consumers in the district where the water from the reservoir enters the pipe network.

■ ■ ■

7–16 GROUNDWATER TREATMENT

Well supplies normally yield cool, uncontaminated water of uniform quality for municipal use. If necessary, a groundwater may require processing to remove toxic contaminants or to improve aesthetic quality. The common nontoxic minerals in groundwater are hardness, that is, calcium and manganese and iron and magnesium. In agricultural regions, groundwater may be contaminated with nitrate from infiltration of fertilizer and from surface application of pesticides. Synthetic chemicals may be found in groundwater from improper disposal of industrial wastewaters. Arsenic, radionuclides, and other inorganic chemicals may originate from natural geological formations.

The flow diagrams in Figure 7–25 illustrate common groundwater treatments. Dissolved iron and manganese in well water oxidize when contacted with air to form tiny rust particles that discolor the water. As shown in Figure 7–25*a*, removal is performed by oxidizing the iron and manganese with chlorine or potassium permanganate and removing the precipitates by filtration.

Precipitation softening, as shown in Figure 7–25*b*, is used to remove excess calcium and manganese. Lime and soda ash, if necessary, are mixed with well water and the resulting precipitate is

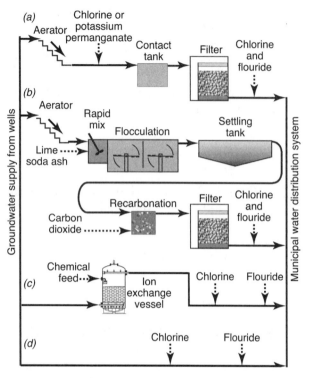

Figure 7–25

Flow diagrams of typical groundwater treatment systems. (a) Iron and manganese removal. (b) Precipitation softening. (c) Ion exchange. (d) Disinfection and fluoridation.

removed. Carbon dioxide is applied to stabilize the softened water prior to final filtration. Aeration is a common first step in the treatment of groundwater to strip out dissolved gases and add oxygen.

The ion exchange process diagramed in Figure 7–25c uses an insoluble resin in a vessel to remove ionized contaminants, such as nitrate and arsenic. The operating cycle consists of passing feedwater through the resin to undergo the exchange followed by backwashing, downflow of brine for regeneration to displace the contaminate ions, and rinsing to remove excess regenerated brine to waste.

The simplest treatment illustrated in Figure 7–25d is chlorination for disinfection and fluoridation to establish the optimum fluoride concentration for dental health. Deep-well supplies may be chlorinated to provide residual protection

against potential contamination in the water distribution system. In the case of shallow wells not under the direct influence of surface water, chlorination can both disinfect the groundwater and provide residual protection.

7–17 PRECIPITATION SOFTENING

Hardness in water is caused by the calcium and magnesium ions resulting from water coming in contact with geologic formations. Public acceptance of hardness varies, although many customers object to water harder than 150 mg/l. The maximum level considered for public supply is 300 to 500 mg/l. A moderately hard water is generally defined as 60 to 120 mg/l. Hardness interferes with laundering by causing excessive soap consumption and may produce scale in hot-water heaters and pipes. To a considerable extent these disadvantages have been overcome by the use of synthetic detergents and the lining of pipes in small hot-water heaters. Industries generally pretreat boiler water to prevent scaling.

Precipitation softening uses lime (CaO) and soda ash (Na_2CO_3) to remove calcium and magnesium from solution. In addition, lime treatment has the incidental benefits of bactericidal action, removal of iron, and aid in clarification of turbid surface waters. Carbon dioxide can be applied for recarbonation after lime treatment to lower the pH by converting the excess hydroxide ion and carbonate ion to bicarbonate ion.

Lime is sold commercially in the forms of quicklime and hydrated lime. Quicklime, available in granular form, is a minimum of 90 percent CaO, with magnesium oxide being the primary impurity. A slaker is used to prepare quicklime for feeding in a slurry containing approximately 5 percent calcium hydroxide. Powdered, hydrated lime contains approximately 68 percent CaO and may be prepared by fluidizing in a tank containing a turbine mixer. Lime slurry is written as $Ca(OH)_2$ in chemical equations. Soda ash is a grayish-white powder containing at least 98 percent sodium carbonate.

Carbon dioxide is a clear, colorless, odorless gas used for recarbonation in stabilizing lime-softened water. The gas is produced by burning a fuel, such as coal, coke, oil, or gas. The ratio of fuel to air is carefully regulated in a CO_2 generator to provide complete combustion. Gas under pressure from the combustion chamber is forced through diffusers immersed in a treatment basin. Several manufacturers produce recarbonation systems for generating and feeding carbon dioxide.

The chemical reactions in precipitation softening are

$$CO_2 + Ca(OH)_2 = CaCO_3\downarrow + H_2O \qquad (7\text{–}19)$$

$$Ca(HCO_3)_2 + Ca(OH)_2 = 2CaCO_3\downarrow + 2H_2O \qquad (7\text{–}20)$$

$$Mg(HCO_3)_2 + Ca(OH)_2 = CaCO_3\downarrow \\ + MgCO_3 + 2H_2O$$

$$MgCO_3 + Ca(OH)_2 = CaCO_3\downarrow + Mg(OH)_2\downarrow$$

$$Mg(HCO_3)_2 + 2Ca(OH)_2 \\ = 2CaCO_3\downarrow + Mg(OH)_2\downarrow + 2H_2O \qquad (7\text{–}21)$$

$$MgSO_4 + Ca(OH)_2 = Mg(OH)_2\downarrow + CaSO_4 \quad (7\text{–}22)$$

$$CaSO_4 + Na_2CO_3 = CaCO_3\downarrow + Na_2SO_4 \quad (7\text{–}23)$$

Lime added to water reacts first with any free carbon dioxide, forming a calcium carbonate precipitate, Eq. 7–19. Next, the lime reacts with any calcium bicarbonate present, Eq. 7–20. In both of these equations, one equivalent of lime combines with one equivalent of either CO_2 or $Ca(HCO_3)_2$. Since magnesium precipitates as $Mg(OH)_2$ ($MgCO_3$ being soluble), two equivalents of lime are needed to remove one equivalent of magnesium bicarbonate, Eq. 7–21. Noncarbonate hardness (calcium and magnesium sulfates or chlorides) requires addition of soda ash for precipitation. Equation 7–23 shows that one equivalent of soda ash removes one equivalent of calcium sulfate. However, magnesium sulfate needs both lime, Eq. 7–22, and soda ash, Eq. 7–23.

The calcium ion can be effectively reduced by the lime additions defined in the equations that raise the pH of the water to approximately 10.3. But precipitation of the magnesium ion demands a higher pH and the presence of excess lime in the amount of about 35 mg/l CaO (1.25 meq/l) above the stoichiometric requirements. The practical limits of precipitation softening are 30 to 40 mg/l of $CaCO_3$ and 10 mg/l of $Mg(OH)_2$ as $CaCO_3$; the theoretical solubility of these compounds is considerably less, being approximately one-fifth of these limits.

A major advantage of precipitation softening is that the lime added is removed along with the hardness taken out of solution. Thus, the total dissolved solids in the water are reduced. When soda ash is added, Eq. 7–23, the sodium ions remain in the finished water along with the accompanying anion, either sulfate or chloride; however, noncarbonate hardness requiring the addition of soda ash is generally a small portion of the total hardness in a raw water. The chemical reactions of lime–soda ash softening can also be used to estimate the quantity of solids produced in waste sludge.

Recarbonation is used to stabilize lime-treated water, reducing its scale-forming potential. Carbon dioxide neutralizes excess lime, precipitating it as calcium carbonate. Further recarbonation converts carbonate to bicarbonate.

$$Ca(OH)_2 + CO_2 = CaCO_3\downarrow + H_2O \qquad (7\text{–}24)$$

$$CaCO_3 + CO_2 + H_2O = Ca(HCO_3)_2 \qquad (7\text{–}25)$$

Excess Lime Softening

Excess lime treatment is used to remove calcium and magnesium hardness to the practical limit of about 40 mg/l. Lime and soda ash dosages are estimated by using the chemical reactions, plus the excess lime addition needed to precipitate the magnesium. A general flow scheme for two-stage excess lime softening is given in Figure 7–26. After excess lime addition, the water is flocculated and settled to remove $CaCO_3$ and $Mg(OH)_2$ precipitates. Subsequent recarbonation is done in two stages. In the first step, carbon dioxide is added to lower the pH to approximately 10.3 and convert the dissolved excess lime into solid $CaCO_3$ (Eq. 7–24) for removal by flocculation and sedimentation. Soda ash, if needed, is also applied in this stage to precipitate noncarbonate calcium

Figure 7–26
Schematic flow diagram for a two-stage excess lime softening plant.

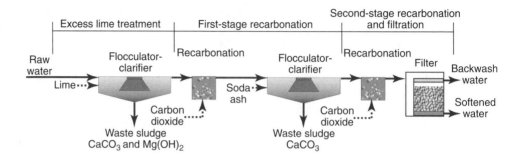

Figure 7–27
Schematic flow diagram for a single-stage calcium carbonate softening plant.

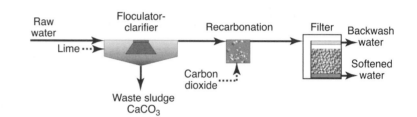

hardness. In the second step, the pH is further reduced to the range of 8.5 to 9.5 to convert most of the remaining carbonate ion to bicarbonate ion (Eq. 7–25) to stabilize the water against scale formation during filtration and in the distribution piping.

Selective Calcium Carbonate Removal

Selective calcium carbonate removal may be used to soften a water low in magnesium hardness, less than 40 mg/l as $CaCO_3$. A magnesium hardness greater than 40 mg/l is not recommended because of the possible formation of hard magnesium silicate scale in high-temperature (180°F) services. The usual processing scheme is lime clarification with single-stage recarbonation followed by filtration, as shown in Figure 7–27. Enough lime is added to precipitate the desired reduction in calcium hardness without adding excess lime for magnesium removal. Soda ash may or may not be required, depending on the extent of noncarbonate hardness. If the $CaCO_3$ precipitate does not settle satisfactorily, a small amount of polymer or alum can be added to aid in flocculation. Recarbonation is usually performed to reduce scale formation on the filter media and to produce a stable softened water.

Split-Treatment Softening

Split treatment consists of dividing the raw water into two portions for softening in a two-stage system, illustrated by the flow diagram in Figure 7–28. The larger portion is given excess lime treatment in the first stage by using a flocculator-clarifier or in-line mixing and sedimentation basins. Soda ash is added to the second stage, where the split flow is blended with the treated water. Excess lime used to force precipitation of magnesium in the first stage now reacts with the calcium hardness that was by-passed around the lime treatment. Thus, the excess lime is used in the softening process instead of being wasted at the expense of carbon dioxide neutralization. Recarbonation is not customarily required, but it may be desirable in the treatment of some waters for stabilization. Since hardness levels of 80 to 100 mg/l are generally considered acceptable, split treatment can result in considerable chemical savings. Lime and recarbonation costs are lower than for excess lime treatment, while the advantage of reducing magnesium hardness to less than 40 mg/l is still possible. The amount of magnesium in the finished water can be calculated by multiplying the by-pass flow times the magnesium concentration in the raw water plus the lime-treated flow times 10 mg/l divided by the total raw water flow. The quantity of flow

Figure 7–28

Schematic flow diagram for a split-treatment softening plant.

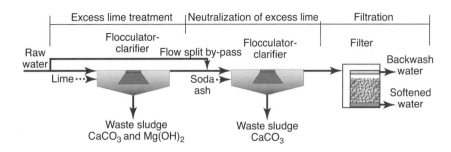

Solution

COMPONENT	mg/l	EQUIVALENT WEIGHT	meq/l
CO_2	8.8	22.0	0.40
Ca^{++}	40.0	20.0	2.00
Mg^{++}	14.7	12.2	1.21
Na^+	13.7	23.0	0.60
Alk	135	50.0	2.70
$SO_4^=$	29.0	48.0	0.60
Cl^-	17.8	35.0	0.51

split around the first stage is often determined by the level of magnesium desired in the softened water.

Softening surface waters may not permit by-passing flow without treatment to remove turbidity. In this case, the split leg is frequently treated by using a coagulant, such as alum. Additionally, coagulant aids or activated carbon or both may be applied to second-stage blending.

■ **EXAMPLE 7–15**

Water defined by the following analysis is to be softened by excess lime treatment. Assume that the practical limit of hardness removal for $CaCO_3$ is 30 mg/l and that of $Mg(OH)_2$ is 10 mg/l as $CaCO_3$.

$$CO_2 = 8.8 \text{ mg/l}$$

$$Ca^{++} = 40.0 \text{ mg/l}$$

$$Mg^{++} = 14.7 \text{ mg/l}$$

$$Na^+ = 13.7 \text{ mg/l}$$

$$Alk(HCO_3^-) = 135 \text{ mg/l as } CaCO_3$$

$$SO_4^= = 29.0 \text{ mg/l}$$

$$Cl^- = 17.8 \text{ mg/l}$$

(a) Sketch a meq/l bar graph, and list the hypothetical combinations of chemical compounds in solution.
(b) Calculate the softening chemicals required, expressing lime dosage as CaO and soda ash as Na_2CO_3.
(c) Draw a bar graph for the softened water before and after recarbonation. Assume that half the alkalinity in the softened water is in the bicarbonate form.

(a) From the meq/l bar graph drawn in Figure 7–29a, the hypothetical combinations are $Ca(HCO_3)_2$, $Mg(HCO_3)_2$, $MgSO_4$, Na_2SO_4, and NaCl. Cacium hardness = 2.00 × 50 = 100 mg/l as $CaCO_3$ and magnesium hardness = 1.21 × 50 = 60.5 mg/l.

COMPONENT	meq/l	APPLICABLE EQUATION	LIME, meq/l	SODA ASH, meq/l
CO_2	0.40	7–19	0.40	0
$Ca(HCO_3)_2$	2.00	7–20	2.00	0
$Mg(HCO_3)_2$	0.70	7–21	1.40	0
$MgSO_4$	0.51	7–22 & 7–23	0.51	0.51
			4.31	0.51

(b) The required lime dosage equals the amount needed for the softening reactions plus 35 mg/l CaO of excess lime to precipitate the magnesium.

$$\text{Dosage} = 4.31 \times 28 + 35 = 156 \text{ mg/l CaO}$$

$$\text{Dosage of soda ash} = 0.51 \times 53$$

$$= 27 \text{ mg/l Na}_2CO_3$$

(c) A hypothetical bar graph for the water after addition of softening chemicals is shown in

Figure 7–29

Bar graphs of raw and treated waters in milliequivalents per liter for Example 7–15. *(a)* Raw water prior to any treatment. *(b)* Water after excess lime softening, but before recarbonation and filtration, *(c)* Lime–soda ash softened water following recarbonation and filtration.

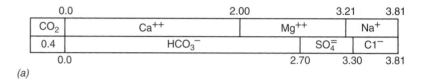

(a)

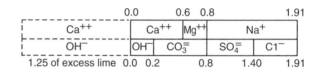

(b)

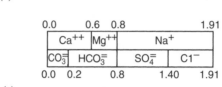

(c)

Figure 7–29*b*. The dashed box to the left of zero is the excess lime (35 mg/l CaO = 1.25 meq/l) added to increase the pH high enough to precipitate the $Mg(OH)_2$. The 0.6 meq/l of Ca^{++} (30 mg/l as $CaCO_3$) and 0.2 meq/l of Mg^{++} (10 mg/l as $CaCO_3$) are the practical limits of hardness reduction. The 1.11 meq/l of Na^+ is the sum of the sodium originally in the water (0.60 meq/l) plus the amount increased by soda ash addition (0.51 meq/l). Alkalinity consists of 0.20 meq/l of OH^- associated with $Mg(OH)_2$, and the remainder is 0.60 meq/l carbonate ion. The $SO_4^=$ meq/l and Cl^- meq/l values are unchanged by the softening process.

Recarbonation converts the excess lime to calcium carbonate precipitate, Eq. 7–24, which is removed by second-stage sedimentation and filtration. Further carbon dioxide addition converts $CO_3^=$ to HCO_3^-, Eq. 7–25. Figure 7–29*c* illustrates the composition of the stabilized, finished water with a total hardness of 40 mg/l.

■ ■ ■

■ **EXAMPLE 7–16**

Consider the softening by selective calcium carbonate removal of a raw water with a bar graph as drawn in Figure 7–30*a*. Calculate the required lime dosage as CaO, and sketch the softened-water bar graph after recarbonation and filtration.

Solution

The only hypothetical combination involving calcium is 2.0 meq/l of $Ca(HCO_3)_2$; therefore, no soda ash is needed and the lime required is 2.0 meq/l (Eq. 7–20), which equals 56 mg/l of CaO.

The softened-water bar graph has 0.6 meq/l of calcium hardness (the practical limit of 30 mg/l), and the total alkalinity is 0.8 meq/l, which is 0.6 meq/l from the practical limit and 0.2 meq/l associated with magnesium in the raw water bar graph. The degree of recarbonation determines the relative amounts of carbonate and bicarbonate anions. The other ions in the softened water are the same as in the raw water.

■ ■ ■

■ **EXAMPLE 7–17**

Consider precipitation softening by split treatment of the raw water, described in Example 7–15. Assume that 40 percent of the flow is by-passed, and that 60 percent is given excess lime treatment in the first stage. Compute the chemicals required and the hardness of the finished water.

Solution

Flow passing through the first stage is processed in the same fashion as in the excess lime treatment described in the solution to Example 7–15.

Figure 7–30

Bar graphs of raw and treated waters in milliequivalents per liter for Example 7–16. (a) Raw water prior to any treatment. (b) Water after selective calcium carbonate removal, recarbonation, and filtration.

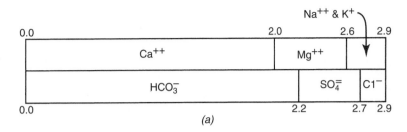

(a)

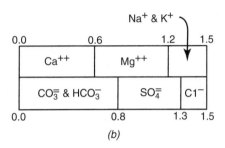

(b)

However, the chemical additions are reduced to 60 percent, since only that fraction of the raw water is being treated.

$$\text{Lime dosage} = 0.60 \times 156 = 94 \text{ mg/l CaO}$$
$$\text{Soda ash} = 0.60 \times 27 = 16 \text{ mg/l Na}_2\text{CO}_3$$

Figure 7–29a is the bar graph of the by-passed water, and Figure 7–29b is the analysis of the first-stage effluent. These two are blended in the second stage where the excess lime reacts with the untreated water. The amount of excess hydroxide in the mixed flow streams is equal to

$$0.60(1.25 + 0.20) = 0.87 \text{ meq/l}$$

The other components of interest in the blended water are

$$\text{CO}_2 = 0.40 \times 0.40 = 0.16 \text{ meq/l}$$
$$\text{Ca(HCO}_3)_2 = 0.40 \times 2.00 = 0.80 \text{ meq/l}$$

First, the carbon dioxide is eliminated by the excess lime:

$$0.87 \text{ meq/l} - 0.16 \text{ meq/l} = 0.71 \text{ meq/l}$$

The balance of the hydroxide ion reacts with calcium bicarbonate, reducing it to

$$0.80 \text{ meq/l} - 0.71 \text{ meq/l} = 0.09 \text{ meq/l}$$

Final calcium hardness equals this remainder plus the limit of calcium carbonate removal,

$$(0.09 \text{ meq/l} + 0.60 \text{ meq/l})50$$
$$= 35 \text{ mg/l as CaCO}_3$$

Magnesium hardness in the finished water is

$$(0.40 \times 1.21 \text{ meq/l} + 0.60 \times 0.20 \text{ meq/l})50$$
$$= 30 \text{ mg/l as CaO}_3$$

Total hardness in the softened water is 35 + 30 = 65 mg/l.

■ ■ ■

7–18 IRON AND MANGANESE REMOVAL

Ferrous iron (Fe^{++}) and manganous manganese (Mn^{++}) are soluble, invisible forms that may exist in well waters or anaerobic reservoir water. When exposed to air, these reduced forms slowly transform to insoluble, visible, oxidized ferric iron (Fe^{+++}) and manganic manganese (Mn^{++++}). The rate of oxidation depends on pH, alkalinity, organic content, and presence of oxidizing agents. If not removed in treatment, the brown-colored oxides of iron and manganese create unaesthetic conditions and may interfere with some water uses.

Preventive measures may sometimes be used with reasonable success. Addition of sodium hexametaphosphate, while not preventing oxidation of the metal ions, may keep them in suspension and thus moving through the system without creating accumulations that periodically cause badly discolored water. The success of this treatment is very difficult to predict, since it depends on the concentrations of iron and manganese, the level of chlorine residual established for disinfection, and the time of passage through the distribution system. The latter is established by the extent of the distribution system, pipe sizes in the network, and location and volume of storage reservoirs.

Reduced iron in water promotes the growth of autotrophic bacteria in distribution mains (Section 3–1). Periodic flushing of small distribution pipes may be effective in removing accumulations of rust particles; however, elimination of iron bacteria is generally difficult and expensive. Biological growths are particularly obnoxious when they decompose in the piping system, releasing foul tastes and odors. Heavy chlorination of isolated sections of water mains followed by flushing has been effective in some cases. The only permanent solution to iron and manganese problems is removal by treatment of the water.

Aeration, Sedimentation, and Filtration

The simplest form of iron oxidation in treatment of well water is plain aeration. A typical tray-type aerator has a vertical riser pipe that distributes water on top of a series of trays from which it then drips and splatters down through the stack. The trays frequently contain coke or stone contact beds that develop and support oxide coatings that speed up the oxidation reactions.

$$\underset{\text{Soluble iron}}{\text{F}^{++}(\text{ferrous})} \ + \ \text{oxygen} \rightarrow \underset{\text{Insoluble iron}}{\text{FeO}_x(\text{ferric oxides})}$$

$$(7\text{–}26)$$

Manganese cannot be oxidized as easily as iron, and plain aeration is generally not effective. Sometimes increasing the pH to about 8.5 with lime or soda ash enhances the oxidation of manganese on coke beds coated with manganic oxides; however, high removals are not assured.

Inefficient removal of manganese can cause serious problems with postchlorination for establishing a residual in the distribution system. Manganese not taken out by the aeration-filtration process may be oxidized in the injector mechanism of a solution-feed chlorinator and may clog the unit; or in the distribution piping, the oxidizing effect of the chlorine residual may create a staining water.

Aeration, Chemical Oxidation, Sedimentation, and Filtration

This is the sequence of processes, illustrated in Figure 7–25a, that is the common method for removing iron and manganese from well water without softening treatment. Preliminary aeration strips out dissolved gases and adds oxygen. Iron and manganese are chemically oxidized by free chlorine residual, Eq. 7–27, or by potassium permanganate, Eqs. 7–28 and 7–29, at rates of oxidation much greater than dissolved oxygen. When chlorine is used, a free available residual is maintained throughout the treatment processes. The specific dosage required depends on the concentration of metal ions, pH, mixing conditions, and other factors. Potassium permanganate is a dark purple crystal or powder available commercially at 97 or 99 percent purity. Theoretically, 1 mg/l of potassium permanganate oxidizes 1.06 mg/l of iron or 0.52 mg/l of manganese. In actual practice, the amount needed is often less than this theoretical requirement. Permanganate oxidation may be advantageous for certain waters, since its rate of reaction is relatively independent of pH.

$$\underset{\text{Soluble ions}}{\text{Fe}^{++} + \text{Mn}^{++} + \text{oxygen}} \ \xrightarrow[\text{Residual}]{\text{Free chlorine}} \ \underset{\text{Insoluble metal oxides}}{\text{FeO}_x{\downarrow} + \text{MnO}_2{\downarrow}}$$

$$(7\text{–}27)$$

$$\underset{\substack{\text{Ferrous}\\\text{bicarbonate}}}{\text{Fe (HCO}_3)_2} + \underset{\substack{\text{Potassium}\\\text{permanganate}}}{\text{KMnO}_4} \ \longrightarrow \ \underset{\substack{\text{Ferric}\\\text{hydroxide}}}{\text{Fe(OH)}_3{\downarrow}} + \underset{\substack{\text{Manganese}\\\text{dioxide}}}{\text{MnO}_2{\downarrow}} \quad (7\text{–}28)$$

$$\underset{\substack{\text{Manganous}\\\text{bicarbonate}}}{\text{Mn(HCO}_3)_2} + \underset{\substack{\text{Potassium}\\\text{permanganate}}}{\text{KMnO}_4} \ \rightarrow \ \underset{\substack{\text{Manganese}\\\text{dioxide}}}{\text{MnO}_2{\downarrow}}$$

$$(7\text{–}29)$$

Figure 7–31

Removal of iron and manganese from groundwater by the manganese greensand process employing a dual-media pressure filter with manganese-treated greensand.

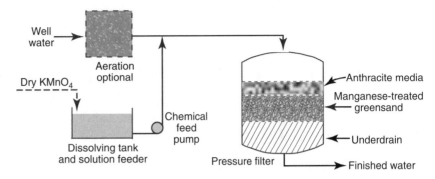

Effective filtration following chemical oxidation is essential, since a significant amount of the flocculent metal oxides are not heavy enough to settle by gravity. Iron and manganese, carried over to the filter, coat the media with oxides that enhance filtration removal. Practice has shown that new filters pass manganese until the grains are covered with oxide that develops naturally during filtration of manganese-bearing water.

Manganese Greensand Process

Manganese greensand is a natural greensand coated with manganese oxide that removes soluble iron and manganese from solution. After the manganese greensand becomes saturated with metal ions, it is regenerated using potassium permanganate. A continuous-flow system is illustrated in Figure 7–31. Permanganate solution is applied to the water ahead of a pressure filter that contains a dual-media anthracite and manganese greensand bed. The iron and manganese oxidized by the permanganate feed are removed by the upper filter layer. Any ions not oxidized are captured by the underlying manganese greensand layer. If surplus permanganate is inadvertently applied to the water, it passes through the coal medium and regenerates the greensand. When the bed becomes saturated with metal oxides, it is backwashed to remove particulate matter from the surface layer and to regenerate the greensand with potassium permanganate.

7–19 WATER STABILIZATION

Corrosive water can cause dissolution of iron, releasing ions into solution that precipitate as ferric oxides. This results in internal pitting and tuberculation of iron pipe and steel water tanks that can ultimately lead to perforation of the pipe or tank wall. Shutoff valves in the distribution pipe network can be made inoperable by corrosion and scale formation. In addition to economic loss, internal corrosion can degrade drinking water quality, creating a health risk to consumers. While dissolved iron causes only unaesthetic discoloration, dissolution of lead from service connections and household plumbing is a serious health hazard, particularly to children. In contrast to iron, lead is not a natural contaminant and is rarely found in either groundwaters or surface waters. Lead is referred to as a corrosion by-product because it is dissolved into corrosive water from lead pipe and solder joints in copper piping.

Corrosion of Iron

Ferrous metal when placed in contact with water results in an electric current caused by the reaction between the metal surfaces and existing chemicals in the water. The basic electrode reactions are represented by Eqs. 7–30 and 7–31. At the anode, iron goes into solution as the ferrous ion, and at the cathode, when dissolved oxygen is present, hydroxide ion is formed.

$$\text{Anode:} \quad \text{Fe} \rightarrow \text{Fe}^{++} + 2 \text{ electrons} \qquad (7\text{–}30)$$

$$\text{Cathode:} \quad 2 \text{ electrons} + \text{H}_2\text{O} + \tfrac{1}{2}\text{O}_2 \longrightarrow \text{OH}^-$$
$$(7\text{–}31)$$

Ferrous (Fe^{++}) iron is further oxidized to the ferric (Fe^{+++}) state in the presence of oxygen and precipitates as insoluble ferric hydroxide, or rust, Eq. 7–32.

$$2Fe^{++} + 5H_2O + \tfrac{1}{2}O_2 \rightarrow Fe(OH)_3\downarrow + 4H^+ \quad \textbf{(7–32)}$$

The rate of corrosion is controlled by the concentration and diffusion rate of dissolved oxygen to the surface of the iron. Diffusion of oxygen to the anode is limited somewhat by the physical barrier of rust forming on the surface. The result of pipe corrosion is pitting and scaling of the surface, referred to as tuberculation, discoloration of the water, and eventual failure of the pipe.

The best way to protect a ductile-iron pipe against internal corrosion is by lining with a thin layer of cement mortar placed during manufacture of the pipe. In addition to cement-mortar lining, a calcium carbonate film can be maintained by chemical stabilization of the water to protect the interior of pipes.

The formation of a protective film of calcium carbonate on the interior of water pipes depends on control of the following chemical reaction:

$$CO_2 + CaCO_3 + H_2O \rightleftharpoons Ca(HCO_3)_2 \quad \textbf{(7–33)}$$

Excess free carbon dioxide dissolves $CaCO_3$, while less-than-equilibrium carbon dioxide forms $CaCO_3$ scale. For stability of a water, the pH must be compatible with carbonate–carbon dioxide equilibrium at the measured water temperature. In actual practice, a calculated oversaturation of 4 to 10 mg/l of $CaCO_3$ is recommended in treatment of water to eliminate aggressive carbon dioxide. The minimum calcium and alkalinity values should be in the range of 40 to 70 mg/l as $CaCO_3$, depending on the concentrations of other ions in solution. Finally, the water must contain sufficient dissolved oxygen, normally 4 to 5 mg/l, to oxidize the nonprotective ferrous hydroxide to ferric hydroxide. Ferric hydroxide is precipitated along with the calcium carbonate scale.

Cathodic Protection

Cathodic protection is a means of counteracting corrosion by preventing the dissolution of iron as shown in Eq. 7–30. Either a sacrificial galvanic anode, composed of a metal higher in the electromotive series, or an electrolytic anode energized by an external source of direct current may be used to protect ferrous structures. Magnesium or zinc galvanic anodes provide a battery action with iron such that they corrode and are sacrificed while the iron structure they are connected to is protected from dissolution. This type of system is common to small hot-water heaters.

Large steel waterworks structures are given cathodic protection using electrolytic anodes that are made anionic by application of a direct current. Anodes may be of many different metals, for instance, graphite, carbon, platinum, aluminum, iron, or steel alloys. They are energized by connecting to the positive terminal of a direct current power supply, usually a rectifier, while the structure being protected is connected to the negative terminal. The impressed current attracts electrons to the iron, preventing ionization and, hence, corrosion.

Cathodic protection is used to prevent corrosion of the interior surfaces of steel water-storage tanks by suspending electrolytic anodes in the stored water. Under some conditions, galvanic anodes are employed in lieu of, or in combination with, the rectifier, depending on the tank condition and chemical characteristics of the water. Buried tanks are protected from exterior corrosion by placing the anodes in the ground surrounding the tank. Except in unusual cases, cathodic systems are not used to protect water distribution piping because of the high cost.

Corrosion of Lead Pipe and Solder

The main sources of lead in tap water are from solder containing half lead and half tin used in joining copper pipes and lead goosenecks that were used to connect galvanized-iron service lines to water mains. The highest lead concentration is in water that has been in contact with solder or lead pipe in a service connection for several hours; the concentration in flowing water is usually undetectable. Therefore, first-flush sampling in the morning is recommended in monitoring for compliance with the allowable maximum contaminant level in drinking water. Lead goosenecks have not been used for decades since the introduction of copper pipe for service connections and, more recently, plastic pipe. Since 1986, lead-free solder has been required in the installation and repair of plumbing for drinking water. As lead solder joints age, the rate of dissolution of lead reduces substantially in 5 years, and old lead goosenecks are often replaced. Therefore, to some extent, lead contamination of tap water is being resolved by the natural aging of solder joints and by replacement of lead pipe. The alternative to removing lead from water lines and plumbing is to reduce corrosivity of the water by chemical treatment.

Lead corrosion control is complicated because the chemical reactions are only partially understood, the physical characteristics of protective surface films are not completely known, and the influence of galvanic action associated with soldered joints is unclear. The pH of a water appears to be a major factor in corrosion of lead. Formation of protective surface films of lead carbonates (for example, $PbCO_3$) that depend on pH and alkalinity appear to be effective in reducing lead dissolution. Calcium carbonate deposition that is effective in reducing iron dissolution may provide only limited effectiveness since the $CaCO_3$ scale can be incomplete and slough off periodically, exposing solder and the surface of lead pipe. Despite this, with hard well water naturally high in calcium and magnesium, service connections containing lead goosenecks and copper pipe joined with lead solder often show negligible first-flush lead concentration.

Treatment of water by orthophosphate addition has been shown to reduce lead dissolution by formation of protective films of lead orthophosphate deposits [for example, $Pb_5(PO_4)_3OH$]. The effectiveness and rate of deposition depend on the phosphate concentration, alkalinity, pH, and temperature. Although the formation is slow, orthophosphate addition with pH control has proved effective in field and laboratory tests. The effectiveness of polyphosphates is controversial. The chemical differences between orthophosphate and polyphosphates in the formation of a protective film is unclear.

7–20 GROUNDWATER CHLORINATION

The Safe Drinking Water Act states:

> . . . the Administrator (of EPA) shall also promulgate national primary drinking water regulations requiring

disinfection as a treatment technique for all public water systems, including surface water systems and, as necessary, (for) groundwater systems.[5]

The first two of three specific categories of drinking water are surface water and groundwater under the direct influence of surface water, which are both covered by the Surface Water Treatment Rules. The third category is groundwater not under the direct influence of surface water, for which the need for disinfection is currently determined based on state and local requirements.

The EPA Proposed Ground Water Rule[6] promulgated in 2003 does not require groundwater systems to practice disinfection. The rule addresses risks through a multiple-barrier approach that relies on five major components:

1. Periodic sanitary surveys of ground water systems requiring the evaluation of eight elements and the identification of significant deficiencies;
2. Hydrogeologic sensitivity assessments to identify wells sensitive to fecal contamination;
3. Source water monitoring for systems drawing from sensitive wells without treatment or with other indications of risk;
4. Corrective actions for significant deficiencies and fecal contamination (by eliminating the source of contamination, correcting the significant deficiency, providing an alternative source water, or providing a treatment which achieves at least 99.99 percent (4 log) inactivation of viruses).
5. Compliance monitoring to insure disinfection treatment is reliably operated where it is used.

Common groundwater supply systems are illustrated in Figure 7–32. In the top diagram, the groundwater is chemically treated by softening or iron and manganese removal followed by filtration; both of these processes can inactive viruses. Subsequent chlorination can enhance disinfection and establish a residual in the distribution system. If chemical treatment and filtration are not necessary, the groundwater may be supplied without disinfection to the distribution system provided the well field is protected against fecal contamination. Nevertheless, chlorine is often applied in larger communities to establish a residual in the

Figure 7–32

Common groundwater supply systems. *(a)* Groundwater from a well field is chemically treated for softening or iron and manganese removal, filtered, and chlorinated prior to distribution.
(b) Groundwater from a well field not requiring treatment is pumped directly to the distribution system with or without chlorination.
(c) Individual wells pump groundwater directly into the distribution pipe network with or without chlorine addition.

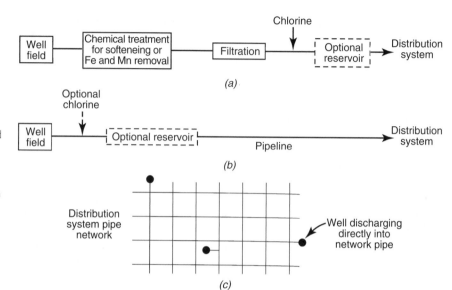

distribution system (Figure 7–32*b*). Community water systems can also receive groundwater from individual wells dispersed throughout the pipe network (Figure 7–32*c*). In communities with dispersed wells, discharges from individual wells cannot be combined into a central supply for treatment. Such reconstruction would require complete disruption of the distribution system because the network pipe sizes are only large enough to distribute dispersed inputs. Nevertheless, chlorine can be applied at individual wells to establish a residual.

Natural disinfection of groundwater results from filtration through surface soils and slow, laminar flow through the aquifer. Filtration through the vadose zone (unsaturated soils above the water table) removes large pathogens, such as *Giardia* cysts and *Cryptosporidium* oocysts. Viruses are removed by inactivation with time and adsorption on the soil grains. Bacteria are removed by both processes. Enteric viruses are of primary concern because of their submicroscopic size. Testing for the presence of pathogenic viruses at low concentration and identifying their species involve complex and difficult procedures. Furthermore, no indicative biological organism or substance has been identified to reliably confirm or deny the presence of enteric viruses.

Evaluation criteria to demonstrate the suitability of natural disinfection include setback distance,

hydrogeologic features, groundwater particle travel time, and virus travel time. The first two are the easiest criteria to apply. Setback distance is the horizontal distance from a well to the nearest potential source of virus contamination, such as a sanitary sewer, septic tank drainage field, or landfill. The specific distance is based on empirical data and professional practice rather than scientific analyses. Common hydrogeologic features include depth of the well screen below the ground surface, thickness of the overlying unsaturated zone, and existence of subsurface impermeable confining layers. Well construction and operating data can often identify these features. Groundwater and contaminant travel times are much more difficult to determine. Although computer models for groundwater flow are available, the accuracy of calculated results for a specific site are difficult to verify.

Outbreaks of waterborne disease attributed to contaminated well water in community water systems are extremely rare. In the absence of significant incidence of waterborne outbreaks, the theory of endemic disease resulting from public water supplies has been proposed. (In this context, *endemic* means constantly present at a significant level affecting individuals within a community rather than a large number of persons; not an epidemic.) Nevertheless, no epidemiological evidence has been established to implicate groundwaters without disinfection treatment in public water supply systems in spreading endemic infectious disease. Almost all of the transmission of enteric viruses in the United States is through person-to-person transmission, fomites, and other close contact. The most effective routes of transmission are in droplets from the sneeze of an infected person or ocular inoculation by rubbing one's eye with a contaminated finger, not by direct ingestion.

Contamination of water in distribution piping, however, has occurred in community systems due to improper operation and physical deterioration. The best protection against contamination is a cross-connection control program to prevent backflow or back-siphonage from high-risk service connections (for example, hospitals, nursing homes, and mortuaries) and enforcement of a plumbing code for residential, commercial, and industrial buildings. In case of an unprotected outlet or poor pipe joint, back-siphonage is prevented by maintaining adequate positive pressure in the supply mains to prevent reversal of flow. The risk of backflow is remote in a properly designed and operated water distribution system. A chlorine residual is not a substitute for a well-maintained system with backflow protection.

Groundwater systems without filtration may add chlorine to establish a residual in the distribution system of 0.2 to 0.4 mg/l. While this level of chlorine residual may inactivate low numbers of susceptible microorganisms in backflow of contaminated water entering a pipe network, the disinfecting action is very limited for resistant microorganisms. Coliform bacteria are inactivated by free chlorine $C \cdot t$ values of less than 0.1 (mg/l) $\cdot$ min; most viruses are many times more resistant, and protozoal cysts are extremely resistant, requiring high $C \cdot t$ values. The degree of protection of health provided by maintaining a chlorine residual in a distribution system supplied by groundwater is very difficult to assess. (A chlorine residual in systems supplied by surface waters is important to suppress the growth of numerous species of heterotrophic bacteria that survive treatment and regrow in the pipe network.) If the water in a portion of the system is contaminated with only 0.1 percent of wastewater containing low concentrations of bacteria and virus pathogens, a chlorine residual of 0.2 to 0.4 mg/l

provides some protection. This residual, however, is no guarantee of protection if the percentage of wastewater is greater or concentrations of pathogens are high. The chlorine demand of the contaminating wastewater can rapidly neutralize the chlorine residual, allowing the pathogens to survive.

A major disadvantage of residual chlorination of uncontaminated groundwater entering a distribution system is interference with coliform testing as an indicator of fecal pollution entering the system. If the water is unchlorinated, the presence of coliforms is a warning of possible backflow contamination. Conversely, if the supply has a chlorine residual, the coliforms can be inactivated and not indicate the presence of more persistent viruses and protozoal cysts. The water utility personnel may not promptly detect that a low level of contamination from backflow is occurring in the absence of positive coliform tests.

■ EXAMPLE 7–18

Disinfection of the groundwater from a well field is required for virus inactivation of 99.9 percent (3.0 log). The proximity of the river that recharges the well field does not provide adequate natural disinfection based on site evaluation by the state authority. The transmission main from the well field to the first customer is 2.0 miles, the velocity of flow at peak pumping capacity is 3.0 ft/sec, water temperature is 5°C, and pH is 7.5. If chlorine is applied at the well field, is disinfection adequate before the water arrives at the first customer?

Solution

From Table 7–5 in Section 7–15, the $C \cdot t$ for 3.0 log virus inactivation by free chlorine at 5°C is 6.

The travel time in the transmission main equals

$$\frac{2.0 \text{ mi} \times 5280 \text{ ft/mi}}{3.0 \text{ ft/sec} \times 60 \text{ sec/min}} = 57 \text{ min}$$

$$C = \frac{6 \text{ (mg/l)} \cdot \text{min}}{57 \text{ min}} = 0.11 \text{ mg/l}$$

However, if the state authority were to require the water entering the distribution system to have a residual of 0.2 mg/l, the chlorine dosage would have to be increased. The actual $C \cdot t$ would then be 0.2 mg/l $\times$ 57 min equal to 11 (mg/l) $\cdot$ min, which is more than adequate for 99.99 percent (4.0 log) virus inactivation.

■■■

7–21 ION EXCHANGE

In the ion exchange process, an insoluble resin removes ions of either positive charge or negative charge from solution and releases other ions of equivalent charge into solution with no structural changes in the resin. For the exchange process to occur, the salts must be ionized in solution separated into cations with positive charge and anions with negative charge. The exchange resins are organic polymer beads usually in the size range of 0.3 to 1.2 mm in diameter with a specific gravity of 1.3 to 1.4.

In water treatment, ion exchange resins can be used to remove specific anions such as nitrate, fluoride, arsenic and other contaminants or cations such as calcium and magnesium. The operating cycle of a typical fixed bed consists of the following four steps:

- Service cycle where feedwater passes down through the resin bed to undergo the exchange reaction;
- Upflow backwashing with product water to prepare the resin for regeneration;
- Downflow of brine for regeneration to displace the exchanged ions from the resin; and
- Rinsing the ion exchange resin to remove excess regenerating brine to waste.

Cation Exchange Softening

This process is used in residential and commercial softeners to reduce hardness in a pubic water supply. Equation 7–34 is the reaction that takes place, with R representing the anionic component of the resin. Calcium and magnesium cations are adsorbed, and an equivalent amount of sodium ions is released into solution. When the exhausted bed is regenerated, the reverse exchange takes place. As shown in Eq. 7–35, a strong sodium

chloride solution displaces the bivalent hardness ions and regenerates the resin with sodium.

$$\begin{matrix} Ca^{++} \\ Mg^{++} \end{matrix} + Na_2R \rightarrow \begin{matrix} CaR \\ MgR \end{matrix} + Na^+ \qquad (7\text{–}34)$$

$$\begin{matrix} CaR \\ MgR \end{matrix} + NaCl \xrightarrow[\text{NaCl}]{\text{excess}} Na_2R + \begin{matrix} Ca^{++} \\ Mg^{++} \end{matrix} \qquad (7\text{–}35)$$

The milliequivalents bar graph of a water after cation exchange softening has the same number of milliequivalents as the raw water, with all of the Ca^{++} and Mg^{++} ions replaced by Na^+ ion.

Homeowners with ion exchange softeners should be aware that the concentration of sodium in the softened water increases in proportion to the hardness ions removed. In some household installations, a separate faucet for drinking water is installed in the kitchen, by-passing the softener. Or, if feasible, the softener could process only the hot-water supply allowing all cold-water appliances and faucets to be directly connected to the public water supply.

The maximum sodium concentration in drinking water for persons on a sodium-restricted diet, which is usually 2000 mg of sodium per day, is 100 mg/l. For a severely restricted diet of 500 mg per day, the maximum concentration recommended by the American Heart Association is 20 mg/l. Cation exchange softening of a water with a hardness of 218 mg/l in a household softener produces a water containing 100 mg/l of sodium. Softening a water with a hardness of 43 mg/l produces a water containing 20 mg/l of sodium.

7–22 ANION EXCHANGE FOR NITRATE REMOVAL

The most effective exchangers for nitrate removal are strongly basic anion resins, referred to as "selective" nitrate resins because they are less selective for multivalent anions such as sulfate.[7] Because of this characteristic, even through sulfate anions are still adsorbed, adsorption of sulfates does not cause massive dumping of nitrates when these resins are depleted as is the case with ordinary resins. Sodium chloride is effective in regenerating both selective and ordinary resins. Reactions for nitrate removal and regeneration are shown in Eq. 7–36.

$$RCl + \begin{matrix} SO_4^= \\ NO_3^- \end{matrix} \underset{\substack{\text{regeneration} \\ \text{with NaCl}}}{\overset{\text{nitrate removal}}{\rightleftharpoons}} \begin{matrix} RSO_4 \\ RNO_3 \end{matrix} + Cl^- \quad (7\text{–}36)$$

The resin beds are susceptible to plugging and loss of efficiency caused by mineral deposits, particulates, and bacterial growths. Thus, the groundwater being treated must be free of silt, iron, and manganese and nonscaling. If a bed is fouled with contaminants, cleaning the bed by backwashing is not feasible since the sand, metal oxides, and most other contaminants have greater resistance to fluidization than the resin beads. Aging of strongly basic resins results in decreased capacity and increased leakage as the sites on the resin beads change from strongly basic to a mixture of weakly and strongly basic sites. The life of a resin should be several years. The factors influencing resin life are type of resin, chemical characteristics of the water being treated, operating temperature, regeneration temperature, regeneration level (salt applied per unit bed volume), water feed rate, and bed depth.

Sketched breakthrough curves for an anion nitrate-selective resin bed during filtration are shown in Figure 7–33. The concentrations of bicarbonate, sulfate, and nitrate anions in the feedwater are shown on the right edge of the diagram, where all of the anions are breaking through the bed. The sulfate anion is removed in preference to all other anions. The bicarbonate is removed initially and then slowly increases in leakage through the resin bed until the concentration is unchanged from the feedwater. Nitrate anions continuously leak through the resin bed at a very low concentration throughout initial filtration. After depletion of the bed starts, the leakage of both sulfate and nitrate increases. The maximum practical operating range is from the start of filtration to this point of leakage.

The common arrangement for a nitrate-removal plant treating groundwater from several wells consists of three ion exchange vessels, a NaCl brine tank for regeneration, a storage tank for spent brine and backwash water, and piping for filtration and regeneration. Figure 7–34 is a cutaway of a typical

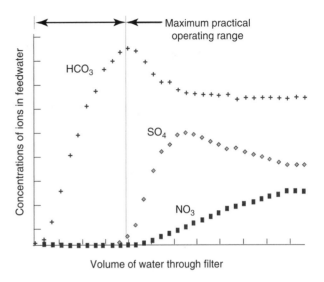

Figure 7–33
Typical breakthrough curves for a nitrate-selective resin. Feedwater concentration of ions increases with volume of water treated.

ion exchange pressure vessel that allows pumping directly through the exchangers into the municipal water distribution pipe network with storage capacity. Feedwater enters the operating vessels at the top through an inlet distributor, passes down through the resin beds, and leaves through the underdrain manifolds at the bottom of the vessels. The depth of the resin beds is 2 ft to 5 ft, supported on underdrain beds. The depth for a particular plant is established by allowable head loss through the beds and by expanded bed depth during backwashing, which is 50 to 75 percent. The three ion exchange vessels are operated in rotational sequence with two units always in service. This provides treated water with more uniform chemical characteristics since the lag unit is treated water in the first half of the operating range and is the lead unit in the second half of the operating range. Three vessels are also necessary in case one is taken out of service for maintenance. Since the nitrate-nitrogen concentration is reduced to 1 to 2 mg/l, well below the maximum contaminant level of 10 mg/l, the treated water is blended with untreated water prior to discharge to the distribution system.

The regeneration cycle consists of three stages: filling the resin bed with brine, slow rinsing, and backwashing. After the vessel is taken out of service, diluted brine is pumped into the ion

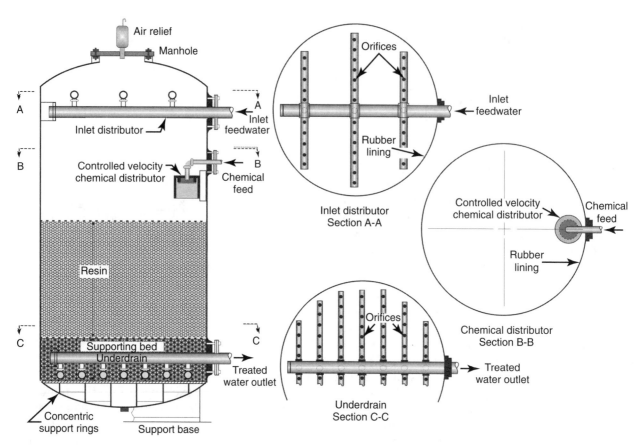

Figure 7–34
Cutaway of a typical ion exchange vessel.

exchange vessel through the chemical distributor located above the resin bed. The dose is about 1.5 bed volumes. The downward flow of brine, with limited mixing with the water in the vessel above the bed, exits through the underdrain manifold and discharges to waste.

Slow-rinse water, which is treated water, enters the vessel at the top through the same distributor used for the untreated water supply during filtration. It exits through the underdrain to waste.

Backwashing is started by upflow of treated water through one-half of the underdrain manifold on one side, followed by a pause to allow the resin to settle. The same is done through the other half of the underdrain manifold and a second time through the first half. The wash water exits out the inlet distributor at the top of the vessel. This backwashing technique declassifies the resin bed, preventing the resin beads from settling into graded

layers with the smallest beads on top and largest beads on the bottom. If this stratification were allowed to occur repeatedly after backwashing, the large beads at the bottom would remain in the nitrate form after bed regeneration and cause increased nitrate in the product water when the bed is returned to service. Wastewater flows usually total 3 to 4 percent of the water processes through the resin beds, which is about 50 percent rinse water, 35 percent backwash water, and 15 percent dilute brine. The amount of sodium chloride for regeneration is commonly 3 to 4 lb per 1000 gal of water processed through the resin. Wastewater disposal options are discharge to the municipal sanitary sewer, land application, discharge to evaporation ponds, or pumping into deep injection wells.

The major disadvantages of anion exchange treatment for nitrate removal are high operating costs and the problem of brine wastewater disposal.

7–23 ARSENIC REMOVAL

The maximum contaminant level of arsenic established by the EPA has been lowered from 50 µg/l to 10 µg/l. Arsenic occurs widely in the earth's crust in the form of insoluble complexes with iron and sulfides and soluble arsenic in water by weathering of rocks, burning fossil fuels, smelting ores, and manufacturing. Nevertheless, for most people, the most significant exposure to arsenic is through food. Although present in some surface waters—mostly mine drainage—the major drinking water source is groundwater. Arsenic in soluble form occurs primarily as arsenites (AsO_3) and arsenates (AsO_4). In oxygen-deprived groundwater, the arsenite species are usually 20 to 50 percent of the total arsenic.

The options for arsenic compliance include switching to a low-arsenic surface-water source; blending existing groundwaters from individual wells or new well fields; and removing arsenic by treatment. The most common large-scale treatment technologies are coagulation and filtration, iron removal, and lime softening. The critical need is for new treatment techniques, such as anion exchange resins, for small communities with several individual wells supplying water to the distribution system. The financial burden will be greatest for communities with arsenic-contaminated wells distributed throughout the pipe network. If individual groundwater treatment processes are not feasible, the alternative solution is to make major changes to the pipe network to connect the individual well supplies at a common location to facilitate treatment and then redistribute the treated water. The development of new technologies is increasing rapidly. The prudent approach for a community is to carefully review options as they become available and consider local pilot-plant studies since the mineral content of each water source is unique. Any existing technology may be improved or made obsolete. Other potential processes to remove arsenic are activated alumina adsorption, iron oxide-coated sand filtration, and nanofiltration/reverse osmosis.

Soluble forms of arsenic can be removed by strongly basic anion exchange resins. Proper pretreatment is necessary to ensure all the arsenic is oxidized to the arsenate form, which can be performed by oxidation with chlorine. The basic anion exchange resins can be regenerated in a single step with a brine of common salt. However, anion resins have a tendency to dump arsenic in preference for sulfate, so the process may not be appropriate for small, unmonitored systems. The presence of high levels of sulfate interfere with and impair the removal efficiency of arsenic in any form, as the sulfate ion is removed along with the arsenic and therefore increases the exchange load. Another concern in ion exchange is disposal alternatives for the waste brine, which is generally limited to land application or landfill.

Many small public groundwater systems, particularly in western United States, have arsenic concentrations greater than the MCL of 10 µg/l. These small communities—often with aging populations—with arsenic-contaminated wells do not have the financial resources to install complex adsorptive/absorptive treatment systems or to employ personnel for operation and maintenance. The need to lower the MCL of arsenic in these public water supplies has caused the EPA to revive exemptions and to delay compliance for small systems based on the affordability of compliance. The EPA is encouraging states to approve arsenic exemptions for systems unable to meet the new arsenic standard. Some primacy states are granting exemptions of the arsenic rule that can allow up to an extra 9 years for compliance for systems with fewer than 3300 people. The EPA is also reconsidering introducing small-system variances for systems serving fewer than 10,000 people to allow installation of treatment systems that are affordable but incapable of meeting the MCL of 10 µg/l. Affordability criteria for arsenic variance are to be prepared with a general guideline of a maximum triple MCL of 30 µg/l, which is within a range of safety reflecting uncertainty factors and safety margins considered in setting the MCL of 10 µg/l. Furthermore, the Safe Drinking Water Act would likely be expanded to authorize variances in small systems for other contaminants with low MCLs that are difficult to meet. New regulations would be required to identify "available and affordable" variance technologies for systems serving 10,000 or fewer people. Furthermore, the EPA would be required to determine that any variance technologies identified would be "protective of public health." Of concern in the EPA's introduction of affordability criteria for variances is that this would result in an authorization of a two-tiered framework, with poorer communities receiving a lesser-quality water than defined by MCLs.

7–24 DEFLUORIDATION

Community water supplies that exceed 5 mg/l must either install defluoridation facilities or change to a low-fluoride source. Although few treatment plants have been built, the preferred treatment technique is filtration through granular activated alumina, which is a semicrystalline inorganic adsorbent.[7] Defluoridation with activated alumina is similar to an ion exchange process but much more complex. The process involves four steps: treatment, backwash, regeneration, and neutralization. For removal of the fluoride, the activated alumina is regenerated with a strong solution of NaOH so that the entire bed is in a pH range of 12.5 to 13 as a result of the caustic regeneration solution. The sequence for regenerating an exhausted bed is backwash with raw water to expand the bed, upflow regeneration with 1 percent NaOH, upflow rinse, and a final downflow.

Fluoride removal is strongly pH dependent with optimum removal at pH 5.0. The bed after caustic regeneration is neutralized with pH feedwater acidified with sulfuric acid. Fluoride will not be removed during initial neutralization, and the acidified feedwater is discharged to waste. As the pH in the bed decreases, the feedwater is adjusted to pH 5.5 and maintained until the bed becomes exhausted. Low-fluoride water produced during the early part of the run can be blended with high-fluoride water produced during the latter stages to yield a combined water with an acceptable level of fluoride. Most activated alumina defluoridation systems have at least two beds.

7–25 MEMBRANE FILTRATION

The most common membrane processes for the treatment of a potable water supply are microfiltration, ultrafiltration, and reverse osmosis. Microfiltration and ultrafiltration use physical straining to remove colloidal and particulate contaminants to meet drinking water regulations for removal of microbial pathogens such as *Giardia* cysts and *Cryptosporidium* oocysts. The reverse osmosis and nanofiltration processes use semipermeable membranes to separate dissolved salts, organic molecules, and metal ions.

The application of different membrane systems is based on pore size for removal of contaminants. The fine particle range of 0.1–10 μm (0.1–10 microns) encompasses turbidity-producing particles, *Giardia* cysts (8–18 μm by 5–16 μm), *Cryptosporidium* oocysts (4–6 μm), and large bacteria. The membrane process in the pore size range of approximately 0.7–7 μm is referred to as *microfiltration* (MF). The molecular range of 0.001–0.1 μm includes microorganisms, colloids, and high-molecular-weight compounds. The membrane process in the pore size range of approximately 0.008–0.8 μm is *ultrafiltration* (UF). The smallest range of 0.0001–0.001 includes removal of aqueous salts, dissolved organic compounds, and metal ions. The membrane processes in this range include *nanofiltration* (NF), approximately 0.005–0.008 μm, and *reverse osmosis* (RO), 0.0001–0.007 μm.

Figure 7–35 summarizes the filtration processes in water treatment for the removal of contaminants. Under the upper scale, in microns (10^{-6} m), the horizontal bars for specific contaminants can be correlated to their size in microns and to the membrane filtration process that can strain them from solution.

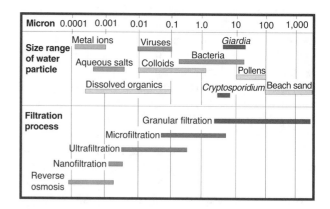

Figure 7–35

Specific contaminants related to their size in microns and membrane filtration processes and granular-media filtration. (Courtesy of the Water Quality Improvement Center, Yuma, Arizona.)

The selection of a membrane process is determined by the treatment desired, such as turbidity reduction, disinfection, removal of organic compounds, softening, desalination, or specific ion removal (for example, arsenic or nitrate). The cost of membrane treatment increases with the decreasing size of the contaminants to be removed. The major factors in operating costs are pretreatment and posttreatment for membrane separation and operating pressure to force water through the membrane. Microfiltration and ultrafiltration can be operated at feed pressures of 60 psi or less; reverse osmosis is operated in the range of 150–800 psi, depending on recovery of product water.

The minimum pretreatment prior to membrane processing requires removal of solids to prevent plugging of the membrane pores. Membrane systems are normally completely automated to control feed rate and frequency of backwashing. In the treatment of raw water, self-cleaning screens with fine openings are often all that is necessary to protect MF and UF membranes. RO membranes are more susceptible to plugging, scaling, and fouling by dissolved organic and inorganic compounds, chemical precipitates, and bacterial growths. When feedwater is of poor quality, RO membranes may require a sequence of pretreatment processes and frequent chemical cleaning of the membranes in addition to backwashing, and membrane life may be significantly reduced.

Disposal of membrane waste streams is a primary consideration in determining the feasibility of a membrane process. The quantity of waste depends on reject concentrate, backwash water, and membrane-cleaning water. Concentrate and pretreatment wastewaters depend upon recovery of permeate (product water) that can be as low as 30 percent or as high as 98 percent, depending on the system. MF and UF membranes are usually backwashed by flushing with permeate on a timely basis to remove accumulated contaminates, depending on raw water quality and the type of membrane system. An RO plant desalting brackish groundwater or seawater generates a concentrate waste containing nearly all the raw water contaminants, including an extremely high salt content. Disposal of wastewaters include the following: discharge to sewer system, ocean discharge, land application, evaporation ponds, and deep-well injection. For surface water or ocean disposal, the Clean Water Act requires a National Pollutant Discharge Elimination (NPDES) permit.

7–26 MICROFILTRATION AND ULTRAFILTRATION

Microfiltration and ultrafiltration are used to treat surface waters of moderate turbidity to produce potable water. These membranes are often hollow fibers bundled into a pressure vessel with one end potted with an epoxy resin. In the UF module shown in Figure 7–36, the feedwater flows inside the fibers and the filtrate is collected from inside the pressure vessel by a central core tube. Concentrate is collected from the ends of the hollow fibers and discharged to waste. The hollow-fiber membranes are available in two sizes, 0.8 mm inside diameter and 1.2 mm, for higher-turbidity feedwater. The flux rates are 15 to 75 gal/ft$^2 \cdot$ d (0.6 to 3.1 m^3/m$^2 \cdot$ h). At the maximum feedwater pressure of 73 psi maximum transmembrane pressure of 30 psi, the filtrate flow is 11–30 gpm (2.7–6.9 m^3/h). In the backwash mode, filtrate flows in reverse from the central core tube inside the pressure vessel through fibers to flush out the contaminants to waste. The water backwash cycle is every 15–60 min for 30–60 sec at 35 psi. Chemical enhanced backwash frequency is a minimum of 1 or 2 times per day for a duration of 1 to 10 min.

Microfiltration and ultrafiltration systems can also be constructed with hollow-fiber membranes that filter water from outside-in in a transverse flow from the pressure vessel through the walls to the inside of the membranes. In transverse flow, contaminants in the feedwater build up on the outside of the hollow-fiber membrane surfaces and clean filtrate flows out of the inside. The advantage is the lower pressure drop for a given flow and number of fibers because of the greater surface area on the outside of the fibers than on the inside. The transmembrane pressure increases as the contaminant load increases, and a set-point air-water backwash is automatically initiated. Pulses of compressed air from inside of a hollow-fiber membrane dislodge contaminants from the outside of the membrane surface. The contaminants are then carried out of the pressure vessel with either feedwater or filtrate as backwash water. Using air in the backwash sequence keeps

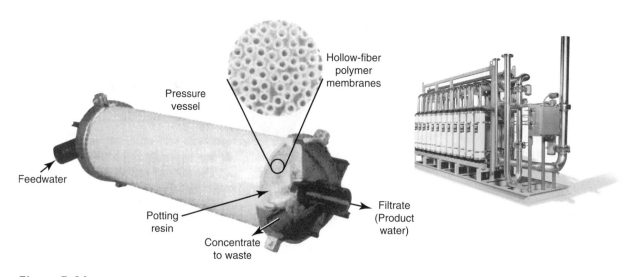

Figure 7–36

Ultrafiltration hollow-fiber module for removal of pathogens, including *Giardia* cysts and *Cryptosporidium* oocysts, and turbidity-producing particles and a Hydranautics treatment plant.

(Courtesy of Hydranautics, a Nitto Denko Corp.)

fibers from coming together to impede cleaning. Other advantages demonstrated by the transverse-flow system are up to 95 percent recovery of feedwater and a significant reduction in the quantity of backwash water production from using air-water backwash.

Figure 7–37 illustrates the Memcor® continuous microfiltration-submerged process. Each cell is self-contained with peripheral equipment including filtrate suction pumps, air blowers, a chemical system for cleaning membranes, and a control system. As shown in Figure 7–37b, the cell is a tank containing a large number of modules filled with hundreds of thousands of individual hollow fibers that filter the water from outside-in. This transverse flow of water from the cell to the inside of the membrane fibers is drawn by vacuum through suction pumps into the filtered water tank. Membrane modules are grouped together in manifolds (Figure 7–37c) that are submerged under the water level in the cell and attached to the filtrate (clean water) manifold.

Combined air-and-water backwash reduces the quantity of the wash water required for flushing the modules, resulting in a higher percentage of filtrate recovery, which is a major advantage of outside-in microfiltration. At backwash initiation, the influent valve to the cell is closed and

filtration is continued until the water level within the cell is lowered to slightly above the top of the modules. Then, air is introduced at the base of the modules to loosen the dirt and debris from the surface of the membrane fibers. To facilitate scouring inside the modules, air is introduced within the base through an air connection in the bottom manifold. As air scour continues outside of the fibers, clean filtered water is fed back to the inside of the fibers to assist in flushing the debris from the outside of the membrane fibers. Upon completion of the air-and-water backwash, the cell is drained to remove the dirty water and debris and refilled with feedwater water. The entire backwash sequence takes less than 3 min.

Membrane fouling is the primary cause of reduced productivity in microfiltration. Chemical cleaning is required for restoration, commonly by using a caustic chemical mixed with surfactants, or dilute citric acid, or sodium hypochlorite. The control system continuously measures and displays the transmembrane pressure of a cell to indicate the need for cleaning. The control system can also perform integrity testing of the membrane barrier, cartridge construction, O-ring seals, and potential cross-connections. If a cell fails the integrity test, maintenance personnel can locate the leakage by introducing air into the

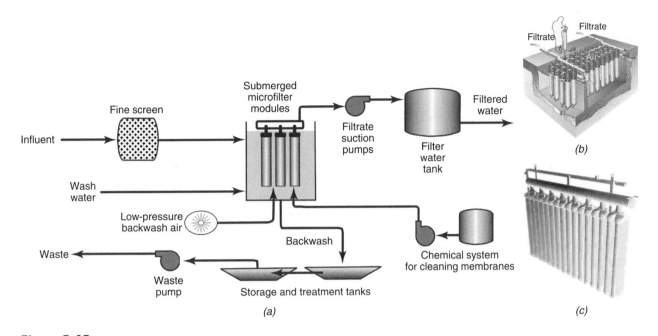

Figure 7–37

Memcor® continuous microfiltration-submerged (CMF-S) process. (a) Cell containing membrane modules and peripheral equipment including influent screen, filtration suction pumps, blowers for air scour, wash-water tank, chemical membrane cleaning system, and backwash processing. (b) Modules in a cell are grouped together in several manifolded assembles. (c) A manifold with suspended microfiltration modules.
(Courtesy of US Filter/Memcor.)

microfiltration system and observing the leaking module by means of the release of air bubbles. The control system also monitors operation of the system, for example, the flow of filtrate based on automatic or manual production command.

■ EXAMPLE 7–19[8]

Until 1997, the public water supply for Marquette, MI was chlorinated water from Lake Superior. The lack of processing, other than chlorination, did not meet the Surface Water Treatment Rule, which states that filtration is required unless the water supplier can demonstrate its drainage basin is fully controlled and the raw water quality meets stringent criteria. Under this rule, filtration was required even though the raw water turbidity averaged between 0.2 and 0.3 NTU (with a historical maximum of 5–6 NTU) from an intake on a rock bottom at a depth of 60 ft and 3000 ft from the shoreline. The problem for the city was lim-

ited building space at the existing pumping station on the shore of the lake. Also, the site was surrounded by private residences and a museum, and the ground surface was bedrock.

Solution

A conventional filtration plant was not a viable option. Microfiltration was recommended because it could, along with chlorination, meet the required $C \cdot t$ (disinfectant concentration · time of contact), only minimal amounts of chemicals were required, and better water quality could be produced compared to conventional filtration.

Two manufacturers provided microfiltration equipment for a 6-month testing period to demonstrate the effectiveness of microfiltration during critical cold-water months. System A was microfiltration using an outside-in flow path (transverse flow) and air-water backwash. System B was microfiltration using an inside-out flow path and water backwash. The pilot study monitored filtrate water quality by particle counts in the critical range of *Giardia* cysts and *Cryptosporidium* oocysts, turbidity, transmembrane pressure, and general operating parameters, such as backwash production and cleaning frequency. Results of the particle removal in the range of 2 to 10 μm in log base 10 was 3.3 for system A and 3.8 for system B, and the turbidity was consistently in the range of 0.03–0.05 NTU for both systems. Both pilot units provided excellent filtrate water quality. Nevertheless, the air-water backwash of system A was more effective in cleaning the hollow fibers and more effective in controlling transmembrane pressure. For systems A and B, the backwash production in percentage of filtered water was 8 percent and 40 percent, respectively, and the average cleaning frequencies expressed in days were 22 d and 2.4 d. Because of these considerations, system A was selected.

The full-scale micofiltration plant was designed based on the following: maximum membrane flux rate of 125 $1/m^2 \cdot h$, based on demonstrated pilot rates; average daily flow of 11,000 m^3/d (2.8 mgd); 8 frames with 90 modules per frame for a total inside membrane area of 10,800 m^2 (116,000 ft^2) with one frame out of service; the plant-rated flow of 26,000 m^3/d (7.0 mgd) with one frame out of service, which is a flux rate of 100 $1/m^2 \cdot h$; and maximum backwash design of 3000 m^3/d (0.80 mgd).

Self-cleaning strainers (screens) were installed as pretreatment for microfiltation. Air-water backwash is discharged under a NPDES (National Pollutant Discharge Elimination System) permit without treatment to Marquette Harbor. Backwash from chemical cleaning of the membranes (sodium hydroxide and surfactant) is neutralized with acid before discharge to the sanitary sewer. The filtered water is chlorinated as it flows into the ground-level storage reservoir and rechlorinated as it is pumped into the distribution system. All operations are highly automated using computer-based control software.

■ ■ ■

7–27 REVERSE OSMOSIS

The normal osmotic flow of water through a semipermeable membrane, separating fresh water and saline water, is for the fresh water to pass through the membrane to dilute the saline water on the opposite side. Reverse osmosis is the forced passage of water through a semipermeable membrane against the natural osmotic pressure to separate dissolved solids, causing the saline water to increase in salt concentration. Under high pressure in the RO process, water from brackish groundwater or seawater passes through the membrane, leaving the dissolved solids behind by blocking passage of salt ions. The osmotic pressure of seawater is about 350 psi; brackish groundwater, having a lower salt concentration, has a significantly lower pressure to overcome for reverse osmosis. Reverse-osmosis operating pressures are typically in the range of 350–800 psi for satisfactory operation. The rate of water transfer depends primarily on the magnitude of the applied pressure, characteristics of the membrane, and the difference in salt concentrations of the saline feed water and the product water.

The spiral-wound module, shown in Figure 7–38, is constructed of large membrane sheets covering

Figure 7–38

Spiral-wound module for reverse osmosis.

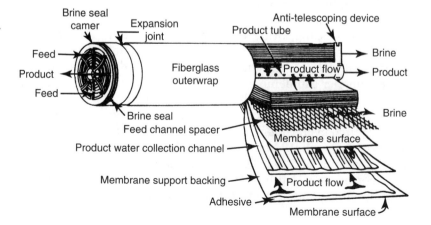

both sides of a porous backing material that collects the permeate (product water). The membranes are sealed in pairs on the two long edges and one end to form an envelope enclosing the permeate collector. The other end of the membrane envelope is connected to a central perforated tube, which receives and carries away the permeate from the collectors. Several of these membrane envelopes with mesh spacers between them for brine flow are rolled up to form a spiral-wound module. Saline water enters the end of the module through voids between the membrane envelopes provided by the mesh spacers. Under high pressure, water is forced from the brine in the spacer voids through the membranes and conveyed by the enclosed porous permeate collectors to the perforated tube in the center of the module. Reject brine discharges from the spacer voids at the outlet end of the module.

A hollow-fiber module, shown in Figure 7–39, is a pressure vessel containing a very large number of microfiber membranes densely packed in a U-bundle with their openings secured in an end block of the module. The hollow fibers with the appearance of coarse thread have an outside diameter of 85 to 100 μm and an inside diameter of 42 μm. Because of their small diameter and relatively thick wall, these tubes can withstand the high reverse-osmosis pressure required to force water from the surrounding brine into the hollow

cores of the fibers. Saline water enters the module through a central perforated feed tube and flows radially through the fiber bundle toward the outer shell of the cylinder. Under high pressure, water enters the hollow fibers and exits from their open ends at the discharge end of the module. Reject brine arriving at the outer shell of the cylinder is collected by a flow screen and conveyed from the module.

A basic reverse-osmosis system consists of pretreatment units, pumps to provide high operating pressure, posttreatment tanks and appurtenances for cleaning and flushing, and a disposal system for reject brine. The saline water fed to RO modules must be clear of suspended solids and free of organic matter and excessive hardness, iron, and manganese to prevent fouling and scaling of the membranes. As diagramed in Figure 7–40, pretreatment commonly includes granular-media and cartridge (25-μm mesh size) filtration, acidification, and addition of a scale inhibitor. In large plants, the granular-media filtration may be preceded by chemical coagulation for turbid waters or precipitation softening for hard waters. If a water contains dissolved organic substances, the common pretreatment is filtration through granular activated carbon followed by chlorination to inhibit biological growths. Reducing the pH to 6 or less by feeding sulfuric acid converts a portion of the bicarbonate ion to carbon dioxide gas, making

Figure 7–39

Hollow-fiber module for reverse osmosis.
(Courtesy of Permasep Products, E. I. du Pont de Nemours & Co.)

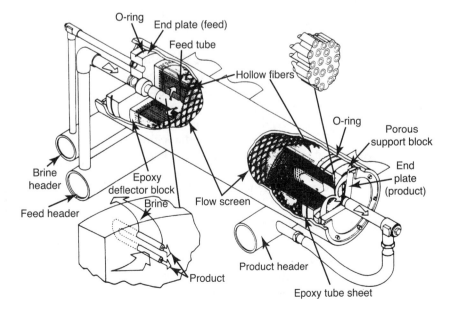

Figure 7–40

Flow diagram of a basic reverse-osmosis system.

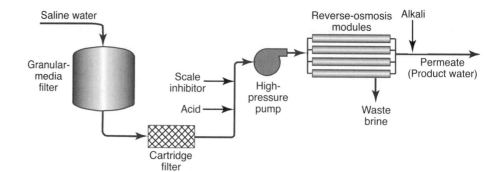

the water corrosive to reduce the probability of calcium carbonate scale and oxidation of iron and manganese. Sodium hexametaphosphate, or some other inhibitor, is applied to avoid chemical scaling. Posttreatment is necessary to stabilize the permeate, since the carbon dioxide formed in acidification can pass through the membranes with the product water. Aerators are used to remove the carbon dioxide, and lime or soda ash is added to provide final pH adjustment. Periodic cleaning of the membrane surfaces is required to maintain high water-transfer efficiency. Modules are flushed with acid rinses and cleaning agents to remove any buildup of metal ions, salt precipitates, organic matter, or biological growths.

The quantity of reject brine varies from 10 to 30 percent of the saline water feed depending on the method of operation. A typical plant that is operating on a feed of 2000 mg/l total dissolved solids can economically achieve about 75 percent conversion; that is, 100 gal of feed produces 75 gal of water with a brine rejection of 25 gal containing 6000 to 7000 mg/l of dissolved solids. A higher percentage of recovery is possible at reduced hydraulic loading or higher pressures; however, the higher dissolved solids content in the rejected water has a greater potential for forming scale on the membranes. The costs involved in disposing of a large volume of reject water are a critical economic and environmental problem associated with reverse-osmosis treatment at inland sites.

■ EXAMPLE 7–20

The groundwater supply for a desert community had increased in salinity beyond the recommended total dissolved solids for drinking water. A reverse-osmosis plant was constructed to process brackish groundwater to drinking water quality. In order to minimize consumption of the potable water, a separate pipe network was constructed to convey the processed water to kitchens of residences and public facilities requiring a potable supply. The untreated brackish groundwater was satisfactory for all indoor uses other then drinking and food preparation. Private lawns and gardens and public parks and landscaping along roads were irrigated with the brackish water.

Solution

The highest-quality groundwater for desalting was pumped from two 400-ft-deep wells with well screens and casings constructed of noncorrosive materials. The groundwater's concentration of total dissolved solids was 2600–2700 mg/l with a pH of 6.8–7.2. Since the well water was free of silt, iron, and manganese and low in silica, the only pretreatment required, as a precaution, was granular-media filters with air backwash.

Figure 7–41 illustrates the primary units in the RO processing at this plant. In Figure 7–41a, booster pumps on the left deliver the filtered groundwater into the plant from a groundwater reservoir. On the right are three cartridge filters to remove particles larger than 5 μm (0.005 mm) in size; the average life of the replaceable filter elements is 12 weeks. Shown in Figure 7–41b are in-line baffle-plate mixers to apply 150 mg/l of sulfuric acid to acidify the brackish water to pH 5.8 to prevent $CaCO_3$ scale formation and 10 mg/l of hexametaphosphate to prevent $CaSO_4$ scale formation. Figure 7–41c is a picture of five RO units in the filtration gallery. In Figure 7–41d, to the left of

(a) *(b)* *(c)*

(d) *(e)*

(f) *(g)* *(h)*

Figure 7–41

Reverse-osmosis processing of a brackish groundwater. *(a)* Influent booster pumps on the left and cartridge filters on the right. *(b)* Chemical feeders to mix sulfuric acid and hexametaphosphate into the feed of brackish groundwater. *(c)* Five independent RO units in the gallery. *(d)* Each RO unit has a high-pressure feed pump and 13 permeator modules. *(e)* On the left is piping for backwashing the modules and chemical cleaning as necessary; on the right are turned-down pipes to convey product water and reject brine to larger pipes in a recessed gallery. *(f)* Countercurrent packed bed to air strip CO_2 from the product water. *(g)* Chemical feeders apply soda ash, fluorosilicic acid, and calcium hypochlorite to the finished water. *(h)* RO module cut open to view hollow fibers.

the RO modules and behind the dark-colored control panel is the high-pressure pump to boost the feed pressure to 370 psi. Each module is 8 in. in diameter and 48 in. long. The central hose of each module discharges the permeate (product water). The RO unit has 13 modules—9 in the first stage and 4 in the second stage. The feedwater applied to the first stage produces 50 percent permeate and 50 percent brine. Brine from the first stage is applied to the second stage to again produce 50 percent permeate and 50 percent brine, which is rejected. Therefore, the total recovery of product water is 75 percent of the feedwater, and 25 percent of the reject brine. Figure 7–41e shows an array of piping on the side of the RO unit for backflushing the modules with product water and selected chemical solutions to remove scale and other contaminants from the membrane surfaces. The turned-down pipes in the rear of the RO unit convey product water and reject brine to larger pipes in a recessed pipe gallery.

The product water is very corrosive from the CO_2 formed from acidification and low pH. Figure 7–41f is an outdoor degasifier with a countercurrent packed bed to air strip CO_2, raising the pH from 5.8 to about 7.0. In Figure 7–41g, the following chemicals are added to the degasified water: approximately 10 mg/l of soda ash to stabilize the water by raising the pH to 8.2–8.5, fluorosilicic acid to increase the fluoride ion concentration from 0.2 mg/l to the optimum of 0.8 mg/l, and calcium hypochlorite to establish a chlorine residual of 0.9 mg/l.

The brackish groundwater treated has a total dissolved solids concentration of 2600–2700 mg/l, alkalinity of about 210 mg/l, calcium hardness of 540 mg/l, and high salinity with approximately 1500 mg/l of NaCl. The finished water after complete RO processing has a total dissolved solids concentration of about 210 mg/l, alkalinity of 80–100 mg/l, and calcium ion concentration of 7 mg/l, which is a stable, noncorrosive, healthy drinking water.

The hollow-fiber modules in this RO plant are as described in Figure 7–39. Pictured in Figure 7–41h is a module cut open to show the hollow fibers that are so fine that they look like threads clustered together. The picture on the left shows the endplate feed and epoxy deflector block slide down the feed tube. The picture on the right shows the block that supports the open ends of the hollow fibers from which the permeate is collected.

■ ■ ■

7–28 DISTILLATION OF SEAWATER

Typical seawater has a salinity (total dissolved solids) of 35,000 mg/l, of which 30,000 mg/l is NaCl. The generally accepted quality standards for drinking water are 500 mg/l of total dissolved solids and 200 mg/l of chloride. Distillation is cost-competitive for desalination of feedwater with a high salt content since the process operates virtually independently of the influent solids concentration. Moreover, a product purity of less than 100 mg/l is easily attained.

Distillation involves heating feedwater to the boiling point and then to steam to form water vapor, which is then condensed to yield a salt-free water. The principal commercial process is multistage flash distillation, which means that a series of evaporation–condensation units are employed to obtain multiple reuse of the energy content of the heated steam. There may be as many as 15 to 25 stages.

Multistage flash distillation is illustrated schematically in Figure 7–42. Seawater entering

Figure 7–42

Multistage flash distillation.

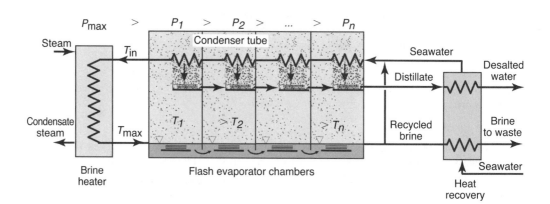

the plant is initially heated in a heat recovery unit in which the hot desalted product water and waste brine discharge are cooled. The warmed seawater is then blended with recycled brine and passed through a series of condenser tubes in the evaporator chambers for further heating. In the process of condensing the steam to distillate (the desalted water), the temperature of the brine is increased in stages to temperature T_{in} as it enters the brine heater. In this unit, the temperature of the brine feed is raised by external thermal energy to just below the saturation temperature T_{max} under a pressure P_{max}. The hot brine is then discharged through the series of stages, each at a reduced pressure compared with the previous stage. The pressure $P_{max} > P_1$, $P_1 > P_2$, and so forth. Because of the reduction in pressure, a portion of the heated feed flashes to vapor in each stage to obtain equilibrium with the vapor condition prevailing in each individual stage. This results in a temperature drop in each stage (for example, T_{max} drops to T_1, and T_1 drops to T_2), arriving at the minimum temperature T_n in the last stage. The total temperature drop is usually from a T_{max} of 250°F to a T_n of 100°F. The fraction of recirculating brine that can be flashed with each cycle is restricted to 0.10–0.15. Therefore, to produce the desired rate of distillate requires a minimum brine recirculation rate in the range of 10, which is 6.6 times the production rate of desalted water. The brine wasted from the recycling feed line may contain approximately 70,000 mg/l from a seawater input of 35,000 mg/l.

The controlling parameters for output of desalted water and energy consumption are the temperature drop allowed in each stage, the difference between the brine inlet temperature at the first stage and the discharge temperature at the last (overall flash range), and the heat transfer coefficients of the stages. The principal unavoidable heat losses result from imperfect heat transfer by the condenser tubes and heat exchangers. Other losses include poor venting, resulting in vapor blanketing of the condenser tube surfaces, and tube fouling as a result of scale formation. Energy consumption in distillation is always well in excess of the ideal theoretical minimum. For a particular installation, efficiency is related to design factors, such as the number of stages.

7–29 SOURCES OF RESIDUALS IN WATER TREATMENT

The characteristics and quantities of water treatment plant residuals vary greatly depending on the water source and kinds of treatment processes. The most common plants are coagulation–filtration systems to remove turbidity and pathogens and precipitation softening to reduce hardness. Water treatment residuals are derived from sedimentation and filtration of chemically conditioned water. Surface supplies yield wastes containing colloidal matter removed from the raw water and chemical flocs; groundwater processing precipitates are mineral with little or no organic material. Sludges vary widely in composition depending on the character of the water source and chemicals added during treatment.

Coagulation Residuals

A typical method of handling a turbid river supply includes presedimentation for reduction of settleable solids, alum coagulation, and filtration for removal of colloids, plus the addition of activated carbon for taste and odor control and chlorination for disinfection. The presedimentation deposit is silt plus detritus; settling-basin sludge is a mixture of inert material, organic solids, and chemical precipitates, including metal hydroxides; and filter backwash water contains floc from agglomerated colloids and unspent coagulant hydroxides. Lake and reservoir waters are often dosed with alum and flocculation aids, plus activated carbon. Settlings during the summer may include significant quantities of algae.

Surface-water wastes are highly variable, owing to changes in raw water quality. High turbidity during spring runoff and periods of high rainfall results in a decreased percentage of aluminum hydroxide solids. The result is a precipitate that settles better and is easier to dewater. Water temperature changes affect algal growth in surface supplies, the rate of chemical reactions in treatment, and filterability of the sludge. In designing a waste-handling system, changes in raw water quality and accompanying variations in sludge characteristics must be investigated by long-term studies of daily and seasonal records.

Sludge production from surface-water treatment can be estimated from chemical additions and raw water characteristics. Based on empirical data, 1 mg of commercial alum applied as a coagulant produces 0.44 mg of aluminum precipitate, which is nearly double the 0.26 theoretical amount (Eq. 7–5). The observed relationship between turbidity of the raw water, expressed in nephelometric turbidity units (NTU), and weight of impurities removed varies from 0.5 to 2.0 mg/NTU, with an average value of 0.74. The following equation for estimating the production of dry sludge solids from alum coagulation is based on these data:

Total sludge solids (lb/mil gal)

$$= 8.34(0.44 \times \text{alum dosage} + 0.74 \times \text{turbidity}) \quad \textbf{(7–37)}$$

The majority of sludge solids settle by gravity in the sedimentation basins; the remainder are removed by subsequent filtration. The nonsettleable portion depends on the kinds of impurities in the water and the chemicals applied in coagulation. The sludge withdrawn from sedimentation basins is 1 to 2 percent, and the solids content of filter wash water is less than 0.05 percent. Gravity thickening of settled sludge and backwash water from alum coagulation can be increased in a clarifier–thickener to 2 to 6 percent.

Lime–Soda Ash Softening Residuals

Softening sludge produced in treating groundwater contains calcium carbonate and, to a lesser extent, magnesium hydroxide. Aluminum hydroxides and other coagulant aids may be present if they are added in water processing. The quantity of $Mg(OH)_2$ depends on magnesium hardness in the raw water and the softening process employed. In general, dry solids are 85 to 95 percent $CaCO_3$. The residue is stable, dense, inert, and relatively pure, since groundwater does not contain colloidal inorganic or organic matter. Calcium carbonate compacts readily, whereas magnesium hydroxide, like aluminum hydroxide, is gelatinous and does not consolidate as well nor dewater as easily. Slurry wasted from flocculator–clarifiers (upflow units) has a solids content in the range of 2 to 5 percent. Stoichiometrically, from Eq. 7–38, 3.6 lb of calcium carbonate is precipitated for each pound of

lime applied. However, in actual practice, the dry solids yield from softening is closer to 2.6 lb/lb of lime applied, owing to incomplete chemical reaction, impurities in commercial-grade lime, and precipitation of variable amounts of magnesium.

$$CaO + Ca(HCO_3)_2 = 2CaCO_3 + H_2O \quad \textbf{(7–38)}$$

The quantity of dry residue produced in softening can be estimated as calculated in Example 7–21 by using the precipitation-softening Eqs. 7–19 to 7–23.

Iron and Manganese Residuals

Iron and manganese oxides removed from groundwater by aeration and chemical oxidation are flocculent particles with poor settleability. The amount of sludge produced in the removal of these metals is relatively small without simultaneous precipitation softening. The majority of hydrated ferric and manganic oxides pass through sedimentation tanks to be trapped in the filters and appear in the dilute backwash water.

Filter Wash Water

Filter backwash is a relatively large volume of wastewater with a low solids concentration of 100 to 400 mg/l. The exact amount of water used in backwash is a function of the type of filter system, cleansing technique, and quality and source of the raw water being treated. Generally, 2 to 3 percent of the water processed in a plant is used for filter washing. The fraction of total waste solids removed by filtration depends on the efficiency of the coagulation and sedimentation stages, type of treatment system, and characteristics of the raw water. The amount may be a substantial portion—say, 30 percent of the dry solids resulting from treatment.

■ EXAMPLE 7–21

A surface-water treatment plant coagulates a raw water having a turbidity of 9 units by applying an alum dosage of 30 mg/l. Estimate the total sludge solids production in pounds per million gallons of water processed. Compute the volume of sludge from the settling basin and filter backwash water using 1.0 percent solids concentration in the sludge and 500 mg/l of solids in the wash water. Assume that 30 percent of the total solids is removed in the filter.

Solution

Applying Eq. 7–37,
Total sludge solids

$$= 8.34(0.44 \times 30 + 0.74 \times 9) = 166 \text{ lb/mil gal}$$

$$\text{Solids in sludge} = 0.70 \times 166 = 116 \text{ lb/mil gal}$$

$$\text{Solids in backwash water} = 0.30 \times 166$$

$$= 50 \text{ lb/mil gal}$$

$$\text{Volume} = \frac{\text{sludge solids (lb)}}{\text{solids fraction} \times 8.34 \text{ (lb/gal)}}$$

$$\text{Sludge volume} = \frac{116}{\dfrac{1.0}{100} \times 8.34} = 1390 \text{ gal/mil gal}$$

$$\text{Wash-water volume} = \frac{50}{\dfrac{500}{1,000,000} \times 8.34}$$

$$= 12,000 \text{ gal/mil gal}$$

■ ■ ■

■ EXAMPLE 7–22

Calculate the residue produced in the excess lime softening of the water defined in Example 7–14.

Solution

COMPONENT IN WATER			PRECIPITATE PRODUCED	
FORMULA	meq/l	APPLICABLE EQUATION	$CaCO_3$ meq/l	$Mg(OH)_2$ meq/l
CO_2	0.40	7–19	0.40	0
$Ca(HCO_3)_2$	2.00	7–20	4.00	0
$Mg(HCO_3)_2$	0.70	7–21	1.40	0.70
$MgSO_4$	0.51	7–22 & 7–23	0.51	0.51
Excess lime	1.25		1.25	0
			7.56	1.21
Minus the practical limits (solubility)			−0.60	−0.20
			6.96	1.01

$$\text{Residue} = CaCO_3 + Mg(OH)_2$$
$$= 6.96 \times 50 + 1.01 \times 29.2 = 378 \text{ mg/l}$$

■ ■ ■

7–30 SELECTION OF PROCESSES FOR WATER TREATMENT RESIDUALS

The pollution of surface waters and groundwaters from the discharge of residuals from water treatment is controlled by state regulations under authorization from the Environmental Protection Agency. In general, clarified waters such as settled backwash water from filters and overflow from solids separation processes can be discharged to flowing waters, provided that in-stream water-quality standards are not violated. After separation of supernatant, the sludge is often disposed of by land application. Water treatment plants may also discharge to a sanitary sewer or directly to a wastewater plant, provided that the wastewater complies with pretreatment requirements for industrial wastewater.

Water treatment processes can be modified to change the characteristics and reduce the quantities of residuals. Polymers as coagulant aids lower the required dosages of alum and auxiliary chemicals. The result is less sludge that is easier to dewater because of the reduced hydroxide precipitate content. Polymers also enhance presedimentation of turbid river waters, thus controlling the carry-over of solids to subsequent chemical coagulation. The addition of specially manufactured clays can be used to aid flocculation of relatively clear surface supplies by producing a denser floc that settles more rapidly. Coagulant aids with alum should be considered as a cost-effective technique for modifying sludge-handling processes at surface-water plants. Groundwater softening plants can also change their mode of operation to lessen the volume of waste sludge. Emphasis is placed on preventing magnesium hydroxide precipitation because it inhibits dewaterability. Also, limiting hardness reduction produces less solid material while still supplying a moderately soft water acceptable to the general public.

The common processes for storage, treatment, and disposal of water treatment sludge are listed in Table 7–7. Each waterworks is unique in that local conditions and existing facilities tend to dictate the techniques applied in processing and disposing of waste solids.

Modern clarifiers equipped with mechanical scrapers discharge sludge at regular intervals,

TABLE 7–7

Processes for Storage, Treatment, and Disposing of Water Treatment Sludge

Storage prior to processing
  Sedimentation basins
  Separate holding tanks
  Flocculator–clarifier basins
Thickening prior to dewatering
  Gravity settling
Chemical conditioning prior to dewatering
  Polymer application
  Lime addition to alum sludge
Mechanical dewatering
  Centrifugation
  Pressure filtration
Air drying
  Shallow lagoons
  Sand drying beds
Disposal of dewatered solids
  Codisposal in municipal solid-waste landfill
  Burial in dedicated landfill
Disposal of liquid and dewatered sludge
  Spread on agricultural land
  Application on dedicated surface-disposal site

usually daily. For storage, separate holding tanks can be installed to accumulate this slurry prior to dewatering. Filter backwash is often stored in clarifier–flocculators that serve as both temporary holding tanks and wash-water settling basins. Equalization and settling are generally the only prerequisites if sludge is discharged to a sewer for processing at the municipal wastewater treatment plant.

A typical system for thickening coagulation waste and filter wash water is shown schematically in Figure 7–43. The two primary sources of waste are sludge from the clarifier, following chemical coagulation, and wash water from backwashing filters. The latter is discharged to a clarifier holding tank for gravity separation of the suspended solids and flow equalization. Settled solids consolidate to a sludge volume less than one-tenth of the wash-water volume. Supernatant is withdrawn slowly and recycled to the plant inlet. After a sufficient portion has been drained, the holding tank is able to receive the next backwash

surge. Settled sludge from both the wash-water tanks and in-line clarifiers are given second-stage consolidation in a clarifier–thickener. Polymer is normally added to enhance the capture of solids, overflow is recycled, and thickened sludge is withdrawn for further dewatering and disposal.

Recycled wash-water supernatant and recycled overflow without treatment can return pathogenic microorganisms (for example, *Cryptosporidium* oocyts and *Giardia* cysts), organic precursors, and disinfection by-products to the influent raw water. The EPA Filter Backwash Recycling Rule has two major provisions. One is that traditional and direct filtration plants must recycle to a location at the head of the plant from which recycle flows pass through all of the treatment processes. The other provision is that traditional and direct filtration plants that recycle must notify the state regulatory agency in writing, report aspects of their recycle practices, and maintain records of these practices. This rule applies to all recycled flows, including filter backwash water, supernatant from sludge thickening, and wastewater from sludge dewatering.

Gravity thickening is the simplest and least expensive process for consolidating waste sludges. Water treatment residuals from both sedimentation and filter backwashing can be compacted effectively by gravity separation (Figure 7–43). The tank of a gravity thickener resembles a circular clarifier except that the depth–diameter ratio is greater and the hopper bottom has a steeper slope. The performance of thickeners in the processing of water treatment plant wastes varies with the character of the water being treated and the chemicals applied. Alum sludge from surface-water coagulations settles to a density in the range of 2 to 6 percent solids. Coagulation-softening mixtures from the treatment of turbid river waters thicken by gravity approximately as follows: alum-lime sludge, 4 to 10 percent; iron-lime settlings, 10 to 20 percent; alum-lime filter wash water, about 4 percent; and iron-lime backwash, up to 8 percent. The density achieved in gravity thickening relates to the calcium-magnesium ratio in the solids, quantity of alum, nature of impurities removed from the raw water, and other factors. Calcium carbonate residue from groundwater softening consolidates to 15 to 25 percent solids. In most cases, special studies have to be

Figure 7–43

Sketch of a water treatment plant showing the thickening and dewatering processes for alum sludge.

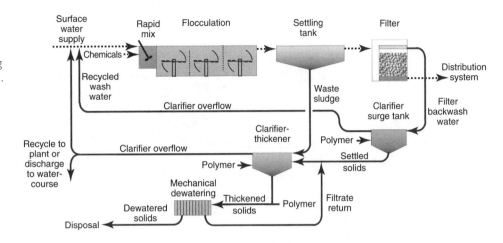

conducted at a particular waterworks to determine the settleability of solids in the waste sludge and wash water. Flocculation aids are used to improve clarification in most cases.

Mechanical dewatering of chemical sludge can be accomplished by centrifugation or plate-and-frame pressure filtration and in some instances by belt filter pressure filtration. A major advantage of a solid-bowl decanter centrifuge is operational flexibility. Machine variables, such as speed of rotation, allow a range of moisture content in the discharged solids, varying from a dry cake to a thickened slurry. Feed rates, sludge solids content, and chemical conditioning are also variables that influence performance. Polymers or other coagulants are normally applied with the slurry feed to enhance the capture of solids. Lime sludge compacts readily, producing a firm cake; alum sludge does not dewater as readily and discharges with a toothpaste consistency suitable for either further processing or transporting by truck to a disposal site. Removal of solids by centrifugation varies over a broad range, depending on operating situations and chemical conditioning; in general, solids recovery and density of cake are related to polymer dosage.

A plate-and-frame filter press is particularly advantageous for dewatering alum sludge if a high solids concentration in the filter cake is desired. Aluminum hydroxide wastes are often conditioned with lime to improve their filterability. The filter medium is precoated with either diatomaceous earth or fly ash before applying sludge solids. Precoating protects against blinding of the filter cloth by fines and ensures easy cake discharge without sticking. With proper chemical conditioning, alum sludge can be pressed to a chunky solid. Belt filter presses can dewater alum sludge to 15 to 20 percent and lime sludge to 50 percent or greater with polymer conditioning.

Lagooning is an accepted method for dewatering, thickening, and temporary storage of waste sludge where suitable land area is available. The diked pond area needed relates to the character of the sludge, climate, design features such as underdrains and decanters, and method of operation. Clarified overflow may be returned to the treatment plant, particularly if filter wash water is directed to the lagoons without prior thickening. Sludge from lime softening consolidates to about 50 percent solids after drying by evaporation and can be removed by a scraper or dragline and hauled to land burial. Alum sludge dewaters and dries more slowly, to a density of only 10 to 15 percent. Although the surface may dry to a hard crust, the underlying sludge turns to a viscous liquid upon agitation. This slurry must be removed, usually by dragline, and spread on the banks to air dry prior to hauling. Freezing enhances the dewatering of alum sludge by breaking down its gelationous character. Air drying at small water plants can be done on sand beds with tile underdrains. Repeated sludge applications over a period of several months can be done in depths up to several feet. Dewatering action is by drainage and air drying, although the operation may include decanting supernatant. Dried cake is removed by either hand shoveling or mechnical

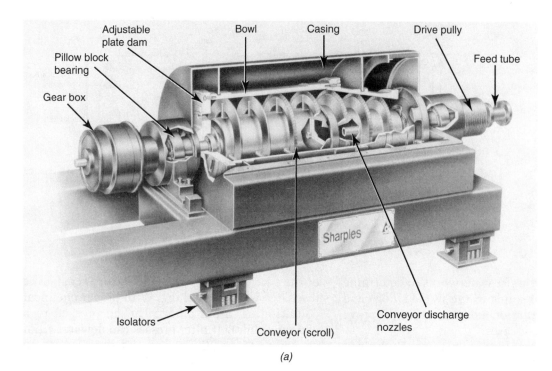

(a)

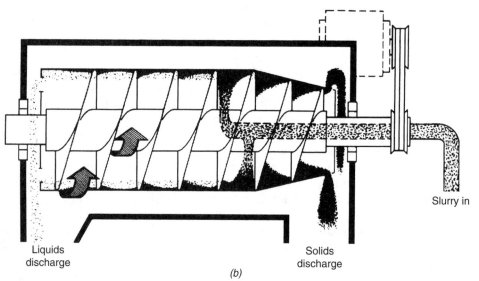

(b)

Figure 7–44

Solid-bowl scroll centrifuge. (a) Cutaway view with components labeled. (b) Schematic diagram showing separation of the feed slurry into liquid discharge and solids discharge.

(Courtesy of Alfa Laval Sharples®, Alfa Laval Separation Inc.)

means. Neither lime nor alum sludge make a good, stable landfill.

7–31 DESCRIPTION OF CENTRIFUGATION

The centrifuge commonly used in dewatering water and wastewater sludges is the solid-bowl scroll type illustrated in Figure 7–44. The unit consists of a rotating solid bowl in the shape of a cylinder with a cone section on one end and an interior rotating screw conveyer. Feed slurry, entering from the center, is held against the bowl wall by centrifugal force. Settled solids are moved by the conveyer to one end of the bowl while clarified effluent discharges at the other end. The conical bowl shape at the solids discharge end enables the conveyer to move the solids out of the liquid for drainage before being discharged.

A major advantage of the solid-bowl centrifuge is operational flexibility. Machine variables include pool volume, bowl speed, and conveyer speed. The depth of liquid held against the bowl wall can be controlled by an adjustable plate dam at the discharge end of the bowl. Adjusting the pool volume changes the drainage deck surface area of the solids discharge section. Bowl speed affects gravimetric forces on the settling particles, and conveyer speed controls the solids retention time. The driest cake product results when bowl speed is increased, pool depth is at the minimum allowed, and differential speed between the bowl and conveyer is the maximum possible. The flexibility of operation allows a wide range of moisture content in the solid discharge, varying from a dry cake to a thickened liquid sludge. Feed rates, solids content, and prior chemical conditioning are also variables that influence performance. The removal of solids can be enhanced by adding polymers or other coagulants with the slurry feed. The scroll centrifuge also operates with little surveillance.

Centrifuges are being used to dewater treatment plant sludges drawn from settling tanks or gravity thickeners. Lime sludges compact readily, producing a cake with 50 percent solids from a feed of 5 to 10 percent. Thus, centrifuges are used to dewater lime-softening sludges in preparation for recalcination. Although aluminum hydroxide

sludges do not dewater as readily as lime precipitates, centrifugation with polymer addition to aid flocculation provides a viscous, liquid, solids discharge suitable for either further processing or disposal. Sludge containing about one-half aluminum hydroxide can be thickened to 10 to 15 percent solids; approximately one-quarter-hydrate slurry can be thickened to 20 to 25 percent. Removal of solids from the feed ranges from 50 to 95 percent, depending on operating conditions and polymer dose. Holding all other variables constant, the percentage of solids recovery and density of the thickened discharge are directly related to polymer dosage.

7–32 DESCRIPTION OF PRESSURE FILTRATION

The plate-and-frame filter press is the type of pressure filter adopted for dewatering waste chemical sludges. It consists of a series of recessed plates with cloth filters and intervening frames held together to form enclosed filter chambers. These chambers are filled with dewatered solids by pumping in sludge under high pressure, forcing the water out through the cloth filters. At the end of the feed and pressure cycles, the plates are separated to remove the sludge cake. Plate-and-frame filters are capable of producing a dry, rigid cake with no visible water and a clear filtrate very low in suspended solids. The two designs of filter presses are the fixed-volume press and the variable-volume diaphragm press.

The variable-volume diaphragm press is designed for completely automatic operation by forcefully discharging the cakes after the frames open and washing the filter cloths before the press closes for another cycle. Figure 7–45a is an overhead view of a press installation. The steps in the operation are chemical conditioning of the raw sludge if necessary, precoating the filter media, pressure sludge dewatering, and discharge of the sludge cakes. The filter media are precoated by feeding a water suspension of diatomaceous earth. Precoating prevents the binding of the filter cloth with sludge particles and facilitates discharge of the cake at the end of the filter cycle. The automatic operation of a variable-volume diaphragm press is diagramed in Figure 7–45b. Sludge is pumped into

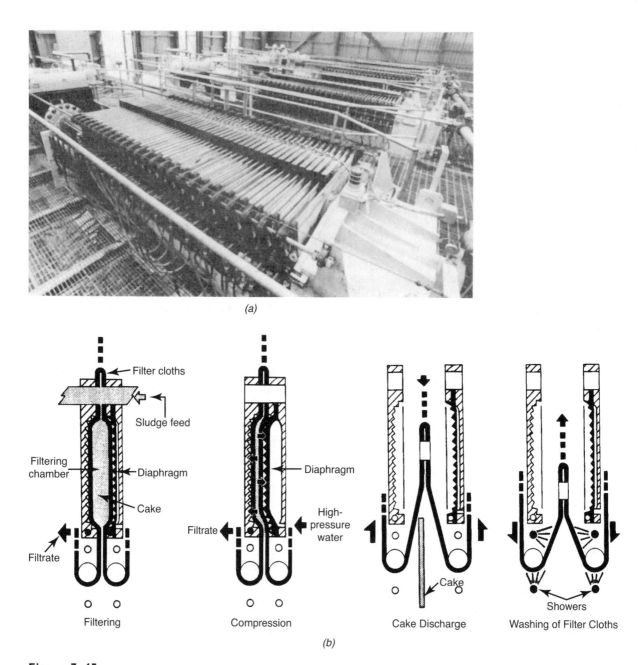

Figure 7–45

Variable-volume diaphragm filter press for dewatering waste sludge (a) Installation of four presses. (b) Diagrams of the four steps in automatic operation of a variable-volume diaphragm press.

(Courtesy of Ingersoll-Rand, IMPCO Division as licensed by Ishigaki Industry Company Limited.)

the recessed chambers at a pressure of 100 psi (690 kN/m²) for 10 to 20 min, allowing filtrate to drain out through the cloths on both sides of the chamber. After the sludge feed is shut off, water under high pressure is pumped into the space between the diaphragm and one side of the plate body. This compression is held for 15 to 30 min to dewater the accumulated solids to the desired cake dryness. At the end of the compression period, the sludge feed and filtrate lines are blown out with compressed air. The press then automatically opens, and the cakes are released by pulling the filter cloths down around small-diameter rollers. The cloths are washed on both sides as they retract up to their original position. The yield for a diaphragm press is more than twice the yield of a fixed-volume press in terms of mass of solids dewatered per unit area of filter cloth per unit of time.

Pressure filtration appears to be particularly advantageous for thickening alum sludges where a high solids content is most desirable. With lime or polymer conditioning, aluminum hydroxide sludges can be pressed to a solids content of 30 to 40 percent, which is adequate for hauling by truck.

7–33 DISPOSAL OF DEWATERED SLUDGE

Dewatered sludge can be buried with household refuse in a municipal landfill. The methods of codisposal can be mixing wet sludge cake with either solid waste or soil. In a sludge–solid-waste operation, sludge is deposited on top of a layer of solid waste at the working face of the landfill and mixed as thoroughly as possible. Using bulldozers and compactors, each mixed layer is then spread, compacted, and covered with a layer of soil. The application of wet sludge should not exceed the absorption capacity of the refuse. To prevent difficulty, a recommended design bulking ratio (weight of refuse to weight of wet sludge) should be established. Codisposal regulations for municipal landfills define water treatment plant residuals as nonhazardous wastes.

Monofills, where water treatment residuals as semisolids are spread and covered with soil, are commonly used as dedicated burial sites. Trenches can be excavated, with a layer of natural stratum or compacted soil between the top of the groundwater or bedrock and the deposited sludge. The sludge mass is usually dumped from the transport trucks directly into the trenches and on-site equipment is used for trench excavation and covering.

Surface disposal is the distribution of semisolid or wet sludge on the land for final disposal or as a beneficial soil amendment, for example, lime waste on acidic soils. In a sludge–soil mixture operation, the sludge is mixed with soil and placed as intermediate layers between layers of sludge and as a final cover. A final cover of mixed sludge and soil (depending on the chemical characteristics of the mixture) can promote growth of vegetation over the field area to minimize wind and water erosion.

REFERENCES

1. *Recommended Standards for Water Works.* Health Research Inc., Health Education Services Division, P. O. Box 7126, Albany, NY 12224, 2003.

2. *Handbook of Public Water Systems,* 2nd Ed., Chapter 1. HDR Engineering, Inc., Omaha, NE, published by John Wiley & Sons, Inc., 2001.

3. U.S. Environmental Protection Agency, Interim Enhanced Surface Water Treatment Rule. *Federal Register,* 63:241:69478 (Dec. 16, 1998).

4. *Guidance Manual for Compliance with the Filtration and Disinfection Requirements for Public Water Systems Using Surface Water Sources.* Office of Drinking Water, U.S. Environmental Protection Agency, Washington, DC, 1991.

5. Amendment to paragraph (8) of section 1412(b) [42 U.S.C. 300g-1(b)(8)] of the Safe Drinking Water Act, SEC. 107, GROUND WATER DISINFECTION (August 1, 1996).

6. *Proposed Ground Water Rule.* Office of Water, EPA 815-F-00-003, April 2000 (www.epa.gov/safewater/gwr.html) and *Fact Sheet: Ground Water Rule* (www.epa.gov/safewater/gwr/gwrfs.html).

7. *Handbook of Public Water Systems,* 2nd Ed., Chapter 18. HDR Engineering, Inc., Omaha, NE, published by John Wiley & Sons, Inc., 2001.

8. W. A. Kelley and R. A. Olson, "Selecting MF to Satisfy Regulations," *J. Am. Water Works Assoc.* 91, No. 6 (1999):52–63.

PROBLEMS

7–1 Compare the contaminants commonly removed in municipal water treatment from surface waters (Section 7–1) and groundwaters (Section 7–16) by tabulating them in two lists. Why does treatment of river water generally require the most extensive and flexible processing?

7–2 The first-order kinetics of an oxidation reaction were analyzed by the addition of chlorine to the water in laboratory experiments. Samples were withdrawn from the completely mixed reaction vessel at 1-min intervals for determining the concentration of the unoxidized metal ions. The data for one test were as follows: $t = 0$ min, $C_0 = 0.70$ mg/l; $t = 1.0$, $C = 0.47$; $t = 2.0$, $C = 0.32$; $t = 3.0$, $C = 0.26$; and $t = 4.0$ min, $C = 0.19$ mg/l. Plot a graph of $\log_e C_0/C_t$ versus t and determine the value of k.

7–3 A flocculator designed to treat 10 mgd is 48 ft long, 52 ft wide, and 12 ft deep. Compute the detention time and horizontal flow-through velocity. Do these values satisfy the *Standards for Water Works*? (*Answers* $t = 32$ min > 30, yes; $v = 1.5$ ft/min = 1.5, yes)

7–4 Locate the rapid-mixing chambers, flocculation basins, and settling tanks in Figure 7–4a. If this plant as pictured had been designed for 30 mgd (15 mgd for each side) based on the *Standards for Water Works*, what would be the volumes of the rapid mixers, flocculation basins, and sedimentation tanks?

7–5 Calculate the detention time and overflow rate for a sedimentation basin with a volume of 1.0 mil gal and surface area of 12,500 sq ft treating 8.0 mgd. (*Answers* 3.0 hr, 640 gpd/sq ft)

7–6 Based on laboratory testing, the settling velocity of alum floc is 0.0014 ft/sec in water at 10°C. Convert this settling velocity to units of overflow rate in gallons per day per square foot. What is the retention time in quiescent water in hours for alum floc to settle from the surface of the water to a depth of 10 ft?

7–7 Two rectangular clarifiers—each 30 ft long and 15 ft wide with a water depth of 10 ft—settle 0.40 mgd following alum coagulation. The effluent channels have a total weir length of 60 ft. Calculate the detention time, horizontal velocity of flow, and rate of flow over the outlet weirs. Do these values meet the *Standards for Water Works*?

7–8 Two rectangular clarifiers, each 27 m long, 5.0 m wide, and 3.8 m deep, settle 6000 m³/d. Total effluent weir length is 50 m. Calculate the detention time, overflow rate, and weir loading. (*Answers* 4.1 h, 22 m³/m² · d, and 120 m³/m · d)

7–9 Calculate the diameter and depth of a circular clarifier for a design flow of 3800 m³/d based on an overflow rate of 0.00024 m/s and a detention time of 3 h.

7–10 What diameter circular clarifier and side water depth are needed for a 15,000 m³/d flow based on a maximum overflow rate of 16 m³/m² · d and detention time of 4.0 h?

7–11 The flow of water through the long rectangular sedimentation tank was analyzed by injecting a pulse of a dye tracer in the influent. The concentrations of dye in the effluent at time intervals of 20 min starting from injection of the dye were as follows:

TIME (min)	CONCENTRATION (µg/l)	TIME (min)	CONCENTRATION (µg/l)
0	0 (start)	120	0.78
20	0.00	140	0.52
40	0.40	160	0.25
60	0.75	180	0.08
80	1.42 (peak)	200	0.02
100	1.20	220	0.00

Draw a residence time distribution curve as shown in Figure 7–7b, locate the theoretical detention time of 120 min, and calculate and plot the mean residence time (centroid). (*Answer* $t_{50} = 96$ min)

7–12 The flow through a sedimentation tank was studied by injecting a pulse of dye tracer in the influent. The concentration of the dye in the effluent at time intervals of 15 min and 30 min starting from injection of the tracer input are listed in the following table.

TIME (min)	CONCENTRATION (µg/l)	TIME (min)	CONCENTRATION (µg/l)
0	0	135	88.0
15	0	150	78.2
30	0	180	55.2
45	3.5	210	33.5
60	16.5	240	20.0
75	46.5	270	12.1
90	72.0	300	7.5
105	89.0	330	4.6
120	95.0	360	2.6

Draw a residence time distribution curve as shown in Figure 7–7b and locate the mean residence time (centroid).

7–13 Flocculator-clarifiers similar to the one illustrated in Figure 7–10 are proposed for precipitation in lime–soda ash softening of a groundwater. The inside diameter of the tank is 40 ft and sidewater depth is 10 ft. The cone-shaped skirt is 12 ft in diameter at the water surface and 24 ft in diameter at the bottom, which is at a depth of 8.0 ft below the water surface. The design flow for each flocculator-clarifier is 750,000 gpd. Does this proposed design meet the *Standards for Water Works* for detention times based on total volume, mixing and flocculation time, and upflow rate based on the open-water surface area? (The cone-shaped shirt can be considered to be the geometric form of a conical basin illustrated in the appendix.)

7–14 The settling velocity of calcium carbonate floc formed during flocculation in a flocculator-clarifier (Figure 7–10) is 2.1 mm/s. If the detention time in the settling zone is 1.0 hr and upflow rate is 1.75 gpm/sq ft, what is the minimum depth of water required to ensure removal of the floc by gravity settling?

7–15 How does the flow scheme for direct filtration differ from the conventional flow scheme shown in the upper diagram of Figure 7–1? What limitations of raw water quality are recommended for adoption of direct filtration?

7–16 What is the function of the rate-of-flow controller in filter operation?

7–17 In a gravity filter, how can the head-loss gauge record a loss of 9 ft if the depth of water above the filter surface is only 4 ft?

7–18 A filter run is terminated as a result of either one of two conditions. State the two conditions.

7–19 What is the principal advantage of a dual-media anthracite-sand filter compared to a single-medium sand filter? After backwashing, why do the media stratify, with the anthracite coal overlying the sand? (Refer to Figure 7–13.)

7–20 List the kinds of filter underdrain systems that allow cleaning by (a) air scour followed by water backwash and (b) concurrent air-and-water scour followed by water backwash.

7–21 Describe the steps used to clean the filter illustrated in Figure 7–14 using separate air scour followed by water backwash.

7–22 Since powdered chemicals cannot be applied directly, how are they applied to the water being treated?

7–23 Results of a jar-test demonstration on alum coagulation are listed in the following table. (Jar testing is discussed in Section 2–11.) Jars 1 through 5 contained a clay suspension in tap water; jar 6 was a clay suspension in distilled water.

Jar	Alum Dosage	Floc Formation
1	0 mg/l	None
2	5 mg/l	Smoky
3	10 mg/l	Fair
4	15 mg/l	Good
5	20 mg/l	Heavy
6	20 mg/l	None

(a) What is the lowest recommended dosage for treating this water? (b) Why didn't the clay suspension in jar 6 destabilize?

7–24 Commercial aluminum sulfate (alum) reacts with natural alkalinity or with the chemicals of lime and soda ash if added to water deficient in alkalinity. Based on Eqs. 7–5, 7–6, and 7–7, calculate the milligram-per-liter amounts of alkalinity, lime as CaO, and soda ash as Na_2CO_3 that react with 1.0 mg/l of alum. (*Answers* 0.50 mg/l, 0.28 mg/l, 0.53 mg/l)

7–25 Alum is applied at a dosage of 20 mg/l in coagulating a surface water. How much natural alkalinity is consumed? What is the change in pH? (*Answers* 10 mg/l, decreases)

7–26 A dosage of 40 mg/l of alum is added in coagulating a water. (a) How many milligrams per liter of alkalinity are consumed? (b) What changes take place in the ionic character of the water? State the ions that change forms and express the amounts in milligrams per liter. (c) What is the effect on the pH of the water? [*Answers* (a) 20 mg/l; (b) $SO_4^=$ increases by 19 mg/l, HCO_3^- reduces by 24 mg/l; (c) decrease in pH resulting from CO_2 formed]

7–27 A dosage of 30 mg/l of alum and a stoichiometric amount of soda ash are added in coagulation of a surface water. What changes take place in the ionic character of the water as a result of this chemical treatment?

7–28 Coagulation of a soft water requires 40 mg/l of alum plus lime to supplement the natural alkalinity for good floc formation. It is desired to react only 10 mg/l (as $CaCO_3$) of the natural alkalinity. What dosage of lime is required?

7–29 The removal of *Giardia* cysts from a soft, cold, low-turbidity water requires 15 mg/l of alum plus 0.10 mg/l of anionic polymer. (a) How many milligrams per liter of natural alkalinity are consumed in the coagulation reaction? How much CO_2 is released by this reaction? (b) What is the stoichiometric dosage of soda ash to react with the 15 mg/l of alum? This reduces loss of alkalinity but still produces carbon dioxide. How much CO_2 is released by this reaction? (c) Would a stoichiometric dosage of lime slurry be better than soda ash? Suggest a reason why lime slurry would not be used. How can CO_2 be removed from water?

7–30 A surface water is coagulated with a dosage of 30 mg/l ferrous sulfate and an equivalent dosage of lime. (a) How many pounds of ferrous sulfate are needed per million gallons of water treated? (b) How many pounds of hydrated lime are required assuming a purity of 70 percent CaO? (c) How many pounds of $Fe(OH)_3$ sludge are produced per million gallons of water treated?

7–31 Write the reaction between ferric chloride and lime slurry that results in precipitation of $Fe(OH)_3$. For a dosage of 40 mg/l $FeCl_3$: (a) What is the required stoichiometric addition of lime expressed as CaO? (b) How many pounds of $Fe(OH)_3$ are produced per million gallons of water treated? [*Answers* (a) 20.7 mg/l CaO, (b) 220 lb/mil gal]

7–32 A dose of 35 mg/l of $FeSO_4 \cdot 7H_2O$ and an equivalent amount of lime are added in coagulation. What are the changes that take place in the ionic character of the water as a result of this chemical treatment?

7–33 A 20.0-mg amount of sodium carbonate is added to 1.0 l of water, resulting in a pH change from 7.2 to 7.4. How much is the alkalinity increased? What is the ionic form of the alkalinity increase?

7–34 In taste and odor control, what is usually the most effective chemical? Why is heavy prechlorination of surface waters likely to be undesirable?

7–35 A river water supply is treated by the sequence of unit operations and chemical additions in the following list. State in a few words the purpose or purposes for each unit process and chemical application.

(a) Presedimentation with polymer addition
(b) Mixing and flocculation with additions of alum and polymer
(c) Addition of activated carbon
(d) Sedimentation
(e) Granular-media filtration
(f) Postchlorination

7–36 The processing scheme for treatment of lake water consists of (1) rapid mixing and flocculation, (2) sedimentation, (3) filtration, and (4) clear-well storage. The chemicals available for treatment are activated carbon, alum, chlorine, fluorosilicic acid, and polymers. At what principal and alternate locations in the processing scheme are these chemicals likely to be added?

7–37 What categories of chemicals are included in synthetic organic chemicals? In treatment of surface waters, how effective are conventional coagulation processes for removal of SOCs and the addition of powdered activated carbon for taste and odor control? What are the limitations in using aeration as a method for removal of VOCs from contaminated well water?

7–38 What dosage of commercial fluorosilicic acid is needed to increase the concentration of fluoride ion from 0.3 mg/l to 1.0 mg/l? Use fluorosilicic acid data from Table 7–2 and express the answer in milligrams per liter and pounds per million gallons.

7–39 A fluoridation installation consists of a feed pump that applies fluorosilicic acid with 30 percent purity to a water supply directly from the shipping drum placed on a platform scale. The recorded weight loss of the drum is 82 lb in processing 2.3 mil gal of water. Calculate the fluoride ion concentration added to the water in milligrams per liter.

7–40 A sodium silicofluoride solution is prepared by dissolving 5.0 kg of 98 percent commercial powder in 95 kg of water. Calculate the feed rate of this solution into a water being treated to increase the fluoride ion content 0.80 mg/l. Express your answer in grams of solution applied per cubic meter of water treated.

7–41 A community with a population of 10,000 has a groundwater supply with a natural fluoride ion concentration of 0.2 mg/l. The annual water consumption is 550 mil gal, and the annual average maximum daily air temperature is 65°F. What fluoride concentration do you recommend maintaining in the distribution system? If fluoride ion is applied using commercial fluorosilicic acid with a purity of 25 percent, calculate the pounds of commercial acid required to fluoridate the community water supply for 1 year.

7–42 Do granular calcium hypochlorite and liquid sodium hypochlorite provide the same disinfecting ions as dissolution of chlorine gas?

7–43 Results of a chlorine demand test on a raw water are as follows:

SAMPLE NUMBER	CHLORINE DOSAGE (mg/l)	RESIDUAL CHLORINE AFTER 10-MIN CONTACT (mg/l)
1	0.20	0.19
2	0.40	0.36
3	0.60	0.50
4	0.80	0.48
5	1.00	0.20
6	1.20	0.40
7	1.40	0.60
8	1.60	0.80

Sketch a chlorine demand curve as shown in Figure 7–17. What is the breakpoint dosage, and what is the chlorine demand at a dosage of 1.2 mg/l?

7–44 How many pounds of available chlorine are contained in 1.0 gal of sodium hypochlorite with a strength of 15 percent? How many gallons would be required for a dosage of 0.6 mg/l to 6.0 mil gal of water? (*Answers* 1.25 lb/gal, 24 gal)

7–45 How many grams of dry hypochlorite powder with 70 percent available chlorine must be added to 400 liters of water to make a 1.0 percent solution? (*Answer* 5700 g)

7–46 Groundwater from a deep well is pumped at 400 gpm directly into the pipe network of a city. During summer, the well pump operates an average of 18 hr per day with chlorine feed controlled by pump operation. The following chlorination systems are being considered to apply 0.50 mg/l of chlorine to the pump discharge: (a) Liquid chlorine from a 100-lb pressurized cylinder through a solution feed chlorinator. (b) Sodium hypochlorite solution with 10 percent available chlorine fed from a 200-gal storage tank by a diaphragm pump. (c) Powdered calcium hypochlorite with 70 percent available chlorine dissolved in water to 15 percent available chlorine in a 50-gal tank and fed by a diaphragm pump. Calculate the quantity of chlorine applied per 18-hr day. Calculate the number of days each system can feed chlorine before requiring renewal. Which system would you recommend for a cold climate? For a warm climate?

7–47 In treating 100,000 m^3 of water, 80 kg liquid chlorine is applied. Compute the dosage in milligrams per liter.

7–48 Why are pressure-regulating and vacuum-compensating valves needed on a solution feed chlorinator?

7–49 Review the written description and illustration of the positive-displacement electronic diaphragm pump (Figure 7–20). What is the purpose of the four-function valve? How is the rate of chemical feed controlled?

7–50 A new main is disinfected with water containing 50 mg/l of chlorine by feeding a 1.0 percent solution of chlorine to the water entering the pipe. (a) How many pounds of dry hypochlorite powder, containing 70 percent available chlorine, must be added to 40 gal of water to make a 1.0 percent solution? (b) At what rate should this 1.0 percent solution be applied to the water entering the pipe to provide a concentration of 50 mg/l?

7–51 A new water storage tank with a volume of 950 m^3 must be disinfected by filling with chlorinated water for 6 hr. The chlorine dosage applied to the filling water is to be 70 mg/l. How many kilograms of commercial hypochlorite should be applied to the 950 m^3 of water if the commercial powder contains 70 percent available chlorine?

7–52 What is the health risk of trihalomethanes (THMs) and haloacetic acids (HAA5) in drinking water? What is the origin of THMs in treated water? If the finished water from a river water treatment plant contains excessive concentrations of THMs and HAA5 during the time of spring runoff, what remedial actions can be taken to reduce by-product formation?

7–53 What is ozone and how is it generated and applied in water treatment? List the possible applications of ozonation in water treatment. Can the use of ozone eliminate the application of chlorine?

7–54 Why is granular-media filtration preceded by chemical coagulation considered essential in the disinfection of surface waters?

7–55 What is the disease in humans caused by *Giardia lamblia* and *Cryptosporidium* species? In what manner do these protozoa infect humans and how are they transmitted to other humans by water? What are other modes of transmission? Describe the waterborne sources of these organisms. (Refer to Section 3–7.)

7–56 Define the meaning of $C \cdot t$ product. In addition to C and t, what factors influence the rate of chemical disinfection? What kind of microorganism is most readily inactivated by free chlorine? What kind is most difficult to inactivate?

7–57 With reference to Table 7–4, list the chemical disinfectants in order of most effective to least effective. Include in the list the $C \cdot t$ values for each chemical for 90 percent (1.0 log) inactivation of *Giardia lamblia* cysts at 10°C and pH 7.

7–58 From Table 7–3, what is the $C \cdot t$ value for 99.9 percent (3.0 log) inactivation of *Giardia* cysts at a free chlorine residual of 1.0 mg/l, temperature of 10°C, and pH 7.5? How does a water temperature increase to 20°C affect the $C \cdot t$? A decrease to 5°C? (*Answers* 134 (mg/l) · min, times 1/2, times 4/3)

7–59 A step-dose tracer test was conducted on the clear well of a water treatment plant. A constant fluoride dosage of 2.0 mg/l was added at the inlet starting at time zero, and the fluoride concentrations in the outflow were measured and recorded every 5 or 10 min as listed in the following table. Calculate the C/C_0 values, plot C/C_0 versus t, and determine t_{10}. (*Answer* t_{10} = 22 min)

t (min)	c (mg/l)	t (min)	c (mg/l)	t (min)	c (mg/l)
0	0	25	0.42	55	1.53
5	0	30	0.70	65	1.65
10	0	35	0.88	75	1.84
15	0	40	1.15	85	1.82
20	0.10	45	1.28	95	1.85

7–60 What are the requirements specified by the EPA Surface Water Treatment Rule for (a) inactivation of *Giardia* and viruses, (b) turbidity of the filtered water, and (c) chlorine residual in the water entering the distribution system.

7–61 A treatment plant with coagulation, flocculation, sedimentation, and filtration produces a filtered water with a turbidity less than 0.3 NTU, pH 7, and temperature of 5°C at peak hourly flow of 3.0 mgd. The filtered water is chlorinated in a baffled reservoir with the residence-time distribution as drawn in Figure 7–22. What is the required disinfection by free chlorine of the filtered water? [*Answers* 25 (mg/l) · min, 90 min, 0.28 mg/l]

7–62 The filtered water at peak hourly flow from a plant with coagulation, flocculation, and filtration has a pH of 7.5 and temperature of 15°C. The t_{10} in the clear well is 22 min, followed by transmission through a pipeline for 4000 ft at a velocity of 5 ft/sec before entering the distribution system. What free chlorine residual is required in the water at the outlet of the clear well and pipeline? (The loss of residual concentration is likely to be negligible during the transmission time in the pipeline.)

7–63 The analysis of a water is as follows:

$$Ca^{++} = 3.7 \text{ meq/l} \qquad HCO_3^- = 4.0 \text{ meq/l}$$

$$Mg^{++} = 1.0 \text{ meq/l} \qquad SO_4^= = 1.2 \text{ meq/l}$$

$$Na^+ = 1.0 \text{ meq/l} \qquad Cl^- = 1.0 \text{ meq/l}$$

$$K^+ = 0.5 \text{ meq/l}$$

(a) Draw a milliequivalents-per-liter bar graph, list the hypothetical combinations of chemical compounds, and calculate the total hardness. (b) Calculate the chemical dose needed for excess lime softening. (c) Draw a bar graph for the finished water after two-step precipitation softening by excess lime treatment with intermediate and final recarbonation, assuming 60 percent of the alkalinity is bicarbonate. Assume the practical limits of hardness removal for calcium to be 30 mg/l and magnesium to be 10 mg/l. (d) Calculate the chemical dose required for selective calcium carbonate removal to produce a finished water with a hardness of 120 mg/l. [*Answers* (a) 6.2 meq/l, 235 mg/l; (b) 175 mg/l CaO, 37 mg/l Na₂CO₃; (c) 3.0 meq/l; (d) 81 mg/l CaO]

7–64 The analysis of a river water in a semiarid region is as follows:

Calcium = 78 mg/l	Alkalinity = 126 mg/l
Magnesium = 30 mg/l	Sulfate = 282 mg/l
Sodium = 106 mg/l	Chloride = 96 mg/l
Potassium = 5 mg/l	pH = 8.1

Total dissolved solids = 680 mg/l

(a) Draw a milliequivalents-per-liter bar graph. What forms of alkalinity are present in the water? (b) The recommended limit of total dissolved solids in drinking water is 500 mg/l. Can precipitation softening reduce the total dissolved solids from 680 mg/l to less than 500 mg/l?

7–65 Settled water after excess-lime treatment, before recarbonation and filtration, contains 35 mg/l of CaO excess lime in the form of hydroxyl ion, 30 mg/l of $CaCO_3$ as carbonate ion, and 10 mg/l as $CaCO_3$ of $Mg(OH)_2$ in the form of hydroxyl ion. First-stage recarbonation (Eq. 7–24) precipitates the excess lime as $CaCO_3$ for removal by sedimentation, and second-stage recarbonation (Eq. 7–25) converts a portion of the remaining alkalinity to bicarbonate ion. Calculate the carbon dioxide needed to neutralize the excess lime and convert 60 percent of the alkalinity in the finished water to the bicarbonate form. Assume an excess of 20 percent of the calculated CO_2 is required to account for unabsorbed gas escaping from the recarbonation chamber.

7–66 Calculate for the following water analysis the lime dosage needed for softening by selective calcium removal. What is the finished water hardness?

Ca^{++} = 63 mg/l	CO_3^- = 16 mg/l
Mg^{++} = 15 mg/l	HCO_3^- = 189 mg/l
Na^+ = 20 mg/l	$SO_4^=$ = 80 mg/l
K^+ = 10 mg/l	Cl^- = 10 mg/l

7–67 Analyze split-treatment softening for the following water analysis. Provide excess lime treatment for two-thirds of the raw water flow and by-pass one-third of the flow. Draw a bar graph of the raw water and finished water after sedimentation and filtration.

$$CO_2 = 15 \text{ mg/l as } CO_2$$
$$Ca^{++} = 60 \text{ mg/l}$$

$$Mg^{++} = 24 \text{ mg/l}$$
$$Na^+ = 46 \text{ mg/l}$$
$$HCO_3^- = 200 \text{ mg/l as } CaCO_3$$
$$SO_4^= = 96 \text{ mg/l}$$
$$Cl^- = 35 \text{ mg/l}$$

7–68 Draw a milliequivalents-per-liter diagram and list the hypothetical combinations for the following groundwater data:

$$\text{Calcium hardness} = 175 \text{ mg/l}$$
$$\text{Magnesium hardness} = 40 \text{ mg/l}$$
$$\text{Sodium ion} = 14 \text{ mg/l}$$
$$\text{Potassium ion} = 4 \text{ mg/l}$$
$$\text{Alkalinity} = 200 \text{ mg/l}$$
$$\text{Sulfate ion} = 29 \text{ mg/l}$$
$$\text{Chloride ion} = 14 \text{ mg/l}$$
$$pH = 7.7$$

(a) Calculate the chemical doses needed for excess lime softening. Draw a bar graph for the finished water after two-stage precipitation softening by excess lime treatment with intermediate and final recarbonation. Assume that half the alkalinity in the finished water is in the bicarbonate form. (b) Calculate the lime dosage required for selective calcium carbonate removal. Draw a bar graph for the finished water. Is this softening process recommended for this water? (c) Calculate the chemical doses for split-treatment softening with 65 percent of the flow given excess lime treatment and 35 percent by-passed to second-stage blending. Estimate the hardness of the finished water.

7–69 A groundwater supply has the following analysis:

$$\text{Calcium} = 94 \text{ mg/l}$$
$$\text{Magnesium} = 24 \text{ mg/l}$$
$$\text{Sodium} = 14 \text{ mg/l}$$
$$\text{Bicarbonate} = 317 \text{ mg/l}$$
$$\text{Sulfate} = 67 \text{ mg/l}$$
$$\text{Chloride} = 24 \text{ mg/l}$$

Calculate the quantities of lime and soda ash for excess lime softening and the carbon dioxide reacted for neutralization by two-stage recarbonation. Assume the practical limits of hardness removal for calcium to be 30 mg/l and magnesium to be 10 mg/l, and three-quarters of the final alkalinity is converted to bicarbonate. Sketch the bar graphs for the raw and finished waters. [*Answers* 240 mg/l CaO, 80 mg/l Na_2CO_3, 45 mg/l CO_2; finished hardness = 10 (Mg) + 30 (Ca) = 40 mg/l]

7–70 Compute the quantities of chemicals for split treatment of the water described in Problem 7–67 based on a magnesium hardness in the finished water of 40 mg/l and maximum total hardness of 100 mg/l. [*Answers* excess lime softening of two-thirds of the raw flow; 160 mg/l CaO, 53 mg/l Na_2CO_3; finished hardness = 40 (Mg) + 60 (Ca) = 100 mg/l]

7–71 The ionic character of a groundwater is defined by the following hypothetical combinations:

$$Ca(HCO_3)_2 = 3.0 \text{ meq/l}$$

$$Mg(HCO_3)_2 = 0.4 \text{ meq/l}$$

$$MgSO_4 = 0.8 \text{ meq/l}$$

$$Na_2SO_4 = 0.2 \text{ meq/l}$$

$$NaCl = 0.6 \text{ meq/l}$$

Draw a milliequivalents-per-liter bar graph. Is the hardness of this water considered excessive? What process do you recommend for precipitation softening of this water?

7–72 A groundwater treatment plant, processing a water high in both iron and manganese, uses tray-type aeration, contact time in a settling basin, sand filtration, and postchlorination. What problem would you anticipate with this system?

7–73 A small community has used an unchlorinated well water supply containing approximately 0.3 mg/l of iron and manganese for several years without any apparent iron and manganese problems. A health official suggested that the town install chlorination equipment to disinfect the water and provide a chlorine residual in the distribution system. After initiating chlorination, consumers complained about water staining washed clothes and bathroom fixtures. Explain what is occurring due to chlorination.

7–74 The iron and manganese removal process for the well supply of a small community is mechanical aeration, the addition of potassium permanganate followed by detention in a contact tank, pressure filtration, and post chlorination. The construction specifications called for manganese-treated greensand; however, the actual filter medium provided was plain sand. Customers often complained that the treated water caused staining of bathroom fixtures and laundry. The common response of the plant operator was to increase the chemical dosage, which did not seem to improve the situation. The operator even tried prechlorination of the water in combination with potassium permanganate addition, but that appeared to increase the staining characteristics of the treated water. Discuss the most probable cause of the poor-quality finished water and your recommendations for improvement.

7–75 The most common treatment for removal of iron and manganese from groundwater is aeration, chemical oxidation, and filtration. What is the purpose of aeration? Name the oxidation chemicals and their formulas. Why is the majority of manganese dioxide and iron oxide removed in the sand filter rather than in the sedimentation tank?

7–76 In the corrosion of iron, what equation represents pitting and what equation represents scaling? What manufacturing process protects the interior of ductile-iron pipe against corrosion?

7–77 How does cathodic protection prevent internal corrosion of a steel water tank? How does cathodic protection influence the corrosion equations?

7–78 What is the health risk of dietary intake of lead? (Refer to Section 5–3.) Why are first-flush samples collected from consumers' faucets to assess lead contamination of drinking water? What are proposed methods of reducing excessive lead in drinking water?

7–79 Why is satisfactory groundwater disinfection based on inactivation of enteric viruses rather than *Giardia* cysts? Define the meaning of natural disinfection of groundwater. What is a variance?

7–80 Groundwater is conveyed to a community from a well field through a 16-in.-diameter pipeline 4200 ft in length. The peak hourly pumping rate is 2000 gpm, and the water temperature is 5°C. Because of the location, the supply has been designated as vulnerable to fecal contamination, and the state has specified a minimum disinfection level of 99.9 percent (3.0 log) virus inactivation. What free chlorine residual is required in the water at the outlet of the pipeline? [*Answers* 6 (mg/l) · min, 22 min, 0.3 mg/l]

7-81 A groundwater treatment plant adds chlorine to oxidize iron and manganese for removal by filtration. The groundwater temperature during peak summer usage is 10°C, and the pH is 7.8. Based on a tracer study, the t_{10} time from chlorine addition in the contact tank through filtration is 1.8 min. The t_{10} time in the clear well during peak pumpage is 8.0 min. The chlorine residual in the filter effluent is 0.8 mg/l, and in the clear well effluent it is 0.4 mg/l. Is the disinfection adequate for 3.0 log inactivation of viruses?

7-82 Assume cation exchange softening is used to remove the hardness ions from the water described in Example 7-16. Sketch a milliequivalents-per-liter bar graph for the finished water. What would be the disadvantages of this treatment process for a municipal water supply?

7-83 What is your comment relative to the following statement: "Fluoridation of a municipal water supply is of no benefit to consumers who have water softeners in their homes because the fluoride ion is taken out by the ion-exchange resin"?

7-84 What is the health hazard associated with excessive nitrate concentration in a public water supply?

7-85 Assume the anion-exchange process is used to remove nitrate from a water with an ionic character as given in Problem 7-69. What would be a major problem of this treatment process?

7-86 Explain the meanings of inside-out flow and outside-in (transverse flow) in microfiltration. In Example 7-19, why was system A with outside-in (transverse flow) with air-water backwash preferred to system B using an inside-out flow path?

7-87 Why is the passage of water through a membrane for removal of dissolved salts called *reverse* osmosis?

7-88 Describe how the feedwater and permeate (product water) flow through a spiral-wound module for reverse osmosis.

7-89 How does the addition of acid prevent scale formation on reverse-osmosis membranes?

7-90 The quality characteristics of a deep-well water in an arid region with a hot climate are

Calcium hardness	= 270 mg/l
Magnesium hardness	= 180 mg/l
Sodium ion	= 138 mg/l
Iron ion	= 0.40 mg/l
Manganese ion	= 0.15 mg/l
Total dissolved solids	= 830 mg/l
Temperature	= 27°C
pH	= 8.1
Alkalinity	= 120 mg/l
Sulfate ion	= 110 mg/l
Chloride ion	= 344 mg/l
Fluoride ion	= 1.4 mg/l
Nitrate nitrogen	= 15 mg/l

Draw a milliequivalents-per-liter bar graph and compare the concentrations of chemicals with the primary and secondary drinking water standards (Tables 5-1, 5-2, and 5-3). (a) Is the water quality satisfactory for a municipal supply without treatment? Discuss any quality problems with respect to health and aesthetic standards. (b) One recommendation is to treat the water by lime-soda ash softening to improve the quality. Will this processing provide a satisfactory quality for a public water supply? Consider each of the quality parameters of concern. What are the advantages and disadvantages of precipitation softening of this water? (c) Propose a treatment scheme that will provide a water quality adequate to meet both health and aesthetic standards. Sketch a flow diagram showing the unit processes, chemical additions, and sources of wastes for disposal. (d) Sketch an approximate bar graph of the treated water.

7-91 Estimate the sludge solids produced in coagulating a surface water having a turbidity of 12 turbidity units with an alum dosage of 40 mg/l. What is the sludge volume if the settled waste sludge and filter backwash water are concentrated to 1000 mg/l of solids in a clarifier-thickener? (*Answers* 180 lb/mil gal, 22,000 gal/mil gal)

7-92 Lime added to water containing calcium hardness results in precipitation of calcium carbonate. If 126 mg/l of lime with a purity of 78 percent CaO is added to a hard water, how many milligrams per liter of $CaCO_3$ are formed? (*Answer* 350 mg/l)

7–93 Calculate the waste sludge produced in the excess lime treatment described in Problem 7–69 in units of pounds of dry solids per million gallons of water processed and gallons per million gallons assuming a settled solids concentration of 10 percent. (*Answers* 5900 lb/mil gal, 70 gal/mil gal)

7–94 Calculate the quantities of waste sludges produced in softening the water described in Problem 7–66. Assume that the solids concentration of the gravity-thickened sludge from the excess lime and split-treatment processes is 8 percent and the solids concentration from selective calcium carbonate removal is 12 percent.

7–95 Extraction of iron and manganese from groundwater is often performed by the process sequence of aeration, chemical oxidation, sedimentation, and filtration. In which unit operation is the majority of the metal oxides removed?

7–96 Which of the following waste sludges is the easiest to dewater: $CaCO_3 + Mg(OH)_2$ precipitate, alum sludge, or $CaCO_3$ slurry?

7–97 Which of the items listed are advantages of pressure filtration relative to centrifugation?
(a) Filtration is a continuous-flow process.
(b) The sludge can be dewatered to a solid cake.
(c) Hydroxide sludges can be dewatered with relative ease.
(d) Filtration is more adaptable for dewatering prior to recalcining.

Operation of Waterworks

Management of a water utility requires extensive knowledge of distribution systems and water processing; refer to Chapters 6 and 7, respectively. Water quality (Chapter 5) and testing for both chemical and biological characteristics are also prerequisite to systems control. This chapter discusses elements of operation and maintenance of water distribution systems, operational control of treatment plants, and management of waterworks.

8–1 DISTRIBUTION SYSTEM INSPECTION AND MAINTENANCE

Fire Hydrants

Hydrant maintenance is the responsibility of the water utility; however, the fire department may assist in hydrant inspections. Use of hydrants as a source of water for street cleaning and construction work must be restricted and controlled. After each use, a hydrant should be inspected to be sure

it is in operating condition in case of fire. A stock of repair parts, particularly replacement internal valves, is necessary to make immediate repairs. Although the frequency of periodic inspection varies among utilities, a yearly maintenance schedule is common. The exact inspection procedure depends on the type of hydrant. The most common is a center-stem dry-barrel fire hydrant (Figure 6–32) equipped with a progressive drain. Rather than opening and closing instantaneously, the drain operates progressively as the main valve opens and closes. During use and inspection, the hydrant should be fully opened so that the drain is fully closed. Water seeping through a partially closed drain can saturate the drain field and lead to washing away the soil around the hydrant. In cold climates, drainage is essential to prevent damage from water freezing and expanding in the barrel.

The first step in inspecting a hydrant is to remove an outlet-nozzle cap and check the barrel for water by inserting and retracting a plumb bob. The presence of water means that either the main valve is leaking or the groundwater level is high.

A listening device is used to check for leakage, and if necessary the main valve is repaired. If groundwater is high, a plug can be placed in the drain to preclude groundwater. Water in the barrel must then be pumped out after hydrant use. Drainage from an unplugged barrel can be checked after flushing. Upon closing the main valve, the drainage rate should be rapid enough to create a noticeable suction on a hand placed over the opening of one open nozzle. Leaks from joints and nozzle caps are detected by opening the main valve with the caps in place. To prevent damage, air is vented from the barrel while filling. Compression packings can be lubricated and tightened or gaskets replaced. For the center-stem hydrant, a screw or fitting in the operating nut indicates the need for manual lubrication, since some brands have automatic lubrication. In either case, the lubrication should be checked. Finally, all of the nozzle caps are inspected and threads cleaned and lubricated and screwed on sufficiently tight to prevent removal by hand.

Valves

Shutoff valves are installed in a distribution system to isolate pipe sections for maintenance or to repair a break. A preventative maintenance program ensures proper functioning of valves when they are needed. Without periodic inspection and operation, they may become inoperable, covered by repaving of streets or paving of driveways, or difficult to locate because of inaccurate mapping. Valves can also be partially or fully closed, obstructing water flow. A good maintenance program includes verifying locations, inspecting the condition of valve boxes, and opening and closing the valves. Depending on local conditions, valves in trunk and feeder mains are inspected at least once a year and others in the pipe network once every 1 to 3 years. Incrustation of internal moving parts can prevent full closure and, in extreme cases, prevent movement of the disk without risk of damage. Periodic opening and closing loosens the interfering scale.

Preventative maintenance involves location, inspection, and operation. Location of the valve box is confirmed by checking the measurements given on the map. Next, the valve box cover is removed and the interior is inspected with a flashlight.

Finally, the valve is turned fully closed and a count of the number of turns required is recorded. By listening with a leak detector, the tightness of closure is checked. To avoid failure of the stem, the applied torque should not exceed the recommended limit specified by the manufacturer. Manual turning, a slow and arduous process, can be done more economically by power equipment. Portable units are powered by compressed air or electricity. Power turning is particularly valuable for repeated operation of a valve to remove internal scale.

The common defects requiring repair are a broken stem, need for packing, and internal incrustation. Repair of a valve is a difficult task requiring isolation of that section of the pipe network, excavation, and removal of the valve. Either the damaged valve can be replaced or, if the overall condition is good, new parts can be installed on-site.

Cross-Connection and Backflow Preventers

A water utility is responsible for protection of its potable water supply from contamination resulting from cross-connection and back siphonage.[1] Therefore, a program for control of cross-connections is essential, including inspection of customer facilities that have the potential of contaminating the water supply with hazardous substances and pathogens. Examples are hospitals, mortuaries, chemical industries, irrigation systems, and reuse systems not intended for potable consumption.

Many cross-connection threats are the result of potential back siphonage from a sink hose into a contaminated fluid or from new construction that was not properly disinfected. With the advent of chemical fertilizer feed to irrigation and dual use of potable and recycled water, pressure adequate to overcome the potable water system and contaminate water distribution has become more common. There are several documented cases where cross-connections have been responsible for contamination of the distribution system. The potential risk is dynamic because piping systems are continuously being modified, extended, and altered.

Backflow preventers (Section 6–9) installed in service lines require periodic testing to ensure proper operation. Since enforcement of a cross-connection-control program involves health agencies, plumbing inspection, water suppliers, and owners, inspection of backflow preventers may be done by water utility personnel, a private testing agency, or another organization.

Cross-Connection Control

Several cross-connection-control requirements are associated with recycled-water distribution systems. Requirements typically include use of purple-colored plastic pipe or purple tape to distinguish recycled water pipe from potable water pipe in white or blue plastic pipe or black ductile iron or grey concrete pipe. All recycled-water appurtenances, such as vaults, boxes, and irrigation system components, must be labeled or preferably colored purple. No hose bibs are allowed on recycled-water piping. Recycled-water areas must be identified with regularly placed signs indicating the use of recycled water and a notification of non-potable water. Buildings and residential areas that are dual plumbed for potable and recycled water must maintain an annual cross-connection inspection program that verifies the connection by isolating each system and checking for flow in the other. If recycled water is connected with the potable water piping, then shutting off the potable water should result in no water, if not cross-connected, or recycled water, if connected.

Water Loss

Water loss accounts for the raw water quantity received and the finished water delivered to homes. Water losses at the treatment plant result from inaccurate metering, leakage from tanks, water associated with sludge disposal, and water wasted with filter backwash.

In the distribution system, authorized consumption includes billed metered consumption, including meterings that overrecord water use. Unbilled authorized consumption includes unbilled metered consumption due to billing errors or nonpayment and unbilled unmetered consumption due to meter underrecording.

Water losses may be grouped into two categories: apparent losses and real losses. Apparent losses include connections to the water system that are unauthorized and unmetered. Most apparent losses are related to inaccurate metering. Accurate water metering is basic to customer billing, collection of rate fees, and data for comparing water-supply quantities with consumption to estimate water losses in the system. Household water meters are under the control of the water utility. Underregistration is common as meters age. Many utilities have an ongoing program of repair for commercial and industrial meters and replacement of residential meters. Meters are sold with a useful life of 15–20 years, but several utilities have documented reductions in metering accuracy beginning at year 10 to 12. For many utilities, a reduction in revenues of 10 percent justifies meter repair or replacement. Billed consumption typically accounts for 85 percent of the water delivered, although values vary greatly with the age of the distribution system and historic construction practices.

Real losses are associated with leakage in the water distribution system, leaks in storage tanks, or leaks in the service connection before the customer meter. Real losses typically occur during unmetered uses for hydrant flushing, sampling, and fire protection. Real losses should be monitored and limited to the extent possible. Many utilities monitor the water used in flushing programs, water sent to drain during sampling, and other nonconsumptive uses to better evaluate losses due to leakage.

The objectives of leak detection are to locate and repair small defects in a pipe network before failures occur and to reduce water loss from the system. As water under pressure exits a crack or small hole, sound waves in the audible range are emitted by the pipe wall and surrounding soil. Electronic detectors, capable of amplifying sound waves and filtering out unwanted background noise, can locate and isolate leaks. With experience, the person operating the detector can also estimate the rate of leakage. The sound transmitted by the pipe wall can be heard by listening at hydrants, main valves, and curb valves. In conducting a survey, areas of water loss are identified by listening at these direct contact points. An approximate location is established by evaluating the intensity from different locations. Water impacting the soil and circulating in a cavity creates lower-frequency waves that have limited transmission through the ground. Through the use of a surface microphone, the leak can be located with greater precision. The nature of the sound emitted by escaping water depends on water pressure in the main, pipe material and size, soil conditions, and configuration of the opening. Consequently, locating and estimating the severity requires operator training and experience.

Pipes and joints are commonly repaired by placing an external cover over the leak. Repair sleeves are cylindrical halves that are fitted around the pipe or joint and bolted together. At a joint, a solid insert may be placed inside with half of the length in each pipe and a split-coupling clamped around the outside.

Tapping of Mains

Employees of a water utility make the service connections to mains for building contractors. A tapping machine drills and taps a hole through the pipe wall under pressure, hence the name "wet" tapping. The machine then either inserts a corporation stop for small-diameter connections or permits closure of an outlet gate valve for large connections. Dry tapping can be performed before a main is filled with water or if the pipe is removed from service.

The tapping machine for small connections is clamped against a saddle that matches the machine to the outside diameter of the pipe. A gasket is placed under the saddle, and the proper tapping tool is selected for the pipe material. The three diagrams in Figure 8–1 illustrate the three main steps in making a house service connection. First, the hole is drilled into the pipe with the combined drill bit and threading tool, and then the drilled hole is threaded with the same tool. After the tool is retracted within the drill-and-tap unit, it is swung aside, placing the corporation stop in position for screwing into the hole.

Large service connections require installation of a sleeve and gate valve. After the pipe exterior has been cleaned, a tapping sleeve in cylindrical halves is clamped around the pipe and sealed by split gaskets held against the sleeve ends by rings placed around the pipe and bolted to the sleeve. One-half of the tapping sleeve has a hole with a short piece of flanged pipe. A gate valve is bolted to the sleeve pipe, and a drilling machine is connected to the valve. The cutting bit, extending through the open gate valve, drills a hole through the pipe wall. After the hole has been completed, the cutter is retracted and the tapping gate valve closed. The drilling machine is removed, and the service line is attached to the flange of the gate valve.

Figure 8–1

Tapping a water main for a service connection. (a) Drilling a hole through the pipe wall. (b) Threading the drilled hole. (c) Inserting the corporation stop. (Courtesy of Mueller Co., A Grinnell Company.)

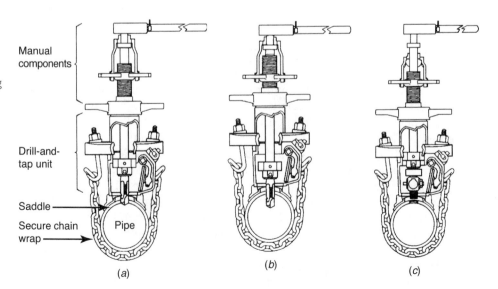

Manual components

Drill-and-tap unit

Saddle

Secure chain wrap

Pipe

(a)   (b)   (c)

8–2 DISTRIBUTION SYSTEM TESTING

Fire Flow Tests

These tests are important in determining the efficiency and adequacy of a distribution system in transmitting water, particularly during days of high demand, and are used to measure the amount of water available from hydrants for fire fighting. The required rate of flow for fire fighting must be available at a specified residual pressure. The minimum is 20 psi if fire department pumpers are used to supply hose streams. If hoses are to be connected directly to hydrants, a residual pressure of 50 to 75 psi is needed, depending on local structural conditions.

Flow tests consist of discharging water at a measured rate of flow from one or more fire hydrants and observing the corresponding pressure drop in the mains through another nearby hydrant. Figure 8–2a illustrates a flow test in which hydrants 1, 2, 3, and 4 are discharging while the residual pressure drop is read at hydrant R. The residual hydrant is chosen so that the hydrants flowing water are between it and the larger feeder mains supplying water to the area. The number of hydrants used for discharge, and the rates of flow, should be such that the drop in pressure at the residual hydrant is not less than 10 psi. For single-main testing, the layouts and hydrant selection are shown in Figure 8–2b.

The typical test procedure is as follows: One of the outlet caps on the residual hydrant is replaced with a pressure gauge, the hydrant valve is opened to exhaust air from the barrel, and the initial water pressure reading is recorded. The surrounding hydrants are then opened, and discharge is measured using pitot gauges. A pitot tube centered in the flow stream from a hydrant nozzle measures the pressure exerted by the velocity of the flowing water. The quantity of discharge from a hydrant nozzle can be calculated from the pitot pressure reading by using Eq. 8–1. Since it is difficult to establish exactly the rate of discharge needed to produce a specific residual pressure, the flow that would be available at a given residual pressure—for example, 20 psi—must be computed from test data by using Eq. 8–2.

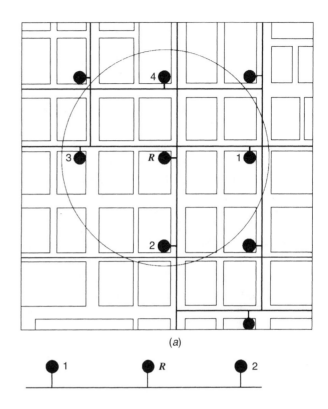

(a)

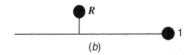

(b)

Figure 8–2

Diagrams showing selection of fire hydrants for flow tests to determine the amount of water available for fire fighting. Numbered hydrants discharge water with the resulting pressure drop measured at hydrant R. (a) Typical group hydrant flow test. (b) Layouts for single-main flow tests.

$$Q = 29.8Cd^2(p)^{1/2} \qquad (8\text{–}1)$$

where Q = discharge, gallons per minute
C = coefficient, normally 0.90
d = diameter of outlet, inches
p = pitot gauge reading, pounds per square inch

$$Q_R = Q_F \frac{H_R^{0.54}}{H_F^{0.54}} \cong Q_F \left(\frac{H_R}{H_F}\right)^{1/2} \qquad (8\text{–}2)$$

where Q_R = computed discharge at the specified residual pressure, gallons per minute

Q_F = total discharge during test, gallons per minute

H_R = drop in pressure from original value to specified residual, pounds per square inch

H_F = pressure drop during test, pounds per square inch

Flow test results show the strength of a distribution system but not necessarily the degree of adequacy of the entire waterworks. Consider a system supplied by pumps at one location and having no elevated storage. If the pressure at the pump station decreases during the test, this is an indication that the distribution system is capable of delivering more than the pumps can provide at their normal operating pressure, and the value for the drop in pressure measured during the test must be corrected. It is equal to the actual drop obtained in the field minus the drop in discharge pressure at the pumping station. If sufficient pumping capacity is available at the station and the discharge pressure can be maintained by operating additional pumps, the water system as a whole is able to deliver the computed quantity. If, however, additional pumping units are not available, the distribution system is capable of delivering the computed amount, but the water system as a whole is limited by the pumping.

The corrections for pressure drops in tests on systems with storage are generally estimated from a study of all the pressure drops observed on recording gauges at the pumping station. The corrections may be very significant for tests near the pumping station while decreasing to zero for remote tests.

Hydrant Test

The water flow from fire hydrant testing is typically discharged to the street and to the storm water system. The chlorine residual in drinking water is toxic to fish and must be dechlorinated prior to disposal. One gram of ascorbic acid (vitamin C) will neutralize 1 mg/l of chlorine in 100 gal. Figure 8–3 shows a typical configuration using vitamin C supply, flow, and pressure monitoring. Immediately following the fire hydrant is the dechlorination mixing chamber. Flow circulates through a tank containing vitamin C tablets back into the hydrant discharge stream. The remote

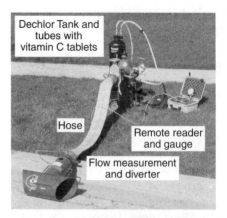

Figure 8–3

Photo of fire hydrant test assembly to measure flow, pressure, and dechlorinate the water prior to disposal in the storm sewer.

(Photo of Hose Monster® equipment courtesy of Hydro Flow Products, Inc.)

data-logging equipment and pressure gauge monitor pressure and flow. The hydrant must maintain a minimum of 20 psi pressure at the designed fire flow. Dechlorinated flow is discharged to the curb, then to the storm drain.

Unidirectional Flushing

Water main flushing is used to remove stagnant water from dead-end lines and scour debris that may have settled over years of service. Water main debris may include sand from water wells, precipitates from water treatment, and iron particles from old cast-iron pipe. The deposits contribute to water quality degradation, color, tastes, odors, and bacterial growth. Unidirectional flushing requires knowledge of the water distribution system to close valves in a specific order to create higher water velocities in one direction toward the open hydrant. The goal is to attain scouring velocities of 5 to 6 ft/sec. The duration of flushing is often determined by visual inspection of water color.

Distribution Network Analysis

Flows and pressures collected during distribution system testing can be used to calibrate computer simulation models of the piping network (Section 4–7). Networks may consist of several hundred pipes, but not all pipes are typically included, to save computational effort. The skeleton model of the system must be detailed enough so that the system can be accurately analyzed and calibrated with existing data.

The first step is to obtain a map of the distribution system, including pipes, sizes, lengths, and location of storage tanks, pumping stations, and wells. Pipe intersections, called nodes, are numbered to separate distinct pipe segments. For each segment, the starting and ending node number, elevation, pipe size, length, and friction coefficient are entered. Nodes are established for water demands, storage tanks, and pumping stations. Nodes are also created at points of data collection that represent flow and pressure readings within the system.

Conditions within the water distribution system are calculated based on the data provided by the user. The input assumptions can be adjusted to make the model simulate results from the real system. Comparison of the computed results with actual field pressure and flow readings improves predictive accuracy under changed conditions. Alternative conditions to be checked include maximum hour demands, fire demands, and loss of storage. When major system expansion is planned, the water system models will not only show the impact on the existing system but assist in sizing future pipelines.

■ EXAMPLE 8–1

Calculate the discharge at a residual pressure of 20 psi based on the following flow test. Hydrants 1, 2, 3, and 4, as illustrated in Figure 8–2a, were all

discharging the same amount of water through $2\frac{1}{2}$-in.-diameter hose nozzles. The pitot tube pressure reading at each hydrant was 16 psi. At this flow, the residual pressure at hydrant R dropped from 90 psi to 58 psi.

Solution

Using Eq. 8–1,

$$Q = (4)(29.8)(0.90)(2.5)^2(16)^{1/2} = 2680 \text{ gpm}$$

From Eq. 8–2,

$$Q_R = 2680 \frac{(90-20)^{0.54}}{(90-58)^{0.54}} = 4090 \text{ gpm}$$

or

$$Q_R = 2680\left(\frac{(90-20)}{(90-58)}\right)^{1/2} = 3970 \text{ gpm}$$

■ ■ ■

8–3 CONTROL OF WATERWORKS OPERATIONS

Water Treatment Process Control

The quality of the raw water supply, chemical processes in treatment, and physical facilities for flocculation, sedimentation, and filtration determine the complexity of controlling treatment plant operation. A high degree of automation, once influenced by the size of the plant, has become commonplace with low-cost computer control systems for small package plants. The control systems allow dial-in capabilities whereby the operating personnel can telephone the system remotely to determine plant operation and modify control strategies. On-line turbidity meters are used to automatically adjust chemical dosage and signal calls to backwash. Local vendors or equipment manufacturers provide service contracts for inspection and replacement of defective components. Even so, sufficient manual controls must be provided to operate equipment in the event of instrument or computer failure. Central control allows local control panels for complex operations like filter control to be coordinated with other filters for flow distribution, backwash sequencing, and emergency operation.

Single-station controllers, programmable logic controllers, and supervisory control and data acquisition (SCADA) systems represent increasing capabilities for automation. A single station controller allows the operator to enter a setpoint value and control such functions as water level and chemical addition. Programmable logic controllers perform the same function but are able to control more than one operation, and the controller action can be changed by altering the software program within the controller. Alarms can be generated from the controller but are typically a separate function. Operators are required to maintain records for run hours, weights, temperature, and process condition. Supervisory control and data acquisition systems use a personal computer for data storage and display and tie the entire control system together with a computer network. From the computer, an operator can start and stop equipment, change chemical feed rates, adjust pump controls, and graph flows, pressures, and weights. Alarms, run time, on/off status, flows, and weights can be recorded, graphed, and formatted for report generation.

Filtration control screens for a 120-mgd surface-water plant are pictured in Figure 8–4. The central computer was installed in an existing control panel, and an old strip charge recorder remains in use to the left of the monitor, Figure 8–4a. Each rectangle shown on the screen represents one of 24 filters. Each filter has a flow meter and modulating control valve in the pipe leaving the filter. The south filter basins, shown at the bottom of the screen, were out of service and placed in manual operation at the time of the photo. The values displayed below each north filter basin rectangle are for head loss through each filter, auto or manual operation, discharge valve position expressed in percent open, and flow through each filter.

The filters were originally programmed to operate using a constant flow rate through the filter. The computer uses a flow-rate value (setpoint) entered by an operator to compare with the value from each flow meter in the filter effluent. Each valve is modulated to a position allowing more or less flow until the flow meter value and opera-

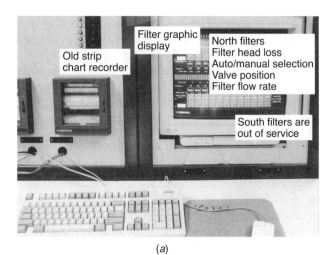

(a)

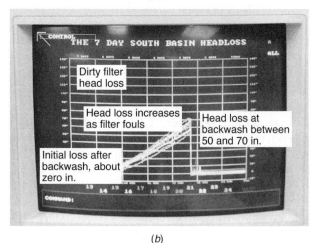

(b)

Figure 8–4

Central computer and display for the supervisory control and data acquisition system at a surface-water treatment plant. (a) Schematic display of the filter gallery. (b) Graphs of filter head versus time for south basin filters.
(San Juan Water District. Granite Bay, CA.)

tional setpoint match. Each valve is only partly open when the filter is clean to keep flow rate low, then as the filter fouls and flow begins to decrease, the controller opens the valve to maintain a constant rate. However, plant staff modified the operation to a declining rate, where the valves are kept at a constant position, about 35 percent open, and the flow decreases as the filter fouls. The change was made to keep the valves from "hunting" to match a flow value and because solids breakthrough is less likely if the flow rate declines as the filters foul.

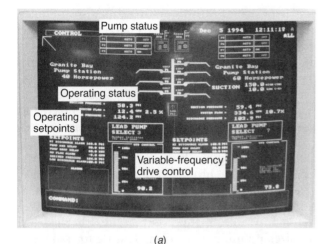

(a)

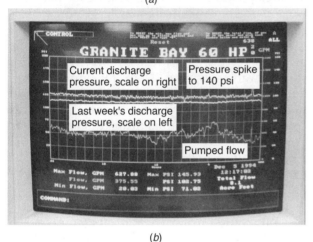

(b)

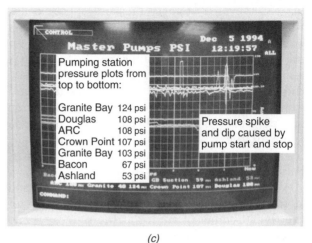

(c)

Figure 8–5
Schematic diagram and graphs of pumping station control.
(a) Schematic diagram and control values for a pumping
station. (b) Graphs of last week's discharge pressure on
left and current flow and discharge pressure on right.
(c) Graphs of eight pumping stations and their respective
pressure readings.
(San Juan Water District, Granite Bay, CA.)

Head loss for each filter is graphed in Figure
8–4b. From the keyboard, an operator can select
auto or manual operation, open and close the discharge valve, and control the process. The computer sequences backwash, controls valves, and
monitors filter operation. Twenty-four on-line turbidity meters record effluent water quality and
alarm based on operator-defined values. Operators
take grab samples of the raw water, sedimentation
basin effluent, filter influent, and filter effluent for
particle count analysis twice per day.

This water plant computerized control permits
operation with a minimum of labor: Three operators work 24 hours on with 48 hours off. One operator is always at the plant for 24 hours (16 hours
working, 8 hours sleeping in the control building).
During the day, the manager and chief operator
assist with plant responsibilities. Two full-time
maintenance staff are supplemented with contracts for all maintenance requirements.

The distribution system serves more than
180,000 connections. Control of the distribution is
monitored at each pumping station using remote
terminal units. The units transmit and receive
information from the central computer at the water
plant by radio communication link. Figure 8–5a
shows a graphic of the pump schematic layout and
control values for a pumping station. Operators can
modify the sequence of pump operation, turn
pumps on and off, and change operating setpoints
from the central computer. The system has only
one elevated storage tank; therefore, system pressure is maintained by pumps operating with
variable-speed controllers, as displayed on the
screen. Control of the pump speed and number of
pumps running is based on an operating pressure,
set at a minimum of 102 psi. A time delay prevents
pumps from starting and stopping due to pressure
fluctuation. Figure 8–5b shows a graph of the last
week's pressure on the left axis and current flow
and discharge pressure on the right axis. The start
and stop of another pump caused the pressure to
spike to 145 psi. Pressure variations are common in
small systems when pumping into a network with

limited elevated storage. Pressure spikes occur when water demand is just enough to start the next pump, but once started, the pump is too large for the demand and pressure builds rapidly in the system before the pump is turned off. Hydropneumatic and small storage tanks help reduce pressure variations by providing storage in the system when pumping exceeds demand. Figure 8–5c shows a graph of the eight pumping stations and their respective pressure readings. For the most part, pump control maintains system pressure within a 10-psi range. Some spikes occur as pumps start and stop and as the instrumentation tries to "hunt" for the right number of pumps at the right speed to meet rapid demand changes.

Water Quality

A quality program is essential to the operation of any water treatment plant. For surface supplies it is mandatory, but even if processing well water is limited to iron and manganese removal, adequate laboratory control is necessary for overall operation. Some small plants routinely carry out comprehensive sampling and testing programs, while others rely completely on state or local health agencies for surveillance. But even most small waterworks now find it within their ability to train operators in many of the common chemical and bacteriological tests by using proprietary kits and reagents.

Availability of complete laboratory reports is valuable for public relations and in assessing system vulnerability. The assessment is based on previous monitoring results, user population characteristics, proximity to sources of contamination, surrounding land uses, protection of water sources, and historical system operation and maintenance data.

All states require analysis of finished water for coliform bacteria; the number of tests required is based on the population served and the number of service connections, or by special circumstance. For each positive total coliform test, at least three repeat samples and testing for both total and fecal coliform are required. Repeated testing is required until no coliforms are detected in three consecutive samples. Public notification is required when two samples are positive in the same month, any repeat samples are positive, or any fecal coliform

test is positive. Minimum analysis also includes general minerals, physical characteristics, and inorganic chemicals. General mineral analyses include bicarbonate, carbonate, hydroxide alkalinity, calcium, chloride, copper, foaming agents, iron, magnesium, manganese, pH, sodium, sulfate, specific conductance, total dissolved solids, total hardness, and zinc. Physical characteristics include color, odor, and turbidity. Inorganic chemical analysis is typical for aluminum, arsenic, barium, cadmium, chromium, lead, mercury, nitrate, selenium, and fluoride. Any chemical concentration exceeding the maximum contaminant level listed in Table 5–1 presents a health risk and requires public notification. Testing for radioactive and organic chemicals may be required every 2 to 4 years and monthly to quarterly for chemicals showing positive results or suspected to be present. Testing is typically more frequent for surface-water supplies than for groundwater. Public notification is required when the initial finding of a radioactive or inorganic chemical is confirmed with two additional samples and the average concentration exceeds the maximum contaminant level. Additional testing may be required for unregulated chemicals as required by the local Health Department.

The sampling of water sources includes general minerals, physical tests, inorganic chemicals, organic chemicals, and radioactivity. Tests exceeding the maximum contaminant levels are not a compliance violation, but may require more frequent monitoring and review of alternative sources. Sampling and testing of waste streams are required to comply with discharge requirements. Concentrations of metals and radioactive materials increase in the sludge and may exceed disposal limits.

Other laboratory testing is related to control of chemical treatment, troubleshooting problems in the distribution system, and handling customer complaints on quality. Shipments of treatment chemicals, if purchased under specifications, should be routinely tested, and penalties should be assessed for deficiencies. A penalty clause in the contract supported by an approved laboratory analysis can protect the plant against inferior material.

The smallest plant laboratory may contain only a turbidimeter, a chlorine residual colorimeter, a pH meter, and glassware for hardness and alkalinity determinations. A completely equipped laboratory for a metropolitan area includes infrared, ultraviolet, and atomic absorption spectrophotometers; a gas chromatograph; an amperometric titrator; and a conductance meter, plus the common ovens and glassware. Incubators, an autoclave, special glassware, and culture media are needed for bacteriological work. If a nuclear power plant is located upstream, some radiochemical surveillance is necessary and a proportional counter and scintillation counter may be needed. Laboratories serving large water systems are usually supervised by a graduate chemist or chemical engineer. Additional professionals and two or three laboratory technicians may also be employed, depending on the extent of the testing program. Laboratory personnel may also serve part-time or full-time in plant supervision. Shift operators can be trained to do routine tests like those for turbidity, alkalinity, and chlorine residual. Standard coliform counts can also be handled by operators who have had the proper training. The advantage of laboratory-trained operators is that operating shifts cover the entire 24-hr day whereas laboratory personnel typically work only an 8-hr day. The capability for around-the-clock surveillance is important in detecting sudden changes in raw water quality.

Laboratory jar tests and past experience are both used to determine optimum chemical dosages for coagulation and lime softening processes. Problems may be anticipated by careful testing of the raw water so that variations in the treatment process can be made promptly. Past treatment records are helpful; for example, efficient lime softening can be constantly evaluated by maintaining a record of lime feed to hardness removed. The ratio of CaO applied to hardness precipitated may range from 0.8 to 1.2 depending on water temperatures and retention time, but it should be fairly uniform during any one season of the year.

Taste and odor problems in surface-water supplies can easily be the primary concern in water-quality control for several months of the year. The specific causative compounds are often dissolved organics that are exceedingly difficult, if not impossible, to identify. The standard chlorine demand test gives an indication of overall water quality. The determination of the threshold odor number is another means of evaluating water quality. The test is conducted by a panel of observers who attempt to detect odor in successively diluted samples. An odor number is calculated from the number of dilutions necessary for the last traces of odor to disappear. Although a useful tool, the test lacks reproducibility, and many instances have been reported in which treated water with a low threshold odor number has been responsible for numerous customer complaints. Odor dilution, it seems, is not always linear, and rather offensive samples may occasionally respond well to only one or two dilutions.

Biological contamination may be associated with wells under the influence of surface water due to poor location, deterioration creating leaks, or unusual circumstances such as low pressure or vacuum in the piping. Contamination in surface water is more likely because of the potential for direct runoff containing animal feces, sewer wastewater, and flooding. Surface-water treatment requires filtration and disinfection because chlorine alone is inadequate to kill harmful microorganisms. The American Water Works Association recommends that operators maintain a turbidity of 0.1 NTU to reduce the risk of biological organisms surviving disinfection. Although the EPA standards require less than 0.5 NTU in 95 percent of the monthly measurements with single spikes less than 5 NTU and tests for total coliform are negative, *Cryptosporidium* oocysts may pass through filtration and disinfection.[2] Conditions adequate to affect public health typically occur under conditions where a spike in contamination is complemented by poor coagulation, problems with filter operation, and high flows that increase effluent turbidity. Water treatment plant operators must be aware of operating conditions that increase the chance of passing biological contamination through the system. Increased risks in the distribution system occur when peak flows scour and release high concentrations of contaminants from the biofilm in the distribution pipes.

When tests for total and fecal coliform are positive, a boil-water advisory may be given to customers, warning them of a potential biological waterborne contamination. A boil-water order is typically issued following confirmation of the contamination and/or investigation of potential risk. An investigation by the water system operator of the source water quality, treatment plant operation, and water distribution system should be conducted to evaluate unusual conditions that may have created conditions for contamination. Source water quality considerations include evaluation of the deviations in source water quality; the vulnerability of source water contamination by reviewing land use activity in the watershed; and unusual events such as sewage spills, heavy runoff, floods, droughts, and other circumstances. Treatment evaluation should start with a review of operating records, especially filter performance (turbidity and particle counts) and disinfection performance (chlorine dose, residual, contact time, and coliform test results at the plant and within the distribution system). Plant records should be reviewed for failure or interruption in process operation, spikes in flow, shortened filter run times, and changes in chemical addition or operational practice. A review of the integrity of the distribution system includes an evaluation of recent construction activity around the water mains, leaks or breaks, reports of low pressure or stagnant water, evidence of peak flows, flushing practices, and a test of the chlorine residual throughout the system. Epidemiologic evidence may be an increase in sales of over-the-counter diarrhea medication and an increase in gastrointestinal illness. Epidemiologic information suggesting an outbreak as waterborne should be used when considering a boil-water order even if water-quality data is not yet available. The boil-water notice should include instructions to bring drinking water to a full boil for at least 1 minute.

Sanitary Survey

A sanitary survey is a physical review of the watershed, water supply, treatment plant, distribution system, and operation and maintenance records.[3] The purpose of the survey is to identify all areas that represent a risk to water quality and delivery. The watershed and water supply are described in terms of geography, climate, hydrology, geology, soils, vegetation, and demography. The treatment facility is described in

terms of its emergency procedures, backup power capability, ability to remove variable influent characteristics, and equipment reliability. The distribution system should have adequate valves for isolation, pipes in a looped or grid system for multiple-feed capability, wells constructed according to code, and storage tanks designed for resistance to natural disasters and sufficient capacity.

The survey, conducted by a sanitary engineer, should assess potential contamination in the entire watershed or aquifer. Activities and sources of contamination within the watershed include waste management, agriculture, herbicide and insecticide use, wild fires, livestock and grazing, mining, landslides, fish and wildlife, urban development, body- and non-body-contact recreation, road building, use, and maintenance. Contamination potentials can be addressed using plant operational procedures modified to treat specific problems, such as control of taste and odor constituents. An emergency response plan should address actions to be taken for major accidents within the watershed, treatment contingencies, and distribution system reliability.

In the United Kingdom, legislation requires continuous monitoring of *Cryptosporidium* occysts if they are found to be present in the raw water supply.[4] The cause-and-effect relationship between the presence of *Cryptosporidium* and illness is particularly difficult to quantify because outbreaks have occurred when levels were below the regulated limit and, in several cases, high counts have been detected with no apparent illness in the community.

8–4   RECORDKEEPING

Efficient management of a water utility is founded on detailed records of physical facilities, operation, and maintenance. A distribution system is documented by data describing the pipe network and its hydraulic characteristics, and water treatment records provide data on the operation and maintenance of the plant and water quality. Records of business affairs include meter readings, billing accounts, material and equipment purchases, insurance, and personnel records.

The operations of a distribution system are documented by water consumption, pressure gauge records, pump station discharge flow and pressure, and levels of storage tanks. These data are essential for the calibration of system distribution models.

Distribution system documentation includes maps and detailed information about valves, hydrants, and maintenance histories. Separate maps are generally maintained for the overall distribution system and individual service connections. Distribution system maps show streets, mains, valves, hydrants, storage tanks, pumping stations, other features, and location information from at least two permanent reference points. The maps can be maintained within an automated mapping or computer-aided design system. Data for each valve, fire hydrant, storage tank, and pump are retained on an individual record card or contained within an automated, preventive-maintenance data base. The data base contains information about the equipment, such as the equipment number, type, size, model, and manufacturer; frequency and tasks for preventive maintenance; and record of inspection and maintenance. Geographic information systems integrate the mapping graphic data and the nongraphic data base information within a single software program. The program can be used to display data about an object, such as a fire hydrant selected from the map, or generate a map of all valves that are scheduled for a preventive-maintenance inspection. Automated programs maintain inventories for valves, meters, supplies, and spare parts.

Treatment plant recordkeeping encompasses water quality and quantity data and plant operations. Some tests on the treated water are required by regulatory agencies to show that the quality meets drinking water standards. Daily tests include coliform analyses and usually one or more chemical parameters. Periodic analyses are performed for inorganic and organic chemicals. For process control, the laboratory testing is defined by the treatment, for example, jar tests for turbidity removal or hardness for softening. Also, the dosages and quantities of chemicals applied in treatment are recorded. Quantity and flow data include the amount of influent water, finished water, filter rates, water loss by backwashing of filters, and sludge wastage. If the source is groundwater, recorded well data are hours of operation, pumping rate, static and pumping water levels, discharge pressure, power consumption, and maintenance. Programs are available for scheduling preventative maintenance, inventory control, process control calculations, plotting water consumption data, and other purposes. Since stored data can be printed without difficulty at any time, records are immediately available for analysis and writing reports.

An annual report summarizes laboratory analyses, operational tests, maintenance, business, chemical dose, and cost data. An annual report distributed to each customer should provide specific information on concentrations of microbiological contaminants, minerals, physical agents, inorganic chemicals, organic chemicals, and radioactivity present in the water supply. The report should compare laboratory results with regulated contaminant and recommended public health levels and information on compliance with drinking water standards.

Public notification is required for procedural and water-quality failures. Procedural failures include violating monitoring requirements, testing procedures, or reporting requirements. Notice of procedural problems is important but distinct from water-quality failures in the frequency and content of messages. Water-quality failures result from exceeding a maximum contaminant level, treatment technique, or a compliance schedule.

Public notification should contain identification of the agency issuing the notice, description of what the notice is about, clear statement of the health significance of the violation, description of the event that triggered the notification, description of any precautions that consumers should take, description of action being taken to correct the problem, and listing of names and telephone numbers for more information.

8–5 WATER CONSERVATION

An important economic advantage of reducing water demand is extending the use of a waterworks without the expansion of facilities. Lower consumption also slows depletion of a limited water supply. Conservation programs can be directed toward either consumer consumption or water losses in the distribution system.

Water conservation, oriented toward reducing consumer usage, is based on public education, installation of water-efficient plumbing fixtures, and institution of water rates that encourage conservation. Although an essential adjunct, an education program without other incentives is not likely to be effective with the passage of time. Plumbing fixtures that use less water are effective; however, the most successful applications have been in new construction where the community has either a limited source of supply or difficulty in wastewater disposal, or both.

High-Efficiency Plumbing Fixtures

The best management practice for instituting indoor residential water conservation is to install high-efficiency plumbing fixtures in new construction and retrofit old fixtures in existing residences, often subsidized by the municipal water provider.

To determine the impacts of high-efficiency retrofits in single-family homes, three studies were undertaken, in Seattle, Washington (1999–2000), East Bay Municipal Water District (EBMWD) service area; California (2001–2003); and in Tampa, Florida. Data from The Tampa Water Department Residential Water Conservation Study (2002–2004)[5] are discussed here. In the study, 26 single-family homes in the service were selected from customers willing to participate who had not installed retrofits and who had an average daily per capita usage higher than 60 gpd. The basic methodology was to establish in 2 weeks the baseline water-use data. Next, the homes were retrofitted with high-efficiency toilets, clotheswashers, showerheads, and faucets. About a month after the retrofit, 2 weeks of water-use data were collected, and a second set of postretrofit data was obtained about 6 months later.

By fitting a data logger strapped to the residential water meter, flow trace analysis for end uses of water is possible. The logger provides precise flow data for each 10-sec interval. A computer with custom signal-processing software can trace the flow with sufficient precision to identify the individual flow signatures of each type of appliance and plumbing fixture in the residence. Each flow trace is disaggregated into its component end uses, that is, toilets, showers, clotheswashers, dishwashers, baths, faucets, and leaks. With this computer technology, precise data on where water is used inside a residence can be collected from the water meter by a nonintrusive flow trace analysis.

Figure 8–6 compares the preretrofit and postretrofit indoor per capita water use percentages. Overall,

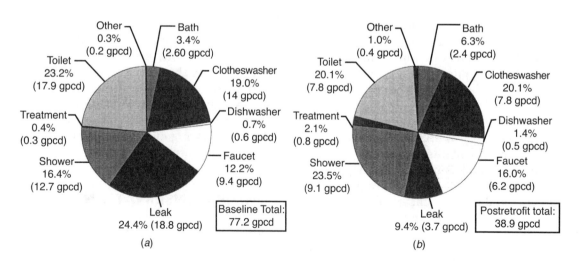

Figure 8–6

Tampa Water Department Residential Water Conservation Study comparing preretrofit and postretrofit indoor water use percentage including leakage. (*a*) Baseline average per capita indoor water use. (*b*) Average per capita indoor water use after installing high-efficiency plumbing fixtures.
(Courtesy of Aquacraft, Inc.)

indoor water usage decreased from 77.2 gpcd to 38.9 gpcd (292 to 147 l/person · d), a reduction of 38.3 gpcd (145 l/person · d), which is 49.6 percent. Leakage decreased from 18.8 gpcd to 3.7 gpcd (71 to 14 l/person · d), from 24.4 percent to 9.4 percent. Toilet leaks, primarily flapper leaks, are the single largest contributor to household leakage. Showers became the largest water use (23.5 percent) followed by toilets and clotheswashers (each 20.1 percent) and faucets (16.0 percent). Baths and dishwashers showed no significant changes.

Data from the two earlier studies are as follows: In Seattle, the preretrofit was 63.6 gpcd (241 l/person · d) and the postretrofit was 39.9 gpcd (151 l/person · d) for 37.3 percent reduction in indoor water usage. For the EBMUD, the preretrofit was 60.4 gpcd (229 l/person · d) and the postretrofit was 43.9 gpcd (166 l/person · d) for 39.0 percent reduction.

In Tampa, outdoor water use can occur in almost any month of the year, depending on weather. Seasonal variations in metered consumption of 986 homes provide an indication of the amount of water used indoors and outdoors. In 2002, annual indoor water usage accounted for 90 percent of the total usage.

Higher Water Rates

Charging a higher price for water can reduce consumption, but unless the public understands the need for restricting usage or for additional revenue, the utility can lose public support. Domestic consumption is difficult to lower below some minimum amount, as determined by normal household uses, without replacing plumbing fixtures. On the other hand, lawn watering can be reduced by instituting an escalating rate schedule that penalizes the homeowner for overirrigation and wastage. Commercial and industrial users are much more likely to reduce their normal consumption as a result of higher prices. With economic incentive, they can often justify modifying processes to save water and develop reuse systems.

Water Audit

Water conservation by a utility is initiated by conducting a water audit to detect leaks to reduce the quantity of water that cannot be accounted for by metered sales. In addition to conserving water, the potential benefits of conducting a water audit are increasing revenue by replacing meters that underregister and preventing serious damage that can result from uncorrected leaks. As a general rule, good performance is when the metered sales are at least 85 to 90 percent of the metered supply. The major phases of a water audit are flow measurement and leak detection.

The survey of a system begins with checking the accuracy of the master meters that measure flow into the distribution system. The common instrument for measuring flow in a pipeline is a pitot tube flow meter and recorder. With an accuracy of ±2 percent, a wide range of flows can be measured in different size pipes. The meter can be temporarily installed through a 1-in. corporation stop. Although testing residential meters during an audit is not practical, the municipal program of maintenance and replacement is assessed if underregistration is suspected. Many communities remove and rebuild meters on a cycle of 8 to 10 years. Since small meters tend to underregister with age, the metered sales are likely to be less than the actual water delivered.

8–6 WATER RATES

A water utility must receive sufficient revenue to cover the costs of adequate service. Basic revenue requirements are for operation and maintenance expenses; debt service, including interest and stipulated reserves; utility extensions and improvements; and plant replacement for perpetuation of the system. The total revenue collected not only should reflect recent cost experience but also should recognize anticipated future costs during the nominal period for which rates are being established, for example, 5 years.

Modern water rates recognize the costs of supplying the amount of water consumed, the rate of use or demand for use, and the expense involved in maintaining the customer account. Recognizing different classes of customers is reasonable in measuring the costs of serving these diverse groups. Charges should be commensurate with the service rendered. For example, a customer with a high peak rate of use in comparison with average consumption would require larger-capacity pumps, pipes, and other facilities than a customer who has a comparable total consumption but uses water continuously at essentially a constant rate.

Determining water rates involves the following major areas of study: the amount of revenue required for the water utility; allocation of this annual revenue to the requirements of basic cost functions, which in turn are attributed to customer classes in accordance with respective class requirements for service; and design of water rates that recover from each customer class the respective costs of providing service. The latter, of course, requires an evaluation of water use for each class of customer based on local conditions that are influenced by climate, industrial demand, and other factors.

One technique for allocating the cost of service to cost functions is referred to as the commodity-demand method. This separates the three primary cost functions of demand, consumption, and customer costs. Demand costs are associated with providing facilities to meet peak rates of use. They include capital charges and operating costs on the part of the treatment works designed to meet peak requirements. Consumption costs include power, chemicals, and other incremental expenses that vary with the quantity of water produced. Customer expenses are associated with serving consumers irrespective of the amount of water used, such as meter reading, billing, accounting, and maintenance on meters and services.

The next step is to ascertain customer classes on the basis of hourly or daily demand characteristics, although some special categories may be based on unusual annual requirements. Typically, the three principal classes are residential, commercial, and industrial. In any given system, there may be users with unusual water-use characteristics that should be given separate consideration, for example, large institutions such as a university.

System costs for demand, consumption, and customer services are now related to the chosen customer classes to establish cost responsibility. Each customer's demand expense is based on his or her fair share of the total system demand. Consumption costs are allocated according to total annual use, and customer costs are allocated on the basis of number of metered services. Finally, a rate schedule is fixed to collect sufficient revenue to meet expenses of the waterworks.

The following is a hypothetical method for billing domestic potable water usage to encourage conservation. Consumption is billed on an inclining block system with the lowest single-family block between 0 to 1000 cu ft/month costing less than $1 per day, 1000 to 2000 cu ft/month, 2000–3000 cu ft/month, and the highest block, over 4000 cu ft/month, costing up to $5 per day. High-efficiency retrofits in a single-family home with four residents would amount to about 600 cu ft/month within the first block for indoor usage. The highest block would include extensive outdoor usage for lawn irrigation and perhaps significant indoor leakage, usually from the toilets. In addition to billing for consumption, a service fee for meter reading and accounting is about $5 per month.

8–7   SECURITY

The waterworks control system requires the physical security of SCADA, and control components should be kept in locked, access-controlled areas, including tightly controlled access to remote locations. Network wiring should be encased in conduits and not be accessible to unauthorized persons. When the SCADA operating system is connected to external networks, access controls, such as firewalls and router packet filters, are recommended to be secure from hackers. System backups and restoration should be available for disaster recovery.

Water treatment plant security is required to ensure safe drinking water and fire protection. A physical protection system must consider potential contamination of water sources, whether it is a reservoir, river, or well field. Although unlikely, since a large quantity of a contaminant would be required, unsubstantiated claims of contamination may be made to the detriment of the water utility. A plan of action should be available to ensure that any claim of contamination, real or false, can be proved or disproved.

The perimeter of the property should have barriers to control vehicle and pedestrian entry. The recommended security is a 7- to 8-ft chain-link fence topped with three strands of barbwire with a 20-ft clear zone outside the fence line. The most secure vehicle gate is a remotely controlled one-piece sliding gate. Electric substations should be fenced and locked. Electronic security systems can provide early warning if an intruder enters the plant property or a building. A wide variety of intrusion detection sensors are available, including passive or active, covert or visible, line-of-sight or terrain-following, volumetric or line detection. The fence can be equipped with vibration sensors or fiber-optic cables that send an alarm if someone tampers with the fence or cuts a fiber-optic cable. To deter any further action, alarms triggered by intruders can be audible. All of these are secondary systems to prevent outsider intrusions, but an uninvited insider is the most difficult to defend against. Installing electronic access devices, such as a card key or proximity card reader, gives management a record of who used a card and when it was used to open a specific door. Using electronic access devices can limit access to critical areas. Closed-circuit television systems with visible cameras and warning signs posted on the perimeter fence can deter intruders or insiders who are in restricted areas, outside or inside. The cameras can provide recorded evidence of acts of sabotage and help determine if the cause of the alarm is human or animal. Digital closed-circuit television technology has improved significantly, resulting in declining costs for cameras and recorders. Digital video recorders produce better images and are easier to integrate with the overall security system.

Chemical deliveries to the plant site should follow a carefully planned protocol. Prior notification of the delivery, a photo ID and the name of the driver making the delivery, and checking the manifest against the purchase order are important security steps. Thorough background checks, including any criminal record, should be conducted on prospective employees. The use of specific uniforms and photo ID badges for all water treatment plant employees should be considered.

REFERENCES

1. U.S. EPA, *Cross-Connection Control Manual*, EPA 816-R-03-002, February 2003.
2. *Cryptosporidium* and Water: A Public Health Handbook. Atlanta, Georgia: Working Group on Waterborne Cryptosporidiosis, Centers for Disease Control and Prevention, National Center for Infectious Diseases, Atlanta, Georgia, 1997.
3. U.S. EPA, *Guidance Manual for Conducting Sanitary Surveys of Public Water Systems; Surface Water and Ground Water Under Direct Influence (GWUDI)*, EPA 815-R-99-016, April 1999.
4. Drinking Water Inspectorate, Drinking Water 2000 Annual Report. London: The Inspectorate; 2001. http://www.dwi.gov.uk/pubs/annrep00/42.htm
5. "Tampa Water Department Residential Water Conservation Study, Impacts of High Efficiency Plumbing Fixture Retrofits in Single-Family Homes," Aquacraft, Inc. Water Engineering and Management, 2709 Pine St., Boulder, Colorado 80302. www.aquacraft.com/Publications

PROBLEMS

8–1 List the major steps in the inspection of fire hydrants and valves. Why is regular opening and closing of valves important?

8–2 Why is cross-connection important? List some of the ways water systems are protected from potential contamination.

8–3 What are the sources of water loss and what can a utility engineer do about water losses?

8–4 A fire flow test was conducted within the distribution system using three hydrants with 2-in. nozzles, as shown in Figure 8–2b. The pitot tube pressure at all hydrants was 12 psi, while the pressure drop at hydrant R went from 50 to 43 psi. Compute the discharge at 20 psi. Is the discharge adequate to fight a residential fire with 1,500 gpm?

8–5 The total discharge during a fire flow test is 800 gpm at a test pressure of 27 psi. The original pressure was 45 psi and the specified residual is 20 psi. What is the computed discharge for the specified pressure?

8–6 A 3400-ft transmission main, 12 in. in diameter, conveys water from storage to the distribution system. Fire flow to meet residential and commercial requirements is 2500 gpm. The booster pumps at the storage tanks have a capacity of 2500 gpm at 63 psi. At $C = 100$, is there adequate residual pressure (see Section 4–3 for the Hazen Williams formula)?

8–7 In general terms, what steps does each of the following perform: single station controller, programmable logic control, and central computer? How does the computer screen in Figure 8–4 help the operator control the filters?

8–8 What are the sources of poor quality of a water supply?

8–9 What are the water quality barriers from the raw water supply through treatment and distribution? How do multiple barriers ensure a greater degree of public health?

8–10 What should be considered before a boil-water notice is issued?

8–11 What should be considered in a sanitary survey?

8–12 What should be included in an annual report for a water utility?

8–13 What should be accomplished in evaluating water conservation? What measures could be taken?

8–14 Describe the logging-computer technology used in The Tampa Water Department Residential Water Conservation Study. What was the average percentage decrease of water usage after retrofitting the single-family homes with high-efficiency toilets, clothes washers, shower-heads, and faucets?

8–15 What are the key elements of security to consider in protecting waterworks?

8–16 What costs are used to determine water rates?

Wastewater Flows and Characteristics

Domestic or sanitary wastewater refers to the liquid discharge from residences, business buildings, and institutions. Industrial wastewater is discharge from manufacturing plants. Municipal wastewater is the general term applied to the liquid collected in sanitary sewers and treated in a municipal plant. In addition, interceptor sewers direct dry weather flow from combined sewers to treatment. Unwanted infiltration and inflow can also enter the collector pipes. A schematic of the system is given in Figure 9–1.

Storm runoff water in most communities is collected in a separate storm sewer system, with no known domestic or industrial connections, and is conveyed to the nearest watercourse for discharge often without treatment. Rain water washes contaminants from roofs, streets, and other areas, and although the pollutional load of the first flush may be significant, the total amount from separated storm-water systems is relatively minor compared with other wastewater discharges. Several large cities have a combined sewer system where both storm water and sanitary wastewaters are collected in the same piping.

Dry weather flow in the combined sewers is intercepted and conveyed to the treatment plant for processing, but during storms, flow in excess of plant capacity is by-passed directly to the receiving watercourse. This can constitute significant pollution and a health hazard in cases where the receiving body is used for a drinking water supply. One solution is to separate the combined sewers, but the cost in large cities would be prohibitive, although this can be done efficiently where only a few combined sewers exist in a municipal system.

9–1 DOMESTIC WASTEWATER

The volume of wastewater from a community varies from 50 to 250 gal per capita per day (gpcd) depending on sewer uses. A common value used for domestic wastewater flow is 120 gpcd (450 l/person · d), which assumes that the residential dwellings have modern appliances, such as automatic washing machines, and accounts for some infiltration and inflow into the manholes and sewer pipe. The organic matter contributed per person per day in

Figure 9–1

Sources of municipal wastewater in relation to collector sewers and treatment. The figure also shows reuse for irrigation, residential, and industrial purposes, effluent disposal to surface water, and separate disposal for industrial gray water.

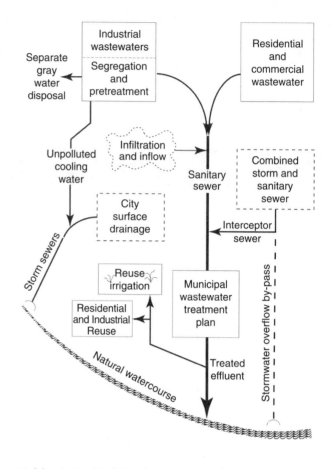

domestic wastewater is approximately 0.24 lb (110 g) of suspended solids and 0.20 lb (90 g) of BOD in communities where a substantial portion of the household kitchen wastes is discharged to the sewer system through garbage grinders. In selection of data for design, the quantity and organic strength of wastewater should be based on actual measurements taken throughout the year to account for variations resulting from seasonal climatic changes and other factors. The average values during the peak month may be used for design. Excluding unusual infiltration and inflow, the average daily sanitary wastewater flow during the maximum month of the year is commonly 20 to 30 percent greater than the average annual daily flow. Excluding seasonal industrial wastes, the average daily BOD load from sanitary wastewater during the maximum month is greater than the annual average by 30 percent or more in small plants (less than 0.5 mgd) and less than 20 percent in large plants (greater than 50 mgd).

Estimated wastewater flows from residential dwellings and other establishments are listed in Table 9–1. Mobile homes and hotels generate less wastewater than residences, since they have fewer appliances. The quantity and strength of wastewater from schools, offices, factories, and other commercial establishments depend on hours of operation and presence of eating facilities. Although cafeterias do not provide a great deal of flow, the wastewater strength is increased materially by food preparation and cleanup.

The common value for sanitary wastewater of 120 gpcd includes residential and commercial wastewaters plus reasonable infiltration, but excludes industrial discharges and reductions in flow associated with water conservation. In some areas, water conservation has reduced flow sufficiently to effect an increase in BOD and suspended solids concentrations. Flow values ranging between 80 and 100 gpcd and waste load concentrations must be selected with an understanding of local water, food preparation, and waste habits. Characteristics of this wastewater prior to treatment, after settling, and

TABLE 9–1

Approximate Wastewater Flows for Various Kinds of Establishments and Services

Type	Gallons per Person per Day	Pounds of BOD per Person per Day
Domestic wastewater from residential areas		
Large single-family houses	120	0.20
Typical single-family houses	80	0.17
Multiple-family dwellings (apartments)	60 to 75	0.17
Small dwellings or cottages	50	0.17
Domestic wastewater from camps and motels		
Luxury resorts	100 to 150	0.20
Mobile home parks	50	0.17
Tourist camps or trailer parks	35	0.15
Hotels and motels	50	0.10
Schools		
Boarding schools	75	0.17
Day schools with cafeterias	20	0.06
Day schools without cafeterias	15	0.04
Restaurants		
Each employee	30	0.10
Each patron	7 to 10	0.04
Each meal served	4	0.03
Transportation terminals		
Each employee	15	0.05
Each passenger	5	0.02
Hospitals	150 to 300	0.30
Offices	15	0.05
Drive-in theaters, per stall	5	0.02
Movie theaters, per seat	3 to 5	0.02
Factories, exclusive of industrial and cafeteria wastes	15 to 30	0.05

TABLE 9–2

Approximate Composition of Average Sanitary Wastewater (mg/l) Based on 120 gpcd (450 l/person · d)

Parameter	Raw	After Primary Settling	After Biological Treatment
Total solids	800	680	530
Total volatile solids	440	340	220
Suspended solids	240	120	30
Volatile suspended solids	180	100	20
Biochemical oxygen demand	200	130	30
Inorganic nitrogen as N	22	22	24
Total nitrogen as N	35	30	26
Soluble phosphorus as P	4	4	4
Total phosphorus as P	7	6	5

following conventional biological processing are given in Table 9–2. Total solids, residue on evaporation, include both dissolved salts and organic matter; the latter is represented by the volatile fraction. BOD is a measure of the wastewater strength. Sedimentation of a typical domestic wastewater diminishes BOD approximately 35 percent and suspended solids 50 percent. Processing, including secondary biological treatment, reduces the suspended solids and BOD content more than 85 percent, volatile solids 50 percent, total nitrogen about 25 percent, and phosphorus only 20 percent.

The surplus of nutrients in the treated effluent indicates that sanitary wastewater has nitrogen and phosphorus in excess of biological needs. The generally accepted BOD/N/P weight ratio required for biological treatment is 100/5/1 (100 mg/l BOD to 5 mg/l nitrogen to 1 mg/l phosphorus). Raw sanitary wastewater has a ratio of 100/17/3; after primary settling the ratio is 100/23/5, and thus there is abundant nitrogen and phosphorus for microbial growth. (The exact BOD/N/P ratio needed for biological treatment depends on the process and availability of the N and P for growth; 100/6/1.5 is often used for unsettled sanitary wastewater, and 100/3/0.7 is used where the nitrogen and phosphorus are in soluble forms.) Another important wastewater characteristic is that not all of the organic matter is

biodegradable. Although a substantial portion of the carbohydrates, fats, and proteins are converted to carbon dioxide by microbial action, a waste sludge equivalent to 20 to 40 percent of the applied BOD is generated in biological treatment.

Loadings on treatment units are often expressed in terms of pounds of BOD per day or pounds of solids per day, as well as quantity of flow per day. The relationship between the parameters of concentration and flow is based on the following conversion factors: 1.0 mg/l, which is the same as 1.0 part per million parts by weight, equals 8.34 lb/mil gal, since 1 gal of water weighs 8.34 lb or 62.4 lb/mil gal, since 1 cu ft of water weighs 62.4 lbs. These relationships are defined by the following equations:

$$\text{Pounds of } C = \text{concentration of } C \text{ (mg/l)} \times Q(\text{mil gal}) \times 8.34 \quad (9\text{–}1)$$

or

$$\text{Pounds of } C = \text{concentration of } C \text{ (mg/l)} \times Q(\text{mil cu ft}) \times 62.4 \quad (9\text{–}2)$$

where C = BOD, SS, or other constituent, milligrams per liter

Q = volume of wastewater, million gallons or million cubic feet

$$8.34 = \frac{\text{lb/mil gal}}{\text{mg/l}}$$

$$62.4 = \frac{\text{lb/mil cu ft}}{\text{mg/l}}$$

Calculations in Example 9–1 show that 120 gal of the sanitary wastewater as described in Table 9–2 contain 0.20 lb of BOD and 0.24 lb of suspended solids; Examples 9–2 and 9–3 illustrate applications of Eqs. 9–1 and 9–2.

■ EXAMPLE 9–1

Sanitary wastewater from a residential community is 120 gpcd, containing 200 mg/l BOD and 240 mg/l suspended solids. Compute the pounds of BOD per capita and pounds of SS (suspended solids) per capita.

Solution

Using Eq. 9–1,

$$\text{BOD} = 200 \text{ mg/l} \times 0.000{,}120 \text{ mil gal}$$

$$\times 8.34 \frac{\text{lb}}{\text{mil gal} \times \text{mg/l}} = 0.20 \text{ lb}$$

$$\text{SS} = 240 \text{ mg/l} \times 0.000{,}120 \text{ mil gal}$$

$$\times 8.34 \frac{\text{lb}}{\text{mil gal} \times \text{mg/l}} = 0.24 \text{ lb}$$

■ ■ ■

■ EXAMPLE 9–2

Industrial wastewaters (Table 9–4) have a total flow of 2,930,000 gpd, BOD of 21,600 lb/day, and suspended solids of 13,400 lb/day. Calculate the BOD and SS concentrations.

Solution

From the relationship in Eq. 9–1,

$$\text{BOD concentration} = \frac{21{,}600 \text{ lb/day}}{2.93 \text{ mil gal/day} \times 8.34}$$

$$= 880 \text{ mg/l}$$

$$\text{SS concentration} = \frac{13{,}400 \text{ lb/day}}{2.93 \text{ mgd} \times 8.34}$$

$$= 550 \text{ mg/l}$$

■ ■ ■

■ EXAMPLE 9–3

An aeration basin with a volume of 300 m³ contains a mixed liquor (aerating activated sludge) with a suspended solids concentration of 2000 mg/l (g/m³). How many kilograms of mixed-liquor suspended solids are in the tank?

Solution

$$\text{MLSS} = \frac{2000 \text{ g/m}^3 \times 300 \text{ m}^3}{1000 \text{ g/kg}} = 600 \text{ kg}$$

■ ■ ■

9–2 INDUSTRIAL WASTEWATERS

Industries within municipal limits ordinarily discharge their wastewater to the city's sewer system after pretreatment. Uncontaminated cooling water is directed to the storm sewer. In this joint processing of wastewater, the municipality accepts responsibility for the final treatment and disposal. The majority of manufacturing wastes are more amenable to biological treatment after dilution with domestic wastewater; however, large volumes of high-strength wastes must be considered in sizing a municipal treatment plant.

A sewer code, user fees, and separate contracts between an industry and the city can provide adequate control and sound financial planning while accommodating the industry through joint treatment. Pretreatment at the industrial site must be considered for wastewaters having strengths or characteristics significantly different from sanitary wastewater. Consideration should be given to modifications in industrial processes, segregation of wastes, flow equalization, and reduction of waste strength. Process changes, equipment modifications, by-product recovery, and in-plant wastewater reuse can result in cost savings for both water supply and wastewater treatment. Modern industrial plant design dictates segregation of separate waste streams for individual pretreatment, controlled mixing, or separate disposal. The latter applies to both uncontaminated cooling water that can be discharged directly to surface watercourses and toxic wastes that cannot be adequately processed by the municipal plant and must be processed or disposed of by the industry. Certain industrial discharges, such as dairy wastes, can be more easily reduced in strength by treatment in their concentrated form at the industrial site or directly hauled to the plant for digestion. Oil and grease retained in restaurant traps must be hauled for disposal or digestion. Others, like metal-plating wastes, require pretreatment for the removal of toxic metal ions. If reuse of the municipal wastewater is planned, rather stringent controls on industrial discharges are needed, since many of the substances in manufacturing wastes are only partially removed by conventional treatment and will interfere with water reuse.

The EPA pretreatment program identified standards for enforcing three general types of discharge: prohibited discharges, categorical discharges, and local pretreatment limits.[1] Prohibited discharges are national standards applicable to all industrial dischargers regardless of whether the treatment plant has an NPDES permit. The standards protect against compounds that pass through conventional treatment and are intended to promote safety, especially as related to biosolids. Categorical standards are stated in the NPDES permit for publicly owned treatment plants. These are concerned with specific process wastewaters and particular industrial categories and apply both to the collection system and to the treatment plants themselves. Local limits are developed to address specific needs and capabilities of individual treatment plants and impose discharge limitations to protect receiving waters.

The characteristics of four selected industrial wastewaters are listed in Table 9–3 for comparison with the sanitary wastewater described in Table 9–2. BOD concentrations range from 5 to 20 times greater than for domestic wastewater. Total solids are also greater but vary in character from colloidal and dissolved organics in food-processing wastewaters to predominantly inorganic salts, such as the chlorophenolic waste. Suspended solids concentration relative to BOD is important when considering conventional primary sedimentation and secondary biological treatment. Settling of the synthetic textile wastewater with a suspended solids to BOD ratio of 2000 mg/l to 1500 mg/l would be as effective as clarifying a sanitary wastewater with a ratio of 240/200, but settling a milk-processing wastewater with a suspended

TABLE 9–3

Average Characteristics of Selected Industrial Wastewaters

	MILK PROCESSING	MEAT PACKING	SYNTHETIC TEXTILE	CHLOROPHENOLIC MANUFACTURE
BOD, mg/l	1,000	1,400	1,500	4,300
COD, mg/l	1,900	2,100	3,300	5,400
Total solids, mg/l	1,600	3,300	8,000	53,000
Suspended solids, mg/l	300	1,000	2,000	1,200
Nitrogen, mg N/l	50	150	30	0
Phosphorus, mg P/l	12	16	0	0
pH	7	7	5	7
Temperature, °C	29	28	—	17
Grease, mg/l	—	500	—	—
Chloride, mg/l	—	—	—	27,000
Phenols, mg/l	—	—	—	140

solids to BOD ratio of 300 mg/l to 1000 mg/l would remove very little organic matter. In addition to high strength and settleability, particular consideration must be given to nutrient content, grease, and toxicity. Food-processing wastes generally contain sufficient nitrogen and phosphorus for biological treatment, but discharges from chemical and materials industries are deficient in growth nutrients. The chlorophenolic waste in Table 9–3 could not be discharged to a sewer without extensive reduction in phenol; the limit set by sewer ordinances is in the range of 0.5 to 1.0 mg/l. Metal-finishing wastes are pretreated to remove oil, cyanide, chromium, and other heavy metals such that the pretreated discharge has fewer contaminants than domestic wastewater. Each municipality should have an inventory of industrial wastewaters being discharged to the sanitary sewer system, as is illustrated in Table 9–4.

In this city the major wastewater contributors are food-processing industries. The manufacturing wastewaters from rubber products, metal working, and carpet weaving have strengths comparable to, or less than, domestic wastewater.

Industrial wastewaters expressed in terms of quantity of flow and pounds of BOD are relatively meaningless to the general public. Therefore, the quantity and strength can be related to the number of persons that would be required to contribute an equivalent quantity of wastewater. Hydraulic and BOD population equivalents, based on average sanitary wastewater, are 120 gpcd and 0.20 lb BOD per person per day, respectively. In addition to equivalent populations, it is desirable to express the quantity of wastewater produced per unit of raw material processed or finished product manufactured. Examples 9–4 and 9–5 illustrate wastewater production and equivalent population calculations.

TABLE 9–4

Results from a Municipal Industrial Wastewater Survey Listing Discharges to the Sanitary Sewer in a City with a Population of 145,000

	FLOW (gpd)	BOD (mg/l)	BOD (lb/day)	SUSPENDED SOLIDS (mg/l)	SUSPENDED SOLIDS (lb/day)	COD (mg/l)	GREASE (mg/l)
Meat processing	1,200,000	1,300	13,000	960	9,600	2,500	460
Soybean oil extraction	478,000	220	880	140	560	440	—
Rubber products	189,000	200	310	250	390	300	—
Ice cream	138,000	910	1,050	260	300	1,830	—
Cheese	110,000	3,160	2,900	970	890	5,600	—
Metal plating	108,000	8	7	27	24	36	—
Carpet mill	103,000	140	120	60	51	490	—
Candy	97,700	1,560	1,270	260	210	2,960	200
Motor scooters	93,500	30	23	26	20	70	—
Potato chips	90,400	600	450	680	510	1,260	—
Flour	83,100	330	230	330	250	570	—
Milk processing	65,100	1,400	760	310	170	3,290	—
Industrial laundry	50,000	700	290	450	190	2,400	520
Pharmaceuticals	40,700	270	91	150	50	390	160
Chicken hatchery	35,300	200	59	310	90	450	—
Luncheon meats	20,900	270	47	60	10	420	—
Soft drinks	16,000	480	64	480	64	1,000	—
Milk bottling	12,700	230	24	110	12	420	—
Totals	2,930,000		21,600		13,400		

■ EXAMPLE 9–4

A dairy processing about 250,000 lb of milk daily produces an average of 65,100 gpd of wastewater with a BOD of 1400 mg/l. The principal operations are bottling of milk and making ice cream, with limited production of cottage cheese. Compute the flow and BOD per 1000 lb of milk received, and the equivalent populations of the daily wastewater discharge.

Solution

Flow per 1000 lb of milk

$$= \frac{1000 \text{ lb}}{250,000 \text{ lb/day}} \times 65,100 \text{ gpd} = 260 \text{ gal}$$

BOD per 1000 lb of milk

$$= \frac{0.0651 \text{ mil gal/day} \times 1400 \text{ mg/l} \times 8.34}{250 \text{ thousands of lb/day}}$$

$$= 3.0 \text{ lb}$$

BOD equivalent population

$$= \frac{0.0651 \text{ mil gal/day} \times 1400 \text{ mg/l} \times 8.34}{0.20 \text{ lb BOD/person/day}}$$

$$= 3800 \text{ persons}$$

Hydraulic equivalent population

$$= \frac{65,000 \text{ gal/day}}{120 \text{ gal/person/day}} = 540 \text{ persons}$$

■ ■ ■

■ EXAMPLE 9–5

A meat-processing plant slaughters an average 500,000 kg of live beef per day. The majority is shipped as dressed halves with some production of packaged meats. Blood is recovered for a salable by-product, paunch manure (undigested stomach contents) is removed by screening and hauled to land burial, and process wastewater is settled and skimmed to recover heavy solids and some grease for inedible rendering with other meat trimmings. After this pretreatment, the waste discharged to the municipal sewer is 4500 m³/d containing 1300 mg/l BOD. Calculate the BOD waste per 1000 kg LWK (live weight kill) and the equivalent populations of the daily wastewater flow.

Solution

BOD per 1000 kg LWK

$$= \frac{4500 \text{ m}^3/\text{d} \times 1300 \text{ mg/l}}{500 \text{ thousands of kg/d} \times 1000 \text{ g/kg}}$$

$$= 11.7 \text{ kg}$$

BOD equivalent population

$$= \frac{4500 \text{ m}^3/\text{d} \times 1300 \text{ mg/l}}{90 \text{ g BOD/person} \cdot \text{d}} = 65,000 \text{ persons}$$

Hydraulic equivalent population

$$= \frac{4500 \text{ m}^3/\text{d} \times 1000 \text{ l/m}^3}{450 \text{ l/person} \cdot \text{d}} = 10,000 \text{ persons}$$

■ ■ ■

9–3 INFILTRATION AND INFLOW

Infiltration is groundwater entering sewers and building connections through defective joints and broken or cracked pipe and manholes. Inflow is water discharged into sewer pipes or service connections from such sources as foundation drains, roof leaders, cellar and yard area drains, cooling water from air conditioners, and other clean-water discharges from commercial and industrial establishments. In comparison to storm sewers, sanitary lines are small, being sized to handle only domestic and industrial wastewaters plus reasonable infiltration. Excessive infiltration and inflow can create several serious problems, including surcharging of sewer lines with backup of sanitary wastewaters into house basements, flooding of street and road areas, overloading of treatment facilities, and by-passing of pumping stations and treatment works.

The quantity of infiltration water entering a sewer depends on the number and condition of the pipe and pipe joints, groundwater level, manholes and structures, and construction practices. Use of sealants and waterproofing along with tighter construction practices reduces the quantity of infiltration. The actual inflow and infiltration rates for a system may be determined by comparing wet weather versus dry weather flows. Monitoring flow within the sewer system helps locate trouble areas and can distinguish between inflow and infiltration by comparing sewer flow during storm events and groundwater levels. For planning purposes, infiltration allowances are used to set a standard for sewer design and construction practices. Often these values are adjusted to reflect local climate and soil conditions or to determine a policy allocating the cost of sewer construction versus treatment. As a planning estimate for new construction, infiltration flow may be estimated in terms of gallons (liters) per acre per day or as a percentage of the per capita flow. Values range from 700 to 1500 gal/day/acre (6,600 to 14,100 l/d/ha). Percentages range from 3 to 5 percent of the peak hourly domestic flow rate, or approximately 10 percent of the average. A more accurate estimate of infiltration is based on pipe diameter and length. Infiltration estimates range from 200 to 500 gal/day per in. diameter per mile of pipe (185 to 465 l/d/cm-dia./km). Infiltration rates vary with groundwater level, pipe material, and materials used for manholes, siphons, and sewer structures. Many agencies prefer lower values to improve design and construction.

Infiltration rates increase with time as soil movement and deterioration create cracks in manholes and pipe, opens joints, and affects waterproofing systems. Correction of infiltration conditions in existing sewer systems involves evaluation and interpretation of sewer flow conditions to isolate and determine the source and rate of excessive infiltration. Many cities are finding that, following significant expenditure in the collection system, significant infiltration occurs in the service connection between the sewer and the home. Present techniques to reduce infiltration include lining manholes and sewer pipe, replacing pipe where joints have failed, and grouting or sealing surrounding soils. All of these alternatives are costly and should be evaluated in relation to the cost of treatment.

Inflow is the result of a deliberate connection of an extraneous water source to the sanitary sewer system. Although unwanted storm water or drainage should be discharged to a storm sewer, the sanitary sewer system may be more convenient. Direct connections include roof downspouts, foundation drains, basement drains and sumps, and driveway, yard, and street drains. Perforated manhole covers or covers that have been removed in flooded areas may also be a significant source of inflow. Municipal sewer ordinances should explicitly restrict direct connections that create inflow. As a planning estimate for new construction, inflow is not expected; however, some municipalities make allowances for inflow ranging from zero, where inflow is included with the 120 gal/capita-day (460 l/d/capita), to 5 gpcd (19 l/d/capita). Most sources of inflow as the result of a direct connection can be readily identified by smoke testing. Smoke introduced into the sewer system between isolated manholes will exit at the top of connected downspouts and into the home where direct connections have been made. Connections to foundation drains and sump discharges are harder to locate. By eliminating the connection with the sanitary sewer, inflow will be reduced.

■ EXAMPLE 9–6

Calculate the infiltration and compare this quantity to the average daily and peak hourly domestic wastewater flows for the following:

Sewered population = 24,000 persons
Average domestic flow = 100 gpcd
Peak hourly domestic flow = 240 gpcd
Infiltration rate = 500 gpd/mile/in. of pipe diameter
Sanitary sewer system:
    4-in. building sewers = 36 miles
    8-in. street laterals = 24 miles
    10-in. submains = 6 miles
    12-in. trunk sewers = 6 miles

Solution

Infiltration (gpd)

$$= \text{rate} \left(\frac{\text{gal}}{\text{day} \times \text{miles} \times \text{in.}} \right) \times \text{dia (in.)}$$
$$\times \text{length (miles)}$$

$$= 500(4 \times 36 + 8 \times 24 + 10 \times 6 + 12 \times 6)$$

$$= 234,000 \text{ gpd}$$

$$\text{Average domestic flow} = 24,000 \times 100$$

$$= 2,400,000 \text{ gpd}$$

$$\frac{\text{Infiltration}}{\text{Average domestic flow}} = \frac{234,000}{2,400,000} \times 100$$

$$= 9.8 \text{ percent}$$

Peak hourly domestic flow

$$= 24,000 \times 240 = 5,760,000 \text{ gpd}$$

$$\frac{\text{Infiltration}}{\text{Peak hourly flow}} = \frac{234,000}{5,760,000} \times 100$$

$$= 4.1 \text{ percent}$$

∎∎∎

9–4   MUNICIPAL WASTEWATER

As shown in Figure 9–1, the flow in sanitary sewers is a composite of domestic and industrial wastewaters, infiltration and inflow, and intercepted flow from combined sewers. Collector sewers must have hydraulic capacities to handle maximum hourly flow, including domestic and infiltration, plus any additional discharge from industrial plants. New sewer systems are usually designed on the basis of an average daily per capita flow of 100 to 120 gal (400 to 480 liters), which includes normal infiltration. However, pipes must be sized to carry peak flows that are often assumed to be 400 gpcd (1500 l/person · d) for laterals and submains when flowing full, 250 gpcd (950 l/person · d) for main, trunk, and outfall sewers; and in the case of interceptors, collecting from combined sewer systems, 350 percent of the average dry weather flow. Peak hourly discharges in main and trunk sewers are less than the maximum flows in laterals and submains, since hydraulic peaks tend to level out as the wastewater flows through a pipe network picking up an increasing number of connections.

Various equations have been developed to determine peaking factors for wastewater flow.

Some equations use population and others use flow as a denominator. The following equation was empirically derived based on flow:

$$PF = \frac{2.5}{Q^{0.145}} \qquad \text{with a maximum of 4} \qquad \textbf{(9–3)}$$

where PF = peaking factor, dimensionless
   Q = flow in million gallons per day

The flow pattern from a separate sanitary sewer system to the wastewater treatment plant is illustrated in Figure 9–2a. The hourly flow rates shown range from less than 40 percent to about 190 percent. The lowest flows occur in early morning (3–4 A.M.) and peak near midday. Flows typically drop sharply and rise again between 6 and 7 P.M. in residential communities. Hourly flow rates from small communities typically range from 20 to 250 percent of average daily flows and from 50 to 200 percent for larger cities.

The BOD and ammonia concentrations in wastewater vary with the time of day in a pattern that may not follow flow variation (see Figure 9–2b). Ammonia closely follows flow variations, but, for this community, BOD strength rises in the afternoon and continues through the evening. Waste strength is greatest as peak noon flows subside and are lowest during the night when activity is at its lowest and slow pipe velocities permit settling of solids within the collection system. Ammonia and BOD concentration patterns vary greatly among facilities due to the time travel from the city to the treatment plant, slope of the sewers, and relative contribution of residential, commercial, and industrial activity. If both flow and BOD concentration variations are known, the time-BOD loading on a treatment plant can be calculated and plotted as shown in Figure 9–2b. Knowledge of influent hydraulic and BOD loadings is essential in evaluating the operation of a treatment plant.

The quantity and characteristics of wastewater fluctuate with season of the year and between weekdays and holidays. Summer discharges frequently exceed winter flows by 10 to 20 percent, and industrial contributions are reduced on Sundays. Hourly fluctuations in large cities are

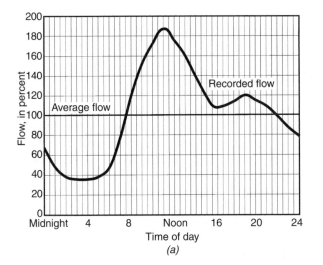

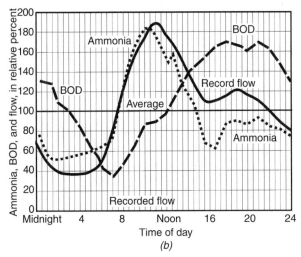

Figure 9–2
Wastewater flow and strength variations for a typical medium-sized city. (a) Typical diurnal flow pattern with peak flows at noon and 6. (b) Variation in flow along with concentrations of BOD and ammonia.

modified in comparison with small towns because of the diversity of activities and operations that take place throughout the 24-hr day. Large volumes of high-strength industrial waste contributions can distort typical flow and BOD patterns by accentuating the peak hydraulic and BOD loadings during operational hours. Excessive infiltration and inflow, while diluting wastewater strength, can have considerable impact on a treatment facility by increasing both the average and

peak flows during periods of high rainfall. All of these factors must be considered in assessing the wastewater flow and strength variations for a particular community.

■ EXAMPLE 9–7

The sanitary and industrial waste from a community consists of domestic wastewater from a sewered population of 7500 persons; potato-processing waste of 30,000 gpd containing 550 lb of BOD; and creamery wastewater flow of 120,000 gpd with a BOD concentration of 1000 mg/l. Estimate the combined wastewater flow in gallons per day and BOD concentration in milligrams per liter.

Solution

SOURCE	FLOW IN GALLONS PER DAY	BOD IN POUNDS PER DAY
Domestic	$7500 \times 120 = 900,000$	$0.20 \times 7500 = 1500$
Potato	$= 30,000$	$= 500$
Creamery	$= 120,000$	0.120×1000 $\times 8.34 = 1000$
Total	$1,050,000$	3050

$$\text{BOD concentration} = \frac{3050 \text{ lb/day}}{1.05 \text{ mil gal/day} \times 8.34}$$
$$= 348 \text{ mg/l}$$

■ ■ ■

■ EXAMPLE 9–8

A city with a sewered population of 145,000 has an average wastewater flow of 18.9 mgd with an average BOD of 320 mg/l. An inventory of the industrial wastewaters entering the sanitary sewer system is given in Table 9–4. (a) Compute the equivalent populations for this municipal wastewater flow that includes both sanitary and industrial wastewaters. (b) Determine the per capita contribution of sanitary wastewater flow and BOD based on the city's population excluding the industrial wastewaters. (c) Calculate the peak flow factor and peak flow.

Solution

(a) For the municipal wastewater,

$$\frac{\text{Hydraulic equivalent}}{\text{population}} = \frac{18{,}900{,}000 \text{ gpd}}{120 \text{ gpcd}}$$

$$= 158{,}000$$

$$\frac{\text{BOD equivalent}}{\text{population}} = \frac{18.9 \text{ mgd} \times 320 \text{ mg/l} \times 8.34}{0.20 \text{ lb/person/day}}$$

$$= 252{,}000$$

(b) Per capita contributions excluding industrial wastewaters are

$$\text{Sanitary flow} = \frac{18{,}900{,}000 - 2{,}930{,}000}{145{,}000}$$

$$= 110 \text{ gpcd}$$

$$\text{Sanitary BOD} = \frac{18.9 \times 320 \times 8.34 - 21{,}600}{145{,}000}$$

$$= 0.20 \text{ lb/person/day}$$

(c) Peak flow based on the average flow is

$$PF = \frac{2.5}{Q^{0.145}} = \frac{2.5}{18.9^{0.145}} = 1.6$$

Peak hydraulic flow $= 1.6 \cdot 18.9 = 30.2$ mgd

∎∎∎

9–5 COMPOSITE SAMPLING

Proper sampling techniques are vital for accurate testing in evaluation studies. To be representative of the entire flow, samples should be taken where the wastewater is well mixed. An instantaneous grab sample represents conditions at the time of sampling only and cannot be considered to represent a longer time period, since the character of a wastewater discharge is not stable. A composite sample is a mixture of individual grabs proportioned according to the wastewater flow pattern. Compositing is commonly accomplished by collecting individual samples at regular time intervals, for example, every hour on the hour, and by storing them in a refrigerator or ice chest; coincident flow rates are read from an installed flow meter or are determined from some other flow recording device. A representative sample is then integrated by mixing together portions of individual samples relative to flow rates at sampling times.

Composite samples representing specified time periods are tested to appraise plant performance and loadings. Weekday specimens collected over a 24-hr period are most common. Average daily BOD and suspended solids data are used to calculate plant loadings, and mean influent and effluent concentrations yield treatment efficiencies. Integrated samples during the period of peak flow, usually 8 to 12 hr depending on influent variation, allow determination of maximum loadings on treatment units.

Sampling data are tabulated in a format convenient for calculating the portions of individual grab samples to be combined into a composite. Flow meter readings (either in terms of flow rate or total flow) are recorded to provide a flow rate for each grab sample. The time interval between collection of samples should be no greater than 2 hr and preferably 1 hr. The compositing time period and frequency of sampling defines the number of grab-sample portions to be combined. The total volume of the composite sample desired depends on the kinds and number of laboratory tests to be performed. The volume of each grab sample collected must be adequate to provide the maximum portion required at maximum flow. The portion of sample needed per unit of flow is designated as the multiplier in the following equation:

$$\text{Multiplier} = \frac{\text{volume of composite sample desired}}{\text{average flow rate} \times \text{number of portions}} \quad (9\text{--}4)$$

Table 9–5 illustrates a general format for tabulating flow data for calculating the portions of individual grab samples to be combined into a composite. In this case, the flow meter registered only total flow and not the rate of flow. Therefore, readings taken at even hours were used to calculate the average flow during the 2-hr interval. Individual grab samples with a volume of 1 liter were collected at odd hours. After completion of the 24-hr flow recording and sampling period, the following values were calculated: average hourly flow rates for each 2-hr period between meter readings, a multiplier using Eq. 9–4 as shown at the bottom of the table, and composite portions for each of the 12 individual samples. If the flow meter had registered the rate of flow, the readings and samples could have been collected simultaneously at the odd hours. A graphical presentation of the compositing process is shown in Figure 9–3, based on the data in Table 9–5, with the wastewater flow line drawn through plotted flow rates and the vertical bars drawn to represent the composite portions.

TABLE 9–5

General Format for Recording the Compositing of a Wastewater Sample from Individual Grab Samples

Time (hr)	Flow (m^3)	Meter Readings (m^3)	2-hr Flow (m^3)	Average Flow (m^3/hr)	Multiplier[a] ($ml/m^3 \cdot hr$)		Composite Portion (ml)
1000	(2) 69616	(1) 69338	278				
1100				139	$\times$ 3.77	$=$	524
1200	(3) 69954	(2) 69616	338				
1300				169	$\times$ 3.77	$=$	637
1400	(4) 70355	(3) 69954	401				
1500				201	$\times$ 3.77	$=$	758
1600	(5) 70732	(4) 70355	377				
1700				189	$\times$ 3.77	$=$	713
1800	(6) 71026	(5) 70732	294				
1900				147	$\times$ 3.77	$=$	554
2000	(7) 71351	(6) 71026	325				
2100				163	$\times$ 3.77	$=$	615
2200	(8) 71652	(7) 71351	301				
2300				151	$\times$ 3.77	$=$	569
2400	(9) 71903	(8) 71652	251				
0100				126	$\times$ 3.77	$=$	475
0200	(10) 72040	(9) 71903	137				
0300				69	$\times$ 3.77	$=$	260
0400	(11) 72113	(10) 72040	73				
0500				37	$\times$ 3.77	$=$	139
0600	(12) 72308	(11) 72113	195				
0700				98	$\times$ 3.77	$=$	370
0800	(13) 72512	(12) 72308	204				
0900				102	$\times$ 3.77	$=$	385
1000		(13) 72512					
			Totals	1591			5999

[a]Multiplier $= \dfrac{\text{volume of composite sample desired}}{\text{average flow rate} \times \text{number of portions}}$

Average flow rate $= 1591/12 = 133$ m^3/hr

Volume of composite sample desired $= 6000$ ml

Multiplier $= \dfrac{6000}{133 \times 12} = 3.77$ ml/m$^3 \cdot$ hr

Figure 9–3
Wastewater flow pattern and composite portions plotted from the compositing data in Table 9–5.

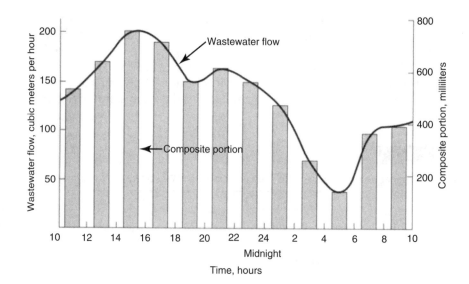

■ EXAMPLE 9–9

Hourly samples were taken of the wastewater entering a treatment plant. The recorded flow pattern is given in Figure 9–3. Tabulate the portions to be used from the hourly grabs to provide composite samples for the 24-hr duration and during the period of maximum 8-hr loading, between 9 A.M. and 5 P.M. The composite sample volumes needed for laboratory testing are approximately 2500 ml.

Solution

The portion of sample needed per unit of flow to provide a 2500-ml composite for the 24-hr and 8-hr periods is calculated using Eq. 9–3.

$$\text{Multiplier for 24-hr period} = \frac{2500 \text{ ml}}{720 \text{ gpm} \times 24}$$
$$= 0.15 \text{ ml/gpm}$$

$$\text{Multiplier for 8-hr period} = \frac{2500 \text{ ml}}{1000 \text{ gpm} \times 8}$$
$$= 0.30 \text{ ml/gpm}$$

Calculations for the portions of hourly samples to be used in compositing are tabulated to create a single porportional composite for the daily and maximum periods.

TIME	FLOW (gpm)	PORTIONS OF HOURLY SAMPLES IN MILLILITERS FOR: 24-HR COMPOSITE	8-HR COMPOSITE
Midnight	490	0.15 × 490 = 74	
I A.M.	420	0.15 × 420 = 63	
2 A.M.	360	0.15 × 360 = 54	
3 A.M.	310	0.15 × 310 = 47	
4 A.M.	290	0.15 × 290 = 43	
5 A.M.	310	0.15 × 310 = 46	
6 A.M.	390	0.15 × 390 = 58	
7 A.M.	560	0.15 × 560 = 84	
8 A.M.	620	0.15 × 620 = 93	
9 A.M.	900	0.15 × 900 = 135	0.3 × 900 = 270
10 A.M.	1040	0.15 × 1040 = 156	0.3 × 1040 = 310
II A.M.	1130	0.15 × 1130 = 170	0.3 × 1130 = 340
Noon	1160	0.15 × 1160 = 174	0.3 × 1160 = 350
I P.M.	1120	0.15 × 1120 = 168	0.3 × 1120 = 340
2 P.M.	1060	0.15 × 1060 = 159	0.3 × 1060 = 320
3 P.M.	1000	0.15 × 1000 = 150	0.3 × 1000 = 300
4 P.M.	950	0.15 × 950 = 143	0.3 × 950 = 290
5 P.M.	910	0.15 × 910 = 136	
6 P.M.	870	0.15 × 870 = 130	
7 P.M.	810	0.15 × 810 = 121	
8 P.M.	760	0.15 × 760 = 114	
9 P.M.	690	0.15 × 690 = 103	
10 P.M.	630	0.15 × 630 = 94	
II P.M.	540	0.15 × 540 = 81	
Total composite sample volumes		2596 ml	2520 ml

■ ■ ■

9–6 EVALUATION OF WASTEWATER

The most common laboratory analyses for defining the characteristics of a municipal wastewater are BOD and suspended solids. BOD and flow data are basic to the design of biological treatment units, and the concentration of suspended solids relative to BOD indicates to what degree organic matter is removable by primary settling. Occasionally, comprehensive testing to yield the type of information given in Table 9–2 should be performed. Treatability as well as wastewater characteristics can be defined from these data. In addition, it may be advantageous to record wastewater temperature, pH, COD, alkalinity, ammonia, phosphorus, color, grease, and presence of specific heavy metals. Selection of these tests depends on the industrial wastes contributed to the sewer system.

Comprehensive BOD testing, as described in Section 3–10, is extremely vital, since the BOD test relates more closely than any other to the operation of a biological treatment system. Besides the 5-day value and k-rate, the ratios of BOD to COD and BOD to volatile solids indicate biodegradability of waste organics. The presence of inhibiting or toxic substances, resulting from industrial wastewaters, is often indicated by increasing BOD values with increasing dilution, lag periods at the beginning of properly seeded tests, and erratic test results. To be significant, laboratory analyses must be correlated with plant operations that prevailed at the time of sampling. Wastewater flow, day of the week, weather and rainfall, and abnormal wastewater discharges caused, for example, by the breakdown of pretreatment facilities at a major industry are essential for future interpretation of recorded testing data. Also, it may be advantageous to record sludge production, equipment performance, chemical and power usage, operational problems, and information on industrial wastewater flows.

Occasionally, a biological process treating municipal wastewater is afflicted by unknown substances that reduce efficiency by inhibiting microbial activity, for example, bulking of activated sludge in an aeration system. If the plant operates satisfactorily under the same loadings at other times, the problem is most likely related to an industrial wastewater being discharged to the sewer system. Often laboratory testing of the industrial wastewater, or comprehensive tests on the municipal wastewater, can pinpoint the difficulty. However, if results do not prove adequate, a treatability study using a small laboratory biological treatment unit is warranted.

Wastewater treatability studies are particularly appropriate for evaluating proposed industrial wastewater discharges from new industries that plan to use the municipal plant for treatment. Although most wastewaters are more amenable to biological treatment after dilution with domestic wastewater, they should be subjected to laboratory analyses prior to permitting their discharge to a municipal sewer. This is especially true if the projected volume is substantial in comparison to the sanitary flow and if the type involved has a history of creating treatment problems. Management of the proposed industry should be required to furnish precise information as to the quantity of wastewater flow, including anticipated flow patterns both daily and weekly, wastewater characteristics with anticipated hourly and daily variations from the norm, and proposed pretreatment facilities.

A laboratory apparatus for studying aerobic treatability is pictured in Figure 9–4. Wastewater is pumped from a refrigerated supply bottle into the aeration chamber; air is supplied through a porous stone diffuser at the bottom. Aerated mixed liquor flows through a connecting tube to the clarifier for gravity separation. Clear supernatant is the effluent, while the settled solids are returned to the aeration cylinder by air lifting. The aeration period and BOD loading should be similar to those used in the full-scale treatment system being simulated. Treatability of the applied wastewater is measured in terms of the efficiency of BOD or COD removal, settleability of the mixed liquor, and microscopic examination of the activated sludge biota. The unit should be operated on a feed of both pure industrial wastewater and a mixture with domestic wastewater. When treated alone, the industrial wastewater may require neutralization and/or addition of inorganic nitrogen and phosphate for nutrient balance. Joint treatment should be conducted at several different ratios of industrial to domestic wastewater to include the dilution anticipated in the actual municipal system.

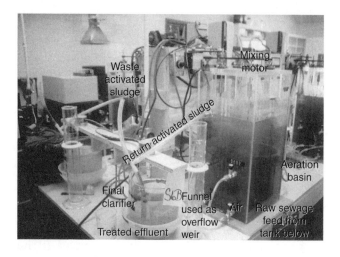
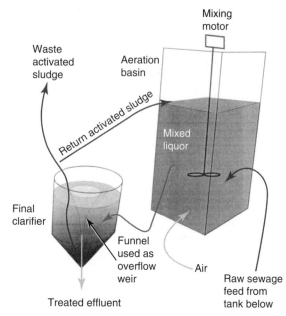

Figure 9–4

Laboratory apparatus for aerobic treatability studies on industrial and municipal wastewaters.

REFERENCE

1. U.S. EPA, *Pretreatment Standards & Limits*, http://cfpub.epa.gov/npdes/pretreatment/pstandards.cfm

PROBLEMS

9–1 Using the values in Table 9–1, estimate the daily wastewater flow and pounds of BOD produced by a recreational area consisting of a luxury resort with 30 employees and 200 guests per day having a restaurant serving lunch and dinner. The plant also serves a small hotel with 20 rooms, 5 employees, and a maximum capacity of 40 persons. Also on the property are 15 single-family homes and 20 apartments for the employees. Calculate the maximum gallons per person per day and pounds of BOD. Calculate the concentration of BOD in the wastewater.

9–2 Using the information listed in Table 9–2, calculate the removal efficiency for suspended solids and BOD after primary settling and after biological treatment.

9–3 For a flow of 12 mgd, use the values listed in Table 9–2 to calculate the pounds of suspended solids, BOD, total nitrogen, and total phosphorus in the raw wastewater.

9–4 What are the general types of pretreatment enforcement as identified by the EPA for publicly owned facilities?

9–5 Using the values in Table 9–3, calculate the pounds of BOD, suspended solids, nitrogen, and phosphorus from 50,000 gpd of milk-processing waste. If the sewer ordinance limits BOD to 500 mg/l, suspended solids to 400 mg/l, nitrogen and phosphorus to 35 mg/l, would the discharge be in compliance?

9–6 Using the values in Table 9–2 and in Table 9–4, what is the population equivalent for meat processing, that is, how many people does it take to generate waste equivalent to the meat-processing waste in Table 9–4?

9–7 A new community of 500 homes is being planned on $\frac{1}{4}$-acre lots with an average of five people per home. Using average infiltration values, what is the estimate of infiltration flow? Using 100 gpcd, what is the percentage of infiltration to wastewater flow and how does that compare with typical percentages?

9–8 An area of 3.4 acres has 1200 ft of 8-in., 800 ft of 10-in., and 430 ft of 16-in. sewer pipe. Calculate the maximum infiltration based on acres and piping. Using 1/3-acre sites and four people per residence at 120 gpcd, calculate the average flow. Calculate the average dry weather and wet weather flows.

9–9 A population of 125,000 has an average daily flow of 80 gpcd. What is the flow peaking factor and peak flow?

9–10 A pumping station is being sized for peak-hour wet-weather flow. For a community of 35,000 people, using 100 gpcd on 78,000 acres, calculate the average dry-weather flow, the peak dry-weather flow, maximum infiltration, and maximum inflow. If the pumping station has five pumps with one out of service, what is the size of each pump in gallons per minute? (*Note*: Remember to calculate the peak wastewater flow from the average while including storm flow.)

9–11 An industrial park has 75 percent industrial flow containing 850 mg/l BOD, 30 mg/l nitrogen, and no phosphorus. The toilet facilities contribute 25 percent of the flow which contains normal domestic-strength wastewater. Calculate the BOD/N/P ratio of the wastewater. Does the wastewater contain adequate nutrients?

9–12 For the synthetic textile waste in Table 9–3 what is the BOD/N/P ratio? Can this waste alone be biologically treated? If 1/3 raw wastewater is added what is the ratio and can it be treated?

9–13 Values of potato chip, milk-processing, and soft-drink wastewater are listed in Table 9–4 for flow, BOD, and suspended solids. The sewered population is 12,500. What is the resulting BOD and suspended solids concentration at the

treatment plant and what is the population equivalent of the industrial wastewater?

9–14 Hourly readings and samples, listed here, were taken of wastewater entering a treatment plant. What is the low, peak, and average flow? Tabulate the portions of the 24-hour composite sample volume of about 3000 ml for BOD testing.

Time	Q (mgd)	Time	Q (mgd)	Time	Q (mgd)
800	17.5	1600	16	2400	10.2
900	21.8	1700	16	100	7.3
1000	24.7	1800	16.7	200	5.8
1100	26.9	1900	17.5	300	5.5
1200	25.5	2000	16.7	400	5.5
1300	23.3	2100	16	500	5.8
1400	20.4	2200	14.6	600	7.3
1500	17.5	2300	13.1	700	11.6

9–15 Set up the format for tabulating the following flow data for calculating the proportion of grab samples to be combined into a composite, as in Table 9–5. The desired final quantity is 2500 ml.

Time	Q(mgd)
600	0.97
900	2.91
1200	3.4
1500	2.33
1800	2.23
2100	2.14
2400	1.55
300	0.74

9–16 Describe the requirements for comprehensive BOD testing industrial wastewaters.

Wastewater Collection Systems

Sewers are underground conduits for conveying wastewaters by gravity flow from urban areas to points of disposal. The earliest drainage systems, constructed in the 16th- and 17th-century cities, were to carry storm runoff from built-up areas, protecting them against inundation. Privies and cesspools were used for disposal of human excreta, and household wastes were often thrown on the streets. Although this created deplorable sanitary conditions, such cities as London and Philadelphia prohibited discharge of household wastes to storm drains as late as 1850. The development of steam-driven pumps and cast-iron pipes for pressure water distribution, however, led to indoor plumbing and flush toilets. Soon cesspools were banned and waterborne wastes were piped to the storm drains, converting them to combined sewers. Although this water carriage system improved conditions within the city, untreated wastes were now being directed to surface watercourses.

Combined sewers were constructed in many cities of the United States prior to 1900 without recognizing the need for segregation and treatment of domestic and industrial wastewaters. Although these systems still exist in older municipalities, separate sewers have become common during the 20th century. Storm sewers carry only surface runoff and other uncontamined waters to natural channels directly, while sanitary sewers convey domestic and industrial wastewaters to treatment works for processing prior to their disposal. Where combined drain pipes exist, intercepting sewers have been constructed to collect the dry-weather flow from a number of transverse outfalls and to transport it to a treatment plant; wet-weather flow in excess of treatment plant capacity is still routed directly to disposal in many cases.

10–1 STORM SEWER SYSTEM

Surface waters enter a storm drainage system through inlets located in street gutters or depressed areas that collect natural drainage. Cooling water from industries and groundwater seepage that enters footing drains are pumped to the storm sewer, since the pipes are usually set too shallow

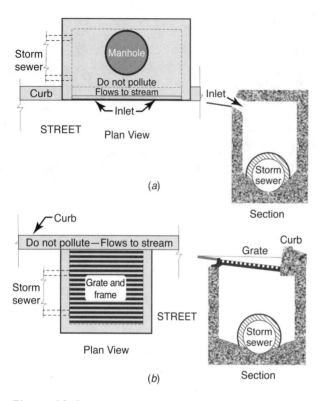

Figure 10–1

Storm-water inlets for street drainage. (*a*) Curb inlet. Water flows under curb into storm-drain drop structure. (*b*) Gutter inlet allows water to enter through a grated cover on the side of the street.

for gravity flow; furthermore, direct connections would be subject to backflow when the pipe surcharges. Figure 10–1 illustrates two common types of storm-water inlets for streets. The curb inlet has a vertical opening to catch gutter flow. Although the gutter may be depressed slightly in front of the inlet, this type of inlet offers no obstruction to traffic. The gutter inlet is an opening covered by a grate through which the drainage falls. The disadvantage is that debris collecting on the grate may result in plugging of the gutter inlet. Combination inlets composed of both curb and gutter openings are also common. Street grade, curb design, and gutter depression define the best type of inlet to select; nevertheless, minimizing interference with traffic and eliminating plugging often take precedence over hydraulic efficiency.

Catch basins under street inlets are connected by short pipelines to the main storm sewer

located in the street right-of-way. Manholes are placed at curb inlets, intersections of sewer lines, and regular intervals to facilitate inspection and cleaning. Pipeline gradients follow the general slope of the ground surface such that water entering can flow downhill to a convenient point for discharge. Sewer pipes are set as shallow as possible to minimize excavation while providing 2 to 4 ft of cover above the pipe to reduce the effect of wheel loadings. Sewer outlets that terminate in natural channels subject to tides or high water levels are equipped with flap gates to prevent backflooding into the sewer system. Backwater gates are also used on combined sewer outfalls and effluent lines from treatment plants where needed.

The rational method, described in Section 4–11, is used to calculate the quantity of runoff for sizing storm sewers. Climatic conditions are incorporated by using local rainfall intensity-duration formulas or curves. In dry regions, sewers may be placed only in high-value districts; streets and roadside ditches serve as surface drains in sparsely populated areas. On the other hand, in regions of the country having intense thunderstorm weather, lined open channels are often found to be more economical than large buried conduits; sewers leading from small drainage areas terminate in grassed or concrete-lined ditches that discharge to surface watercourses.

The flowing-full velocities used in the design of storm sewers are a minimum of 3.0 ft/sec (0.9 m/s) and a maximum of about 10 ft/sec (3.0 m/s). The lower limit is set so that the lines are self-cleaning to avoid deposition of solids, and the upper limit is fixed to prevent erosion of the pipe by grit transported in the water.

A major difference in the design approach to sanitary and storm sewers is that the latter are assumed to surcharge and overflow periodically. For example, a storm drain sized on the basis of a 10-year rainfall frequency presumes that one storm every 10 years will exceed the capacity of the sewer. Sanitary sewers are designed and constructed to prevent surcharging. Where backup of sanitary sewers does occur, it is more frequently attributable to excess infiltration of groundwater through open pipe joints and unauthorized drain connections. A second easily recognizable difference between sanitary and storm sewers is the pipe sizes that are needed to serve a given area. As illustrated in Example 10–1, storm drains are many times larger than the pipes collecting domestic wastewater. Consequently, only a small amount of infiltrating rain water results in overloading domestic sewers.

Circular concrete pipe is commonly used for storm sewers. Nonreinforced concrete pipe is available in sizes up to 24 in. in diameter in lengths of 3 or 4 ft. Reinforced pipe has steel embedded in the concrete to give added structural strength. Circular pipe is manufactured in diameters from 12 to 144 in. and in laying lengths that range from 4 to 12 ft. Elliptically shaped and arch-type concrete pipe are also manufactured for special applications. Various types of joints are available for connecting pipe sections. The one selected depends on construction conditions, pipe size, manufacturer, and whether the pipe is reinforced or nonreinforced. The single-rubber-ring gasket joint is commonly used in concrete sewer lines because of economy, ease of construction, and satisfactory performance. In manufacture, the pipe ends are carefully cast with a space left for the rubber ring. With the gasket in place and lubricated, the spigot end is pushed into the bell to seat the joint properly.

■ EXAMPLE 10–1

(a) What is the maximum population that can be served by an 8-in. sanitary sewer laid at minimum grade using a design flow of 400 gpcd and a flowing-full velocity of 2.0 ft/sec? (b) Compute the diameter of a storm drain to serve the same population based on population density = 30 persons per acre, coefficient of runoff = 0.40, 10-yr rainfall frequency curve (Figure 4–30), a duration (time of concentration) = 20 min, and a velocity of flow = 5.0 ft/sec.

Solution

(a) Based on Table 10–1, Q flowing full at a velocity of 2.0 ft/sec for an 8-in.-diameter pipe is 310 gpm (0.7 cu ft/sec). The maximum population that can be served

$$= \frac{310 \text{ gpm} \times 1440 \text{ min/day}}{400 \text{ gal/day}} = 1100 \text{ persons}$$

TABLE 10–1

Minimum Slopes for Various Sized Sewers at a Flowing-Full Velocity of 2.0 ft/sec and Corresponding Discharges[a]

SEWER DIAMETER (in.)	MINIMUM SLOPE (ft/100 ft)	FLOWING-FULL DISCHARGE (cu ft/sec)	(gpm)
8	0.33	0.7	310
10	0.25	1.1	490
12	0.19	1.6	700
15	0.14	2.4	1080
18	0.11	3.5	1570
21	0.092	4.8	2160
24	0.077	6.3	2820
27	0.066	8.0	3570
30	0.057	9.8	4410
36	0.045	14.1	6330

[a]Based on Manning's formula with $n = 0.013$.

(b) Drainage area $= \dfrac{1100 \text{ persons}}{30 \text{ persons per acre}}$

$= 36$ acres

From Figure 4–30, duration = 20 min, $I = 4.2$ in./hr. Using Eq. 4–24, $Q = 0.40 \times 4.2 \times 36 = 60$ cu ft/sec.

Based on $Q = VA$, for $Q = 60$ cu ft/sec and $V = 5.0$ ft/sec, the pipe diameter is 42 in. Therefore, a 48-in.-diameter storm sewer is needed to drain the housing area of 1100 persons that can be served by an 8-in.-diameter sanitary sewer.

■ ■ ■

Storm-water management programs are designed to reduce pollutants carried with the drainage from entering local waterways. Pollutants of interest include sediment, trash, metals, organics, pesticides, nutrients, oil, and grease. Spills, if not removed by cleaning, may run off with storm water. Specific activities and areas of interest include maintenance grounds, parking lots, fueling areas, outdoor storage, and loading facilities. Residential development may contribute fertilizer, pesticides, and smaller quantities of other pollutants. Source control is critical to reducing storm-water pollution. Outdoor facilities must be designed to contain spills, eliminate areas of soil erosion, and control the flow and discharge of storm flows. In addition, the application rates of irrigation water, fertilizer, and pesticides must be controlled to reduce the potential for runoff.

Source control measures include sweeping and cleaning; controlling litter and erosion; spill, leak, and overflow control; limits on the use of chemicals; and monitoring for illegal dumping. In general, source control falls under the category of good housekeeping and application rates.

Storm-water retention ponds are used to attenuate the storm flow rate and provide a quiescent reservoir for the retention of pollutants. Heavier solids settle and become incorporated into the bottom of the pond. Trash accumulates and must be collected and removed. Organics, pesticides, nutrients, oils, and grease receive a degree of treatment in the biological environment created in the pond and by filtration and decomposition in the soil floor. The design features of a storm-water retention point are important to pollution reduction and pond performance. Increased settling area allows heavier solids to settle before reaching the permanent water pool, where lighter solids settle. Shallow edges along the permanent pool provide habitat for aquatic plants and wetlands. These plants biologically uptake nutrients and provide habitat for other organisms. Controlled permeability allows groundwater recharge while maintaining the permanent pool.

10–2 SANITARY SEWER SYSTEM

Sanitary sewers transport domestic and industrial wastewaters by gravity flow to treatment facilities. A lateral sewer collects discharges from houses and carries them to another branch sewer and has no tributary sewer lines. Branch or submain lines receive wastewater from laterals and convey it to large mains. A main sewer, also called a trunk or outfall sewer, carries the discharge from large areas to the treatment plant. A force main is a sewer through which wastewater is pumped under pressure rather than by gravity flow.

Design flows for sewer systems are based on population served using the following per capita quantities: laterals and submains 400 gpcd (1500 l/person · d), main and trunk 250 gpcd (950 l/person · d), and interceptors 350 percent of the estimated average dry-weather flow. These figures include normal infiltration and are based on flowing-full capacity. Excluded are industrial wastewaters and excessive infiltration. Sewer slopes should be sufficient to maintain self-cleansing velocities; this is normally interpreted to be 2.0 ft/sec (0.60 m/s) when flowing full. Table 10–1 lists sewer size, minimum slope for 2 ft/sec, and the corresponding quantity of flow. Slopes slightly less than those listed may be permitted in lines where the design average flow provides a depth of flow greater than one-third the diameter of the pipe. Where velocities are greater than 10 ft/sec (3.0 m/s), special provision must be made to protect the pipe and manholes against displacement by erosion and shock hydraulic loadings.

Sanitary sewers are placed at sufficient depth to prevent freezing and to receive wastewater from basements. As a general rule, laterals placed in the street right-of-way are set at a depth of not less than 11 ft (3.3 m) below the top of the house foundation. To provide economical access to the sewer after street construction, service connections are generally extended from laterals to outside the curb line at the time of sewer placement. An alternative is to place the sanitary sewer behind the curb on one side of the street, making it readily accessible for service connections on that side. Where soil conditions permit, pipe connections from the opposite side of the street are accomplished by excavating working pits on each side and by jacking the house sewer into position for connection to the lateral on the opposite side. In jacking, pipe sections are pushed beneath the roadway by hydraulic jacks. For hard soils, boring machines are used to cut an opening through which the pipeline is pushed. The cutter head operates in front of the first pipe section, and an auger pulls the excavated material out through the pipe to the jacking pit.

Ease of maintenance dictates many of the design criteria for wastewater collection systems. The minimum recommended size for laterals is 8-in. (200-mm) diameter. Manholes located at regular intervals allow access to the pipe for inspection and cleaning. Pipes laid on too flat a grade require periodic flushing and cleaning to remove deposited solids and to prevent pipe plugging.

Sewers less than 24 in. should be laid on a straight line between manholes, although in recent years curves have been permitted to follow the curvature of the streets. The degree of curvature is determined by the length of the pipe sections and the flexibility of the joints. Curved sewer lines can be cleaned with a high-pressure jet without damaging the pipe interior. Pumping stations are used in a collection system only when continuing flow by gravity is impractical—because of their high cost and potential maintenance problems.

Sewer Plan and Profile

The drawing in plan and profile for a sanitary sewer is illustrated in Figure 10–2. The plan view is drawn over the profile and often an aerial photo or survey topography is included as a background to the sewer plan. The plan shows land features such as streams and roads along with the routes of underground piping. The profile is drawn to an exaggerated vertical scale, 10 times greater than the horizontal scale, so that elevations and vertical separations can be clearly shown. It shows the ground surface, sewer line with slope and diameter, manholes, connecting

sewers, all other underground utilities, elevation at the pipe invert at critical points, and special features, such as inverted siphons.

New sewer lines and water mains must be installed so as to ensure adequate separation to prevent any possible passage of polluted water from the sewer into the potable water supply. The Department of Health in most states requires a minimum horizontal separation of 10 feet (3 m) and a minimum vertical separation of 18 inches (46 cm). When separation cannot be maintained or the water line crosses under the sewer, the Department of Health must be contacted to determine alternative construction methods. Requirements may include specific locations of sewer joints, encasement of the sewer, or limits on the type of sewer and joint material to prevent leaks in the vicinity of the water line.

Inverted Siphons

A siphon is a depressed sewer that drops below the hydraulic gradient to avoid an obstruction, such as a stream, railway cut, or depressed highway. The design incorporates provisions for maintenance-free operation and minimum hydraulic head loss.

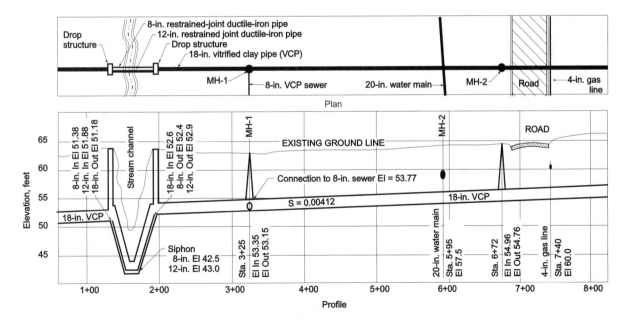

Figure 10–2

Sewer plan and profile showing land features, buried piping, and sewer design details. A stream crossing is conducted with a siphon consisting of two differently sized pipes to limit solids deposition in the siphon. The profile shows the sewer to be installed at a constant slope of 0.00412 with in and out invert elevations at each structure.

Since a depressed sewer acts as a trap, the velocity of flow in the pipes should be greater than 3 ft/sec (0.9 m/s) to prevent deposition of solids. This is accomplished by constructing an inlet splitter box that directs flow to two or more siphon pipes placed in parallel. For example, in Figure 10–2, flow in the 18-in.-diameter sewer is carried by two depressed pipes. During low flow in the main sewer, all of the wastewater is directed to the 8-in. pipe. When the depth of flow in the main sewer exceeds 6 in., the wastewater stream is split in the inlet box and is directed to both siphon pipes; with the main sewer flowing half full, the velocity in both is greater than 3 ft/sec. In addition to control of flow, the inlet and outlet structures provide access for cleaning the sewer lines.

Where a waterway, railroad, road, or other structures do not allow a trench to be open-cut and the pipe installed from the ground surface, pipe may be installed using trenchless technologies. Sewer pipe like that shown in Figure 10–3 may be installed using horizontal auger boring or microtunneling methods. Both techniques begin with the construction of a drilling and receiving pit. Horizontal auger boring uses an auger machine to turn an auger within a carrier pipe (see Figure 10–4a). As the casing is pushed into the soil, a cutting head and flighted auger remove the soil and hollow out the hole. The casing supports the soil as the auger pro-

gresses. The auger boring machine places a jacking pressure on the carrier pipe while simultaneously rotating the auger. After the pipe connects to the receiving pit, the auger is removed, the pipe is installed on skids, and the ends are sealed with concrete. Auger boring machines are capable of installing pipe from 8 to 60 in. (200 to 1500 mm) in diameter. Microtunneling machines have a cutting-style tunneling face with computerized directional control to lead the pipe through the hole (see Figure 10–4b). Smaller machines are used to install pipe between 10 to 28 in. (250 to 700 mm) in diameter. As the cutting face removes the soil, it is crushed and mixed with a slurry fluid (typically bentonite and water) to allow pump and pipe transport of the removed material and decrease friction on the pipe inserted behind the machine. The flushing fluid is conveyed into the center of the machine through a hollow driveshaft. The pipe is pushed in behind to fill the void created by the tunneling machine. Many machines use a laser measuring system to determine direction of travel and to report on location.

Manholes

Most manholes are circular in shape with an inside diameter of 4 ft, which is considered sufficient to perform sewer inspection and cleaning (Figure 10–5). For small-diameter pipes, the manhole

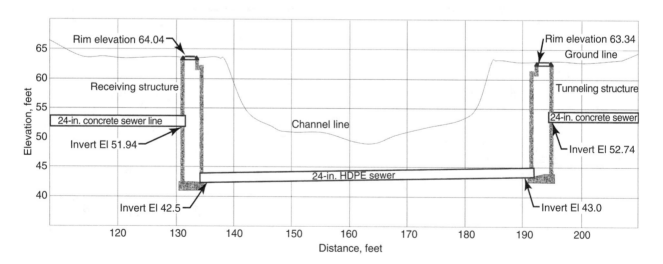

Figure 10–3

Profile of a sanitary sewer river crossing. The HDPE pipe was installed behind a microtunnel machine. Construction began with installation of tunneling and receiving structures. The concrete sewer was installed from the surface in an open-cut trench.

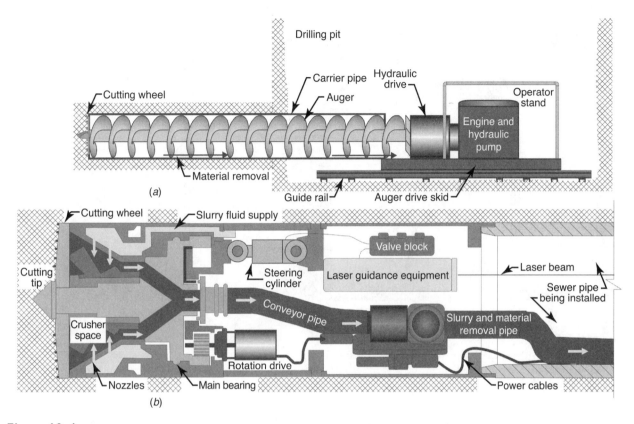

Figure 10–4

Illustrations of trenchless sewer pipe installation. (*a*) Horizontal boring machine in the drilling pit installing a carrier pipe as a casing for the sewer pipe to be inserted after removal of the auger. Material is cut by the leading wheel and augered into the drilling pit for removal. (*b*) Microtunnel machine with sewer pipe trailing the hole. Cuttings are crushed, mixed with a slurry material, and pumped out of the tunnel. The sewer pipe is jacked behind the machine into position. The microtunnel machine is removed in a receiving pit.

(Illustration of microtunnel model AVN250 courtesy of Herrenknecht AG.)

is usually constructed directly over the centerline of the sewer. For very large sewers, access may be provided on one side with a landing platform for the convenience of introducing cleaning equipment. Manhole frames and covers are usually cast iron with a minimum clear opening of 21 in. (54 cm). Solid covers are used on sanitary sewers; open-type covers are common on storm sewers. Steps or ladder rungs are placed for access. Walls may be constructed of precast concrete rings, concrete block, brick, or poured concrete.

Wastewater flow is conveyed through the manhole in a smooth U-shaped channel formed in the concrete base. Where more than one sewer enters a manhole, the flowing-through channels should be curved to merge the flow streams. If a sewer changes direction in a manhole without change of size, a drop of 0.05 to 0.10 ft (1.5 to 3.0 cm) is provided in the manhole channel to account for head loss. When a smaller sewer joins a larger one, the bottom of the larger pipe should be lowered sufficiently to maintain uniform flow transition. An approximate method for achieving this is to place the 0.8 depth point of both sewers at the same elevation; an alternative technique is to have the pipe crowns at the same height.

A drop manhole is used when it is necessary to lower the elevation of a sewer in a manhole more than 24 in. (Figure 10–5*b*). This construction is needed to protect a man entering the structure and to eliminate the nuisance created by solids splashed onto the walls. The tee pipe section extending past the drop into the manhole allows access to the sewer line for cleaning tools.

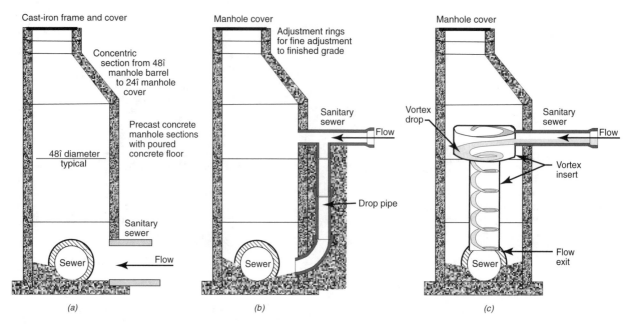

Figure 10–5

Sewer manholes: (*a*) Typical sewer manhole. The concentric section allows easier access for maintenance. (*b*) Drop manhole structures are typically used when the sewer vertical separation is 24 in. or greater. (*c*) Vortex inserts reduce odors generated by falling sewage by reducing velocity and aerating sewage in the drop structure.

Manhole drops over 6 ft (2 m) create excessive corrosion and odor as a result at the drop. Vortex manhole inserts, like that shown in Figure 10–5*c*, reduces the velocity during the drop and reduce corrosion and odor generation.

Manholes should be placed at all changes in sewer grade, pipe size, or alignment; at all intersections; at the end of each line; and at distances not greater than 400 ft (120 m) for sewers 15 in. or less, and 500 ft (150 m) for sewers 18 to 30 in. In some municipalities, a spacing of 300 ft (90 m) is considered the maximum for smaller-sized sewers. Distances of greater than 500 ft may be used where the pipe is large enough to permit walking in the sewer.

Manholes constructed in high groundwater should be waterproofed to prevent excessive infiltration. Waterproofing is applied to the exterior surfaces prior to backfilling the excavation.

Service Connections

House sewers are laid on a straight line and grade using 4-in. or 6-in. (100- or 150-mm) pipe. The preferred minimum slope is 2 percent, or $\frac{1}{4}$ in./ft,

although slopes as shallow as $\frac{1}{8}$ in./ft are occasionally used. In some housing developments, the setback of dwellings from the street dictates the slope of the connection. Pipe trenches for laying the service line should be straight and should be excavated to the required slope. Where the soil is suitable for pipe support, the natural floor of the trench can be shaped to support the barrel of the pipe. Crushed stone or coarse sand may be used for bedding when fill is required to provide uniform support. High-quality pipe, watertight joints, and good workmanship are required to minimize infiltration and root penetration.

The service connection to a sanitary sewer is made through a tee branch turned upward 45° or more from the horizontal, as shown in Figure 10–6, so that backflooding does not occur when the collecting sewer is flowing full. For a deep sewer, the tee connection and riser pipe are often vertical and may be encased in concrete to prevent damage during backfilling.

Building vents attached to sewer drains provide a connection between the air in sewer pipes and the atmosphere. Plumbing traps on toilet and sink drains prevent backup of sewer gases into the building

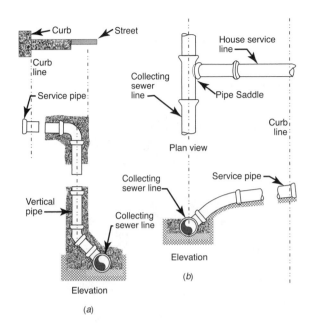

Figure 10–6

Typical sanitary sewer service connection. (*a*) Connection to a deep sewer. (*b*) Connection to a shallow sewer.

interior, while the fall of wastewater down the stack draws air into the pipe. Oxygen in the sewer atmosphere reduces production of hydrogen sulfide, and ventilation carries off volatile gases that may originate from illegal disposal of flammable liquids. In some instances where natural aeration is not sufficient because of long sewer lines with few service connections, forced ventilation is installed to draw air out of the sewer, exhausting it to a high stack or some deodorizing process.

10–3 MEASURING AND SAMPLING OF FLOW IN SEWERS

Monitoring sewer use requires measuring of flow and composite sampling at critical points in the collection system. Evaluation of peak flow periods defines the unused capacity of pipes and indicates infiltration-inflow problems. Sampling stations on industrial waste discharges are necessary to regulate sewer use and to determine flow and strength data to establish user fees.

The estimated quantity of flow in a sewer can be computed from the measured depth of flow using the Manning formula if the slope of the pipe

is known (Section 4–8). Another approximate method is to multiply the measured wetted cross-sectional area of flow times the velocity obtained by a current meter or the use of dyes or floats.

Installation of a flume is necessary for accurate flow measurement and automatic composite sampling. The Palmer-Bowlus flume with a trapezoidal throat cross section, as illustrated in Figure 10–7, can be inserted into an existing sewer pipe with adequate slope. (Palmer-Bowlus flumes have various cross-sectional shapes; however, the standard among commercial manufacturers is the trapezoidal shape.[1]) The trapezoidal section with a flat bottom (Figures 10–7*a* and *b*) is preferred for circular pipes because it has the least constriction through the critical flow area and minimum head loss. The flume is essentially a restriction in the channel to produce a critical flow of higher velocity in the throat.

The upstream height of water level, *H*, above the throat during free flow is related to the quantity of flow. Nevertheless, dimensions for Palmer-Bowlus flumes are not standardized to yield a mathematical flow equation as Parshall flumes are. Instead, a rating curve, or tabulated data, is provided by the manufacturer to relate height of upstream head with rate of flow. The ideal location for measuring water level is at an upstream distance from the flume entrance of one-half *D* (pipe diameter or channel width). This exact location is not essential provided it is upstream from the upper transition section. The flow in the upstream channel should be smooth, and downstream a shooting flow should be evident to indicate free discharge. The flume itself should be level, but a minimum pipe or channel slope of the downstream section is necessary to maintain critical flow through the throat to prevent submergence of the flume. The required critical flow occurs if the downstream depth of flow is less than 85 percent of the upstream depth (Figure 10–7*a*).

Palmer-Bowlus flumes are constructed for temporary installation in the half section of a sewer in a manhole. They are prefabricated of fiberglass, reinforced plastic, and stainless steel in sizes from 4 to 30 in. in diameter to match the inside diameters of sewer pipes. For a permanent installation, the flume is built to be embedded in poured concrete. Sharp-crested weirs with triangular, rectangular, and trapezoidal openings can be used for flow measurement and composite sampling. Installation in a manhole, however, is difficult and generally

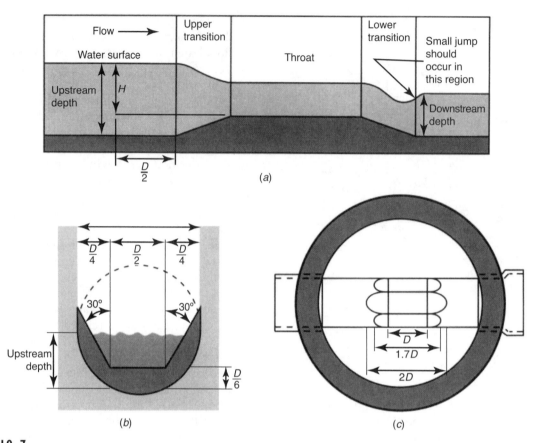

Figure 10–7

Palmer-Bowlus flume with a trapezoidal throat cross section.[1] (a) Free-flowing flume with the depth H above the throat of the flume as the index of discharge. (b) Dimensional configuration of the trapezoidal throat cross section of the flume with D equal to pipe diameter. (c) Plan view of a manhole with a Palmer-Bowlus flume installed in the sewer pipe.

(Courtesy of *Isco Open Channel Flow Measurement Handbook*, Isco, Inc., Environmental Division.)

not suitable, since settleable solids deposit in the approach channel behind the weir and stringy and floating solids collect on the weir plate.

The open-channel flow meters in Figure 10–8 use an ultrasonic sensor, submerged probe, or bubbler to measure water level or depth. The meters accommodating the three different probes all have a liquid crystal display (LCD) and keypad for programming. Mathematical conversions for level-to-flow rate are stored in the computer memory for the common kinds of flumes and weirs and the Manning open-channel flow equation. For special flow-measuring devices, sets of data points for levels and flow rates or a mathematical equation can be entered into the memory of the flow meter for conversion of the measured water levels to flow rates. The backlighted LCD continuously displays water level, flow rate, and cumulative flow in addition to its use as a pro-

gramming screen. A built-in dot matrix printer plots water level or flow rate data and total flow data. When a sampler is connected to the meter, the time and bottle number of each sample is recorded. Upon command, or at selected time intervals, summary reports can be printed. The flow meter program can also be printed. Data stored in the internal memory of the meter can also be printed as graphs, reports, and summaries using a computer and software available from the meter manufacturer. The meters can be operated on AC power with built-in backup battery power or by a portable rechargeable battery pack.

The ultrasonic sensor (Figure 10–8a) transmits a sound pulse that is reflected from the surface of flowing water. The elapsed time between sending a pulse and receiving an echo determines the water level. A built-in temperature probe compensates

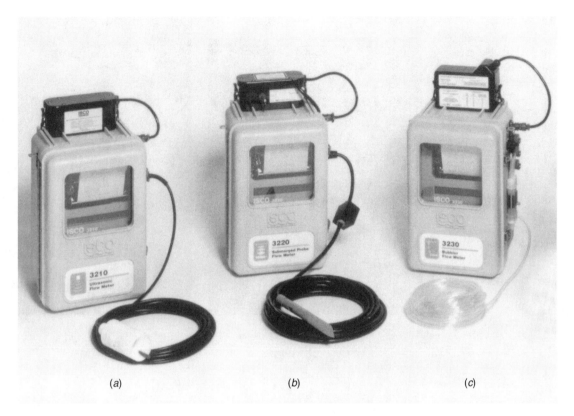

(a)   (b)   (c)

Figure 10–8

Open-channel flow meters have level-to-flow rate conversions for most weirs and flumes; visual display to show water level, flow rate, and cumulative flow; internal memory to store data; and a device to measure water level or depth of flow using (a) an ultrasonic sensor, (b) a submerged probe, or (c) a bubbler. (Courtesy of Isco, Inc., Environmental Division.)

for changes in air temperature. Since the sensor does not contact the water, it is positioned above the measuring location using a floor mount or suspension bar. The ultrasonic sensor is suitable for wastewater-level measurements upstream from flumes or weirs where the flow is smooth, since it is not influenced by suspended solids or grease. The submerged probe (Figure 10–8b) is a differential pressure transducer to measure the depth of the water. To minimize obstruction of the flow stream, the probe has a streamlined, low-profile shape. A venting system compensates for changes in atmospheric pressure. By using a special mounting ring or strap, the probe can be installed in a pipe, manhole invert, or open channel. Most prefabricated flumes are available with an integral recess for mounting a probe. The submerged probe is suitable for wastewater depth measurements and has the advantage of no interference from floating debris, grease, or

foam. The bubbler tubing from the flow meter (Figure 10–8c) is attached to a rigid tube with the outlet end submerged in the flow channel. The meter has an internal air compressor to force a metered flow of air through the submerged bubble tube. By measuring the pressure needed to force air bubbles out of the tube, the depth of the water is accurately measured. The bubbler can be used in wastewater applications, since its operation is unaffected by turbulence, floating or suspended solids, or temperature fluctuations.

The variable-gate open-channel flow meter illustrated in Figure 10–9 is a portable meter for measuring flow in 4-, 6-, or 8-in.-diameter pipes. The major advantage is no flume or weir is needed, and no calibration is required. The metering insert is installed in the upstream pipe of a manhole. A stainless steel expansion ring inside the pipe holds the insert in position, and a rubber

Figure 10–9
Variable-gate open-channel flow meter. (*a*) Variable-gate metering insert placed in a wastewater pipe with the flow meter on the left and automatic composite sampler on the right. (*b*) The insert with flow passing under the variable gate. (*c*) The operating principle of the insert is that the gate position and upstream water depth determine the rate of flow.
(Courtesy of Isco, Inc., Environmental Division.)

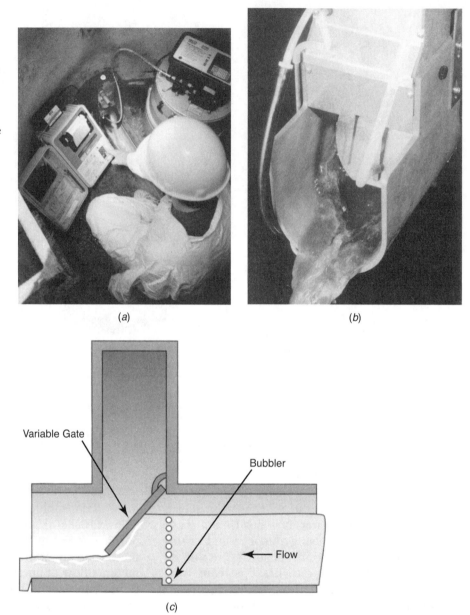

(*a*)

(*b*)

Variable Gate

Bubbler

Flow

(*c*)

bladder seals the pipe, routing flow through the insert. The flow meter has a programming display and keypad, built-in printer, and internal memory to store data that can be printed as graphs and reports by computer. As shown in Figure 10–9*c*, the insert has a pivoting gate under which the wastewater flows. The gate creates an upstream water level measured by a bubbler and adjusts automatically in response to changing flow rates. Together, the gate position and upstream level determine the rate of wastewater flow. To main-

tain accuracy, periodically the bubble tubing is purged and the gate is opened to allow any settled solids to flush through.

The portable wastewater sampler in Figure 10–10 has a controller with an LCD and keypad for programming the desired sampling setup and an attached peristaltic pump, which are protected by a cover (not shown). Programmed sampling can be at uniform or nonuniform time intervals for single-bottle sequential collection, multiple-bottle or single-bottle compositing, split sampling, or

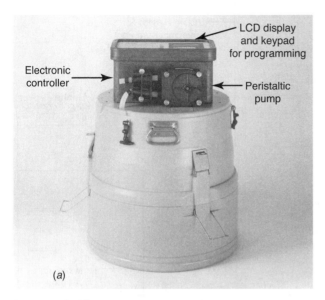

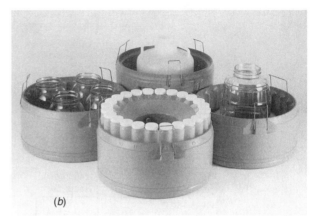

Figure 10–10
Portable wastewater sampler. (*a*) Sampler with cover removed to display the programmable controller and peristaltic pump. (*b*) Collection bottle options for sequential or composite sampling.
(Courtesy of Isco, Inc., Environmental Division.)

flow-paced collection. The peristaltic pump draws a programmed volume of sample through a flexible suction tube with a strainer on the submerged end. Both before and after sampling the pump reverses to air purge the suction line to minimize cross-contamination between samples. After withdrawal, a rotating distributor arm moves to the programmed position above a collection bottle. The base of the sampler (Figure 10–10*b*) has bottle options for sequential sampling in 24 bottles or 4 bottles and composite sampling in single large bottles. Sample collection in proportion to flow (composite sampling) is controlled by a wire attachment that receives electronic signals from a flow meter.

Sampling stations for industrial wastewaters range from a standard manhole to a separate chamber that is large enough to house flow-recording and automatic-sampling equipment. A manhole on the service sewer to a commercial business is generally adequate, provided neither flow measurement nor composite sampling is required. Frequently, for a small enterprise, the quantity of wastewater flow can be determined from water consumption recorded by the supply meter and grab sampling is adequate for monitoring wastewater quality.

For moderate-sized industries and commercial businesses, a sampling station similar to Figure 10–11 is appropriate. The sewer passes through a reinforced concrete chamber of sufficient size to conveniently set up a flow meter and sampler. The usual flow measuring device is a Palmer-Bowlus flume installed in the sewer pipe. A slightly elevated bench and stilling well are used to accommodate measurement of the water level upstream from the flume. A portable sampler is used to collect composite or sequential wastewater samples automatically at programmed time intervals. The intake hose is placed in the flow downstream from the flume, and the sampler is set on the bench above. For large industries, completely automated systems are essential for continuous monitoring of wastewater discharge. The flow measuring system consists of a Parshall or Palmer-Bowlus flume with flow recorder and totalizer, and the sampler is often programmed to collect composite samples continuously over defined time intervals.

10–4 SEWER PIPES AND JOINTING

Circular sewer pipe is manufactured with inside diameters from 4 to 144 in., at intervals of 2 in. for pipes from 4 to 12 in. in diameter, increases of 3 in. between 12 and 36 in. in diameter,

Figure 10–11

Sampling chamber to monitor industrial wastewater discharge to a municipal sewer where only intermittent analyses are required. The measuring flume is permanently installed, but the portable flow meter and automatic sampler are needed to perform composite sampling.

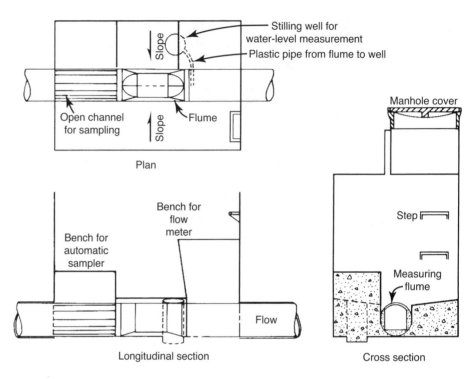

and 6-in.-diameter intervals from 36 to 144 in. Maximum sizes vary with materials; for example, the maximum clay pipe is 42 in. in diameter. The physical characteristics essential for sewer pipe are durability for long life; an abrasion-resistant interior to withstand the scouring action of wastewater carrying gritty materials; impervious walls to prevent leakage of water; and adequate strength to resist failure or deformation under backfill and traffic loads. Joints should be durable, easy to install, and watertight to prevent leakage or entrance of roots.

The most important chemical characteristics of a pipe material are resistance to dissolution in water and to corrosion. Pipe surfaces must be able to withstand both electrochemical and chemical reactions from the surrounding soil and wastewater conveyed in the pipe. Figure 10–12 illustrates the process of crown corrosion in sanitary sewers. Bacterial activity in anaerobic wastewater produces hydrogen sulfide gas, particularly in warm climates when sewers are laid on flat grades. Hydrogen sulfide absorbed in the water condensed on the crown is converted to sulfuric acid by aerobic bacterial action. If the pipe is not chemically resistant, the acid deteriorates it and eventually results in collapse of the crown. The most effective preventative

measure is to select a pipe material resistant to corrosion, such as vitrified clay or plastic. When reinforced concrete pipe is needed for larger sizes, an interior protective coating of coal tar, vinyl, or epoxy should be considered. Generation of hydrogen sulfide in a sewer can be reduced by placing the pipe on as steep a gradient as possible and by ventilating if necessary. Corrosion of the pipe bottom is caused by disposal of acidic industrial wastewaters. This problem is best solved by limiting the discharge of acid wastes to the municipal sewer system. In concrete pipe, corrosion-resistant liners, such as PVC sheets, may be placed in the crown of the sewer for protection.

Vitrified clay is the most common material used in sanitary sewer pipe. It is manufactured of clay, shale, or combinations of these materials that are pulverized and mixed with a small amount of water. The moistened clay is extruded through dies under high pressure to form the barrel and socket, dried, and then fired in a kiln for vitrification. Vitrified clay pipe (VCP) is manufactured in standard strength to 36 in. (900 mm) in diameter and in extra strength to 42 in. (1050 mm) in diameter. Laying lengths range from 2 to 7 ft depending on diameter. Curves, elbows, and branching fittings,

Figure 10–12

Environmental conditions leading to crown corrosion occur in sanitary sewers as a result of flat grades and warm wastewaters.

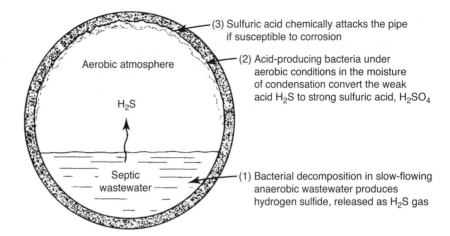

(3) Sulfuric acid chemically attacks the pipe if susceptible to corrosion

(2) Acid-producing bacteria under aerobic conditions in the moisture of condensation convert the weak acid H_2S to strong sulfuric acid, H_2SO_4

Aerobic atmosphere

H_2S

Septic wastewater

(1) Bacterial decomposition in slow-flowing anaerobic wastewater produces hydrogen sulfide, released as H_2S gas

including single and double tee and wye shapes, are available in most pipe sizes.

Bell-and-spigot vitrified clay pipes are connected by compression joints with seals of flexible plastic to prevent leakage of wastewater, infiltration of groundwater, and penetration of tree roots. The joint illustrated in Figure 10–13a has bonded polyester liners in the bell and on the spigot end with a separate compression ring. The adjoining ends compress the O-ring in the annular space between the liners to form a tight seal. The joint in Figure 10–13b has bonded polyurethane elastomer seals in the bell and on the spigot end. To connect the pipes, the seals are lubricated and the spigot is pushed into the bell of the already installed pipe using a pry bar or pipe puller. A wooden block is placed in the bell of the pipe being pushed into place to prevent damage. In addition to bell-and-spigot pipe, plain-end pipe is manufactured in 6- through 18-in. diameters with polyurethane elastomer seals on both ends. A polyvinyl chloride sleeve, or collar, is attached to one pipe end in the factory as a coupling. The seal on the plain end is lubricated and pushed into the sleeve, compressing the polyurethane seal for a tight fit. All of these compression joints allow limited deflection without leaking. The allowable deflection per foot of pipe length is $\frac{1}{2}$ in. for pipe diameters of 4 to 12 in. (42 mm/linear m), $\frac{3}{8}$ in. for 15 to 24 in., $\frac{1}{4}$ in. for 27 to 36 in., and $\frac{3}{16}$ in. for 39 to 42 in.

The plastic pipe used most frequently in sewer systems is made of polyvinyl chloride (PVC) or polyethylene (PE). PVC pipe is produced in two strength classifications in sizes from 4 to 12 in. A few manufacturers make pipe up to 30 in. in diameter. The standard length is 20 ft with other lengths available. PVC pipe sections have a deep-socket bell end to accommodate a chemical weld joint. Solvent is spread both on the inside of the bell and on the plain end before they are pushed together. PE pipe is joined by softening the aligned faces of the ends in a suitable apparatus and pressing them together under controlled pressure. PVC pipe is used for building connections and branch sewers; the popular application of PE pipe has been for long pipelines, often laid under adverse conditions, for example, in swamp or underwater crossings.

Precast concrete sewer pipe may be obtained in many sizes with several kinds of joints. The choice of a particular type of pipe and joint depends on application, location, and conditions for installation. Nonreinforced concrete pipe is available in 4- to 24-in. diameters in 3- or 4-ft lengths. It is manufactured in standard-strength and extra-strength grades. The bell and spigot ends are normally joined by using a rubber ring gasket. Circular reinforced concrete pipe, produced in a size range from 12 to 108 in., is available in five classes based on strength. Precast conduits are also manufactured in elliptical and arch-type shapes. Tongue-and-groove joints, which are

Figure 10–13

Bell-and-spigot vitrified clay sewer pipe connected by compression joints.

(*a*) Polyester liners bonded in the bell and spigot ends have an annular space to secure a rubber O-ring for a tight seal.
(Courtesy of The Logan Clay Products Company.)

(*b*) Ployurethane elastomer seals in the bell and on the spigot compress when the ends are pushed together for a tight seal.
(Courtesy of Pacific Clay Products, Inc.)

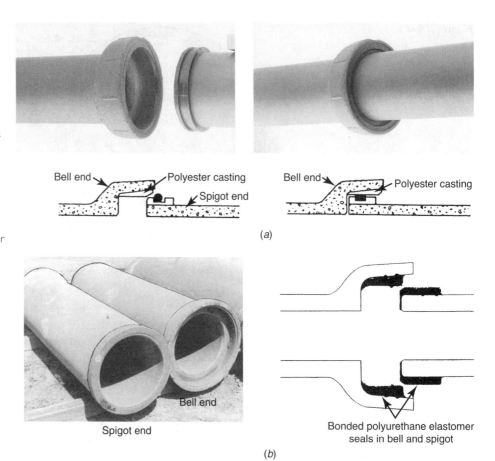

common in reinforced pipe, use a mastic compound or rubber gasket to form a watertight seal.

Concrete pipe is used extensively in storm sewer systems for its abrasion resistance, availability in large sizes, high crushing strength, and, generally, lower cost relative to pipes of other materials. Since concrete is an alkaline material subject to attack by acids, it should not be used for small-sized sanitary sewers where hydrogen sulfide production can cause internal corrosion. However, reinforced concrete pipe is installed for sanitary trunk sewers where the diameter exceeds the vitrified clay pipe sizes available. Here the installation must be protected by control of acidic, high-temperature, and high-sulfate wastewaters by adding chemicals to control biological growth, maintaining flushing velocities and adequate ventilation, and installation of pipe lining if necessary. Epoxy or plastic lining may be cast into the concrete pipe during manufacture, or bitumastic or coal-tar epoxy may be painted on the concrete pipe surfaces after installation.

Ductile-iron pipe is used for force mains, inverted siphons, inside pumping stations and treatment plants, when proper separation from water mains cannot be maintained, and where poor sewer foundation conditions exist. Other pipe materials for special applications in wastewater collection systems are smooth-wall and corrugated steel pipe, bituminized fiber, and reinforced resin pipe.

10–5 BEDDING AND BACKFILL

Bedding and backfill varies with native soil conditions, pipe material, local conditions, and sometimes with bury depth. Bedding and backfill materials must be able to be placed in a readily compacted state under, around, and over the pipe.

Various trench sections are shown in Figure 10–14. Bedding is critical to pipe integrity,

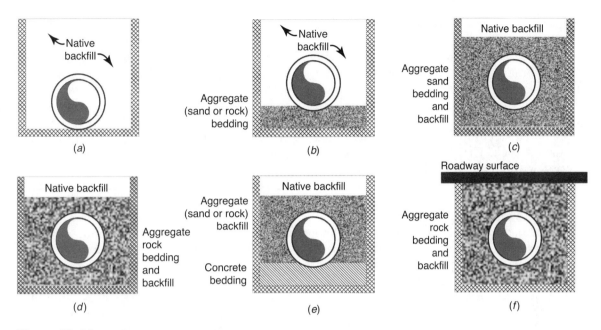

Figure 10-14

Common types of bedding for sewer pipe installations. (*a*) Class D bedding consists of placing the pipe on undisturbed soil and backfilling with native material. The bedding does not provide proper support unless the native soil is sand or gravel. (*b*) Class C bedding consists of sand or rock placed and compacted under the sewer pipe prior to placement. (*c*) Class B bedding consisting of a sand bedding up to the centerline or over the top of the pipe. (*d*) Class CS bedding and backfill of crushed rock. (*e*) Class A-I concrete cradle bedding used to enhance pipe support and mitigate poor soils that cannot support aggregate bedding. (*f*) Backfill extended to below the pavement to avoid settlement that may occur with native soil.

and when placed in the bottom of the trench, it provides a level compacted surface for pipe support. If piping is simply laid by placing the pipe barrels on the flat native trench bottom, as illustrated in Figure 10–14*a*, the pipe may not be able to support the overburden or surface loads. The native soil under the pipe may settle, allowing pipe joints to separate or creating a high point load that may crack the pipe, causing leaks. To provide firm support for the pipe, a bedding of aggregate material consisting of crushed rock or well-graded sand (Figure 10–14*b* and *c*) is used. Bedding is placed in the trench and compacted prior to pipe installation. In some cases, the underlying soil is inadequate to support even backfill. In those cases, concrete may be used to support the pipe and bridge over areas of poor soil (see Figure 10–14*e*).

Backfill must be capable of flowing under the pipe in a compacted state. Native materials used as backfill are typically sandy loam or loam materials that can be readily compacted, as in Figure 10–14*a* and *b*. Most native soils do not readily flow and leave poorly compacted material or voids under the pipe. Backfill using granular material of well-graded sand, gravel, or crushed rock, as in Figures 10–14*c* through *f*, with little effort can be placed under and around the pipe in a compacted state without voids. This type of bedding and backfill provides the maximum uniform pipe support. Warning tape and tracer lines may be installed in the trench on the backfill above the pipe to warn future excavators of the buried pipeline. Tracer lines are common on PVC and other nonmetallic pipelines, allowing easy identification of the pipe location. After bedding is placed over the top of the pipe, the balance of the trench is filled with native soil, which is compacted as it is placed. Trench backfill is typically compacted to 85 to 95 percent of maximum compaction for that soil. Optimum moisture content is critical to obtaining compaction and soil may need to be dried or wetted, depending upon its condition. In many areas, it is advisable to extend the aggregate backfill up to roads or structures to eliminate any settlement that may occur with the use of native material above the pipe and under the road (see Figure 10–14*f*).

Bedding and backfill materials may vary with the pipe being installed. Clay, concrete, and concrete-coated steel pipe are typically installed with crushed rock. Crushed rock reaches full compaction without much work. The angular portions of the rock interlock and do not slip under load. Polyethylene-encased ductile-iron, painted steel, and PVC pipe may be scratched or penetrated by crushed rock; therefore, pea gravel and sand are preferred. Placement requires greater care, but materials also reach full compaction without much work. Water is typically used to help sand flow under the pipe, but direct jetting is discouraged.

10–6 SEWER INSTALLATION

Sewer installation requires knowledge of appropriate excavation requirements, installation needs, and compaction requirements. Considerations include environmental requirements, shoring, dewatering, and field testing of placed material.

A backhoe is commonly used for trench excavation, since it can serve also for backfilling and lifting pipe sections. Trenches can be excavated with vertical walls in stiff clayey soils, but sloping sides are often necessary in less cohesive soils. Any excavation over 5 ft deep requires shoring, bracing, sloped sidewalls, or other provisions for worker protection against the hazard of cave-ins and pipe installation. Safety plans and trench requirements are specified by the U.S. Department of Labor, Occupational Safety and Health Administration (OSHA) and state and local ordinances. Excavation support systems must withstand earth pressure, unrelieved hydrostatic pressure, utility load, and traffic loads to prevent cave-ins and allow safe construction around adjacent structures. The most popular method of excavation is to open-cut the trench with side slope, although there are a number of factors to consider when assessing the stability of excavation walls, including soil behavior, groundwater, traffic, trench stockpile, and other forces adjacent to the trench. Over the last 20 years, empirical evidence has shown that when sloped properly, excavation walls remain safely stable, the angle of the slope depending on the properties of the soil. OSHA classifies materials and defines a maximum allowable slope for excavation in a hierarchy from stable to less stable as listed in Table 10–2. A quick method of determining if a soil is type A or C is to squeeze a handful, taking care that the soil is not saturated with water. If the soil falls away, it is granular in nature. If the soil is moldable and breaks into a couple of pieces as it falls, it has

TABLE 10–2

Soil Types and Slopes Required for Open Trench Excavation[2]

Type of Soil		Compressive Strength		Horizontal/ Examples	Slope[*]	
		ton/sq ft	kPa		Vertical (ft/ft)	Angle
Rock	Stable rock			Unfractured rock	Vertical	90
A	Cohesive	>1.5	>144	Clay, silty clay, sandy clay	¾ : 1	53
B	Cohesive	>0.5	>48	Silt, silty sandy loam, dry unstable rock	1 : 1	45
C	Granular	<0.5	<48	Gravel, sand, loamy sand	1½ : 1	34

*Limited to a total vertical excavation of 20 feet.

cohesive properties. Pocket penetrometers have a direct-reading gauge to measure the unconfined compressive strength. Pushed into the soil, the indicator sleeve displays results in either tons per sq ft or kPa. Penetrometer results have error ranges of ±20 to 40 percent, but are useful in distinguishing between soil types. The most accurate method of soil classification is to submit samples to a laboratory for determination of gradation, plasticity, and unconfined compressive strength.

Figure 10–15 depicts sewer construction in unstable soils. Trench boxes have the advantage of being moved along the trench with the backhoe used to install the pipe. Figure 10–15 shows the combination of trench box and side slope excavation.

The 54-in.-diameter pipe sections are reinforced concrete precast in 12-ft lengths. On the right side of the top picture is a pipeline connected to dewatering wells spaced alongside the trench to lower the water table below the bottom of the excavation. A backhoe is being used to excavate the trench ahead of a steel trench box in which the pipe sections are placed. This bottomless steel safety shield consists of side walls connected together by heavy braces so that the distance between the walls is the trench width. The purpose of the box is to provide a safe working space for laying the pipe without having to set shoring. In the top picture, a section of pipe has already been placed in the space protected by the safety shield, and as the trench excavation is extended, the backhoe is used to pull the box forward. Figure 10–15b shows the workers leveling the bedding on the trench bottom with hand shovels. In the bottom picture, a pipe section is being lowered into the box by a crane. The workers in the trench place the rubber gasket on the spigot of the pipe and guide the section into position to join the buried pipe. The bedding shown is similar to that shown in Figure 10–14c. Each pipe joint is tested for tightness immediately after placement with a special circular expander that presses against the walls of the pipe on both sides of the joint, forming an enclosure that is filled with compressed air to test for leakage. Alignment of pipe sections is done using a laser beam.

A laser system is used to establish grade and alignment in the installation of sewer pipes. The laser beam also serves as a guide in excavation of the trench to proper slope. The laser unit illustrated in Figure 10–16a projects a pencil-thin ray of red light through the pipe opening onto a target for position reference. After rough leveling to within ±5° using foot-screws, the laser unit self-levels and maintains this position unaffected by external environmental factors, such as temperature changes and vibration caused by earth-moving or compaction equipment. The desired grade and alignment are then set by touch controls on the rear of the unit. As each pipe is positioned, a target with a bull's-eye is placed in the end of the pipe to check the slope and alignment. Targets are centered horizontally and adjusted vertically for different size pipes.

As each pipe is installed, the target is placed in the end of the pipe so the pipe can be shifted into position by bringing the laser spot on target center. A laser attached to a calibrated vertical pole can also be set outside the pipe supported on a platform or inside a large-diameter pipe on a T-bar (Figures 10–16c and 10–16d). The laser beam cannot be projected through the pipe if the sewer is laid to follow the curvature of a street or if refraction of the beam is caused by fumes or thermal stratification of the air in the pipe. Instead, the laser is mounted on a stand and directed toward a target attached to a grade rod with a foot that is placed on the invert at the end of each pipe to maintain the proper grade. A hand-held remote (shown at the bottom of Figure 10–16c) can be used to turn the alignment of the laser beam to the correct angle for each length of pipe for the correct curvature of the pipeline.

Dewatering is necessary to prevent excessive groundwater from entering the trench and disturbing the pipe bedding and weakening the trench walls. In some cases, a pocket can be dug for a pump at the end of the trench for removal of groundwater. Where groundwater level is high, shallow wells are installed to lower the water level below the excavation. Permits are required to dispose of water without damage to adjacent property or pollution of waterways or drainage ditches. Prior to disposal, filtering may be required to remove sand and silt from the water.

To reduce migration of water through the porous trench material along the pipeline, seepage barriers, also called cut-off walls, are placed at about 1000-ft intervals along the pipeline. Barriers

(a)   *(b)*

Figure 10–15

Construction of sanitary sewers using vitrified clay and reinforced concrete pipe. (*a*) Trench sidewalls sloped back and benched for safe excavation. Pipe is being installed on a bedding of crushed rock. Native backfill will be used above the rock to bring the trench back to ground level. (*b*) Backhoe excavating the trench ahead of a safety shield or trench box in which pipe sections are lowered. Workers level the sand bedding on the bottom of the excavation.

(Photos courtesy of Gladding McBean Company, a division of Pabco building products, LLC.)

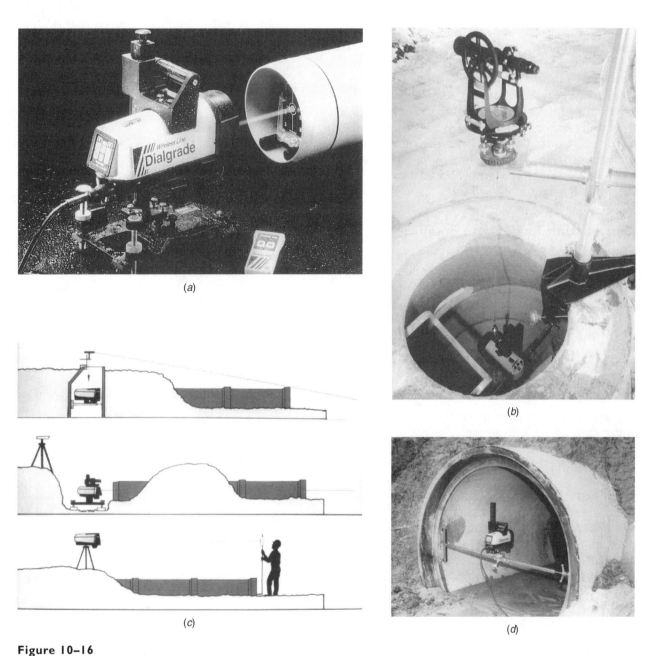

Figure 10–16

Laser system for installation of sewer pipe.[3] (a) The laser unit projects a ray of red light through the pipe to a target and has a dial for setting the desired sewer grade. (b) A transit mounted over the laser unit in a manhole is used to set the laser beam on the desired alignment. (c) The laser can be set in a manhole, on a platform to project through the pipe, or on a stand to project over the pipe. (d) Laser unit attached to a T-bar in a large storm sewer.
(Courtesy of Spectra-Physics Laserplane, Inc.)

extend at least 1 ft beyond the backfill material into the native soil.

Environmental considerations apply to excavation in environmentally sensitive areas, protection of excavated material, and runoff control. Environmental requirements may limit excavation to certain seasons when sensitive species are not impacted. Construction during the rainy season includes covering excavated material with plastic tarps and installation of straw materials to prevent silt and soil from running off into local waterways.

Compaction Testing and Equipment

Bedding and backfill materials are granular-type soils of crushed rock, gravel, or sand and readily compact with little work. In contrast, most native soils are classified as cohesive, consisting of a combination of silts and clays that will mix with water to form a paste that can be plastic and sticky. The compaction of silty clay soils varies with moisture—too little and compaction will be inadequate, and too much creates water-filled pockets that weaken when a load is applied. The highest density (greatest compaction) is at an optimum water content for a given soil. Testing of cohesive materials placed in the excavation is necessary to verify compaction and limit future soil settlement.

Laboratory tests are used to determine the optimum water content and maximum density for a given soil. Field measurements are used to evaluate the effectiveness of the contractor's use of compaction equipment. The most popular laboratory test is ASTM D1557, often called the Modified Proctor Test. The test determines the maximum density for the soil and the effects of moisture on soil density. It is conducted by first oven drying the soil sample and adding water to obtain a known moisture content as a percent of the soil dry weight. The wet soil is compacted in five layers using a 10-lb hammer with 25 blows per layer. The compacted soil is weighed and dried to determine the final moisture content. The process is repeated after changing the water content. The results are graphed as shown in Figure 10–17. The data show that the maximum dry density for this sample is 12 lb/cu ft at an optimum moisture content of 11 percent.

Engineers typically require the contractor to compact material placed in the trench to 95 percent

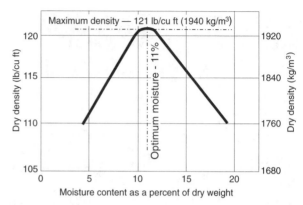

Figure 10–17

Results from a Modified Proctor Test on a cohesive soil in a laboratory. Soil is repeatedly wetted and dried to determine the relationship between dry density and moisture content. The results are used to determine the maximum density and optimum moisture content.

of the maximum density when used as a backfill around the pipe, 85 percent of maximum density above the pipe in open areas, and 90 percent above the pipe and under roads and structures. The higher compaction requirements limit the differential compaction over the trench relative to compaction of surrounding soils. The four devices used for field density testing are shown in Figure 10–18. The sand cone and balloon density meter (Figure 10–18a and b) are used by making a small hole in the compacted soil. The volume of the hole is determined by filling the hole with a measured amount of dry sand or fluid enclosed in a balloon to determine the exact volume. The soil removed is

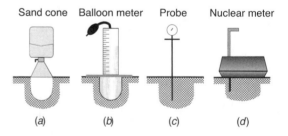

Figure 10–18

Devices for the field measurement of soil compaction. (a) A measured quantity of sand is released to determine the volume of removed soil. (b) A balloon is inflated with water to determine the volume of removed soil. (c) A probe with a resistance meter is pressed into the soil. (d) Radiation sent through the soil is used to determine density.

weighed, dried, and weighed again to determine the weight and percent of moisture. Soil density is calculated by dividing the dried weight of compacted soil in pounds by the cubic footage of the hole. This field measurement is then compared with the laboratory results of the Modified Proctor Test and the percentage of maximum density or the relative density of the compacted soil is calculated.

■ EXAMPLE 10–2

The Modified Proctor Test results are shown in Figure 10–17. Two field tests were made along the trench using a sand cone. Sample #1: The sand volume is 170 cu in. and laboratory tests determined a wet weight of 11.8 lb and a dried soil weight of 11.0 lb. Sample #2: Volume = 156 cu in. with a wet weight of 10.26 lb and a dry weight of 9.25 lb. Determine the density of the contractor's backfill and if the compaction meets the engineer's 90 percent compaction requirement. Has the contractor provided adequate compaction?

Solution

Sample #1

Dry density of the compacted material
= 11.0 lb/(170/1728) = 111.8 lb/cu ft
Percent of maximum density = 111.8/121 · 100
= 92% > 90%
The field test result meets the engineer's requirement for compaction.
The moisture is $((11.8 - 11.0)/11.0 - 100 = 7.3\%)$
The moisture is not optimal and additional compaction work will not improve the compaction at this moisture content.

Sample #2

Dry density of the compacted material
= 9.25 lb/(156/1728) = 102.5 lb/cu ft
Percent of maximum density = 102.5/121 · 100
= 85% < 90%
The field test result does not meet the engineer's requirement for compaction.
The moisture is $((10.26 - 9.25)/9.25 \cdot 100 = 11\%)$, which is optimal. The compaction work is inadequate and additional work is required to increase the density.

■ ■ ■

The dial compaction probe shown in Figure 10–18c is pressed into the soil to a depth of 36 to 48 in. (1 to 1.3 m). The resistance to penetration is shown on the dial as the rod moves down through the soil under steady and even pressure. The resulting soil density measurement should be considered approximate. A nuclear density meter, depicted in Figure 10–18d, is a direct-reading meter to determine density using a radioactive isotope source and detector. Photons generated by the isotope are absorbed more in dense soil than in loose soil. The meter can also determine the water content of the soil. The field density values are used to compare with the Proctor values and calculate the relative density.

Bedding and backfill is placed in layers, or lifts, of up to 12 in. and compacted using the equipment shown in Figure 10–19. Limiting the backfill layer depth ensures complete compaction throughout the vertical trench profile. Compaction equipment includes vibratory plates designed to consolidate granular material, rammers (also called sheep's foot), rollers, and wheels attached to backhoes to consolidate cohesive materials. Vibratory plates, as shown in Figure 10–19a, have two eccentric weights powered by an engine to develop a high-speed vibratory force. Vibratory plates are used for granular soils such as bedding and backfill. Vibrators are not recommended for native cohesive soils. Rammers (Figure 10–19b) obtain compaction from a small engine powering a large piston held between two springs. The forward incline propels the machine forward as it jumps. Rammers are good in narrow trenches with native cohesive soil. Small motorized rollers (see Figure 10–19c) are operated by wire from a tethered control panel. The weight of the equipment and the vibration generated by the motor create consolidation. Ride-on rollers (Figure 10–19d) are used for larger areas of soil consolidation and wide areas of granular consolidation. A variety of rollers are designed to be attached to backhoes and rolled over the trench backfill, and various attachments like that shown in Figure 10–19e are available for backhoes and excavators. The weight of the equipment plus the roller action are used to consolidate soils.

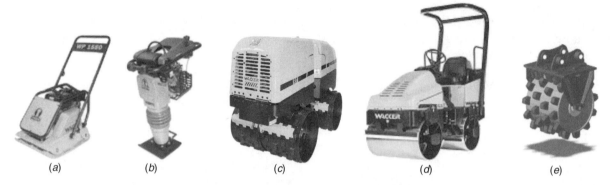

Figure 10–19

Equipment used to compact trench materials. (*a*) Vibratory soil plate used to consolidate pipe bedding and granular materials. (*b*) Vibratory rammer or sheep's foot used to consolidate small areas in narrow trenches and between the pipe and trench wall. (*c*) Trench roller operated by a joystick tethered to the machine. (*d*) Ride-on roller for large trenches. (*e*) Roller attachment to backhoe, attaches in place of the bucket to roll along the pipe trench. (Equipment photos courtesy of Wacker Corporation.)

10–7 SEWER TESTING

New sewers are tested for watertightness by monitoring infiltration, measuring exfiltration with the pipeline full of water, and air testing.

Exfiltration testing, the reverse of measuring actual infiltration, is used mainly in dry areas where the groundwater table is below the pipe crown. Filling the pipe with water and observing the loss during a specified time period is a positive method, since it subjects the sewer and manholes to a known water pressure. Excessive pressures can have destructive results in lower sections of a sewer; however, testing of sections between manholes has few risks. The maximum hydrostatic head applied is usually 10 ft, and the water is allowed to stand for at least 24 hr before measuring exfiltration. This allows the pipe and joint material to become saturated with water and permits the removal of entrapped air. The maximum allowable exfiltration specified varies from 100 to 500 gal/in. of diameter/mile/day. Table 10–3 gives infiltration and exfiltration rates for various sizes of sewer pipes.

Low-pressure air testing is applicable to sewer lines constructed of vitrified clay pipe[4] or combinations of clay and other pipe materials. Although

TABLE 10–3

Conversion of Infiltration or Exfiltration Rates from Gallons per Inch of Diameter per Mile per Day to Gallons per Hour per 100 ft for Various Sizes of Sewer Pipes

DIAMETER OF SEWER (in.)	*FILTRATION RATES IN GAL/HR/100 FT FOR THE FOLLOWING RATES IN GAL/IN. DIAMETER/MILE/DAY*				
	100	200	300	400	500
8	0.63	1.3	1.9	2.5	3.2
10	0.79	1.6	2.4	3.2	4.0
12	0.95	1.9	2.8	3.8	4.7
15	1.2	2.4	3.5	4.7	5.9
18	1.4	2.8	4.3	5.7	7.1
21	1.7	3.3	5.0	6.6	8.3
24	1.9	3.8	5.7	7.6	9.5
27	2.1	4.3	6.4	8.5	10.7
30	2.4	4.7	7.1	9.5	11.8
36	2.8	5.7	8.5	11.4	14.2
42	3.3	6.6	10.0	13.3	16.6
48	3.8	7.6	11.4	15.2	19.0

Figure 10–20

Low-pressure air testing to determine the structural Integrity of newly constructed sewer lines.
(Courtesy of Cherne Industrial, Inc.)

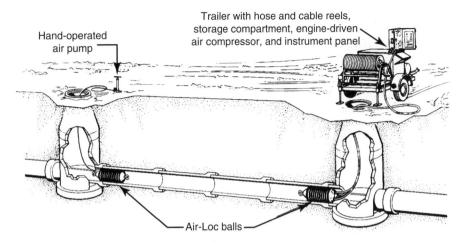

Hand-operated air pump

Trailer with hose and cable reels, storage compartment, engine-driven air compressor, and instrument panel

Air-Loc balls

air leakage cannot be correlated to water leakage, the air test does demonstrate the structural integrity of a sewer line. A well-constructed sewer will exhibit some loss of air pressure even if watertight; however, significant air loss is caused by damaged pipe or defective joints. To reduce the passage of air through the walls, the interior of the sewer line is wetted before conducting the air test.

For the air test, a section of sewer between manholes is isolated as shown in Figure 10–20 by plugging both ends and tightly capping any service connections. Air pressure is applied, raised to 4.0 psi, and held for 2 to 5 min while all plugs are checked for leakage and the temperature of the air stabilizes. The air supply is then disconnected and the time required for the pressure to drop from 3.5 psi to 2.5 psi is measured with a stopwatch. If the 1.0-psi drop does not occur within the calculated test time, the sewer line has passed the test. If the pressure drops more than 1.0 psi during the test time, the sewer has failed the test. The test times required per 100 ft of sewer for pipes of various diameters are given in Table 10–4. If the section of sewer line tested includes more than one size of pipe, the test time is calculated for each diameter and the total of these times is the test time for the section.

Manholes are tested using a vacuum test after completion of pipe testing. The manhole to be tested is isolated from sewer lines by installing plugs at least 8 in. into the sewer pipes. The plugs must be inflated past the manhole/pipe gasket to avoid interference with pipe infiltration. The manhole is considered to pass the test if the

TABLE 10–4

Minimum Test Times per 100 Feet of Sewer Line for Low-Pressure Air Testing[5]

Pipe Diameter (in.)	Test Time (min)	Pipe Diameter (in.)	Test Time (min)
		21	3.0
4	0.3	24	3.6
6	0.7	27	4.2
8	1.2	30	4.8
10	1.5	33	5.4
12	1.8	36	6.0
15	2.1	39	6.6
18	2.4	42	7.3

(ASTM C828-06 Low-Pressure Air Test of Vitrified Clay Pipe Lines. © ASTM International. Reprinted with permission.)

time for the vacuum reading to drop from 10 to 9 in. of Hg meets or exceeds the values indicated in Table 10–5.

If the test fails, the manhole may be repaired with a nonshrinkable grout or other appropriate sealant material.

■ EXAMPLE 10–3

A 400-ft length of 8-in. vitrified clay sewer was air tested for structural integrity. The time measured for the air pressure to drop from 3.5 to 2.5 psi was 6.5 min. Did the sewer line pass this low-pressure air test?

TABLE 10–5

Minimum Test Times for Various Manhole Depths and Diameters

| Depth (ft) | Diameter, in. | | | | |
| | 48 | 54 | 60 | 66 | 72 |
	Time, in seconds				
8	20	23	26	29	33
10	25	29	33	36	41
12	30	35	39	43	49
14	35	41	46	51	57
16	40	46	52	58	67
18	45	52	59	65	73
20	50	53	65	72	81
22	55	64	72	79	89
24	59	64	78	87	97
26	64	75	85	94	105
28	69	81	91	101	113
30	74	87	98	108	121

(ASTM C1244 Standard Test Method for Concrete Sewer Manholes by the Negative Air Pressure (Vacuum) Test Prior to Backfill. © ASTM International. Reprinted with permission.)

Solution

From Table 10–4, the minimum test time per 100 ft of 8-in. pipe is 1.2 min. Therefore,

$$\text{Test time} = \frac{400 \text{ ft} \times 1.2 \text{ min}}{100 \text{ ft}} = 4.8 \text{ min}$$

The measured time of 6.5 min is greater than this, so the sewer line passes the test.

■ ■ ■

10–8 LIFT STATIONS IN WASTEWATER COLLECTION

Lift stations are required to elevate and transport wastewater in collection systems when continuation of gravity flow is no longer feasible. In flat terrain, sewers en route to a treatment plant may increase in depth to the point where it is impractical to continue gravity flow. Here, a pumping station can be installed to lift the wastewater to an intercepting sewer at a higher level, or the pumps can discharge to a force main conveying waste-water under pressure to the treatment plant. New subdivisions or industries locating on the edge of an existing sewer system may find that the pipe is too shallow to receive sewer connections from basement level. In this case, a small lift station can pump wastewaters from the area to the main sewer.

Pumping stations are expensive to install, require periodic inspection and maintenance, and strain public relations if they malfunction and allow backup of domestic wastewaters into residences or businesses and create sewer system overflows. For these reasons, they are avoided wherever possible by incorporating sewer planning and treatment plant location into municipal land-use zoning and comprehensive planning.

Pneumatic ejectors are commercially available for pumping flows not exceeding about 100 gpm, for example, for several homes or a commercial building. They have the important advantage of being able to handle unscreened domestic wastewater without clogging. (Although low-capacity pumps can be employed, the smallest centrifugal pump designed to pass 3-in. solids and operate with reasonable freedom from clogging has a rated capacity greater than 100 gpm.)

Wet-well pumping stations greater than 100 gpm typically use nonclog centrifugal pumps installed in or over a wet well (see Figure 10–21). Submersible pumps and self-priming pumps reduce the cost of installation by avoiding the added cost of a wet well and dry well. Submersible pumps (Figure 10–21a) have the motor in a watertight enclosure coupled with the pump. The pump is held by the pump base plate on the floor of the wet well. Rails and a lifting chain are used to remove the pump for inspection and maintenance. Pumps start and stop using mercury switches enclosed in a teardrop-shaped plastic casing. The units tip up when submerged, activating the switch. Some units are fitted with air bubblers for continuous water-level monitoring. The discharge check valve and isolation gate valve are mounted above ground or in a separate belowground vault. Pumps can be installed in a contractor-built wet well or as a package assembly in a fiberglass wet-well enclosure.

Self-priming centrifugal pumps, like that shown in Figure 10–21b, are popular because access to the pump and motor is aboveground and

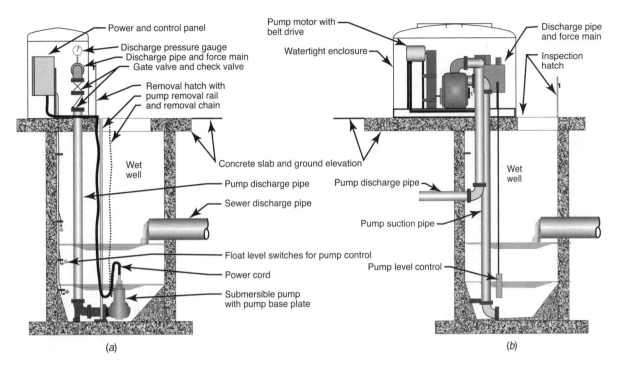

Figure 10–21

Typical collection system pumping stations for lifting wastewater from a deep sewer to an intercepting sewer or to the wastewater treatment plant headworks via a discharge pipe and force main. (a) Typical submersible pumping station with submersible pump mounted on the floor of the wet well. A rail system and chain are used for removal. Check and gate valves may be placed in an underground vault or aboveground as shown. (b) Self-priming centrifugal pump mounted over the wet well. Access to the pump is easier because the pump is aboveground. A belt-drive motor allows the pump speed to be adjusted as flows increase over time. An access hatch is provided for inspection and wet well cleaning. (Courtesy of Gorman-Rupp Company.)

they do not have to be removed for inspection from the sump like submersible pumps. The pump impellers offer greater capability for handling solids, and an access cover allows rag removal in the volute at the impeller. Pumps may be direct-drive, but are often belt-driven to allow flexibility in setting pump speed and adjusting pump capacity. Discharge check and isolation gate valves are installed aboveground for access. A hatch is installed over the wet well for access and cleaning. Factory-built duplex pumping stations are available on a common skid and enclosure.

REFERENCES

1. Isco, A Teledyne Technologies Company, Water and Wastewater Monitoring Products. http://www.isco.com/products/productswm.asp

2. *OSHA Technical Manual*, Section V, Chapter 2, "Excavations: Hazard Recognition in Trenching and Shoring." U.S. Department of Labor, Occupational Safety & Health Administration. http://www.osha.gov/dts/osta/otm/otm_v/otm_v_2.html#8

3. *Concrete Design Manual*, American Concrete Pipe Association, Irving, TX. http://www.concrete-pipe.org/manual/START.pdf

4. *Engineering Manual*, National Clay Pipe Institute, Lake Geneva, WI. http://www.ncpi.org/engineer.htm

5. American Society for Testing and Materials (ASTM). http://www.astm.org/

PROBLEMS

10–1 What is the purpose of storm sewers as distinct from sanitary sewers? How are combined sewers different from storm and sanitary sewers?

10–2 How is storm-water pollution managed?

10–3 If 500 homes generate an average of 480 gpd/home with a peaking factor of 3.5, what size sewer pipe would be required, based on Table 10–1?

10–4 What is the basis for the minimum slopes listed in Table 10–1?

10–5 Why is the design of a sanitary sewer based on 400 gpcd when the wastewater flow is 80–120 gpcd?

10–6 A development has 520 homes, followed by 180 homes along the access road, with another 440 homes to the sewer main. Homes are expected to contribute a peak flow of 750 gpd/home. Size the pipes in each segment.

10–7 What is the maximum number of homes that can be served by a 24-in. sewer with 4.5 people per home? Peaking factor is 3.5 and flow = 120 gpcd.

10–8 Why are multiple pipes used in inverted siphons and what is the criteria for sizing? If the average flow is 8 mgd, low flow is ½ the average, and peak flow is given by Eq. 9–3, then what would be the appropriate size of the siphon pipes?

10–9 What are the requirements for separation between potable water pipes and sewer pipes?

10–10 What methods are used for trenchless construction? How are the methods similar and different?

10–11 What are the types of typical manholes and why or when are they used?

10–12 What is the purpose of installing plumbing traps on house drains? Why is a vertical vent pipe installed? How are sanitary sewers in the street vented?

10–13 How are flumes placed in a sewer for proper flow measurement? How is flow measured?

10–14 Why do sewer pipes corrode? What is the origin of and the conditions for sewer pipe corrosion? What pipe materials are used to resist sewer corrosion?

10–15 How are industrial sewers monitored?

10–16 Why is clay pipe preferred for sanitary sewer service and concrete pipe used for storm-water flow?

10–17 How are pipe joints sealed to prevent leakage? How are minor changes in pipe direction made?

10–18 For a vitrified clay pipe to be installed in a poor soil, what types of bedding materials are appropriate (refer to Figure 10–14)?

10–19 For a PVC water line installed in a clay soil, what type of bedding and material are appropriate?

10–20 The native soil is a clayey-silt. What type of backfill should be considered for the following pipe: PVC, clay, concrete, and polyethylene-encased ductile-iron pipe? What additional considerations should be given for backfill under pavement?

10–21 The contractor wants to open-cut a pipe trench. When a handful of the soil is squeezed, no water drops from the sample. As the hand is opened, the soil falls away quickly without clumping. What is the safe trench side slope?

10–22 The contractor has started a trench. A handful of soil is squeezed and the soil molds, then breaks and falls away in chunks. Just based on this information, what is the minimum safe trench slope? What additional information would be needed to reduce the volume of excavation?

10–23 How is sewer pipe installed to be at a constant slope?

10–24 How is trench compaction tested? What equipment is used to test compaction and how is it used?

10–25 A modified Proctor Test of a soil sample resulted in an optimum moisture content of 12 percent with a dry weight density of 111 lb/cu ft. Field testing of 173 cu in. of soil resulted in a wet weight of 10.8 lb and a dry weight of 9.65 lb. What are the field dry weight, field density, and percent of optimum density? If the specifications require 90 percent of optimum compaction, is compaction adequate? What changes can be made to enhance compaction?

10–26 A modified Proctor Test resulted in an optimum moisture content of 8 percent with a dry density of 128 lb/cu ft. Field testing using a sand cone resulted in the following: Volume = 166 cu in. with a wet weight of 11.6 lb and a dry weight of 11.2 lb. What are the field dry weight, field density, and percent of optimum density? If the specifications require a minimum of 95 percent, is compaction adequate? What changes can be made to enhance compaction?

10–27 What compaction equipment is used for granular as opposed to cohesive soils?

10–28 A 1000-ft segment of 12-in. pipe had an exfiltration loss of 13 gal in 30 min. Calculate the exfiltration rate in gal/hr/100 ft and determine the filtration rate. If the limit is 200 gal/in. dia./mile/day, are the sewer losses acceptable?

10–29 What is the maximum allowable exfiltration flow for a 200-ft-long, 8-in.-diameter sewer if the limit is 100 gal/in. dia./mile/day? If the

pipe is tested by low-pressure air, what would be the minimum testing time for a pressure drop from 3.5 to 2.5 psi?

10–30 A 250-ft segment of 8-in. clay pipe was tested by air. If the pressure drop from 3.5 to 2.5 psi occurred in 180 sec, was the pipe installation acceptable?

10–31 The air testing of a standard 20-ft-deep, 48-in. manhole resulted in a pressure drop of 1 in. Hg in 2.2 min. Is the manhole acceptable?

10–32 What are the common types of sewage pumping stations? What are the major differences between them?

Wastewater Processing

Conventional mechanical wastewater treatment is a combination of physical and biological processes designed to remove organic matter and solids from solution. The earliest mechanical method was plain sedimentation in septic tanks. Imhoff tanks used by municipalities were two-story septic tanks that separated the upper sedimentation zone from the lower sludge digestion chamber by a sloping bottom with a slot opening. Solids settling in the upper portion of the tank passed through the slot into the bottom compartment, from which the digested sludge was periodically withdrawn for disposal. Today, primary sedimentation and sludge processing are performed in separate treatment tanks. Sludge is usually processed in separate digestion tanks and mechanically dewatered prior to disposal.

Primary sedimentation of municipal wastewater has limited effectiveness, since less than half of the wastewater organic content is typically settleable. The initial attempt at secondary treatment involved chemical coagulation at the primary clarifier to improve settleability of the wastes. Although this provided considerable improvement, the heavy chemical dosages resulted in high cost, and dissolved organics were still not removed. The first major breakthrough in secondary treatment occurred when it was observed that the slow movement of wastewater through a gravel bed resulted in rapid reduction of organic matter and BOD. This process, referred to as trickling filtration, was developed for municipal installations starting in about 1910. A more accurate term for a trickling filter is *biological bed*, since the process is microbial oxidation of organic matter by slimes attached to the stone rather than a straining action.

A second major advancement in biological treatment took place when it was observed that biological solids, developed in polluted water, flocculated organic colloids. These masses of microorganisms, referred to as activated sludge, rapidly metabolized pollutants from solution and could be subsequently removed by gravity settling. In the 1920s, the first continuous-flow treatment plants were constructed using activated sludge to remove BOD from wastewaters. (Biological systems are also discussed in Section 3–11.)

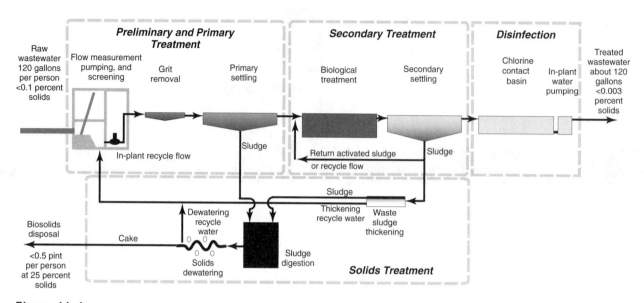

Figure 11–1

Schematic of a conventional municipal wastewater treatment plant. Liquid treatment consists of preliminary treatment, primary sedimentation, biological treatment, final sedimentation, and disinfection prior to discharge. Solids treatment consists of digestion of primary and thickened secondary solids, mechanical dewatering, and land disposal.

The pictorial diagram in Figure 11–1 summarizes the processes applied in conventional municipal wastewater treatment. Preliminary steps include influent flow measurement, screening to protect the pumps from large solids, pumping as needed to lift the wastewater above ground level, and grit removal to protect mechanical equipment from abrasive wear. Primary treatment removes heavier solids from the wastewater by sedimentation and scum that floats to the surface. Primary clarifiers are also used to store and thicken sludge, but their main function is to reduce the organic load on secondary treatment processes. Secondary treatment is the biological conversion of organic material (BOD) into solid biomass for removal in the secondary clarifier. Several treatment alternatives exist, including fixed media (trickling filters) and suspended growth (activated sludge). Following secondary settling, disinfection of the effluent reduces the risk of disease and biological activity for use in-plant, irrigation, or as required for discharge to a receiving stream. Excess microbial growth settled out in the secondary clarifier is wasted and may be thickened prior to digestion. The figure shows that waste sludges from primary settling and secondary treatment are anaerobically digested for stabilization prior to dewatering. Plants without primary clarifiers use aerobic digestion for secondary sludges. Some plants use a mechanical process to extract water directly from raw sludge after chemical conditioning. Ultimate disposal of dewatered solids may be by landfill, incineration, or land application if biologically stabilized.

The overall process of conventional treatment can be viewed as the conversion of soluble matter to an organic solid and solids thickening; pollutants removed from solution are concentrated in a small volume convenient for ultimate disposal. The contribution of raw sanitary wastewater is about 120 gal/person (400 l) with a total solids content of less than 0.1 percent, 240 mg/l suspended solids, and 200 mg/l BOD. During treatment, microorganisms consume organic material, thus converting the BOD into a settable biological solid. Liquid waste sludge withdrawn from primary and secondary processing amounts to approximately 0.5 gal/person (2 l), with a solids content of 5 percent by weight (Figure 11–1). This is further concentrated to a handleable material by mechanical dewatering; the extracted water is returned for reprocessing. Cake from dewatering amounts to less than 0.5 pint/person (0.25 l) with a 25 percent solids concentration. This type of physical-biological scheme is effective in reducing

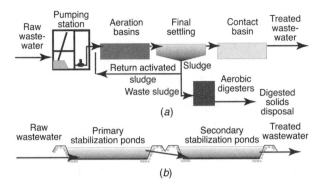

Figure 11–2

Processing diagrams of systems for treatment of small wastewater flows. (a) Biological processing without primary sedimentation. (b) Natural facultative stabilization ponds.

the organic content of wastewater and accomplishes the major objective of BOD and suspended solids removal. Dissolved salts and other refractory pollutants are removed to a lesser extent. Referring to Table 9–2, 50 percent of total volatile solids, 70 percent of total nitrogen, and 70 percent of total phosphorus remain in the effluent after secondary biological treatment. Advanced wastewater treatment processes are needed to remove these refractory contaminants. Orthophosphate can be precipitated by chemical coagulation, nitrogen may be reduced by biological nitrification and denitrification, and activated carbon removes refractory soluble organics.

Completely mixed aeration without primary sedimentation (Figure 11–2a) is popular for treatment of small wastewater flows, for example, from subdivisions, villages, and towns. The size of aeration basins ranges from factory-built metal tanks with diffused aeration to accommodate the waste flow from a few hundred persons to mechanically aerated concrete-lined basins, such as an oxidation ditch, to serve a town with a population of several thousand. Elimination of primary settling dramatically affects the character of waste sludge. Instead of a septic primary settled sludge with relatively high solids content, the waste is aerobic and much more voluminous, with solids in the range of 0.5 to 2 percent. Therefore, the solids handling system for the flow scheme in Figure 11–2a consists of aerobic digesters (aerated sludge holding tanks) and hauling of the stabilized liquid sludge for spreading on farmland. Larger

plants may reduce the volume of sludge by gravity or mechanical thickening prior to digestion and mechanical dewatering.

Hundreds of villages and commercial establishments in rural areas use stabilization ponds for wastewater treatment (Figure 11–2b). The organic loading applied to lagoons is very low, and the liquid retention time is long, normally more than 90 days. The natural biological processes that stabilize the waste are sketched in Figure 3–19. In dry climates the evaporation rate may equal or exceed the liquid loading so that the ponds provide complete retention of the wastewater. Stabilized wastewater may be used for irrigation or, where lagoons overflow to streams, discharge of the treated wastewater may be screened to remove algae and meet effluent discharge requirements.

11–1 CONSIDERATIONS IN PLANT DESIGN

A basic schematic diagram positioning fundamental design considerations relative to the boundary of a treatment plant is shown in Figure 11–3. The desired effluent quality is specified for the design engineer by a government pollution control authority. Because effluent standards are specific for each receiving water, they must be established for each treatment plant based on the location of discharge. Based on the input of raw wastewater flow and characteristics and the output of required effluent quality, the design engineer can recommend alternative wastewater treatment systems. Because the waste sludge produced varies with the kind of treatment system, the design engineer

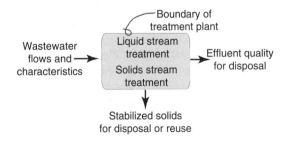

Figure 11–3

Fundamental considerations for the design or expansion of wastewater treatment and sludge processing.

needs to know the desired method of sludge disposal or reuse. Processes for sludge stabilization and dewatering are then selected based on both the wastewater treatment system and method of sludge disposal or reuse.

Effluent Quality

The technology-based standards for secondary treatment of municipal wastewater are a maximum monthly average BOD of 30 mg/l, suspended solids of 30 mg/l, oil and grease of 10 mg/l, and pH between 6.0 and 9.0. Disinfection of a wastewater effluent is required where necessary to protect public health. Where chlorine is used for disinfection, dechlorination may be required to eliminate any chlorine residual prior to discharge. A common quality standard for fecal coliform bacteria is a geometric mean not to exceed 200 per 100 ml. The national pollutant discharge elimination system (NPDES) permit may be more restrictive for these conventional pollutants and include other pollutants to meet water quality-based standards. Effluent standards for discharge to surface waters and the NPDES program are discussed in Section 5–5. Effluent standards for restricted and unrestricted irrigation and other reuse applications are presented in Chapter 14.

Design Loading

The wastewater quantity used for sizing basins and equipment varies with the nature of the equipment and whether the process is hydraulic or loading limited. The selection of flow and load values must take into account hourly, daily, and seasonal variations. For example, in warm climates summer monthly average flows are often 20 to 30 percent greater than the annual average. Saturdays and Sundays, when industries are not in operation, generally have lower flows. The selected design flow includes normal infiltration/inflow that occurs during the wet season. Storm runoff collected in combined sewers draining to the plant must also be considered. Where few data are available, the engineer may increase the quantity of flow from areas with older sewers by about 25 percent. This value is referred to as the wet-weather design flow.

Flow and load values are typically expressed in terms of peak hour, maximum day, maximum average month, and annual average. Further, flows may be referred to as average dry weather and average wet weather. The loading (pounds of BOD or suspended solids) is the same; the quantity of flow changes the concentration. Loading may be designated by season; for example, separate values may be listed for increased industrial and agricultural activity. Peak hour typically refers to the peak observed for a 1-hr period during a storm event or diurnal loading peak. The flow value is important when sizing pumps, pipes, and hydraulically limited equipment. The loading value is important in sizing aeration and digestion equipment to meet peak demands. The maximum day values include wet-weather flow and peak industrial loading impacts. The maximum average month values for flow, BOD, and suspended solids are used in sizing basins because effluent requirements are also expressed in monthly average limits. Values for flow and loading are taken from the same month. Average annual values are calculated from the average of the monthly average values. The values used in design require considerable judgement so that equipment is not under- or oversized.

Table 11–1 lists the design criteria for the treatment processes shown in Figure 11–1. Maximum and minimum hourly hydraulic flows are applicable for sizing certain units, for example, lift pumps. Normally, low flows range from 20 to 50 percent of the average daily, and maximum rates are 200 to 250 percent. Sometimes these daily extremes are confusing, since they may be expressed in units of million gallons per day while referring to flows that occur only over a 1-hr period. For example, a specification may state that an aeration system is planned for a maximum monthly average flow of 5 mgd and a peak flow of 10 mgd. This means that the unit has a capability of treating 5 mil gal during one day at the maximum month conditions with a flow variation such that the peak hourly rate does not exceed 10 mgd. This should not be interpreted as the system is able to process 10 mil gal in a 24-hr time period. All process and equipment components must be checked against average, minimum, and peak conditions. The range from maximum to minimum may exceed the range of the equipment and require multiple units for complete coverage. Many plants designed for seasonal peaks remove units from service during off-peak periods.

TABLE 11–1

Typical Design Criteria for the Treatment Processes Shown in Figure 11–1

PROCESS	LOADING
Flow measurement	Peak hourly flow
Bar screen	Peak hourly flow
Pumps	Peak hourly flow
	Min. hourly flow
Grit chamber	Max. monthly flow
	Peak hourly flow
Primary settling	Max. monthly flow
Biological treatment	Max. monthly BOD loading
	Check peak hourly BOD loading
Final settling	Max. monthly flow
Disinfection	Peak hourly flow
Thickening	Max. daily sludge flow
	Check max. solids loading
Digestion	Max. monthly volatile solids load
	Check max. monthly sludge flow
Dewatering	Max. sludge flow
	Check max. solids loading
Land application	Max. nutrient loading (sludge)
	Max. hydraulic loading (water)

The organic content of a municipal wastewater is defined by the concentrations of BOD and suspended solids. Design organic loadings on a treatment plant are expressed in terms of average pounds per weekday. Values given in the units of milligrams per liter must be converted to pounds per day, using the flow associated with the concentration measurement. Peak flow is not typically associated with peak concentration and may or may not be the peak loading value when converted to pounds. Other significant parameters to consider in design loadings are ammonia nitrogen, irregular flow, high strength, and the presence of toxins in industrial wastewaters.

Design Parameters

Standards for the design of treatment units are expressed by a variety of terms. Hydraulic criteria are expressed in terms of rates (flow per unit length or flow per unit area) and detention times (flow per unit volume). Sedimentation basins are sized on weir loading rate, overflow rate, and detention time. The weir loading rate is the effluent flow over the weir divided by the weir length, expressed in gallons per day per linear foot of weir. The overflow rate is the effluent flow divided by the surface area of the tank, expressed in units of gallons per day per square foot. Detention time in hours is calculated by dividing the tank volume by influent flow, expressed in units of hours.

Organic loadings on treatment units are stated in terms of pounds of suspended or volatile solids per unit volume and pounds of 5-day BOD per unit volume. Loading on an aeration basin is commonly expressed as pounds BOD applied per 1000 cu ft of tank volume per day. A biological filter loading uses the same units, except the volume refers to the quantity of media rather than liquid volume. The aeration period (hydraulic detention time) in an activated-sludge process is equal to the volume of the aeration basin divided by the flow of raw wastewater (hours). Since biological filters do not contain a liquid volume, hydraulic loading is presented as the amount of wastewater applied per unit of surface area, for example, million gallons per day per acre (mgad). Digester loading is expressed in pounds of volatile suspended solids per 1000 gallons per day.

The specific numerical values of design parameters used by an engineer in sizing treatment units are frequently recommended in published standards. For example, the *Recommended Standards for Wastewater Facilities, Policies for the Design, Review, and Approval of Plans and Specifications for Wastewater Collection and Treatment Facilities, A Report of The Wastewater Committee of the Great Lakes—Upper Mississippi River Board of State and Provincial Public Health and Environmental Managers* are used by regulatory agencies in the ten states surrounding the Great Lakes in reviewing plans for sanitary facilities. Increasingly, states have established unit process design standards when wastewater or biosolids are reused or are applied in public areas. Even so, many design specifications are not listed and must be selected by the engineer based on experience or on the recommendations of equipment manufacturers. These may not be readily available in the published literature, and for a particular treatment plant the criteria applied must be obtained from the design engineer or developed based on actual plant performance.

Construction drawings show plant layout, flow schemes, and tank dimensions, but typically do not list design loadings or show proprietary equipment. Contract specifications describe quality of workmanship, materials, and equipment to be installed and include performance requirements for mechanical units, such as pumps and aerators. The preliminary study report, written prior to preparation of plans and specifications, discusses conditions relative to plant design and includes loadings, flows, and alternative treatment processes. Planning, design, and construction are evolutionary processes. Values in the predesign report must be compared against the as-built drawings, because the designs may be modified between the time the preliminary report is written and plant construction. The design engineer is required to write operation manuals that include design loadings and effluent quality requirements and is typically the best source of design information.

11-2 PRELIMINARY TREATMENT

Flow measurement, screening, pumping, and grit removal are normally the first steps in processing a municipal wastewater. Chlorine solution or ferric chloride may be added to raw wastewater for odor control and to improve settling characteristics of the solids. The arrangement of preliminary units varies but the following general rules apply. A Parshall flume is typically located first and ahead of screens and prior to the introduction of in-plant recycle flows. Bubbler or ultrasonic meters measure the water level, which is converted into flow using flume equations. Screens protect pumps and prevent large solids from fouling subsequent units. With variable-speed pumps, a magnetic flow meter in the discharge pipe or a flume may be placed on the discharge side of the pumps. Grit removal reduces abrasive wear on mechanical equipment and prevents the accumulation of sand in tanks and piping. Although ideally grit should be taken out ahead of the lift pumps, grit chambers located aboveground are far more economical and offset the cost of pump maintenance. Two possible arrangements for preliminary units are illustrated in Figure 11–4. The upper scheme is typical of a large plant; the lower sequence is common in small systems.

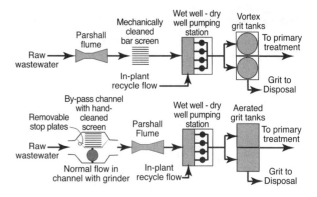

Figure 11–4

Typical arrangements of preliminary treatment units in municipal wastewater processing: flow measurement, screening, sewage pumping, and grit removal. The lower sequence is common for smaller plants.

Flow Measuring

All treatment plants are required to monitor influent wastewater flow. The best system is a Parshall flume that is equipped with an automatic flow recorder and totalizer (refer to Section 4–10). The advantages of a flume are low head loss and smooth hydraulic flow to prevent deposition of solids. Some existing treatment plants have magnetic flow meters in the discharge piping from the influent pumping station, but these are not recommended because of in-plant recycle flows and potential operation and maintenance problems.

Screens, Fine Screens, and Shredders

Mechanically cleaned screens have clear bar openings of between $\frac{1}{2}$ and $2\frac{1}{4}$ in. As illustrated in Figure 11–5, collected solids are removed from the bars by a travelling rake that lifts them to the top of the unit. Here the screenings may be fed to a compactor for dewatering prior to discharge to a portable receptacle for hauling to land burial. Inclined bar screens with chain-driven rakes are also manufactured. Screens with finer openings manufactured of high-strength plastic are popular for removing greater quantities of paper and plastics. Screens are manufactured as a continuous fabric; openings of 3 to 6 mm ($\frac{1}{8}$ to $\frac{1}{4}$ in.) are common throughout Europe.

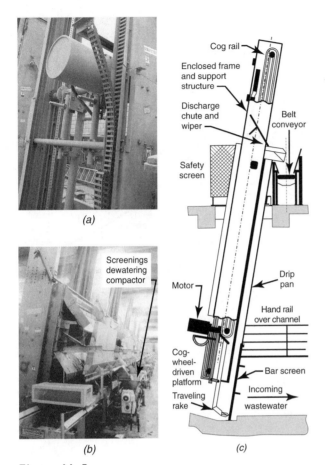

(a)

(b)   *(c)*

Figure 11–5

Climber-style mechanically cleaned bar screen. *(a)* Photo of drive at the top of the bar screen showing motor and cog wheels on the fixed plate between side rails. *(b)* Photo from behind the unit where screenings travel on a belt conveyor prior to compression dewatering and final disposal in a landfill. *(c)* Drawing of the bar screen. Screen openings are $\frac{3}{4}$ in. No moving parts are submerged in wastewater. The bottom portion of the bar screen is shown in Figure 11–11b.

Figure 11–6

Photo of 2.5-mm rotary screen. Wastewater enters the rear, flows through the screen drum, and out the end in a pipe to secondary treatment. Solids collected on the drum are removed by a doctor blade and discharged into a container for land disposal.
(Photo courtesy of Parkson Corporation)

Screens with openings of 1 to 3 mm ($\frac{1}{32}$ to $\frac{1}{8}$ in.) are required to remove nonbiodegradable fibrous material and hair from the wastewater. Removal of this material is important when sludge may be recycled for public use and preceding biological treatment in a membrane unit. The rotary screen shown in Figure 11–6 is externally fed such that wastewater is directed under then up over the length of the screen and through the screen and into a pipe for secondary treatment. Sticky solids collected on the drum are removed with a doctor

blade and backwashed using pressure nozzles. Sometimes the spray water is heated to remove grease that may blind the screen openings. The unit shown has openings of 2.5 mm ($\frac{3}{32}$ in.) with a capacity of 7800 gpm (1.8 m³/h). Other fine-screen units are stationary and are installed in the channel flow. Fine-screen units may be installed as part of preliminary treatment in lieu of bar screens or in series with a bar screen protecting preliminary treatment and a fine screen protecting the membranes used in secondary treatment. Placing progressively smaller screens in series reduces the amount of organic material removed at each screening step.

The volume of screenings removed from wastewater varies with the quantity of material in influent wastewater and exponentially with decreased screen size. Plant operators do not normally keep track of the quantity of screenings, therefore data on screening volumes are not readily available. The quantity of screening shown in Figure 11–7 is a rough estimate of the solids removal used to size screening conveyance equipment. Figure 11–7a

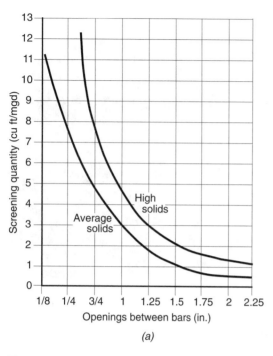

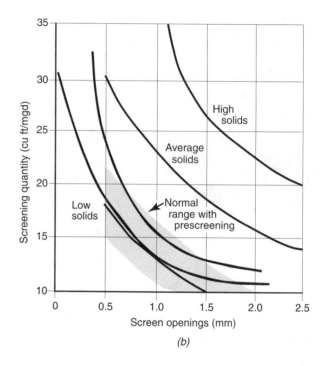

Figure 11–7

Estimates of screening quantity in cu ft/mgd for coarse and fine bar screens based on screen openings and ranges of solids. (a) Screening quantities for bar screen openings from $\frac{1}{8}$ through $2\frac{1}{4}$ in. Lines represent potential screening quantities for average and high solids loadings. (b) Screening quantities for fine screen openings between 1 and 2.5 mm for low, average, and high solids loadings. The grey area highlights the screening quantity when a bar screen is used ahead of the fine screen for prescreening.
(Courtesy of Parkson Corporation)

shows an estimate of screenings removal under average and peak or high solids conditions. Bar sizes range from $\frac{1}{8}$ in. through $2\frac{1}{4}$ in., and the associated screenings removal ranges from 11 ft³/mgd to about 1 ft³/mgd. Removal in fine screens (Figure 11–7b) of 1 to 2.5 mm are shown for low, average, and high solids removal. Screenings removal ranges from 35 to 10 ft³/mgd. Fine screens following coarse screens tend to have removals in the low solids range, as highlighted in gray in Figure 11–7b. Such screening values are used only as a gross approximation of potential screenings removal.

Following removal, screenings contain free water and a varying quantity of organic material. Many landfills place limitations on free water and on the percentage of organic material. Figure 11–8 is an illustration of equipment used to wash and dewater screenings. Spray nozzles break up and wash away organic materials while the auger

and compression pipe squeeze the water out, resulting in a plug of compressed solids suitable for landfill disposal.

A shredder or grinder, used only in small plants, cuts solids in the wastewater passing through the device to about $\frac{1}{4}$ in. (6 mm). Such devices are installed directly in a flow channel and are provided with a by-pass so that the section containing the shredder can be isolated and drained for machine maintenance. A grinder is not often used with a mechanically cleaned bar screen; nevertheless, if installed, it is placed at some point after the screen or in the sludge piping. In a small municipal plant, a shredder can be placed in the flow stream. For this type of arrangement (Figure 11–4), the by-pass channel contains a hand-cleaned bar screen for emergency use. The channels are equipped with stop gates so that wastewater can be directed through the fixed screen while maintenance is performed on the shredder.

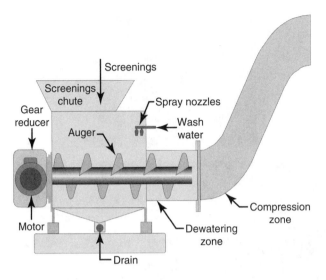

Figure 11–8
Screenings treatment unit to wash organic material and dewater screenings prior to disposal. Screenings enter the chute, are washed, dewatered, and compacted into a solid plug suitable for landfill disposal.
(Drawing adapted from screw wash press manufactured by Parkson Corporation.)

Grit Chambers

Grit includes sand and other heavy particulate matter, such as seeds and coffee grounds, which settle from wastewater when the velocity of flow is reduced. If not removed in preliminary treatment, grit in primary settling tanks can cause abnormal abrasive wear on mechanical equipment and sludge pumps, can clog pipes by deposition, and can accumulate in sludge holding tanks and digesters. Grit chambers are designed to remove particles equivalent to a fine sand, defined as 0.2-mm-diameter particles with a specific gravity of 2.7, with a minimum of organic material included. A variety of systems are employed depending on the quantity of grit in the wastewater, the size of the treatment plant, and the amount of money allocated to installation and operation. Standard chambers include channel-shaped settling tanks, aerated units with hopper bottoms, and forced vortex tanks. Separated grit may be further processed in a screw-type grit washer or cyclone separator.

The earliest grit chambers consisted of two or more long, narrow channels constructed in paral-

lel with space for grit storage. A proportional weir placed at the discharge end controlled the flow such that the horizontal velocity was maintained at about 1.0 ft/sec independent of the quantity of flow. This allowed the grit to settle while providing a scouring velocity to flush through organic suspended solids. Buckets on a continuous chain scrape material from the bottom into a receptacle, thus avoiding hand cleaning.

The most popular type of grit chamber in smaller plants is a hopper-bottomed tank with the influent pipe entering on one side and an effluent weir on the opposite. The chamber is small, with a detention time of approximately 1 min. at peak hourly flow, and is often mixed by diffused aeration to keep the organics in suspension while grit settles out. Solids are removed from the hopper bottom by an air-lift pump, screw conveyor, or centrifugal pump.

Grit clarifiers, sometimes called detritus tanks, are generally square, with influent and effluent weirs on opposite sides. A centrally driven scraper arm pushes the grit into a hopper on the bottom for removal by a screw auger. The clarifier is a shallow tank with a short detention time or a deeper aerated chamber to improve grit separation while freshening the raw wastewater.

Forced-vortex grit units use a mechanically induced centrifugal force to enhance grit capture. Figure 11–9 shows the grit chamber and mechanical paddle used to create the vortex. As the vortex induces solids toward the center, the rotating paddles increase the velocity at the center enough to suspend the lighter organic material and return it to the wastewater stream. Grit settled in the hopper becomes dense and must be scoured with air and fluidized with air and water to be pumped from the lower grit hopper.

Grit processing through dewatering is critical to actual grit removal and protection of equipment. Grit dewatering equipment includes augers, small vortex units installed over augers, and specialty grit classification and dewatering devices. Augers convey the grit up an inclined trough using a helical spiral. Water flows down as grit is conveyed up. The equipment becomes less efficient with age and deterioration of the auger and trough. The equipment in Figure 11–10 is a classifying unit followed by a dewatering screen

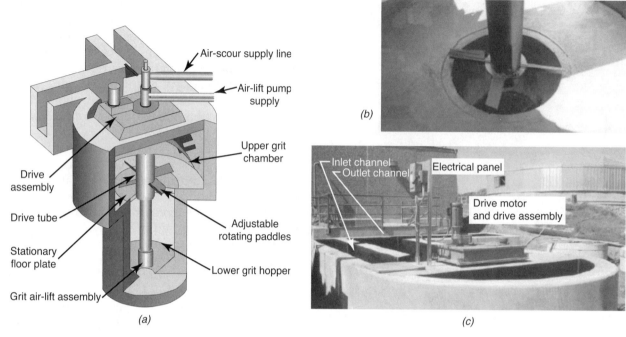

Figure 11–9

Forced vortex unit for removing grit. (a) The vortex suspends organic solids while grit settles in the lower chamber. A grit pump is used to remove settled grit to be washed and dewatered prior to placement in a dumpster for disposal in a landfill. (b) Photo of the upper grit chamber showing the rotating paddles. (c) External view of a grit unit showing the inlet and outlet channels, drive assembly, and electrical panel.
(Courtesy of Parkson Corporation)

for grit processing. The classifying unit uses a vortex to spin the grit to the outside of the free vortex area and down to the grit underflow cone. Wastewater along with organic material flows up through the center discharge orifice and back for treatment. The concentrated grit must again be fluidized by a smaller quantity of water prior to dewatering. The dewatering equipment uses a slow-moving cleated belt to permit water to escape from the grit. The grit slurry enters a quiescent pool maintained by an overflow weir at the lower end of the unit. Dewatering occurs as the grit is carried up to the top of the unit where it is scraped off of the belt cleats into a disposal hopper.

The quantity of grit varies depending on local sewer condition, age, and soil carried with infiltration. Like screenings, grit quantities are not monitored by plant operators. Grit tends to settle in the sewer and the grit load to the treatment plant abruptly increases when storm flows increase sewer velocities, conveying additional grit. Equation 11–1 gives an approximation of the first-flush grit loading used to size grit removal equipment. This concentration is the quantity of grit that can enter the plant, regardless of the type of grit removal equipment. Using a Q equal to the peak day flow produces the grit load used to size grit processing equipment.

$$\text{Grit} = 670Q\left(\frac{\text{PDWW}Q}{\text{AD}Q} - 1\right) \qquad \textbf{(11–1)}$$

where
$$\begin{aligned}
\text{Grit} &= \text{quantity of grit (lb/day)} \\
Q &= \text{wastewater flow (mgd)} \\
\text{PDWW}Q &= \text{peak day wet-weather flow (mgd)} \\
\text{AD}Q &= \text{average daily flow (mgd)}
\end{aligned}$$

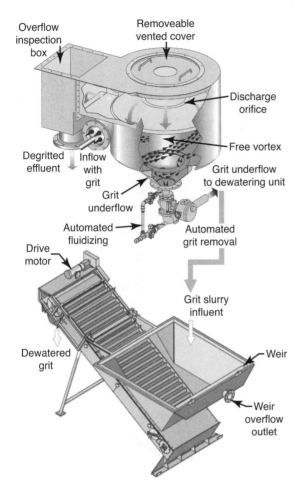

Figure 11–10

Illustration of grit classification and dewatering equipment. Grit separation occurs under a forced vortex with grit settling into a collection cone and grit-free flow forced up to an overflow. Grit dewatering allows free water to flow through cleated steps as grit is conveyed to discharge. (Illustration of grit equipment courtesy of Eutek® Systems™, Inc.)

11–3 PUMPING STATIONS

Pumping stations at large treatment plants have wet and dry chambers separated by a common wall, as shown in Figure 11–11. Raw wastewater, after measurement and screening, flows into the wet well, where the pipe intakes are positioned near the bottom of the chamber to prevent deposition of solids. Automatic controls governing pump operation maintain the water surface between preset levels. At the high mark all pumps are running, except standby units; at the minimum level the controls shut off the last pump before it can draw air into the pump suction. In addition to the influent wastewater, drain and recycle lines return in-plant wastewaters to the wet well. These include supernatant from digesters, filtrate from sludge thickening and dewatering, underflow from secondary clarifiers in trickling filter plants, and building drains throughout the plant.

The effective capacity of a wet well for constant-speed pumps is based on the number of pump starts and the capacity of the smallest pump. For variable-speed pumping stations, the wet well must only be large enough to maintain proper pump control. A small wet well keeps the wastewater mixed and moving through at sufficient velocity to reduce the accumulation of settled and floating solids. Yet, if the detention time is too short, on-and-off cycling increases mechanical wear on the pumps and the temperature of the drive motors. For best performance, with any combination of inflow and pumping, the cycle of operation for each pump should not be more than 6 starts per hour, or about 10 min, and a maximum detention time of wastewater in the wet well of 30 min. To meet these conditions, the selection of individual pumps and operating water levels is coordinated in the design of a wet well.

Centrifugal wastewater pumps located in the dry well can be mounted with either horizontal or vertical driveshafts. Placing motors above the dry well or in waterproof enclosures has the advantages of accessibility for maintenance and protection in case the dry well is accidentally flooded. A separate sump pump is installed in a dry well, but its capacity is limited to removal of leakage or floor drainage that enters the chamber. Pump housings are set below the low-water level in the wet well for automatic priming. Suitable valves are placed on the suction and discharge lines of each pump so that individual pumps can be removed for maintenance. A check valve is essential in the discharge line, between the shutoff valve and the pump, to prevent wastewater from flowing back through the pump when it is not in operation.

Independent mechanical ventilation is required for the dry and wet chambers, even if the latter are not covered. Switches for operation are marked and located conveniently, and if the ventilation system is intermittent rather than continuous, the electrical

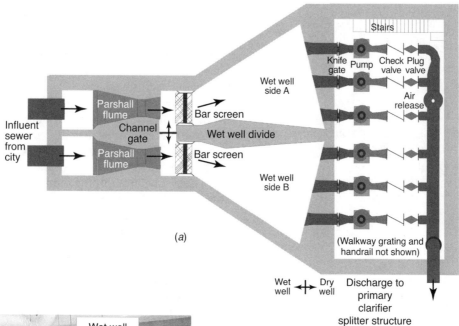

(a)

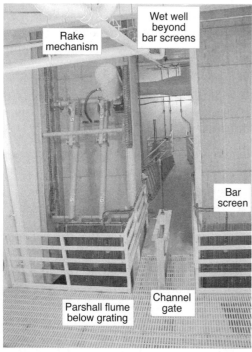

(b)

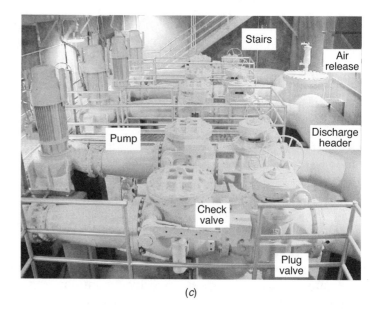

(c)

Figure 11–11

Drawing of a raw wastewater pumping station. (a) Plan view showing separate wet well and dry well. The walls of the wet well are coated with a plastic lining material to prevent corrosion of the concrete walls. The wet well contains flow measurement devices, bar screens, and a split wet well. During low flows, each half of the wet well can be removed from service for cleaning. The dry well contains pumps, valves, and discharge piping. (b) A photo of the bar screens and view toward the wet well. (c) A photo of the dry well, pump motors, pumps, and a check valve and plug valve on each discharge. An air release valve on the discharge allows any entrained air to escape. (Photos taken at the Water Conservation Plant in Visalia, California.)

switches are interconnected with the pit lighting system. Newer designs include sensors that detect the presence of an explosive atmosphere or the absence of oxygen in the wet well. A safe means of access by either a stairway or caged ladder, with rest landings at least every 10 ft, is needed for dry wells and wet wells containing mechanical equipment. Alarm systems are installed to notify plant personnel in cases of power or pump failure or hazardous environment. Emergency power is supplied by using two incoming power lines or an engine-driven generator with automatic switching equipment to transfer the load from one utility system to a standby source in the event of failure.

Centrifugal pumps for lifting raw wastewater are designed for ease of cleaning or repair and are provided with open impellers to reduce the risk of clogging. Most are made with hand holes that can be removed for access to the interior. Freedom from clogging depends on impeller type, casing shape, and operating clearance between the two. Impellers are generally made with only two or three vanes, set between two parallel plates so that any object entering the pump passes through.

The number of pumps installed depends on the station capacity and range of flow. The total pumping capacity is equal to or greater than the maximum rate of inflow with the largest pump out of service. To meet the variations of flow, the minimum number of pumps is three or more of the same or varying capacities. In small stations with a peak flow of less than 1.0 mgd, two units may be installed, with each being able to meet the maximum inflow. The pumps may be either constant-speed or variable-speed. The characteristics of centrifugal pumps are discussed in Section 4–4, and system head curves are explained in Section 4–5.

Constant-Speed Pumps

A typical pumping station lifts wastewater from the wet well for discharge into an open tank, which is usually a grit chamber or distribution structure. The pumping head is determined by the static head (water level at the discharge minus the wet-well level) plus the dynamic head (the friction head loss in the discharge piping). Because the system curves include only the head losses in the piping from the common header to the discharge tank, the pump characteristic curves are modified to account for losses in the individual pump suction and discharge lines. These modified curves are then considered the pump head-discharge curves. The curves sketched in Figure 11–12 are for

Figure 11–12

Head-discharge curves for three constant-speed pumps.

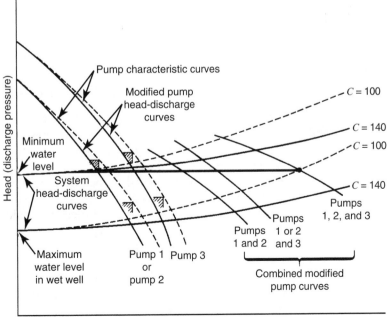

a station with four pumps (all of equal size). The pump characteristic curves have been modified, and the system head-discharge curves bracket the minimum-maximum operating range.

Pumps must operate from today's low flow when the pipes are new to the future maximum flow when the pipes are old. Pumps are selected such that their optimum efficiency point is closest to the normal operating flow and head. In Figure 11–12, two average system head curves were drawn, where the $C = 140$ lines represent when the discharge piping is new and $C = 100$ represents 20-year-old ductile-iron pipe. When pumping is at maximum capacity, the wet-well water level is at the maximum elevation and all three pumps are operating. The maximum resistance to flow is along the lower system head-discharge curve labeled $C = 100$. Therefore, the operating point is at the intersection of the combined pump curve for pumps 1, 2, and 3 and the $C = 100$ system curve. A horizontal line drawn through this intersection locates the operating points on the modified pump curves. Then, by projecting vertically to the pump characteristic curves, the pumping rates at maximum anticipated head are located, as indicated by the shaded angles. At minimum pumping capacity, the wet-well water level is at the maximum elevation and the pumps are operating individually against the lower system head curve at a $C = 140$. Pump operating points are where the modified pump curves cross the $C = 140$ curve. These are projected vertically to locate the minimum pumping rates indicated by the shaded angles on the pump characteristic curves.

Variable-Speed Pumps

The graphical analysis of a typical pumping station with a combination of two variable-speed pumps and one constant-speed pump is shown in Figure 11–13. All three pumps, plus a third variable-speed standby pump, have identical characteristic curves, so they can be operated in parallel in alternating combinations. The speed reduction to discharge at the desired minimum capacity is 87 percent of full speed. The first variable-speed pump starts when the water level in the wet well reaches Start 1 (Figure 11–13a). If the pumping rate exceeds inflow, the water level drops to the minimum level and the pump stops,

Stop 1. On the other hand, if the inflow continues to rise, the pump speed increases to 100 percent. As pumping discharge continues to rise, pump number 2 starts (Start 2) and numbers 1 and 2 operate in parallel at the same speed. When the constant-speed pump number 3 starts, the discharge of the two variable-speed pumps share in pumping the remaining discharge. Finally, at maximum design flow all three pumps are operating at full speed. The straight-line chart in Figure 11–13b relates pump discharge to pump speeds and water levels in the wet well. Because the most efficient pump operation is at the highest speed, keeping a minimum number of pumps in operation is an economic advantage. Thus, with diminishing flow rates, pumps are stopped as soon as possible. To prevent pumps from cycling on and off during inflow reduction, pump speeds are slowed to less than 100 percent at a discharge slightly lower than the inflow rate. The slight rise in water level is then handled by an increase in speed of the operating pumps rather than by restarting the stopped pump. For example, when pump number 3 is stopped, variable-speed pumps numbers 1 and 2 are operating at about 94 percent of full speed. Then as the level rises, these two pumps increase in speed to a value somewhat less than maximum to match pump number 3's stop capacity. The objective is to have a sufficient capacity overlap between stopping and starting of pumps to prevent fluctuation between the starting pumping rate and the stopping rate.

Screw Pumps

A screw pump is a rotating helix conveyor that pushes wastewater up an inclined trough to a higher elevation, as shown in Figure 11–14. The main advantage is the wide range of flows that can be efficiently pumped at a constant speed of rotation. Also, when used to pump return activated sludge, the reduced turbulence compared to centrifugal pumps prevents breakup of the biological floc. Disadvantages include high power consumption at low flows and release of odors due to turbulence. The filling level (Figure 11–14) is the depth in the influent chamber when a screw pump operates at full capacity, best efficiency, and highest power consumption. If the level rises, the

Figure 11–13

Operating diagrams for a wastewater pumping station with two variable-speed pumps and one constant-speed pump controlled by wet-well level. (*a*) System head curve with superimposed diagram of pump operator. (*b*) Straight-line chart relating station discharge to pump speeds and water levels in the wet well.

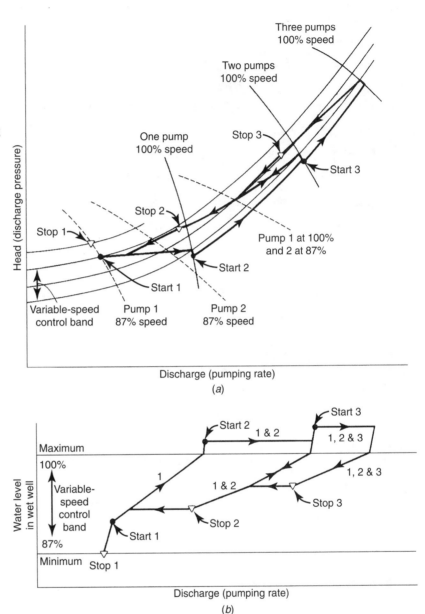

capacity remains the same but efficiency decreases. If the level drops below the filling point, capacity and efficiency are reduced. Some screw pumps are enclosed in pipes to increase pumping efficiency and allow increased pump speeds.

The practical lift heights of screw pumps range from 6 to 30 ft (2 to 10 m), with screw diameters increasing from 12 to 144 in. (0.3 to 3.7 m) in proportion to height. The diameter is related directly to pumping capacity. The best

speed is the highest rate of rotation possible without having wastewater overflow the screw shaft into a lower chamber. Screws are manufactured with one, two, and three flights (single, double, and triple helices). A single flight is a 360° helix in a length of one diameter along the shaft. The discharge capacity of a screw increases with the number of flights. For the same diameter, the capacity of a two-flight screw is approximately 80 percent of that of a three-flight screw.

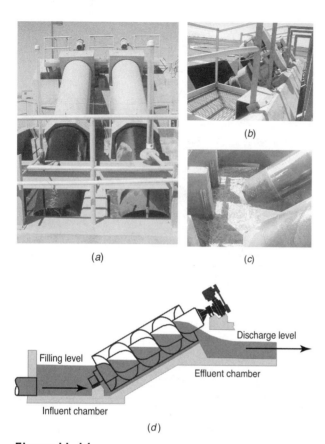

(a)

(b)

(c)

(d)

Figure 11–14

A pumping station using screw pumps to lift wastewater. (a) Photo of enclosed screw pumps. (b) Access to motor drive at top of pump. (c) Influent chamber in operation. (d) Diagram showing screw pump in section. (Photos taken at the Wastewater Treatment Plant in Woodland, California.)

Labels in diagram (d): Filling level, Influent chamber, Discharge level, Effluent chamber

11–4 SEDIMENTATION

Settling, often called clarification, is performed in rectangular or circular tanks where the wastewater is held quiescent to permit particulate solids to settle out of suspension. To prevent short-circuiting and hydraulic disturbances in the tank, flow enters behind a baffle to dissipate inlet velocity. Overflow weirs, placed near the effluent channel, are arranged to provide a uniform effluent flow. The discharge of floating materials with the liquid overflow is prevented by placing a baffle in front of the weir. A mechanical skimmer collects and deposits the scum in a pit outside the tank. Settled sludge is slowly moved toward a hopper in the tank bottom by a collector arm. Clarifiers following activated-sludge aeration may be equipped with hydraulic pickup pipes for rapid sludge return.

The criteria for sizing settling tanks are weir loading rate, overflow rate (surface settling rate), tank depth at the side wall, and detention time. Surface settling rate is defined as the average daily overflow divided by the surface area of the tank, expressed in terms of gallons per day per square foot (cubic meters per square meter per day), Eq. 11–2. The area is calculated by using the inside tank dimensions, disregarding the central stilling well or inboard weir troughs. The quantity of overflow from a primary clarifier is equal to the wastewater influent, since the volume of sludge withdrawn from the tank bottom is negligible. However, secondary settling tanks may have recirculation that draws liquid from the tank bottom, for example, recirculation of activated sludge, in which case the effluent flow is equal to the influent minus the recycle flow. For these, effluent flow is used to calculate overflow, and the influent (which includes recirculation) is used to calculate detention time.

$$V_o = \frac{Q}{A} \qquad (11\text{--}2)$$

where V_o = overflow rate (surface settling rate), gallons per day per square foot (cubic meters per square meter per day)

Q = design daily flow, gallons per day (cubic meters per day)

A = total surface area of tank, square feet (square meters)

Detention time is computed by dividing tank volume by influent flow expressed in hours, Eq. 11–3. Numerically, it is the time that would be required to fill the tank at a uniform rate equivalent to the design flow. The depth of a tank is taken as the water depth at the side wall measuring from the tank bottom to the top of the overflow weir; this excludes the additional depth resulting from the slightly sloping bottom in both circular and rectangular clarifiers. Clarifier depth varies for primary, intermediate, biological filter, and activated-sludge secondary clarifiers. The detention time represents a balance between overflow rate and clarifier depth. Effluent weir loading is equal to the average daily quantity of overflow divided by the total weir length, expressed in gallons per day per linear foot (cubic meters per meter per day).

$$t = 24 \frac{V}{Q} \qquad (11\text{–}3)$$

where t = detention time, hours
V = tank volume, million gallons (cubic meters)
Q = design daily flow, million gallons per day (cubic meters per day)
24 = number of hours per day

Primary Clarifiers

Settling tanks that receive raw wastewater prior to biological treatment are called primary tanks. Figure 11–15 shows a rectangular tank. Raw wastewater enters through a series of ports near the surface along one end of the tank. A short baffle

dissipates the influent velocity, directing the flow downward. Water moves through at a very slow rate and discharges from the opposite end by flowing over multiple weirs around a collection trough. Settled solids are scraped to a sludge hopper at the inlet end by redwood or plastic flights that are attached to endless chains riding on sprocket wheels. Sludge is withdrawn periodically from the sludge hopper for disposal. The upper run of flights protrudes through the water surface, pushing floating matter to a skimmer placed in front of the weirs. The scum trough is a cylindrical tube with a slit opening along the top. When the trough is manually or mechanically rotated, scum collected on the surface flows through the slot into the tube that slopes toward a scum pit. The length-to-width ratio of rectangular tanks varies from about 3:1 to 5:1. Widths range from 10 ft to 20 ft (3 to 6.1 m) based on the maximum span of the flights. Depths are typically 7 ft to 8 ft (2 to 2.5 m). The bottom has a gentle slope toward the sludge hopper.

Sludge pumps draw sludge from the bottom of the clarifiers at preset time intervals. Where gravity

Figure 11–15

A rectangular primary clarifier. (a) Photo from the influent end of the clarifier showing scum pumps, sludge pumps, and associated piping at the wastewater treatment plant (El Dorado Irrigation District, El Dorado Hills WWTP, California). (b) Longitudinal section of the rectangular clarifier. The sludge collector scrapes settled solids to the hopper at the influent end of the clarifier. The upper run of flights pushes floating matter to a scum trough, where it is removed by pumping. At some plants the sludge pumps are used to pump scum by changing the valves isolating the scum and sludge piping.

(a)

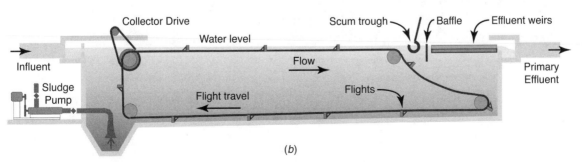

(b)

thickening tanks are not employed, sludge is allowed to accumulate in the clarifiers and is pumped directly to digesters or dewatering equipment. Scum boxes drain to a box connected to the sludge pumps so that the scum can be disposed of with the sludge. Where clarifiers are installed

below ground, sludge pumps would be located in a dry pit that is typically covered to reduce accumulation of debris. In cold climates, pumps are located in building enclosures to protect against freezing.

Views of a circular primary clarifier are shown in Figure 11–16. Raw wastewater enters through

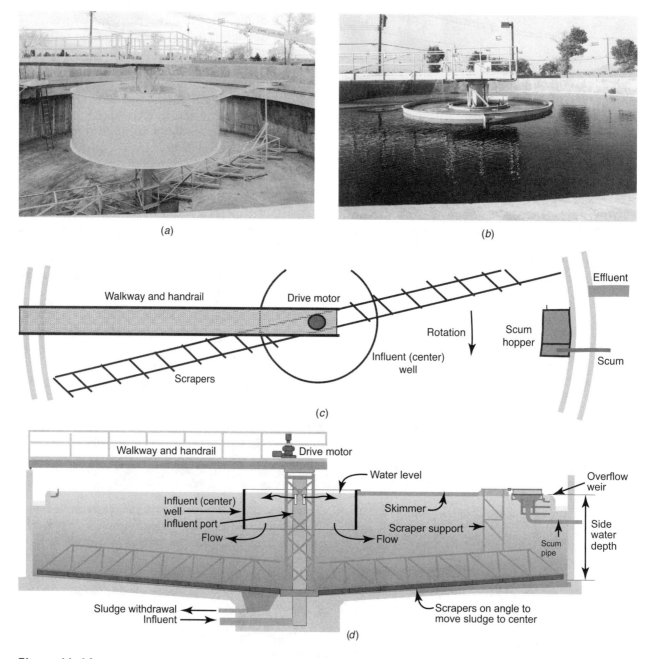

Figure 11–16

Circular primary clarifier with a pier-supported center drive and peripheral effluent weir. (*a*) Photo of empty clarifier. (*b*) Photo of same clarifier filled with wastewater. (*c*) Partial plan view of clarifier. (*d*) Section through clarifier.

ports in the top of a central vertical pipe into a center well baffle, then flows radially to a peripheral effluent weir. The center well directs flow downward to reduce short-circuiting across the top. A very slowly rotating collector arm plows settled solids to the sludge draw-off at the center of the tank. Discharge of floating solids migrating toward the edge of the tank is prevented by a baffle set in front of the weir. A skimmer attached to the arm collects scum from the surface and drops it into a scum box that drains outside the tank wall. Circular tanks are from 30 to 150 ft (9 to 46 m) in diameter, although some are as large as 200 ft (60 m). The center well diameter is 15 to 20 percent of the tank diameter. Side water depths range from 7 to 12 ft (2.1 to 3.7 m), and the bottom slopes are about 8 percent.

Rectangular tanks are popular where space is limited because they can be constructed together using a common wall. Improvements using plastic flights, chain, and stainless steel sprockets have greatly reduced the maintenance required. Circular basins are generally preferred to rectangular tanks in new construction because of improved performance and lower maintenance costs. Bridge- or pier-supported, center-driven collector arms have fewer moving parts than the chain-and-sprocket scraper mechanisms in rectangular tanks. Although inlet turbulence is greater behind the small influent well of a circular clarifier, as the flow radiates toward the effluent weir the wastewater movement slows, thus reducing the exit velocity. Greater weir lengths can be more easily achieved around the periphery of a circular tank than across the end of a rectangular one.

Design criteria for primary clarifiers are listed in Table 11–2. The EPA provides separate values when humus or activated-sludge solids are returned to the primary for removal with the primary solids. These hydraulic loadings achieve 30 to 40 percent BOD removal in settling raw domestic wastewater. The effectiveness of plain sedimentation, of course, depends largely on the character of the wastewater. If a municipal waste contains a large amount of soluble organic matter, BOD removal may drop to less than 20 percent; on the other hand, industrial wastes that contribute settleable solids may increase BOD removal to as high as 60 percent. Hydraulic loading also

TABLE 11–2

Typical Design Criteria for Primary Clarifiers

OVERFLOW RATES—GALLONS/DAY/SQUARE FOOT (gpd/sq ft)

	AVERAGE MONTHLY FLOW	PEAK FLOW
EPA	800–1200	2000–3000
Standards[1]	1000	1500
EPA with secondary solids	600–800	1200–1500

SIDE WATER DEPTH—FEET (ft)

EPA	10–13	
Standards[1]	7	
EPA with Secondary solids	13–16	

WEIR LOADING—GALLONS/DAY/LINEAR FOOT (gpd/ft)

EPA	10,000–40,000	
Standards[1]	10,000	

influences the density of the accumulated sludge. With overflow rates of less than 600 gpd/sq ft (24 m³/m² · d), the settled solids tend to thicken in the bottom of the tank as the collector arm slowly moves through the accumulated sludge. Rates in excess of 800 gpd/sq ft may create hydraulic movements in the tank that inhibit consolidation of the sludge. In addition to a more dilute waste sludge, hydraulic overloading is identified by increased turbidity in the effluent during the period of maximum wastewater flow. Storing sludge in a tank for too long can also upset clarification, particularly if waste is returned to the head of the plant for settling with the raw wastewater. Microorganisms decomposing the waste organics produce gas that makes the solids more buoyant, thus expanding the sludge blanket and reducing the solids concentration. A severe case of detrimental biological activity is characterized by foul odors, floating sludge, and a darkening of the wastewater color.

The liquid depth of mechanically cleaned primary tanks can be as shallow as practicable but not less than 7 ft (2.1 m). Designers sometimes increase the side water depth to provide additional volume for accumulated sludge. Detention time is generally not a specified criterion for sizing primary clarifiers, since it is already defined by overflow rate and depth as related in Eq. 11–4. For example, an overflow rate of 600 gpd/sq ft and depth of 7 ft yield a detention time of 2.1 hr.

$$t = \frac{180 \cdot H}{V_o} \qquad \textbf{(11–4)}$$

where
t = detention time, hours
H = depth of water in tank, feet
V_o = overflow rate, gallons per day per square foot
180 = 7.48 gal/cu ft × 24 hr/day

Weir loading is the hydraulic flow over an effluent weir. For primary tanks, weir loadings are not to exceed 10,000 gpd/foot (125 m³/m · d) for plants of 1 mgd or smaller and preferably not more than 20,000 gpd/foot (250 m³/m · d) for design flows above 1 mgd. These values limit the water velocity approaching the effluent weir to minimize carryover of suspended solids.

Total suspended solids removal in primary clarifiers does not tend to vary directly with overflow rate, but is more related to the characteristics of the influent solids. Practical experience suggests a removal percentage between 70 to 50 percent over an overflow rate range of 600 to 1500 gpd/sq ft (24.4 to 61.1 m³/m² · d). Many clarifiers will maintain a minimum of 50 percent removal efficiency up to 2500 gpd/sq ft (102 m³/m² · d).

Advanced primary removal uses a chemical aid to improve suspended solids removal. The approach is often used to reduce solids loading on subsequent biological treatment processes and temporarily avoid expansion of secondary treatment units. Advanced primary removal is also used to enhance primary treatment performance prior to ocean disposal. Ferric chloride and alum act as coagulant aids to reduce the electrical charge holding solids in suspension. A high-molecular-weight polymer enhances flocculation and solids capture. A dose of 5 to 15 mg/l ferric and 0.1 to 0.15 mg/l polymer will generally increase suspended solids removal from 50 percent to 75 or 90 percent. Polymer doses vary greatly with effectiveness, and overall treatment removal varies because of the differences in solids characteristics, flocculation, and clarifier overflow rates.

■ EXAMPLE 11–1

Two primary settling tanks are 95 ft in diameter with a 9-ft side water depth. Single effluent weirs are located on the peripheries of the tanks. For an average design flow of 10.0 mgd and peak flow of 15.4 mgd, calculate the overflow rate, detention time, and weir loading.

Solution

Calculating surface area and volume,

$$\text{Surface area} = 2\pi r^2 = 2 \times 3.14 \times (95/2)^2$$
$$= 14{,}200 \text{ sq ft}$$
$$\text{Volume} = 14{,}200 \times 7 = 89{,}000 \text{ cu ft}$$
$$= 0.665 \text{ mil gal}$$

.592

From Eq. 11–2,

$$V_o = 7{,}000{,}000/12{,}700$$
$$= 550 \text{ gpd/sq ft (max. monthly } Q\text{)}$$
$$V_o = 14{,}500{,}000/12{,}700$$
$$= 1140 \text{ gpd/sq ft (peak } Q\text{)}$$

By Eq. 11–3,

$$t = (2 \cdot 0.665)/7.0 = 2.3 \text{ hr (max. monthly } Q\text{)}$$

Alternate calculation for t using Eq. 11–4,

$$t = \frac{180 \times H}{V_o} = \frac{180 \cdot 7}{550} = 2.3 \text{ hr}$$

Weir length = circumference of both tanks = $2\pi d$

$$\text{Weir loading} = \frac{7{,}000{,}000}{2 \cdot 3.14 \cdot 90} = 12{,}400 \text{ gpd/ft}$$

■ ■ ■

Intermediate Clarifiers

Sedimentation tanks between trickling filters, or between a filter and subsequent biological aeration, in two-stage secondary treatment are called intermediate clarifiers. The following may be used for sizing intermediate settling tanks: The overflow rate should not exceed 1000 gpd/sq ft $(41 \text{ m}^3/\text{m}^2 \cdot \text{d})$, minimum side water depth is 7 ft, and weir loadings should be less than 10,000 gpd/lin ft for plants of 1 mgd or smaller and should not be over 20,000 gpd/lin ft for larger plants.

Secondary Clarifiers

Settling tanks following biological filters are similar to primary clarifiers illustrated in Figure 11–16. Common criteria for secondary clarifiers of trickling filter plants are overflow rate not exceeding 800 gpd/sq ft $(33 \text{ m}^3/\text{m}^2 \cdot \text{d})$, minimum side water depth of 7 ft, and maximum weir loadings the same as for intermediate tanks, although lower values are preferred.

The purpose of gravity settling following a trickling filter is to collect biological growth, or humus, flushed from filter media. These sloughed solids are generally well-oxidized particles that settle readily. Therefore, a collector arm that slowly scrapes the accumulated solids toward a hopper for continuous or periodic discharge gives satisfactory performance. The depth of accumulated sludge in a trickling filter final is normally a few inches if recirculation flow is drawn from the tank bottom. Even if sludge is drained only twice a day, the blanket of settled solids rarely exceeds 1 ft.

The clarifier in Figure 11–17 is specially designed for activated-sludge systems. Gravity separation of biological growths suspended in the mixed liquor of aeration systems is more difficult. The greater viability of activated sludge results in lighter, more buoyant flocs with reduced settling velocities. In part, this is the result of microbial production of gas bubbles that buoy up the tiny biological clusters. The accumulated microbial floc in a final basin for separating activated sludge may be 1 to 2 ft thick in a well-operating plant. During peak loading periods, the sludge blanket may expand further to occupy one-third to one-half of the tank volume; this is particularly true in high-rate aeration systems. The wastewater flow pattern is the same as that of other circular clarifiers, but the sludge collection system is unique. Suction pipes are attached to and spaced along a V-plow-scraper mechanism rotated by a turntable above the liquid surface. The discharge elevation of the suction riser pipes in the sight well is lower than the water surface in the tank so that sludge is forced up and out of the uptake pipes by water pressure. The flow from each suction pipe is controlled by adjusting the elevation of a slip tube over the riser pipe. The sludge collected in the sight well flows out through a vertical pipe

Figure 11–17
Secondary clarifier designed for use with biological aeration. Activated sludge is withdrawn through suction pipes located along the collector arm for rapid return to the aeration basin.

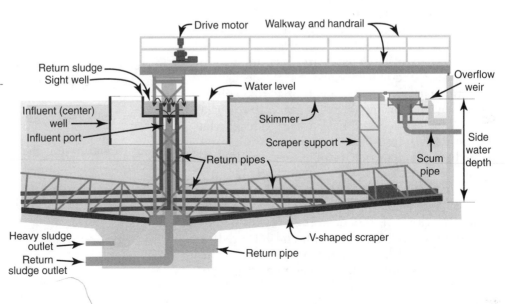

centrally located in the influent wastewater pipe. A seal between the sight well and the wastewater riser allows the well to rotate around with the collector arm without leaking sludge into the tank. The unit illustrated is referred to as a sight well clarifier; other manufacturers produce similar units using such descriptive titles as a rapid-sludge-removal clarifier. Each proprietary unit has features that may be of importance in the operation of a specific biological aeration system. For example, one mechanism separates the sludge collected by the uptake pipes from the heavier solids, which are plowed to a sludge hopper. The collected sludge is returned as activated sludge while the solids scraped from the bottom are wasted.

Rapid uniform withdrawal of sludge across the entire bottom of an activated-sludge clarifier has two distinct advantages. The retention time of solids that settle near the tank's periphery is not greater than those that land near the center; thus, aging of the biological floc and subsequent floating solids due to gas production is eliminated. With a scraper collector, the residence time of settled solids depends directly on the radial distance from the sludge hopper. The second advantage is that the direction of activated-sludge return flow is essentially perpendicular to the tank bottom, rather than horizontal toward a centrally located sludge hopper. Downward flow through a sludge blanket enhances gravity settling of the floc and increases sludge density. This is an important factor when one considers that the return flow may be as great as one-half of the influent flow.

Design parameters for clarifiers in activated-sludge processes take into account the reduced settleability of a flocculent biological suspension. Compared to other wastewater sedimentation tanks, activated-sludge clarifiers are deeper to accommodate the greater depth of settled solids, have a lower overflow rate to reduce carryover of light biological floc, and have longer weir lengths by the installation of an inboard weir channel to reduce the approach velocity of the effluent. Typical overflow rates are 600 gpd/sq ft (24 m^3/m$^2 \cdot$ d) for plants smaller than 1 mgd and 800 gpd/sq ft (33 m^3/m$^2 \cdot$ d) for larger plants. During the peak hydraulic flow of the day, the overflow rate should not exceed 1200 gpd/sq ft and 1600 gpd/sq ft for small and larger plants, respectively. The recommended minimum side water depth is 10 ft

(3.1 m), with greater depths for larger-diameter tanks, for example, 11 ft for 50-ft diameters and 12 ft for 100-ft diameters. Depending on the values selected for overflow rate and depth, detention time is in the range of 2.0 to 3.0 hr. The maximum recommended weir loading is 10,000 to 20,000 gpd/ft (125 to 250 m^3/m $\cdot$ d).

■ EXAMPLE 11–2

Determine the recommended size of a new circular secondary clarifier for an activated-sludge system with a design flow of 26,000 m^3/d with a peak hourly flow of 32,000 m^3/d. Use maximum overflow rates of 33 m^3/m$^2 \cdot$ d at design monthly flow and 66 m^3/m$^2 \cdot$ d at peak hourly flow.

Solution

At design flow, the surface area required for each tank is

$$\text{Area} = \frac{20{,}000 \text{ m}^3/\text{d}}{2 \cdot 33 \text{ m}^3/\text{m}^2 \cdot \text{d}} = 303 \text{ m}^2$$

$$\text{Peak overflow rate} = \frac{32{,}000}{2 \cdot 303}$$
$$= 53 \text{ m}^3/\text{m}^2 \cdot \text{d} < 66 \text{ m}^3/\text{m}^2 \cdot \text{d}$$
$$\text{(OK)}$$

$$\text{Tank diameter} = \left(\frac{4 \cdot 303}{\pi}\right)^{0.5} = 19.6 \text{ m} = 64 \text{ ft}$$

The recommended side water depth for a tank diameter greater than 50 ft is 11 ft. Therefore,

$$\text{Tank side water depth} = 11 \text{ ft} = 3.4 \text{ m}$$

Use an inboard weir channel set on a diameter of 18 m.

$$\text{Weir loading} = \frac{10{,}000 \text{ m}^3/\text{d}}{2\pi \cdot 18 \text{ m}}$$
$$= 88 \text{ m}^3/\text{m} \cdot \text{d} < 125 \text{ m}^3/\text{m} \cdot \text{d} \quad \text{(OK)}$$

■ ■ ■

11–5 BIOLOGICAL FILTRATION

Fixed-growth biological systems are those that contact wastewater with microbial growths attached to the surfaces of supporting media. Where the wastewater is distributed over a bed of crushed rock, the unit is commonly referred to as a trickling filter. With the development of synthetic media used in place of stone, the term *biological tower* was introduced, since these installations are often 14 ft to 20 ft in depth rather than the traditional 6-ft stone-media filter. Another type of fixed-growth system is a rotating biological contractor, where a series of circular plates on a common shaft are slowly rotated while partly submerged in wastewater. Although the physical structures differ, the biological process is essentially the same in all of these fixed-growth systems.

Domestic wastewater sprinkled over fixed media produces biological slimes that coat the surface. The films consist primarily of bacteria, protozoa, and fungi that feed on waste organics. Sludge worms, fly larvae, rotifers, and other biota are also found, and during warm weather sunlight promotes algae growth on the surface of a filter bed. Figure 11–18 portrays the biological activity.

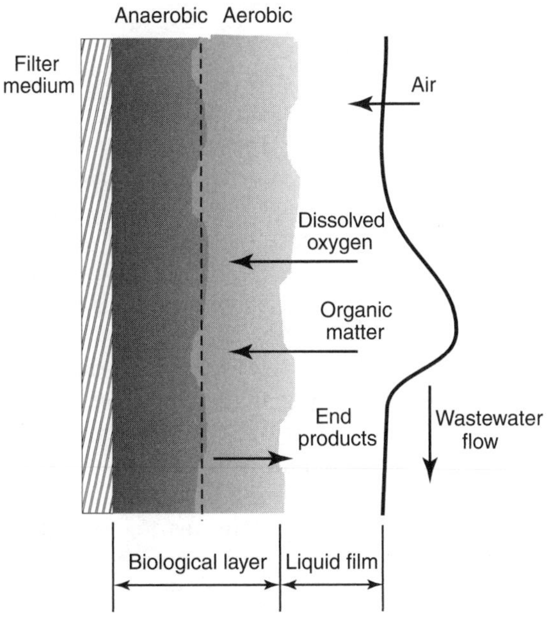

As the wastewater flows over the slime layer, organic matter and dissolved oxygen are extracted, and metabolic end products such as carbon dioxide no, are released. Dissolved oxygen in the liquid is replenished by absorption from the air in the voids surrounding the filter media. Although very thin, the biological layer is anaerobic at the bottom. Therefore, although biological filtration is commonly referred to as aerobic treatment, it is in fact a facultative system incorporating both aerobic and anaerobic activity.

Organisms attached to the media in the upper layer of a bed grow rapidly, feeding on the abundant food supply. As the wastewater trickles downward, the organic content decreases to the point where microorganisms in the lower zone are in a state of starvation. Thus, the majority of BOD is extracted in the upper 2 or 3 ft of a 6-ft filter. Excess microbial growth sloughing off the media is removed from the filter effluent by a secondary clarifier. Organic overload of a stone-media filter, in combination with insufficient hydraulic flow, can result in plugging of passages with biological growth, causing ponding of wastewater on the bed, reduced treatment efficiency, and foul odors from anaerobic conditions.

Stone-Media Trickling Filters

Views of trickling filters are shown in Figure 11–19. The major components are a rotary distributor, underdrain system, and filter media. Influent wastewater is pumped up a vertical riser to a rotary distributor for spreading uniformly over the filter surface. Rotary arms are driven by reaction from the wastewater flowing out of the distributor nozzles. Bed underdrains carry away the effluent and permit circulation of air. Ventilation risers and the effluent channel are designed to permit free passage of air. In some installations, the underdrain blocks empty into a channel between double exterior walls to allow improved aeration and access for flushing of underdrains.

The most common media in existing filters are crushed rock, slag, or field stone that are durable, insoluble, and resistant to spalling. The size range preferred for stone media is 3- to 5-in. diameter. Although smaller stone provides greater surface area for biological growth, the voids tend to plug and limit passage of liquid and air. Bed depths

Figure 11–18

Sketch illustrating the biological process in a filter bed.

Figure 11–19

Photos of stone-media trickling filter.
(a) Cutaway view of trickling filter having concrete side walls. (b) External view of trickling filter without side walls showing tile underdrain. (Photo taken at the El Dorado Hills wastewater treatment plant.)

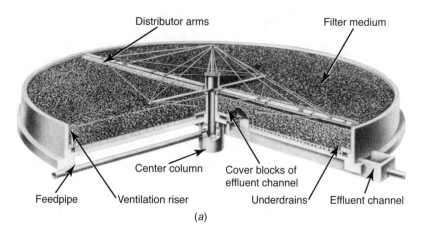

(a)

(b)

range from 5 to 7 ft; greater depths do not materially improve BOD removal efficiency. Stone-media filters in the treatment of municipal wastewaters are always preceded by primary settling to remove larger suspended solids.

BOD load on a trickling filter is calculated using the BOD in the primary effluent applied to the filter, without regard to any BOD contribution in the recirculated flow from the clarifier, Eq. 11–5. Hydraulic loading is the amount of liquid applied to the filter surface including both untreated wastewater and recirculation flows, Eq. 11–6. This surface loading is commonly expressed in units of million gallons per acre of surface area per day (mgad) or gallons per minute per square foot (cubic meters per square meter per day). The recirculation ratio, calculated by Eq. 11–7, is the ratio of recirculated flow to the wastewater entering the treatment plant.

$$\text{BOD loading} = \frac{\text{settled wastewater BOD}}{\text{volume of filter media}} \quad (11\text{–}5)$$

where   BOD loading = pounds of BOD applied per 1000 cu ft per day (grams per cubic meter per day)

settled BOD = wastewater BOD remaining after primary sedimentation, pounds per day (grams per day)

volume of media = volume of stone in the filters, thousands of cubic feet (cubic meters)

$$\text{Hydraulic loading} = \frac{Q + Q_R}{A} \quad (11\text{–}6)$$

where hydraulic loading = million gallons per acre per day (cubic meters per square meter per day)

Q = wastewater flow, million gallons per day (cubic meters per day)

Q_R = recirculation flow, million gallons per day (cubic meters per day)

A = surface area of filters, acres (square meters)

$$R = \frac{Q_R}{Q} \qquad (11\text{--}7)$$

where R = recirculation ratio
Q_R and Q= same as above

Typical loadings for stone-media filters are listed in Table 11–3. Filter plants return sufficient flow from the clarifier hopper to the wet well for removal of accumulated settled solids and to prevent stalling of the distributor arm during low wastewater flow. Also, in-plant recirculation of wastewater increases liquid flow through the filter bed to allow greater organic loading without filling the bed voids with biological growths that would inhibit aeration. Experience has shown that BOD loadings in excess of about 25 lb/1000 cu ft/day (400 g/m$^3 \cdot$d) require a minimum hydraulic flush-

ing of 10 mil gal/acre/day to keep a stone-filled bed open. In addition, BOD removal efficiency is enhanced by passing wastewater through a filter more than once.

Numerous recirculation patterns are applied in existing filter plants. The most popular new design is gravity return of underflow from the secondary clarifier to the wet well during periods of low wastewater flow and direct recirculation by pumping filter discharge back to the filter influent, as diagramed in Figure 11–20. Flow to the filter is shown by the lines in Figure 11–20b. Recycling of clarifier underflow is generally limited to a recirculation ratio of about 0.5, since this is adequate to return settled solids for removal in the primary clarifier and to maintain adequate flow for turning the distributor arm. Yet, by limiting this return flow, the peak outflow rate of the primary clarifier is not increased. Direct recirculation has the advantage of influencing neither the primary nor secondary settling tank; the disadvantage is that a separate pumping station is required.

A two-stage trickling filter consists of two filter-clarifier units in series (Figure 11–21); sometimes the intermediate settling tank is omitted. This type of system is needed to achieve an effluent BOD of 30 mg/l when processing a wastewater in cold climates or stronger-than-average domestic wastewater. Both filters are normally constructed the same size for economy and optimum operation. Generally, several options for recirculation are incorporated in design. For example, in Figure 11–21, underflow from the intermediate and clarifiers is returned to the wet well for solids disposal, while direct recirculation around each filter stage is possible. Direct recirculation may be pumped from the bottom of a settling tank or directly from the influent well (filter discharge).

Several mathematical equations have been developed for calculating the BOD removal efficiency of biological filters based on such factors as depth of bed, kind of media, temperature, recirculation, and organic loading. The exactness of these formulas depends on the filter media having a uniform biological layer and an evenly distributed hydraulic load. These conditions rarely occur in stone-media filters, which may develop unequal biological growths, resulting in short-circuiting of wastewater through the bed. Therefore, general practice has been to use empirical relationships

TABLE 11–3

Typical Loadings for Trickling Filters with a 5- to 7-ft Depth of Stone or Slag Media

	High Rate	Two-Stage
BOD loading		
lb/1000 cu ft/day[a]	30 to 90	45 to 70
lb/acre-ft/day	1300 to 3900	2000 to 3000
Hydraulic loading		
mil gal/acre/day[b]	10 to 30	10 to 30
gpm/sq ft	0.16 to 0.48	0.16 to 0.48
Recirculation ratio	0.5 to 3.0	0.5 to 4.0

[a]1.0 lb/1000 cu ft/day = 16.0 g/m$^3 \cdot$d
[b]1.0 mil gal/acre/day = 0.935 m^3/m$^2 \cdot$d

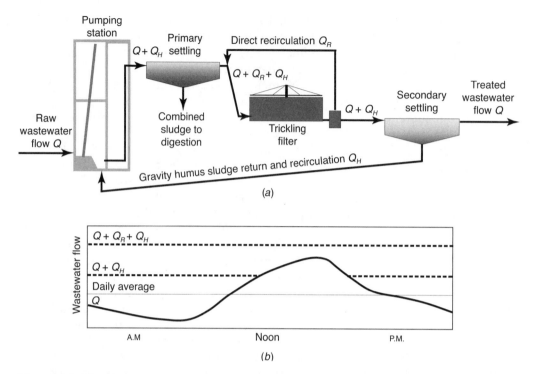

Figure 11–20

Single-stage trickling filter plant. (a) Profile of a single-stage trickling filter showing related wastewater flow diagrams including in-plant recirculation. (b) General flow patterns: Q = wastewater influent flow, $Q + Q_H$ = influent plus humus return from the bottom of the clarifier, and $Q + Q_R + Q_H$ = flow to the filter with direct and indirect recirculation.

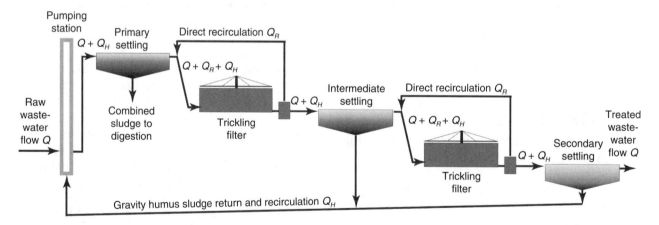

Figure 11–21

Typical flow diagram for a two-stage trickling filter plant with intermediate clarifier.

based on operational data collected from existing treatment plants. One of the most popular formulations evolved from National Research Council (NRC) data that were collected from filter plants at military installations in the United States during the early 1940s. The results, graphed in Figure 11–22, are considered applicable to single-stage stone-media trickling filters followed by a settling tank and treating settled domestic wastewater with a temperature of 28°C (68°F). Efficiency

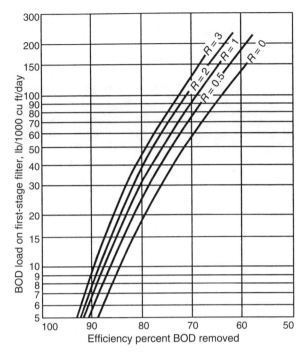

Figure 11–22

Efficiency curves for a single-stage stone-media trickling filter treating domestic wastewater at 20°C, based on National Research Council data.

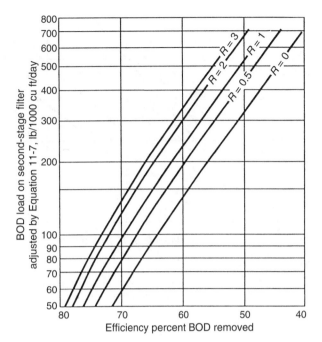

Figure 11–23

Efficiency curves for a second-stage stone-media trickling filter domestic wastewater at 20°C, based on National Research Council data. Adjusted for BOD load for entering diagram as calculated using Eq. 11–7.

is read along the bottom scale below the point where the horizontal load line intersects the curved line with the proper recirculation ratio. For example, at a load of 40 lb BOD/1000 cu ft/day and a recirculation ratio of 0, the efficiency is 74 percent; with the same BOD load and $R = 3$, efficiency is 81 percent.

A second-stage filter is less efficient than the first stage because of the decreased treatability of the waste fraction applied to the second bed. In other words, the most available biological food is taken out first, passing the organics that are more difficult to remove through to the second-stage filter. Based on NRC observations, this effect can be incorporated by increasing the actual load to the second-stage filter as shown in the following relationship:

$$
\frac{\text{Second-stage BOD load adjusted for treatability}}{\text{actual second-stage BOD load}}
$$
$$
= \frac{1}{\left[\dfrac{(100 - \text{percentage of first-stage efficiency})}{100}\right]^2}
$$

$$(11\text{–}8)$$

The second-stage BOD load is calculated by Eq. 11–5, where the settled wastewater is taken as the overflow from the intermediate clarifier. After adjusting for treatability by Eq. 11–8, the efficiency of the second stage can be determined from Figure 11–23. The overall treatment plant efficiency of a two-stage filter system can be calculated by Eq. 11–9:

$$
E = 100 - 100\left[\left(1 - \frac{35}{100}\right)\left(1 - \frac{E_1}{100}\right)\left(1 - \frac{E_2}{100}\right)\right]
$$
$$(11\text{–}9)$$

where E = treatment plant efficiency, percent

   35 = percentage of BOD removed in primary settling

   E_1 = BOD efficiency of first-stage filter and intermediate clarifier corrected for temperature, percent

   E_2 = BOD efficiency of second-stage filter and final clarifier corrected for temperature, percent

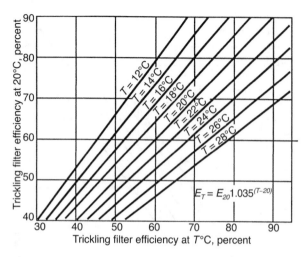

Figure 11–24

Diagram for correcting BOD removal efficiency from NRC data at 20°C to efficiency at other temperatures between 12°C and 28°C.

Figure 11–25

The trickling filter in the background has been covered for odor control. The odor control scrubber to the left of the filter connects with a duct for ventilation. In northern climates, trickling filters are covered to reduce heat loss and maintain removal efficiency during winter conditions. (Photo taken at the Las Vegas Water Pollution Control Facility.)

BOD removal in biological filtration is influenced significantly by wastewater temperature. Filters in northern climates operate at efficiencies about 5 percent or more below the yearly average during winter months. The plot in Figure 11–24 can be used to adjust efficiencies from Figures 11–22 and 11–23 for temperatures above or below 20°C. Filter covers may be required in cold climates to achieve the effluent standard of 30 mg/l BOD or less. An installation of the type shown in Figure 11–25 protects the filters from wind and snow, thus preventing excess cooling of wastewater sprayed over the rocks. Positive ventilation is provided to maintain passage of air through the bed and to dissipate corrosive gases, namely, hydrogen sulfide. Odor control may be required to reduce trickling-filter odors (organic compounds such as mercaptans).

■ EXAMPLE 11–3

The design flow for a two-stage trickling filter plant illustrated in Figure 11–21 is 12 mgd with an average BOD concentration of 220 mg/l. Calculate the unit loadings and treatment plant efficiency at a wastewater temperature of 16°C. Unit processes were sized as follows:

Primary clarifiers	= 4 at 80 ft, 7 ft swd
First-stage filter	= 4 at 110 ft dia., 6 ft deep
Intermediate clarifiers	= 2 at 90 ft, 7 ft swd
Second-stage identical to first	
Secondary clarifiers	= 4 at 80 ft, 8 ft swd

The recirculation pattern is as shown in Figure 11–21: The return to the wet well is 0.5 mgd (average flow) underflow from each clarifier for total Q_R = 0.6 mgd; and direct recirculation around each filter Q_R is 0.8 mgd. Q_R and Q_H are $\frac{1}{2}$ influent Q.

Solution

Primary Settling Tank
V_o includes Q_H average return flow:

$$A = 4\pi (80/2)^2 = 20,100 \text{ sq ft}$$

$$V_o = (12 + 0.5) \cdot 10^6/20,100 = 620 \text{ gpd/sq ft}$$

Given V_o within design limits and BOD removal of 35 percent,

$$\text{BOD of settled effluent} = 0.65 \cdot 220 = 143 \text{ mg/l}$$

First-Stage Trickling Filter

$$\text{Volume} = 4\pi(110/2)^2 \cdot 6 = 228,000 \text{ cu ft}$$

$$\text{BOD load} = \frac{12 \text{ mgd } (220 \text{ mg/l}) (1 - 0.35)8.34}{228 \text{ thousand cu ft}}$$

$$= 62.8 \text{ lb/1000 cu ft/day}$$

$$\text{Hydraulic load} = \frac{12 + 6 + 0.5 \text{ mgd}}{0.873 \text{ acres}}$$

$$= 21.2 \text{ mgad} \qquad \text{(OK)}$$

From Figure 11–22, at 63 lb/1000 cu ft/day and $Q_R = 6.5/12 = 0.54$ or 54 percent at 20°C, efficiency is 74 percent. Efficiency corrected for 16°C from Figure 11–24 is 64 percent.

Intermediate Clarifier

$$A = 2 \times \frac{90^2 \pi}{4} = 12,700 \text{ sq ft}$$

$$V_o = \frac{(12 + 0.5) \cdot 10^6 \text{ gal}}{12,700 \text{ sq ft}} = 980 \text{ gpd/sq ft}$$

$$\text{Effluent BOD} = (1 - 0.64)143 = 51.5 \text{ mg/l}$$

Second-Stage Filter and Secondary Clarifier

$$\text{BOD load} = \frac{12 \text{ mgd} \cdot 51.5 \text{ mg/l} \cdot 8.34}{228 \text{ thousand cu ft}}$$

$$= 22.6 \text{ lb/1000 cu ft/day}$$

From Eq. 11–8,

$$\text{Adjusted BOD load} = \frac{22.6}{(1 - 0.64)^2}$$

$$= 170 \text{ lb/1000 cu ft/day}$$

$$\text{Hydraulic load} = \frac{12 + 6 + 0.5 \text{ mgd}}{0.873 \text{ acres}}$$

$$= 21.2 \text{ mgad}$$

Efficiency from Figure 11–23 at 170 lb/1000 cu ft/day and $Q_R' = 6.5/12 = 0.54$ is 62 percent. Efficiency corrected for 17° from Figure 11–24 is 54 percent.

$$\text{Clarifier area} = 4 \cdot \frac{80^2 \pi}{4} = 20,100 \text{ sq ft}$$

$$V_o = \frac{12 \times 10^6 \text{ gpd}}{20,100 \text{ sq ft}} = 600 \text{ gpd/sq ft}$$

$$\text{Plant effluent BOD} = (1 - 0.54) \cdot 51.5 = 24 \text{ mg/l}$$

Total Plant BOD Removal Efficiency at 17°C

$$E = \frac{(220 - 24)}{220} \cdot 100 = 89 \text{ percent}$$

■ ■ ■

Biological Towers

Several kinds of plastic media are manufactured for trickling filters. The main advantage relative to stone media is the high specific surface (surface area per unit of volume) with a corresponding high percentage of void volume, which permits substantial biological slime growth without inhibiting the passage of air to supply oxygen. Other advantages are uniform media for better liquid distribution, light weight allowing construction of deep filters, chemical resistance, and the ability to treat high-strength wastewaters.

The common kinds of media for shallow filters are random packing, placed like stone media, and high-density cross-flow modules that can be fitted into a circular tank. A typical random packing is small cylinders with perforated walls and internal ribs manufactured of plastic with dimensions of 2 to 4 in. The specific surface is 30 to 40 ft²/ft³ (100 to 130 m²/m³) with a void volume of 91 to 94 percent. The packing is lightweight and placed by dumping in the filter tank on top of the underdrains. Because of the random placement, the filter is effective in distributing the applied wastewater to the media surfaces as it trickles down through the bed.

Modules of packing with various internal configurations are manufactured of polyvinyl chloride

(PVC) in bundles 2 ft wide, 4 ft long, and 2 ft high. The module in Figure 11–26a has corrugated sheets bonded between flat sheets to prevent clear vertical openings and distribute the wastewater over the surfaces of the media. The specific surface varies with manufacturer from 26 to 43 ft^2/ft^3 (85 to 140 m^2/m^3), and the void space is about 95 percent. The cross-flow module in Figure 11–26b is constructed of ridged corrugated sheets with the ridges on adjacent sheets at 45° or 60° angles to each other and bonded together where the ridges contact. As the wastewater flows down through

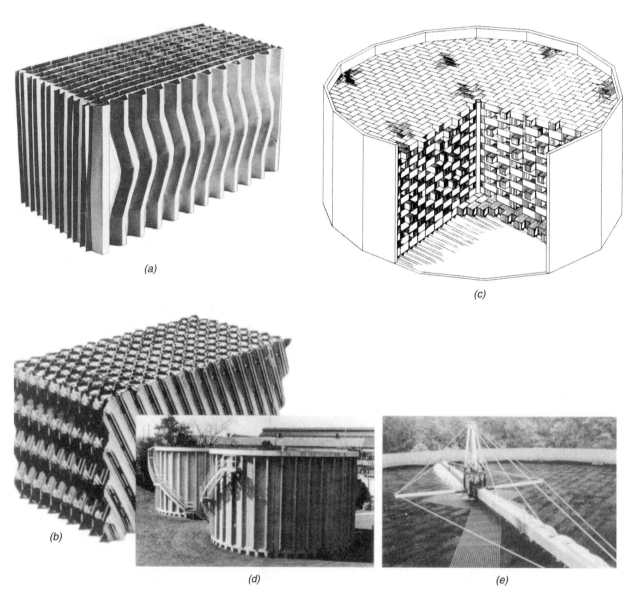

(a)

(b)

(c)

(d)

(e)

Figure 11–26

Biological tower media and construction. (a) Vertical-flow module with corrugated sheets bonded between flat sheets. (b) Cross-flow module with corrugated sheets assembled with adjacent sheets in a crossed pattern. (c) Cutaway view showing the construction of a circular biological tower. (d) Picture of two biological towers. (e) Distributor arm spreading wastewater on surface of tower media.

(Courtesy of TLB Corporation, Newington, CT.)

the media, each contact point permits flow splitting and combining. This cross-flow pattern provides better wetting of the surfaces and slows the downward rate of flow, resulting in increased hydraulic residence time in the bed. The specific surface of low-density cross-flow media is 27 ft²/ft³ (90 m²/m³) and of high-density is 42 ft²/ft³ (140 m²/m³). The strength of the modules is adequate to support the packing with attached wet biological growth in towers of 20 ft in height. The media bundles are stacked to interlock for structural stability and can be cut to fit the edge modules in a circular tower equipped with a rotary distributor (Figure 11–26c, d, and e).

The hydraulic profile of a plant with a biological tower is shown in Figure 11–27. Primary clarification is required to remove settleable and floating solids prior to filtration. Direct recirculation of the tower underflow is blended with the clarified raw wastewater to provide dilution and a greater flow through the media. In this way, treatment is improved by vertical distribution of the BOD load throughout the depth of the bed and passage of the wastewater through the filter more than once.

BOD loadings on biological towers are usually 50 lb/1000 cu ft/day (800 g/m³·d) or greater with surface hydraulic loadings of 1.0 gpm/sq ft (60 m³/m²·d, 0.68 l/m²·s) or greater. The design loading selected for treatment of a particular wastewater depends on BOD concentration, biodegradability, temperature, kind of modules, depth of media, and the ratio and pattern of wastewater recirculation.

Efficiency equations have been developed for plastic filters, since the uniformity of the packing can be defined by the specific surface. However, no universal formulas exist that can precisely describe the removal of organic matter, partially because of the different geometric shapes of the media. The packing configuration influences the residence time of the liquid in the bed, which is in turn related to hydraulic loading and filter depth. Furthermore, removal of organic matter depends directly on its solubility. In theoretical equations, the BOD is the filtered (soluble) BOD, not the total BOD as is normally measured.

The removal of soluble BOD in a filter with random or modular packing based on first-order kinetics is

$$\frac{S_e}{S_0} = e^{-k_{20}A_sD/Q^n} \qquad (11\text{–}10)$$

where S_e = soluble (filtered) BOD in effluent, mg/l

S_0 = soluble (filtered) BOD in effluent, mg/l

e = 2.718 (the Napierian base)

k_{20} = reaction rate coefficient at 20°C, (gpm/ft²)$^{0.5}$ [(l/m²·s)$^{0.5}$]

A_s = specific surface area of media, ft²/ft³ (m²/m³)

D = depth of media, ft (m)

Q = hydraulic loading, gpm/ft² (l/m²·s)

n = empirical flow constant (normally selected as 0.5 for vertical-flow and cross-flow media)

The reaction-rate coefficient is corrected for temperature by the following equation:

$$k = k_{20}\theta^{T-20} \qquad (11\text{–}11)$$

Figure 11–27

Profile of a biological tower with direct recirculation to the tower and indirect recirculation of humus from the bottom of the clarifier to the wet well.

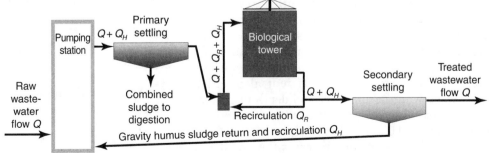

where k = reaction-rate coefficient at temperature T, °C

k_{20} = reaction-rate coefficient at 20°C

θ = temperature coefficient, normally
selected as 1.035

T = wastewater temperature, °C

The BOD concentrations of the applied wastewater before and after dilution with clear recirculation flow are given by

$$S_0 = \frac{S_p + RS_e}{1 + R} \qquad (11\text{--}12)$$

where S_0 = soluble BOD in influent after dilution with recirculated flow, mg/l

S_p = soluble BOD in primary effluent
before dilution with recirculated flow,
mg/l

S_e = soluble BOD in effluent, mg/l

R = recirculation ratio recirculated flow/
primary effluent flow (Q_R/Q_P)

Combining Eqs. 11–10, 11–11, and 11–12 will
result in

$$\frac{S_e}{S_p} = \frac{e^{-k_{20}\theta^{T-20}A_sD/[Q_p(1+R)]^n}}{(1 + R) - Re^{-k_{20}\theta^{T-20}A_sD/[Q_p(1+R)]^n}} \qquad (11\text{--}13)$$

where S_e = soluble (filtered) BOD in effluent,
mg/l

S_p = soluble BOD in primary effluent
before dilution with recirculated
flow, mg/l

Q_p = hydraulic loading of primary effluent
without recirculation flow, gpm/ft^2

These equations are a simplification of the
complex biological-physical interactions that
occur in trickling filters. Values for the reaction-
rate coefficient k_{20} vary with media of different
configurations even though the specific surface
areas are the same. The coefficient also varies
with wastewater treatability and depth of media.
The approximate range for k_{20} values for vertical-
flow media is 0.0008 to 0.0016 (gpm/ft^2)$^{0.5}$ [0.0010
to 0.0020 (l/m^2 · s)$^{0.5}$] and for cross-flow media is

0.0012 to 0.0023 (gpm/ft^2)$^{0.5}$ [0.0017 to 0.0028
(l/m^2 · s)$^{0.5}$]. The k_{20} values for cross-flow media
are greater than those for vertical-flow media,
which is attributed to longer contact time and better wastewater flow distribution. Often manufacturers provide field data to verify reaction-rate
coefficients for their media; however, where feasible, pilot studies using selected media are recommended to determine the k_{20} for design.

■ EXAMPLE 11–4

A single-stage trickling filter plant consists of a
primary clarifier, a trickling filter 70 ft in diameter
with a 7-ft depth of random packing, and a secondary clarifier. The primary effluent has a soluble (filtered) BOD of 100 mg/l. The temperature of
the wastewater is 15°C. The constants for the random plastic media are n = 0.44, k_{20} = 0.0026
(gpm/ ft^2)$^{0.5}$, and A_s = 35 ft^2/ft^3. Calculate the
effluent-soluble BOD assuming indirect recircula-
tion to the wet well of 0.40 mgd.

Solution

Filter area = 3.14 × (35)^2 = 3850 sq ft

$$Q_p = \frac{800{,}000 \text{ gal/day}}{1440 \text{ min/day} \cdot 3850 \text{ sq ft}} = 0.144 \text{ gpm/ft}^2$$

$$R = 0.4/0.8 = 0.5$$

The exponent for e in Eq. 11–13 equals

$$\frac{-0.0026 \cdot (1.035)^{15-20} \cdot 35 \cdot 7}{[0.144 \,(1 + 0.5)]^{0.44}} = -1.05$$

Substituting into Eq. 11–13,

$$\frac{S_e}{S_p} = \frac{e^{-1.05}}{(1 + 0.5) - 0.5 \times e^{-1.05}} = 0.26$$

Soluble effluent BOD = 2.06 × 100 mg/l

= 26 mg/l

■ ■ ■

■ EXAMPLE 11–5

A biological tower has a diameter of 24.3 ft (A = 464 sq ft) and media depth of 20 ft. The packing is cross-flow modules with a k_{20} = 0.0018 (gpm/ft^2)$^{0.5}$, A_s = 42 ft^2/ft^3, and n = 0.50. The settled wastewater flow is Q_p = 0.50 mgd with a BOD = 162 mg/l, S_0 = 80 mg/l, and temperature = 15°C. The recirculation is R = 1.0 for a media wetting rate of 0.75 gpm/ft^2. Calculate the soluble BOD loading and unfiltered BOD loading in pounds per thousand cu ft of packing per day. Calculate the soluble-effluent BOD S_e, in milligrams per liter, and estimate the unfiltered BOD assuming that 50 percent of the effluent BOD is soluble.

Solution

Volume of packing = 464 ft^2 · 20 ft = 9280 ft^3

$$\text{Soluble BOD loading} = \frac{0.50 \text{ mgd} \cdot 80 \text{ mg/l} \cdot 8.34}{9.28}$$

$$= 73 \text{ lb/1000 cu ft/day}$$

The exponent for e in Eq. 11–13 equals

$$\frac{-0.0018 \, (1.035)^{15-20} \cdot 42 \cdot 20}{[0.75(1 + 1.0)]^{0.50}} = -1.20$$

Substituting into Eq. 11–13,

$$\frac{S_e}{S_p} = \frac{e^{-1.20}}{(1 + 1.0) - 1.0 \times e^{-1.20}} = 0.12$$

Soluble effluent BOD = 0.12 · 80 mg/l = 10 mg/l

Effluent BOD = 2 · 10 = 20 mg/l

■ ■ ■

Combined Filtration and Aeration Process

Biological-tower filtration can be combined with second-stage aeration to improve treatment, particularly for municipal wastewaters with variable strength resulting from contributions of high-strength industrial or seasonal wastes. These processes can be operated with various flow patterns, as illustrated in the flow diagram in Figure 11–28.

One method is to operate the filter and aeration tank as independent sequential processes. The biological tower is operated as a roughing filter by direct recirculation for adequate hydraulic loading, with the filtered wastewater passed on to the second stage. The aeration process can be operated as plain aeration to enhance settleability of the biological solids in the filter effluent or as an

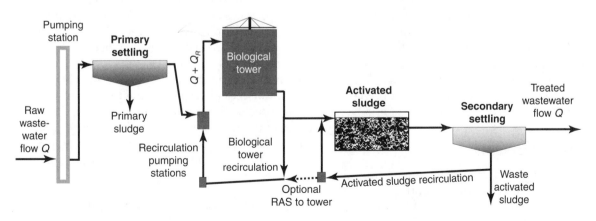

Figure 11–28

Profile of the combined filtration and aeration process with direct recirculation through the tower and recirculation of activated sludge from the bottom of the secondary clarifier. Recirculation of activated sludge to the tower is done where raw wastewater strength is high and the tower contains plastic media.

activated-sludge process by return of settled biological floc from the secondary clarifier.

An alternative flow pattern recirculates a portion of the settled biological floc from the clarifier to the filter influent to form an activated sludge that recycles through both the filter and aeration tank. This process is referred to by several names: activated biological filtration (ABF), activated biofiltration–activated sludge, and trickling filtration–solids contact process. This composite system has the characteristics of both fixed-growth and suspended-growth processes. The mixed liquor in the aeration tank enhances removal of nonsettleable and dissolved BOD in the filter effluent. Because of good process stability, a high-quality effluent can be consistently produced. In a way, the biological filter in this process acts as a static aerator; therefore, by increasing the hydraulic loading by recirculation, the transfer of oxygen is increased. However, if excess sludge solids are recirculated through the biological filter, the oxygen demand exceeds the transfer capability of the media and limits BOD removal.

Equations for the combined filtration and aeration process have not been successfully formulated because of the difficulty in analyzing the system. The following are ranges of various design and operational parameters of these treatment systems: percentage of the total volume (filter media plus aeration tank) that is aeration tank volume, 40 to 75 percent; system BOD loading based on total volume, 20 to 50 lb/1000 cu ft/day (320–800 g/m^3 · d); biological tower loading, 50 to 110 lb/1000 cu ft/day (800–1800 g/m^3 · d); system food-to-microorganism ratio, 0.2 to 1.0 with the median at 0.5; and effluent BOD and effluent suspended solids, 5 to 25 mg/l. Operating personnel at most plants believe the system is better able to absorb shock loads and more stable than the conventional activated-sludge process.

Operational Problems

Two major problems of stone-media trickling filters are effluent quality and odors; both are associated with organic loading, industrial wastes, and cold-weather operation. The average BOD removal efficiency of a single-stage filter plant is about 85 percent. Therefore, to achieve an effluent BOD of 30 mg/l, the raw wastewater must be essentially a domestic waste with a BOD not greater than 200 mg/l. If a municipal wastewater contains significant industrial waste contributions, a two-stage filtration system is necessary to meet the required effluent standards. In northern climates, the temperature of wastewater passing through the bed may be considerably lower in the winter and may adversely influence BOD removal. Covers can be placed over trickling filters to help maintain the temperature during cold weather.

The microbial zone immediately adjacent to the surface of the media is anaerobic and capable of producing metabolic end products that have offensive odors. Reduced compounds formed in treating domestic wastewaters, such as hydrogen sulfide, appear to be oxidized as they move through the aerobic zone with adequate aeration. Forced-air ventilation with an air scrubbing tower to remove the odors from exhausted air may be required to maintain adequate air passage through the bed and to prevent a corrosive atmosphere under the dome.

Biological towers using synthetic media are less susceptible to the operational difficulties of quality control and odors than are stone-media beds. This is primarily attributed to improved aeration and hydraulic distribution of the wastewater. Nevertheless, potential odor problems and the influence of cold weather must be considered in the design and operation of treatment systems that employ biological towers.

Filter flies, *Psychoda*, are a nuisance problem near filters during warm weather. They breed in sheltered zones of the media and on the inside surfaces of the retaining walls. Although wind can carry these small flies considerable distances, their greatest irritation is to operating personnel. Periodic spraying of the peripheral area and walls of the filter with an insecticide is a common method of fly control. Filter snails are a maintenance problem in warm climates. Snails grow on the sidewalls and filter media. Snail shells accelerate the wear on pumps and accumulate in downstream pipes and tanks.

11–6 BIOLOGICAL AERATION

Understanding aeration processes in wastewater treatment requires a greater knowledge of biology than that required for biological filtration.

Figure 11–29

Generalized biological process in aeration (activated-sludge) treatment.

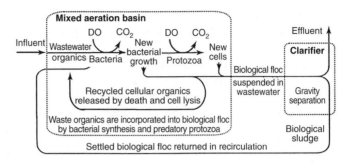

Aeration Tank Loadings

Therefore, the reader is encouraged to review the discussions on bacteria and protozoa presented in Sections 3–1 and 3–2 and population dynamics in Section 3–11.

The generalized biological process that takes place in an aeration system is sketched in Figure 11–29. Raw wastewater flowing into the aeration basin contains organic matter (BOD) as a food supply. Bacteria metabolize the waste organics, producing new growth while taking in dissolved oxygen and releasing carbon dioxide. Protozoa graze on bacteria for energy to reproduce. Some of the new microbial growth dies, releasing cell contents to solution for resynthesis. After the addition of a large population of microorganisms, aerating raw wastewater for a few hours removes organic matter from solution by synthesis into microbial cells. Mixed liquor is continuously transferred to a clarifier for gravity separation of the biological floc and discharge of the clarified effluent. Settled floc is returned continuously to the aeration basin for mixing with entering raw wastewater.

The liquid suspension of microorganisms in an aeration basin is generally referred to as a mixed liquor, and the biological growths are called mixed liquor suspended solids (MLSS). The name *activated sludge* was originated in referring to the return biological suspension, since these masses of microorganisms were observed to be very "active" in removing soluble organic matter from solution. This extraction process is a metabolic response of bacteria in a state of endogenous respiration, or starvation. The activated-sludge process is truly aerobic because the biological floc is suspended in mixed liquor containing oxygen.

An activated-sludge process is defined by the aeration period, BOD loading per unit volume, food-to-microorganism ratio, and sludge age. The aeration period is calculated in the same manner as detention time. Using Eq. 11–3, V is the volume of the aeration tank and Q the quantity of wastewater entering, without regard to recirculation flow. BOD load is usually expressed in terms of pounds of BOD applied per day per thousand cu ft of liquid volume in the aeration tank (grams per cubic meter per day). This can be calculated by Eq. 11–5, where V is the liquid volume of the aeration basins rather than volume of the trickling-filter media.

The food-to-microorganism ratio (F/M) is a way of expressing BOD loading in proportion to the microbial mass in the system. Equation 11–14 calculates the F/M value as BOD applied/day/unit mass of MLSS in the aeration tank.

$$\frac{F}{M} = \frac{Q \times \text{BOD}}{V \times \text{MLSS}} \qquad (11\text{–}14)$$

where F/M = food-to-microorganism ratio, pounds of BOD per day per pound of MLSS (grams of BOD per day per gram of MLSS)

Q = wastewater flow in million gallons per day (cubic meters per day)

BOD = wastewater BOD, milligrams per liter (grams per cubic meter)

V = liquid volume of aeration tank, million gallons (cubic meter)

MLSS = mixed liquor suspended solids in the aeration basin, milligrams per liter (grams per cubic meter)

Some authors express the food-to-microorganism ratio in terms of mass of BOD/day/unit mass of MLVSS (mixed liquor volatile suspended solids).

Sludge age, also referred to as mean cell residence time, is an operational parameter related to the F/M ratio. While liquid retention times (aeration periods) vary from 3 to 30 hr, the residence time of biological solids in a system is much greater and is measured in terms of days. In other words, while wastewater passes through aeration only once and rather quickly, the resultant biological growths and extracted waste organics are repeatedly recycled from the clarifier back to the aeration tank. Several formulations have been developed to compute a value for sludge age. The most common establishes sludge age on the basis of the mass of MLSS in the aeration tank relative to the mass of suspended solids in the wastewater effluent and waste sludge.

$$\text{Sludge age} = \frac{\text{MLSS} \times V}{\text{SS}_e \times Q_e + \text{SS}_w \times Q_w} \quad (11\text{--}15)$$

where Sludge age = mean cell residence time, days

$\quad$ MLSS = mixed liquor suspended solids, milligrams per liter

$\quad V$ = volume of aeration tank, million gallons (cubic meters)

$\quad \text{SS}_e$ = suspended solids in wastewater effluent, milligrams per liter

$\quad Q_e$ = quantity of wastewater effluent, million gallons per day (cubic meters per day)

$\quad \text{SS}_w$ = suspended solids in waste sludge, milligrams per liter

$\quad Q_w$ = quantity of waste sludge, million gallons per day (cubic meters per day)

Sludge age can also be expressed in terms of volatile suspended solids, founded on the argument that the volatile portion of suspended solids is more representative of biological mass. For facilities operating with normal domestic wastewater, the volatile solids fraction is about 70 percent of the total suspended solids.

BOD loading per unit volume and aeration period are interrelated parameters dependent on the concentration of BOD in the wastewater entering and the volume of the aeration tank. (For example, if a 200-mg/l wastewater flows into a tank with an aeration period of 24 hr, the resulting BOD load is 12.5 lb/1000 cu ft/day. If the tank volume is reduced to an aeration period of 8 hr, the BOD load would be tripled to 37.5 lb/1000 cu ft/day). The F/M ratio is an expression of BOD loading relating to the metabolic state of the biological system rather than to the volume of the aeration tank. The advantage of this expression is that pounds of BOD/day/pounds of MLSS defines an activated-sludge process without reference to aeration period or strength of applied wastewater. Two quite different systems may operate at the same F/M ratio. For example, an extended aeration process with a 24-hr aeration period and MLSS concentration of 1000 mg/l would have an F/M of 0.2 for an influent wastewater with a BOD of 200 mg/l. A conventional activated-sludge system with an 8-hr aeration period can treat this same wastewater at the same F/M ratio by operating with a higher MLSS of 3000 mg/l.

A summary of volumetric BOD loadings, F/M ratios, sludge ages, aeration periods, return sludge rates, and typical BOD removal efficiencies is provided in Table 11–4. The concentration of MLSS maintained in an aeration tank varies with the kind of process and the method of operation. In general, less than 1000 mg/l does not provide a low enough F/M for good sludge settleability, and over 4000 mg/l results in loss of suspended solids in the clarifier overflow. The common range is 2500 to 3500 mg/l. The rate of return sludge from the clarifier to the aeration basin is expressed as a percentage of the raw wastewater influent. For example, if the return activated-sludge rate is 30 percent and the raw wastewater flow equals 3.0 mgd, BOD efficiency is calculated by dividing the quantity of BOD removed in aeration and subsequent settling by the raw BOD entering.

Sludge Settleability

The effectiveness of treatment achieved in an aeration process depends directly on settleability of the biological floc that agglomerates and settles by gravity in the clarifier, leaving a clear supernatant

TABLE 11–4

Summary of Loadings and Operational Parameters for Aeration Processes

Process	BOD Loading (lb BOD/day per 1000 cu ft)[a]	MLSS (mg/l)	F/M Ratio (lb BOD/day per lb MLSS)[b]	Sludge Age (days)	Aeration Period (hr)	Return Sludge Rates (percent)	BOD Removal Efficiency (percent)
Conventional	20 to 40	1000 to 3000	0.2 to 0.5	5 to 15	4.0 to 7.5	20 to 40	80 to 90
Step aeration	40 to 60	1500 to 3500	0.2 to 0.5	5 to 15	4.0 to 7.0	30 to 50	80 to 90
Extended aeration	10 to 20	2000 to 8000	0.05 to 0.2	20 and up	20 to 30	50 to 100	85 to 95
High-purity oxygen	120 and up	4000 to 8000	0.6 to 1.5	3 to 10	1.0 to 3.0	30 to 50	80 to 90

[a] 1.0 lb/1000 cu ft/day = 16.0 g/m³ · d

[b] 1.0 lb/day/lb MLSS = 1.0 g/d · g MLSS

for discharge. Conversely, poorly flocculated particles (pin floc) or buoyant filamentous growths that do not separate by gravity contribute to BOD and suspended solids in the process effluent. Excessive carryover of floc resulting in inefficient operation is referred to as sludge bulking. This may be caused by adverse environmental conditions that are created by insufficient aeration, lack of nutrients, presence of toxic substances, or overloading.

Settleability of a biological sludge, under a normal operating environment, depends on the food-to-microorganism ratio and sludge age. Figure 11–30 graphically relates *F/M* values to sludge settleability. Extended aeration systems with long aeration periods and relatively high MLSS concentrations operate in the endogenous growth phase. This yields high BOD efficiency, since starving microorganisms effectively scavenge the organic matter and flocculate readily under quiescent conditions. At the opposite extreme, high-rate aeration employs a high *F/M* ratio, which allows a greater volumetric BOD loading and a reduced aeration period. The resulting activated sludge has reduced settleability that can result in suspended solids in excess of 40 mg/l in the effluent. All of the aeration processes listed in Table 11–4, except for high rate, can consistently produce an effluent with less than 30 mg/l of BOD and suspended solids if properly designed and operated. The *F/M* range between 0.2 and 0.5 appears to be optimum

for achieving the efficiency needed for treating municipal wastewater. Volumetric loadings in this range are great enough to allow aeration periods of 5 to 7 hr, thus permitting economical construction and operating costs.

Effects of Temperature

Effluent quality from well-operated activated-sludge processes in the BOD loading range of 30 to 50 lb BOD/day/1000 ft³ can reliably meet the secondary standards of average maximum BOD of

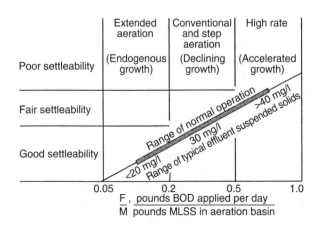

Figure 11–30

Approximate relationship between activated-sludge settleability and operating food-to-microorganism ratio.

30 mg/l and suspended solids of 30 mg/l with the temperature of mixed liquor at 10° to 20°C (50 to 68°F). At loadings lower in the listed range, or mixed liquor temperature in the upper range, and with a low clarifier overflow rate, the effluent quality is more likely in the range of 10 to 20 mg/l suspended solids. Biological activity doubles or halves for every 10 to 15°C temperature change. Therefore, for processes in the loading range of 30 to 50 lb BOD/day/1000 ft³, a decrease or increase in mixed liquor temperature has less influence on changes of effluent quality.

The selection of aeration equipment is to some extent dictated by this relationship between allowable BOD loading and operating temperature. In a cool climate, submerged diffused aeration is common to reduce cooling of the mixed liquor in winter operation. Surface aerators do not create adverse temperature effects. The effectiveness of clarification also fluctuates as wastewater temperature decreases. Settling rates decrease, but the greatest impacts are from thermal stratification within the clarifier that reduce the effective clarification volume and increase short-circuiting. Solids in greater number pass over the weir and into the plant effluent.

Conventional and Step Aeration

These processes are similar to the earliest activated-sludge systems that were constructed for secondary treatment of municipal wastewater. As diagrammed in Figure 11–31, the aeration basin is a long rectangular tank with air diffusers along the bottom for oxygenation and mixing. In a conventional basin, the air supply is tapered along the length of the tank to provide greatest aeration at the head end, where raw wastewater and return activated sludge are introduced. Air is provided uniformly in step aeration, while wastewater is introduced at intervals, or steps, along the first portion of the tank.

Air diffusers are set on the bottom of the tank or attached to pipe headers along one side at a depth of 8 ft or more to provide deep mixing and adequate oxygen transfer. Fine-bubble diffusers, in the form of aeration domes or plates mounted over channels in the floor, are distributed over the entire bottom of the tank to provide uniform vertical mixing. The smaller bubble size and

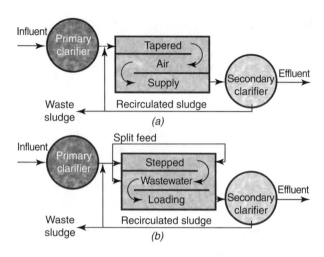

Figure 11–31

Flow schemes for conventional and step-aeration activated-sludge processes. *(a)* Conventional tapered aeration. *(b)* Step feed to aeration basin.

increased floor coverage improve oxygen transfer and decrease air requirements and blower size. In spiral-flow aeration, a large number of diffuser tubes or nozzles are attached to air headers along one side of the tank. A number of different kinds are manufactured, including perforated tubes with or without removable sleeves, jet nozzles, and a variety of air spargers. These produce coarser bubbles than a ceramic diffuser. The air headers are connected to a jointed arm so that the diffusers can be swung out of the tank for cleaning and maintenance.

The plug-flow pattern of long rectangular tanks produces an oscillating biological growth pattern. The relatively high food-to-microorganism ratio at the head of the tank decreases as mixed liquor flows through the aeration basin. Since the aeration period is 5 to 8 hr, and can be considerably greater during low flow, the microorganisms move into the endogenous growth phase before their return to the head of the aeration basin. This starving microbial population must quickly adapt to a renewed supply of waste organics. The process has few problems of instability where wastewater flows are greater than 0.5 mgd; however, because of wide hourly variations in waste loads from small cities, the conventional plug-flow system can experience serious problems of biological instability. This phenomenon was a major factor contributing to the development of completely mixed aeration for handling small flows.

■ EXAMPLE 11–6

A conventional activated-sludge plant without primary clarification operates under the following conditions:

Design flow = 2.14 mgd
Influent BOD = 185 mg/l
Suspended solids = 212 mg/l
Aeration basins: 4 units, 40 ft square × 15.5 ft deep
Mixed liquor suspended solids = 2600 mg/l
Recirculation flow = 1.0 mgd
Waste sludge quantity = 39,000 gpd
Suspended solids in waste sludge = 8600 mg/l
Effluent BOD = 15 mg/l, suspended solids
= 15 mg/l

Calculate the following: aeration period, BOD loading, F/M ratio, suspended solids and BOD removal efficiencies, sludge age, and return activated-sludge rate.

Solution

Aeration basin volume = $4 (40^2) \cdot 15.5 \cdot (7.481/10^6)$

$$= 0.74 \text{ mil gal}$$

Aeration basin d_t = 0.74 mil gal/2.14 mgd $\cdot$ 24

$$= 8.3 \text{ hr}$$

$$\text{BOD load} = \frac{2.14 \text{ mgd} \cdot 185 \text{ mg/l} \cdot 8.34}{4 \cdot 40^2 \cdot 15.5/1000 \text{ cu ft}}$$

$$= 33.3 \frac{\text{lb/day}}{1000 \text{ cu ft}}$$

$$\frac{F}{M} = \frac{2.14 \text{ mgd} \cdot 185 \text{ mg/l} \cdot 8.34}{0.74 \text{ mil gal} \cdot 2600 \text{ mg/l} \cdot 8.34}$$

$$= 0.21 \frac{\text{lb BOD/day}}{\text{lb MLSS}}$$

Suspended solids removal

$$= \frac{212 - 15}{212} \cdot 100 = 93 \text{ percent}$$

BOD removal efficiency

$$= \frac{185 - 15}{185} \cdot 100 = 92 \text{ percent}$$

Suspended solids in the effluent

$$= 2.14 \text{ mgd} \cdot 15 \text{ mg/l} \cdot 8.34 = 268 \text{ lb/day}$$

Suspended solids in waste activated sludge

$$= 0.039 \text{ mgd} \cdot 8600 \text{ mg/l} = 2800 \text{ lb/day}$$

$$\text{Sludge age} = \frac{0.74 \text{ mil gal} \cdot 2600 \text{ mg/l} \cdot 8.34}{2800 + 268 \text{ lb/day}}$$

$$= 5.2 \text{ days}$$

$$\text{Return sludge rate} = \frac{1.0}{2.14} \cdot 100 = 48 \text{ percent}$$

■ ■ ■

High-Purity Oxygen

The high-purity oxygen system differs from aeration by the use of an enriched oxygen supply and a specially compartmented aeration tank. The oxygen supply is generated either by manufacturing liquid oxygen or by producing high-purity gas by adsorption separation from air. Standard cryogenic air separation, involving liquefaction of air followed by fractional distillation to separate the major components of nitrogen and oxygen, is employed at large installations. For the majority of treatment plants, the simpler pressure-swing adsorption system is more efficient. Air is compressed in a vessel filled with a granular adsorbent that takes in carbon dioxide, water, and nitrogen gas under high pressure, leaving a gas relatively high in oxygen. The adsorber is regenerated by depressurizing and purging. Three vessels are installed for continuous flow of high-purity oxygen, with two units alternating between operation and regeneration and the third used as a standby or backup unit.

The aeration tank is divided into stages by means of baffles and covered with a gas-tight enclosure; thus, the liquid and gas phases flow concurrently through the sections (Figure 11–32). Raw wastewater, return activated sludge, and oxygen gas under a slight pressure are introduced to the first stage. Oxygen may be mixed with the

Figure 11–32
Schematic section of a high-purity oxygen aeration system.

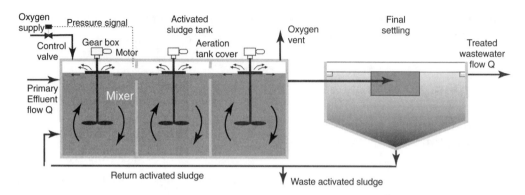

tank contents by recirculation through a hollow shaft to a rotating sparger device, as illustrated in Figure 11–32, or a surface aerator may be installed at the top of the mixer turbine shaft to contact oxygen gas with the mixed liquor. Successive aeration chambers are connected to each other so that liquid flows through submerged ports and head gases pass freely from stage to stage with only a slight pressure drop. Exhausted waste gas is a mixture of carbon dioxide, nitrogen, and about 10 to 20 percent of the oxygen applied. Effluent mixed liquor is settled in a rapid sludge-return clarifier, and activated sludge is recirculated to the aeration tank.

Several advantages are attributed to high-purity oxygen aeration compared with air systems. As is shown in Table 11–4, high efficiency is possible at increased BOD loads and reduced aeration periods. Disposal of waste sludge is easier because of the higher solids content and smaller quantity produced; the settled sludge generally contains 1 to 2 percent solids. Effective odor control is easier to achieve with oxygen aeration, covered tanks, and the reduced volume of exhaust gases.

■ EXAMPLE 11–7

A high-purity oxygen aeration system, as diagramed in Figure 11–32, is being considered for treatment of a combined domestic and industrial wastewater. Since the combined wastewater is high in soluble BOD and low in suspended solids, primary clarification is not included in the processing scheme. The design flow is 3000 m³/d with an average BOD of 300 mg/l. From previous experience in the aerobic treatment of similar

wastewater, the design F/M is 0.60 g/d of BOD per gram of MLSS and the operating MLSS is 4000 mg/l. The maximum volumetric BOD loading is 2.5 kg/m³·d, minimum aeration period is 2.0 h, and minimum recommended sludge age is 3.0 d. Based on these criteria, determine the volume of the aeration tank.

Solution

Rearranging Eq. 11–13,

$$V = \frac{Q \cdot \text{BOD}}{F/M \cdot \text{MLSS}} = \frac{3000 \text{ m}^3/\text{d} \cdot 300 \text{ mg/l}}{0.6 \text{ g/d} \cdot \text{g} \cdot 4000 \text{ mg/l}}$$

$$= 375 \text{ m}^3$$

$$\text{BOD loading} = \frac{3000 \text{ m}^3/\text{d} \cdot 300 \text{ g/m}^3}{375 \text{ m}^3 \cdot 1000 \text{ g/kg}}$$

$$= 2.4 \text{ kg/m}^3 \cdot \text{d} \qquad \text{(OK)}$$

$$\text{Aeration period} = \frac{375 \text{ m}^3 \cdot 24 \text{ h/d}}{3000 \text{ m}^3/\text{d}}$$

$$= 3.0 \text{ h} > 2.0 \text{ h} \qquad \text{(OK)}$$

Assuming excess suspended solids production of 0.4 g of MLSS per gram of BOD applied,

$$\text{Estimated sludge age} = \frac{375 \text{ m}^3 \cdot 4000 \text{ g/m}^3}{0.4 \cdot 3000 \text{ m}^3/\text{d} \cdot 300 \text{ g/m}^3}$$

$$= 4.2 \text{ d} > 3 \text{ d} \qquad \text{(OK)}$$

■ ■ ■

Extended Aeration

The most popular application of this process is in treating small flows from schools, subdivisions, trailer parks, and villages. The cross-sectional diagram of a common design is given in Figure 11–33. Aeration basins may be cast-in-place concrete or steel tanks fabricated in a factory. Continuous complete mixing is either by diffused air or by mechanical aerators, and aeration periods are 24 to 36 hr. Because of these conditions, as well as low BOD loading, the biological process is very stable and can accept intermittent loads without upset. For example, a unit serving a school may receive wastewater during a 10-hr period each day, for only 5 days a week.

Clarifiers for small plants are conservatively sized with low overflow rates, ranging from 200 to 600 gpd/sq ft (8 to 24 m³/m² · d) and long detention times. Sludge is returned to the aeration chamber by an air-lift pump for positive process control (Figure 11–33). Sludge that floats to the surface of the sedimentation chamber is returned to the aeration tank by a skimming device attached to the air-lift pump return.

Usually no provision is made for wasting of excess activated sludge from small extended-aeration plants. Instead, the mixed liquor is allowed to increase in solids concentration over a period of several weeks and then is removed directly from the aeration basin. This is performed by allowing the suspended solids to settle in the tank with the aerators off and then pumping the concentrated sludge from the bottom into a vehicle for hauling away. The MLSS operating range varies from a minimum of 1000 mg/l to a maximum of about 10,000 mg/l. In treating domestic wastewater under normal loading, the mixed liquor concentration increases at the rate of approximately 30 to 50 mg/l suspended solids per day.

Larger extended-aeration plants consisting of an aeration tank, clarifier, and aerobic sludge digester are used by small municipalities. The aeration unit may be a concrete tank with diffused aeration, a lined earthen basin with mechanical aerators, or a channel with horizontal rotor aeration. Final clarifiers are usually separate circular concrete tanks with mechanical sludge collectors. Aerated holding tanks are used to digest and store waste-activated sludge prior to disposal. The oxidation ditch plant illustrated in Figure 11–34 is popular because of its high efficiency and easy operation. The aeration tank is usually an elongated oval with a liquid depth of 4 to 16 ft. Depending on the size of the plant, the outside channel walls may be sloping or vertical, and the median may be either a wide strip with sloping walls or a dividing wall with flow-guide baffles in the channel. The rotor aerator oxygenates the mixed liquor while

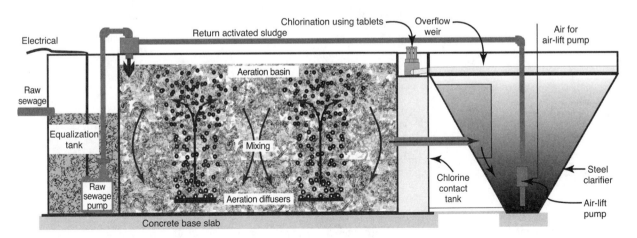

Figure 11–33

Cross-sectional diagram illustrating the operation of a typical small extended-aeration plant with diffused aeration and an air-lift pump for return of settled solids and scum to the aeration tank. Raw sewage equalization and effluent chlorination are built into the same structure as the aeration tank.

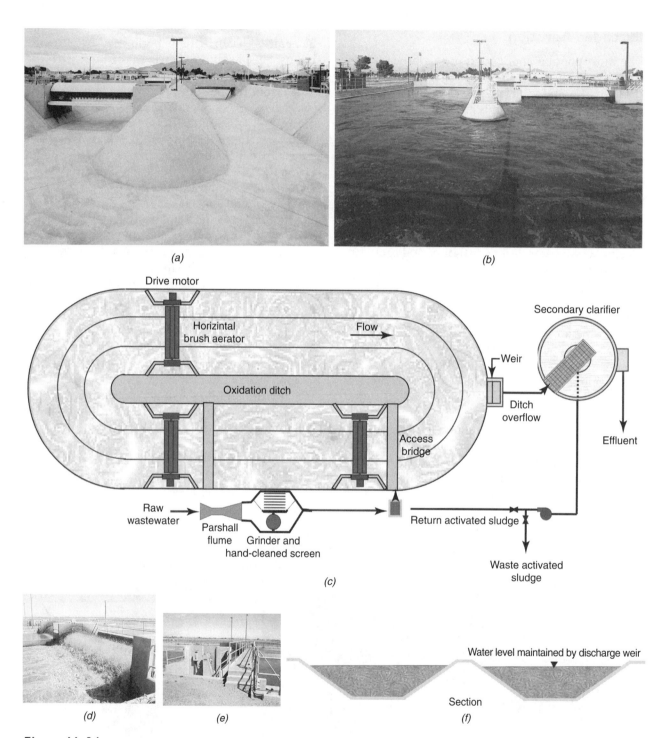

(a)

(b)

(c)

(d)

(e)

(f)

Figure 11–34

Extended-aeration process employing a horizontal mechanical aerator mounted over an aeration channel. (*a*) Photo of an empty oxidation ditch. (*b*) Photo of the same basin in operation at BOMO Water Reclamation Facility in Las Vegas, Nevada. (*c*) Plan view showing associated processes including headworks (flow measurement and screening), final clarifier, and sludge return. (*d*) Photo of a brush aerator in operation. (*e*) Photo of motor drive and access walkway. (*f*) Section through oxidation ditches showing sloped side walls. Vertical walls are preferred due to ease of construction.

propelling it around in the channel. The aerator pictured in Figure 11–34 is a 42-in.-diameter, high-capacity rotor with blades mounted on a 14-in.-diameter shaft. In most cases, the BOD loading on larger extended-aeration systems is less than 20 lb/1000 cu ft/day (320 g/m^3 · d), and the aeration period is greater than 12 hr.

■ EXAMPLE 11–8

A small extended-aeration plant without sludge-wasting facilities is loaded at a rate of 10.5 lb BOD/1000 cu ft/day with an aeration period of 24 hr. The measured suspended-solids buildup rate in the aeration tank is 40 mg/l/day. What percentage of the raw influent BOD is converted and retained as MLSS? If the MLSS concentration is allowed to increase from 2000 mg/l to 6000 mg/l before wasting solids, how long would this buildup take?

Solution

Because the aeration period is 24 hr,

$$\frac{\text{BOD load/day}}{\text{liter of tank}} = 10.5 \, \frac{\text{lb BOD/day}}{1000 \, \text{cu ft}}$$

$$\cdot \frac{\text{mg/l}}{62.4 \, \text{lb/1,000,000 cu ft}}$$

$$= 168 \, \text{mg/l}$$

$$\frac{\text{MLSS buildup}}{\text{BOD applied}} = \frac{40 \, \text{mg/l/day}}{168 \, \text{mg/l/day}} \cdot 100$$

$$= 24 \, \text{percent}$$

$$\text{Buildup time} = \frac{6000 \, \text{mg/l} - 2000 \, \text{mg/l}}{40 \, \text{mg/l/day}}$$

$$= 100 \, \text{days}$$

■ ■ ■

Aeration and Oxygen Transfer

Oxygen is supplied to the mixed liquor in an aeration tank by dispersing air bubbles through submerged diffusers or by entraining air into the liquid by mechanical means. Air diffusers are porous plates, tubes, or nozzles attached either to

Figure 11–35
Photo of centrifugal blowers providing air to a diffused aeration system. The blowers are driven by electric motors and engines. Engines may operate on natural or excess digester gas.

(a)   (b)

Figure 11–36
Underwater photos of aeration diffuser systems.
(a) Coarse-bubble diffuser (b) Fine-bubble diffuser.
(Courtesy of Sanitaire Water Pollution Control Corp.)

air piping on the bottom of the tank or to pipe headers that can be lifted out of the tank. Centrifugal blowers like those shown in Figure 11–35 provide compressed air to the diffusers. Coarse-bubble devices are orifices or nozzles designed so that the discharged air is broken up into bubbles and dispersed in the surrounding liquid (see Figure 11–36a). Fine-bubble devices are porous materials that release air as fine bubbles, as shown in Figure 11–36b. Although each kind of diffuser has individual features, coarse-bubble nozzles are noted for maintenance-free operation, but fine-bubble diffusers have the advantage of higher oxygen-transfer efficiency. Mechanical aerators are horizontal paddle, vertical turbine, and vertical-turbine draft tube. A horizontal rotor rotates partially submerged in an aeration channel. A vertical turbine may be a surface unit or completely submerged with compressed air supplied under the rotating blade. For deep mixing, a vertical turbine may be in a draft tube so that liquid from the bottom of the tank is drawn up through the tube and discharged at the surface.

Oxygen transfer is a two-phase process, as diagrammed in Figure 11–37. First, gaseous oxygen is dissolved in the wastewater by diffused or mechanical aeration. Then, the dissolved oxygen is taken up by the microorganisms in metabolism of the waste organic matter. If the rate of oxygen utilization exceeds the rate of dissolution, the dissolved oxygen in the mixed liquor is depleted.

The oxygen demand in aeration processes must account for the BOD and ammonia demand. The BOD demand is estimated by the difference between the BOD entering the aeration basin minus the BOD in the plant effluent and BOD in the waste solids, plus any ammonia demand, as shown in the following equation:

$$O_2 \text{ demand} = Q(BOD_i - BOD_e) \cdot 8.34$$
$$- (\text{waste TSS})(1.4)$$
$$+ (\text{BOD in recycle and}$$
$$\text{return flows})$$
$$+ 4.6Q(NO_3)8.34 \qquad \textbf{(11–16)}$$

where O_2 demand = carbonaceous oxygen demand, lb/day

Q = wastewater flow to the aeration (without recycle), mgd

BOD_i, BOD_e = influent BOD to aeration, effluent BOD from the secondary clarifier, mg/l

waste TSS = from the secondary clarifier, 1.4 lb BOD/lb TSS, lbs/day
BOD in recycle and return flows are unique for every facility. As an initial estimate, the BOD in the waste TSS and BOD in the return flow are considered equal and cancel each other out.

NO_3 = ammonia converted to nitrate, 4.6 lb BOD/lb NO_3, mg/l

To satisfy the demand, aeration equipment must provide sufficient oxygen for normal and peak demands. Aeration equipment must meet the peak demand, including peak seasonal and peak day demands. Many engineers use a peaking factor of 2.0 to account for BOD and flow variation; however, a review of actual peaking is preferred.

The oxygen transfer values for mechanical aeration devices are published by the manufacturer and represent the rate in pure water at standard temperature and pressure. The capacity is based on tests conducted in accordance with standards for the measurement of oxygen transfer in clean water. Clean water transfer rates reported for various types of aerators are listed in Table 11–5. These rates do not account for wastewater characteristics, residual oxygen concentration, and the increased pressure at the depth air is released. The actual oxygen transfer rate in wastewater is represented by the following equation:

$$OTR = SOTR \cdot \alpha F \cdot \theta^{T-20} \cdot \left(\frac{\beta \cdot C_s - C_t}{C_\infty} \right) \quad \textbf{(11–17)}$$

where OTR = O_2 demand corrected for wastewater, lb/hp-hr, kg/kWh, or lb/hr

SOTR = O_2 clean water at standard temperature and pressure, lb/hp-hr, kg/kWh, or lb/hr

Figure 11–37

Two-phase oxygen transfer in activated-sludge aeration. Gaseous oxygen is first converted to dissolved oxygen and is taken up by the biological floc.

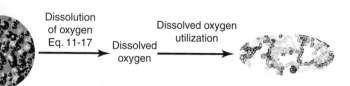

Dissolution of oxygen Eq. 11-17

Dissolved oxygen utilization

Dissolved oxygen

Air bubbles

Biological cells

TABLE 11–5

Typical Clean Water Oxygen Transfer Rates for Typical Mechanical Aeration Equipment

Mechanical Aerator Type	Figure Reference	Oxygen Transfer Efficiency (%)	SOTR (O_2 lb/hp-hr)	SOTR (O_2 kg/kWh)[a]
Fine-bubble membrane diffusers (total floor coverage)[b]	11–33b	20–39	4.0–7.7	2.4–4.7
Coarse-bubble diffusers[b]	11–33a	10–15	2.0–2.9	1.2–1.8
Jet aerators[b]	13–19	22–27	4.0–5.0	2.4–3.1
Fixed mechanical aerators	11–29	10–18	2.0–3.5	1.2–2.1
Floating aerators	11–40c	10–18	2.0–3.6	1.2–2.2
Rotor brush aerators	11–31d	10–18	2.0–3.5	1.2–2.1

[a] 1.0 lb/hp-hr = 0.61 kg/kWh
[b] Oxygen transfer efficiency at a 15-ft submergence, typical values are fine bubble = 2.0%, coarse bubble = 0.75%, jet aerators = 1.6% per foot of submergence.

αF = ratio of the wastewater to clean water transfer rate, dimensionless. Based on the transfer rate value from tests of a specific diffuser under field conditions divided by the transfer rate in clean water; the factor includes wastewater effects and aerator fouling. Alpha values vary greatly depending on solids retention time, MLSS concentration, and wastewater characteristics. Low alpha values range from 0.3 to 0.4 in activated sludge plants that are not nitrifying; fully nitrified systems have higher values ranging from 0.45 to 0.75. Membrane fouling factors range from 0.7 to 0.9. In the absence of field testing, typical αF values range between 0.5 and 0.6 for fine-bubble systems and between 0.7 to 0.8 for coarse-bubble aeration tanks.[1]

θ = dimensionless correction for temperature = 1.024

T = wastewater temperature, °C

β = saturation modifier, dimensionless = process water C/clean water C, range between 0.9 to 0.95.

C_s = dissolved oxygen saturation, mg/l; concentration at standard temperature and pressure, multiply C_s by (barometric pressure at job site / barometric pressure at mean sea level) to adjust for elevation.

C_t = dissolved oxygen, mg/l; concentration residual in the basin

C_∞ = dissolved oxygen concentration, mg/l; corrected for pressure of diffuser in basin

$$C_\infty = C_s(1 - 0.325 \cdot \text{diffuser depth}/D_c) \quad \textbf{(11–18)}$$

where C_∞ and C_s are defined above

D_C = 10.34 if diffuser depth is in meters, and D_c = 34 if diffuser depth is in feet

The energy required to meet the oxygen demand can be calculated as

$$\text{hp} = \frac{O_2 \text{ demand (lb/day)}}{O_2 \text{ supply (lb/hp} \cdot \text{hr)} \cdot 24}$$

$$\text{kWh} = \frac{O_2 \text{ demand (lb/day)}}{O_2 \text{ supply (lb/kWh} \cdot \text{hr)} \cdot 24} \quad \textbf{(11–19)}$$

The air supply required for diffused aeration systems can be calculated as follows:

$$Q_{\text{air}} = \frac{Q_{O2}}{\text{OTE} \cdot \rho_{\text{air}} \cdot f_{O2} \cdot 24 \cdot 60}$$

$$= \frac{Q_{O2}}{\text{QTE} \cdot 0.0174 \cdot 1400} \quad \textbf{(11–20)}$$

where Q_{air} = air flow rate, cu ft/min
Q_{O2} = oxygen transfer rate, lb/day
OTE = oxygen transfer efficiency listed in Table 11–5, percent
ρ_{air} = lb/cu ft, density of air = 0.075
f_{O2} = percent of oxygen in air = 0.232

The number of aeration units is based on incremental size for air-flow capacity and horsepower and is unique for each manufacturer. Since biological activity is just as great at low levels and the transfer rate from air to dissolved oxygen increases with decreasing concentration, it is logical to operate a system as close to critical minimum dissolved oxygen as possible. Operation of the air compressors at reduced capacity, or even turning off one blower on weekends, may be feasible to conserve electrical energy with no adverse effects on the biological process. Automatic control of blowers using dissolved oxygen probes has increased with improvements in control systems and the reliability of dissolved oxygen meters.

■ EXAMPLE 11–9

Floating aerators of 25 hp each are used for BOD reduction in the first cell of a high-rate pond system. The flow is 0.5 mgd with an influent BOD of 180 mg/l and an effluent BOD of 50 mg/l. There is no recycle flow and no sludge removal. The cell does not nitrify and there is a residual DO of 5 mg/l. Average water temperature is 15°C. Calculate the oxygen demand, total motor horsepower, and number of aerators.

$\alpha = 0.3$
$\beta = 0.9$
F = no fouling

Solution

$C_\infty = C_s = 10.2$ mg/l (see Table 2–5)

Floating aerators provide 2.8 lb O_2/hp-hr (Table 11–5).

O_2 demand = $Q(BOD_{in} - BOD_e)8.34$

$= 0.5(180 - 50)8.34 = 540$ lb/day

$= 1080$ lb/day   with a 2.0 peaking factor

$$OTR = 2.8 \cdot 0.3 \cdot 1 \cdot (1.024)^{15-20} \cdot \left(\frac{0.9 \cdot 10.2 - 2}{10.2} \right)$$

$$= 2.8 \cdot 0.135 = 0.375 \text{ lb } O_2\text{/hp-hr}$$

$$hp = \frac{O_2 \text{ demand (lb/day)}}{O_2 \text{ supply (lb/hp-hr)} \cdot 24}$$

$$= \frac{1080}{0.135 \cdot 24} = 121 \text{ hp}$$

or 5 aerators at 25 hp each.

■ ■ ■

■ EXAMPLE 11–10

A fine-bubble aeration system is used to remove BOD and fully nitrify the wastewater. The temperature of the wastewater is 18°C, the diffuser depth is 18 ft, the chloride concentration is < 2000 mg/l, and there is a 2.0 mg/l dissolved oxygen residual. What are the oxygen demand, total horsepower requirement, and air-flow requirement using an influent flow of 3.5 mgd, 140 mg/l BOD primary effluent, 15 mg/l in the plant effluent, and 23 mg/l ammonia?

$\alpha = 0.65$
$\beta = 0.95$
$F = 0.75$

Solution

$C_s = 9.5$ mg/l (see Table 2–5)

$C_\infty = C_s(1 - 0.325 \cdot \text{diffuser depth}/D_c)$

$C_\infty = 9.5 \cdot (1 - 0.325 \cdot 18/34) = 9.5 \cdot 0.83$

$= 7.9$ mg/l

O_2 demand = $Q(BOD_{in} - BOD_e)8.34$

$+ 4.6 \cdot Q \cdot NH_3 \cdot 8.34$

$= 3.5(140 - 15)8.34$

$+ 4.6 \cdot 3.5 \cdot 23 \cdot 8.34 = 6{,}740$ lb/day,

or 13,500 lb/day with a peaking factor of 2.0

$$OTR = STOR \cdot 0.65 \cdot 0.75 \cdot (1.024)^{18-20}$$

$$\cdot \left(\frac{0.95 \cdot 9.5 - 2}{7.9} \right) = STOR/0.413$$

OTR = 13,500 lb/day/0.413 = 32,700 lb/day

Fine-bubble oxygen transfer rate

$$= 8.3 \text{ lb } O_2/\text{hp-hr}$$

$$\text{hp} = 32,700/8.3/24 = 164$$

$$Q_{air} = \frac{Q_{O2}}{OTE \cdot \rho_{air} \cdot f_{O2} \cdot 1440}$$

$$= \frac{13,500}{0.36 \cdot 0.0174 \cdot 1440} = 1,500 \text{ cfm}$$

■ ■ ■

Empirical Design of Processes

Activated-sludge systems treating municipal wastewater are commonly designed on the basis of operating experience of existing plants using the same aeration processes. Local conditions, such as sewer infiltration and the effect of industrial wastewaters, are taken into account, from experience, by the designer. Climatic conditions, particularly wastewater temperature and seasonal variations in flow, also have an influence on the selection of loading parameters used in sizing aeration tanks. For instance, to produce an effluent of less than 30 mg/l BOD in a moderate climate, the typical design values for conventional secondary aeration on essentially domestic wastewater are a volumetric BOD loading of 40 lb/1000 cu ft/day (640 g/m^3 · d) and a minimum aeration period of 6 hr. Based on this aeration tank sizing, the process can be operated at an F/M of 0.2 to 0.3 lb BOD/day/lb MLSS with a mixed-liquor suspended solids concentration of approximately 2500 to 3000 mg/l.

Operation and Control

Aeration processes are regulated by the quantity of air supplied, the rate of activated-sludge recirculation, and the amount of sludge wasted, which, in turn, controls the MLSS, F/M ratio, and sludge age. A properly designed system treating domestic wastewater experiences few problems if the system is operated in a steady-state condition. Namely, the quantity of wastewater applied each day is approximately the same, aeration mixing and

dissolving oxygen concentration are relatively steady, and excess sludge is wasted continuously in quantities necessary to maintain a proper food-to-microorganism ratio. The latter may require reduced sludge wasting on weekends when the organic load on the plant is less.

When a plant is operated considerably under the design BOD load, nitrifying bacteria can convert ammonia to nitrate in the aeration basin. During subsequent detention in the clarifier, the nitrate can be used as an oxygen source under anaerobic conditions, releasing nitrogen gas that buoys up biological floc. The best solution is to increase sludge wasting to deplete the nitrifying bacterial populations and to decrease the air supply to reduce the dissolved oxygen concentration, provided that these control measures do not reduce the carbonaceous BOD removal efficiency.

Laboratory tests for monitoring activated-sludge treatment are dissolved oxygen, MLSS concentration, and effluent BOD and suspended solids. Influent BOD and flow are needed to calculate organic loading, the F/M ratio, and the aeration period. The concentration of solids in the return activated sludge, the effluent quality from the clarifier, and probing to determine the depth of the sludge blanket in the clarifier yield information to establish the proper recirculation rate for optimum process efficiency and maximum solids concentration in the waste sludge.

Industrial wastes are frequently the source of the operational instability of aeration processes. Shock loads of high-strength wastewater can deplete dissolved oxygen and unbalance the biological system. Toxic wastes in sufficient amounts interfere with microbial metabolism, and high-carbohydrate wastes may cause nutrient deficiency. All of these result in loss of MLSS in the clarifier overflow, thus decreasing treatment efficiency and shifting the F/M ratio by this unintentional wasting of microorganisms.

Section 9–6 discusses techniques to evaluate the treatability of wastewater. Hydraulic shock loads resulting from excessive inflow and infiltration to the sewer system can be just as detrimental as toxins or organic overload. High rates of overflow from a clarifier, even for short time periods, can carry over a substantial portion of the viable activated sludge needed in the system.

Several days may be required to rebuild the density of mixed liquor and to return to a proper operating F/M ratio. Determining the cause of poor settleability of an activated sludge involves investigating biological, chemical, and physical factors. First, the operating parameters of volumetric and F/M BOD loadings, MLSS concentration, and sludge age should be calculated and compared to the recommended values for the particular aeration process.

Microscopic examination and visual observation of the settling characteristics of the MLSS can reveal filamentous growths and poorly agglomerated floc. Nitrification-denitrification is evidenced by sludge bulking in the clarifier and testing of the mixed liquor for nitrate concentration. Chemical factors that can cause poor biological growth are insufficient dissolved oxygen, lack of nutrients, presence of toxic substances, and low temperature. Laboratory tests can be used to establish the adequacy of dissolved oxygen and availability of nitrogen and phosphorus nutrients. Since lowering the temperature of the mixed liquor decreases the rate of biological activity, cold temperature has an effect similar to increasing the BOD loading. Finally, the physical characteristics of the aeration system and clarifier should be considered. Excessive agitation due to underloading or overly aggressive mechanical mixing can cause shearing of biological floc into tiny particles with reduced settleability. Ineffective final clarification can result from an inadequate rate of returning sludge, excessive overflow rate, hydraulic turbulence, short-circuiting, or a faulty collector mechanism.

11–7 MATHEMATICAL MODEL FOR COMPLETELY MIXED AERATION

The principles of biological kinetics as defined by Monod in the batch growth of pure cultures (Section 3–12) have been applied in developing a mathematical model for biological aeration. The two key assumptions are that the kinetics reactions of pure cultures are the same as those of mixed cultures and that growth and substrate metabolism in a batch culture simulates BOD or COD removal in a completely mixed, continuous-flow aeration tank.

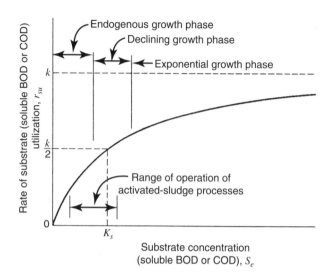

Figure 11–38

Rate of substrate utilization versus substrate concentration in the aeration tank (Eq. 11–31). The constant K_s is equal to the concentration of substrate S when the rate of substrate utilization r_{su} equals one-half the maximum rate k.

The basic mathematical relationship between rate of substrate utilization and substrate concentration in the declining growth phase is depicted in Figure 11–38 as a hyperbolic function. The shape of this curve is identical to Monod's equation graphed in Figure 3–26. For good activated-sludge settleability, aeration processes operate at a low substrate concentration (low F/M ratio) in the declining and endogenous growth phases. The unique mathematical feature of a hyperbolic function is that the constant K_s is equal to the substrate concentration when the substrate utilization is equal to half the maximum rate of substrate utilization. The equation for the curve in Figure 11–38 is

$$r_{su} = k\,\frac{XS_e}{K_s + S_e} \qquad (11\text{–}21)$$

where r_{su} = rate of substrate utilization, milligrams per liter of soluble BOD or COD per day

k = maximum rate of substrate utilization, milligrams per liter of soluble BOD or COD per milligram per liter of volatile MLSS per day

X = concentration of biomass, milligrams per liter of volatile MLSS

S_e = concentration of substrate surrounding the microorganisms, milligrams per liter of soluble BOD or COD

K_s = saturation constant, equal to substrate concentration when K_s equals $k/2$, milligrams per liter of soluble BOD or COD

The flow scheme shown in Figure 11–39 is used in deriving the mathematical equations for the completely mixed, activated-sludge process. The symbols of the diagram are

Q = rate of influent wastewater flow, cubic meters per day

Q_w = rate of excess sludge wasting from the aeration tank, cubic meters per day

$Q - Q_w$ = rate of effluent flow, cubic meters per day

R = recirculation ratio (Q_R/Q)

RQ = rate of activated-sludge recirculation

$Q(1 + R)$ = rate of flow from aeration tank

V = volume of aeration tank, cubic meters

X = concentration of biomass in aeration tank, milligrams per liter of volatile MLSS

X_R = concentration of biomass in recirculated activated sludge, milligrams per liter of volatile SS

X_e = concentration of biomass in effluent

S_o = concentration of substrate in influent wastewater, milligrams per liter of soluble BOD or COD

S_e = concentration of substrate in effluent flow, recirculating sludge, and aeration tank

The following relationships use the symbols shown in the flow scheme of Figure 11–39.

$$E = \frac{(S_o - S_e)100}{S_o} \qquad (11\text{–}22)$$

where E = efficiency of soluble BOD or COD removal, percent

$$\theta = \frac{V}{Q} \qquad (11\text{–}23)$$

where θ = aeration period, days

$$F/M = \frac{QS_o}{VX} = \frac{S_o}{\theta X} \qquad (11\text{–}24)$$

where F/M = food-to-microorganism ratio, grams of soluble BOD or COD applied per day per gram of volatile MLSS in the aeration tank

$$\theta_c = \frac{VX}{Q_w X + (Q - Q_w)X_e} \qquad (11\text{–}25)$$

where θ_c = sludge age (mean cell residence time), days

When the activated-sludge process in Figure 11–39 is operated under steady-state conditions, the rate of biomass growth equals the rate of biomass losses in the effluent and waste sludge.

$$r_g V = Q_w X + (Q - Q_w)X_e \qquad (11\text{–}26)$$

Substituting the endogenous rate of growth from Eq. 3–25 in Section 3–12 for r_g, Eq. 11–25 becomes

$$\frac{1}{\theta_c} = Y\frac{r_{su}}{X} - k_d \qquad (11\text{–}27)$$

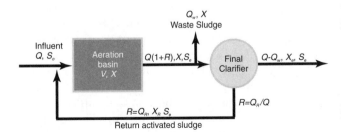

Figure 11–39

Flow schematic for a completely mixed activated-sludge process.

where Y = growth yield, milligrams per liter of biomass increase per milligram per liter of substrate utilized

Inserting U as the symbol for the specific substrate utilization rate r_{su}/X,

$$\frac{1}{\theta_c} = YU - k_d \qquad (11\text{–}28)$$

Also, at steady state, the substrate utilization rate r_{su} and specific substrate utilization rate U are constant values defined by the following equations:

$$r_{su} = \frac{S_o - S_e}{\theta} \qquad (11\text{–}29)$$

$$U = \frac{r_{su}}{X} = \frac{S_o - S_e}{\theta X} \qquad (11\text{–}30)$$

where U = specific substrate utilization rate per day.

Substituting the last term of Eq. 11–30 into Eq. 11–27 for r_{su}/X and V/Q for θ, solving for aeration tank volume yields

$$V = \frac{\theta_c YQ(S_o - S_e)}{X(1 + k_d \theta_c)} \qquad (11\text{–}31)$$

The biological kinetics relationship given by Eq. 11–21 is now introduced to derive an equation incorporating the saturation constant K_s. By setting Eq. 11–29 equal to Eq. 11–21, dividing by X, inverting and linearizing, and substituting in U as defined in Eq. 11–30, the resulting relationship is

$$\frac{1}{U} = \left(\frac{K_s}{k}\right)\left(\frac{1}{S_e}\right) + \frac{1}{k} \qquad (11\text{–}32)$$

Based on observed growth yield, Y_{obs} (Eq. 3–31), the production of excess biomass in the waste-activated sludge is calculated from Eq. 11–32.

$$P_x = \frac{YQ(S_o - S_e)}{1 + \theta_c k_d} \qquad (11\text{–}33)$$

where P_x = volatile solids in the waste sludge, grams per day.

The assumed operating conditions of the aeration process in the derivation of these equations are as follows:

1. Complete mixing in the aeration tank.
2. Steady-state flows, biomass concentrations, and substrate concentrations.
3. Soluble substrates, filtered BOD or COD.
4. Excess activated sludge is wasted from the aeration tank, rather than the sludge recirculation line.
5. Substrate concentration in the aeration tank equals substrate concentration in the effluent.
6. Biological activity takes place only in the aeration tank.

Kinetic Constants

To apply the mathematical equations for completely mixed aeration, numerical values for the following constants are required:

Y = growth yield, milligrams of volatile suspended solids per milligram of soluble BOD or COD

k_d = biomass decay coefficient, per day

K_s = saturation constant, milligrams per liter of soluble BOD or COD

k = maximum rate of substrate utilization, (milligrams of soluble BOD or COD per milligram of volatile suspended solids) per day

A wastewater treatability study using a bench-scale activated-sludge unit is required to determine these constants, unless full-scale study data are available. The operating conditions in the laboratory are the same as those assumed in deriving the mathematical equations. To collect sufficient data, the experimental runs are conducted at several sludge ages in the range of 3 to 20 days under steady-state conditions including continuous uniform flow rates, complete mixing in the aeration tank, and constant substrate loading and sludge wasting. The resulting operating conditions should be constant θ, volatile MLSS, F/M, and θ_c. Temperature, pH, and dissolved oxygen concentration are held constant throughout the series of tests.

The operating time for each test period should be at least twice the sludge age to ensure a steady-state analysis. From the average of daily measurements of flow rates, soluble BOD or COD, and volatile suspended solids concentrations, the specific substrate utilization rate U is calculated by Eq. 11–28, and the sludge age θ_c is calculated by Eq. 11–25. Values for $1/\theta_c$ and U for each test are plotted as shown in Figure 11–40a. Equation 11–28 is the equation of a straight line drawn through these plotted points. The slope of the line is the growth yield Y, and the intercept with the vertical axis is the biomass decay coefficient k_d. Values of $1/U$ are then plotted against $1/S_e$, as shown in Figure 11–40b. Based on Eq. 11–32, the slope of a straight line drawn through the data is $K_s \neq k$ and the intercept with the vertical axis is $1/k$.

The general ranges of magnitude for kinetic constants for completely mixed activated-sludge aeration are listed in Table 11–6.

Process Design

The aeration process designed using the kinetics model must be completely mixed activated sludge, since the application of Monod's growth kinetics in a batch culture to those of a continuous-flow system is valid only with complete mixing. Other ideal conditions assumed during derivation of the mathematical model must be compensated for in selecting design parameters. Steady-state conditions do not exist in operation of an actual treatment plant; therefore, allowance must be made for diurnal and random variations in wastewater flow and loading. Since most wastewaters contain suspended solids, the oxygen demand of the wastewater is from both suspended and soluble organic matter, whereas the kinetics equations are founded on soluble BOD or COD. To correlate total to soluble oxygen demand, BOD or COD tests can be conducted on both unfiltered and filtered samples during the treatability analyses. Finally, sludge settleability must be considered in the selection of design parameters to ensure good separation of the biological suspended solids. Design criteria for the secondary clarifiers are very important.

The first step in aeration design is to select the desired concentration of effluent soluble BOD, based on the allowable total BOD. The criterion for aeration loading can be either selection of a sludge age (Eq. 11–25) or food-to-microorganism ratio (Eq. 11–24). Sizing an aeration tank based on either of these parameters requires selection of a design flow and operating volatile MLSS. Knowing these data, the aeration tank volume can be calculated by Eq. 11–31.

The design sludge age for a conventional loading rate, to produce an effluent of less than 30 mg/l of total BOD, is in the range of 5 to 15 days. The selection of an operating volatile MLSS concentration is based on settleability of the activated sludge

Figure 11–40
Method of plotting laboratory data to determine kinetic constants from wastewater treatability study. (a) Graph of Eq. 11–27. (b) Graph of Eq. 11–31.

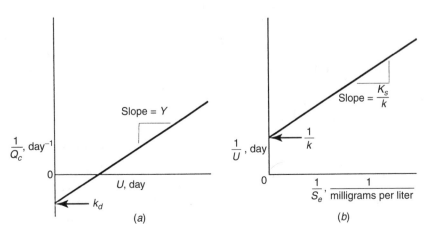

TABLE 11-6

Typical Ranges for Kinetic Constants for Completely Mixed Activated-Sludge Aeration Treating Municipal Wastewater at a Temperature of Approximately 20°C

Constant	Units of Expression	Range
Y	mg VSS/mg BOD	0.4–0.8
Y	mg VSS/mg COD	0.3–0.4
k_d	per day	0.04–0.08
K_s	mg/l of BOD	25–100
K_s	mg/l of COD	25–100
k	per day	4–8

as well as food-to-microorganism loading. A low design value results in an extended aeration period, and an excessive MLSS concentration results in poor gravity separation of the activated sludge in the secondary clarifier. In general, a conventional activated-sludge process operates in the MLSS range of 1500 to 3000 mg/l with 70 to 80 percent of the suspended solids being volatile. Therefore, a typical design value in a volatile MLSS concentration is between 1200 and 2400 mg/l.

■ EXAMPLE 11-11

The aeration tank for a completely mixed aeration process is being sized for a design wastewater flow of 2.0 mgd. The influent total BOD is 130 mg/l with a soluble BOD of 90 mg/l. The design effluent total BOD is 20 mg/l with a soluble BOD of 7.0 mg/l. Recommended design parameters are a sludge age of 10 days and volatile MLSS of 1400 mg/l. Selection of these values takes into account the anticipated variations in wastewater flows and strengths. The kinetic constants from a bench-scale treatability study are $Y = 0.60$ mg VSS/mg soluble BOD and $k_d = 0.06$ per day.

Solution

Using Eq. 11–31, the required aeration tank volume is

$$V = \frac{10 \cdot 0.60 \cdot 2.0(90 - 7)}{1400(1 + 0.06 \cdot 10)}$$

$$= 0.44 \text{ mil gal}$$

From Eq. 11–23,

$$\theta = \frac{0.44 \cdot 24}{2.0} = 5.3 \text{ hr}$$

From Eq. 11–24,

$$F/M = \frac{90}{(5.3/24)1400}$$

$$= 0.29 \frac{\text{lb soluble BOD/day}}{\text{lb volatile MLSS}}$$

Using Eq. 11–33, the excess biomass per day is

$$P_x = \frac{0.60 \cdot 2.0(90 - 7)8.34}{1 + 10 \cdot 0.06}$$

$$= 519 \text{ lb VSS/day}$$

■ ■ ■

11-8 STABILIZATION PONDS

Stabilization ponds, also called lagoons or oxidation ponds, are generally employed as secondary treatment in rural areas or as polishing ponds for mechanical treatment plants. Although they serve only about 7 percent of the population, thousands of lagoon installations exist in the United States, with about 90 percent in communities of less than 10,000 population. Ponds are classified as facultative, tertiary, aerated, or anaerobic according to the type of biological activity that takes place in them. Pretreatment includes flow measurement and screening. Each of these kinds of ponds is discussed in the following paragraphs.

Facultative Ponds

These are the most common lagoons employed for stabilizing municipal wastewater. The bacterial reactions include both aerobic and anaerobic decomposition, hence the term *facultative pond*.

Figure 11–41 illustrates the basic biological activity. Waste organics in suspension are broken down by bacteria, releasing nitrogen and phosphorus nutrients and carbon dioxide. Algae use these inorganic compounds, along with energy from sunlight, for growth, releasing oxygen to solution. Dissolved oxygen is in turn taken up by the bacteria, thus closing the symbiotic cycle. Oxygen is also introduced by reaeration through wind action. Settleable solids decomposed under anaerobic conditions on the bottom yield inorganic nutrients and odorous compounds, for instance, hydrogen sulfide and organic acids. The latter are generally oxidized in the aerobic surface water, thus preventing their emission to the atmosphere.

Bacterial decomposition and algae growth are both severely retarded by cold temperature. During winter when pond water is only a few degrees above freezing, the entering waste organics accumulate in the frigid water. Microbial activity is further reduced by ice and snow cover, which prevents sunlight penetration and wind reaeration. Under this environment, the water can become anaerobic, causing odorous conditions during the spring thaw, until algae become reestablished. This may take several weeks depending on climatic conditions and the amount of waste organics accumulated during the cold weather.

Operating water depths range from 2 to 5 ft, with 3 ft of dike freeboard above the high-water level (Figure 11–41). The minimum 2-ft depth is needed to prevent growth of rooted aquatic weeds, but exceeding a depth of 5 ft may create excessive odors because of anaerobiosis on the bottom.

Typical construction is two or more shallow pools with flat bottoms enclosed by earth dikes (Figure 11–42). Wastewater enters through an inlet division box, flows between cells by valved cross-connecting lines, and overflows through an outlet structure. Inlet lines, controlled by stop gates in the division box, discharge near the pond centers. Operating water depth is managed by a valving arrangement in the discharge structure. Connections between cells permit either parallel or series operation. Dikes are constructed with relatively flat side slopes to facilitate grass mowing and reduce slumping of the earth wall into the lagoon. Often the inside slopes along the waterline are protected by stone rip-rap to prevent erosion by wave action. If the soil is pervious, the pond bottoms should be sealed with bentonite clay or lined with plastic to prevent groundwater pollution. The area should be fenced to keep out livestock and discourage trespassing.

BOD loadings on stabilization ponds are expressed in terms of pounds of BOD applied per day per acre of water surface area (grams of BOD per day per square meter), or sometimes as BOD equivalent population per acre. The maximum allowable loading is about 20 lb of BOD per acre per day ($2.2 \text{ g/m}^2 \cdot \text{d}$) in northern states to minimize odor nuisance in the spring of the year. In those climates where ice coverage does not prevail, higher organic loadings may be used; for example, in the South and Southwest a loading of 50 lb of BOD per day per acre ($5.6 \text{ g/m}^2 \cdot \text{d}$) is practical. These loadings are 0.9 to 1.3 lb of BOD per thousand cu ft per day for a 5-ft water depth, substantially less than the volumetric loads applied to aeration and filtration units. The retention time of the wastewater in lagoons is 3 to 6 months depending on the applied load, depth of the wastewater, evaporation rate, and loss by seepage.

The two-cell lagoon in Figure 11–42a can be operated in either parallel or series. The flow pattern for series operation is shown by the arrows. In larger installations, a minimum of three cells is

Figure 11–41

Schematic of a faculative stabilization pond showing the basic biological reactions of bacteria and algae. (Also refer to Figure 3–19.)

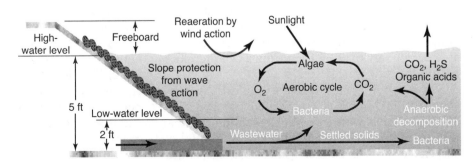

Figure 11–42

Typical stabilization pond arrangements. (a) Two-cell lagoon that can be operated in either parallel or series. (b) Three-cell lagoon with primary ponds, operated either in parallel or series, followed by a secondary pond to provide additional storage capacity and controlled discharge.

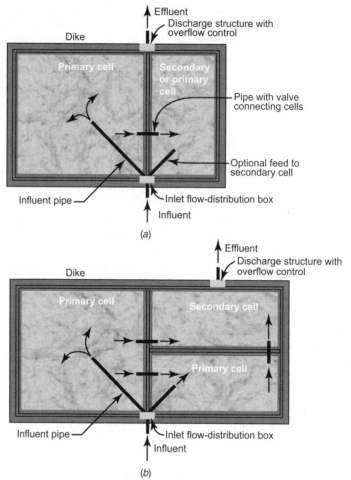

recommended with a secondary pond that cannot receive raw wastewater. The arrangement illustrated in Figure 11–42b allows parallel or series operation of the primary cells, which operate in series with the secondary cell. This flow pattern reduces short-circuiting. The secondary cell also provides additional storage capacity and permits isolation of the effluent wastewater before discharge. In sizing of ponds, the secondary cell is not included in BOD loading calculations; however, the volume is considered in calculating retention time of the wastewater. The water depth in a secondary pond is often increased to 8 ft, since odors are not as likely to be emitted from a cell that does not receive raw wastewater.

Stabilization ponds treating domestic wastewater produce an effluent with a BOD of less than 30 mg/l during warm-weather operation. Nevertheless, meeting an effluent standard of 30 mg/l of suspended solids is unlikely, since algae suspended in the water generally contribute 50 to 100 mg/l. In some instances, the algae population can be reduced in lightly loaded ponds by series operation. If an effluent is unacceptable for discharge to a watercourse, an alternative may be land disposal by irrigation of agricultural crops near the lagoon site.

■ EXAMPLE 11–12

Stabilization ponds for a town of 3000 population are constructed as shown in Figure 11–42b, with the larger cell having 14 acres and the two smaller cells each having 7 acres. The average daily wastewater flow is 0.24 mgd containing 450 lb of BOD; this is equivalent to 80 gpcd and 0.15 lb of BOD per person per day, or 225 mg/l BOD. (a) Calculate the BOD loadings based on the total area of primary cells. (b) Estimate the number of days of winter storage available between the 2-ft and 5-ft water levels assuming an evaporation and seepage loss of 0.10 in. of water per day.

Solution

(a) BOD load on the primary pond area is $450/21 = 21.4$ lb per day per acre. The equivalent population load is $21.4/0.17 = 125$ persons per acre.

(b) Storage volume:

$$(5 \text{ ft} - 2 \text{ ft}) \cdot 28 \text{ acres} \cdot 43,560 \text{ sq ft per acre}$$

$$= 3,660,000 \text{ cu ft} = 27.4 \text{ mil gal}$$

Water loss:

$$\frac{0.10 \text{ in./day}}{12 \text{ in./ft}} \cdot 28 \text{ acres} \cdot 43,560 \frac{\text{sq ft}}{\text{acres}}$$

$$\cdot 7.48 \frac{\text{gal}}{\text{cu ft}} = 76,000 \text{ gpd}$$

Storage time available:

$$\frac{27.4 \text{ mil gal}}{0.24 \text{ mgd} - 0.076 \text{ mgd}} = 170 \text{ days}$$

■ ■ ■

Tertiary Ponds

These units, also referred to as maturation or polishing ponds, serve as third-stage processing of effluent from activated sludge or trickling-filter secondary treatment. Stabilization by retention and surface aeration reduces suspended solids, BOD, fecal microorganisms, and ammonia. The water depth is generally limited to 2 or 3 ft for mixing and sunlight penetration. BOD loads are less than 15 lb of BOD per acre per day ($1.7 \text{ g/m}^2 \cdot \text{d}$), and detention times are relatively short at 10 to 15 days.

Aerated Lagoons

Completely mixed aerated ponds, usually followed by facultative ponds, are used for first-stage treatment of high-strength municipal wastewaters and for pretreatment of industrial wastewaters. The basins are 10 to 12 ft deep and are aerated with pier-mounted or floating mechanical units (Figure 11–43). The aerators are designed to provide mixing for suspension of microbial floc and to supply dissolved oxygen. The biological process does not include algae, and organic stabilization depends on the mixed liquor that develops within the basin, since no provision is made for settling and returning activated sludge. BOD removal is a function of aeration period, temperature, and nature of the wastewater. The common

Figure 11–43
Photos of a floating aerator. (*a*) Motor and propeller assembly. (*b*) Complete aerator with float assembly and intake bell. (*c*) Floating aerator in operation. (Photos courtesy of Aqua-Aerobic Systems, Inc.)

(a)

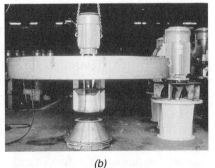

(b)

(c)

mathematical relationship for reduction of BOD in a completely mixed aerated pond is as follows:

$$\frac{\text{Effluent BOD}}{\text{Influent BOD}} = \frac{1}{1 + (k \cdot t)} \qquad \textbf{(11–34)}$$

where k = reaction-rate constant to base e, per day
t = aeration period, days

The value of k depends primarily on the biodegradability of the waste organics and temperature of the wastewater. At 20°C, k values for different wastewaters occur in a wide range from 0.3 to over 1.0 per day; the value for a particular wastewater must be experimentally determined. Design aeration periods are normally in the range of 3 to 8 days depending on the degree of treatment desired and wastewater temperature during the cold season of the year. For example, aerating a typical municipal wastewater for 5 days at 20°C provides about 85 percent BOD reduction; lowering the temperature to 10°C reduces the efficiency to approximately 65 percent. Oxygen transfer by surface aerators in wastewater lagoons can be computed from the manufacturer's oxygen transfer rate in clean water as follows:

$$R = R_o \frac{\beta C_s - C_t}{C_\infty} \cdot 1.024^{T-20}(\alpha) \qquad \textbf{(11–35)}$$

Problems of odors and low efficiency result when aerated lagoons are improperly designed or poorly operated. Thorough mixing and adequate dissolved oxygen ensure odor-free operation. If the aeration equipment is inadequate, deposition of solids and reduced oxygenation can result in anaerobic decomposition that leads to foul odors. Pretreatment and control of industrial wastes are required, since large inputs of either biodegradable or toxic wastes can upset the process. Infiltration entering the sewer collection system during wet weather can have a deterimental effect on an aerated lagoon by reducing the aeration period and flushing microbial floc out of the basin. Where infiltration is a problem, diverting a portion of wet-weather flow around the aerated lagoon to the second-stage facultative ponds can prevent

undesirable hydraulic loading on the aerated basins. In the winter, the aerators should be adjusted and windbreaks should be set up to reduce cooling of the lagoon water.

■ EXAMPLE 11–13

An aerated lagoon with a depth of 10 ft and liquid volume of 2.10 mil gal is equipped with four surface aerators of 10 hp each. The manufacturer's oxygen transfer rating is 2.5 lb of oxygen per horsepower hour. The influent is 0.30 mgd of combined domestic and industrial wastewater with an average BOD of 600 mg/l, which is equal to 1500 lb/day. Based on treatability studies, the wastewater characteristics are an alpha coefficient of 0.9 and a beta coefficient of 0.8. The temperature extremes of the aerating wastewater are 10°C in the winter and 25°C in the summer. If the desired BOD removal is a minimum of 75 percent, is the aeration capacity adequate? Assume 1.0 lb of oxygen is required per pound of BOD removed and that the dissolved oxygen concentration in the aerating wastewater should be a minimum of 2.0 mg/l.

Solution

Detention time = 2.10 mil gal/0.30 mgd

= 7.0 days   (OK)

Desired minimum BOD removal and oxygen

transfer = 1500 lb/day · 0.75 = 1130 lb/day

Rate of oxygen transfer required

= 1130/24 = 47 lb of oxygen/hr

From Table 2–5, C_s at 10°C = 11.3 mg/l

Oxygen transfer by four 10-hp aerators at 10°C using Eq. 11–35 is

$$R_{10°C} = 4 \cdot 10 \cdot 2.5 \frac{0.8 \cdot 11.3 - 2.0}{11.3} \cdot 1.024^{10-20}(0.9)$$

= 44 lb/hp-hr   (OK)

From Table 2–5, C_s at 25°C is

$$R_{25°C} = 4 \cdot 10 \cdot 2.5 \frac{0.8 \cdot 8.4 - 2.0}{8.4} \cdot 1.024^{25-20} (0.9)$$

$$= 57 \text{ lb/hp-hr} \qquad\qquad (\text{OK})$$

■■■

11–9 EFFLUENT DISINFECTION

The purpose of disinfecting wastewater effluent is to protect public health by inactivating pathogenic organisms, including enteric bacteria, viruses, and protozoans. Satisfactory disinfection of a secondary effluent is defined by a maximum geometric mean concentration test for fecal coliform bacteria of 200 per 100 ml after 30 consecutive days and 400 per 100 ml after 7 consecutive days. The number of coliform organisms in effluent from trickling filters, activated-sludge systems, and stabilization ponds is in the range of 100,000 to 10,000,000/100 ml. The safety of effluent disposal by dilution in surface waters after chlorination is based on the argument that reduction of fecal coliforms from 10^6/100 ml to 200 to 400/100 ml eliminates the great majority of bacterial pathogens and inactivates large numbers of enteric viruses. The most popular disinfectants are chlorine (gas and liquid) and ultraviolet radiation.

Chlorination

Chlorine is the most economical and widely used chemical for disinfection. The chlorine dosage needed for disinfection depends on wastewater pH, presence of interfering substances, temperature, and contact time. Applications of 8 to 15 mg/l provide adequate disinfection in well-designed units with a minimum contact time of 20 to 30 min. (Chlorine chemistry is discussed in Section 7–12 and chlorination equipment is shown in Figures 7–18 and 7–19.)

Chlorine contact consists of rapid initial mixing, creating a high initial kill at the point of chlorine contact and in the plug-flow basin for at least 30 min at peak flow providing for increased disinfection, as shown in Figure 11–44. The rapid-mixing chamber uses a mechanical mixing device to disperse chlorine solution throughout the wastewater. Adding the solution to an open channel results in very little dispersion and in poor chlorination efficiency because the flow is usually stratified. The most efficient tank for contact is a long, narrow chamber that approaches plug-flow conditions. To create plug flow conditions, tanks are divided by internal walls to create an area 10 ft wide by 10 ft deep and as long as required for the

Figure 11–44
Photo of a chlorine contact basin. (*a*) Rapid-mix tank showing high-energy mixing at the point where chlorine solution is injected. (*b*) Long, narrow channels within the basin create plug-flow conditions for chlorine contact. Three basins are sized for 20 min of contact prior to discharge. One basin is sized at 90 min for reuse and in-plant water. (Photos taken at the city of Las Vegas Water Pollution Control Facility.)

(*a*)

(*b*)

detention time. Dye testing of the tank reveals dead spaces and short-circuits that yield a more accurate measure of the tank volume.

Precise measurements and control of chlorine residual are extremely important for efficient and economical operation. An automatic residual monitoring and feedback control can prevent low concentrations and inadequate disinfection as well as excessive chlorination, resulting in discharge of an effluent toxic to aquatic life. Some regulatory agencies have specified maximum chlorine residuals in undiluted effluent of 0.0 to 0.5 mg/l to prevent potential toxicity in the receiving stream. Dechlorination may be required to detoxify the discharge after disinfection has been achieved. The least expensive and most effective method is by adding sulfur dioxide (gas form) or sodium metabisulfite (liquid form). Feed systems for sulfur dioxide and sodium metabisulfite are similar to the respective chlorine systems.

Wastewater contains significant quantities of microorganisms and may contain pathogens even after a high degree of treatment. Chlorination is commonly used for disinfection, but chlorine also reacts with organic compounds to generate by-products, called trihalomethanes, that may themselves by considered a health hazard. Regulatory limits placed on chlorination by-products may be sufficiently restrictive as to prohibit the use of disinfection using chlorine. Alternatives include[2] disinfection using ozone, chlorine dioxide, potassium permanganate, chloramines, hydrogen peroxide, and ultraviolet (UV) radiation. UV disinfection has the advantage of providing disinfection without a chemical agent. UV light may be used in conjunction with hydrogen peroxide to enhance oxidation and disinfection. Chlorine may be used with UV disinfection as a backup disinfectant and when a disinfectant residual is needed.

Ultraviolet Disinfection

The UV spectrum lies between X rays and visible light (see Figure 11–45) and is subdivided into four regions. UV light in the wavelengths between 200 and 300 nm has the greatest effect on microorganisms with the least degradation in water. UV radiation inhibits organisms from reproducing by damaging the DNA. During normal reproduction,

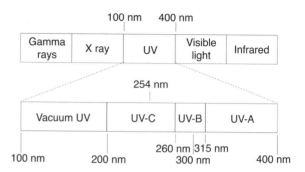

Figure 11–45

Portion of the electromagnetic spectrum from gamma rays through infrared with wavelengths in nanometers. Highlighted is the UV spectrum from 100 to 400 nm.

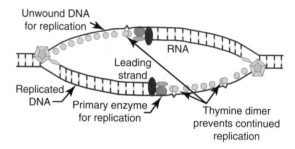

Figure 11–46

Diagram illustrating DNA reproduction inhibited by a thymine dimer caused by UV radiation. DNA strands unwind for synthesis. Enzymes used for replication are blocked by being unable to remove the link holding the thymine together.

the strands of the double helix separate, allowing DNA to be read and replicated (see Figure 11–46). UV absorption causes portions of the DNA structure (cytosine and thymine) to link together (called dimers) and inhibiting reproduction. UV radiation also damages viruses and bacteria in their spore and cyst forms. Where viral destruction is required, pretreatment with hydrogen peroxide or posttreatment with chlorine may be used.

Most bacteria do not have enzymes necessary to repair UV damage. Bacteria may undergo a process of photo repair in sunlight. Photons striking damaged DNA may unlock the dimers, allowing reproduction. Research studies have not shown reactivation of *Giardia*, and although *Cryptosporidium* may undergo some repair, infectivity is not restored. Coliform and *Shigella* have exhibited some photo reactivation, but viruses have not. UV effectiveness is measured by the response of organisms to UV light as the log of viable organisms divided by the number of organisms following exposure.

$$\text{Log inactivation} = \log_{10}(N_o/N) \quad \textbf{(11–36)}$$

where N_o = Concentration of reproductive organism prior to UV exposure

N = Concentration of reproductive organisms after exposure

UV lamps are specially manufactured mercury lamps under a low pressure and high pressure. Figure 11–47a shows the relative DNA absorbance over a UV wavelength from 200 to 300 nm. The figure also shows the UV dose for low- and high-pressure lamps over the same wavelength. Low-pressure lamps have a higher biocide efficiency, because nearly all of the lamp output is at the highly absorbant wavelength of 254 nm. Due to lower power requirements, lamp life is increased; however, more lamps are needed for a given UV dose. Medium-pressure lamps radiate over a broader DNA absorbance range, requiring fewer lamps, and have a lower overall energy cost for biocide effectiveness. However, higher operating temperatures accelerate lamp fouling. Lamps are typically oriented on a swing arm or removable array to allow easy removal for additional cleaning and maintenance. Wastewater organisms, as shown in Figure 11–47b, have a varied response to exposure[3], ranging from *E. coli* to the more difficult to inactivate *Bacillus subtilis* endospores. Inactivation of *E. coli* to a 4 $\log_{10}$ inactivation level is rapid with a UV dose of 10 mJ/sq cm. In contrast, *B. subtilis* requires a dose of 80 mJ/sq cm to achieve the same 4 $\log_{10}$ inactivation. In addition, the dose response is different for any specific wastewater, because of compounds in that water that interfere with UV effectiveness.

Figure 11–48 shows a small UV disinfection system with one arm of bulbs raised for cleaning. Several factors impact the efficiency of inactivation. Clumps of microorganisms (floc) shield some organisms from the UV dose. Compounds such as hardness (calcium, manganese), alkalinity, iron, and aluminum foul the external face of the lamps, and solids in the water absorb the UV radiation.

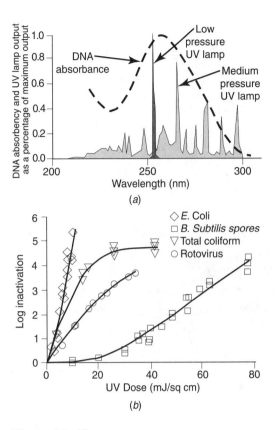

Figure 11–47

The figures show (*a*) DNA absorbency and (*b*) inactivation doses for various bacteria. UV lamp output for low-pressure and medium-pressure wavelengths are plotted with DNA absorbance for comparison. Medium-pressure lamps are more effective because they cover more of the DNA absorbance range. UV does response (*b*) indicates the relative ease of *E. coli* versus *B. subtilis* spore inactivation.

Medium-pressure lamps operate at higher temperature and are more susceptible to deposits and fouling. Fouling may occur in hours or over several months. Cleaning systems include lamp sleeve wipers and chemical solutions, including citric and phosphoric acid, to remove material adhering to the lamp. Mechanical wipers may consist of stainless steel brushes mounted on collars or cleaning rings that move along the lamps. Wiper frequencies are variable, but may be as frequent as four times per hour. UV lamp systems should be field-tested to determine the effectiveness of cleaning systems and dose response prior to final sizing and design.

Figure 11–48

Photo of ultraviolet disinfection equipment. The glow results from lamps in the effluent channel. One rack of bulbs is raised for cleaning. Hypochlorite is used as standby disinfectant. (Photo taken at the Wastewater Treatment Plant in Grants Pass, Oregon.)

11–10 INDIVIDUAL HOUSEHOLD DISPOSAL SYSTEMS

Approximately one-fourth of the homes in the United States are located in unsewered areas and must rely on individual treatment. The quantity of wastewater is from 60 to 100 gpcd, with BOD concentrations varying from 200 to 400 mg/l. The number of bathrooms, automatic washing machines, and garbage grinders, as well as the number of residents, define the magnitude and character of the wastewater.

The most popular system is the septic tank and absorption field because of its low cost and the desirable characteristic of underground disposal of effluent (Figure 11–49). A septic tank is an underground concrete box sized for a detention time of approximately 2 days. With garbage grinders and automatic washers, the recommended minimum capacity is 750 gal (2.8 m³) for a two-bedroom house, 900 gal for three bedrooms, 1000 for four bedrooms, and 250 for each additional bedroom. Inspection and cleaning ports must be accessible for maintenance, usually by removing about 1 ft of earth cover. Inlet and outlet pipe tees prevent clogging of the drains with scum that accumulates on the liquid surface. The functions of a septic tank are settling of solids, flotation of grease, anaerobic decomposition of accumulated organic matter, and storage of sludge. Retaining large solids is essential to prevent plugging of the percolation field.

An absorption field, where the majority of the biological stabilization takes place, consists of looped lateral trenches 18 to 24 in. wide and at least 18 in. deep. Drain tile or perforated pipe in an envelope of gravel is used to distribute the wastewater uniformly over the trench bottom. Organics are decomposed in the aerobic-facultative environment of the bed, and water seeps down into the soil profile. Air enters the drain tile through the backfill covering of the trenches and by ventilation through the house plumbing stack. The percolation area required depends directly on soil permeability, and for a four-bedroom dwelling the

Figure 11–49

A typical septic tank–absorption system for disposal of household wastewater. The tank provides sedimentation, scum accumulation, anaerobic digestion, destruction of organic matter, and sludge storage. The absorption field permits aerobic-facultative decomposition of organics and seepage of the wastewater.

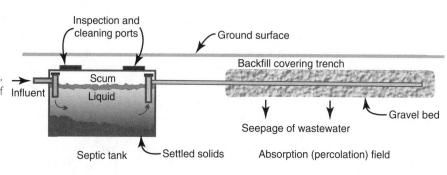

area needed ranges from 300 to 1300 sq ft (28 to 121 m^2). The trench area for a particular location is determined by subsurface soil exploration and percolation tests. Most state environmental control agencies and county health departments have guidelines for installations based on local conditions.

The most frequent complaints in operation of septic tank–absorption field systems concern plumbing stoppages and odorous seepage through the ground surface. Tanks fill with nonbiodegradable solids and must be pumped out every few years. When cleaning them, leaving a small amount of the black-colored digesting sludge in the tank ensures adequate bacterial seeding to continue solids digestion. Introduction of chemical or enzymatic conditioners have not been shown to be of any significant value in reviving or increasing bacterial activity in a septic tank.

11–11 CHARACTERISTICS AND QUANTITIES OF WASTE SLUDGES

The purpose of wastewater settling and biological aeration is to remove organic matter and to concentrate it in a much smaller volume for ease of handling and disposal. As illustrated in Figure 11–1, the solids of 0.5 pint/person amount to about 0.6 gal of total sludge or about 5000 gal of sludge per 1.0 mil gal (5 m^3 per 1000 m^3) of wastewater treated. The cost of facilities for stabilizing, dewatering, and disposal of this concentrate is about one-third of the total investment in a treatment plant. Operating expenses in sludge handling may amount to an even larger fraction of the total plant operating costs, depending on the system employed. For these reasons, a properly designed and efficiently run sludge disposal system is essential.

The quantity and nature of the sludge generated relate to the character of the raw wastewater and processing units employed. Primary settling produces an anaerobic sludge of raw organics that are being actively decomposed by bacteria. Therefore, these solids must be handled properly to prevent emission of obnoxious odors. In comparison with secondary biological waste, primary sludges thicken and dewater readily because of their fibrous and coarse nature. The following formula can be used to estimate the raw solids that are removed by plain sedimentation:

$$W_p = f \cdot SS \qquad (11\text{--}37)$$

where W_p = raw primary sludge solids, pounds of dry weight per day (grams per day)

f = fraction of suspended solids removed in primary settling (f is about 0.5 for domestic wastewater)

SS = suspended solids in unsettled wastewater, pounds per day (grams per day) concentration of suspended solids in unsettled wastewater in mg/l · wastewater flow in mil gal/day · 8.34 lb/mil gal per mg/l (concentration of suspended solids in mg/l · flow in m^3/d)

Waste from aeration is flocculated microbial growths with entrained nonbiodegradable suspended and colloidal solids. It is relatively odor-free because of biological oxidation, but the finely divided and dispersed particles make it difficult to dewater. Excess activated-sludge solids from aeration processes and humus from biological filtration can be estimated by Eq. 11–38, which relates solids production to BOD load. The coefficient K depends on the process food-to-microorganism ratio shown in Figure 11–50. Although this formulation is reasonable for domestic wastewater, calculated values may differ considerably from real sludge yields when treating municipal discharge that contains a substantial portion of industrial or food-processing waste.

$$W_s = K \cdot BOD \qquad (11\text{--}38)$$

where W_s = biological sludge solids, pounds of dry weight/day (grams/day)

K = fraction of applied BOD that appears as excess biological solids from Figure 11–50 assuming about 30 mg/l of suspended solids in the plant effluent

BOD = BOD in applied wastewater after primary sedimentation, pounds/day (grams/day)

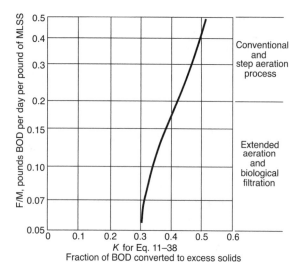

Figure 11–50

Hypothetical relationship between food-to-microorganism ratio and coefficient K in Eq. 11–38 for calculating excess activated-sludge production in biological aeration of wastewater, based on an effluent suspended solids of approximately 30 mg/l.

The total sludge solids production in a conventional treatment plant with primary sedimentation and secondary aeration is equal to the sum of the values calculated by Eqs. 11–37 and 11–38.

$$W_{ps} = W_p + W_s \qquad (11\text{–}39)$$

where W_{ps} = total sludge solids from primary sedimentation and secondary biological aeration.

The sludge solids production for an activated-sludge system treating unsettled wastewater, for example, extended aeration without primary sedimentation, will be less than W_{ps} from Eq. 11–39 but more than W_s from Eq. 11–38. The quantity of solids produced can be estimated based on the influent BOD of unsettled wastewater, disregarding the suspended solids, by increasing the value determined from Figure 11–50 by 100 percent. Thus, for aeration systems without primary clarifiers,

$$W_{as} = 2.0 \cdot K \cdot BOD \qquad (11\text{–}40)$$

where W_{as} = biological sludge solids from activated-sludge processing without primary sedimentation, pounds of dry weight per day (grams per day)

K = value from Figure 11–50

BOD = plant influent BOD without primary sedimentation, pounds per day (grams per day)

Design and operation of a sludge disposal system are based on volume of the wet sludge as well as the dry solids content. Once the dry weight of the solids has been determined, the volume of wet sludge can be calculated using Eq. 11–41 by knowing the percentage of solids, or water content. This formula assumes a specific gravity for the wet sludge of 1.0, which is sufficiently accurate for normal computations. For example, a slurry with 10 percent organics has a specific gravity of about 1.02.

$$V = \frac{W}{\left(\dfrac{s}{100}\right) \cdot 8.34} = \frac{W}{\left(\dfrac{100 - p}{100}\right) \cdot 8.34} \qquad (11\text{–}41)$$

where V = volume of wet sludge, gallons
W = weight of dry solids, pounds
s = solids content, percent
p = water content, percent

Equation 11–41 in metric units is

$$V = \frac{W}{\left(\dfrac{s}{100}\right)} = \frac{W}{\left(\dfrac{100 - p}{100}\right)} \text{ (SI)} \qquad (11\text{–}42)$$

where V = volume of wet sludge, liters
W = weight of dry solids, kilograms

Typical solids concentrations in raw waste sludges are primary sludge only, 4 to 6 percent; primary sludge plus humus from secondary filtration, 4 to 5 percent; primary plus thickened waste-activated sludge from secondary aeration, 4 to 5 percent; and unthickened waste-activated sludge only, 0.5 to 1.5 percent. In all of these wastes, the percentage of volatile solids is about 70 percent of the total dry solids. For a given quantity of dry solids, the percentage concentration dramatically influences the sludge volume. Doubling the solids concentration reduces the wet volume by half; conversely, the amount would double if the slurry were diluted to half the original percentage of solids. For example, using Eq. 11–41, the volume of wet sludge containing 10 lb of dry solids at 2 percent concentration is 60 gal. If thickened to 4 percent solids, the volume becomes 30 gal, and a further concentration to 8 percent reduces the quantity to 15 gal. With this concept in mind, it is easier to understand why dilute secondary sludges are more difficult to process and handle than primary wastes.

■ EXAMPLE 11–14

A domestic wastewater with 200 mg/l BOD and 220 mg/l of suspended solids is processed in a trickling filter plant. (a) Using Eqs. 11–37 and 11–38, estimate the quantities of sludge solids from primary settling and secondary biological filtration per million gallons of wastewater treated. (b) How many pounds of dry solids are produced relating to the BOD equivalent population load on the plant? (c) What is the volume of wet sludge per million gallons of wastewater if underflow from the clarifiers returns filter humus to the plant influent and the combined sludge drawn from the primaries has a solids content of 5.0 percent?

Solution

(a) Assuming 50-percent suspended solids reduction from settling the raw wastewater, and substituting into Eq. 11–37,

$$W_p = 0.5 \cdot 220 \cdot 1.0 \cdot 8.34$$

$$= 917 \text{ lb dry solids/mil gal}$$

From Eq. 11–38, assuming 35 percent BOD removal in the primary and an estimated K of 0.34 from Figure 11–50,

$$W_s = 0.34 \cdot 0.65 \cdot 200 \cdot 1.0 \cdot 8.34$$

$$= 369 \text{ lb dry solids/mil gal}$$

(b) For 1.0 mgd with a BOD of 200 mg/l,

$$\text{BOD equivalent population} = 1.0 \cdot 200 \cdot 8.34/0.2$$

$$= 8340 \text{ persons}$$

Sludge production per capita

$$= \frac{W_p + W_s}{\text{population}} = \frac{917 + 369}{8340}$$

$$= 0.15 \text{ lb dry solids/person}$$

(*Note:* The value of 0.20 lb of dry solids per person per day has often been used as a conservative estimate of solids yield in trickling filter plants treating domestic wastewater.)

(c) Substituting into Eq. 11–41 with $W_p + W_s$,

$$V = \frac{917 + 369}{\left(\dfrac{5.0}{100}\right)8.34} = 3080 \frac{\text{gal of sludge}}{\text{mil gal of wastewater}}$$

■■■

11–12 SLUDGE PUMPS

High-viscosity sludges are primary sludges and mixtures of primary and waste-activated sludges, scum from clarifiers, and raw and digested sludges that have been thickened by gravity, flotation, or centrifugation. Positive displacement allows metering of the quantity of sludge pumped with respect to time. Also, heavy solids can be moved at a slow rate without clogging the pump chambers.

A thickened-sludge pumping station is shown at the primary clarifer in Figure 11–15. The progressing cavity pumps are positive-displacement pumps that discharge at a uniform rate of flow, shown in cutaway in Figure 11–51. The chamber is an abrasion-resistant rubber stator with a double-helix opening, and the moving element is a steel single-helix rotor driven by an electric motor. When the single-helix rotor revolves within the double-helix stator, cavities are formed that fill with sludge and progress toward the outlet end. High-pressure sensors on the pump discharge immediately stop the pump to prevent damage to the pump and piping.

Sludge Pumping Headloss

The Hazen-Williams calculation for head loss is based on a fluid in turbulent flow (refer to Section 4–3). As the solids in sludge increase, the fluid

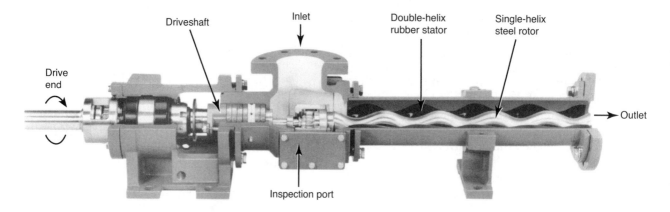

Figure 11–51

Cutaway view of a progressing cavity pump used for pumping sludge.
(Courtesy of Moyno, Moyno® is a registered trademark of Robbins & Myers. Inc.)

becomes increasingly thicker, changing the fluid characteristics and increasing the velocity required for the fluid to change from laminar to turbulent flow. In water flow, velocities of 5 to 6 ft/sec are used as an economic balance between pipe size and head loss. To prevent clogging and for ease of cleaning, sludge lines are rarely sized less than 6 in. (150 mm) and velocities less than 2 ft/sec are common.

The Reynolds number for turbulent flow is calculated using Eq. 11–43, and for laminar flow using Eq. 11–44. The values are important in determining when there is sufficient velocity gradient to shear the fluid and create turbulent flow. Flow is generally regarded as turbulent if $R_e > 4000$ and laminar if $R_e < 2300$. Flow may be considered in transition between these values.

$$R_e = \frac{d \cdot v \cdot \rho}{12 \cdot \eta} \qquad \textbf{(11–43)}$$

where R_e = dimensionless Reynolds number
d = pipe diameter in inches
v = fluid velocity in ft/sec
ρ = density in lb/cu ft
η = fluid viscosity in lb/ft-sec

$$R_e = \frac{3 \cdot (d/12) \cdot v^2 \cdot \rho}{3 \cdot \eta + 16(d/12)S_y} \qquad \textbf{(11–44)}$$

where units are listed in Eq. 11–45.

Figure 11–52 provides a comparison between the head loss of water and that for digested, primary, and thickened waste-activated sludge. Water has a shear stress of zero; therefore, at even the smallest amount of energy, water will flow. Sludge will not flow until a threshold amount of pressure or yield stress is applied. That stress is different for different sludges because of differences among the solids in each sludge. Even when sludge is moving, the amount of energy required is greater than for

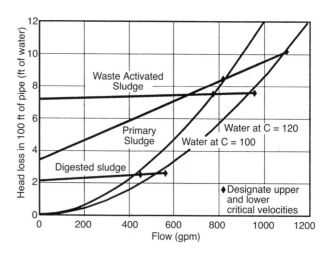

Figure 11–52

Head loss versus flow for 100 ft of 6-in. pipe. Curves are plotted for waste-activated, digested, and primary sludges at 3.5 percent, water at $C = 100$, and water at $C = 120$ with no solids.

water as defined by the coefficient of rigidity. Digested sludge tends to be the easiest to pump because the solids are small and less "sticky." Primary sludge contains a high concentration of fibrous material that remains difficult to pump even at high velocities. Waste-activated sludge is the most difficult to pump because the pili (cell hairs) and polymer tie the organisms together into a semisolid floc that resists becoming turbulent.

The Bingham plastic model is a good predictor of sludge head loss in laminar flow. The equation may be written as follows:

$$H/L = \frac{16S_y}{3wd/12} + \frac{\eta v}{w(d/12)^2} \qquad \textbf{(11–45)}$$

where d = diameter of pipe, in inches

S_y = shear stress at the yield point where sludge begins to flow, pounds per square foot

η = coefficient of rigidity, pounds per foot per second

H = head loss measured in feet of water height

L = length of pipe, feet

v = average velocity, feet per second

w = weight of water, 64.4 lb/cu ft

The Hazen-Williams equation is used when flow is turbulent. The velocity where sludge becomes turbulent may be bounded by the following equations for lower critical and upper critical velocities:

$$V_{lc} = \frac{1000\eta + 103\sqrt{94\eta^2 + (d/12)^2 S_y \rho}}{(d/12)\rho} \qquad \textbf{(11–46)}$$

$$V_{uc} = \frac{1500\eta + 127\sqrt{140\eta^2 + (d/12)^2 S_y \rho}}{(d/12)\rho} \qquad \textbf{(11–47)}$$

where V_{lc} = lower critical velocity, feet per second

V_{uc} = upper critical velocity, feet per second

ρ = density of sludge = W weight of water, pounds per cubic foot

Values for yield stress and coefficient of rigidity versus sludge solids concentration are plotted in Figure 11–53a and 11–53b, respectively. Lines

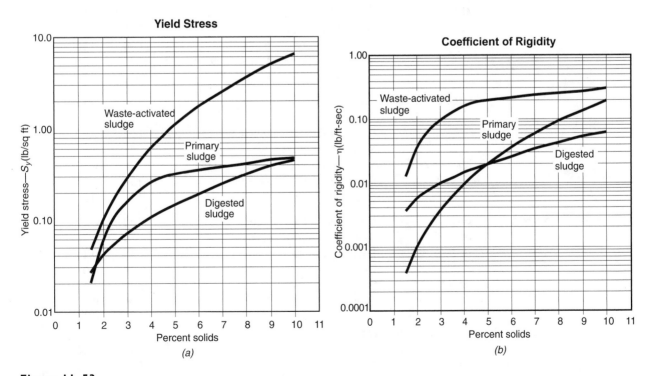

Figure 11–53

Coefficients for calculating the head loss of sludges at various solids concentrations. (a) Values for yield stress. (b) Values for coefficient of rigidity.

representing theoretical values are drawn for raw primary sludge, digested sludge, and thickened waste-activated sludge.

Head loss in fittings may be estimated by substituting an equivalent length of straight pipe or by the velocity loss coefficient K. For laminar flow, the values in Table 4–1 are not appropriate. A second K valve (K_1) is required to account for losses during laminar flow; refer to Table 11–7. The values are used in Eqs. 11–44 and 11–48 to estimate the head loss of laminar sludge flow.

$$K = K_1/R_e + K_\infty(1 + 1/d) \qquad (11\text{–}48)$$

where d = pipe diameter in inches
K_1 and K_∞ are listed in Table 11–7.

■ EXAMPLE 11–15

Sludge recirculation of 150 gpm is required for digester heating. Digesting sludge concentrations average 3.5 percent. Determine the piping head

TABLE 11–7

Values for K_1 and K_∞, Using the Two-K Method for Head Loss in Fittings

		FITTING TYPE	K_1	K_∞
Elbows	90°	Standard ($R/D = 1$), screwed	800	0.4
		Standard ($R/D = 1$), flanged, welded	800	0.25
		Long-radius ($R/D = 1.5$), all	800	0.20
	45°	Standard ($R/D = 1$), all	500	0.20
		Long-radius ($R/D = 1.5$), all	500	0.15
	180°	Standard ($R/D = 1$), screwed	1000	0.60
		Standard ($R/D = 1$), flanged, welded	1000	0.35
		Long-radius ($R/D = 1.5$), all	1000	0.30
Tees	Used as elbow	Standard, screwed	500	0.70
		Long-radius, screwed	800	0.40
		Standard, flanged or welded	800	0.80
		Stub-in-type branch	1000	1.00
	Run thru tee	Standard, screwed	200	0.50
		Standard, flanged or welded	150	0.10
		Stub-in-type branch	100	0.00
Valves	Gate	Full line size, 100% open	300	0.10
	Ball	Reduced trim, 90% open	500	0.15
	Plug	Reduced trim, 80% open	1000	0.25
		Lift	2000	10.0
	Check	Swing	1500	1.50
		Tilting-disk	1000	0.50
	Globe, standard		1500	4.00
	Globe, angle or Y-type		1000	2.00
	Diaphragm, dam type		1000	2.00
	Butterfly		800	0.25
In	Pipe entrance		160	0.5
Out	Pipe exit		0	1.0

loss in 1000 feet of 6-in. pipe. Piping contains the following fittings: one check valve, ten 90° elbows, 4 tees, and 5 plug valves. Determine the fitting head loss.

Solution

Using Figure 11–53 and a digested sludge concentration of 3.5 percent, $S_y = 0.10$, $\eta = 0.012$, and $\rho = 64.4$. The velocity in 6-in. pipe is 1.69 ft/sec.

(a) Check if the flow is laminar using Eqs. 11–46 and 11–47:

$$V_{lc} = \frac{(1000 \cdot 0.012) + 103\sqrt{94(0.012)^2 + 0.5^2(0.10)64.4}}{0.5 \cdot 64.4}$$
$$= 4.4 \text{ ft/sec}$$

$$V_{uc} = \frac{(1500 \cdot 0.012) + 127\sqrt{140(0.012)^2 + 0.5^2(0.10)64.4}}{0.5 \cdot 64.4}$$
$$= 5.6 \text{ ft/sec}$$

(b) Calculate laminar head loss using Eq. 11–45:

$$H/1000 = \frac{16 \cdot 0.10}{3 \cdot 64.4 \cdot 0.5} + \frac{0.012 \cdot 1.69}{64.4 \cdot 0.5^2}$$
$$= 0.018 \text{ ft water/ft}$$
$$H = 18 \text{ ft of water}$$

Note: The head loss for water using the Hazen-Williams Eq. 4–8 would have predicted the following result:

$$h_L = 0.002083(1000)\left(\frac{100}{100}\right)^{1.85} \frac{150^{1.85}}{6^{4.8655}}$$
$$= 3.6 \text{ ft of water}$$

or 5 times less than the head loss of sludge. The K_1 and K_∞ totals for the fitting listed are

	No.	K_1	K_∞
Check Valve	1	1500	1.5
90° elbows	10	8000	2.5
Tees	4	3200	3.2
Plug valves	5	5000	1.25
		17700	8.45

$$R_e = \frac{3 \cdot (6/12) \cdot (1.69)^2 \cdot 64.4}{3 \cdot 0.012 + 16(6/12)0.10} = \frac{276}{0.036 + 0.8}$$
$$= 330 \quad (<2300 \text{ laminar flow})$$

$$K = 17700/330 + 8.45(1 + 1/6) = 54 + 10 = 64$$

$$\text{Fitting } H_f = K\frac{v^2}{2 \cdot g} = 64\frac{(1.69)^2}{64.4} = 3 \text{ ft}$$

■ ■ ■

11–13 SLUDGE THICKENING

Sludge thickening is used to increase sludge solids to at least 4 percent and reduce the volume required for anaerobic digestion. Aerobic secondary sludge in the range of 0.5 to 2 percent is suitable for aerobic digestion but is dilute for anaerobic digestion. Primary solids may be removed from the primary clarifier at a high rate to prevent carryover and thickened in a separate tank. Many plants cothicken primary and secondary solids, often with mixed results. Gravity thickening is used for primary and trickling filter solids and is typically performed without chemical addition; however, ferric chloride may be added to reduce odors and enhance settling. Mechanical thickening is typically used for activated sludge with polymer addition to enhance flocculation and thickening.

Gravity Sludge Thickening

Gravity thickening is performed in circular settling tanks that are equipped with scraper arms having vertical pickets. Sludge withdrawn from primary clarifiers or sludge blending tanks is applied to the gravity thickener through a central inlet well. Overflow containing the nonsettleable fraction is returned to the head of the plant for reprocessing and the thickened sludge is drawn from the tank bottom. Unit loadings are normally in the range of 6 to 12 lb of solids per square foot of tank bottom per day (30 to 60 kg/m^2 · d). Ferric chloride may be added to help reduce hydrogen sulfide generation in the thickener and anaerobic digester. In handling domestic wastes, the under-flow solids concentration is often double that of the applied sludge. For example, expected under-flow solids content would be 6 percent for a mixture of primary and activated sludge and 8 percent for primary plus filter humus. The precise performance of a thickener is difficult to predict because of the varying character of waste sludges. Depending on the characteristics of the sludge, solids capture varies from about 80 to 95 percent. Coagulant chemicals may be used to improve the capture of suspended solids and density of the thickened waste.

Thickening of Waste-Activated Sludges

Waste-activated sludges are not generally amenable to gravity settling. Some extended aeration sludges are sufficiently endogenously respired to thicken by gravity without significant degasification or bulking. Gravity thickening of activated sludge from an extended aeration plant without primary clarification can increase solids from 0.75 percent to 3.5 percent. Activated sludges are more effectively thickened using mechanical equipment such as dissolved air flotation, gravity belt, or centrifuge thickeners, where sludges can be thickened from 0.5 percent to 4 to 6 percent.

Dissolved Air Flotation

Dissolved air flotation is achieved by releasing fine-air bubbles that attach to sludge particles and cause them to float. Figure 11–54 is a flow diagram for a typical unit. Small units tend to be rectangular and fabricated using steel. Larger units are circular and manufactured in steel or concrete. Waste-activated sludge enters the bottom of the flotation tank, where it is merged with recirculated flow that contains compressed air. A portion of the clarified effluent is pressurized in a separate retention tank under an air pressure of approximately 60 psi to force air into solution. On pressure release, air dissolved in the recirculated flow forms fine bubbles attached to the suspended solids. The process underflow is returned to wastewater treatment, and the overflow, discharged by a mechanical skimming device, is the thickened sludge.

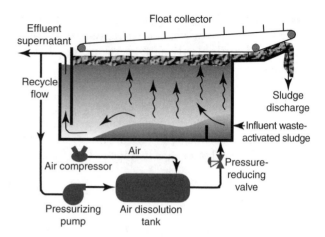

Figure 11–54

Schematic diagram of a dissolved air floatation tank for thickening of waste-activated sludge.

Flotation thickeners used for waste-activated sludge normally produce a concentrate of approximately 4 percent solids with a solids recovery of 85 percent. Polymers or other flocculents may be added to increase the capture of solids to 95 percent or more. Chemical treatment, however, may not increase the concentration of the thickened sludge. Practical loading rates for waste-activated sludge are in the range of 2 to 4 lb of solids per square foot (10 to 20 kg/m^2 · d) of flotation area per hour. Because the characteristics of biological floc are not consistent, even in the same treatment plant, the response to chemical flotation aids and solids loadings must be determined in each individual case.

Gravity Belt Thickeners

A gravity belt thickener consists of a continuously moving permeable fabric belt that passes over a horizontal surface, allowing free water to drain from the sludge while retaining the solids on the fabric. The process is illustrated in Figure 11–55. Polymer is required to liberate free water and agglomerate the solids. Thickeners are rectangular and vary in belt width from 1 to 3 m. Sludge enters a flocculation tank where the polymer interacts with solids. The inlet chute uniformly distributes the sludge across the width of the belt. Rows of plows assist in liberating

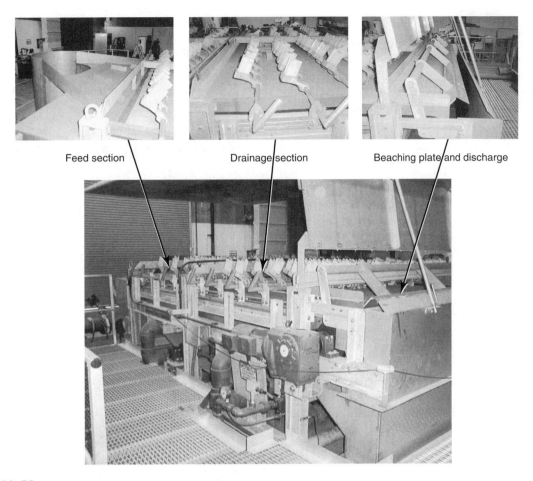

Feed section   Drainage section   Beaching plate and discharge

Figure 11–55
Gravity bell used to thicken waste-activated sludge. Polymer addition agglomerates floc and allows free water to drain through a continuously moving plastic fabric. Wash water keeps the fabric clear of trapped solids. (Photos taken at the Sacramento Regional Wastewater Treatment Plant in Elk Grove, California.)

additional water by opening drainage passages. Free water drains through the belt as the belt travels the horizontal length of the thickener. A pan beneath the belt collects the filtrate and directs flow to a drain. A beaching plate at the end of the belt causes a backrolling action and may be used to increase dewatering. Dewatered sludge falls into a small hopper at the end of the belt for pumping. Some sludge clings to the fabric and is scraped off by a doctor blade. Other particles drain into the weave of the fabric, decreasing the capacity to pass water. A high-pressure water spray washes the belt to remove polymer and sludge from the fabric on the belt return. The speed of

the belt can be controlled to increase or decrease the rates of flow.

Unit loading is normally in the range of 100 to 250 gpm (400 to 1000 l/m) per meter of belt width. Based on an influent feed between 0.5 to 1.0 percent, dewatered solids range from 4 to 6 percent, and 6 to 8 percent when using the beaching plate. Polymer in the range of 3 to 10 lb/ton (1.5 to 4.5 kg/ton) results in a solids capture between 90 and 98 percent. The wash-water requirement is 150 gpm at 50 psi. Rotary drum thickeners function similarly to the gravity belt, where free water from a flocculated sludge drains through a wire stainless steel screen.

Centrifuge Thickening

Centrifuge thickening machines are similar to dewatering centrifuges. For the same size machine, thickening machines can accept flow rates of 400 to 600 gpm (1500 to 2300 l/m) at 0.5 to 2 percent solids and produce thickened sludge in the range of 4 to 8 percent.

■ EXAMPLE 11–16

A conventional aeration plant treats 7.7 mgd of municipal wastewater with a BOD of 240 mg/l and suspended solids of 200 mg/l. The sludge flow pattern is shown in Figure 11–1 with a gravity belt thickener to concentrate the excess activated sludge. Primary and thickened activated sludge are pumped separately to the anaerobic digester. Estimate the quantities and solids contents of the primary, secondary, and mixed sludges. Assume the following: 50 percent SS removal and 35 percent reduction in BOD in the primary, a raw sludge concentration of 4 percent, an operating F/M ratio of 1:3, a waste-activated sludge concentration of 15,000 mg/l, and a 5 percent solids concentration from the gravity belt thickener.

Solution

Applying Eqs. 11–37 and 11–41 for the primary sludge,

$$W_p = 0.50 \cdot 200 \text{ mg/l} \cdot 7.7 \text{ mgd} \cdot 8.34$$
$$= 6410 \text{ lb/day}$$

$$V_p = \frac{6410}{\left(\dfrac{100 - 94}{100}\right) 8.34} = 12,800 \text{ gal/day}$$

From Figure 11–50, for an F/M ratio of 1:3, the K is 0.48. Substituting in Eqs. 11–38 and 11–41,

$$W_s = 0.48 \cdot 0.65 \cdot 240 \text{ mg/l} \cdot 7.7 \text{ mgd} \cdot 8.34$$
$$= 4800 \text{ lb/day}$$

$$V_s = \frac{4800}{(1.5/100)8.34} = 38,400 \text{ gal/day}$$

After gravity belt thickening, assuming 98 percent solids capture, the waste-activated sludge volume is

$$V_s = \frac{4800 \cdot 0.98}{(5.0/100)8.34} = 11,300 \text{ gal/day}$$

The blended sludge, after mixing primary and thickened secondary, has the following characteristics:

$$W_{ps} = W_p + W_s = 6410 + 4800 = 11,200 \text{ lb/day}$$
$$V_{ps} = V_p + V_s = 12,800 + 11,300$$
$$= 24,100 \text{ gal/day}$$
$$s = \frac{100 \cdot W_{ps}}{8.34 \cdot V_{ps}} = \frac{100 \cdot 11,200}{8.34 \cdot 24,100} = 5.57 \text{ percent}$$

■ ■ ■

■ EXAMPLE 11–17

Primary sludge containing trickling filter humus is gravity thickened in a circular tank with a diameter of 12.0 ft and side water depth of 10.0 ft. The applied sludge is 2600 gpd with 4.5 percent solids, and the thickened sludge withdrawn is 1300 gpd at 7.5 percent solids. The blanket of consolidating sludge in the tank has a depth of 3.0 ft. For odor control and to enhance thickening, 45,000 gpd of treated wastewater is pumped to the tank along with the sludge to increase the overflow rate. Calculate the solids loading on the tank bottom, the solids capture, the overflow rate, and the estimated time of solids retention in the tank.

Solution

$$\text{Tank area} = \pi(12.0 \text{ ft})^2/4 = 113 \text{ sq ft}$$

$$\text{Solids loading} = \frac{2600 \text{ gpd} \cdot 0.045 \cdot 8.34}{113 \text{ sq ft}}$$
$$= 8.6 \text{ lb/sq ft}$$

$$\text{Solids capture} = \frac{1300 \text{ gpd} \cdot 0.075 \cdot 8.34}{2600 \text{ gpd} \cdot 0.045 \cdot 8.34}$$
$$= \frac{815 \text{ lb/day}}{976 \text{ lb/day}} = 0.83 = 83\%$$

$$\text{Overflow rate} = \frac{(45,000 + 2600 - 1300) \text{ gpd}}{113 \text{ sq ft}}$$

$$= 410 \text{ gpd/sq ft}$$

Estimated solids in tank

$$= 3.0 \text{ ft} \cdot 113 \text{ sq ft} \cdot 0.075 \cdot 62.4 = 1590 \text{ lb}$$

$$\text{Solids retention time} = \frac{1590 \text{ lb} \cdot 24 \text{ hr/day}}{813 \text{ lb/day}}$$

$$= 47 \text{ hr}$$

■■■

11–14 REGULATORY REQUIREMENTS FOR TREATMENT AND DISPOSAL OF SEWAGE SLUDGE

Treatment and use of sewage sludge are regulated in 40 CFR Part 503, which was issued in 1993 under the authority of the Clean Water Act (CWA).[4] The rule was designed to encourage recycling by land application rather than disposal in a land- or monofill. Land application has been long used for biosolids and has been further encouraged by the limitations and costs associated with landfill disposal and incineration. The regulations cover numerical limits for pathogen reduction performance as well as technology-based requirements.

Some bacteria form protective shells, viruses remain dormant outside of the host, protozoa cyst, and helminth ova are resistant to the competitive pressures of other organisms during treatment in aerobic and anaerobic conditions. The survivability of pathogens in soil and on plants was studied and is

listed in Table 11–8. *Salmonella* sp. and fecal coliform are good indicators of the degree to which pathogenic bacteria are reduced by treatment. These bacteria are readily found in higher concentrations and are at least as hardy as pathogenic organisms. Untreated sewage sludge generally contains approximately 100 million MPN/g of fecal coliform.

Insects that do not cause disease but spread infection are often referred to as vectors, for example, mosquitoes, flies, fleas, and ticks. Untreated and partially treated sewage sludge attracts vectors because of the warm, moist, and biologically rich environment. Control of vectors is important to limiting the spread of infection.

11–15 REQUIREMENTS FOR CLASS B BIOSOLIDS

Class B biosolids receive significant treatment but are not considered pathogen-free. Requirements may be met by

1. Monitoring the fecal coliform content
2. Use of processes to significantly reduce pathogens
3. Using equivalent processes to significantly reduce pathogens
4. Site restrictions for land application

Monitoring fecal coliform requires that seven samples over a 2-week period have less than 2 million MPN/g of dry-weight biosolids. Given an average of 100 million MPN/g in untreated sludge, processes achieving a Class B biosolid are required to achieve a 1.7 log removal. A high standard deviation suggests sampling or laboratory analysis variability or may indicate inconsistent reduction

TABLE 11–8

Survivability of Pathogens in Soil

| PATHOGEN | SOIL | | PLANTS | |
	ABSOLUTE MAXIMUM	COMMON MAXIMUM	ABSOLUTE MAXIMUM	COMMON MAXIMUM
Bacteria	1 year	2 months	6 months	1 month
Viruses	1 year	3 months	2 months	1 month
Protozoa cyst	10 days	2 days	5 days	2 days
Helminth ova	7 years	2 years	5 months	1 month

during treatment. A log standard deviation under 0.3 is acceptable. Each sample should be taken at the same point, handled consistently, and analyzed within 24 hr. In addition, treatment requirements for vector control must be met.

Table 11–9 lists processes to significantly reduce pathogens, including aerobic and anaerobic digestion, air drying, composting, and lime stabilization. Equivalent processes must be able to consistently reduce coliform to 2 million MPN/g under a variety of conditions that may be site-specific. These processes include a variety of proprietary equipment and arrangements promoted by individual manufacturers.

Aerobic Digestion

Waste sludge can be stabilized by long-term aeration that biologically destroys volatile solids. The aerobic digestion process was developed specifically to handle excess activated sludge from aerobic treatment plants without primary clarifiers (Figure

TABLE 11–9

Processes to Significantly Reduce Pathogens Creating Biosolids Meeting Requirements for a Class B Grade[5]

TREATMENT PROCESS	CLASS B REQUIREMENTS	VECTOR CONTROL REQUIREMENTS
AEROBIC DIGESTION		
Feed alternatives: Continuous, no supernatant removal Continuous, with supernatant removal Continuous, with batch sludge removal Batch or staged digesters	40 days at 68°F (20°C) or 60 days at 58°F (15°C)	VSS reduction $\geq$ 38% or when SOUR $\leq$ 1.5 mg at 68°F (20°C) or if additional VSS destruction $<$ 15% over 30 days
ANAEROBIC DIGESTION		
Feed alternatives: Continuous, no supernatant removal Continuous, with supernatant removal Continuous, with batch sludge removal Batch or staged digesters	**Standard rate (no mixing, little or no heating)** 60 days at 68°F (20°C) **High rate (continuous mixing, controlled temperature)** 15 days at 95 to 131°F (35 to 55°C)	VSS reduction $\geq$ 38% or additional VSS destruction $<$17% over 40 days; SOUR test is not appropriate
AIR DRYING		
Air drying in open air on sand or sludge drying beds	Air drying for a minimum of 3 months with 2 of 3 months having an air temperature $>$32°F (0°C)	VSS reduction $\geq$ 38%
COMPOSTING		
Using wood hips, bark, sawdust, rice hulls, or additive to absorb moisture and increase air flow in sludge pile	Windrows or in vessel mixing achieving sustained elevated temperatures	Sludge pile must achieve temperatures of $>$104°F (40°C) for 14 days with an average of 113°F (45°C)
LIME STABILIZATION		
Using hydrated lime $Ca(OH)_2$ or quicklime CaO	Lime added to raise the pH of sludge to 12 for 2 hr	Sludge pH must be $\geq$ 12 for 2 hr and $\geq$ 11.5 for 24 hr

11–2a). Early attempts to anaerobically digest this sludge failed because of the low solids concentration and aerobic nature of the waste. High water content, in the range of 98 to 99 percent, also prevented economical dewatering by mechanical means without prior thickening. Furthermore, because most of the aeration plants served small communities, large investments for mechanical thickening equipment and anaerobic digestion could not be justified for waste sludge processing. Consequently, aerated holding tanks were introduced to stabilize and store mixed liquor drawn from the aeration tank. Aerobic digesters are single or multiple tanks equipped with diffused or mechanical aerators. Gates or pipes are provided to draw off supernatant after gravity settling of the suspended solids with the aerators shut off. Figure 11–56 shows two aerobic digesters used to treat septage collected from household, industrial, and forest-service septic tanks.

Design standards for aerobic digestion vary with the character of waste sludge and method of ultimate disposal. The general guideline for stabilizing waste-activated sludge, with a suspended solids concentration of about 1.0 percent, is a maximum volatile solids loading of 0.04 lb/cu ft/day ($640 \text{ g/m}^3 \cdot \text{d}$) and an aeration period in the range of 200 to 300 degree-days, which is computed by multiplying the digesting temperature in °C times the sludge age. However, EPA requires at least 800 degree days to meet class B requirements. Volatile solids and BOD reductions at these loadings are in the range of 30

to 50 percent. In small plants, the aerobic digester volume is often 2 to 3 cu ft per design population equivalent of the treatment plant. This provides a loading in the range of 0.01 to 0.02 lb VS/cu ft/day ($160 \text{ to } 320 \text{ g/m}^3 \cdot \text{d}$). Aeration of waste-activated sludge creates a digested sludge that resists gravity thickening. By quiescent setting without air mixing, the solids concentration can usually be gravity thickened to approximately 2.0 percent solids. The maximum concentration with careful withdrawal of supernatant is not likely to exceed 2.5 percent. This poor settleability frequently causes problems in disposing of the large volume of sludge produced. The common methods of disposal are liquid spreading on farmland, lagooning, and drying on sand beds. If drying beds are used, they must be deep enough to allow application of 3 ft or more of digested sludge so that the dried cake will be 1 to 2 in. thick.

An aerobic digester is normally operated by continuously feeding raw sludge, with intermittent supernatant and digested sludge withdrawals. The digesting sludge is continuously aerated during filling and for the specified digestion period after the tank is full. Aeration is then discontinued to allow the stabilized solids to settle by gravity. Supernatant is decanted and returned to the head of the treatment plant, and a portion of the gravity-thickened sludge is removed for disposal. In practice, aeration and settlement may be a daily cycle with feed applied early in the day and clarified supernatant decanted later in the day. Digested solids are withdrawn when the sludge in the tank does not gravity thicken to provide a supernatant of adequate clarity.

Figure 11–56

Photo of aerobic digesters for treatment of septage at the Union Mine Disposal Site, El Dorado County, California. Air on the right digester has been turned off to allow solids to settle and removal of supernatant. The septage hauling truck discharges to a grit removal unit.

Anaerobic Digestion

The purpose of sludge digestion is to convert bulky, odorous sludges to a relatively inert material that can be rapidly dewatered without obnoxious odors. The bacterial process, as summarized in Eq. 11–49, consists of two successive processes that occur simultaneously in digesting sludge. The first stage consists of breaking down large organic compounds and converting them to organic acids along with gaseous by-products of carbon dioxide, methane, and trace amounts of hydrogen sulfide. This step is performed by a variety of facultative bacteria operating in an environment devoid of oxygen. If the process were to stop there, the accumulated acids would lower the pH and would inhibit further decomposition by "pickling" the remaining raw sludge. For digestion to occur, second-stage gasification is needed to convert the organic acids to methane and carbon dioxide. Acid-splitting methane-forming bacteria are strict anaerobes and are very sensitive to environmental conditions of temperature, pH, and anaerobiosis. In addition, methane bacteria have a slower growth rate than the acid formers and are very specific in food supply requirements. For example, each species is restricted to the metabolism of only a few compounds, mainly alcohols and organic acids, while carbohydrates, fats, and proteins are not available as energy sources.

$$\text{Organic} \xrightarrow[\substack{\text{Acid-forming} \\ \text{bacteria}}]{\substack{CO_2, CH_4 \\ H_2S}} \text{Organic acids} \xrightarrow[\substack{\text{Acid-splitting} \\ \text{methane-forming} \\ \text{bacteria}}]{} \substack{CH_4 \\ \text{and} \\ CO_2} \qquad (11\text{–}49)$$

The stability of the digestion process relies on the proper balance of the two biological stages. A buildup of organic acids may result from either a sudden increase in organic loading or a sharp rise in operating temperature. In either case, the supply of organic acids exceeds the assimilative capacity of the methane-forming bacteria. This imbalance results in decreased gas production and eventual drop in pH, unless the organic loading is reduced to allow recovery of the second-stage reaction. Digesters may generate foam as a result of overfeeding. An accumulation of toxic substances from industrial wastes, such as heavy metals, may also inhibit the digestion reaction. The exact cause of digester problems is often difficult to determine. Monitoring volatile solids loading, total gas production, volatile acids concentration in the digesting sludge, and percentage of carbon dioxide in the head gases are the methods most frequently employed to give advance warning of pending failure. These measurements can also indicate the most probable cause of difficulties. Gas production should vary in proportion to organic loading. Volatile acids content is normally stable at a given loading rate and operating temperature. The percentage of carbon dioxide should also remain relatively constant. Monitoring digestion by pH measurements is not recommended, since a drop in pH does not precede

failure but announces that it has occurred. Table 11–10 lists the general operating and loading conditions for anaerobic digestion.

TABLE 11–10

General Operating and Loading Conditions for Anaerobic Sludge Digestion

Temperature	
Optimum	98°F (36.7°C)
General operating range	85°–99°F (29°–37°C)
pH	
Optimum	7.0 to 7.1
General limits	6.7 to 7.4
Gas production	
Per pound of volatile solids added	8–12 cu ft (230–340 l)
Per pound of volatile solids destroyed	16–18 cu ft (450–510 l)
Gas composition	
Methane	65 to 69 percent
Carbon dioxide	31 to 35 percent
Hydrogen sulfide	trace to 80 mg/l
Volatile acids concentration	
General operating range	200 to 800 mg/l
Alkalinity concentration	
Normal operation	2000 to 3500 mg/l
Volatile solids loading	
Conventional single stage	0.02–0.05 lb VS/cu ft/day[a]
First-stage high rate	0.05–0.15 lb VS/cu ft/day
Volatile solids reduction	
Conventional single stage	50 to 70 percent
First-stage high rate	50 percent
Solids retention time	
Conventional single stage	30 to 90 days
First-stage high rate	15 to 20 days
Digester capacity based on design equivalent population	
Conventional single stage	4 to 6 cu ft/PE[b]
First-stage high rate	0.7 to 1.5 cu ft/PE

[a] 1.0 lb/cu ft/day = 16,000 g/m³·d
[b] 1.0 cu ft = 0.0283 m³

Single-Stage Digestion

A photo of a single-stage fixed-cover anaerobic digester is shown in Figure 11–57. The photo also shows ancillary equipment associated with digester heating: boiler, heat exchanger, and sludge recirculation piping. Raw sludge is pumped into the tank through feed pipes. Mixing pumps discharge at nozzles within the digester to keep the contents from stratifying. Without mixing, sludge separates, with a scum layer on top, a middle zone of supernatant (water of separation) underlain by actively digesting sludge, and a bottom layer of digested concentrate. A limited amount of mixing is also provided by withdrawing digesting sludge, passing it through a sludge heater, and returning it through the inlet piping. Supernatant is withdrawn from any one of a series of pipes extended from the supernatant box. Digested sludge is taken from the tank bottom for dewatering. High-rate digesters are completely mixed, the contents do not tend to separate or develop a clear supernatant, and the entire contents of the digester must be dewatered.

For digesters designed with floating covers, the cover floats on the sludge surface, and liquid extending up the sides provides a seal between the tank wall and the side of the cover. Gas rising out of the digesting sludge is collected in the gas dome and is burned as a fuel in the sludge heater; often the excess is wasted to a gas burner. The cover can rise vertically from the landing brackets to near the top of the tank wall, guided by rollers around the circumference to keep it from binding. The volume between the landing brackets and the fully raised cover position is the amount of storage available for digested sludge; this is approximately one-third of the total volume.

Digestion in a single-stage floating-cover tank performs the functions of volatile solids digestion, gravity thickening, and storage of digested sludge. When sludge is pumped into the digester from the primary settling tanks, the floating cover rises, making room for the sludge. Unmixed operation permits daily drainage of supernatant equal to approximately two-thirds of the raw sludge feed. Being high in both BOD and suspended solids, the withdrawn water is returned to the inlet of the treatment plant. Periodically, digested sludge is removed for dewatering and disposal. In large plants, digested sludge may be dewatered mechanically; however, in small installations it is frequently spread in liquid form on farmland or is

Figure 11–57

Photo of a single-stage fixed-cover anaerobic digester. The support equipment includes the heating system (hot-water boiler and sludge heat exchanger), gas withdrawal piping, sludge feed, and withdrawal piping (El Dorado Irrigation District, El Dorado Hills WWTP, CA).

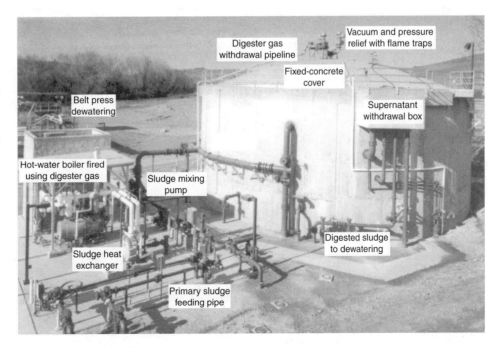

dried on sand beds and hauled to land burial. Weather often dictates the schedule for land disposal, and, consequently, substantial digester storage volume is required in northern climates. In a typical operation, the cover is lowered to the landing brackets in the fall of the year to provide maximum storage volume for the winter.

Fixed-cover digesters, where sludge is withdrawn as the digested sludge is displaced by the raw feed sludge, maintain a constant volume. Fixed-cover digesters require holding tanks, sludge lagoons, or other locations where displaced digested sludge can drain. Because the volume is constant and the cover is fixed, these digesters can be mixed by roof-mounted turbine mixers, externally mounted pumps, and gas mixing in draft tubes. Turbine, roof-mounted mixers are very efficient at mixing the entire tank contents. Rags can be removed by reversing the mixing direction. External mixing pumps can be mounted in draft tubes inside or outside of the digester or in a pump piped to the digester. Figure 11–57 shows mixing pumps mounted outside of the digester tank. Pump mixing is also very effective but may require

multiple discharge points for large digesters. Gas mixing induces a flow within the draft tube to provide mixing. Mixing requirements may be expressed in terms of power input or turnover time. Typical values for power are 0.2 to 0.3 hp/1000 cu ft (0.005 to 0.008 kW/m³). No allowance is made for the efficiency of converting power into mixing. Turnover time is calculated by taking the volume of the digester divided by the mixing flow rate. Typical designs are based on turnover rates of 30 to 60 min.

Two-Stage Digestion

In this process, two digesters in series separate the functions of biological stabilization from gravity thickening and storage, as shown in Figure 11–58. The first-stage high-rate unit is completely mixed and heated for optimum bacterial decomposition. These systems are available for installation in either fixed- or floating-cover tanks. By using a floating cover, digested sludge does not have to be displaced simultaneously with raw sludge feed, as is required with a fixed-cover tank. In either case,

Figure 11–58

Two-stage anaerobic digestion is performed by two tanks in series. (a) The first-stage tank on the left is completely mixed for optimum digestion. The second-stage with a gas dome cover is for gravity thickening and storage of digested sludge. (b) Photo of two-stage digesters at the Northeast Wastewater Treatment Facility in Lincoln. Nebrasaka.

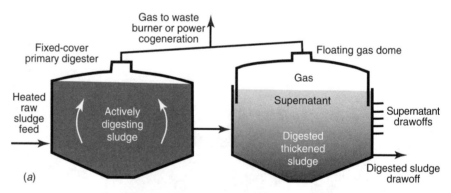

(a)

(b)

however, the sludge cannot be thickened in a high-rate process because continuous mixing does not permit formation of supernatant. Actually, the discharged sludge has a lower solids concentration than the raw feed because of the conversion of volatile solids to gaseous end products. The second-stage digester must have either a floating cover or gas dome and have provisions for withdrawing supernatant. The unit is often unheated, depending on the local climate and degree of stabilization accomplished in the first stage. By minimizing hydraulic disturbances in the tank, the density of the digested sludge and clarity of the supernatant are both increased. Two-stage digestion may be advantageous in some plants, while conventional operation may be better in others. The determining factors include the size of the treatment plant, flexibility of sludge-handling processes, method of ultimate solids disposal, storage capacity needed, and the interrelated element of climatic conditions. For large plants with a number of digesters, series operation provides better utilization of digester capacity, but for small plants with limited supervision, the conventional operation is frequently more feasible.

Sizing of Digesters

Historically, conventional single-stage tanks have been sized on the basis of population equivalent load on the treatment plant. Heated digester capacity for a trickling filter plant processing domestic wastewater was established at 4 cu ft (0.11 m³) per capita of design load. For primary plus secondary activated sludge, the total tank volume requirement was increased to 6 cu ft (0.17 m³) per capita. These values are still used as guidelines for sizing conventional digesters for small treatment works.

Total digestion capacity can be calculated for conventional single-stage operation using Eq. 11–50. Application of this formula requires knowing the characteristics of both the raw and digested sludges.

$$V = \frac{V_1 + V_2}{2} T_1 + V_2 \times T_2 \qquad \textbf{(11–50)}$$

where V = total digester capacity, gallons (cubic meters)
V_1 = volume of daily raw sludge applied, gallons per day (cubic meters per day)

V_2 = volume of accumulated digested sludge, gallons per day (cubic meters per day)
T_1 = period required for digestion, days (approximately 30 days at a temperature of 85° to 90°F or 30° to 35°C)
T_2 = period of digested sludge storage, days

The volume needed for the high-rate unit in a two-stage digestion system is based on a maximum volatile solids loading and minimum detention time. For new designs, the generally adopted maximum allowable loading is 0.08 lb VS/cu ft/day (1300 g/m³·d) and the minimum liquid detention time is 15 days. At these loadings and a temperature of 95°F, volatile solids reduction should be 50 percent or greater. No specific design criteria are established for second-stage tanks in high-rate systems, since thickening and digested sludge requirements depend on local sludge disposal procedures.

Start-up of Digesters

Anaerobic digestion is a difficult process to start because of the slow growth rate and sensitivity of acid-splitting methane-forming bacteria. Furthermore, the number of these microorganisms is very low in raw sludge compared with acid-forming bacteria. The normal procedure for start-up is to fill the tank with wastewater and to apply raw sludge feed at about one-tenth of the design rate. If several thousand gallons of digesting sludge from an operating digester are used as seed, the new process can be operational in a few weeks. However, if only raw sludge is available, developing the biological process may take months. Careful additions of lime with raw sludge are helpful in maintaining the pH near 7.0, but erratic dosage can result in sharp pH changes detrimental to the bacteria. After gas production and the volatile acids concentration have stabilized, the feed rate is gradually increased by small increments to full loading. Daily monitoring of this process involves plotting the daily gas production per unit of raw sludge fed, percentage of carbon dioxide in the head gases, and concentration of volatile acids in the digesting sludge.

■ EXAMPLE 11–18

An anaerobic sludge digester operating at a feed rate of 0.06 lb VS/cu ft/day has good gas production and volatile solids reduction. The external sludge heater fails and is not repaired for about 6 weeks. During heater shutdown, the same feed rate is maintained and a noticeable decrease in gas production occurs as the temperature of the digesting sludge drops from 95° to 75°F. When the heater is returned to service, the digester temperature is rapidly elevated back to 95°F with a sudden initial increase in gasification. However, after a few days of renewed operation, gas production decreases sharply, volatile acids are measured at about 2000 mg/l, and the pH of the digesting sludge begins to drop. Describe what has happened to the anaerobic biological process.

Solution

Initially, the acid-forming and methane-forming populations are in balance, and, because they are operating near optimum temperature, the majority of the volatile solids are being converted to gaseous end products. When the heater fails and the temperature of the digesting sludge drops, a new equilibrium of populations is established at a reduced gasification efficiency. Therefore, raw organic matter accumulates in the digesting sludge. These dormant volatile solids suddenly become available to the bacteria when the temperature is raised back to 95°F. The sharp increase in gas production is related primarily to carbon dioxide produced when first-stage bacteria convert volatile solids to organic acids. The acid-splitting methane formers attempt to respond to the increased organic acid supply. However, their populations have been reduced by the lack of food during low-temperature operation. Furthermore, the methane bacteria, being more sensitive, are inhibited by the acid conditions; hence, total gas yield decreases. The solution is to drop the temperature of the digesting sludge back to 75°F and to increase it slowly, perhaps 1 degree per week, to allow the methane formers to adjust gradually to the increasing production of organic acids.

■ ■ ■

■ EXAMPLE 11–19

Calculate the capacity required per population equivalent for a single-stage floating-cover digester based on the following: 0.24 lb of solids contributed per person, 4.0 percent solids content in raw sludge, 7.0 percent solids concentration in digested sludge, 40 percent total solids reduction during digestion, 30-day digestion period, and 90-day storage period for digested sludge.

Solution

Using Eq. 11–41, the daily volumes of raw sludge produced and digested sludge accumulated are

$$V_1 = \frac{0.24 \text{ lb/person/day}}{(4.0/100)8.34} = 0.72 \text{ gal/person/day}$$

$$V_2 = \frac{0.60 \cdot 0.24 \text{ lb/person/day}}{(7.0/100)8.34}$$

$$= 0.25 \text{ gal/person/day}$$

Then, the digester capacity by Eq. 11–50 is

$$V = \frac{0.72 + 0.25}{2} \cdot 30 + 0.25 \cdot 90 = 37 \text{ gal}$$

$$= 5.0 \text{ cu ft}$$

(*Note:* This calculation verifies the single-stage digester capacity based on design equivalent population listed in Table 11–10.)

■ ■ ■

■ EXAMPLE 11–20

A trickling filter plant has two floating-cover digesters each with a total capacity of 30,000 cu ft with 20,000 cu ft below the landing brackets. The remaining 10,000 cu ft in each tank is the volume between the lowered and fully raised cover positions (storage volume). The sludge piping is arranged so that the digesters may be operated in parallel as conventional digesters or in series as a two-stage system. One digester is equipped with gas mixing to serve as a first-stage high-rate process. Daily raw sludge production is 6400 gal, containing 2800 lb of dry solids that are 70 percent volatile. Digested sludge is 8.0 percent solids, and the process converts 60 percent of the volatile solids to gas. (a) For conventional single-stage operation with half of the waste applied to each tank, calculate the volatile solids loading, supernatant produced, and number of days of digested sludge storage available. (b) For two-stage operation, compute the volatile solids loading on the first tank, solids content of the digested sludge leaving the high-rate digester, and available sludge storage in the system.

Solution

(a) For conventional single-stage operation,

$$\text{VS applied to each tank} = (0.70 \cdot 2800)/2$$
$$= 980 \text{ lb/day}$$

$$\text{Loading with covers lowered} = \frac{980 \text{ lb VS/day}}{20{,}000 \text{ cu ft}}$$

$$= 0.049 \text{ lb VS/cu ft/day}$$

$$\text{Loading with covers raised} = \frac{980 \text{ lb VS/day}}{30{,}000 \text{ cu ft}}$$

$$= 0.033 \text{ lb VS/cu ft/day}$$

$$\frac{\text{Supernatant return}}{\text{day}} = \frac{\text{raw sludge volume}(V_1)}{\text{day}}$$
$$- \frac{\text{digested sludge volume}(V_2)}{\text{day}}$$

$$V_1 = 6400 \text{ gal}$$

$$V_2 = \frac{\text{inert solids} + \text{remaining VS}}{(\text{solids content}/100)8.34}$$
$$= \frac{0.3 \cdot 2800 + 0.4(0.7 \cdot 2800)}{(8.0/100)8.34}$$

$$= 2400 \text{ gal}$$

Supernatant return $= 6400 - 2400 = 4000$ gal/day

Digested sludge storage available

$$= \frac{2 \cdot 10{,}000 \text{ cu ft} \cdot 7.48 \text{ gal/cu ft}}{2400 \text{ gal/day}} = 62 \text{ days}$$

(b) For two-stage operation,

$$\text{VS applied to high-rate tank} = 0.7 \cdot 2800$$
$$= 1960 \text{ lb/day}$$

Loading with covers lowered

$$= (20{,}000 \cdot 7.48)/6400 = 23 \text{ days}$$

With the cover raised, the volume is 30,000 cu ft.

$$\text{Loading} = 0.065 \text{ lb VS/cu ft/day}$$

$$\text{Detention time} = 35 \text{ days}$$

The solids content in digested sludge leaving the first-stage high-rate digester (Eq. 11–41) is given by

$$s = \frac{0.3 \cdot 2800 + 0.4(0.7 \cdot 2800)}{6400 \cdot 8.34/100} = 3.0 \text{ percent}$$

(*Note:* The raw sludge is 5.2 percent solids. The percentage of solids reduces to 3.0 percent in high-rate digestion, since volatile solids are converted to gas while no supernatant is withdrawn to thicken the digested sludge.

Digested sludge storage, available only in the second-stage tank, is equal to 31 days. This reduced storage capacity is a disadvantage of high-rate digestion compared with parallel operation, which permits 62 days storage.)

■ ■ ■

Air Drying Beds

Open-air drying of anaerobically digested sludge has been used since the practice of sludge digestion was started. The standard design has been a bed consisting of a 12-in. sand layer underlain by graded gravel surrounding tile or perforated pipe underdrains. Large beds are partitioned by concrete walls, and a pipe header from the digesters with gated openings allows application of sludge independently to each cell. Seepage collected in the underdrains is returned to the plant wet well for treatment with the raw wastewater. Cleaning dried sludge cake from the beds is a laborious job. The cake has to be shoveled and taken off in wheelbarrows for loading in a truck. Attempts to use mechanical equipment result in excessive loss of sand and disturbance of the gravel underdrain.

In modern design, drying beds are constructed to permit the use of tractors with front-end loaders to scrape up the sludge cake and dump it into a truck. The arrangement of drying beds with paved surfaces to allow mechanical removal of dried cake is illustrated in Figure 11–59. The main features are: (a) watertight walls extending 18 to 24 in. above the surface of the bed; (b) an end opening in the wall, sealed by inserting planks, for entrance of a front-end loader; (c) centrally located drainage trenches filled with a coarse sand bed supported on a gravel filter with a perforated pipe underdrain; (d) paved areas on both sides of the trenches with a 2 to 3 percent slope for gravity drainage; and (e) a sludge inlet at one end and supernatant drawoff at the opposite end. The dimensions of the individual beds are a width of 20 ft and length of 100 ft. The operating procedure is to apply digested sludge to a depth of 12 in. or

more, draw off supernatant after the solids settle, and allow the sludge to dry. A well-digested sludge forms a thoroughly dry black cake 3 to 5 in. in thickness and with cracks resulting from horizontal shrinkage.

The sizing of sludge beds in design is difficult because of the multitude of factors that influence drying time. These include climate and atmospheric conditions of temperature, rainfall, humidity, and wind velocity; sludge characteristics of degree of stabilization, grease content, and solids concentration; depth and frequency of sludge application; and condition of the drying bed sand and underdrain piping. Typical criteria for the sizing bed area are 1 to 2 sq ft per BOD design equivalent of the treatment plant, or solids loadings of 20 lb/sq ft/yr (100 kg/m^2 · y) in northern states and up to 40 lb/sq ft/yr in southern states. Drying time ranges from several days to weeks depending on gravity drainage of water from the wet sludge and suitable weather conditions for evaporation.

Figure 11–59

Open-air beds for drying digested sludge with paved surfaces to allow mechanical removal of dried cake. Water is separated by decanting supernatant, draining to trenches, and evaporation. (Courtesy of HDR Engineering, Inc.)

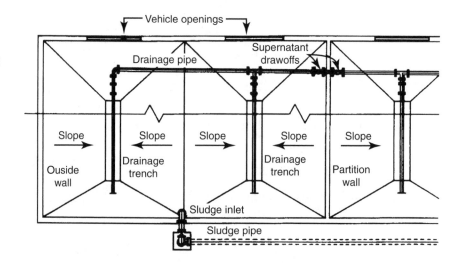

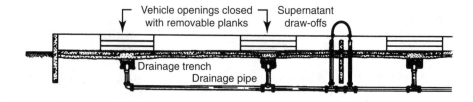

Composting

The most common method of composting includes periodically mixed windrows, a static aerated pile, and processing within an enclosed vessel. Windrows and piles may be either exposed to the atmosphere or sheltered under a roofed structure. The choice of process and sheltering is based on climate, environmental control, availability of a binding agent, and economic considerations. Static pile composting to further reduce pathogens requires that sludge temperatures of 131°F (55°C) or higher are maintained for 3 days. Windrow piles must maintain 131°F (55°C) or higher for 15 days and be turned a minimum of five times during that period.

Composting is the stabilization of moist organic solids by natural biological processes when the organic matter is placed in piles that allow ventilation. In addition to digesting putrescible organic matter, the objectives of sludge composting are to destroy pathogenic organisms and reduce the mass and volume of waste. The optimum moisture content for a compost mixture is 50 to 60 percent. Less than 40 percent limits the rate of decomposition, and over 60 percent is generally too wet for adequate ventilation. The volatile solids reduction and water loss during composting are 50 percent or more. For efficient stabilization and pasteurization, the temperature in a compost pile should rise to at least 104°F (40°C) for 14 days, but not above 176°F (80°C).

Composting temperature is influenced by moisture content, degree of aeration, size and shape of the pile, and climatic conditions, particularly air temperature and rainfall. The finished compost is a friable humus with a moisture content less than 40 percent. Although too low in nutrients to be considered a fertilizer, compost is an excellent soil conditioner. For example, when mixed with soil, the added humus content increases the capacity for retention of water.

Sludge cake for composting is usually raw organic solids dewatered using polymer as a conditioning chemical, although partially digested sludge may also be composted for additional stabilization. Dewatered cake alone, commonly having a solids content of 20 to 30 percent, is too wet and too compact to have adequate porosity for aeration of the interior of a pile. If it is not mixed with another substance, the pile of sludge cake will become a dense mass with a wet, anaerobic interior and a dried external crust. Therefore, the dewatered cake is mixed with either an organic amendment like dried manure, straw, or sawdust or a recoverable bulking agent, which is usually wood chips, to reduce the unit weight and increase air voids.

Compost piles in the windrow system are arranged in long parallel rows. These windrows are remixed and turned at regular intervals to restructure the compost, ensuring uniform biological stabilization. Depending on the equipment used for turning, the shape varies in height and width. The common dimensions are 4 to 8 ft in height and 8 to 12 ft in width. Figure 11–60 pictures a self-propelled windrow composting machine straddling a triangular pile 6 ft high and 12 ft wide that is being restructured. The mixing is done by a rotating agitator mounted near the ground between the side plates located just inside the wheels.

Lime Stabilization

Hydrated lime ($Ca(OH)_2$) or quicklime (CaO) may be added to undigested liquid sludge, digested sludge, or dewatered sludge cake to raise the pH and destroy pathogens. Lime stabilization does not reduce the volatile solids concentration, but at a pH greater than 11, biological activity stops. If

Figure 11–60

Windrow composting machine for mixing and piling compost in triangular-shaped rows.

the pH drops below 10.5, bacteria may regrow, causing rapid decomposition and generating a high degree of hydrogen sulfide and organic odors. Stabilization reduces bacterial and viral pathogens by more than 99 percent, but has little impact on hardy species of helminth ova.

11–16 REQUIREMENTS FOR CLASS A BIOSOLIDS

Class A biosolids require a further reduction of pathogens to a point below detectable limits. Requirements (see Table 11–11) are met by

TABLE 11–11

Processes to Further Reduce Pathogens Meeting Requirements for Class A Biosolids[5]

TREATMENT PROCESS	CLASS A REQUIREMENTS	PROCESS REQUIREMENTS
COMPOSTING		
Aerated pile (similar to windrow with forced air ventilation under pile), mixed to enhance treatment In-vessel mixing Windrows, mixed to enhance treatment	Aerated pile or in vessel mixing achieving ≥ 131°F (55°C) for 3 days; Windrows mixing achieving ≥ 131°F (55°C) for 15 days with 5 turnings of the pile	Sufficient moisture must be maintained in the compost pile to support biological activity, typically 45–60% moisture
HEAT DRYING		
Flash dryers—hot gases interact with pulverized sludge Spray dryers—atomized sludge is sprayed into a drying chamber with hot gases Rotary dryers—drum mixes sludge with hot gases Steam dryers—disks or paddles used to mix sludge with steam	Sludge dried to 90% solids	Unstabilized dry solids may become odorous if they become wet
THERMOPHILIC AEROBIC DIGESTION		
Mixed with air or oxygen to maintain aerobic conditions Batch fed no more than once per day	10 consecutive days at 131 to 140°F (55 to 60°C)	Digested biosolids must be tested for fecal coliform or *Salmonella* sp. at the time of use
Beta Ray and Gamma Ray Radiation	Dose 1.0 megarad at 68°F (20°C)	To meet vector attraction requirements, sludge requires VSS reduction or pH adjustment or land incorporation
Pasteurization	Sludge temperature maintained >158°F (70°C)	To meet vector attraction requirements, sludge requires VSS reduction or pH adjustment or land incorporation

1. Processing under high temperatures
2. Radiation
3. Chemical disinfection
4. Drying.

Class A biosolids requirements limit *Salmonella* sp. to less than 3 MPN/4 g total dry weight, enteric viruses to less than 1 PFU/4 g, and viable helminth ova to less than 1 ova/4 g. One of the vector attraction requirements must also be met, typically through a process that meets Class B requirements. Class A biosolids may be used without restriction. Evidence of compliance with Class A requirements may be established with a fecal coliform value of less than 1000 MPN/g or *Salmonella* sp. value of less than 3 MPN/4 g of total dry-weight solids. Using 100 million MPN/g in the untreated sludge, Class A treatment requires a 5 log removal.

Heat Drying

Mechanical dewatering is used as a pretreatment prior to drying biosolids. Dryers may be horizontally or vertically oriented (see Figure 11–61). In the vertical multistage tray dryer shown, sludge is dried and pelletized in a single process. The horizontal, hollow trays are heated by thermal oil in a recirculating loop. The heat needed for drying can be generated in an oil heater or by heat exchange with another heat source. An external dosing hopper is used to hold dry sludge pellets. Some of the pellets are recirculated through a coater, where dry sludge pellets are mixed with incoming dewatered sludge, forming a thin layer of wet sludge on the pellet. These are fed into the distribution cone at the top of the feeding and heating section. A raking assembly connected to a central rotating shaft moves the drying pellets over the tray to the outer edge where they fall onto the next tray and are

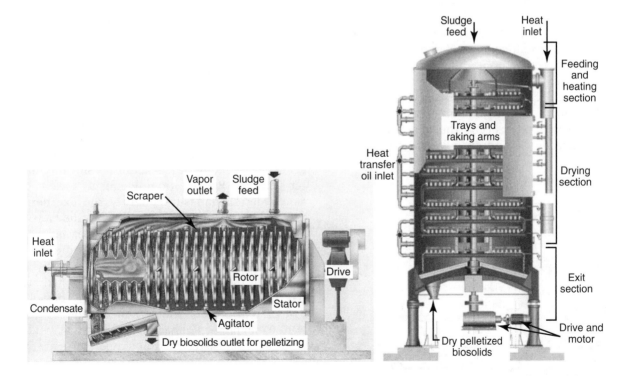

Figure 11–61

Cutaway illustrations showing (*a*) disk dryer and (*b*) multiple tray dryer. A multiple-tray dryer is an indirect drying system using heat-transfer oil that circulates through a series of hollow trays. Vapor removed from the dryer is passed through a vapor condenser. Dried biosolids are pelletized and bagged and sold as fertilizer or as fuel pellets. (Courtesy of Keppel Seghers, Inc.)

pushed back to the center. Sludge dries as it falls from tray to tray in the drying section. Dried pellets collected at the bottom are transported to the top where smaller pellets are separated and recirculated. Finished pellets are cooled to less than 105°F to prevent spontaneous combustion and avoid absorption of humidity to prevent pellets from sticking together. Pellets are between 2 and 5 mm and contain 95% solids. The pellets meet Class A biosolids requirements and may be bagged and sold for fertilizer or burned as a pellitized fuel.

Thermophilic Aerobic Digestion

Autothermal thermophilic aerobic digestion (ATAD) is a high-temperature aerobic process producing a Class A biosolid. Normal operating temperatures range from 132° to 149°F (56° to 65°C) and are typically maintained from 4 to 20 hours to pasteurize the sludge. The metabolic activity generates sufficient heat to raise the temperature without an external heat source. Processing is accomplished in a single-stage insulated treatment tank with a detention time of 10–15 days. Support equipment consists of a mixing pump, jet aeration system, blower, and control system (see Figure 11–62). Sludge is fed daily, creating a feast-and-famine cycle of biological activity. Sufficient oxygen supply is critical to maintain an aerobic environment and prevent the odors associated with anaerobic conditions. Foam control, as a

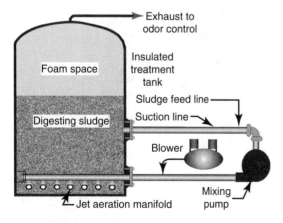

Figure 11–62
Autothermal thermophilic aerobic digestion consisting of an insulated treatment vessel, mixing pump, and blower.

consequence of protein degradation, is important to normal operation. An oxidation–reduction potential (ORP) sensor is used to monitor the treatment process and control the air supply. Volatile solids destruction ranges from 45 to 55 percent and the result can be readily dewatered, producing a cake with 20 to 25 percent solids.

Pasteurization

Pasteurization occurs when sludge is heated to 70°F (158°C) or higher for 30 min. Sludge may be heated using heat exchangers, steam injection, or a combination of lime addition and external heating.

Figure 11–63 illustrates a pasteurization process as pretreatment to anaerobic digestion. Cold solids are initially heated using the pasteurized solids. This recovers excess heat and reduces the solids temperature from the pasteurization to a point closer to the digester operating temperature. A separate boiler and heat exchanger is used in conjunction with a mixed reaction vessel to provide the heat and detention time required for pasteurization. Pasteurization provides the pathogen reduction, while anaerobic digestion provides the volatile solids reduction to meet the Class A requirements for volatile solids reduction and vector control.

A combination of anaerobic digestion with lime addition to the sludge cake and supplemental heating can also achieve the pasteurization requirements for time and temperature. The reaction of lime and the water remaining in the sludge cake generates an exothermic reaction, raising the sludge temperature. Preconditioning enhances the hydration reaction and subsequently the exothermic reaction rate. From the thermofeeder, the biosolids will be transferred into a thermoblender and mixed with quicklime. The sludge-lime mixture is then augered into the pasteurization vessel to fully complete the 70°C for 30 min time-temperature requirements for producing Class "A" biosolids. The pasteurized biosolids are then ready for distribution. Supplemental heat, often electrical, is added to ensure that pasteurization temperatures are met.

11–17 DEWATERING

During the sludge dewatering process, water is removed from sludge in order to concentrate the solids present. Mechanical assistance, such as

Figure 11–63

Schematic of pasteurization preceding anaerobic digestion designed to meet the treatment requirements for Class A biosolids. The resulting solids may be spread as a liquid or dewatered prior to land application.

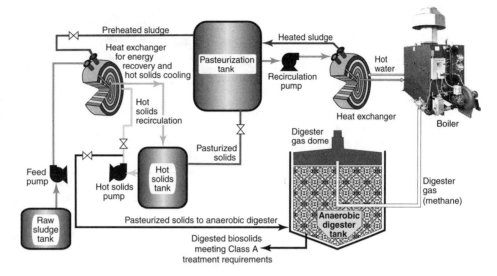

centrifugation, vacuum, or pressure, increases the drainage rate and amount of water released from sludge. However, the best dewatering performance is reached only by optimizing both chemical treatment and mechanical equipment operation.

Pressure Filtration

The belt filter press and the plate-and-frame filter press are pressure filters. Although both are used in dewatering wastewater sludges, the belt filter press is much more popular because it is available in small sizes, uses polymer for chemical flocculation of the sludge, is lower in cost, and less expensive to operate.

Plate-and-frame presses, described in Section 7–32, are used primarily to dewater chemical sludges. They are large machines that use ferric chloride and lime or polymer for conditioning of organic sludges prior to dewatering. Plate-and-frame presses are noted for producing a very compact and dry cake; hence, they can be economically justified for dewatering wastewater sludges prior to incineration or landfilling or for long-haul shipping.

Description of a Belt Filter Press

A belt filter press has two continuous porous belts that pass over a series of rollers to squeeze water out of the sludge layer compressed between the belts. Figure 11–64 is an operational diagram and photograph of a belt press. After addition of polymer to flocculate the solids, the wet sludge is applied at a uniform rate to the upper belt at the beginning of the gravity drainage zone—similar to the gravity belt thickener. The belt is supported in this zone on an open framework to allow water that separates by gravity to drain through the porous belt into a collection pan. After about half of the water is removed in the gravity zone, the sludge drops onto the lower belt and is gradually compressed between the two belts as they come together in the cake-forming wedge zone. The two belts then pass over a series of rollers for high-pressure dewatering by squeezing the water out of the sludge layer through the belts. The dewatering rollers may be plain or perforated cylinders constructed of corrosion-resistant material; the belt tension, alignment, and drive rollers have a rubber coating to prevent slippage of the belt. The cake is scraped from the belts by doctor blades held against both belts. Features of belt filter presses vary with different manufacturers.

The schematic layout of a belt filter press system is shown in Figure 11–65. A complete, separate system of auxiliary equipment is recommended for each press in a multiple-press installation, although large facilities will have common polymer feed and cake conveyance to carry away dewatered sludge from several machines. The

Figure 11–64

Operational diagram and photograph of a belt filter press with two continuous belts for gravity and pressure dewatering with uniform-diameter rollers.

(Courtesy of Ashbrook Corporation.)

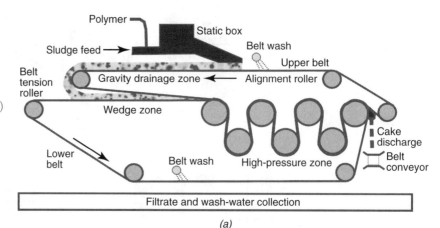

(a)

(b)

sludge feed must be applied at a uniform rate. A variable-speed drive for the sludge feed pump is necessary, and a pumping capacity of 25 to 50 percent greater than the design hydraulic loading is recommended. Auxiliary equipment for centrifuge and plate-and-frame presses is similar, particularly with respect to the polymer feed system.

Polymer and Auxiliary Systems

The polymer preparation and feed equipment is designed with flexibility to allow for variations in feed rate and use of different kinds of polymers. Dry polymer has greater than 95 percent active solids and must be carefully wetted and aged to

activate (uncoil) the polymer strands. Emulsion polymer is 25 to 50 percent active polymer solids packed in a carrier oil. When mixed with water, the polymer uncoils rapidly and may be fed directly from the blending unit or aged prior to use. Aging is recommended for optimal polymer performance. Typically, dry and emulsion polymers are diluted with water, aged at a concentration of 0.5 to 1.0 percent total solids, and postdiluted for a concentration of 0.2 to 0.5 percent. Dry and emulsion polymers can be purchased in a variety of molecular weights. High-molecular-weight polymers are typically used for sludges. Mannich polymer is made of polymer fragments chemically bonded together to make longer strands. Packed in

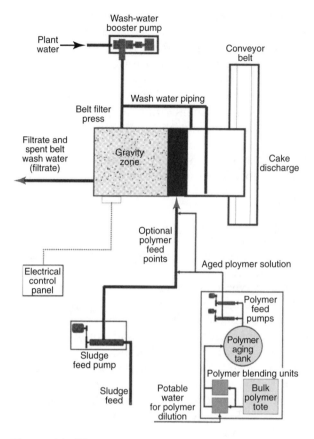

Figure 11–65

Schematic layout showing the major components of a belt filter press system including a polymer feed system.

water, the polymer typically contains 2 to 8 percent active polymer solids and only requires additional dilution to 0.2 to 0.5 percent for feeding. Potable water is used for dilution and aging because solids in the treated water will begin to tie up the polymer and because polymer is sensitive to high chlorine concentrations. Treated effluent may be used for postdilution. The polymer metering pumps must be positive displacement with a variable-speed drive.

A calibration column mounted at the pump suction allows the operator to determine actual pumping rates and make adjustments. For cleaning the belts, either potable water or clean plant effluent can be used. Relatively clean water is required to reduce clogging of spray nozzles. An easily cleanable strainer with a fine-mesh element should be installed in the wash-water supply line. Wash water is required for all sludge dewatering

equipment and the general dewatering area for housekeeping purposes. The electrical control panel should provide automatic start-up and shutdown. On a preprogrammed schedule, the automatic system initiates and monitors each step of the cycle to provide start-up, operation, automatic sludge feed and polymer feed controls, discharge cake conveyance, washing, and shutdown. Alarm conditions should shut down the sludge and polymer feed in case of problems.

Operation of a Belt Filter Press

The main operating parameters of a belt filter press are hydraulic or solids loading and polymer dosage. Hydraulic loading is expressed in gallons per minute of sludge feed per meter of belt width (cubic meters per meter per hour). Solids loading is expressed as pounds of total dry solids feed per meter per hour (kilograms per meter per hour). Polymer dosage is calculated as pounds applied per ton of total dry solids in the feed sludge (kilograms per tonne). The key performance parameters are solids recovery, cake dryness, and wash-water consumption. Solids recovery is calculated as the total solids in the feed sludge minus the suspended solids in the wastewater (filtrate plus wash water), divided by the total solids in the feed sludge. Cake dryness is expressed as the percentage of dry solids by weight in the cake. For easy comparison with hydraulic sludge loading, wash-water consumption and wastewater discharge are expressed in units of gallons per minute per meter of belt width (cubic meters per meter per hour). Belt presses can dewater both thickened and dilute sludges that are either raw or digested. The sludge characteristics of greatest importance are solids concentration, nature of the solids, and prior biological or chemical conditioning. When the solids concentration is less than 4 percent, a press is limited to a hydraulic capacity essentially independent of solids concentration. Most manufacturers suggest a maximum hydraulic loading of 50 to 60 gpm per meter of belt width (11.4 to 13.7 $m^3/m \cdot h$). With solids concentration greater than about 6 percent, the capacity of a press is restricted by solids loading. The nature of the solids influences both polymer flocculation and the degree of dewatering. Fibrous solids that exist in primary sludge are much easier to dewater than fine biological solids in waste-activated sludge. Polymer is applied to flocculate the sludge solids, forming aggregates to allow release of water with dosage depending on the sludge characteristics and sludge feed rate. The final sludge concentration is also related to belt speed and belt tension. If sludge loading is too high, the gravity drainage zone does not release sufficient water, resulting in extrusion of fine solids through the belt fabric and from between the belts at the edges. Excessive belt tension can also cause extrusion of solids. In actual plant operation, the optimum solids loading is based on the economy of operation, considering the two major costs of polymer consumption and hours of operation. Adjusting process variables at a given sludge loading is done by selecting initial settings for polymer dosage, belt speed, and belt tension, and then readjusting these three settings to achieve the desired cake dryness and solids recovery with minimum polymer dosage.

Sizing of Belt Filter Presses

The design of a sludge dewatering facility considers the size of the treatment plant, desired flexibility of operations, and anticipated conditions of dewatering, which depends on the quantity of sludge production, design feed rate, and

operating time. Filter presses are manufactured with belt widths in the range of 0.5 to 3.0 m, with the most common sizes between 1.0 and 2.5 m. Unless adequate sludge storage is provided, at least two presses should be installed so that one press can be operated while the other is out of service for maintenance to change belts or repair mechanical components, such as roller bearings. In a small plant, the operating time may be 7 hr per day, whereas in larger plants the presses may be operated for a period of 15 to 23 hr per day. These schedules allow 1 hr for start-up and shutdown.

Typical operating parameters for filter pressing of wastewater sludges are listed in Table 11–12. The solids loadings account for both the feed solids concentration and hydraulic loading. For example, using the average values for anaerobically digested primary sludge, a 5 percent feed at 50 gpm/m yields a solids loading of 1250 lb/m/hr. Also, the percentage of cake solids decreases and polymer dosage increases with low feed solids concentration and for those sludges containing waste-activated sludge. The data in Table 11–12 should be used only to estimate the capacity of presses required in the design. For a given manufacturer's press, operating experience at other installations dewatering a sludge of similar characteristics is considered. In general, press capacity should be prudently selected

TABLE 11–12

Typical Operating Parameters for Belt Filter Press Dewatering of Polymer Flocculated Wastewater Sludges

Type of Sludge	Feed Solids (percent)	Hydraulic Loading (gpm/m)[a]	Solids Loading (lb/m/hr)[b]	Cake Solids (percent)	Polymer Dosage (lb/ton)[c]
Anaerobically digested primary only	4 to 6	40 to 60	1000 to 1600	20 to 30	3 to 8
Anaerobically digested primary plus waste-activated	2 to 5	40 to 60	500 to 1000	15 to 26	8 to 14
Aerobically digested without primary	1 to 3	30 to 45	200 to 500	12 to 18	8 to 14
Raw primary and waste-activated	3 to 6	40 to 50	800 to 1200	18 to 26	4 to 10
Thickened waste-activated	3 to 5	40 to 50	800 to 1000	14 to 20	6 to 8
Extended aeration waste-activated	1 to 3	30 to 50	200 to 600	12 to 22	8 to 14

[a] 1.0 gpm/m = 0.225 m^3/m·h
[b] 1.0 lb/m/hr = 0.454 kg/m·h
[c] 1.0 lb/ton = 0.500 kg/tonne

for new installations to account for the probable inaccuracy in performance.

Sizing belt filter presses for an existing treatment plant can be done most reliably by conducting field tests using a narrow-belt machine. Most manufacturers have a full-scale 0.5- or 1.0-m press enclosed in a mobile trailer that is representative of their larger machines. During the preliminary design investigation, a rented trailer unit can be used to determine dewaterability of the sludge and to establish testing criteria for the performance specifications. After the press facility has been sized and designed, the press manufacturer can be selected based on both competitive bidding and qualification testing using trailer units. Field testing reduces the risk in design by demonstrating that the selected manufacturer's press can achieve the results required by the performance specification. Acceptance testing after construction to evaluate installed presses is still conducted to ensure compliance with the specifications.

■ EXAMPLE 11–21

A belt filter press with an effective belt width of 1.5 m dewaters anaerobically digested sludge at a sludge feed rate of 70 gpm. The polymer dosage is 6.0 gpm containing 0.20 percent polymer by weight, and the wash-water usage is 50 gpm. Based on laboratory analyses, total solids in the feed sludge are 4.0 percent, total solids in the cake are 35 percent, and suspended solids in the wastewater (filtrate, polymer feed, and wash water) are 1800 mg/l. Calculate the hydraulic feed rate, solids loading rate, polymer dosage, and solids recovery. Compare these values with those listed in Table 11–12.

Solution

Hydraulic loading

$$70 \text{ gpm}/1.5 \text{ m} = 47 \text{ gpm/m}$$

(Table 11–12, 40–60 gpm/m)

Solids loading

$$47 \text{ gpm/m} \cdot 60 \text{ min/hr} \cdot 0.040 \cdot 8.34$$
$$= 940 \text{ lb/m/hr}$$

The range of solids loadings from Table 11–12 is 1000 to 1600 lb/m/hr.

Polymer dosage

$$\frac{6.0 \text{ gpm} \cdot 60 \text{ min/hr} \cdot 0.0020 \cdot 8.34}{1.5 \text{ m} \cdot (940 \text{ lb/m/hr})/(2000 \text{ lb/ton})}$$

$$= 8.5 \text{ lb/ton}$$

This polymer dosage is greater than the 3-to 8-lb/ton range given in Table 11–12; however, flocculation efficiencies of polymers vary considerably among different brands. Selection of the best polymer is based on cost per ton of cake solids, not the weight per ton.

The quantity of filtrate can be estimated by subtracting the calculated cake volume from the sludge feed. Assuming a specific gravity of 1.05, the volumetric flow of sludge cake using Eq. 11–41 with the factor for specific gravity in the denominator is

$$V = \frac{70 \text{ gpm} \cdot 0.040 \cdot 8.34}{0.35 \cdot 8.34 \cdot 1.05} = 7.6 \text{ gpm} = 8 \text{ gpm}$$

Flow of filtrate = sludge feed − cake volume

$$= 70 - 8 = 62 \text{ gpm}$$

Wastewater flow = filtrate + polyfeed
+ wastewater

$$= 62 + 6 + 50 = 118 \text{ gpm}$$

Solids in wastewater

$$= \frac{118 \text{ gpm} \cdot 60 \text{ min/hr} \cdot 0.0018 \cdot 8.34}{1.5 \text{ m}}$$

$$= 71 \text{ lb/m/hr}$$

Solids recovery

$$= \frac{940 \text{ lb/m/hr} - 71 \text{ lb/m/hr}}{940 \text{ lb/m/hr}} = 0.92 = 92\%$$

■ ■ ■

Centrifugation

Centrifuges are employed for dewatering raw, digested, and waste-activated sludges, although they are not as common as belt filter presses in small and medium-sized facilities. Centrifuges are more costly than belt filter presses but are economical for large facilities, where they are operated 24 hours per day, 5 to 7 days per week. Centrifuges are enclosed, thus reducing the release of odors. A discussion of centrifugation is presented in Section 7–31, and the common type employed is illustrated in Figure 7–44. Centrifuges are manufactured in many sizes, from small 50-gpm machines to medium-sized machines accepting flow rates of 150 to 300 gpm and large machines designed for 500 gpm. Centrifuges are hydraulically limited below 6 percent solids and solids limited at greater concentrations.

Typical operation produces a cake with 15 to 30 percent solids concentration, depending on the character of the sludge. Without chemical conditioning, the solids capture is in the range of 50 to 80 percent; proper chemical pretreatment can increase solids recovery to 80 to 95 percent. Polymers, ferric chloride, and lime may be used as chemical conditioners. The choice between centrifugation and pressure filtration is based on performance and economics. The latter involves both initial cost of installation and operating costs, which include labor, chemicals, and power.

Screw Press

A screw press consists of an auger screw within a screen. Solid/liquid separation is accomplished by gradually reducing the volume available for the sludge, thus increasing the pressure and dewatering the sludge as it travels from the inlet to the outlet end of the press. The reduction in volume is achieved by using a tapered shaft that increases in diameter (see Figure 11–66). The screen contains punched holes less than $\frac{1}{8}$ in. in diameter and is welded to a support frame. As pressures increase, an additional screen with about 1-in. openings is attached for support. Polymer is critical to the flocculation of the sludge and to limiting the solids from being extruded through the screen openings. Solids capture efficiency ranges between 88 and 95 percent.

At the end of the production cycle, wash water is used to remove any solids buildup, improving restoring efficiency and reducing the odor potential. The entire unit is enclosed to contain water and sludge spray through the screen and contain odors and operates more like a belt press, squeezing solids against the screen, in an orientation resembling a centrifuge. The unit requires significantly less power than a centrifuge.

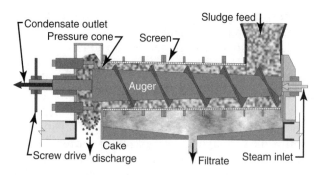

Figure 11–66

Cross-sectional diagram illustrating the operation of a screw press for dewatering waste activated sludge. Sludge mixed with polymer enters the feed hopper and is augered to the cake discharge chute. Water is released through the screen to the filtrate hopper. Steam may be added to heat the sludge and create a Class A biosolid. (Courtesy of FKC Co., Ltd./Fukoku Kogyo Co., Ltd.)

As an option, steam and lime may be introduced to improve solids dewatering and achieve a Class A biosolid quality. Steam at about 30 to 50 psi is injected into the driveshaft to heat the tapered shaft and flights and transfer heat to the biosolids. The addition of lime improves the sludge dewatering characteristics and aids in temperature generation within the sludge. The resulting cake is 30 to 40 percent solids.

11–18 BIOSOLIDS DISPOSAL

The ultimate disposal of biosolids is based on the degree of treatment and the class as defined under federal statute 40 CFR Part 503. Using these regulations, Figure 11–67 shows the questions leading to the acceptable disposal alternatives and practices, with disposal options ranging from unrestricted use for biosolids of "exceptional quality," landfill for biosolids that are not Class B, and a range of limitations for the use of Class B biosolids.

Incineration

Mechanical dewatering is used as pretreatment prior to the burning of waste solids. Incineration converts solids into an inert ash that can be disposed of easily. If dewatered to approximately

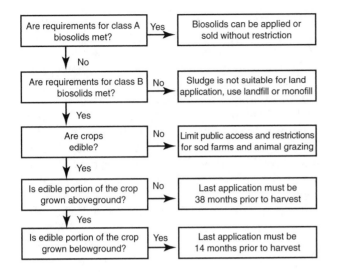

Figure 11–67
Decision tree for the ultimate disposal of biosolids, adopted from 40 CFR Part 503.

35 percent solids, the process is usually self-sustaining without supplemental fuel, except for initial warm-up and heat control. It is generally preferable to burn raw rather than digested sludge because of its higher heat value. The multiple-hearth furnace has proved to be successful for both sludge drying and incineration. The furnace consists of a circular steel shell containing several hearths arranged in a vertical stack and a central rotating shaft with rabble arms. Sludge cake is fed onto the top hearth and is raked slowly in a spiral path to the center. Here, it drops to the second level, where it is pushed to the periphery and drops, in turn, to the third hearth, where it is again raked to the center. The two upper levels allow for evaporation of moisture, the middle hearths burn the solids, producing temperatures exceeding 1400°F, and the bottom zone cools the ash prior to discharge. The hollow central shaft is cooled by forced air vented out the top. A portion of this preheated air from the shaft is piped to the lowest hearth and is further heated by the hot ash and combustion as it passes up into the furnace. The air then cools as it gives up its heat to dry the incoming sludge. The countercurrent flow of air and sludge solids optimizes combustion efficiency. Ash and water vapor are carried out with the combustion gases. A cyclonic wet scrubber is used to remove ash from the exhaust gases.

Finally, the ash is separated from the scrubber water in a cyclone separator.

Agricultural Land Application

The term *biosolids* is used to designate sludge that has been properly treated for land application. The majority of wastewater biosolids are disposed of on land, with about three-quarters being used as soil conditioner and the remainder buried in a landfill or used in landfill cover soil. Wastewater biosolids are spread on agricultural land as a soil conditioner and fertilizer. Liquid disposal of digested biosolids is popular where acceptable sites are located within convenient hauling distance. Land application of sewage sludge is regulated by federal statute in 40 CFR Part 503. The regulation covers sewage sludge and biosolids sold or given away in bulk, bags, or other containers for application to agricultural land, such as cropland, pastures, rangelands, forests, reclamation sites, and public contact sites (parks, lawns, and home gardens). The rule gives states the discretion of administering more restrictive application standards.

Table 11–13 lists the 10 regulated pollutants for land application. Ceiling limits are established to prevent the land application of high concentrations of pollutants. Annual loading rates limit the amount of a pollutant applied in a 12-month

TABLE 11–13

Federal Limits for Land Application of Sewage Sludge Listed in 40 CFR Part 503

Pollutant	Ceiling Concentration Limits (mg/kg)	Annual Loading Rate Limits (kg/hectare/yr)
Arsenic	75	2.0
Cadmium	85	1.9
Chromium	3000	150
Copper	4300	75
Lead	840	15
Mercury	57	0.85
Molybdenum	75	
Nickel	420	21
Selenium	100	5.0
Zinc	7500	140

period. Management practices limit application during wet weather and floods or on frozen or snow-covered sites. Sites must have runoff control to prevent sludge from entering wetlands or other waters. In addition to chemical limits, application rates to agricultural lands, forests, or public contact sites must be limited to agronomic uptake rates to the site.

Standards for pathogen reduction and vector attraction limits are established to protect public health and the environment. Biosolids are divided into Class A and Class B categories based on the extent of pathogen reduction. Class A sludge can be applied to land in areas open to the public without any pathogen-related restrictions. Class B sludges are processed to a point where health problems are unlikely; however, because pathogens may be present, site restrictions must limit human and animal contact. Class B biosolids cannot be sold or given away in bag or container form.

Site restrictions are imposed for harvesting, animal grazing, and public contact where Class B biosolids are applied. Food crops with harvested parts aboveground require 14 months from the last application and 38 months where the harvested parts are below the ground surface. Food, feed, and fiber crops where edible parts do not touch the soil surface require the last application to be 30 days prior to harvest. Animals must not graze for 30 days after application. Public contact

is restricted for 1 year where potential public exposure is high, such as at parks or ball fields. Where public exposure is low, for example, farms and private property, access must be restricted for 30 days.

Liquid disposal of digested sludge is most applicable to small plants that are located in rural areas and are equipped with aerobic or anaerobic digesters to biologically stabilize the sludge before spreading. Even though hauling costs may be high, the alternatives of solid-liquid separation on drying beds or mechanical dewatering are often more expensive. The liquid sludge can be applied on the surface by a vehicle equipped for spreading or injected underneath the surface using chisel plows (Figure 11–68a). Spreading from fixed or portable irrigation nozzles (Figure 11–68b) may be practiced where odor and insects are not problems.

Subsurface injection of liquid digested sludge is the most environmentally acceptable method, since the sludge is incorporated directly into the soil, reducing exposure to the atmosphere. At large plants, digested sludge is dewatered by belt filter presses to reduce the weight and cost of hauling. The sludge cake is stored in piles on an open site and placed on cropland at appropriate times using a spreader with back beaters that throw the solids outward from the rear of a wagon or truck-mounted box. Disk cultivators then mix

(a)

(b)

Figure 11–68

Methods of land application for liquid sludge. (a) Subsurface injection using a chisel plow and tractor. Sludge is fed in a hose pulled behind the equipment. (b) Surface application of liquid sludge using a portable nozzle.

the sludge solids with the surface soils before planting a crop.

The environmental concerns of sludge disposal on agricultural land are pollution of groundwater and surface water, contamination of the soil and crops with toxic substances, and transmission of human and animal diseases. Normally, laboratory analyses of a sludge are conducted for solids concentration; nitrogen, phosphorus, and potassium contents; presence of the heavy metals cadmium, copper, nickel, lead, and zinc; and selected organic compounds, such as polychlorinated biphenyls. For protection of human health, the preferred crops are fodder grasses, feed grains, and fiber crops that are not in the human food chain. Grasses are considered acceptable provided cattle are restricted from grazing for a specified period after sludge application. Feed grains for animal consumption, like corn, are fertilized by tilling sludge solids into the soil before the planting of crops. To ensure adequate environmental control, the sludge, soil, groundwater, and runoff are monitored for fecal coliforms, nutrients, and contaminants. Cadmium is the heavy metal of greatest concern, since it can be taken up by crops and therefore has the potential of entering the human food chain. Despite this, plant uptake from soils containing cadmium does not always occur. The pH of the soil and the kind of plant species are important considerations. The highest amounts of cadmium occur in fibrous roots and plant leaves; thus, potential high-risk food plants for humans are leafy vegetables like lettuce and spinach. Fodder crops and cereal grains appear to present the least risk because of their low concentration of cadmium. The nitrogen content of a sludge often determines the allowable rate of application. The concern is pollution of groundwater with nitrate, which is a health-related contaminant in drinking water standards.

Agronomic application rate refers to the distribution of sludge application on a dry-weight basis such that the amount of nitrogen and phosphorus does not exceed the needs of the crop. This limits the amount of nitrogen and phosphorus that is available to pass below the root zone and into the groundwater. Limits on phosphorus have been established in Michigan and Florida on a case-by-case basis. The Florida Department of Environmental Protection lists the established

crop nitrogen demand for the purpose of determining agronomic rates. Nitrogen demand during the growing season ranges from 100 lb/acre for peanuts, soybeans, and alfalfa to 200 lb/acre for grasses and 400 lb/acre for hay or silage.

■ EXAMPLE 11–22

A 10-mgd municipal activated sludge treatment plant operating at an F/M of 0.25 and digestion with a 50 percent VSS destruction is dewatered to a cake containing 30 percent solids with 34 lb/ton plant-available nitrogen and 140 lb of phosphorus as P_2O_5. The site is used to grow corn with a total estimated crop nitrogen demand of 250 lb/acre and phosphorus demand of 50 lb/acre. Tests of the soil indicate a concentration of 60 lb/acre nitrogen and 5 lb/acre phosphorus.

Solution

Refer to Figure 11–50 for sludge solids production. Assuming 200 mg/l BOD and a 70 percent volatile suspended solids fraction, the total solids are calculated as follows:

$$Solids = (10 \cdot 200 \cdot 2 \cdot 0.45) \cdot 0.3$$
$$+ (10 \cdot 200 \cdot 2 \cdot 0.45) \cdot 0.7 \cdot 0.5$$
$$= 1170 \text{ lb/day} \quad or \quad 1170 \cdot 365/2000$$
$$= 214 \text{ ton/yr}$$

$$Nutrient\ load = 214 \cdot 34 = 7{,}300 \text{ lb}$$
$$nitrogen\ and$$
$$214 \cdot 140 = 30{,}000 \text{ lb phosphorus}$$

$$Crop\ demand = 250 - 60$$
$$= 190 \text{ lb/acre nitrogen}$$
$$and \quad 50 - 5$$
$$= 45 \text{ lb/acre phosphorus}$$

Acres required based on nitrogen loading:
$$7300/190 = 40 \text{ acres}$$

Acres required based on phosphorus loading:
$$30000/45 = 670 \text{ acres}$$

Agronomic application based on phosphorus uptake requires sludge to be distributed over a much larger area, and supplemental nitrogen will be required to meet plant needs.

■ ■ ■

Landfill or Monofill

Dewatered sludge may be disposed of in a landfill or monofill specifically designed for sludge. Burial can be the final disposal for dried digested sludge and dewatered raw sludge cake. Since these pose health problems, the landfill sites are designed to prevent groundwater and surface-water pollution. Many cities bury wastewater sludge in their sanitary landfill along with municipal refuse. The basic operations include spreading, compacting, and covering the wastes daily with excavated soil. Many locations incorporate dewatered sludge into the daily cover for disposal.

11–19   ODOR CONTROL

Odors result from the generation of hydrogen sulfide, release of organic compounds, and other vapors. Decomposition results in mercaptans, organic sulfates, and amines. The odor of hydrogen sulfide can be detected at concentrations above 0.005 ppm. At 10 ppm, the concentration has toxic effects on the respiratory system. The greatest odors are evident at the headworks (raw sewer, screenings, and grit), thickeners, sludge storage, and processing. Primary clarifiers, in-plant pumping stations, and tricking filters are secondary sources of odor. Ventilation rates range between 6 and 20 air changes per hour. Odor-control equipment includes adsorption systems, biological systems, and wet scrubbers. Because of the high moisture content and corrosion potential of hydrogen sulfide, blowers and piping are manufactured out of corrosion-resistant materials such as plastic or fiberglass.

Adsorption Systems

The most common media for adsorption systems is granular activated carbon, which adsorbs organic and inorganic compounds. Granular media are supported from a screen within the tank. Blowers or fans may be used prior to or following the tank. As air passes through the media, odor compounds are adsorbed on the carbon's extensive surface area. When the carbon is saturated, the filter no longer removes odors and must be regenerated. New carbon and carbon regeneration are costly.

Biological Systems

Organic and inorganic compounds can be reduced by biological activity. At some plants, odorous air is piped to the inlet of the aeration blowers and uses the activated sludge system to remove odor compounds. The system is effective, but issues associated with corrosion must be addressed because most blowers are not designed for such humid/corrosive conditions. Biological trickling filters dedicated to treating foul odors have been used, with mixed results. Biological beds consist of sand or compost media with perforated distribution piping, as shown in the diagram in Figure 11–69. The media remove odors from the air stream while biological activity in the media reduces the compounds. The beds must be watered to remain moist to support biological growth. The media decompose slowly and are typically replaced every 10 years or so.

Wet Scrubbers

Wet scrubbers use chemically treated water to dissolve and reduce odor compounds. Sodium hydroxide is used to lower the pH, increasing solubility, and sodium hypochloride is used to

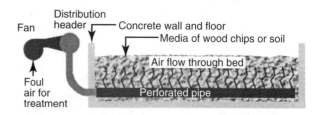

Figure 11–69

Biological odor control bed. Air is diffused under a bed of wood chips or sand. Odor-causing compounds are removed by adsorption and destroyed by biological activity. Sprinklers maintain a moisture content suitable for biological growth.

Figure 11–70

Photo of wet scrubber packed tower, fan, and chemical feed. Air is forced up through the tower, while chemically treated recirculation water flows down to dissolve and oxidize odorous compounds.

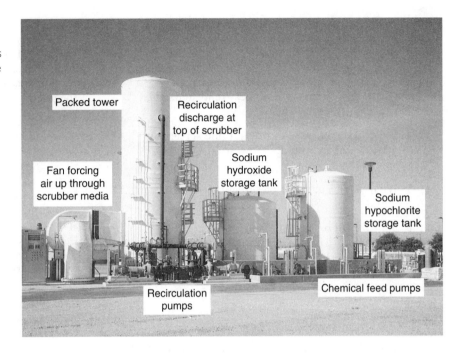

chemically reduce the compounds. A wet scrubber system is shown in Figure 11–70. The scrubber system consists of a scrubber tower (open or packed with plastic media), a chemical feed, a wet well at the bottom of the scrubber tower, recirculation pumps, and a spray system at the top of the scrubber. Air is forced up the scrubber as the water falls down, scrubbing the liquid. A constant stream of fresh water is required to prevent the buildup of salts in the recirculation water. Acid cleaning is periodically necessary to remove deposits from the scrubber media.

REFERENCES

1. *Recommended Standards for Wastewater Facilities, Policies for the Design, Review, and Approval of Plans and Specifications for Wastewater Collection and Treatment Facilities*, A Report of The Wastewater Committee of the Great Lakes—Upper Mississippi River. Board of State and Provincial Public Health and Environmental Managers. Albany, NY: Health Research, Inc., Health Education Services Division, 1997.

2. *Alternative Disinfectants and Oxidants Guidance Manual*, U.S. Environmental Protection Agency, EPA 815-R-99-014, April 1999.

3. Linden, Karl G., et al., "Effectiveness of UV Irradiation for Pathogen Inactivation in Surface Waters," *2002 Progress Report: Effectiveness of UV Irradiation for Pathogen Inactivation in Surface Waters*. Duke University, University of North Carolina at Chapel Hill, EPA Grant Number: R829012, 2002.

4. *CFR Title 40: Protection of Environment*, Chapter 1, Subchapter O—Sewage Sludge (Parts 501–503), U.S. Environmental Protection Agency. http://www.epa.gov/epahome/cfr40.htm

5. *Control of Pathogens and Vectors in Sewage Sludge*, U.S. Environmental Protection Agency, EPA 625/R-92/013, 2003.

PROBLEMS

11–1 Conventional wastewater treatment has been described as a process converting BOD to solids and solids thickening. Describe the overall process of wastewater treatment, and compare the flow volumes and solids concentrations for influent, effluent, and solids stream (Figure 11–3) with the discussion in Figure 11–1.

11–2 What are the individual unit processes within preliminary treatment, primary treatment, secondary treatment, disinfection, and solids treatment?

11–3 A treatment plant has a minimum flow of 2 mgd, max. monthly flow of 4 mgd, and a peak flow of 8.5 mgd. What values would be used to size the pumping station, primary treatment, secondary treatment, disinfection, and solids treatment? What number best represents the treatment capacity and why?

11–4 What is the purpose of bar and fine screens in wastewater treatment? How are they similar and different?

11–5 What is the purpose of screening treatment units?

11–6 If a 6.5 mgd treatment plant is designed with a bar screen ($\frac{3}{4}$ in. clear spacing) followed by a fine screen (1-mm clear spacing), what would be the average and high screening solids estimate for each unit and their total?

11–7 What is the purpose of grit removal, and what equipment is used?

11–8 Why is grit processing important to grit removal?

11–9 A treatment plant receives an average flow of 930,000 gpd with a peak-day wet-weather flow of 1.2 mgd. Calculate the quantity of grit the processing units must handle at peak flow.

11–10 A pumping station like that shown in Figure 11–11 receives a total average flow of 10 mgd and a peak flow of 18 mgd. All pumps are the same size, and one is standby. Size the pumps, and determine how many pumps operate at average conditions.

11–11 The pumping station shown in Figure 11–11 is designed for future conditions. The ultimate wet-weather capacity is 60 mgd with one pump out of service. Flow measurement at the plant currently is 32 mgd peak, 15 mgd average day, and 6.2 mgd minimum. Size the pumps. How many pumps should be installed to meet existing flow? If pumps at the slowest speed operate at 50 percent of total capacity, will one pump run at the minimum flow?

11–12 A small pumping station serves 150 homes with a peak flow of 850 gpd per home. Initial development will be 60 homes. Size constant-speed pumps and size the wet-well operating volume assuming that two pumps will be operating and one is standby in the future.

11–13 A pump has the following characteristics: 34 ft at zero flow, 28 ft at 500 gpm, 25 ft at 1500 gpm,

and 18 ft at 2500 gpm. The wet well operates at a relatively constant elevation of 1020 ft. The discharge flows through an 6-in. line for 540 ft to a splitter box at elevation 1035. Use $C = 100$ and Eq. 4–8 to calculate the system head curve. Plot the pump curve and identify the operating point.

11–14 A pumping station wet well operates between 540- and 550-ft elevation. The pump curve is defined by the following points: 80 ft at zero flow, 78 ft at 200 gpm, 65 ft at 800 gpm, and 50 ft at 1200 gpm. The pump discharge contains an equivalent of 50 ft of 6-in. pipe. The discharge pipe is 120 ft long and terminates at a splitter box, elevation 570. Using $C = 100$, plot the pump curve and the corrected pump curve. Plot the pump discharge curves at each wet-well elevation and for $C = 100$ and $C = 140$. What is the pump flow at the low and high wet-well elevations for new and old pipe?

11–15 The operating characteristics of a screw pump are shown in the following table. Plot the water level in the influent chamber on the horizontal axis versus capacity and efficiency on the vertical axis. Use a second axis in the same graph to plot efficiency. When the water level is 60 percent of total, what is the pump operating capacity? Over what range of pump capacity is the efficiency greater than 70 percent?

Water Level in Influent Chamber (in.)	Pump Capacity (gpm)	Pump Efficiency (percent)
0	0	0
12	60	45
16	120	54
23	240	63
28	360	68
32	480	71
35	600	73
38	720	74
43	960	75
48	1200	75

11–16 What is the purpose of primary sedimentation, and how are units sized?

11–17 A treatment plant has three primary clarifiers at 60 ft in diameter with side water depths of 9 ft. The inboard weirs are set at 54 ft. The average daily flow is 4.6 mgd and the peak flow is 8.7 mgd. Calculate the overflow rates, detention time, and weir loading. Using the calculations, are the clarifiers adequately sized? What are the overflow rates with one unit out of service?

11–18 A primary clarifier is 65 ft in diameter and receives a peak flow of 1.2 mgd containing 230 mg/l TSS. Calculate the overflow rate, and based on the information in the text, estimate the percentage of TSS removal and effluent TSS concentration. If ferric and polymer are added to improve primary removal, what is the estimated removal and effluent TSS concentration.

11–19 A circular clarifier is designed to handle 2.5 mgd at average day conditions and 5.5 mgd at peak flow. Using the average overflow rates, calculate the size of the clarifier and detention time based on an average depth of EPA values.

11–20 How are primary clarifiers and secondary clarifiers similar and different?

11–21 A treatment plant operates with two 45-ft-diameter final clarifiers at an average flow of 3.4 mgd and a peak flow of 5.2 mgd. What are the overflow rates? Are they adequate? Can one unit be removed from service at average flow?

11–22 A treatment plant consists of eight rectangular final clarifiers, each having dimensions of 18 ft × 100 ft with a 16-ft side water depth. Inboard weirs have a total length of 110 ft each. For an average flow of 6 mgd and a peak flow of 25 mgd, calculate the overflow rate, detention time, and weir loading. Are the clarifiers adequately sized for the flow, and if not, why not?

11–23 A rock trickling filter operates with an influent of 2.8 mgd, 215 mg/l BOD, and 240 mg/l suspended solids. The primary clarifier removes 50% of the influent solids and 35% of the BOD. The two filters are 60 ft in diameter and 5 ft deep. Calculate the loading rate. Significant odors are present and a third filter of 60 ft is being considered. If it is constructed, what would be the revised loading?

11–24 Referring to Problem 11–23, advanced primary treatment is being considered until the new filter is built. What chemicals should be added to increase suspended solids removal. Settling tests showed that BOD is removed in the same proportion as suspended solids. What is the resulting loading to the two filters? Would the change in treatment decrease the odor problems?

11–25 A single-stage rock-media trickling filter plant consists of two 160-ft-diameter by 7-ft-deep

units using the flow schematic shown in Figure 11–20. The design flow rate is 10.5 mgd with a BOD of 205 mg/l and suspended solids of 200 mg/l. The primary clarifier removes 25% of the BOD, and the filter recirculation pumps operate at 9000 gpm. Calculate the BOD and hydraulic loading. The filter is uncovered and wastewater temperatures drop to 14.8°C. What is the resulting BOD effluent? If the filters are covered and the temperature increases to 18.5°C, what is the resulting effluent BOD?

11–26 A two-stage rock-media trickling filter plant, as shown in Figure 11–21, operates with an influent flow of 4.8 mgd with a BOD of 225 mg/l and suspended solids of 240 mg/l. The primary clarifier removes 28% of the BOD. The first-stage and second-stage trickling filters (two each) are the same size, at 120 ft in diameter and 6.5 ft deep. Recirculation pumps are 2400 gpm each. Calculate the loading, and determine the BOD removal at each filter and for secondary treatment.

11–27 An industrial treatment plant has a strong wastewater with 450 mg/l BOD (the soluble portion is 250 mg/l) and 280 mg/l suspended solids. The biological tower treatment process (Figure 11–27) consists of high-density, cross-flow packing media with dimensions of 480 sq ft by 20 ft high. The packing has a specific area of 42 ft^2/ft^3, $n = 0.5$, and $k_{20} = 0.0035$ $(gpm/ft^2)^{0.5}$. The primary clarifier effluent is 0.8 mgd at 18°C. The recirculation ratio is 1.0. Calculate the effluent soluble BOD at 20°C, then correct the calculation for actual temperature. Calculate the soluble and total effluent BOD assuming that the effluent has 0.54 soluble BOD/total BOD.

11–28 An extended aeration plant consists of three oxidation ditches without primary clarification. Each ditch has a volume of 2.0 mil gal. The average annual flow is 6.0 mgd, the maximum flow is 7.8 mgd, and the BOD is 240 mg/l. The MLSS is maintained at 1800 mg/l. Calculate the liquid detention time, BOD loading, and F/M ratio. Refer to Figure 11–30 and determine if sludge will settle properly.

11–29 A conventional aeration plant has a flow of 12 mgd with a BOD of 190 mg/l. The primary clarifier removes 25 percent of the BOD. The 10 aeration basins are 40 ft wide, 100 ft long, and 15 ft deep. Determine the liquid detention time, loading rate, F/M ratio (with an MLSS of

2350 mg/l), sludge age (with 5330 lb/day waste sludge Eff. SS = 20 mg/l. Refer to Figure 11–30 and determine if sludge will settle properly.

11–30 The design maximum monthly flow for a conventional activated-sludge plant is 2.75 mgd with a BOD of 195 mg/l and a suspended solids of 205 mg/l. Calculate the diameter and side water depth for two primary clarifiers. Using the loading and operational parameters listed in Table 11–2, recommend the required volume for the aeration tanks. Calculate the diameter and side water depth for two secondary clarifiers, F/M, and aeration period.

11–31 How does wastewater temperature impact treatment?

11–32 What are the advantages and disadvantages of high-purity oxygen treatment?

11–33 Compare the operation and control for activated-sludge aeration versus biological filtration systems.

11–34 A trailer park is being planned with 140 units at an occupancy of 2.0 people per unit. Wastewater flow is 80 gpcd with 170 mg/l BOD and 190 mg/l suspended solids. Size the extended aeration basin and clarifier (see Figure 11–33) using the minimum detention time and overflow rates. Calculate the average F/M using an MLSS of 1000 mg/l. Calculate the BOD loading. Compare values with those listed in Table 11–4.

11–35 An oxidation ditch has a flow of 11.5 mgd, an influent BOD of 200 mg/l, and 35 mg/l of ammonia, and effluent of 10 mg/l. Wastewater nitrifies year-round at a temperature of 16°C. The two ditches (Figure 11–34) have four brush aerators each operating at average conditions (Table 11–5). Use an αF of 0.5 and β of 0.9. The residual DO is 4.0. Calculate the values for oxygen demand, oxygen transfer, and hp for each brush.

11–36 A fine-bubble aeration basin has a flow of 21.5 mgd, an influent BOD of 180 mg/l, 31 mg/l of ammonia, and effluent of 10 mg/l. During the winter, the wastewater temperature of 14.8°C does not nitrify. During the summer, wastewater temperatures increase to 18.5°C and operators cannot avoid nitrification. The residual DO is controlled to maintain 2.0 mg/l. Diffusers are 15 ft deep. Use average oxygen transfer values for αF and β. Calculate (1) winter

and (2) summer values for oxygen demand, oxygen transfer, hp, and air-flow rate.

11–37 An activated sludge plant uses coarse-bubble aeration at a depth of 15 ft. The flow is 6.5 mgd with an influent BOD of 220 mg/l (primary clarifier removal is 35%). Alpha, beta, and DO are like those in Example 11–10. Wastewater temperature is 16°C. Calculate the hp and air flow required.

11–38 A laboratory study was performed on a soluble wastewater to determine the kinetic constants for the mathematical model presented in Section 11–8. The aeration chamber had a volume of 10 l mixed by a bubbler stone. The wastewater was fed at a rate of 38 l/day. Mixed liquor was removed daily. What is the liquid detention time? Determine the values for Y, k_d, k, and k_s. From the laboratory data for each run, calculate θ_c, $1/\theta_c$, r_{su}, U, $1/U$, and $1/S_e$. Plot these values as illustrated in Figure 11–40 to determine the kinetic constants.

11–39 A completely mixed activated-sludge plant is being designed for a wastewater flow of 3.0 mgd. The BOD of 225 mg/l is 200 mg/l soluble. The effluent soluble BOD requirement is 7 mg/l (70% of the total 10 mg/l). For sizing, the mean cell residence time is to be 8 days at an MLSS of 2500 mg/l. Laboratory analysis resulted in $Y = 0.6$ lb VSS/lb BOD, $k_d = 0.06$ per day, $k = 5.0$ per day, and $k_s = 60$ mg/l. Calculate the aeration basin size (use a safety factor of two times larger than what is calculated) and the waste sludge rate.

11–40 Describe the biological process in facultative ponds. Describe the biological process associated with aerated lagoons versus activated-sludge aeration.

11–41 Based on a loading of 25 lb BOD/day/acre, what pond area is needed for an average flow of 0.65 mgd with 165 mg/l BOD? Calculate the minimum water depth for a retention time of 90 days with a net loss of 0.7 in./week.

11–42 Stabilization ponds with a total surface area of 16 acres receive a wastewater flow of 0.2 mgd with a BOD of 230 mg/l. Calculate the BOD loading and winter storage available between the 2-ft and 5-ft depths assuming a net water loss of 0.05 in./day during the winter months.

11–43 An aerated lagoon with a 10-ft depth and liquid volume of 175,000 cu ft treated an average of

0.2 mgd of wastewater with a BOD of 450 mg/l. The anticipated temperature extremes of the aerating wastewater range from 10° to 30°C. The two 25-hp surface aerators were recommended for adequate mixing. Based on laboratory treatability studies, the wastewater has the following characteristics: k_{10} = 0.68 per day, C_t = 2.0, α = 0.9, β = 0.8, and an oxygen utilization rate of 1.0 lb oxygen/lb BOD. Is the lagoon adequately sized? Are the aerators adequate? Will the lagoon meet an effluent BOD requirement of 50 mg/l?

11-44 An aerated pond receives 0.375 mgd with a BOD of 220 mg/l. How many 10-hp aerators should be used?

11-45 What is the primary reason for disinfecting the wastewater effluent? Write the chemical reaction that takes place when chlorine is mixed with wastewater effluent. Why is a chlorine residual detrimental to a receiving stream? What chemicals are used to neutralize the chlorine residual?

11-46 Why is chlorination conducted in long, narrow tanks?

11-47 A treatment plant has a maximum monthly flow of 14.6 mgd. Calculate the size of the chlorine contact tank and recommend dimensions for the tank. For a 5% chlorine concentration, calculate the number of pounds per day used and the number of days a 1-ton storage tank will last.

11-48 How does ultraviolet light provide disinfection?

11-49 UV provides an effluent of 2.2 MPN/100 ml total coliform. If the influent has 4,400 MPN/ 100 ml, calculate the log removal and determine the UV dose. What is the removal of *B. Subilis* spores?

11-50 Describe the treatment for individual household disposal systems.

11-51 A 5.2-mgd activated-sludge plant with 200 mg/l BOD and 220 mg/l suspended solids consists of a primary clarifier, aeration basin (dt = 8 hr), and secondary clarifier. The primaries removal is 48% solids and 32% BOD. The aeration basin operates in a step feed at an MLSS of 3000 mg/l. Calculate the sludge solids and flow rate where the primary solids concentration is 5.5% and the waste-activated sludge is 0.5%.

11-52 Using a per capita contribution of 120 gpcd of wastewater containing a BOD of 0.2 lb and a suspended solids of 0.24 lb, calculate the quantity of sludge solids and volume of waste sludge produced per person using a biological filter plant and two different activated-sludge plants. The plants are as follows:

1. Biological filtration
   The primary clarifier removes 50% solids and 35% BOD with 5% sludge solids concentration. The k-value for the filter is 0.31 and the effluent solids are 30 mg/l.
2. A conventional activated sludge plant
   The plant has a primary clarifier that removes 50% solids and 30% BOD with 5% sludge solids concentration and 8-hour detention time with 1500 mg/l MLSS. The effluent solids are 20 mg/l. The sludge is a blend of primary at 5% and thickened waste-activated sludge at 4% solids.
3. An extended aeration plant
   There is no primary clarifier. The basin has 24 hr of detention time and 1200 mg/l. Sludge wasted from the secondary clarifier contains 1% solids.

11-53 For digested, primary, and waste-activated sludge at 5% solids, calculate the upper and lower critical velocities. For a 6-in pipe, 800 ft long, plot the head loss from zero flow to 600 gpm at 100-gpm increments. Label the upper and lower critical velocities, as in Figure 11–52.

11-54 Sludge must be pumped from the primary clarifier to the anaerobic digester at 500 gpm at 4.5% solids, from the secondary clarifier to the thickener at 500 gpm at 0.5% solids, from the thickener to the anaerobic digester at 100 gpm at 5% solids, and from the anaerobic digester to dewatering at 350 gpm at 3.5% solids.

All pipe is 8-in.-diameter. Distances and fittings are as follows:
   Primary sludge: 850 ft, 14 plug valves, 1 check valve, 8 flow-through tees, 24 standard 90° elbows.
   Secondary clarifier sludge: 420 ft, 8 plug valves, 1 check valve, 3 flow-through tees, 8 standard 90° elbows (use C = 100).
   Thickened sludge: 1200 ft, 14 plug valves, 1 check valve, 9 flow-through tees, 18 standard 90° elbows.
   Digested sludge: 520 ft, 6 plug valves, 1 check valve, 3 flow-through tees, 14 standard 90° elbows. Calculate the head loss in the pipe and fittings.

11-55 A plant with 20 mgd and 215 mg/l suspended solids maintains a primary clarifier removal of

60% by overdrafting sludge solids (at 22%) to a gravity sludge thickener. Determine the size of the total area of thickening required at the average loading rate of 9 lb/sq ft. Calculate the diameter of two equally sized tanks.

11–56 An oxidation ditch without primary clarification receives 28 mgd of 185 mg/l BOD and 200 mg/l suspended solids (dt = 24 hr). Sludge is wasted to maintain an MLSS of 2500 mg/l at a concentration of $\frac{3}{4}$% solids. Calculate the waste solids in lb/day and flow rate. How many thickening units are required at a capacity of 250 gpm/unit, operating 8 hr/day? If sludge is thickened to 4.5% with 0.95% capture, calculate the solids and flow in the thickened sludge and filtrate.

11–57 What are the regulatory requirements and treatment processes to attain Class B biosolids? And where can these biosolids be used?

11–58 Aerobic digesters have a total capacity of 50,000 cu ft. The average daily sludge flow is 32,000 gal per day with 1.5% solids of which 63% are volatile. Calculate the sludge detention time and VS loading rate, and determine whether the requirements for Class B biosolids are met at 15%.

11–59 Sludge from an oxidation ditch is aerobically digested. The waste sludge volume is 5250 lb/day at 1.2% solids, and 65% are volatile. Wastewater temperature is 15.5°C. Calculate the sludge flow rate, and size the digester volume to meet Class B requirements? What is the minimum aerobic digester volume? Calculate the degree days.

11–60 Raw sludge at a rate of 180 gpm is pumped for similar to three anaerobic digesters (40-ft diameter, 20-ft depth) containing 5% solids that are 65% volatile. Sludge does not settle and is not withdrawn. The digesters maintain 42% volatile solids destruction. Calculate the liquid detention time, loading rate, and final solids percentage.

11–61 Two-stage anaerobic digesters have two fixed-cover primaries and two floating-cover secondary digesters. All digesters are 35 ft in diameter with a 22-ft liquid depth. Floating covers have a 10-ft operating range. Sludge is fed at a rate of 500 gpm for 5 min every hour. Sludge contains 4.5% solids that are 62% volatile. An average of 50% volatile solids is destroyed. Weather requires 3 months of storage. Calculate the

loading rate, treated solids percentage, and storage. Is storage adequate?

11–62 A drying bed in a northern state needs to be sized for a rock-media trickling filter plant. Primary sludge flow is 1000 lb/day and secondary sludge flow is 400 lb/day. Recommend a size for the sludge drying beds. Determine the number of beds based on the general dimensions given in the text.

11–63 Describe the process of sludge composting. How is pathogen reduction accomplished?

11–64 Describe the process of lime treatment. How is pathogen reduction accomplished?

11–65 What are the regulatory requirements and treatment processes to attain Class A biosolids? And where can these biosolids be used?

11–66 Contrast the operating difference between aerobic digestion and ATAD. How are the processes similar and different?

11–67 Describe the two methods for pasteurization, and explain how each achieves the requirements of Class A biosolids treatment.

11–68 A belt filter press test unit, housed in a mobile trailer, was used to test the dewaterability of anaerobically digested sludge. The 0.5-m unit operated at a feed rate 20 gpm and a polymer dosage of 2 gpm containing 0.2% polymer by weight. Wash-water flow was 17 gpm and contained 2000 mg/l suspended solids, and the filtrate was 19 gpm with 500 mg/l. The feed solids of 3.5% are thickened to a cake of 32%. Calculate the hydraulic loading, solids loading, polymer dosage, and solids recovery.

11–69 A 2.0-m belt press dewaters 150 gpm of sludge at 4% solids. The polymer dose is 6.4 gpm containing 0.5% polymer by weight. The wash-water consumption is 60 gpm. Cake solids are 24%, and the suspended solids in the wash water is 1400 mg/l. Calculate the hydraulic loading rate, solids loading rate, and polymer dosage and estimate the solids recovery.

11–70 A centrifuge dewaters sludge at a feed of 250 gpm containing 2.6% solids. Active polymer addition is 22 lb/ton of sludge (polymer is 45% active). The cake solids are 26% and the centrate contains 1720 mg/l solids. Calculate the pounds of solids applied, the removal efficiency, the pounds of active polymer, and total pounds of polymer used. If polymer weighs about 9 lb/gal, what is the polymer feed rate in gallons per hour?

11–71 A treatment plant produces 8000 lb/day of total sludge dewatered to 26% solids. The sludge contains 36 lb/ton of plant-available nitrogen. Tests show a residual of 75 lb/acre of nitrogen. How many acres are required for hay growth?

11–72 A two-stage rock-media trickling filter plant consists of preliminary treatment and secondary treatment as shown in Figure 11–21. Chlorinated effluent is discharged to a river. Solids treatment consists of returning the intermediate and secondary clarifier sludge to the primary clarifier for removal. Sludge is digested in two-stage anaerobic digestion and hauled to land application.

Operating conditions:
    Max month design flow = 4 mgd; peak
        flow = 8.1 mgd; minimum flow = 1.8 mgd
    BOD = 185 mg/l, suspended solids = 198 mg/l
    Wastewater temperature = 14.5°C

Effluent treatment requirements:
    BOD = 30 mg/l, suspended solids = 30 mg/l,
        200 MPN/100 ml
Treatment units:
    Influent pumping station with two 48-in.
        screw pumps
    Bar screen with 1-in. openings
    No grit removal
    Primary clarifiers—two at 45-ft diameter,
        7-ft side water depth each
    Intermediate clarifiers—two at 35-ft diameter, 7-ft side water depth each
    Final clarifiers—two at 50-ft diameter, 8-ft
        side water depth each
    First-stage trickling filters—two at 60-ft
        diameter, 6-ft depth each
    Second-stage trickling filters—two at 60-ft
        diameter, 6-ft depth each
    Chlorine contact basin—3 passes in channels 100 ft long, each 10 ft wide and
        10 ft deep
    First-stage anaerobic digester—35 ft in diameter by 12 ft below the landings, 20-ft
        operating volume
    Second-stage anaerobic digester—35 ft in
        diameter by 12 ft below the landings, 10-ft
        operating volume above
Determine the operating characteristics for each of the treatment units and compare against the unit process design parameters listed in the text:
    For the influent pumping station, determine
        the adequacy of the wet-well volume and
        pump sizes.
    For the bar screen, determine the maximum
        screenings removal.
    For the clarifiers, determine the adequacy of
        the overflow rates, weir loading, and
        detention times.
    For the trickling filters, check the BOD and
        hydraulic loading.
    For the chlorine contact basin, determine
        the detention time, chlorine feed rate in
        pounds, and amount of chlorine needed for
        a 30-day supply.
    Calculate the sludge production and
        volume:
    For the anaerobic digesters, determine loading, detention time, and storage time.
    For land application, determine the land area
        required using 35 mg/l nitrogen in the
        liquid sludge.

Will treatment meet effluent requirements?

11–73 A biological filter plant consists of preliminary treatment and secondary treatment as shown in Figure 11–27. Chlorinated effluent is discharged to a river. Solids treatment consists of returning the secondary clarifier sludge to the primary clarifier for removal. Sludge is held in a storage tank and dewatered, lime is added, and the sludge is hauled to landfill.

Operating conditions:
    Max month design flow = 0.12 mgd; peak
        flow = 0.32 mgd; minimum flow =
        0.05 mgd
    BOD = 360 mg/l, suspended solids =
        280 mg/l
    Wastewater temperature = 16.5°C, $n = 0.5$,
        and $k_{20} = 0.0035$ $(gpm/ft^2)^{0.5}$
1 lb BOD = 1 lb suspended solids · 0.7

Effluent treatment requirements:
    BOD = 20 mg/l, suspended solids = 20 mg/l,
        200 MPN/100 ml

Treatment units:
    Influent pumping station wet-well dimensions of 60 in. in diameter, 4-ft operating
        range; two pumps 250 gpm
    Bar screen with 6-mm openings
    No grit removal
    Primary clarifiers—one at 12-ft diameter, 5-
        ft side water depth each

Final clarifiers—two at 16-ft diameter, 7-ft side water depth each

Biological filters—two at 16 ft. in diameter, 15-ft depth each

High-density, cross-flow packing media with a specific area of 46 ft^2/ft^3

Recirculation pumping of 175 gpm

Chlorine contact basin—1 pass in channels 30 ft long, each 3 ft wide and 5 ft deep

Sodium hypochlorite fed from storage tank at 12.5% concentration of chlorine

Belt filter press—three at 0.5 m each, operating 8 A.M.–5 P.M., 5 days a week as sludge is available

Determine the operating characteristics for each of the treatment units and compare against the unit process design parameters listed in the text:

For the influent pumping station, determine the adequacy of the wet-well volume and pump sizes.

For the bar screen, determine the maximum screenings removal.

For the clarifiers, determine the adequacy of the overflow rates, weir loading, and detention times.

For the trickling filters, check the BOD and hydraulic loading.

For the chlorine contact basin, determine the detention time, chlorine feed rate in pounds, and amount of hypochlorite needed for a 30-day supply.

Calculate the sludge production and volume.

For the belt filter press, determine the loading and feed rate.

What is the lime requirement to meet Class B biosolids?

Will the treatment plant meet effluent requirements?

11–74 An oxidation ditch plant consists of preliminary treatment and secondary treatment without a primary clarifier, as shown in Figure 11–34. Chlorinated effluent is discharged to a river. Solids are digested by aerobic digestion and hauled to land application.

Operating conditions:

Max month design flow = 2.4 mgd; peak flow = 4.2 mgd; minimum flow = 1.3 mgd

BOD = 260 mg/l, suspended solids = 210 mg/l, ammonia = 25 mg/l

Wastewater temperature = 17.5°C

Effluent treatment requirements:

BOD = 10 mg/l, suspended solids = 10 mg/l, 200 MPN/100 ml

Treatment units:

Influent pumping station wet-well dimensions 15 × 20 × 12 operating range of 8 ft; five pumps 3000 gpm each

Bar screen with $\frac{3}{4}$ in. openings

Grit removal—none

Final clarifiers—two at 60-ft diameter, 12-ft side water depth each

Oxidation ditch—four with a volume of 80,000 cu ft, four brush aerators in each tank, MLSS = 2000 mg/l, sludge age = 15 days

Recirculation pumping station—six pumps at 300 gpm each

Chlorine contact basin—four passes in channels 20 ft long, 10 ft wide, and 10 ft deep.

Aerobic digesters—four units at 50 ft × 50 ft × 15 ft water level

Determine the operating characteristics for each of the treatment units and compare against the unit process design parameters listed in the text:

For the influent pumping station, determine the adequacy of the wet-well volume and pump sizes.

For the bar screen, determine the maximum screenings removal.

For the clarifiers, determine the adequacy of the overflow rates, weir loading, and detention times.

Will the sludge settle well?

For the oxidation, check the BOD and hydraulic loading. Check the aeration capacity.

For the chlorine contact basin, determine the detention time, chlorine feed rate in pounds, and amount of chlorine needed for a 30-day supply.

Calculate the sludge production and volume.

For the aerobic digesters, determine loading, detention time, and storage time.

For land application, determine the land area required using 30 mg/l nitrogen in the liquid sludge.

Will the treatment plant meet effluent requirements?

11–75 An activated-sludge plant consists of preliminary treatment, primary, and secondary treatment as shown in Figure 11–31a. Effluent is UV disinfected prior to discharge to the river. Solids treatment consists of thickening the secondary clarifier sludge; primary sludge is pumped directly to digestion. Sludge is digesed in two-stage anaerobic digestion and dewatered prior to hauling to landfill.

Operating conditions:

Max month design flow = 10.3 mgd; peak flow = 19 mgd; minimum flow = 5 mgd

BOD = 210 mg/l, suspended solids = 240 mg/l, ammonia = 22 mg/l

Wastewater temperature = 17.5°C

Effluent treatment requirements:

BOD = 10 mg/l, suspended solids = 10 mg/l, 2.2 MPN/100 ml

Treatment units:

Influent pumping station wet-well dimensions 15 × 25 × 10 operating range of 6 ft; five pumps, 4000 gpm on variable-speed drives.

Bar screen with $\frac{1}{2}$ inch openings

Grit removal—two units 40 ft × 12 ft

Primary clarifiers—two at 65-ft diameter, 7.5-ft side water depth each

Final clarifiers—two at 90-ft diameter, 14-ft side water depth each

Aeration basins—eight at 80 ft × 40 ft × 19 ft water depth; the fine-bubble diffusers are at a depth of 18 ft, MLSS = 400 mg/l, sludge age = 15 days.

Recirculation pumping station—six pumps at 3000 gpm each

Blowers—four blowers at 56,000 scfm, 150 hp

UV disinfection—four units operated in parallel

Thickening—gravity belt, 96% solids capture, 4.5% thickened solids

First-stage anaerobic digester—40 ft in diameter by 12 ft below the landings, 20-ft operating volume

Second-stage anaerobic digester—40 ft in diameter by 12 ft below the landings, 10-ft operating volume above

Centrifuge—two at 300 gpm each, 20 hr per day, 5 days per week operation

Sludge pumping distances and fittings are as follows:

Primary sludge: 420 ft, 16 plug valves, 1 check valve, 6 flow-through tees, 18 standard 90° elbows, one entrance and one exit.

Secondary clarifier sludge: 860 ft, 6 plug valves, 1 check valve, 6 flow-through tees, 10 standard 90° elbows, one entrance and one exit

Thickened sludge: 520 ft, 12 plug valves, 1 check valve, 8 flow-through tees, 22 standard 90° elbows, one entrance and one exit

Digested sludge: 1200 ft, 4 plug valves, 1 check valve, 6 flow-through tees, 16 standard 90° elbows, one entrance and one exit

Determine the operating characteristics for each of the treatment units and compare against the unit process design parameters listed in the text:

For the influent pumping station, determine the adequacy of the wet-well volume and pump sizes.

For the bar screen, determine the maximum screenings removal.

For the grit unit, determine the maximum grit removal?

For the clarifiers, determine the adequacy of the overflow rates, weir loading, and detention times.

Will the sludge settle well?

For the activated sludge, check the BOD and hydraulic loading.

For the UV disinfection, what is the power required to meet discharge requirements?

Calculate the sludge production and volume.

For the anaerobic digesters, determine loading, detention time, and storage time.

For the centrifuge press, verify the feed rate and operating conditions.

For sludge pumping, determine the solids concentration, flow, and head loss.

Will the treatment plant meet effluent requirements?

Wastewater Systems Capacity, Management, Operation, and Maintenance

Wastewater systems are generally divided between the wastewater treatment plant and sewer collection system with its associated sewer pipe, manholes, and sewage lift stations. Requirements for treatment performance and operation are listed in the National Pollutant Discharge Elimination System (NPDES) permit regulations or state permit for irrigation and reuse. Permits contain water-quality requirements for discharge, capacity limitations, and requirements for operation and maintenance. Permit conditions for sewer collection systems are focused on reducing the frequency and occurrence of sewer overflows and public notification when overflows occur.

The EPA has estimated that between 23,000 and 75,000 overflows of sanitary sewers occur annually. Untreated wastewater contaminates streams, creeks, beaches, and other water bodies. Sewage can also back up into basements and low-lying areas. The EPA Region IV developed a program to help utilities identify needs for collection and treatment systems. The Management, Operation, and Maintenance program (MOM)

is a comprehensive checklist of self-monitoring activities designed to improve system performance and specifically to reduce sanitary sewer overflows.

The principles for MOM were developed specifically to reduce the frequency of sewer system overflows but, in general, apply to all types of asset management, including treatment facilities and water distribution systems. The general principles include use of best management practices; condition and capacity assessment; O&M planning; and identification of solutions to repair, rehabilitate, or replace equipment and components, thus improving reliability. Management practices are focused on information collection (peak flows, normal flows, loading) to compare with design capacity, planning to coordinate increases in capacity with existing facilities, appropriate use of analytical tools (modeling software), and compliance with ordinances limiting sewer discharges (in particular fats, oils, and grease). Capacity assessment requires an understanding of the system design, current performance, and strategies to increase capacity. O&M planning

463

involves determining how to deal with capacity problems through changes in operation, maintenance, enhancement, or replacement. The all-too-common practice of emergency or reactive maintenance waits until sewers overflow or equipment fails before identifying a problem and determining the appropriate corrective maintenance. Proactive maintenance involves actively monitoring performance and controlling capacity degradation on a regular basis. The need for rehabilitation arises out of deterioration of structures or equipment, the need for capacity beyond the original design, and the inability of processes and facilities to meet regulatory requirements. The decision to repair, rehabilitate, or replace must be determined on a case-by-case basis.

Sewer maintenance requires a thorough knowledge of the layout and appurtenances used in collection systems, covered in Chapter 10. Wastewater flows, infiltration, and inflow are discussed in Chapter 9. Prerequisite information regarding the function of various unit operations and how they relate to each other is presented in Chapter 11. The following sections discuss elements of capacity assurance, management, operation, and maintenance of wastewater systems with an emphasis on sewer collection systems.

12–1 CAPACITY OF WASTEWATER TREATMENT

NPDES permits establish flow and load limitations for treatment. Exceeding these limits generally results in a building moratorium that stops development until the plant is rerated through evaluation or expanded to increase capacity. Upgrades to a treatment plant may be required to meet more stringent effluent quality requirements, but new treatment processes typically do not add to treatment capacity.

Performance Evaluation of Treatment Plants

Compliance inspections by the state regulatory agency or the EPA regional office to ensure implementation of self-monitoring are performed annually, or more frequently if the plant is not in compliance. Flow measurement, sampling,

laboratory testing, and recordkeeping are primary considerations. Flow measurement is evaluated to verify the accuracy of the installed measuring system. For a flume, the procedure is to measure the head and calculate the rate of flow by applying the proper formula or chart. To be acceptable, the simultaneously recorded meter reading must be within plus or minus 10 percent of the calculated flow. Also, maintenance records are reviewed at least once a year to verify calibration of the flume. Sampling and composition procedures (Section 9–5) are reviewed for compliance with the specified frequency, locations of collection, and proper preservation. The laboratory personnel must be trained in analytical and quality control procedures required by the NPDES permit, including tests for conventional parameters, toxic substances, and biomonitoring. Recordkeeping and reporting include detailed documentation of physical facilities, operation, maintenance, and business affairs, and personnel records. In a performance inspection, the inspector observes sample collection, flow measurement, laboratory testing, and reporting. In compliance monitoring, split portions of samples are tested by the regulatory agency to verify the accuracy of the plant laboratory testing.

The primary purpose of regulatory inspections is to verify compliance with the NPDES self-monitoring program. Therefore, the main concerns are influent plant hydraulic and organic loadings, quality of the effluent wastewater, and proper disposal of sludge solids. The evaluation of a treatment plant requires examination of each unit process to study in detail its operation and how the process functions in the overall treatment scheme.

The first step in performance evaluation is to sketch a process flow diagram showing normal operation. Changes in processing to meet unusual flow and load conditions would be noted. Second, the dimensions of all treatment units should be determined from field measurements and design drawings. Sampling points must be carefully selected to ensure the collection of representative portions of flow for compositing. If access ports and flow-measuring devices are not available for all major in-plant flows, physical modifications must be made to permit isolation of each unit process for study. Finally, the laboratory needs to perform routine tests in accordance with

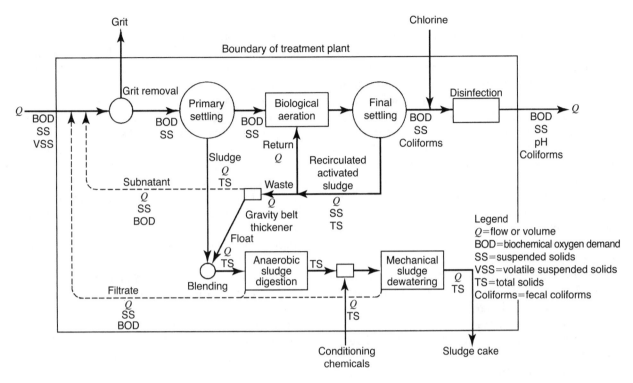

Figure 12–1

Typical process diagram for wastewater treatment by conventional activated sludge showing interrelated unit processes.

established procedures. Special studies and specific processes may dictate analyses of chemical oxygen demand (COD), grease, phosphorus, various forms of nitrogen, sulfide, volatile acids, gas analyses, sludge filterability, biological oxygen uptake, heavy metals, or total organic carbon (TOC).

The process diagram for a typical activated-sludge plant in Figure 12–1 suggests a minimum testing program for plant evaluation. Influent and effluent monitoring routinely requires testing for composite biochemical oxygen demand (BOD), suspended solids (SS) concentrations, and pH. In some locations, tests for fecal coliform count, chlorine residual, phosphorus, ammonium nitrogen, and the presence of heavy metals may be included for evaluation. Also crossing the boundary of the treatment plant are conditioning chemicals imported for thickening and dewatering of sludge and the export of grit and sludge cake to disposal. Recycle flows, including filtrate, subnatant, and drain water have significant impacts on flow and loading and require continuous monitoring.

Examination of Individual Processes

Examination of individual unit operations within a treatment plant requires understanding the following: (1) the purpose of each unit process and its function in the overall treatment scheme, (2) proper procedures for operation, and (3) performance expected under all possible loading conditions, including anticipated minimum and maximum values. Data collection must include composite testing of all influent and effluent flows and documenting the actual method of operation. Of course, the unit should be examined for any physical defects that would prevent satisfactory operation.

The measurements for evaluation of primary sedimentation include flow, BOD and SS concentrations of the influent, BOD and SS concentrations of the effluent, volume of sludge withdrawn, and its total solids content. Overflow rate, weir loading, and detention time can be calculated from the wastewater flow and these values compared with design parameters to indicate the loading status of the tanks. Although the efficiency of primary settling is most often related to BOD and SS removal, the quantity of solids content of sludge withdrawn is equally important as a measure of satisfactory performance. Poor sludge thickening in a sedimentation tank can be the result of either hydraulic overload or poor operating procedure. For example, a voluminous amount of thin sludge results if high pumping rates cause withdrawal of water from above the settled sludge layer.

Secondary biological aeration is examined on the basis of both organic matter removal and characteristics of the waste-activated sludge. To view the aeration-clarification functions only in terms of BOD and SS removals, without regard for concentration of solids in the waste sludge, ignores the thickening function of biological flocculation. The air supply, the rate of activated-sludge recirculation, and sludge wasting can be adjusted to provide the thickest waste sludge possible while maintaining a clear wastewater effluent. Operation at various dissolved oxygen levels, mixed-liquor suspended solids concentrations, and food-to-microorganism ratios with careful surveillance testing can define optimum performance.

Effluent disinfection requires careful analysis to ensure efficient use of chlorine. Poorly operated disinfection units can result in excessive chlorine residual being carried into the receiving watercourse. Chlorine residuals at different dosages can be measured under varying flow conditions and compared with laboratory tests on wastewater samples held for the same contact time. An efficient chlorination system provides rapid initial mixing of the chlorine solution in the wastewater followed by an adequate contact time in a plug-flow basin. Often, automatic residual monitoring and feedback control units are necessary to prevent either inadequate disinfection or excessive dosing.

The operation of a gravity belt thickener is judged on thickened sludge solids concentration and clarity of the filtrate. For a given loading of waste-activated sludge, operating variables for gravity belt thickening are hydraulic loading, solids loading, and polymer dosage. This process has proven to be an effective method of thickening waste sludge, with excellent solids capture commonly greater than 95 percent. Obviously, higher solids concentration in the influent requires reduced

dosage for separation of the suspended solids. Therefore, operation of the biological aeration system directly affects the sludge thickening process. Excess activated sludge is normally discharged from the recirculation line, returning from the bottom of the final clarifier to the head of the aeration basin. If the recycled flow is greater than that necessary for the settled sludge volume in the final clarifier, the settled solids are diluted by pumping overlying clarified wastewater, resulting in a thinner recirculated flow. For overall optimum plant operation, the final clarifier should be used to gravity thicken an activated sludge in order to reduce recirculation pumping and increase the solids content of the return flow and waste sludge.

Overall performance of anaerobic sludge digestion is determined by volatile solids reduction, gas production and composition, and clarity of withdrawn supernatant. The operational controls include temperature of the digesting sludge, mixing in high-rate digesters, rate of raw sludge feed, and solids retention time.

Evaluating the operation of mechanical sludge dewatering normally requires extensive studies because it involves economic considerations related to operation time, chemical dosage, and disposal of sludge cake. Because the quantity of chemical used is related to the solids concentration in the sludge feed, satisfactory dewatering is keyed to the previous unit operations of wastewater treatment and sludge thickening. Performance assessment too often stops with sludge yield and cake solids without regard to the efficiency of solids captured or the quality of filtrate, which is frequently hard to sample. Solids processing is an integral part of plant evaluation.

Careful consideration must be given to all unit operations that return flow back to the head of the plant. Sludge processing may return supernatant from anaerobic or aerobic digesters, overflow from gravity thickeners, underflow from flotation thickeners, centrate from centrifuges, and filtrate from belt thickeners or pressure filters. The return of excessive suspended solids can result in recycling of fine solids within the treatment plant. For example, if insufficient conditioning chemical is applied in sludge dewatering, the filtrate may carry a large amount of solids back to the plant influent. Being primarily colloidal in nature, these solids pass through primary sedimentation for capture in biological aeration. They are then returned in the waste-activated sludge for thickening and dewatering again. Cycling solids can lead to overloading and upset of all systems. However, their presence is normally first noticed in the thinning of the primary sludge and the increased oxygen demand in biological aeration.

Estimating Solids Capture in Sludge Processes

A relatively easy method of estimating solids capture by a sludge thickening or dewatering unit is to measure solids concentrations in the process flows. Figure 12–2 is a hypothetical processing unit for which the following relationships are apparent.

The solids mass balance is

$$M_S = M_F + M_C \qquad (12\text{–}1)$$

where M_S = mass of sludge solids
M_F = mass of filtrate solids
M_C = mass of cake solids

Liquid flow balance is

$$Q_S = Q_F + Q_C \qquad (12\text{–}2)$$

where Q_S = flow of sludge
Q_F = flow of filtrate
Q_C = flow of cake

Without introducing significant error, the specific gravity of all flows can be assumed to be 1.0.

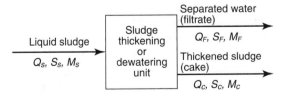

Figure 12–2

Flow diagram of a sludge-processing unit for Eqs. 12–1 through 12–4; Q is the flow rate, S is the solids concentration, and M is the mass of solids.

(For example, a wastewater sludge with 10 percent organic solids has a specific gravity of approximately 1.02.) Hence,

$$M = S \times Q \qquad (12\text{–}3)$$

Combining these equations results in the following:

$$\text{Fraction of solids removal} = \frac{M_C}{M_S} = \frac{S_C(S_S - S_F)}{S_S(S_C - S_F)} \qquad (12\text{–}4)$$

where S = solids concentration.

This equation is easy to apply because no volumetric quantities are necessary.

The solids concentration S can be either total solids (residue upon evaporation) or SS (nonfiltrable residue), with the latter being more common. Testing for SS by the standard laboratory technique of filtration through a glass-fiber filter is feasible for dilute suspensions. Dissolved solids may be a major proportion of the total solids in supernatant or filtrate because of the low SS concentration. Waste flows with high solids contents are, however, difficult to test accurately by laboratory filtration. Therefore, SS analyses of sludge samples are performed by total solids tests and then corrected by subtracting an estimated dissolved solids concentration. This procedure does not create significant error because the filterable solids content usually amounts to less than 5 percent of the total solids in sludge samples. Example 12–1 illustrates the use of this computational technique.

Recordkeeping

Complete and accurate records of all phases of plant operation and maintenance are essential. Too often, extensive testing and flow records are filed daily and are not reviewed until the plant is having serious operational problems or requires expansion. A rigid testing program without a directed purpose wastes labor and, worse yet, does not achieve its objectives. An effective sampling and testing program is one of continuous evaluation. Data collected may be applied to calculate existing hydraulic and organic loadings on all unit processes and to pinpoint problem areas in operations that are related to industrial wastes or excessive infiltration and inflow. Errors or omissions in laboratory testing will be recognized in the process of analyzing accumulated records. One procedure of value in compiling data is to trace suspended solids through the treatment processes, namely, removal from suspension in primary and secondary units, thickening, dewatering, and ultimate disposal. In plant data on sludge production, thickening efficiency, dewatering operations, and other processes are extremely valuable to an engineer in modifying or expanding an existing facility. Effluent quality is of value as a measure of overall plant performance.

■ EXAMPLE 12-1

The performance of a centrifuge dewatering anaerobically digested sludge process was evaluated to determine the capture and removal efficiency for suspended solids. Samples of liquid sludge and cake were tested for solids concentration. The centrate was tested for suspended solids concentration. Results of the laboratory analyses were

Centrifuge feed rate = 200 gpm
Centrifuge feed solids = 35,000 mg/l, or 3.5%
Cake solids = 28%, or 280,000 mg/l
Centrate suspended solids = 1200 mg/l, or 0.125%

$$\text{Removal efficiency} = \frac{28(3.5 - 0.12)}{3.5(28 - 0.12)} = \frac{94.6}{97.6} = 97\%$$

■ ■ ■

12-2 SEWER CAPACITY

Sewer capacity is established during design by the selection of pipe diameter and slope. Design values include planning assumptions for sewage quantities and allocations for infiltration and inflow. Allowances may be made for reduced capacity due to future storm flows or grit deposition in the sewers. New connections require a review of sewer capacity based on the impact of future flows and assessment of current conditions.

Infiltration and Inflow Surveys

The entrance of extraneous waters into sewer systems is of concern for several reasons. These include sewer surcharging during periods of intensive rainfall resulting in flooded basements, overloading of treatment plant facilities, overtaxing of pumping stations, excessive costs in processing diluted wastewater flows, and health hazards from discharging raw wastewater. In the past, problems of excessive infiltration and inflow were often handled by construction of relief sewers that bypassed treatment works and flowed directly to surface watercourses. This practice allowed adequate wastewater treatment 90 to 95 percent of the time and appeared to be the best economical design choice. Currently, the goal is to eliminate these pollutional discharges by requiring treatment of all wastewater flows even during peak periods.

Infiltration results from groundwater entering sewer lines through poor joints and cracks in manholes and sewer pipe; the sources are widespread, and the flow is relatively steady during times of high groundwater levels. Inflow comes from direct connections, such as roof drains, and results in sudden high rates of flow of short duration. The quantity of infiltration-inflow is considered to be the maximum wastewater flow minus the peak domestic and industrial discharge; in other words, it is the difference between peak flow measurements during wet weather and dry weather.

A systematic approach to sewer system evaluation includes identifying the quantity and nature of infiltration-inflow, isolating problem areas, and then determining the most economical corrective measures. The basic alternatives are either to rehabilitate the sewer system to reduce extraneous flows or to extend treatment facilities to handle peak wet-weather flows.

The diagram in Figure 12-3 outlines the general approach applied in infiltration-inflow surveys. The first step is to analyze and identify the magnitude of the problem. The sewer system is divided into areas that drain to key points, such as a manhole or pumping station. Flow tests are then conducted to determine peak discharges. Obviously, measurements must be taken at the proper times during both dry weather and high groundwater and rainfall situations. Evaluation of these data

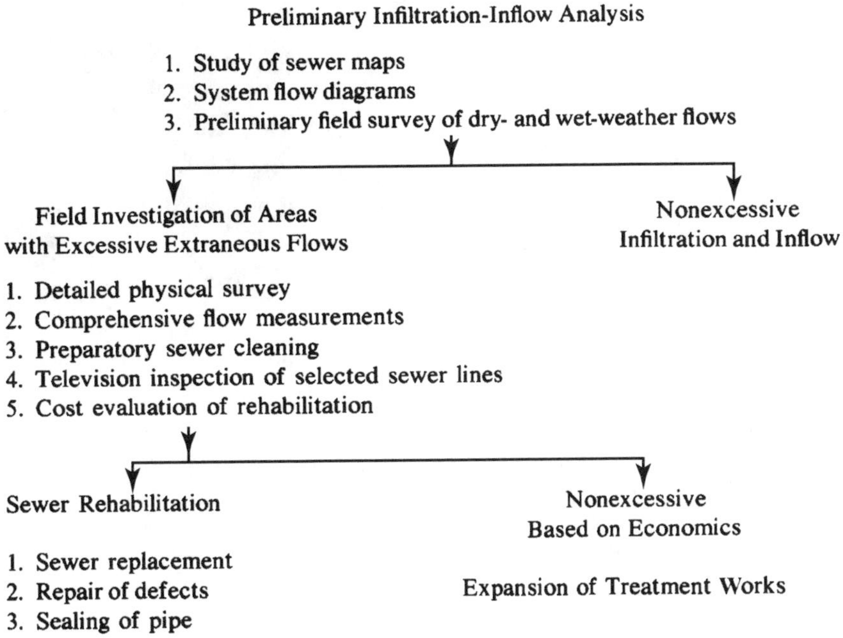

Figure 12–3

Diagram outlining the general steps that are included in infiltration and inflow evaluations of sanitary sewer collection systems.[2]

Preliminary Infiltration-Inflow Analysis

1. Study of sewer maps
2. System flow diagrams
3. Preliminary field survey of dry- and wet-weather flows

Field Investigation of Areas with Excessive Extraneous Flows

Nonexcessive Infiltration and Inflow

1. Detailed physical survey
2. Comprehensive flow measurements
3. Preparatory sewer cleaning
4. Television inspection of selected sewer lines
5. Cost evaluation of rehabilitation

Sewer Rehabilitation

Nonexcessive Based on Economics

1. Sewer replacement
2. Repair of defects
3. Sealing of pipe

Expansion of Treatment Works

reveals whether infiltration and inflow are reasonable. Detailed field investigations are then conducted in those areas where flows are considered excessive. Field investigations start with a survey of physical conditions, which requires descending every manhole in the study area and visually inspecting the sewer lines. This often uncovers deficiencies that are not obvious by merely looking into manholes from the street level. Cross-connections from storm sewers and unauthorized drains attached to sanitary sewers may be detected by using low-pressure smoke or tracer dyes in water. Often, eliminating inflows that are observed during the physical survey materially reduces peak discharges. For this reason, additional flow measurements may be required. Finally, those sewer lines identified as the source of excessive infiltration and inflow are inspected by using inspection television. Preparatory cleaning is required prior to televising. Many municipalities find that a significant amount of infiltration occurs in the service laterals, between the house and the sewer. These laterals are scheduled for replacement even though the service laterals are the property of the homeowner.

The final report of the field survey presents data from investigations, recommendations for a sewer rehabilitation program, and, most important, an economic evaluation of the problem. The expenses of rehabilitation are compared to the costs of expanding the treatment works to process extraneous water. Sewer repair may involve replacing pipelines, repairing structural deterioration or defects, or sealing openings by external grouting or pipe relining. The most economic and best techniques must be determined from local conditions and experience.

Figure 12–4 is a drawing that identifies an area of a sewer system to be rehabilitated. Rehabilitation of the sewers may include slip lining, cured-in-place linings, and pipe bursting. Linings are inserted into the existing pipe and take up some of the pipe diameter. Cured-in-place linings are inserted inside-out using air pressure and cured in place using hot air or water. Pipe bursting is used when an oversized steel pipe bursting head is pulled through the existing clay pipe. The advantage of pipe bursting is that it replaces the pipe with a larger pipe, maintaining or increasing capacity. Manhole repair methods include pressure cementitious grout, hydrophobic polyurethane grout, injectable foam, and surface-applied sealant materials. Laterals and side sewers are generally excavated and replaced. Such complete sewer system rehabilitation results in infiltration and inflow reductions of 70 to 90 percent.

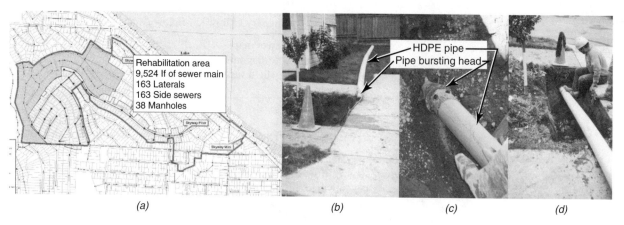

(a)   *(b)*   *(c)*   *(d)*

Figure 12–4

Map of rehabilitation area and photos of pipe bursting for sewer rehabilitation. (*a*) Within rehabilitation area, 100 percent of the sewer mains and manholes were rehabilitated along with laterals and side sewers. (*b*) and (*c*) Photos of steel pipe bursting head and HDPE sewer replacement pipe. (*d*) Bursting head and pipe being pulled through existing VCP sewer. Following installation, the HDPE is connected to the service laterals and cut open at the manholes.

Inspection of Sewers

Collection systems can be inspected for structural soundness by several techniques. Low-pressure smoke detection is an easy method of locating openings that connect to the atmosphere, such as service connections left unplugged from houses that have been removed, cross-connections with storm sewers, and unauthorized drainage connections. Since rats need access to a dry area above a sewer opening, smoke testing is a good way to locate holes accessible to rodents. A large volume of air combined with smoke is blown into a sewer line through a manhole by a mechanical air-smoke blowing machine. Flowing in the reverse direction of the water flow, the smoke follows the sewer pipes to an opening and then rises to the surface through the drain line or cracks in the earth.

Smoke detection simultaneously tests the sewers, house connections, and house plumbing. The smoke is nontoxic and leaves no residue to damage the interior of buildings, although it is irritating to breathe. Before testing, the residents of buildings along a sewer line should be notified to reduce the possibility of alarm. If smoke enters a building, the occupants should be evacuated and the interior ventilated while locating the point of entry.

The best method for inspecting the inside of small-diameter pipes is by closed-circuit television.

After a sewer line has been cleaned, the skid-mounted video camera is pulled through the pipe and a continuous picture is transmitted to a receiver in the service truck. This technique allows visual inspection and location of structural defects. The videotape can be replayed for further examination and retained for a permanent record.

A diagram of the major pieces of equipment in television inspection is shown in Figure 12–5. As shown in Figure 12–5*a* and *b*, the truck is parked near the manhole, a plug is installed in the upstream manhole to block flow, and the camera is tethered by wire. The operator in the van has control over the camera and computer and monitors the camera position and sewer condition. Video from the camera is digitally encoded and stored for future reference. Often, the operator documents the type and severity of pipe damage for future repair by location, type of damage, and severity to prioritize and schedule repair. The camera shown in Figure 12–5*c* and *d* is wheel-mounted and the camera can be raised or lowered to accommodate pipe size. Most tractor transports allow changes in wheel diameter, material, and gear teeth as conditions change. Some trucks can be converted from wheels to treads. Figure 12–5*e* shows the computer, tethered wire, power, communication cables, and pull-type camera. The camera is pulled between manholes

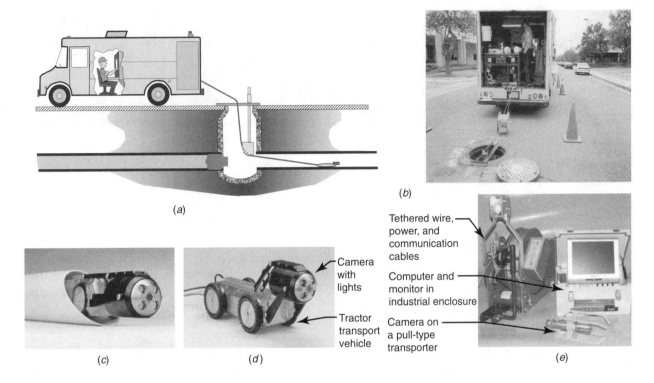

Figure 12–5

Television inspection of sewers. (*a*) Schematic drawing of CCTV van with operator and computer monitor and a camera in the sewer. The upstream portion of the sewer pipe is plugged with an inflatable "pig" to prevent flow. (*b*) Photo of van and wires tethered to the camera. (*c* and *d*) Camera with lights mounted on a self-propelled transport. (*e*) Photo of computer, tethering wire, and skid-mounted camera pulled by a wire through the sewer.
(Photo courtesy of RS Technical Services, Design and Manufacture of Video Pipeline Inspection Equipment)

using a stainless steel cable pulled by a winch. Cameras mounted on tractor transport trucks have become more popular due to the reduced setup required and their ability to explore sewers without being pulled.

Figure 12–6*a* shows a sewer overflow in the street. Sewer cracks like that shown in Figure 12–6*b* allow soil to enter the pipe and allow sewage to leak from the pipe, undermining the integrity of the sewer system. In combination with the weight of the soil, the pipe may collapse and fill with soil, creating a plug. The photo in Figure 12–6*c* shows the buildup of grease in the sewer. Lack of maintenance at restaurant grease traps results in a significant grease load that may lead to sewer plugging. Roots, like those shown in Figure 12–6*d*, obstruct flows and, with sewer solids, lead to sewer plugging. In each of these cases, failures in sewer maintenance lead to

flow restrictions and, with the introduction of sewage solids, will cause the pipe to plug and overflow.

Regulation of Sewer Use

The key purposes of a sewer ordinance are to control discharges to the sanitary sewer to ensure that water-quality standards of the receiving watercourse can be achieved and to establish equitable customer charges for wastewater service.[1] A comprehensive code contains regulations requiring the use of public sewers where available, control of private waste disposal in the absence of public sewers, construction of service connections, control of the quantity and character of wastewaters admissible to municipal sewers, procedures for wastewater sampling and analyses, provisions for the powers and authority of inspectors, an

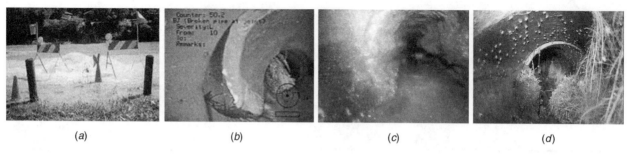

(a)   *(b)*   *(c)*   *(d)*

Figure 12–6

Photo of sewer system overflow and a growing sewer blockage. (*a*) Sewer flow boiling out of a manhole. (*b*) Crack in a sewer allowing groundwater infiltration that may in time cave in, creating a line blockage. (*c*) Buildup of grease within the sewer. (*d*) Roots growing in the sewer pipe through cracks in the pipe or gaps in joints. Eventually, roots and solids will plug the sewer, preventing flow.

enforcement (penalty) clause, other legal clauses, and signatures to validate the document.

The proper use and separation of sanitary and storm sewers are essential in protecting public health and reducing the quantity of inflow. Privies and septic tanks are not permitted where sewer service is available. In the absence of public sewers, regulations are established regarding construction, installation, and monitoring of private household systems to ensure that surface water and groundwater are not contaminated. Unpolluted waters must be excluded from the sanitary collection system and must be directed to storm sewers or a natural drainage outlet where feasible. These include inflow from downspouts, footing drains, surface-water inlets, swimming pools, and cooling water from air conditioners and industrial refrigeration units.

Certain wastes cannot be admitted to sanitary sewers, and others must be carefully controlled by specifying discharge limitations. These can be considered in the following four categories: (1) wastes that create a fire or explosion hazard; (2) substances that impair hydraulic capacity; (3) contaminants that create a hazard to people, the physical sewer system, or the biological treatment process; and (4) the refractory wastes that pass through treatment and result in degradation of the receiving watercourse. Examples of flammable liquids are gasoline, fuel oil, and cleaning solvents. Solids and viscous liquids that create sewer stoppages include ashes, sand, metal shavings, paunch manure, unshredded garbage, grease, and oil. The most common sewer stoppage, however, is related to root growth in sewer lines. This

can be minimized by discouraging the planting of certain trees like elm, poplar, willow, sycamore, and soft maple near sewer lines. As a preventative measure, proper jointing methods and materials should be used when sewers are installed in areas where roots could be a problem.

Uncontrolled industrial wastes may contain corrosive or toxic compounds. For example, sulfur compounds and high-temperature wastewaters can promote bacterial formation of sulfuric acid and lead to sewer crown corrosion. Acid wastewaters cause invert corrosion and, if not sufficiently diluted, may interfere with treatment processes. Toxic metal ions, such as chromium and zinc, and organic chemicals in rather small concentrations can lead to inhibition of the biological activity in aeration and anaerobic digestion. Dissolved salts and color- and odor-producing substances are only partially removed by conventional treatment. For these, protection of the receiving watercourse is best achieved by separate disposal at industrial sites rather than by discharge to the sewer system. Examples are spent brine solutions, dye wastes, and phenols. Where industrial wastes are highly variable, it may be necessary to install equalizing tanks to prevent shock loads on the treatment works. In addition to controlling quality variations by neutralization and dilution, pretreatment by equalization can also provide discharge control to smooth out flow and to prevent shock hydraulic loads.

Flow measuring and sampling of wastewaters entering the sewer system are essential for enforcing a sewer ordinance and establishing equitable fees for sewer use. Each industry is required to install and maintain a suitable sampling station that may range from a simple manhole for grab samples to a structure that includes a flow recorder and automatic sampler. Inspectors must have the authority to enter all properties for the purposes of observation, measurement, sampling, and testing pertinent to wastewater discharge. Facilities for sampling and flow measurement are absolutely essential for instituting a sewer ordinance, and, therefore, their construction must be given primary consideration. For a small enterprise, the cost may involve only the installation of a manhole for sampling. Additional water meters may be needed to determine wastewater discharge if the wastewater is assumed to equal the water supplied minus the consumption for lawn watering or discharge to the storm sewer. Automated sampling stations for large industries, particularly with more than one outlet sewer, may be a substantial investment.

Violation of an ordinance is declared a misdemeanor or sometimes disorderly conduct. Penalties usually involve a fine for each violation plus costs of prosecution. If a user fails to correct an unauthorized discharge, the city is commonly authorized to discontinue sewer or water service or both. Repeated violations may result in revoking such an industry's discharge permit. Any of these actions requires written notice. An ordinance may specify a hearing board to arbitrate differences between the city and an institution aggrieved by a penalty.

An ordinance also includes a section on charges for wastewater service. Sometimes, this is put in a separate document apart from the regulations governing sewer use so that the two subjects can be handled independently at public hearings and during enactment. This approach is particularly practical when instituting a new code, since fees cannot be collected on the basis of wastewater flow or excessive strength until after installation of sampling stations.

Sewer Charges and Revenues

Annual revenues of a wastewater system include costs associated with operation and maintenance of the collection system, treatment plant, and other facilities; principal and interest payments for debt financing of major capital improvements, as well as required reserve payments for general obligation or revenue bonds; and costs of capital additions not debt financed, such as interim replacements, betterments, and minor extensions to physical facilities that are paid for from current revenues. The allocation of revenue requirements must consider all costs of providing service. These vary with each municipality. Payment for service from each customer should be in proportion to use and benefits received. Arriving at a fair, proper, and practical system for raising annual revenue from users according to their responsibility for costs is a complex subject.

Wastewater service charges based on volume and strength reflect the actual physical use of the system to a considerable degree, but raising all revenues on the basis of service does not provide for payment by undeveloped property, for having facilities available whether or not they are used, or for the general community benefit that results from having adequate wastewater facilities. Yet a service charge system yields a relatively stable source of funds that allows for orderly planning, upgrading, and expansion. A major advantage in handling industrial wastes is that volume and strength charges encourage reduction of wastewater loads by in-plant modifications, improved housekeeping, by-product recovery, and water reuse.

The most equitable charge for flow is a uniform rate for all users regardless of quantity, with no tapered schedule of charges. In many cities, household wastewater use is based on water meter readings during the winter when outdoor water use is minimal. Water meters can also be used to determine wastewater flows from small industries and commerical establishments. Large industries install flow recorders that directly measure the quantity discharged to the municipal sewer.

The second component of a service charge is related to wastewater strength. This is generally applied as a surcharge for concentrations of pollutants that are greater than those found in domestic wastewater. The strength parameters often employed are BOD and suspended solids. These must be determined by analyzing composite samples of industrial discharges in accordance with guidelines that are specified in the ordinance. Equation 12–5 is a typical formula for calculating service charges. The surcharge portion is for BOD in excess of 250 mg/l and suspended solids greater than 300 mg/l. These values were used to develop the service charge in Eq. 12–5.

Wasterwater service charge

= charge for volume + surcharge for strength

= $VR_V + V[(BOD - 250)R_B$

$+ (SS - 300)R_S]8.34$ **(12–5)**

where V = wastewater volume, million gallons
BOD = average BOD, milligrams per liter
SS = average suspended solids, milligrams per liter
R_V = charge rate for volume, dollars per million gallons
R_B = charge rate for BOD, dollars per pound
R_S = charge rate for suspended solids, dollars per pound

A sewer fee schedule should stipulate a minimum monthly charge for each property. The amount is often set on a graduated scale that increases with the size of water service to the property. One reason for a minimum bill is to cover the costs for maintaining capacity in the system even though it is not used. This is referred to occasionally as the right-to-service factor. Cities may also have a fee for making service connections to the municipal system. The charge may be small, only offsetting the costs of inspection, or much larger to pay a share of the cost involved in extending the sewer system.

■ **EXAMPLE 12–2**

Calculate the service charge for a dairy wastewater by using Eq. 12–5 based on the following: daily flow = 150,000 gal, average BOD = 910 mg/l, average suspended solids = 320 mg/l, sevice charge for flow = \$450.00/mil gal, and surcharges of 2.38 ¢/lb excess BOD and 1.83 ¢/lb excess SS.

Solution

Substituing into Eq. 12–5,

$$SC = 0.15 \times 450 + 0.15[(910 - 250)0.0238$$
$$+ (320 - 300)0.0183]8.34$$
$$= 67.50 + 20.10 = \$87.60 \text{ per day}$$

■ ■ ■

12–3 MANAGEMENT OF WASTEWATER SYSTEMS

Organizational structure varies significantly among utilities. Many cities and agencies separate the management of the treatment plant from the sewer system because of the specialized nature of the work and geographic change. Wastewater department staff may interact with a variety of other departments, such as engineering, public works, the building department, human resources, safety, customer service, information services, and general services.

The information collected about the utility includes operation and maintenance records, complaint tracking, performance monitoring, construction, and assessment documents. Wastewater treatment operators and maintenance staff rely on as-built drawings and manufacturer equipment manuals for information about processes and equipment. Collection system staff typically maintain their own drawings showing sewer and manhole location. Geographic information systems (GIS) allow information about the system to be attached to an electronic drawing of the sewer. Operations and maintenance staff may query the data base to for example, create a map highlighting sewers that have been cleaned or repaired in the last 6 months and sewers to be cleaned in the next 6 months.

Computer maintenance management systems assist maintenance staff with tracking work orders for predictive, preventative, and requested service calls. Predictive maintenance may be based on vibration or heat (infrared) readings on rotating and electrical equipment. For example, bearing failure may be predicted by detecting an increase in equipment vibration, and electrical breaker failure may be detected by measuring an increase in heat at the connection. Values programmed into the computer automatically trigger a work order to evaluate equipment operating outside of a performance range. Preventive maintenance is triggered by date or running-hour limits. For example, a work order to change engine oil may be automatically initiated every 12 months or 200 hours of operation. Requested service may initiate work orders based on information collected by operating staff or complaint tracking. Work orders may be manually entered and tracked by the system. Maintenance management systems may be used to track inventory, adjust workload, and make decisions about equipment replacement by monitoring the aggregate cost of maintenance.

Supervisory control and data acquisition systems assist operations staff with monitoring and controlling equipment. At a central computer, an operator can monitor what equipment is being operated, total operating time, liquid levels, pressures, temperatures, and other operating details. Alarms indicating equipment failure and operating conditions above or below target values are reported and logged. The information allows operators to identify problems that require attention and maintenance. The information also allows operators to monitor changes in performance over time during normal diurnal changes, storms, and emergency events and long-term trends between seasons and from year to year.

Electronic document management systems assist management, operations, maintenance, engineers, and other users in retrieving information maintained by the utility. Operation and maintenance manuals, documents of construction, regulatory requirements, and utility records may be stored in an electronic format accessed from the agency's intranet. For example, the MOM program is based on defined goals, performance measures, and periodic evaluation that must be in writing. Records must be available to staff and accessible for audit and inspection.

12–4 OPERATING AND MAINTAINING WASTEWATER SYSTEMS

Wastewater treatment plant operators and sewer workers are certified and licensed by the state. The operation of wastewater treatment plants includes putting unit processes and equipment in and out of service, adjusting chemical feed, establishing setpoint values, and controlling the treatment process to create an effluent meeting or exceeding NPDES permit requirements. Sewer systems operation includes pumping stations, control of pretreatment, and flow monitoring to avoid sewer system overflows. All operation includes documentation, inspection during regular rounds, and for most wastewater treatment

plants, operation 24 hours per day 365 days per year. Supervisory control and data acquisition systems assist operators by monitoring equipment and reporting alarms at a single location. The control system may include an automated dialer to notify on-call operators while facilities are unstaffed.

Maintenance staff for wastewater treatment plants varies with the type of equipment but usually includes general maintenance workers, mechanics, electricians, and laborers. Operations or maintenance staff may lubricate equipment, clean filters, and perform routine tasks. Maintenance staff is responsible for maintenance, installations, inspections, troubleshooting, and adjustments of a variety of mechanical equipment, which may include pumps, gates, compressors, conveyors, engines, and rotating equipment. Electrical maintenance includes isolating power, troubleshooting, logic and functional testing, repairs, maintenance of analog and digital equipment, and restoring power.

12–5 SEWER MAINTENANCE AND CLEANING

Serious and expensive sewer problems can result from improper design or poor construction. Adequate slopes to maintain self-cleaning velocities are essential to minimizing maintenance. Selection of a suitable pipe joint is vital to prevent penetration of roots and excessive infiltration. The cutting of tree roots from sewer lines can be an expensive and recurring cleaning process. Groundwater entering joints carries with it soil from around the pipe, which ultimately causes structural failure. In addition to the review of new designs and supervision of construction, building permits should require careful inspection of all service connections before backfilling. It is important to make certain that unused service lines are properly capped when buildings are demolished.

A successful maintenance program operates on a maintenance management or graphic information system and requires keeping effective records. Maps are used to show the location of manholes, flushing inlets, service connections, and other appurtenances. Records should be kept on maintenance performed, with particular

Figure 12–7

Sewer cleaning using a high-pressure water jet. (*a*) Schematic drawing of the system. (*b*) Line drawings of the hose nozzles available for various cleaning applications: (*b1*) Radial nozzles are designed for general cleaning applications in small and medium-sized pipes. (*b2*) Thrust nozzles are lightweight, fast, and forward-traveling and are used to thread TV camera lines or break through blockage. (*b3*) Bayonet nozzles are used for penetrating grease and complete stoppages. (*b4*) Missile nozzles have two banks of rear-propulsion orifices for increased thrust to penetrate difficult stoppages. (*b5*) Double storm radial nozzles allow lead jets to dislodge debris while the second jets flush it to the manhole. (*b6*) Rotating root cutters with a cement removal tool use steel chains that are rotated by high pressure to remove roots and hard deposits. (*b7*) Torpedo nozzles are used to clean the bottom of large-diameter pipes. Their smooth, hardened surface allows them to glide over sand and debris. (*b8*) Sting-ray nozzles are engineered to ride evenly on the bottom of the pipe and include a swivel and rollover bar to eliminate flipover.

(Courtesy of Aquatech, Inc., Cleveland. OH)

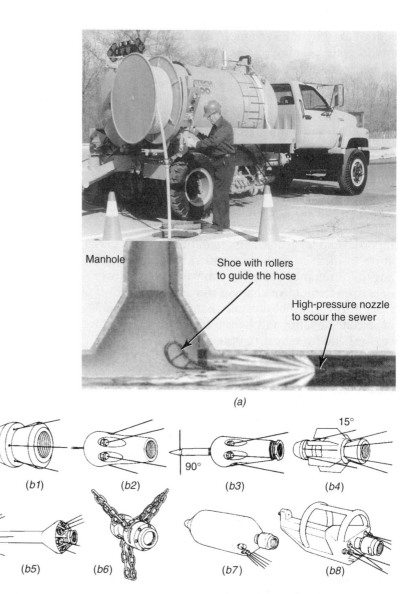

Manhole

Shoe with rollers to guide the hose

High-pressure nozzle to scour the sewer

(*a*)

(*b1*)   (*b2*)   90° (*b3*)   15° (*b4*)

(*b5*)   (*b6*)   (*b7*)   (*b8*)

emphasis on troublesome lines that are known to require more frequent inspection or cleaning. Although large sewers on adequate slopes may never require flushing or cleaning, others must be placed on a regular schedule that may range from every month to once a year. The number of emergency sewer blockages can be materially reduced by such preventative maintenance.

Sewer stoppages are caused chiefly by sand, greasy materials, sticks, stones, and tree roots. Sewer system overflows, like that shown in Figure 12–6*a* are the result of overloaded pipes or blockages caused by pipe cracks leading to cave-ins, grease restrictions, and root plugs (examples of each are shown in

Figure 12–6*b* through 12–6*d*). Common cleaning techniques are hydraulic scouring with high-pressure jets, scraping with mechanical tools, and addition of chemicals. When a clogged sewer line is reported, the manholes are inspected to determine the location. With few exceptions, the sewer pipe is cleaned in the up-slope direction by working in the first dry manhole downstream from the stoppage. This procedure allows working in a clear manhole, and the wastewater flow from upstream helps to flush the sewer line after the blockage is removed.

Sewer cleaning using a high-pressure water jet is performed as shown in Figure 12–7. The truck is positioned with the hose reel over the opening of

Figure 12–8

Sewer cleaning using a high-pressure water jet and vacuum to remove debris. (*a*) The nozzle flushes solids back to a manhole with a temporary dam where the collected solids are removed by a vacuum tube. (*b*) The truck has two tanks. The first one stores solids separated from the wastewater by a cyclone separator. The second tank holds clarified wastewater for continuous cleaning using the high-pressure jet. (Courtesy of Aquatech. Inc., Cleveland, OH)

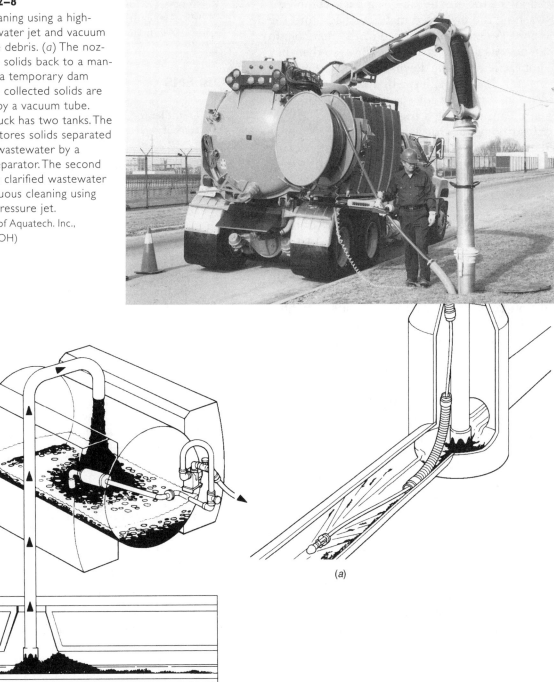

(*a*)

(*b*)

the downstream manhole. After a shoe with rollers is placed into the manhole as a hose guide, the hose with a high-pressure nozzle on the end is inserted through the manhole and into the sewer pipe. Water pumped through the hose is ejected out through orifices on the rear of the nozzle at a pressure up to 2000 psi at a rate of up to 60 gpm with a 400-ft hose (Figure 12–7*b*). The resulting force thrusts the nozzle through clogging materials as the jets of water break up the solids. By slowly withdrawing the nozzle, pipe cleaning is completed by washing the dislodged materials back into the manhole. A dam can be placed in the manhole to retain heavy solids for manual removal so that they are not carried into the downstream sewer pipe.

Several hose nozzles, illustrated in Figure 12–7b, are available for specific applications in sewer cleaning using a high-pressure jet. Radial nozzles (b1, b5) are for general cleaning. Bayonet (b3) and missile (b4) nozzles are for penetrating and clearing sewer stoppages. Nozzles with steel chains rotated by high pressure (b6) are used to remove roots and hard deposits, such as mortar protruding into the pipe through a poorly grouted joint. For large-diameter pipes, nozzles are designed to ride on the invert for removal of sand and debris from the bottom of the pipe (b7, b8).

The sewer cleaning vehicle illustrated in Figure 12–8a (page 477) jet-cleans sewers using recycled water and vacuum removes debris. The hose with a high-pressure nozzle is placed in the clogged sewer pipe, and the vacuum tube is hung down into the manhole. As the water from jet cleaning flows into the manhole, the wastewater and debris are drawn out into an 850-gal cylindrical tank on the truck (Figure 12–8b). The water is transferred to a second 1000-gal tank through a cyclone separator to clarify the wastewater by returning solids to the debris tank. From this tank, the water is recycled through the high-pressure cleaning nozzle. The cylindrical debris tank can be emptied by hydraulically opening a hinged door and tilting the tank to drain out the wastewater and accumulated solids.

Tree roots are extremely troublesome in some sewer systems. Fine roots can be removed by high-pressure jetting and heavier roots by using a root-cutting nozzle. Treatment with chemicals can be effective in removing and controlling the recurrence of root intrusion. After the chemicals are allowed to kill and partially dissolve the roots, the pipe is cleared by hydraulic cleaning.

REFERENCES

1. U.S. Environmental Protection Agency, Region 4, *Management, Operations, and Maintenance Program.* http://www.epa.gov/region4/water/wpeb/momproject/index.html

2. *Wastewater Collection Systems Management,* MOP7. Alexandria, VA: Water Environment Federation, 1999. http://www.wef.org/Home

PROBLEMS

12–1 Describe the purpose of the MOM program for sanitary overflows.

12–2 How is the capacity of a treatment plant established and what information is required?

12–3 The following information is available about a primary sedimentation processing unit:

	V (mgd)	S (mg/l)	M (· 8.34 = lb)
Liquid influent	1.2 mgd	220 mg/l	
Separated water		121 mg/l	
Thickened sludge			1010 lb/day

Complete the table and determine the fraction of solids being removed using Eq. 12–4.

12–4 The following information is available about a gravity belt thickener processing unit with a 90% capture efficiency:

	V (mgd)	S (mg/l)	M (· 8.34 = lb)
Liquid influent	500 gpm	250 mg/l	
Thickened sludge		4000 mg/l	
Recycled flow			

Complete the table and determine the fraction of solids being removed using Eq. 12–4.

12–5 A sewer pipe has been overflowing onto the street during storms. How is the capacity of a sanitary sewer system determined? How is the decision made to repair sewer infrastructure, and what are the common types of repair?

12–6 Flow monitoring in the sewer suggests a direct connection between rain water runoff and sewer flow. What method of evaluation should be used, and what precautions should be taken?

12–7 What wastes are strictly prohibited from a sanitary sewer and why?

12–8 What are the considerations in setting sewer rates?

12–9 What is necessary to maintain sewer capacity? What are the common causes of reduced sewer capacity?

12–10 An industrial discharger produces 1200 gpd with 4200 mg/l BOD and 800 mg/l suspended solids. Sewer changes are $0.35 per gallon, $120 per lb BOD, and $180 per lb suspended solids. Calculate the monthly service charge. Is the discharge acceptable?

Advanced Wastewater Treatment

dvanced wastewater treatment refers to methods and processes that remove more contaminants from wastewater than are taken out by conventional biological treatment. The term may be applied to any system that follows secondary treatment or that includes removal or nitrification in conventional secondary treatment. The expression *tertiary treatment* is often used as a synonym, but the two terms do not have precisely the same meaning. *Tertiary* suggests a third step that is applied after primary and secondary processing. Wastewater reclamation consists of a combination of conventional and advanced treatment processes employed to return a wastewater to nearly original quality, reclaiming the water.

Common advanced wastewater treatment processes remove phosphorus from solution, oxidize ammonia to nitrate (nitrification), convert nitrate to nitrogen gas (denitrification), and remove or inactivate pathogenic bacteria and viruses (disinfection). In water reclamation, the objectives may be expanded to include removal of heavy metals, organic chemicals, inorganic salts, and elimination of all pathogens.

13–1 LIMITATIONS OF CONVENTIONAL TREATMENT

Contamination of municipal wastewater results from human excreta, food preparation wastes, and a variety of organic and inorganic industrial wastes. Conventional wastewater treatment is a combination of physical and biological processes to remove organic matter. For acceptable processing, at least 85 percent of the BOD and suspended solids are removed to produce an effluent of 30 mg/l of BOD and 30 mg/l of suspended solids or less. Some treatment plants use a disinfection process to reduce the numbers of viable bacteria and viruses in the effluent wastewater. Conventional treatment typically results in negligible reduction in ammonia and phosphorus, incomplete disinfection, removal of toxins present in the raw wastewater to varying degrees, and no removal of soluble nonbiodegradable chemicals.

In highly urbanized and environmentally sensitive areas, discharge standards of 30/30 do not adequately protect the environment. Limits for BOD and suspended solids may need to be stricter based on the actual capacity of the receiving water to complete the treatment process. In addition, nutrients, pathogenic organisms and viruses, and soluble metals remain relatively untreated in conventional secondary processes. States have been asked to identify environmentally sensitive streams, rivers, and lakes for a detailed study of maximum pollutant limits, which are then allocated to the dischargers in that area. As the National Pollution Discharge Elimination System (NPDES) permits are renewed, total maximum daily loads (TMDL) are established for each pollutant in terms of pounds, concentration, or both.

The primary pollutional effect of phosphorus in surface waters is eutrophication. Because phosphorus is usually the growth-limiting plant nutrient in natural waters, discharge of wastewater high in soluble phosphates leads to accelerated fertilization. The results in lakes and reservoirs are excessive growth of algae, causing reduced water transparency, depletion of dissolved oxygen, release of foul odors, loss of finer fish species, and dense growths of aquatic weeds in shallow bays. Flowing waters with turbidity that blocks the sunlight necessary for photosynthesis are not subject to eutrophication; however, estuaries and slow-moving rivers with clear water can be adversely affected in the same way as impounded waters.

Un-ionized ammonia is toxic to fish and other aquatic animals. The amount of un-ionized ammonia is based on the pH of the water, since ammonia is converted to the nontoxic ammonium ion with decreasing pH. The criterion for salomonid fish is 0.02 mg/l of un-ionized ammonia, and for tolerant fish species it is 0.08 mg/l. These values are equivalent to total ammonia-nitrogen concentrations of approximately 0.5 mg/l and 5.0 mg/l at pH 8, respectively. As a result of nitrification, ammonia can also result in dissolved oxygen uptake in flowing and impounded waters. Theoretically, 1.0 mg of ammonia nitrogen can exert an oxygen demand of 4.6 mg when converted to nitrate nitrogen. Nitrification of ammonia from a wastewater discharge rarely occurs to this extent in natural waters because of competing reactions, such as algal photosynthesis and environmental conditions adverse to nitrifying bacteria. In oligotrophic waters, the contribution of nitrogen in wastewaters can increase the rate of eutrophication.

Protection of public health is the primary concern regarding pathogens in wastewaters discharged to surface waters used for body-contact recreation and drinking water supplies. Conventional biological treatment removes 99 to 99.9 percent of pathogenic microorganisms in raw wastewater. In addition, effluent chlorination can kill an additional 99.99 percent or more of bacteria and an unknown number of viruses and protozoa. Protozoal cysts and helminth eggs are resistant to the combined chlorine residual formed in wastewater. Although a treated wastewater may be considered satisfactory for discharge if the fecal coliform count is reduced to 200 per 100 ml, effluent should not be considered pathogen free. In addition to resistant pathogens, viruses and bacteria can be harbored and protected from the effects of chlorine in suspended organic matter. Because chlorine residual is toxic to aquatic life, dechlorination by the addition of sulfur dioxide can be used to destroy excess chlorine.

Soluble organic and inorganic chemicals are difficult to remove by biological treatment. Some cause aesthetic problems like foam and color; others are injurious to aquatic life and, in sufficiently high concentrations, prevent safe reuse of receiving water. The chemicals are best removed at their point of origin by instituting an industrial pretreatment program. The traditional approach to detection uses numerous chemical analyses to identify the presence of specific compounds. As the list of potential toxic substances increased, testing for specific compounds became increasingly difficult. Biomonitoring was developed to reduce the cost of broad-spectrum effluent toxicity monitoring.

13–2 SUSPENDED SOLIDS REMOVAL

The inability of gravity sedimentation in secondary clarifiers to remove small particles is the major limitation of conventional wastewater treatment for removal of suspended solids (and associated BOD), phosphorus, and a variety of solids ranging from bacteria to viruses and dissolved organics to dissolved salts. Figure 13–1 displays the range of particle sizes and the corresponding filtration and membrane removal processes required for removal. Tertiary granular filtration can be used to remove

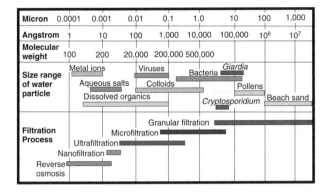

Figure 13–1

Size range of wastewater solids and removal by filtration processes.

relatively large solids and upgrade effluent quality. The filters are similar to those used in water treatment (Section 7–7). Membrane filtration removes particles by physical straining and, depending on media porosity, has the ability to remove bacteria, viruses, and dissolved solids. Increased demands for water reuse and reclamation have increased the use of membrane treatment for removal of solids, bacteria, and as a pretreatment for reverse osmosis. Microfiltration can remove *Giardia* and *Cryptosporidium* cysts. Reverse osmosis can be used to remove viruses and reduce dissolved solids.

Granular-Media Filters

The design of gravity filtration systems must take into account the higher suspended solids content and fluctuating rates of wastewater flow not common in water treatment. The preferred filters are coal-sand dual media or mixed media containing anthracite coal, garnet, and sand. A sand medium alone cannot accommodate the quantity of coarse solids in wastewater because of surface plugging. Multimedia beds allow in-depth filtration and greater solids holding capacity, resulting in longer filter runs. Efficient backwashing requires an auxiliary scour because hydraulic suspension of the media, by upward flow of water through the bed, does not provide adequate cleaning. Either air scrubbing or a rotating agitator improves scouring action. The filters may be either gravity or pressure units, depending on the size of the treatment plant. Chlorination prior to filtration prevents growth within the filter.

Flow schemes for tertiary filtration shown in Figure 13–2 range from traditional filtration with chemical coagulation, flocculation, and sedimentation to direct filtration and use of upflow filtration. Normally, at least two and usually four filter cells are provided for operational flexibility to accommodate diurnal flow variations. The total filter area should be sufficient to allow peak design flow with one unit out of service for backwashing or repair. When a filter is cleaned, the surge of dirty backwash water is collected in an equalizing tank and returned to the plant influent at a constant rate for treatment. Tertiary filtration can reduce suspended solids to 4 mg/l or less; however, small particles, including microoganisms, pass through the granular media. Direct filtration with chemical coagulation and optional flocculation-sedimentation is the most common tertiary filtration process (Figure 13–2b). The secondary effluent from biological processing and final settling is applied directly to granular-media filters. The impurities, including tiny particles, removed from the water are collected and stored in the filter media and removed by backwashing. Although rapid mixing of alum or other coagulant is necessary, subsequent flocculation is reduced to less than 30 min or eliminated. Contact flocculation of the chemically coagulated particles in the wastewater takes place in the granular media of the filter. Polymer as a coagulant aid is usually added just ahead of filtration or in the flocculation tank. Either this tertiary system is sized to process the diurnal flow variation or a flow equalization basin is installed to eliminate or attenuate the diurnal flow pattern. Maintaining a relatively uniform rate of flow allows improved chemical feed control and process reliability. An upflow filter with coarse media may be used in lieu of flocculation and sedimentation to reduce the area and cost of treatment units (Figure 13–2c). The upflow filter, sometimes referred to as a solids contact clarifier, functions to flocculate and remove large solids. Solids removal reduces the loading on the gravity filters, increases filter run times, and reduces effluent solids concentrations. The upflow filter is backwashed using air to fluidize the bed and release solids for removal in the backwash flow. All filtration systems with chemical addition can physically remove protozoal cysts and helminth eggs by filtration and inactivate the remaining

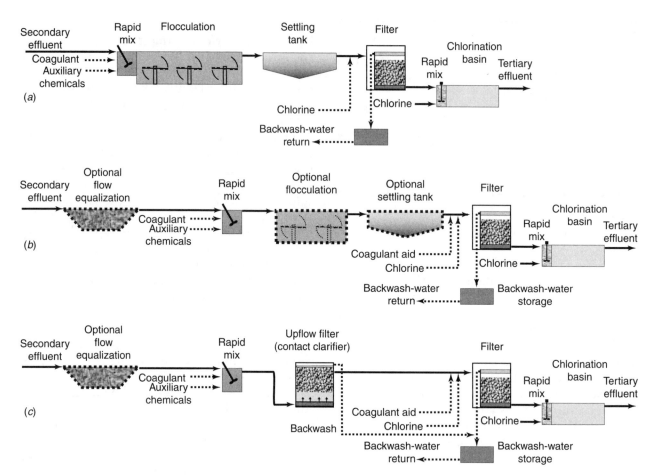

Figure 13–2

Flow diagrams for tertiary filtration of biologically treated wastewater. (a) Traditional filtration for solids removal with rapid mix, flocculation, settling, and filtration, followed by chlorination (for additional coliform and virus removal). (b) Direct filtration with optional equalization, flocculation, and settling. Sedimentation and flocculation are added with increasing influent solids concentrations. Equalization reduces peak flows and filter size. (c) Use of upflow filtration for flocculation and large solids removal. Chemical addition varies with the need for increasing solids removal. Chlorine addition prior to filtration reduces biological growth on the filters.

bacteria and viruses in the clear filtered water by extended chlorine contact.

Hydraulic loading and time of filtration before a bed requires backwashing are interrelated. Design rates in wastewater filtration are generally in the range of 3 to 5 gpm/sq ft (2.0 to 3.3 l/m$^2 \cdot$ s). Proper design of a filtration system depends on providing sufficient bed area. Although removal efficiency is relatively independent of the hydraulic loading rate, too small a surface area causes short filter runs. Excess backwash water cycled to the head of the plant and filter downtime result in inefficient operation. A minimum filtration time of 24 hr between

backwashes is desirable under normal conditions of wastewater flow and suspended solids concentration. Effective suspended solids removal relies on the design of the installation and the characteristics of the applied wastewater. In general, the best effluent quality achievable by plain filtration is 5 to 10 mg/l of suspended solids. If further reduction is desired, chemical coagulation must precede filtration to flocculate the colloidal solids and reduce the effluent suspended solids to less than 5 mg/l.

Figure 13–3 shows an aerial view of the City of Las Vegas filtration complex. The complex, similar to Figure 13–2b, reduces suspended solids from

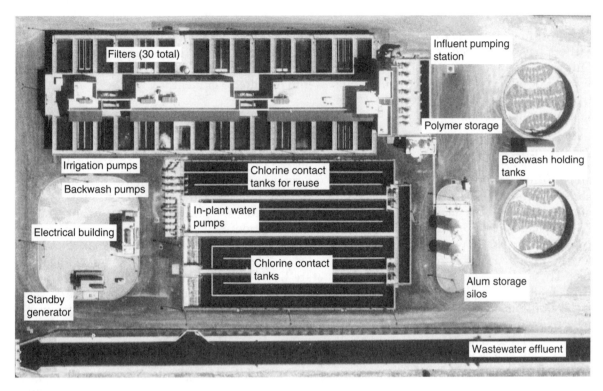

Figure 13–3

Processing arrangement for tertiary filtration for removal of suspended solids and effluent disinfection, City of Las Vegas Water Pollution Control Facility, Nevada.

(Courtesy of the City of Las Vegas, NV.)

about 30 mg/l to an effluent of 3 mg/l at a design flow rate of 99 mgd. Following secondary and tertiary treatment for nitrification, wastewater is pumped, mixed with alum and polymer, and chlorinated prior to filtration. Wastewater flows down through the filters and through the chlorine contact basins to the plant discharge. The 30 filters are each 18 by 28 ft and contain 24 in. of anthracite (1.0 mm) above 12 in. of sand (0.45 mm) with 12 in. of gravel below the sand. Design flow is 5 gpm/sq ft with air and water backwash. The filters operate with a constant water level in a declining-rate mode controlled by a 30-in.-diameter air-operated butterfly valve. Each filter has its own backwash panel, but the backwash sequence for all 30 filters is controlled by a central computer. The chlorine contact basins are baffled into long, narrow channels 12 ft deep, 10 ft wide, and 500 ft long. Three tanks provide 20 min of detention time at peak flow prior to discharge into Las Vegas Wash, and one tank provides 90 min of

detention time prior to reuse for power plant cooling water and golf course irrigation. Backwash holding tanks provide 24 hr of storage for backwash flow, representing 7 percent of the influent flow. Additional equipment includes pumps for backwash, irrigation, and in-plant water, electrical service, an emergency power generator, and chemical storage.

Variations of gravity filtration include pressure, traveling-bridge, and upflow filters. Pressure filters are more typical in smaller treatment plants. The granular media is retained in a pressure vessel, allowing higher heads and longer filter runs, thus reducing the filter size but requiring additional pumping and equipment. Underdrain design and media size are similar to gravity media filters. Traveling-bridge filters were named for the backwash pump and shroud that travel on a bridge and rails along the filter. Filter media depth tends to be shallow at 10 to 12 in., allowing effective backwash. The filter media is divided into horizontal segments that can be covered by the backwash shroud as it travels across the filter. Upflow filters have been used as pretreatment to gravity filters with coarse media for large solids removal and flocculation. The design and media size are coordinated as part of an overall system with subsequent disinfection.

Cloth-Media Filters

Cloth-media filters remove solids at the surface of the fabric by physical straining. The cloth is typically mounted on disks to facilitate backwash. Figure 13–4 depicts the unit in cutaway and in operating sequences of filtration and backwash. Disk units may be installed in concrete, fiberglass, or steel tanks. These filters have lower backwash flow volume and smaller overall size as compared with granular-media filters. Figure 13–5 shows the relationship between influent and effluent turbidity for various filtration types. The data indicate that all filters perform well at low influent loadings, but as

influent turbidity increases, the cloth media maintains a lower effluent turbidity, while traveling-bridge, deep-bed, and upflow granular-media filters begin to release increasing concentrations of solids into the filtered effluent.

Membrane Filters

Membrane filters are used to provide an increased degree of treatment for industrial applications, wastewater reuse, and reclamation. Membrane filters are used in lieu of secondary sedimentation, flocculation, settling tank, and granular filtration. Chlorination may be added for virus removal and to carry a disinfectant residual in the effluent water. Membrane design criteria and applications are listed in Table 13–1. Membranes are categorized by pore size, ranging from microfiltration to ultrafiltration, nanofiltration, and reverse osmosis. Membrane treatment and selection are based on the desired degree of treatment, such as turbidity reduction, disinfection, virus removal, and

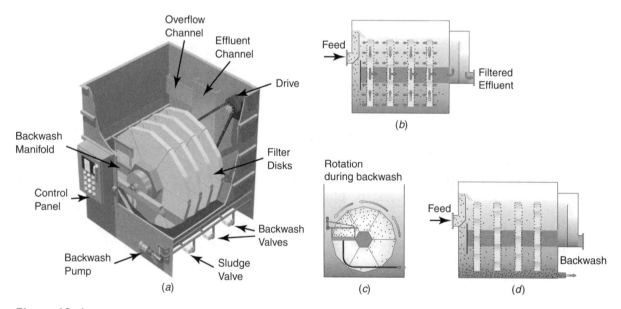

Figure 13–4

Cutaway view and operation of a cloth-filter unit. (a) The cutaway shows cloth-filter modules mounted on a shaft in a water-containing structure of steel. The diagram shows associated equipment for operation. (b) Section showing filter in filtration mode. Wastewater flows outside through the filter fabric, inside the disk, and through a hollow shaft to the effluent channel. (c) Filter disks rotate during backwash mode to remove captured solids. (d) Backwash removes settled solids as well as solids retained on the filter fabric.

(Courtesy of Aqua-Aerobic, Rockford, IL.)

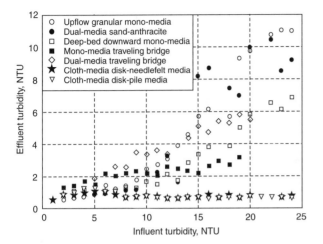

Figure 13–5

Performance comparison of effluent and influent turbidity for various types of granular-media filtration and fabric filter. (Courtesy of Aqua-Aerobic, Rockford, IL.)

desalination. See section 7–26 for microfiltration in water treatment.

Figure 13–6 shows a single ultrafiltration hollow fiber. Fibers are available in two sizes, 0.8 mm (inside diameter) and 1.2 mm with wall thickness of 0.2 to 0.4 mm for higher-turbidity applications. Hollow fibers are constructed from hydrophilic polymers or plastics. Membrane fibers may be

housed in a sealed module or suspended on an open frame installed directly in the activated sludge basin and as shown in Figure 13–7. Modules containing 12,000 fibers may be bundled into an 8-in. pressure vessel with epoxy resin ends to seal the fibers. To reduce fouling, feed water is screened using a 150 μm mesh. Wastewater is pumped to sealed modules that operate under pressure; open modules operate under a vacuum pulled into the center of the fiber suspended in the fluid.

Membranes filter from either outside-to-inside or inside-to-outside. Outside-in has the inherent advantage of having filtration occur on the exterior of the filter membrane. Increasing solids concentrations using inside-out membranes must be controlled to limit high head or plugging as the solids thicken in the membrane. In an outside-in mode, filters operate under a vacuum created within the hollow membrane fiber. Air scour pulsing into the feed water helps keep the membranes clean by driving solids off of the fiber.

Processing may be conducted using a cross-flow mode with recirculation or in a direct-flow filtration (dead-end) mode, as shown in Figure 13–8. Many manufacturers use a combination, beginning with direct-flow from the bottom then direct-flow from the top, to create uniform fouling, and finally cross-flow as the membrane begins to foul or during

TABLE 13–1

Various Design Criteria and Applications for Filtration

Process	Operating Pressures	Recovery (Percent)	Flux	Primary Application
Granular Filtration	10 ft of water (gravity) up to 150 psi (1000 kPa) (pressure)	95–98	5 gpm/sq ft (12 m/d)	Suspended solids removal
Microfiltration	5–40 psi (5–7 psi vacuum)	95–98	0.4–8 gpd/sq ft (1–20 m/d)	Suspended solids and bacteria removal
Ultrafiltration	15–60 psi	80–95	0.2–4 gpd/sq ft (0.5–10 m/d)	Virus removal and RO pretreatment
Nanofiltration	80–200 psi	70–90	0.12–0.4 gpd/sq ft (0.3–1 m/d)	Special applications
Reverse Osmosis	150–600 psi	70–85	0.16–0.33 gpd/sq ft (0.4–0.8 m/d)	Demineralization, TDS removal

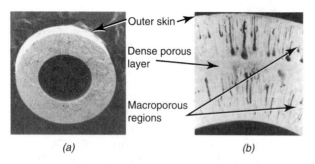

(a)   *(b)*

Figure 13–6

Magnified photos of ultrafiltration fiber membrane.
(*a*) Section through a 0.8-micron fiber. (*b*) Magnification of fiber wall showing fiber elements.
(Courtesy of Pall Advanced Separation Systems, East Hills, NY.)

Mode	Flow Direction
Processing	
Direct-flow	A to C or B to C
Cross-flow	A to B and C
Backwash	
Flush	A to B
Backwash top	C to B
Backwash bottom	C to A
Soak	None
Backwash both	C to A and B
Cleaning	A to B and C

Figure 13–8

Schematic representation of flow patterns illustrated using a single fiber (inside-out operation). Outside-in operation is similar, with the A and C directions reversed and B withdrawn from the module.

(a)

(b)   *(d)*

Figure 13–7

Installation alternatives for membrane filtration fibers.
(*a*) Rack-mounted pressure modules containing ultrafiltration fibers. (Manufactured by U.S. Filter.) (*b*) Contents of a pressure module showing filter fibers and seal rings.
(*c*) Schematic drawing of filter fibers installed directly into an aeration basin. (*d*) Fibers and support structure being removed from an aeration basin.
(Courtesy of Zenon Environmental, Inc., Oakville, Ontario, Canada.)

periods of high feed-water turbidity. Cross-flow generally maintains a minimum fluid velocity of 8.0 ft/sec (2.5 m/s) to scour solids off of the fibers. Backwashing is necessary to remove particles that accumulate on the membranes during normal operation. Backwash (see Figure 13–8) is automated in sequential steps: flushing, backwash, soak, and final rinse. Flushing forces the feed water out of the module to physically remove particles from the surface of the fibers. Backwash uses product water in reverse flow to lift particles off the membrane and is applied to both the top and bottom of the fibers. A chlorine solution (5–10 mg/l) may be added to aid in removal of organic material. A soak cycle allows time for chlorination to disinfect the fibers. A final backwash rinse feeds the top and bottom at the same time to remove any residual chlorine and remaining solids.

Plugging and fouling result from solids (sand, silt, and clay), high-molecular-weight organics, bacterial slimes, metal oxides, and precipitates. If backwashing loses effectiveness, cleaning is required. Cleaning includes removing a module from service and recirculating a low pH solution (typically citric acid) followed by a high pH solution of sodium hydroxide to dissolve organics and precipitates from the membranes.

Module fibers may break, leading to a decrease in performance. A visual bubble test is performed while the module is off-line to verify the integrity of the fibers. If a broken fiber is found, it can be isolated by placing a stainless steel pin inside of the tube at both ends.

Reverse-osmosis equipment and processing for wastewater is similar to water treatment as discussed in Section 7–27. Pretreatment to reduce solids is critical to reverse osmosis. Figure 13–9*a* shows traditional pretreatment processes, including lime precipitation, sedimentation, recarbonation, granular-media filtration, and carbon filtration.

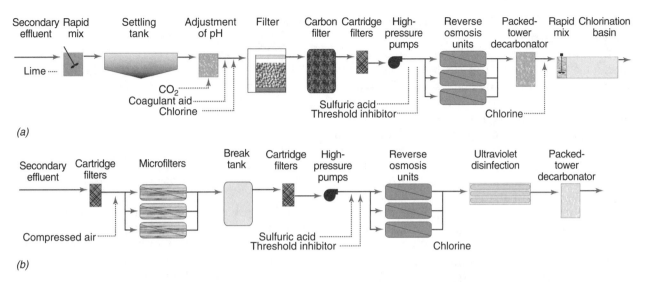

Figure 13–9

Flow diagrams showing pretreatment and reverse osmosis treatment. (a) Schematic of traditional pretreatment using lime addition, sedimentation, carbon dioxide addition (for pH adjustment), granular filtration (solids removal), carbon filters (organic removal), cartridge filters, acid addition, and reverse osmosis. (b) Schematic of membrane filtration used as pretreatment for reverse osmosis.

Lime precipitation is the first step in chemical-physical processing, in which high alkalinity kills bacteria and inactivates viruses. Flocculation and sedimentation are required for solids removal. Recarbonation is used to stabilize the wastewater, and granular-media filtration is used to remove nonsettleable solids. Granular-activated carbon is used to adsorb trace concentrations of a variety of nonpolar organic compounds. Acid addition may be used to reduce the potential for precipitation of calcium and silica compounds. The resulting water is ready for reverse osmosis. Figure 13–9b shows pretreatment using microfiltration or ultrafiltration, with chemical addition as pretreatment. Both processes result in treatment that removes bacteria, viruses, and organic substances and reduces dissolved solids concentrations.

13–3 PATHOGEN REMOVAL

Many infectious diseases are transmitted by fecal wastes, and the pathogens encompass all categories of microorganisms: viruses, bacteria, protozoa, and helminths. The kinds and concentrations of pathogens in domestic wastewater from a community depend on the health of the population.

The typical pathogens excreted in human feces are listed in Table 3–1. (Section 3–5 discusses waterborne diseases.)

Disinfection of biologically treated wastewater requires coagulation and granular-media filtration followed by chlorination with an extended contact time. Although chlorination is effective in killing bacteria and inactivating enteric viruses, these pathogens can be protected in suspended and colloidal solids if the wastewater has not been filtered first for turbidity (solids) removal. Cysts of protozoa and helminth eggs are resistant to chlorine, and they need to be physically removed by effective chemical coagulation and granular-media or membrane filtration. Even though the disinfection of wastewater by tertiary treatment is not as well defined as disinfection of surface waters for potable water supplies (Section 7–15), the principles for removal of pathogens are the same.

In the reuse of wastewater effluents containing enteric viruses, human contact is of major concern. The reuse may be unintentional, for instance, discharge to a dry streambed where children can play, or intentional, as in the case of landscape watering. The following is a brief summary of a significant study on tertiary disinfection for virus removal and its implementation by the Sanitation Districts of Los Angeles County.

The Pomona Virus Study[1] investigated the efficiency of virus removal in several tertiary treatment systems on a pilot-plant scale. The results demonstrated that direct filtration with a low alum dosage of 5 mg/l with polymer aid was as effective as alum coagulation at a dosage of 150 mg/l with conventional in-line mixing, flocculation, and sedimentation units prior to filtration. The rate of filtration was 5 gpm/sq ft, and chlorination was in a plug-flow tank with a residual of 5 to 10 mg/l after a 2-hr contact period. The concentration of coliforms in the tertiary effluent was reduced to less than 2.2 per 100 ml. The efficiencies of the tertiary systems were based on the removal of poliovirus that had been added to the wastewater before processing. The major conclusions of the study were that the majority of virus inactivation occurred during disinfection and that the main function of filtration with alum addition was to remove suspended solids that interfere with effluent disinfection. Also, the concentration of the chlorine residual directly affected effluent virus concentration.

A chlorination chamber must be properly designed for effective disinfection. Rapid mixing is critical to blend the chlorine solution with the wastewater, followed by plug flow through a contact tank that is a long, narrow, serpentine channel. Chlorine dosage is maintained by automatic residual monitoring and feedback control. If dechlorination of the effluent is necessary, a solution of sulfur dioxide can be added at the discharge end of the chlorination chamber.

Tertiary treatment by direct filtration, as diagramed in Figure 13–10, has been implemented at six plants of the Sanitation Districts of Los Angeles County (Long Beach, Los Coyotes, Pomona, San Jose Creek, Whittier Narrows, and Valencia) treating a total of approximately 150 mgd. Secondary effluent from activated-sludge processing is filtered through anthracite-sand dual media in five plants and through granular carbon in one. The plants have deep filter boxes with constant-rate filtration controlled by influent flow splitting, and the backwash systems are combined air-water with wash-water recovery. For chemical coagulation, alum and anionic polymer are added. The chlorine contact time averages approximately

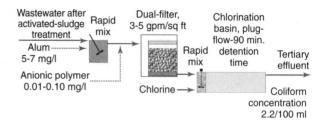

Figure 13–10
Tertiary treatment scheme for virus removal, Sanitation Districts of Los Angeles County, California.

90 min with a final residual of 4 to 5 mg/l, which is dechlorinated before discharge.

The effluent standard for high-quality reclaimed water in California is a median coliform concentration not to exceed 2.2 total coliforms per 100 ml, as determined from the last seven tests, and a maximum concentration not to exceed 23 per 100 ml in any sample and a turbidity limit not to exceed either an average of 2 NTU or more than 5 NTU for 5 percent of the time during a 24-hr period. All six plants complied with these criteria, with a small percentage of samples exceeding the minimum turbidity and total coliform limits. Fecal coliforms were rarely detected. Based on virus sampling over a 10-yr period, only one of 590 samples (each averaging 275 gal of filtered wastewater) was positive for enteric viruses (Coxsackie B3).

13–4 TOXIC SUBSTANCE REMOVAL

Numerous organic chemicals and several inorganic ions, mostly heavy metals, are classified as toxic water pollutants. To qualify, a toxic substance must be hazardous to aquatic life or human health and known to be present in polluted waters. Toxicity to aquatic life may be related either to acute or chronic effects on the organisms themselves or to humans by biological accumulation in seafood. Persistence in the environment and treatability are also important factors. Toxicity to humans may be related to either carcinogenicity or chronic disease through long-term consumption of contaminants. Currently, over 100 substances are listed as priority toxic water pollutants. Some are well-documented toxic substances; for others, there is limited data to support their hazard to the aquatic environment and human health.

The gross toxicity of a raw wastewater is evidenced by upset to the biological processes in wastewater treatment. This is an indication of industrial toxic chemicals being discharged to the sewer collection system. Heavy metals are partially removed by entrapment and adsorption onto settled solids or biological floc; nevertheless, a portion appears in the effluent. For cadmium, chromium, copper, lead, nickel, and zinc, the removal percentages may be up to 70 percent. Very little data are available on the removal of toxic organic compounds. Some may be biologically decomposed and others entrapped in settled solids or volatilized, but many are likely to be only slightly affected and present in the treated wastewater.

The general goal of municipal wastewater treatment with pretreatment of industrial wastewaters is to reduce the discharge of toxic pollutants to an insignificant level. An evaluation of toxicity reduction in a treatment plant is the first step. The objectives are to evaluate plant operation and performance to identify and correct treatment deficiencies causing effluent toxicity, identify the toxic substances in the effluent, trace them to their sources in the wastewater collection system, and implement appropriate remedial measures to reduce effluent toxicity.

The conventional approach to evaluating toxicity is to test raw and treated wastewater samples for specific substances. This can be very costly unless the toxic substances likely to be present can be reduced to a reasonable number. Initial tests should be for those substances related to wastes from industries served by the sewer system. After the toxic pollutants have been quantitatively identified, remedial measures must be taken by pretreatment of the industrial wastewaters at the factory and improved wastewater processing at the treatment plant. The potential deficiencies of this analytical approach through chemical testing are the possible presence of undetected toxic chemicals and combinations of synergistic toxic substances.

The second approach to evaluating toxicity is by biomonitoring of effluent, often referred to as whole-effluent toxicity. Bioassays to determine toxicity are performed by exposing selected aquatic organisms to wastewater effluent in a controlled laboratory environment. In the static test for acute toxicity, effluent is placed in laboratory containers,

the test organisms added, and the containers held in a controlled environment for 24 hr. A control test in prepared water is conducted in parallel for quality assurance of the test organisms and test procedures. An enhanced bioassay method is the flow-through test, where fresh effluent is continuously supplied to the laboratory containers with continuous outflow. The effluent is usually a 24-hr composite sample that has been filtered, aerated, and stabilized at 20°C for testing warm-water species (12°C for cold-water species). Among the several kinds of vertebrates and invertebrates used, the common warm-water test organisms are fathead minnow and water flea (*Daphnia*). The fathead minnow, which is a popular bait fish, feeds primarily on algae and grows to an average length of 50 mm with a normal life span of less than 3 years. *Daphnia* are invertebrates that feed on algae, grow to a maximum length of 4 to 6 mm, and have a life span of 40 to 60 days. From 10 to 20 of each species are tested in several containers under controlled temperature, light intensity, dissolved oxygen, pH, and food supply. The results of a screening test are expressed as percentages of surviving minnows and water fleas.

Definitive bioassays are conducted for longer test periods at effluent dilutions in a geometric series. For chronic toxicity, the recommended test durations are extended up to 7 days by renewing the diluted effluent waters every 24 hr in static tests or continuously in the flow-through tests. The dilution water is either a synthetic water or filtered water from the receiving water away from the point of wastewater discharge. Based on the results, the no-observed-effect concentration (NOEC) is determined for effluent in the diluted test samples. This is compared to the wastewater concentration after initial mixing of the wastewater in the receiving water body. The initial wastewater concentration should be less than the NOEC, with a margin of safety.

Pretreatment of industrial wastewaters is essential to control toxic substances. Since many hazardous substances are biocides, reductions in their concentrations in raw wastewater are essential to effective biological processing. If the evidence of toxicity is less dramatic, their presence in the raw wastewater and the treatment plant effluent may pass unnoticed. A typical list of toxic pollutants includes 13 metals, cyanide, 15 polynuclear aromatic hydrocarbons, 13 aromatics, 10 phenols,

21 aliphatics, 6 phthalates, 2 nitrogen compounds, 2 oxygenated compounds, and pesticides.

13–5 PHOSPHORUS IN WASTEWATERS

The common forms of phosphorus in wastewater are orthophosphates ($PO_4^{\equiv}$), polyphosphates (polymers of phosphoric acid), and organically bound phosphates. Polyphosphates, such as hexametaphosphate, gradually hydrolyze in water to the soluble ortho form; bacterial decomposition of organic compounds also releases orthophosphate. With the majority of compounds in wastewater being soluble, phosphorus is removed only sparingly by plain sedimentation. Secondary biological treatment removes phosphorus by biological uptake; however, relative to the quantities of nitrogen and carbon, the amount of influent phosphorus is greater than necessary for biological synthesis. Consequently, conventional treatment removes only about 20 to 40 percent of the influent phosphorus.

Biological phosphorus is that phosphorus incorporated into the biomass for cell growth. Orthophosphates and polyphosphates result from the decomposition of organic matter and can be removed through biological or chemical processes. Orthophosphate is the most abundant phosphorus species and can be removed from wastewater through biological uptake or chemical precipitation. Polyphosphates are condensed orthophosphates. Metal salts of aluminum or iron are commonly used to precipitate phosphates. Reverse osmosis and other membrane processes will physically remove organic molecules and chemical forms of phosphorus.

Remedial action for phosphorus pollution is the treatment of wastewaters that discharge directly into lakes and rivers or streams that flow into lakes. Several states have adopted effluent standards for phosphorus, which range from 0.1 to 2.0 mg/l as P, with many established at 1.0 mg/l. To protect the lakes and surface waters of the Great Lakes and Chesapeake Bay drainage basins, phosphorus removal has been implemented at many wastewater treatment facilities.[2] Of 526 plants in the Chesapeake Bay drainage basin (Maryland, North Carolina, Pennsylvania, and Virginia), 99 are removing phosphorus.

Phosphorus Removal in Conventional Treatment

The nutrient composition of an average sanitary wastewater based on 120 gpcd (450 l/person · d) is listed in Table 13–2. Phosphorus enters the sewer in the form of soluble and organically bound phosphates. Biological activity in the sewer releases organically bound phosphates into solution that are not removed by plain sedimentation of raw wastewater. The amount of organically bound phosphates released into a soluble form varies with the sewer length, wastewater temperature, and biological conditions.

Based on the tabulated values, the total phosphorus is reduced from 7 mg/l to 6 mg/l by sedimentation. Secondary biological treatment removes phosphorus by biological uptake; however, the amount of phosphorus is surplus relative to the quantity of nitrogen and carbon necessary for synthesis (Section 9–1). In general, the amount of phosphorus in the excess biological floc produced in activated-sludge treatment of a wastewater is equal to about 1 percent of the BOD applied. Based on this, the total phosphorus is further reduced from 6 mg/l to approximately 5 mg/l. As a result, in

TABLE 13–2

Approximate Nutrient Composition of Average Sanitary Wastewater Based on 120 gpcd (450 l/person · d)

Parameter	Raw	After Settling	Biologically Treated
Organic Content (mg/l)			
Suspended solids	240	120	30
Biochemical oxygen demand	200	130	30
Nitrogen Content (mg/l as N)			
Inorganic nitrogen	22	22	24
Organic nitrogen	13	8	2
Total nitrogen	35	30	26
Phosphorus Content (mg/l as P)			
Inorganic phosphorus	4	4	3
Organic phosphorus	3	2	2
Total phosphorus	7	6	5

Table 13–2, the total phosphorus of 7 mg/l in the raw wastewater is reduced to 5 mg/l in the biologically treated effluent.

■ EXAMPLE 13–1

Using the values given in Table 13–2, trace the inorganic, organic, and total phosphorus through a conventional activated-sludge treatment plant. Label the total phosphorus as a percentage of the influent phosphorus equal to 100 percent. Assume a primary clarifier removal of 35 percent BOD and 50 percent solids with 0.9 percent phosphorus. The waste-activated sludge system operates with an F/M of 0.40 and contains 2.0 percent phosphorus in the waste-activated sludge. Filtrate recycles 5 percent of the influent phosphorus.

Solution

Using the influent (raw) values from Table 13–2—influent SS = 240 mg/l, BOD = 200 mg/l, inorganic P of 4 mg/l, and organic P of 3 mg/l—effluent SS and BOD equal 30 mg/l each. At 50 percent SS removal, primary solids are 120 mg/l of the influent flow.

Phosphorus in the primary sludge equals:

$$0.009 \cdot 120 = 1.1 \text{ mg of organic phosphorus}$$

The primary effluent consists of SS = 120 mg/l, BOD = 130 mg/l, inorganic P remains unchanged at 4.0 mg/l plus the recycle P of 0.35 mg/l, and organic P is reduced to 1.9 mg/l for a total primary effluent P concentration of 6.25 mg/l.

From Figure 11–50 at an F/M of 0.40, K equals 0.50:

$$W_s = 0.50 \cdot 130 = 65 \text{ mg of sludge solids}$$

The phosphorus content of the solids in the sludge solids is

$$0.02 \cdot 65 = 1.3 \text{ mg of organic phosphorus}$$

The total P in the primary effluent is 6.25 mg/l into the aeration basin minus 1.3 mg/l (based on the influent flow) in the sludge, for 4.95 mg/l in the plant effluent. The 1.3 mg of P in the underflow is considered to be removed from the inorganic P consumed to provide cell growth, leaving 3.05 mg/l in the effluent. The organic P of the plant effluent is $0.02 \cdot 30$ mg/l SS = 0.6 mg, but an additional organic P of 1.3 mg/l is contained in organic matter that was not settled in the clarifier. The total plant effluent P concentration is 4.95 mg/l, or about 70 percent of the influent concentration.

From these calculations, the balance of the values for BOD, SS, and P can be calculated and are presented in Figure 13–11. If the total influent phosphorus concentration is assigned a value of

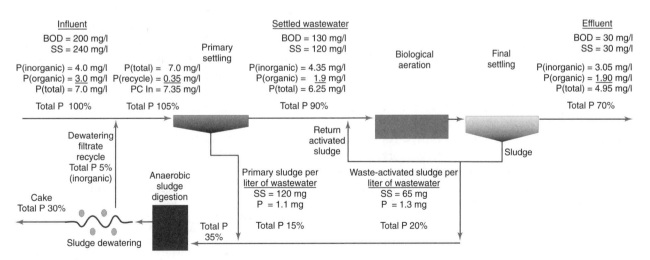

Figure 13–11

Diagram tracing wastewater phosphorus through a hypothetical conventional treatment plant. Of 100 percent influent phosphorus, 70 percent is in the treated wastewater and 30 percent is in the cake of the dewatered digested sludge solids. Diagram annotated with the results from Example 13–1.

100 percent, removal in the primary would be 15 percent and 20 percent in the secondary, resulting in the effluent discharge of 70 percent of the influent phosphorus. Because of phosphorus release under anaerobic conditions, some phosphorus is returned to the head of the plant in the filtrate recycle in the form of ortho (inorganic) phosphorus. Of course, variations in characteristics of the wastewater and different methods of treatment can result in higher or lower removal efficiencies. Still, phosphorus removal in conventional biological treatment systems is generally within the range of 20 to 40 percent.

■ ■ ■

13–6 CHEMICAL-BIOLOGICAL PHOSPHORUS REMOVAL

Chemical precipitation using aluminum or iron coagulants is effective in phosphate removal. Although coagulation reactions are complex and only partially understood, the primary action appears to be the combining of orthophosphate with the metal cation. Polyphosphates and organic phosphorus compounds are probably removed by being entrapped, or adsorbed, in the floc particles. Aluminum ions combine with phosphate ions as follows:

$$Al_2(SO_4)_3 \cdot 14.3H_2O + 2PO_4^{\equiv}$$
$$= 2AlPO_4\downarrow + 3SO_4^{=} + 14.3H_2O \quad \textbf{(13–1)}$$

The molar ratio for Al to P is 1 to 1, and the weight ratio of commercial alum to phosphorus is 9.7 to 1.0. Coagulation studies have shown that alum dosage greater than this is necessary to precipitate phosphorus from wastewater. One of the competing reactions, which accounts in part for the excess alum requirement, is with natural alkalinity, as follows:

$$Al_2(SO_4)_3 \cdot 14.3H_2O + 6HCO_3^{-}$$
$$= 2Al(OH)_3\downarrow + 3SO_4^{=} + 6CO_2 + 14.3H_2O_3$$

$$\textbf{(13–2)}$$

As a result, phosphorus reductions of 75 percent, 85 percent, and 95 percent require alum-to-phosphorus weight ratios of about 13 to 1, 16 to 1, and 22 to 1, respectively. For example, to achieve 85 percent phosphorus removal from a wastewater containing 10 mg/l of P, the alum dosage needed is approximately $16 \cdot 10 = 160$ mg/l, which is substantially greater than the $9.7 \cdot 10 = 97$ mg/l stoichiometric quantity of alum based on Eq. 13–1.

Iron coagulants precipitate orthophosphate by combining with the ferric ion, as shown in Eq. 13–3, at a molar ratio of 1 to 1.

$$FeCl_3 \cdot 6H_2O + PO_4^{\equiv} = FePO_4\downarrow + 3Cl^{-} + 6H_2O_3$$

$$\textbf{(13–3)}$$

Just as with aluminum, a greater amount of iron is required in actual coagulation than this chemical reaction predicts. One of the reactions competing with natural alkalinity is

$$FeCl_3 \cdot 6H_2O + 3HCO_3^{-}$$
$$= Fe(OH)^3\downarrow + 3CO_2 + 3Cl^{-} + 6H_2O \quad \textbf{(13–4)}$$

Provided that the wastewater has sufficient natural alkalinity, ferric salts applied without coagulant aids result in phosphorus removal at Fe to P dosages of 1.8 to 1.0 or greater. This is equivalent to an application of approximately 150 mg/l of commercial ferric chloride for treatment of a wastewater containing 10 mg/l of P. Since the reaction of ferric chloride with natural alkalinity is relatively slow, lime or some other alkali may be applied to raise the pH and supply the hydroxyl ion for coagulation, as follows:

$$2FeCl_3 \cdot 6H_2O + 3Ca(OH)_2$$
$$= 2Fe(OH)_3\downarrow + 3CaCl_2 + 6H_2O \quad \textbf{(13–5)}$$

Ferrous sulfate also forms a phosphate precipitate with an Fe-to-P molar ratio of 1 to 1; the dosages for coagulation are similar to those for ferric salts.

Commercially available iron salts are ferric chloride, ferric sulfate, ferrous sulfate, and waste pickle liquor from the steel industry. The last is the least expensive and most common source of iron coagulants for wastewater treatment in industrial regions. Pickle liquor is variable in composition, depending on the metal treatment process. Ferrous sulfate from pickling with sulfuric acid and ferrous chloride from pickling with hydrochloric acid are the two common waste liquors from metal finishing. Waste liquors have an iron content from 5 to 10 percent and free acid ranges from a low of 0.5 percent to a high of 15 percent. Preparation prior to use includes neutralization and pH adjustment of the liquors with lime or sodium hydroxide.

Chemical-biological treatment combines chemical precipitation of phosphorus with biological removal of organic matter. Alum or iron salts are added prior to primary clarification, directly to the biological process, or prior to final clarification. In a survey of 83 treatment plants using metal salts for phosphorus removal in Wisconsin, Michigan, Ohio, Indiana, Pennsylvania, and Virginia, 32 added chemicals to the primary clarifiers, 24 ahead of the biological process, and 27 to the final clarifier. For all application points, the amount of chemical added is about the same to achieve a specific phosphorus removal. Addition to the primary clarifier enhances both suspended solids and BOD removal, resulting in 75 percent solids and 50 percent BOD removal. Thus, subsequent treatment capacities increase and tend to be hydraulically rather than BOD limited. Alum and ferric sulfate add sulfate molecules to the wastewater, and in high concentrations will increase corrosion and odor problems associated with hydrogen sulfide created under anaerobic conditions in thickeners and digesters.

In activated-sludge aeration, the coagulant can be added to the aerating mixed liquor. Although the resulting chemical-biological floc has fewer protozoa, the efficiency of BOD removal is not adversely influenced. In trickling filtration or rotating biological contactors, the coagulant is usually mixed with the process effluent just prior to final sedimentation. Depending on coagulant dosage, the production of sludge solids in chemical-biological treatment is generally in the range of 20 to 60 percent more than the biological solids produced without chemical addition. In part, this increase in solids production is the result of improved effluent clarification because of chemical coagulation. The volume of sludge, on the other hand, usually increases by a smaller percentage

because of the higher density of the chemical-biological sludge.

Alum addition and filtration are required if phosphorus concentrations below 1.0 mg/l are required. The purpose of this filtration is not necessarily to remove additional phosphorus precipitate, but to remove the phosphorus adsorbed and entrapped in biological floc. Alum is used as a filter aid.

■ EXAMPLE 13–2

Alum is applied in the primary sedimentation tank of a conventional activated-sludge system (diagramed in Figure 13–12) to reduce the total phosphorus in the effluent to 1.3 mg/l. The commercial dose of alum is 90 mg/l. Calculate the alum dose in terms of the weight ratio of alum applied to the phosphorus content of the wastewater and the molar ratio of alum to phosphorus. Calculate anticipated sludge-solids production assuming the following: alum addition to the primary clarifier increases solids removal to 75 percent and BOD removal to 55 percent. The F/M ratio is 0.4, and the average plant effluent is 12 mg/l BOD with 10 mg/l suspended solids. How much has sludge production increased as compared to conventional treatment?

Solution

Alum-to-phosphorus and aluminum-to-phosphorus ratios:

$$\text{Alum dosage} = 90 \text{ mg/l}$$

$$\text{Aluminum dosage} = 90\frac{2 \cdot 27}{600} = 8.1 \text{ mg/l}$$

Phosphorus in the primary influent = 7.35 mg/l (including recycle shown in Figure 13–6)

$$\frac{\text{Alum applied}}{\text{Phosphorus in wastewater}} = \frac{90}{7.35} = 12.2 \text{ mg/l}$$

Molar ratio of Al to P = 8.1/7.35 = 1.1

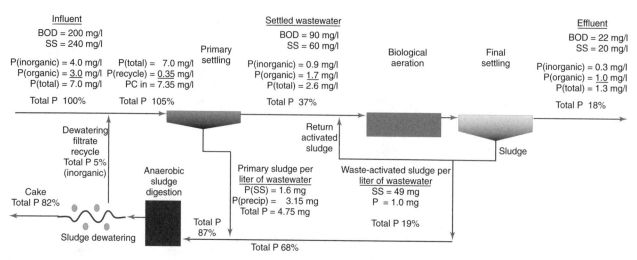

Figure 13–12

Results from Example 13–2. Diagram tracing wastewater phosphorus through a treatment plant using chemical alum addition for phosphorus removal. Of 100 percent influent phosphorus, 18 percent is in the treated wastewater and 82 percent is in the dewatered digested chemical sludge solids.

Sludge-solids production:

SS removed in the primary = $0.75 \cdot 240 = 180$ mg/l
The K value from Figure 11–50 is 0.5 for an F/M of 0.4 and the primary effluent BOD is 90 mg/l (200 mg/l · 45 percent remaining); therefore,
SS in waste activated sludge = $0.4 \cdot 90 = 36$ mg/l based on an effluent of 30 mg/1 SS.
Additional BOD removal creates an additional $0.4(30 - 20) = 4$ mg/l of SS, and improved final clarification captures an additional $30 - 20 = 10$ mg/l.

Total SS in waste-activated sludge = $36 + 4 + 10 = 50$ mg/l

The organic P content of the waste-activated sludge is $0.02 \cdot 50 = 1.0$ mg/l.
The P in the final effluent is 1.3 mg/l total.
Because there is insufficient ortho P to create 1.0 mg in the sludge, additional P was taken from the organic P content. Based on actual plant data, the effluent ortho P was 0.3 mg/l, leaving 1.0 mg/1 of effluent organic P. The inorganic P in the primary effluent is 0.9 mg/l inorganic, 1.7 mg/l organic, and 2.6 mg/l total.
Phosphorus removed in the primary clarifier solids = $180 \cdot 0.009 = 1.6$ mg/l.

Phosphorus removed by alum precipitation is the primary influent phosphorus minus the primary effluent phosphorus minus the P associated with the biological solids, which is

$$7.35 - 2.6 - 1.6 = 3.15 \text{ mg/l}$$

From Eq. 13–1,

$$\text{AlPO}_4 \text{ precipitate} = \frac{(\text{P precipitated})(\text{molecular weight of AlPO}_4)}{\text{molecular weight of P}}$$

$$= (3.15)(122)/(31) = 12 \text{ mg/l}$$

From Eq. 13–1, unused:

$$\text{Alum} = \frac{\text{alum}}{\text{dosage}} - \frac{(\text{P precipitated})(\text{molecular weight of AlPO}_4)}{2 \times \left(\dfrac{\text{molecular}}{\text{weight of P}}\right)}$$

$$90 \text{ mg/1} - (3.15)(600) \neq (2 \cdot 31) = 59.5 \text{ mg/l}$$

The alum not used in phosphorus precipitation reacts with natural alkalinity to precipitate Al(OH)_3. From Eq. 13–2,

$$Al(OH)_3 \text{ precipitate} = \frac{\text{(unused alum)} \times (2 \cdot \text{molecular weight of } Al(OH)_3)}{\text{molecular weight of alum}}$$

$$= (59.5)(2 \cdot 78)/600 = 15 \text{ mg/l}$$

The total sludge-solids production equals the sum of the SS removal in primary sedimentation, SS in waste-activated sludge, $AlPO_4$ precipitate, and $Al(OH)_3$ precipitate. Therefore, solids production per liter of wastewater treated equals

$$180 + 55 + 3 + 15 = 253 \text{ mg/l of the influent flow}$$

Increase in sludge-solids production: The solids production by chemical-biological processing relative to conventional activated-sludge processing as given in Figure 13–12 is $253 - 185 = 68$ mg/l of wastewater treated, which is an increase of 37 percent.

■ ■ ■

13–7 NITROGEN IN WASTEWATERS

The common forms of nitrogen are organic, ammonia, nitrate, nitrite, and gaseous nitrogen. Decomposition of nitrogenous organic matter releases ammonia to solution, Eq. 13–6. Under aerobic conditions, nitrifying bacteria (Eq. 13–7) oxidize ammonia to nitrite and subsequently to nitrate. Bacterial denitrification, Eq. 13–8, occurs under anaerobic or anoxic conditions when organic matter (AH_2) is oxidized and nitrate is used as a hydrogen acceptor, releasing nitrogen gas.

$$\text{Organic nitrogen compounds} \xrightarrow[\text{decomposition}]{\text{bacterial}} NH_3 \text{ (ammonia)} \quad \textbf{(13–6)}$$

$$NH_3 + O_2 \xrightarrow[\text{nitrification}]{\text{aerobic}} NO_3^- \text{ (nitrate)} \quad \textbf{(13–7)}$$

$$NO_3^- + AH_2 \xrightarrow[\text{denitrification}]{\text{anaerobic}} A + H_2O + N_2\uparrow \quad \textbf{(13–8)}$$

Nitrogen in municipal wastewater results from human excreta, ground garbage, and industrial wastes, particularly from food processing. The greatest source of nitrogen in wastewater is from human fecal matter. Organic nitrogen is contained in proteins, amines, nucleic acids, peptides, amino acids, and other organic constituents excreted in fecal matter. Ammonia results from bacterial mineralization of proteins and urea (the principal component in urine), and increases in concentration due to biological activity in the sewer system. Nitrites and nitrates are typically not found in raw wastewater because of the lack of free oxygen for nitrification. Nitrates may be contributed by industrial discharge or from drinking water high in nitrate concentrations.

Approximately 40 percent of influent nitrogen is in the form of ammonia, and 60 percent is bound in organic matter with negligible nitrate. The total nitrogen contribution is in the range of 8 to 12 lb per capita per year (4 to 6 kg/person · y), and the average concentration in domestic wastewater is 35 mg/l.

Ammonia is created by the decomposition of organic matter and can be removed through biological or chemical processes. Ammonification is the process of converting organic nitrogen to ammonia through biological activity under anaerobic conditions. This activity is found in sewers and anaerobic digestion. Ammonia is used for cell production in biological growth. Under strictly aerobic conditions, the process of nitrification can be used to biologically convert ammonia to nitrite and then to nitrate. Under an anoxic (lacking dissolved oxygen) condition, nitrate can be biologically converted to nitrogen gas and removed from the wastewater using the process of denitrification. Ammonia can be removed chemically by stripping ammonia gas after raising the wastewater pH above 10. Full-scale facilities were constructed at Lake Tahoe and Water Factory 21; however, they are no longer used because of the high maintenance associated with air-stripping processes. Ammonia may also be removed using breakpoint chlorination to chemically convert ammonia to nitrogen gas. The process is costly to operate but may be installed as a backup to biological nitrification. In small systems, ion exchange resins can be used to exchange ammonium ions for another cation or nitrate for another anion. Reverse-osmosis and other membrane processes physically remove organic molecules, ammonia, and nitrate.

Nitrogen Removal in Conventional Treatment

The nitrogen compounds in an average sanitary wastewater are listed in Table 13–2. With most of the nitrogen in soluble and colloidal organic forms, the amount removed by primary sedimentation is limited to about 15 percent. Based on the tabulated values, the uptake in subsequent biological treatment is only another 10 percent. In general, the amount of nitrogen in excess biological floc produced in activated-sludge treatment of a wastewater is equal to about 4 percent of the BOD applied. With a total reduction of only 25 percent, the effluent contains 26 mg/l of the original 35 mg/l. Approximately 2 mg/l is organic nitrogen bound in the effluent suspended solids. The remaining 24 mg/l is in the form of ammonia, except when nitrification occurs during aeration. Oxidation of a portion of the nitrogen to nitrate is most likely to occur in an activated-sludge process treating a warm wastewater at a low BOD loading.

■ EXAMPLE 13–3

Using the values given in Table 13–2, trace inorganic and organic nitrogen through a conventional activated-sludge treatment plant. Label the total nitrogen as a percentage of the influent phosphorus equal to 100 percent. Assume a primary clarifier removal of 35 percent of BOD and 50 percent of solids with 4.2 percent organic nitrogen content. The waste activated-sludge system operates with an F/M of 0.40 and contains 6.2 percent nitrogen in the waste-activated sludge.

Processing of the raw sludge can release some of the nitrogen extracted from the wastewater. In this treatment scheme, the sludge is stabilized by anaerobic digestion, and ammonia released from decomposition of the sludge solids is returned to the influent of the treatment plant in the filtrate. Assuming 40 percent of the organic nitrogen in the sludge is converted to ammonia, 10 percent of the original 25 percent is recycled back to the treatment plant. Use a recycle of 10 percent of the influent nitrogen as ammonia (inorganic nitrogen).

Solution

Using the influent (raw) values from Table 13–2, influent SS = 240 mg/l, BOD = 200 mg/l, inorganic N is 22 mg/l (all in the form of ammonia), and organic N is 13 mg/l. Given a recycle contribution of inorganic N equal to 10 percent of the influent N, or 3.5 mg/l, the primary clarifier influent consists of N (ammonia) = 25.5 mg/l and N (organic) = 13 mg/l, for a total of 38.5 mg/l.

At 50 percent SS removal, primary solids are 120 mg/l of the influent flow. Organic nitrogen in the primary sludge equals

$$0.042 \cdot 120 = 5.0 \text{ mg of organic nitrogen}$$

The primary effluent consists of SS = 120 mg/l, BOD = 130 mg/l, inorganic N unchanged at 25.5 mg/l, and organic N reduced to 8 mg/l.

From Figure 11–50 at an F/M of 0.40, K equals 0.50,

$$W_s = 0.50 \cdot 130 = 65 \text{ mg of sludge solids}$$

The nitrogen content of the solids in the sludge solids is

$$0.062 \cdot 65 = 4.0 \text{ mg of organic nitrogen}$$

The organic N of the plant effluent is $0.062 \cdot 30$ mg/l SS = 2.0 mg.

The inorganic N remains in solution through all of the treatment processes.

From these calculations, the balance of the values for BOD, SS, and N can be calculated and are presented on Figure 13–13.

If the influent total nitrogen concentration is assigned a value of 100 percent, removal would be 14 percent in the primary and 11 percent in the secondary, resulting in an effluent discharge of 79 percent of the influent nitrogen. The influent inorganic nitrogen is in the form of ammonia and remains in solution through the plant. Depending on variations in the nitrogen content of the wastewater and methods of water and sludge processing, nitrogen removal in conventional biological treatment systems ranges from nearly zero to 40 percent.

■ ■ ■

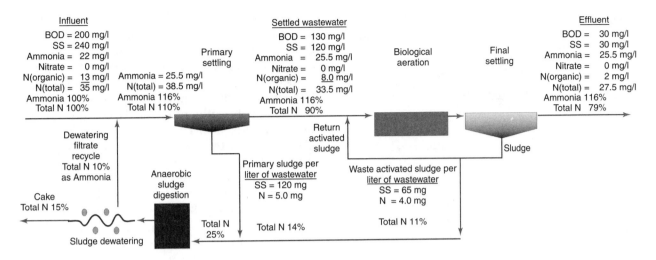

Figure 13–13

Diagram tracing wastewater ammonia and nitrogen through a hypothetical conventional treatment plant. Of 100 percent influent ammonia, 79 percent is in the treated wastewater and 15 percent is in the dewatered digested sludge solids. The diagram is annotated with the results from Example 13–3.

13–8 BIOLOGICAL NITRIFICATION AND DENITRIFICATION

Nitrification of a wastewater is practiced where the ammonia content of the effluent causes pollution of the receiving watercourse. The process does not remove the nitrogen but converts it to the nitrate form (Eq. 13–7). Nitrification-denitrification, which reduces the total nitrogen content, includes conversion of the nitrate to gaseous nitrogen (Eq. 13–8).

Nitrification

Nitrification is usually a separate process following conventional biological treatment. Although nitrification may be possible to perform along with organic matter removal in a single-stage extended aeration unit in a warm climate, two-step treatment is necessary for reliable operation at a reduced wastewater temperature. The conventional biological treatment removes BOD, without oxidation of ammonia nitrogen, to produce a suitable effluent for nitrification. The high ammonia content and low BOD provide greater growth potential for the nitrifiers relative to the heterotrophs. This allows operation of the nitrification process at a greater sludge age to compensate

for lower operating temperature and to ensure that the growth rate of nitrifying bacteria is rapid enough to replace those lost through washout in the plant effluent. Although synthetic-media filtration in biological towers can perform nitrification, the most reliable system is suspended-growth aeration.

Figure 13–14 is the common flow scheme for nitrification following biological treatment of a wastewater for organic matter reduction. After the nitrifying bacteria oxidize the ammonia in the aeration tank, the activated sludge containing high populations of nitrifiers is settled in the final clarifier for return to the aeration tank. Sludge can be wasted, if necessary, to remove excess bacterial growth from the system. Because the rate of oxidation of ammonia is nearly linear, the tank configuration is plug-flow to minimize short-circuiting. The important parameters in bacterial nitrification kinetics are temperature, pH, and dissolved oxygen concentration. The reaction rate is decreased markedly at reduced temperatures, with about 8°C being the reasonable minimum value. Optimum pH is near 8.4, and the dissolved oxygen level should be greater than 1.0 mg/l.

Biological nitrification destroys alkalinity, which can result in lower pH when processing wastewaters of moderate hardness or where alum

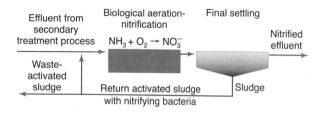

Figure 13–14

Flow diagram for nitrification by suspended-growth aeration following conventional biological treatment.

precipitation has been used for phosphorus reduction in preceding processes. Theoretically, 7.2 lb of alkalinity is destroyed per pound of ammonia nitrogen oxidized to nitrate:

$$2NH_4HCO_3 + 4O_2 + Ca(HCO_3)_2$$
$$= Ca(NO_3)_2 + 4CO_2 + 6H_2O \qquad \textbf{(13–9)}$$

Destroying alkalinity causes the wastewater to become aggressive, attacking the cement at the concrete surface and corroding ferrous metals, such as clarifier mechanisms. Lime or soda ash addition can be used to maintain alkalinity. Although steel weir plates are common in primary and secondary clarifiers, fiberglass weirs are used in clarifiers following nitrification because of the aggressive environment. Steel clarifier mechanisms are typically protected by cathodic protection systems.

Nitrification of 1.0 lb of ammonia expressed as N requires 4.6 lb of oxygen for both the creation of nitrate and alkalinity, Eq. 13–9. Additional oxygen must be included for the carbonaceous BOD that also occurs in the nitrification stage. A minimum dissolved oxygen concentration of at least 2.0 mg/l is typically required to reduce denitrification in the settling tank.

Denitrification in the settled sludge of an activated-sludge secondary clarifier can result in flotation of settled solids and a decrease in clarifier performance. Nitrogen gas produced in the sludge becomes entrapped in the floc, floats to the surface, and escapes over the effluent weir.

Ammonia nitrogen loadings applied to the aeration tank are 10 to 20 lb/1000 cu ft/day (160 to 320 g/m³ · d) with corresponding wastewater temperatures of 10° to 20°C, respectively. For an average wastewater effluent, this is an aeration period of 4 to 6 hr.

Figure 13–15 shows an aerial view of the nitrification complex for the city of Las Vegas. Treatment is designed to reduce 18 mg/l of ammonia nitrogen (21 mg/l total Kjeldahl nitrogen) to 0.8 mg/l of effluent ammonia nitrogen at a flow rate of 46 mgd. Effluent from the secondary trickling filter plant is diverted at the effluent channel and pumped into the aeration basins. The aeration basins operate at 2100 mg/l of MLSS with 4 hr of liquid detention time, 8 days of sludge age, and 50 to 100 percent return sludge flow. Aeration is provided by three electric and two engine-driven centrifugal blowers that each deliver 18,300 cfm to fine-bubble diffusers for oxygen transfer. The final clarifiers are 140 ft in diameter with an overflow rate to 550 gpd/sq ft. The hexagon shape allows for common-wall construction and forms a tunnel containing pumps, piping, and electrical equipment. Waste sludge, stored in aerated tanks, is thickened in centrifuges to 6 percent prior to blending with primary and secondary sludges for anaerobic digestion. Alkalinity lost by prior phosphorus removal using alum at the primary clarifiers is replaced using soda ash for a minimum alkalinity of 100 mg/l.

■ EXAMPLE 13–4

The sludge age and mixed-liquor suspended solids are increased in the aeration basin of a conventional activated-sludge system (diagramed in Figure 13–13) to reduce the ammonia in the effluent to 2.0 mg/l. Calculate anticipated sludge-solids production assuming an F/M ratio of 0.4. The average plant effluent is 22 mg/l BOD with 20 mg/l suspended solids. How much is sludge production increased as compared with a conventional plant?

Solution

Using the influent (raw) values from Table 13–2, influent SS = 240 mg/l, BOD = 200 mg/l, inorganic N is 22 mg/l (all in the form of ammonia), and organic N is 13 mg/l. At 50 percent SS removal, primary solids are 120 mg/l of the influent flow. Organic nitrogen in the primary sludge equals

$$0.042 \cdot 120 = 5.0 \text{ mg of organic nitrogen}$$

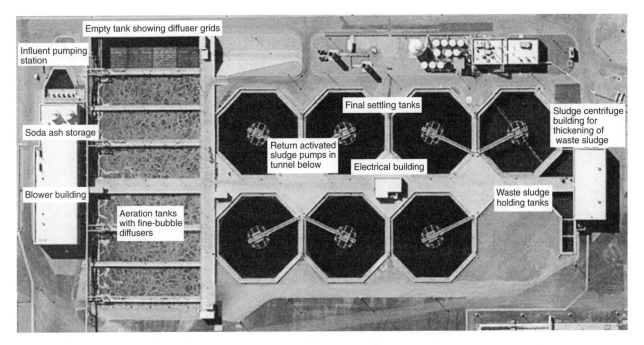

Figure 13–15

Processing arrangement for nitrification by suspended-growth aeration. City of Las Vegas Water Pollution Control Facility, Nevada.

(Courtesy of the City of Las Vegas, NV.)

The primary effluent consists of SS = 120 mg/l, BOD = 130 mg/l, inorganic N unchanged at 22.0 mg/l, and organic N reduced to 8 mg/l.

From Figure 11–50 at an F/M of 0.4, K equals 0.5,

$$W_s = 0.5 \cdot 130 = 65 \text{ mg of sludge solids}$$

Increased BOD removal represents an additional 0.5 (30 − 22) = 4 mg, and increased solids capture represents 30 − 20 = 10 mg, for a total of 79 mg.

The nitrogen content of the solids in the sludge solids is

$$0.062 \cdot 79 = 5 \text{ mg of organic nitrogen}$$

The total N in the plant effluent is 33.5 − 5, or 28.5 mg/l. The organic N of the plant effluent is 0.062 · 20 mg/l SS = 1.2 mg. Effluent ammonia is 2.0 mg/l, resulting in the conversion of 28.5 mg/l of ammonia to nitrate. The total solids production is 120 + 79 = 199 mg, or 14 mg greater than without nitrification, as shown in Figure 13–13. The inorganic N remains in solution through all of the treatment processes.

From these calculations, the balance of the values for BOD, SS, and N can be calculated and are presented on Figure 13–16. If the influent total nitrogen concentration is assigned a value of 100 percent, removal would be 14 percent in the primary and 14 percent in the secondary, resulting in an effluent discharge of 82 percent of the influent nitrogen. However, 25 mg/l of ammonia was converted to nitrate, thus reducing the effluent toxicity.

■ ■ ■

Denitrification

Nitrification results in a reduction in effluent ammonia toxicity but releases higher concentrations of nitrate, a groundwater pollutant. Denitrification is a process that reduces nitrate to nitrogen gas using facultative hetrotrophic bacteria, thus removing nitrate from the plant effluent. Additional benefits include the recovery of 2.86 lb of oxygen per pound of nitrate-nitrogen reduced and the recovery of 3.0 lb of alkalinity per pound of nitrate-nitrogen reduced. An organic carbon source, in AH_2 (Eq. 13–8), is needed to act as a

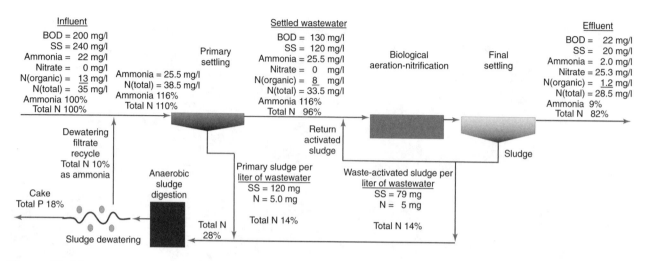

Figure 13–16

Results from Example 13–4. Diagram tracing wastewater ammonia and nitrogen through a treatment plant using nitrification for ammonia removal. Of 100 percent influent ammonia, 9 percent is in the treated wastewater, with the balance of the inorganic nitrogen in the form of nitrate.

hydrogen donor and to supply carbon for biological synthesis. Although any biodegradable organic substance can serve as a carbon source, methanol is common because of its availability and ease of application and because it doesn't leave a residual BOD in the process effluent. As with all other chemical carbon sources, methanol is expensive. In fact, the cost of methanol makes the widespread application of this denitrification process unrealistic in wastewater treatment.

The denitrification reaction between methanol and nitrate is as follows:

$$5CH_3OH + 6NO_3^-$$
$$= 3N_2\uparrow + 5CO_2\uparrow + 7H_2O + 6OH^- \quad \text{(13–10)}$$

Since the process effluent from nitrification also contains dissolved oxygen and nitrite, the total methanol required as a hydrogen donor in denitrification is as given in Eq. 13–10. In addition, methanol is used as a carbon source in bacterial synthesis. Based on approximately 30 percent excess methanol needed for synthesis, the total methanol demand is calculated using Eq. 13–11.

$$CH_3OH = 0.7\ DO + 1.1NO_2\text{-}N + 2.0NO_3\text{-}N \quad \text{(13–11)}$$

$$CH_3OH = 0.9\ DO + 1.5NO_2\text{-}N + 2.5NO_3\text{-}N \quad \text{(13–12)}$$

where CH_3OH = methanol, milligrams per liter

DO = dissolved oxygen, milligrams per liter

$NO_2\text{-}N$ = nitrite nitrogen, milligrams per liter

$NO_3\text{-}N$ = nitrate nitrogen, milligrams per liter

The recommended denitrification system consists of a plug-flow tank with underwater mixers followed by a clarifier for sludge separation and return (Figure 13–17). The level of agitation in the denitrification chambers must keep the microbial floc in suspension but be controlled to prevent undue aeration. Since nitrogen gas released from solution can float the biological floc, the last chamber must strip the nitrogen gas from solution for efficient final clarification. This may be done using an aeration chamber, which also has the advantage of oxygenating the plant effluent, or by a degasifier. The detention time required for denitrification of a domestic wastewater is usually in the range of 2 to 4 hr, depending on nitrate loading and temperature. Since methanol is expensive, denitrification following nitrification is generally

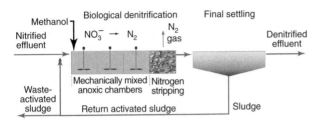

Figure 13–17

Flow diagram for denitrification by anoxic suspended growth following nitrification.

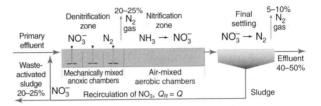

Figure 13–18

Diagram tracing wastewater nitrogen through a hypothetical biological nitrification-denitrification system. Of 100 percent influent nitrogen, 30 to 35 percent is released as nitrogen gas, 20 to 25 percent is organic nitrogen in the waste-activated sludge, and 40 to 50 percent appears in the effluent, primarily as nitrate nitrogen.

performed only where the receiving watercourse is used as a source for the public water supply and an effluent nitrogen concentration of less than 10 mg/l requires strict control.

Biological Nitrification-Denitrification

Unoxidized organic matter can be used as an oxygen acceptor (hydrogen donor) for conversion of nitrate to nitrogen gas. This reaction satisfies a portion of the BOD in a wastewater.

$$\begin{matrix} \text{Unoxidized} & & \text{Oxidized} \\ \text{organic matter} + NO_3^- \rightarrow & \text{organic matter} + N_2\uparrow \\ \text{(BOD)} & & \text{(reduced BOD)} \end{matrix}$$

$$(13\text{–}13)$$

The flow scheme of a biological nitrification-denitrification process requires mixing of raw organic matter with nitrified wastewater; hence, an aerobic zone is needed for nitrification and an anoxic zone for denitrification. (The term *anoxic* means "lacking in dissolved oxygen.")

The plug-flow activated-sludge system, diagramed in Figure 13–18, blends nitrified recirculation flow with raw settled wastewater (primary effluent) in an anoxic zone. In this mechanically mixed chamber—or segregated zone of a long, narrow tank—the biological floc in the return activated sludge uses recirculated nitrate as an oxygen source, releasing nitrogen gas. In the subsequent chambers, mixed by diffused or mechanical aeration, dissolved oxygen is taken up by the biological floc for nitrification of the ammonia in the wastewater. Both the anoxic and aerobic chambers reduce the wastewater BOD—

the anoxic zone by denitrification and the aerobic zone by uptake of dissolved oxygen. The final clarifier contributes to denitrification as the bacterial floc extracts the oxygen from nitrate in the metabolism of the organic solids during sedimentation. The majority of nitrogen removal, however, occurs in the anoxic chamber, and the degree of nitrogen removal is controlled by the rate of recirculated flow. The amount of return nitrate that can be reduced depends on the maximum rate of denitrification possible in the anoxic zone. Too great a recirculation carries nitrate through the anoxic zone into the aerobic chambers, and, similarly, too short an aeration period can reduce the degree of nitrification. Other factors, such as the relative concentration of nitrogen to BOD in the wastewater and temperature, also influence the method of process operation. Figure 13–18 illustrates typical treatment of a settled domestic wastewater at a moderate temperature with a total retention time of approximately 8 hr in the biological chambers. Influent nitrogen is assigned a value of 100 percent. Of this, 30 to 35 percent is converted to nitrogen gas, 20 to 25 percent appears in the waste sludge as organic nitrogen, and 40 to 50 percent is in the plant effluent, primarily as nitrate nitrogen. Thus, this biological nitrification-denitrification process removes 50 to 60 percent of the influent nitrogen. Denitrification ahead of the nitrification zone is advantageous because the BOD in the raw wastewater is used as a carbon source, the oxygen demand of the nitrification zone is reduced, and denitrification recovers alkalinity lost during nitrification.

Combined Biological Phosphorus Removal and Nitrification-Denitrification

Biological nitrification-denitrification begins treatment with an anoxic zone followed by an aerobic zone to remove nitrogen. Starting the process with an anaerobic zone promotes the biological release of organic phosphorus and changes the behavior of bacteria, resulting in enhanced phosphorus uptake during biological aeration. The process is referred to as enhanced biological phosphorus removal (EBPR). Excess phosphorus in the range of 6 to 8 percent, as compared with the typical 2 percent, is removed in the waste-activated sludge. The actual removal depends on operational control of the process, the efficiency of solids removal processes, and sludge processing that limits the return of phosphorus into wastewater treatment.

Figure 13–19a shows a three-stage EBPR biological nutrient removal system for phosphorus and nitrogen. The individual tanks allow for internal recycle and plug-flow conditions. Sizing of the aerobic and anoxic tanks depends on the wastewater characteristics, pH, temperature, dissolved oxygen, and arrangements of the internal recycle streams. Anaerobic zones are commonly sized to 10 and 20 percent of the aeration tank volume. Anoxic zones must bring the dissolved oxygen to zero and provide sufficient time for biological conversion of nitrate to nitrogen gas. They typically range in size from 0.5 to 3.0 hr of detention time. The aerobic zone is the largest and is designed for a detention time of 6 to 24 hr. The treatment facility design shown in Figure 13–19b shows three anaerobic tanks (detention time = 1.0 hr each) preceding three anoxic tanks (detention time = 2.25 hr each) and a final aerobic tank (detention time = 6 hr). The anaerobic and anoxic zones are mixed by submersible propeller mixers to assure complete mixing. The aerobic zone is aerated using fine-bubble aeration from adjustable blowers. Although not required for mixing, the aerobic zone also uses propeller pumps to create a circular velocity of 1 ft/sec, to increase the bubble travel time and oxygen transfer. Other plants use an oval aeration tank similar to an oxidation ditch with horizontal aerators to propel the wastewater around the channel. Mixed-liquor concentrations are maintained between 2500 and 5000 mg/l. Internal recycle is critical to performance. Raw wastewater enters the

first anaerobic tank and is mixed with an anaerobic recycle from the end of the third anaerobic tank. Return activated sludge, entering the first and second anaerobic tanks, supplies new bacteria. The second recycle stream returns nitrate to the first anoxic zone for denitrification. Recycle and return-activated-sludge pumps are sized for 1 to 1.5 times the design flow. The mixing energy required for anoxic zones ranges from 0.25 to 0.8 hp/1000 cu ft. Air requirements for the aeration basins are based on oxygen demand for BOD and ammonia.

Continuous monitoring of dissolved oxygen, oxidation reduction potential (ORP), ammonia, nitrate, and phosphorus aids in process control. Adjustments in recycle and dissolved oxygen change the concentration and biological activity in the respective process phases. Denitrification control is possible by adjusting the return-activated-sludge flow, recycle flow, and feed points into the aeration basin. Control of the return-activated-sludge flow, as with conventional activated sludge, is based on final clarifier performance and depth of the sludge blanket. The recycle flow is adjusted based on treatment performance and dissolved oxygen concentrations. Placement of the feed points and flow splits is based on operating experience. Total phosphorus removal of 70 to over 80 percent and total nitrogen removal in the range of 50 to over 70 percent is possible using a combination of anaerobic, anoxic, and aerobic zones.

Bacteria used for denitrification and phosphorus removal require carbon and volatile fatty acid (VFA) sources that may be insufficient even in normal domestic wastewater. As with two-stage nitrification-denitrification, methanol may be used to supplement a weak wastewater. Phosphorus removal is achieved under aerobic conditions by growing phosphorus-accumulating organisms that rapidly uptake soluble orthophosphate. The preceding anaerobic environment increases the concentration of volatile fatty acids through decomposition of organic matter and causes orthophosphate to be released into solution from facultative organisms. In the aerobic environment, cellular use of volatile fatty acids is critical to the uptake of orthophosphate; however, other organisms, including nitrifiers, also require VFAs for cell growth. Methanol or acetic acid are commercially available and may be used to supplement low-strength wastewater. Anaerobic

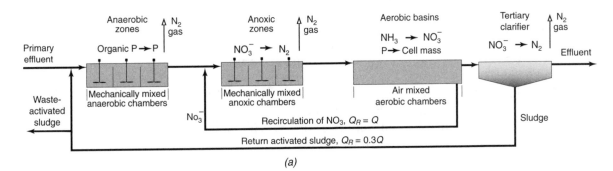

(a)

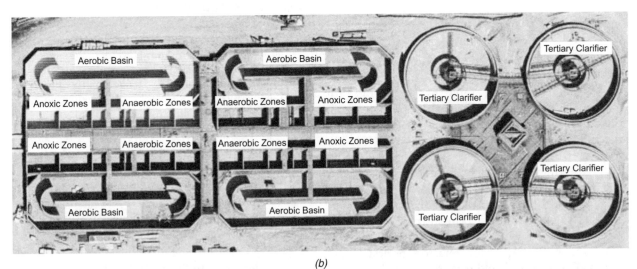

(b)

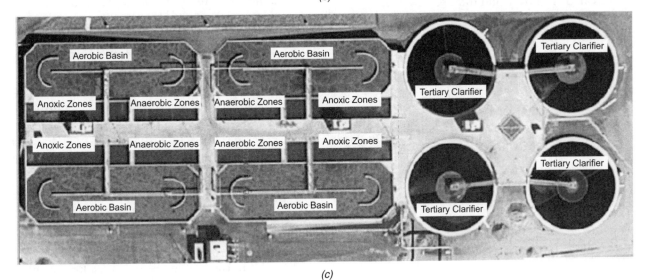

(c)

Figure 13–19

Three-stage biological nutrient removal processes for phosphorus and nitrogen removal. (*a*) A schematic of a diagram showing the order of treatment and side streams. (*b*) Basins under construction. (*c*) Completed basins under operation at the City of Las Vegas Water Pollution Control Facility.

fermentation of primary or return activated sludge produces acetic acid and other VFA by-products. Fermentation tank design must address elutriation of the VFA from the biomass and release of dissolved gases produced under anaerobic conditions for efficient settling. Fermentation of food waste, primary sludge, or RAS may be used in lieu of chemical addition. Chemical phosphorus precipitation may be necessary to remove remaining phosphorus to attain the desired level.

■ EXAMPLE 13–5

An oxidation ditch extended-aeration plant is operated as a biological nitrification-denitrification process. The aeration tank has a volume of 5600 m^3. Based on several months' operating data, the average wastewater flow was $4300 \text{ m}^3/\text{d}$ with the following characteristics: 260 mg/l BOD, 300 mg/l SS, 51 mg/l total N, and 31 mg/l NH_3-N. The average effluent quality was 5 mg/l BOD, 8 mg/l SS, 7 mg/l total N, 4 mg/l NH_3-N, and 2 mg/l NO_3-N. The operating MLSS concentration averaged 4700 mg/l, and the temperature was in the range of 16° to 18°C. Calculate the volumetric BOD loading, F/M ratio, aeration period, BOD removal, SS removal, and total N removal.

Solution

BOD loading

$$= \frac{(260 \text{ mg/l})(4300 \text{ m}^3/\text{d})}{5600 \text{ m}^3} = 200 \text{ g/m}^3 \cdot \text{d}$$

$$F/M = \frac{(260 \text{ mg/l})(4300 \text{ m}^3/\text{d})}{(4700 \text{ mg/l})(5600 \text{ m}^3)} = 0.042 \frac{\text{g BOD/d}}{\text{g MLSS}}$$

$$t = \frac{5600}{4300}24 = 31 \text{ h}$$

$$\text{BOD removal} = \frac{260 - 5}{260}100 = 98\%$$

$$\text{SS removal} = \frac{300 - 8}{300}100 = 97\%$$

$$\text{Total N removal} = \frac{51 - 7}{51}100 = 86\%$$

■ ■ ■

REFERENCES

1. *Pomona Virus Study*, Sanitation Districts of Los Angeles County, California State Water Resources Control Board, Sacramento, CA, 1977.
2. *Retrofitting POTWs for Phosphorus Removal in the Chesapeake Bay Drainage Basin*, U.S. Environmental Protection Agency, Office of Research and Development, EPA, 625/6-87/017.

FURTHER READINGS

Nitrogen Control, U.S. Environmental Protection Agency Office of Research and Development, EPA 625/R–93/010. Washington, DC, September 1993.
Phosphorus Removal, U.S. Environmental Protection Agency, Office of Research and Development, EPA 625/1–87/001. Washington, DC, September 1987.

PROBLEMS

13–1 What are the EPA water-quality standards for secondary treatment? List pollutants not removed by conventional treatment. What environmental pollution is caused by the release of ammonia and phosphorus?

13–2 Referring to Figure 13–1, what filtration technology is required for pathogen removal? Virus removal?

13–3 How is public health protected from infectious diseases?

13–4 What is the limitation of gravity filtration, and what are the results of various filtration methods?

13–5 How does gravity settling and various types of filtration in combination with disinfection impact pathogen removal?

13–6 The relationship between horsepower and pumping is given by the following equation: hp = (gpm × ft of water)/(3960 × pump efficiency). Power, in kilowatt hours (kWh), is equal to 0.749 times hp. Energy costs $0.15 per kWh. For a 1-mgd filtration facility, calculate the annual cost of pumping based on the least operating pressures listed in Table 13–1.

13–7 Wastewater average flow is 3 mgd and 9.6 mgd peak. Filtration must meet unrestricted recycled water requirements. Size multimedia filter beds of equal size with one unit out of service.

13–8 What is effluent toxicity, and how is it measured?

13–9 Refer to Figures 13–1, 13–2, and 13–10. What treatment schemes are appropriate for virus removal? In tertiary treatment for virus removal, why does coagulation precede filtration? What are the effluent water-quality criteria for virus removal?

13–10 Data published by the EPA related suspended solids to fecal coliform in the following table. Plot the data on a log-log scale. For each of the fecal coliform values, list the minimum treatment processes required to achieve the limit.

SUSPENDED SOLIDS (mg/l)	FECAL COLIFORM MPN No./100 ml
3	2.2
10	23
21	100
30	200

13–11 What is the typical phosphorus removal process in primary and secondary treatment? How can phosphorus be removed in primary, secondary, and tertiary treatment?

13–12 If a 10-mgd plant has an influent phosphorus concentration of 7 mg/l, how much alum or ferric feed would be required for removal?

13–13 Alum at a concentration of 100 mg/l is added to the primary clarifier. What chemical and biological solids removal may be expected? What might be the corresponding BOD removal? (*Hint*: refer to Example 13–2.)

13–14 Alum is used at a 12-mgd plant to reduce phosphorus from 5 mg/l to 0.5 mg/l. The alum addition is 90 mg/l; using $210 per ton, calculate the annual alum use and cost.

13–15 Explain methods for evaluating effluent toxicity. How can toxicity be reduced?

13–16 A plant with 12 mgd operates like the plant in Figure 13–6. Calculate the pounds of phosphorus in the effluent. Size alum feed pumps using a commercial grade of alum containing 5 lb of Al per gallon of solution.

13–17 Using Example 13–2, what is the organic and inorganic fraction of sludge solids in mg/l and pounds for a plant flow of 6 mgd? If the organic sludge is 75 percent VSS and anaerobic digestion destroys 50 percent of the VSS, how many pounds of solids are in the final digested

sludge? If dewatering captures 95 percent of the solids and the cake contains 26 percent solids, what are the dry and wet tons?

13–18 What are the chemicals used in chemical phosphorus removal? What are the chemicals used to enhance biological phosphorus removal?

13–19 If the influent phosphorus concentration is 7 mg/l and the effluent target is 0.75 mg/l to maintain an effluent permit limit of less than 1 mg/l, then how much alum must be added? If commercial alum contains 5 lb Al per gal, size the pump and storage tank required for 10 days of storage.

13–20 A treatment plant, like that in Figure 13–12 with a flow of 6.3 mgd, has a BOD of 320 mg/l, SS of 280 mg/l, and phosphorus of 7 mg/l. Keeping the effluent the same at 22 mg/l BOD and 20 mg/l SS, calculate the settled wastewater phosphorus sludge and WAS in concentration and pounds (refer to Example 13–2).

13–21 What are the sources of phosphorus and nitrogen in wastewater? List the forms of nitrogen and phosphorus in influent wastewater. If wastewater generation is 120 gpcd, what is the average contribution of phosphorus and nitrogen in lb per capita day?

13–22 Explain aerobic, anoxic, and anaerobic conditions in terms of the forms of oxygen present. Under which conditions does nitrification take place and why?

13–23 What is the relationship between nitrification and alkalinity?

13–24 Refer to Figure 13–15 for the City of Las Vegas. If the oxygen transfer is 6 percent and air contains 0.0154 lb oxygen per cu ft of air, how much nitrogen can be treated?

13–25 A 4.3-mgd plant contains 32 mg/l nitrate prior to two-stage nitrification-denitrification. How much methanol is required? If commercial methanol is 95 percent pure and costs $1.30/gal, what is the monthly cost of methanol addition?

13–26 A 2.2-mgd plant is designed per Figure 13–18. Using 1 lb soluble BOD = 1 lb methanol, with a soluble fraction of 40 percent of the influent total BOD and using 200 mg/l influent BOD, what is the BOD consumed in the denitrification process?

13–27 A laboratory activated-sludge system was used to evaluate chemical-biological phosphorus removal by applying alum to the aeration chamber. Twelve liters of settled wastewater were applied daily at a constant rate to the aeration chamber, which had a volume of 3.6 l. The aeration tank MLSS was held near 2000 mg/l by wasting 200 to 250 ml of mixed liquor per day. The temperature was 22° to 24°C, pH was 7.4, and SVI varied from 90 to 130 mg/l. The alum solution feed to the aeration tank had a strength of 10.0 mg of commercial alum per milliliter. The following data were collected at various feed rates:

INFLUENT

WASTEWATER FEED (l/d)	ALUM APPLIED (ml/d)	BOD (mg/l)	SS (mg/l)	P (mg/l)	ALK (mg/l)
12	0	150	94	10.3	350
12	70	158	104	10.9	350
12	140	169	114	10.6	340
12	210	173	135	10.4	360
12	390	173	123	9.3	320

EFFLUENT

WASTEWATER FEED (l/d)	ALUM APPLIED (ml/d)	BOD (mg/l)	SS (mg/l)	P (mg/l)	ALK (mg/l)
12	0	6	7	7.9	230
12	70	6	8	5.4	200
12	140	5	11	2.4	160
12	210	10	10	1.0	140
12	390	8	17	0.5	80

Calculate the phosphorus removal and the weight ratio of alum applied to total phosphorus in the influent wastewater for each run. Plot a graph of percentage of phosphorus removal versus the weight ratio of alum to phosphorus. On the same graph, plot the phosphorus removal and alum dosage values given in the text under Eq. 13–2.

Calculate the average BOD and SS removal efficiencies. Suggest some reasons why the effluent BOD and SS values are lower and resulting efficiencies higher than are normally achieved by a full-scale treatment plant. Why did the concentration of alkalinity in the effluent decrease with increasing alum dosage?

13–28 Data for ammonia and alkalinity were collected along the length of an aeration basin 100 ft wide, 18 ft deep, and 168 ft long. The flow rate is 10 mgd. Plot the ammonia and alkalinity versus distance. Calculate the value of k and plot k versus distance.

$$\ln\left(\frac{\text{ammonia at distance } D}{\text{influent ammonia}}\right)$$
$$= -k(\text{MLVSS})1.024^{T-20}t$$
$$t = \frac{\text{area} \cdot \text{distance}}{\text{flow}}$$

DISTANCE (ft)	AMMONIA (mg/l)	ALKALINITY (mg/l)
0	16.11	243
6	16.46	185
12	12	190
18	8	190
24	5.28	188
30	4.48	189
36	3.96	188
42	4.55	191
48	2.93	181
54	2.64	178
60	2.62	178
66	2.59	178
72	2.44	178
78	2.31	179
84	2.44	178
90	2.44	180
96	1.69	178
102	2.01	175
108	1.44	172
114	0.87	172
120	1.05	171
126	1.25	172
132	0.6	177
138	0.8	173
144	0.53	173
150	0.42	174
156	0.4	172
162	0.59	169
168	0.55	170

13–29 For a 5-mgd plant, size the basins for biological phosphorus and nitrogen removal. What chemicals will be necessary to enhance removal?

13–30 Using an ammonia load of 10 lb/100 cu ft (160 g/m³/d), size the basin for a 10-mgd plant using data in Figure 13–16.

13–31 How are phosphorus and nitrogen removed in conventional treatment?

13–32 Redo Example 13–4 and Figure 13–16 for a treatment plant that does not have a primary clarifier.

Water Reuse

The practice of discharging treated wastewater to surface waters and withdrawal downstream is not considered reuse because dilution and separation in time and space allow additional purification to take place. Reuse of wastewater involves the direct application of treated wastewater for agricultural irrigation, urban irrigation, industrial reuses, groundwater recharge, and, in some cases, street cleaning, car washing, and toilet flushing. Reuse applications require a degree of treatment ranging from secondary treatment to advanced tertiary treatment. The degree of treatment increases with the degree of public access and concern for public health. Irrigation of pasture land on private property (with an appropriate perimeter buffer and fence) limits public access and treatment may be only pond systems or mechanical plants with secondary treatment. Irrigation of street medians and residential landscaping exposes the public to reuse water and aerosols from spray irrigation, thus tertiary treatment with high levels of disinfection is required to remove the health risk associated with pathogens in the wastewater. Groundwater recharge requires

a water reclamation degree of treatment to remove organic compounds and dissolved salts and to improve water quality prior to injection.

The decision to develop reuse applications is often economic and may be the result of viewing wastewater from a disposal perspective or as a water resource. From a disposal perspective, the question is: What level of treatment is required for discharge to a surface water? Discharge water-quality requirements depend on the assimilative capacity of the receiving body, discussed in Chapter 5. A measure of indirect reuse occurs when effluents that have been discharged and diluted in a river are withdrawn downstream for irrigation or as a public water supply. In some instances, the treated wastewater represents a significant portion of the drought flow. A stream flow providing a 20:1 dilution of treated wastewater effluent has been deemed protective of public health in some states. Dilution of less than 20:1, or stream assimilative capacity that is deemed inadequate, may require a level of treatment equal to that for recycled water for surface discharge in order to protect public health.

As a disposal alternative, water reuse may be less costly than discharge to surface waters or other alternatives. For example, the reuse system for agricultural irrigation in Tallahassee, Florida, was developed because of eutrophication and water-quality problems caused by surface discharge to a downstream lake. The irrigation system and treatment for reuse was developed in lieu of the higher cost of providing advanced wastewater treatment to remove phosphorus and nitrogen. Only recently have water shortages been a motivating factor in the development of recycled water.

Water reuse supplies may be considered part of an overall water supply during water resources master planning. Recycled water may be used instead of some potable water where additional potable water supplies are more costly or not available, thus freeing an equivalent amount of potable water for other uses. Because of limitations on water supplies in many urban areas of California, Arizona, Florida, and other states, recycled water is considered an alternative source of water supply along with groundwater, surface water, and imported water supplies. In areas where water supplies do not meet the potable water demands, discussions of water reuse and reclamation to supplement potable water supplies continue. For example, groundwater demands in Orange County, California, drew ocean water inland, contaminating the fresh groundwater supply. Beginning in 1976, groundwater injection started to create a hydraulic mount using highly treated reclaimed water to prevent seawater intrusion. This water commingles with other groundwater and is ultimately withdrawn as part of the potable water supply.

Use of recycled water should not adversely affect downstream water rights, degrade water quality, or injure plants, fish, and wildlife. Withdrawal of treatment plant effluent from stream discharge may be limited by the quantity of water necessary to maintain an established habitat along the stream. Treatment requirements for discharge depend on the beneficial uses associated with the stream. Where the beneficial use is for unrestricted recreation, high-quality recycled water treatment may be required for discharge. Additional treatment requirements for the removal of nutrients, metals, salts, and total dissolved solids (TDS) would be based on the assimilative capacity of the stream.

The U.S. Geological Survey[1] maintains water statistics for the trends, sources, and delivery of public water supplies and wastewater releases. In 1995, recycled wastewater totaled 983 mgd, or about 2 percent of the total wastewater release of 41,000 mgd. In the same year, total water sources included 19.25 percent from groundwater, 80.5 percent from surface water, and 0.25 percent from recycled water. The states of Arizona (209 mgd), California (216 mgd), and Florida (271 mgd) accounted for a total of 696 mgd, or about 71 percent of total national recycled water uses in 1995. In Arizona, about 37 percent of all wastewater treated is recycled. Table 14–1 lists withdrawals for groundwater and surface and recycled water from 1950 through 1995. The U.S. Geological Survey reports that "most of the increases in water use from 1950 through 1980 were caused by expansion of irrigation systems and energy development. Development of center-pivot irrigation and plentiful groundwater sources supported groundwater use."

Higher energy prices in the 1970s and a large drawdown in groundwater levels increased the cost of irrigation. In the 1980s, improved agricultural techniques, a downturn in the farm economy, and increased water competition decreased demand and supported a transition from water-supply management to water-demand management.

TABLE 14–1

U.S. Geological Survey Data for Water Withdrawals and Recycled Water Use in Thousands of Million Gallons per Day

Year	Ground	Surface	Recycled
1950	34	150	
1955	48	198	0.20
1960	50	221	0.60
1965	61	253	0.70
1970	69	303	0.50
1975	83	329	0.50
1980	84	361	0.50
1985	74	325	0.58
1990	81	325	0.75
1995	78	324	1.02

Additional decreases can be attributed to an increase in water recycling, improved industrial efficiencies, and changes in regulations to reduce the discharge of pollutants. The overall result was less water being returned to natural systems after use.

The U.S. Geological Survey suggests that water withdrawals will continue to increase as population increases. Higher water prices and active conservation programs may reduce the per capita usage. Irrigators will have increasing difficulty competing with urban and industrial users for available water supplies. Major attention needs to be given to water-management problems to ensure that maximum benefits will be obtained from the nation's water resources.

Figure 14–1 presents the national average for water demands and the reuse by category in California and Florida. The largest water demands are for thermoelectric power generation, agricultural irrigation, public water supply, and industrial uses. In the U.S., agricultural and industrial uses account for 90 percent of the total water withdrawn from streams, lakes, and groundwater. Agricultural irrigation represents 40 percent of total withdrawals. Only 10 percent of water is used in households and contributes to an average wastewater flow of 75 to 80 gpcd. Although the volume of household water that generates wastewater and could be treated to be reused as recycled water is low, this recycled water source is important to arid, urbanized areas. The largest reuse applications in California are for agricultural irrigation, landscape irrigation, and groundwater recharge. Other applications include environmental enhancements, for example, wetlands for animal habitats and recreational impoundments. In Florida, more water is used for recreational impoundments (including urban reuse) than for agricultural irrigation, groundwater recharge, and industrial use. The actual use depends on the availability and relative cost of freshwater sources, proximity of the treatment plant to the proposed use, and relative treatment requirements for disposal. In the year 2000, the state of Florida reported an inventory of 103,660 residential properties, 401 golf courses, 385 parks, and 159 school grounds irrigated with recycled water. Reuse in California and Florida has just about doubled since 1995.

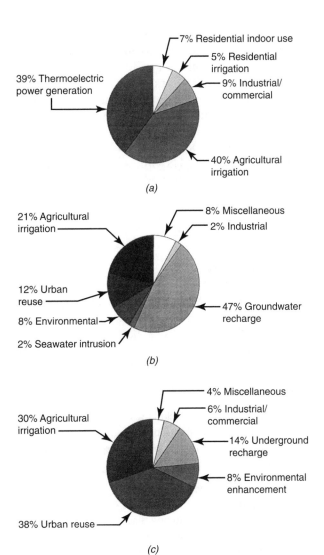

Figure 14–1

Fresh water demand and recycled water use. (a) Total fresh water withdrawal by water-use category in year 2000[1]. (b) Recycled water use in California by use category in year 2002[2]. (c) Recycled water use in Florida by use category in year 2005[3].

(Adapted from U.S. Geological Survey, *Water Use in the United States*, State of California, Department of Water Resources; *Water Facts: Water Recycling*; Florida Department of Environmental Protection, Water Reuse Program, *2005 Reuse Inventory*.)

14–1 WATER QUALITY AND REGULATIONS

The U.S. Environmental Protection Agency regulates treatment plant discharge through the national pollution discharge elimination system (NPDES) permitting process but does not govern water reuse regulations or permitting. Regulations for a wide range of reuse applications have been adopted by individual states, and although many states have thorough lists of reuse applications and corresponding water-quality standards or treatment requirements, no state has regulations that include all potential applications. Arizona, California, Florida, Hawaii, and Texas strongly encourage water reuse and have adopted regulations covering most applications. Regulations in those states specify water-quality requirements

and/or treatment processes necessary to protect public health, maintain water quality, and enhance the environment. Other regulatory requirements include storage, monitoring, runoff, provisions for alternative disposal, and irrigation loading rates.

Table 14–2 lists general reuse applications, treatment processes, and common water-quality requirements compiled from standards developed by individual states. The types of reuse may be grouped into the broad categories of restricted agricultural irrigation, unrestricted agricultural

TABLE 14–2

Type of Recycled Water Use, Treatment, and Water-Quality Requirements Compiled from Regulations Developed by Individual States

TYPE OF RECYCLED WATER USE	TREATMENT	WATER-QUALITY REQUIREMENTS
Unrestricted reuse applications	Classification A+	Turbidity <2 NTU (24-hr) and does not exceed 5 NTU
Irrigation: Food crops, including edible root crops, parks, playgrounds, school yards, residential landscaping, unrestricted-access golf courses	Secondary treatment, filtration using coagulants or polymers, nitrogen removal (denitrification), and disinfection	No detectable fecal coliform 7-day median, single sample <23/100 ml
		Total nitrogen <10 mg/l (5-sample mean concentration)
Industrial: Industrial or commercial cooling or air conditioning that involves use of a cooling tower, evaporative condenser, spraying, or any mechanism that creates a mist		Alternative treatment and blended water must have no detectable enteric virus in 4 of 7 last monthly samples.
Where mist can contact employees or public, chlorine or biocide must be used to treat the cooling circulating water to minimize growth of *Legionella* and other microorganisms	Classification A Secondary treatment, filtration using coagulants or polymers, and disinfection	Turbidity <2 NTU (24-hr) and does not exceed 5 NTU No detectable fecal coliform 7-day median, single sample <23/100 ml
Other: Flushing toilets and urinals, priming drain traps, fire fighting, decorative fountains, commercial laundries, construction, artificial snow making, commercial car washes Restricted reuse applications	(Equal to California Title 22 requirement for disinfected tertiary recycled water) Secondary treatment using microfiltration, nanofiltration, ultrafiltration, or reverse osmosis	Alternative treatment and blended water must have no detectable enteric virus in 4 of 7 last monthly samples. Turbidity <0.2 NTU (24-hr) and does not exceed 0.5 NTU

TABLE 14–2 Continued

TYPE OF RECYCLED WATER USE	TREATMENT	WATER-QUALITY REQUIREMENTS
Irrigation: Cemeteries, freeway landscaping, restricted-access golf courses, ornamental nursery stock and sod farms with access to the general public, pasture for animals producing milk for human consumption, irrigation of nonedible vegetation where public access is controlled	Classification B+ Secondary treatment, nitrogen removal (denitrification) and disinfection Classification B Secondary treatment and disinfection	Fecal coliform <200/100 ml 7-day median, single sample <800/100 ml Total nitrogen <10 mg/l (5-sample mean concentration) Fecal coliform <200/100 ml 7-day median, single sample <800/100 ml Limits to suspended solids may be set at 5 to 10 mg/l prior to disinfection.
Industrial: Industrial or commercial cooling or air conditioning that does not create a mist Other: Industrial boiler feed, soil compaction, mixing concrete, dust control, street cleaning, industrial processes with no contact with workers	(Equal to California requirement of disinfected secondary with more stringent coliform requirements)	Edible crops that are not peeled, skinned, cooked, or thermally processed must not contact recycled water.
Restricted irrigation applications Irrigation: Orchards and vineyards where recycled water does not contact the edible portion of the crop, non-food-bearing trees, fodder and fiber crops, seed crops not eaten by humans, food crops that undergo commercial pathogen-destroying processing, ornamental nursery stock and sod farms (no watering 14 days prior to harvesting) Other: Flushing of sanitary sewers	Secondary treatment in a series of stabilization ponds for at least 20 days, including aeration, and with/without disinfection (Equal to California requirement for undisinfected secondary)	Fecal coliform <1000/100 ml 7-day median, single sample <4000/100 ml BOD and suspended solids may be limited to 40 to 60 mg/l Use may be limited to overland flow and drip irrigation systems. Additional requirements for disinfection may apply if wastewater is sprayed.
Industrial Unlisted type of direct reuse	Treatment is determined on a case-by-case basis for: Industrial wastewater containing sewage Irrigation or processing of any crop used as human or animal food Treatment is determined on a case-by-case basis using the following factors: Risk to public health Degree of public access and exposure Level of treatment necessary to ensure water is aesthetically acceptable Level of treatment to prevent nuisance conditions Specific water-quality requirements for the intended type of direct reuse Means of water application Degree of treatment necessary to avoid violation of surface water-quality or aquifer water-quality standards Potential for improper or untended use Reuse guidelines, criteria, or standards of the U.S. EPA Similar reclamation experience in the U.S.	

irrigation, restricted urban irrigation, unrestricted urban irrigation, industrial reuse, and recharge of aquifers used for groundwater supply. Treatment processes may be linked with water-quality limits because a combination of treatment and water-quality requirements is known to produce reclaimed water of acceptable quality and reduce the number of criteria to be monitored. Secondary treatment does not require primary treatment, but must include oxidation. Biological aeration is typically used because of its superior BOD and solids removal. Chemicals are added to the wastewater for solids coagulation and passed through a bed of filter media at a rate not exceeding 5 gpm/sq ft, resulting in an average turbidity of less than 2 NTU. Other treatment processes, such as membrane filtration, must meet or exceed the turbidity and overall coliform removal requirements. Chlorine contact time must have a CT value (residual chlorine concentration times model contact time) of not less than 450 mg/l · min with a model contact time of at least 90 min at peak flow. Many states require testing and certification of new processes and equipment to establish limits on loading and identify subsequent treatment requirements for reuse applications. For small treatment systems, stabilization ponds are effective for removal of helminth eggs and protozoal cysts with multicelled ponds of greater than 20 days' detention. Fecal bacteria are also significantly reduced in warm ponds with over 25 days' detention. This water would be suitable for restricted agricultural reuse and some industrial applications. As described in Chapter 13, membrane treatment removes suspended solids, bacteria, and cysts. Disinfection may be required for virus removal and to maintain a residual to the point of application. Industrial and other types of reuse may be judged on a case-by-case basis.

Considerable variability exists in setting treatment requirements; for example, some states allow stabilization ponds prior to irrigation, but most states require a minimum of secondary treatment with oxidation and chlorination. Not listed in Table 14–2, but part of most states' treatment requirements, are reliability criteria, including standby power, alarms, standby treatment units, emergency storage, and automated control systems. Alarms include those for a loss of power and the failure of biological, disinfection,

coagulation, or filtration processes. Alarms automatically initiate short-term retention or disposal provisions to avoid contamination of treated recycled water and maintain treated water quality.

Referring to Table 14–2, some states apply letter grades (A through C) to recycled water quality. A plus designation on the letter grade refers to treatment that includes denitrification. Water quality is typically measured using BOD, suspended solids, and/or turbidity and coliform. Coliform is used as a measure of the degree of treatment and as a surrogate measure for disease organisms and viruses. Some states use fecal coliform and others total coliform. Effluent wastewater contains more total than fecal coliform; therefore, total coliform is more restrictive and indicative of a higher degree of treatment. In general, the extent of treatment, degree of disinfection, and coliform limits relate to increasing public exposure.

California Title 22, Chapter 3, Water Recycling Criteria, refers to four levels of recycled water treatment: disinfected tertiary (equal to grade A), disinfected secondary—2.2 total coliform/100 ml (grade B with greater coliform removal), disinfected secondary—23 total coliform/100 ml (grade B with greater coliform removal), and undisinfected secondary (equal to grade C). Disinfected tertiary treatment refers to an oxidized wastewater (secondary treatment) that has been coagulated and passed through natural undisturbed soils or a bed of filter media at a rate that does not exceed 5 gpm/sq ft or does not exceed 2 gpm/sq ft in a traveling-bridge automatic backwash filter. Filter discharge must not exceed a 2 NTU 24-hr average and must not exceed 5 NTU for more than 15 min and never exceed 10 NTU. Wastewater passed through a microfiltration, ultrafiltration, nanofiltration, or reverse-osmosis membrane must not exceed 0.2 NTU for more than 15 min and never exceed 0.5 NTU. The chlorine disinfection process following filtration must provide a CT value (the product of total chlorine residual and modal contact time measured at the same point) of not less than 450 mg/l · min at all times with a modal contact time of at least 90 min, based on peak dry-weather design flow or a disinfection process that, when combined with filtration, has been demonstrated to inactivate and/or remove 99.999 percent of plaque-forming units of F-specific bacteriophage MS2, or polio virus, in the wastewater. The 7-day median (4 of 7 samples) of total coliform bacteria measured in the disinfected effluent must not exceed an MPN of 2.2/100 ml, and the number of total coliform bacteria must not exceed an MPN of 23/100 ml in more than one sample in any 30-day period. No sample shall exceed an MPN of 240/100 ml total coliform bacteria. The modal contact time is that amount of time elapsed between the time that a tracer dye is injected into the influent and the time that the highest concentration of the tracer is observed in the effluent from the chamber.

Nutrient removal during wastewater treatment is not required to meet the water-quality reuse requirements for any state, but has some advantages in water-quality issues. The application of reuse water for irrigation is limited by the nitrogen uptake in crops; wastewater treatment using denitrification allows for greater application rates to reduce soil salinity buildup and may limit excess water disposal requirements. Denitrification also enhances reuse water quality by improving final clarifier performance, resulting in lower turbidity levels. Phosphorus and nitrogen removal reduces, but will not eliminate, algae growth in large storage reservoirs.

Salinity is a concern to all semi-arid and arid areas using recycled water. Salinity increases as water is reused for potable and agricultural purposes and results in salt buildup in soils and water supplies. Increased agricultural production, the use of water softeners, and treatment chemicals lead to increasing salinity in recycled water. Self-regenerating residential water softeners discharge salt brine reject water directly to the sewer system. The chemicals selected for wastewater treatment vary in their contribution to increases in salinity; for example, chlorination using sodium hypochlorite contributes sodium salt as opposed to gas chlorination or ultraviolet disinfection. Physical removal of excess salt in recycled water tends to be limited to water reclamation because of the high cost involved in reverse-osmosis treatment.

Engineering reports serve as a preliminary design report and are required by most states to obtain recycled water permits for new and expanded systems and may be required for existing facilities that violate permit conditions or water-quality standards. Some states allow abbreviated reports for slow-rate irrigation and land application systems. Table 14–3 lists a summary of report requirements compiled from requirements in California, Arizona, and Florida.

14–2 AGRICULTURAL IRRIGATION

Agricultural reuse is advantageous because the wastewater treatment requirements for it are often moderate, wastewater contains plant nutrients and

---

TABLE 14–3

Summary of the Contents of an Engineering Report Required for Recycled Water (Adapted from Arizona,[4] California,[5] and Florida[6] State Requirements)

Land Use Requirements	Legal description of property
	Map showing site with setback distances, present land uses, and planned land uses
	All potable and nonpotable water wells and monitoring wells with associated ownership information
	Surface waters and descriptions regarding their classification, beneficial uses, and distance from site
Soils Information	Soils map showing type
	Physical characteristics of each soil and subsoil layer. Hydraulic conductivity to support design loading and application rates
	Depth to groundwater

Continued

TABLE 14–3 Continued

Hydrogeologic Survey	Data necessary to evaluate the capability of the site's performance, including flood and runoff conditions
	A proposed groundwater monitoring plan
	Direction and rate of existing groundwater movements
	For aquifer storage and recover projects, the report must address existing and future total dissolved solids.
Land Management System	Present and future soil-vegetation management program. Data and other documents to verify crop uptake of recycled water nutrients, evapotranspiration rate, salt tolerances, pollutant toxicity levels, and expected crop yield
	Harvesting frequency, length of operating seasons, application periods and rates, resting and drying periods, and ultimate use of crops
	Plans for storage, reuse, and disposal of reclaimed water or effluents during crop removal, wet weather, control for pests, equipment failure, or other conditions that preclude land application
Project Evaluation	Evaluation of the overall long-term impact of the project on environmental resources in the area: changes in water table, rate and direction of movement of applied reclaimed water, changes in water quality in the area associated with the project
	Evaluation of pretreatment program
	Evaluation of reclaimed water characteristics: physical, chemical, and biological concentrations; flow patterns (average, peak, low flow during wet and dry seasons); treatment unit processes; hydraulic, organic, and nutrient loadings (average, peak, low); concentrations of reclaimed water percolated to groundwater or discharged to surface water; and operating and control strategies
For Projects Involving Groundwater Recharge	Time of travel from the discharge point to removal point for potable purposes
	Groundwater mounding analysis
	Assessment of the impact on water levels, surface-water levels, groundwater quality, surface-water quality, and uses of property in the area
For Projects Involving Industrial Uses of Reclaimed Water	Any additional treatment and disinfection requirements beyond the minimum requirements
	Disposition of any industrial wastewaters originating from the use of reclaimed water

soil amendments, agricultural areas may be adjacent to treatment plants, and income is gained by growing cash crops. In arid regions, wastewater may be the only source of water available. Water not lost in evaporation and transpiration percolates to groundwater or drains to surface impoundments with no discharge from the irrigation site. This is consistent with the 1987 Clean Water Act Amendments for promoting zero discharge. The major contamination problems are percolation of nitrate to groundwater, retention of heavy metals in the soils, and pathogenic hazards to farm workers.

Storage is required to balance supply and demand for irrigation. While the supply of wastewater is continuous, demands for irrigation water depend on the growing season, crops, application rates, and climatic conditions. Storage must ensure an irrigation use without overflow during peak flows or wet weather. The volume is greatest in humid northern regions. Some plants use storage reservoirs for additional treatment, pathogen removal, and the prevention of discharge.

Water distribution is by surface spreading or fixed and moving sprinkling equipment. On flat land, having less than a 1 percent slope, surface irrigation is possible by the ridge-and-furrow method. Water applied to furrows flows by gravity and seeps into the ground. Border-strip irrigation uses parallel soil ridges constructed in the direction of slope. Water introduced between the ridges at the upper end flows down 20- to 100-ft-wide strips. Sprinkling systems may be fixed or center-pivot. Fixed nozzles are attached to headers that are 2 to 6 ft off the ground. The center-pivot system uses nozzles attached to a spray boom that rotates around a central tower with the distribution piping suspended between wheel supports. Runoff collected at the low point is retained or returned for irrigation.

Loading rates are determined by hydraulic, organic, nutrient, and salinity limits. Hydraulic loading rates may be restricted by numeric limits established by the state or by antidegradation restrictions limiting groundwater percolation. Some states limit hydraulic loading to 2.0 to 2.5 inches per week and others up to 4.0 inches per week. Zero percolation to groundwater requires loading be limited to crop uptake and an allowance for evaporation. Background groundwater quality, crop type, and irrigation method are required to help set maximum loading rates. Organic loading rates are based on providing optimum biological activity to maintain aerobic conditions in the upper soil layer. One state limits COD loading to 50 lb/ac/day. Nutrient loading typically includes both phosphorus and nitrogen, where nitrogen is the limiting factor. Nitrate-nitrogen is limited to less than 10 mg/l to protect groundwater and surface water as a drinking water source. High salinity impedes biological activity in the soil, affects plant growth, and runoff may degrade surface water quality. Salt stress in plants is evidenced by the yield of plant parts such as seeds, roots, fruits, or leaves. Moderately tolerant plants exhibit less than 10 percent impact at chloride concentrations up to 400 mg/l. Target chloride concentrations range between 100 and 200 mg/l with peaks of 400 mg/l and 600 mg/l for salt tolerant plants.

Restricted Agricultural Irrigation

Reuse water for irrigation of fodder, fiber, and seed crops presents the least opportunity of human contact and so allows less stringent treatment requirements where buffer areas and restrictions to public access are maintained. Treatment requirements and water-quality requirements vary considerably among states, from lagoon treatment to secondary treatment with disinfection. California requires a minimum of treatment, including primary sedimentation for fodder, fiber, and seed, with disinfection (23/100 ml fecal coliforms) for pasture with milking animals. Florida requires secondary treatment and disinfection with an effluent quality of 20 mg/l BOD, 20 mg/l suspended solids, chlorine residual of 0.5 mg/l, and 200/100 ml fecal coliform limit. Most states require biological secondary treatment (BOD range from 20 to 75 mg/l and suspended solids range from 10 to 90 mg/l), fecal coliform limits (from 23 to 2000/100 ml depending on buffer zone), and buffer areas to protect water supplies, lakes, and human inhabitance. There are additional treatment and disinfection requirements or limitations for dairy animal pasture.

Storage is based on a minimum detention time or water balance requirements. Some states require 0 to 7 days of storage; others require 45 to 150 days of storage that may be used for additional treatment and pathogen removal.

Unrestricted Agricultural Irrigation

Unrestricted irrigation of crops includes both processed food crops and foods eaten raw, with the required quality of water depending on the method of application. Spray irrigation requires a higher degree of wastewater reclamation than surface irrigation. Also, required water quality varies with aboveground and root crops and fruit formation on trees. Although people are warned that the water is not for drinking purposes, site access may not be controlled. While irrigation of food crops is prohibited in some states, it is regulated on a case-by-case basis in others. Colorado and Hawaii regulate the largest number of applications, including processed food, raw food consumption, orchards and vineyards, and root and nonroot crops, using surface and spray irrigation. Other states require a high level of treatment, ranging from secondary treatment with chlorination to treatment for pathogen removal using coagulation, filtration, and disinfection. Secondary treatment with chlorination is not adequate for removal of helminth eggs, cysts, and viruses without adequate storage following treatment.

For foods eaten raw, secondary treatment with chemical addition, coagulation, filtration, and long-detention chlorination results in 2.2/100 ml fecal coliform and turbidity below 2 NTU. Processed food requires secondary treatment, with fecal coliform limits in the range of 23 to 1000/100 ml. Buffer zones are provided between domestic water supplies and adjoining property based on use. Storage required for detention ranges from 5 to 15 days or is based on rainfall and loading rates. Loading rates are based on the percolation of nitrate-nitrogen of less than 10 mg/l, crop uptake of nutrients, water balance, and maximum loading rates of 0.5 to 4.0 in./week.

14–3 URBAN IRRIGATION AND REUSE

Urban applications represent a small quantity of the total reuse except in large urban areas where other options are not available. Urban irrigation includes golf courses, landscaped medians, parks, and, in some locations, front and back yards. Following treatment, reclaimed water is stored for repumping into the distribution and irrigation system. Water degraded during storage may require additional treatment for algae removal and chlorination to maintain a residual. Piping systems are identified by color-coded pipe or by using purple plastic pipe (potable water is blue or white). Cross-connection is prohibited and other connections restricted. Warning signs stating "Irrigation with Reclaimed Wastewater," should be prominently displayed, and hose bibs should be posted with signs stating "Reclaimed Wastewater, Do Not Drink."

Restricted Urban Irrigation

Areas such as golf courses, cemeteries, and highway medians, where public access is restricted and where water is applied only during night hours without airborne drift or surface runoff into public areas, present limited exposure risk. Where transient human activities take place, such as on a golf course, the vegetation should be allowed to dry and excess water should be allowed to soak into the ground before activities begin. Some states require unrestricted reuse limits for golf course irrigation.

California, Colorado, and Hawaii require secondary treatment with chlorination to a fecal coliform limit of 23/100 ml. Florida requires a higher degree of treatment, with coagulation, filtration, and disinfection to a fecal coliform limit of 25/100 ml. Some states allow stabilization pond treatment with a limit of 20 to 30 mg/l suspended solids. Additional treatment may be required at the point of application for solids removal that can plug nozzles and chlorination to prevent growth within the irrigation system.

Storage is based on detention time or water balance. During nonirrigation periods, reclaimed water can be used for other purposes appropriate to the degree of treatment. Loading rates, where specified, are typically less than 2.5 in./week but depend on vegetation and weather conditions.

Unrestricted Urban Irrigation and Reuse

Unrestricted irrigation includes parks, playgrounds, schoolyards, residences, and commercial landscaping. Other urban uses include toilet flushing, fire protection, and construction. Although the water purveyor may be in control of watering times and limit exposure by night watering, public contact is expected and the water may be ingested by children. Unrestricted reuse water must be pathogen free, requiring a high degree of treatment and disinfection.

California requires secondary treatment with coagulation, filtration, and chlorination with a long detention time, less than 2.2/100 fecal coliforms, and turbidity less than 2 NTU. Florida has the same treatment requirement for restricted and unrestricted applications. Other states require at least secondary treatment and disinfection with storage lagoon detention of 15 to 150 days.

14–4 GRAY WATER AND INDUSTRIAL REUSE

Residential gray water is wastewater from the clothes washer, bathtub, shower, or sink that is collected separately from the toilet, dishwasher, and kitchen sink. Industrial gray water refers to wastewater from the manufacturing process that may or may not require treatment for reuse within the manufacturing facility or for irrigation of crops or landscaping on private property. Separation of sanitary facilities and industrial processes is necessary to avoid human pathogen contamination of industrial wastes. Gray water use is typically not considered a reuse, but part of an industry's use of internal recycle flows in a water conservation program.

Industrial water use accounts for about 6 percent of total water demand; however, including power generation (47 percent), the total industrial and power-generation water demand is 53 percent of total water use. Commonly, reuse water is used for cooling tower makeup water, where the water-quality concerns are growths, scaling, corrosion, and foaming. These are common concerns in any water supply and are controlled using physical-chemical pretreatment and biocides. Hardness and salt buildup are controlled using dilution water (adding to the water demands) or by treatment using reverse osmosis. Industrial applications such as those for textiles or pulp and paper manufacturing have specific water-quality requirements depending on the use and limitation on dissolved salts and color. A few states have reuse treatment requirements for common industrial applications. Hawaii,[7] for example, requires that water used for cooling be treated using secondary treatment with coagulation, filtration, and chlorination to 2.2/100 ml fecal coliform and additional disinfectant for *Legionella* and *Klebsiella*. For industrial processes not involved in food production, secondary treatment including chlorination to a fecal coliform limit of 23/100 ml is used. Treatment processes tend to be physical-chemical (solids removal) but

may also include biological treatment processes. Depending on use and water quality, reductions in suspended solids, total dissolved solids, and color may be required. Processes such as disinfection and membrane filtration are the same as those used for wastewater treatment. Overall water quality may require a blending of water sources to meet manufacturing needs.

Water uses requiring wastewater treatment include landscape irrigation, seal water for pumps, wash-down water, polymer dilution, cooling tower makeup, and fire protection. Wastewater treatment facilities fall under industrial reuse requirements because of limited access by the public. Treatment requirements for each use vary from solids removal to high-level disinfection. Small package filters and chlorine contact tanks may be used on a side stream of plant effluent for in-plant needs.

14–5 CONSTRUCTION AND OTHER REUSE APPLICATIONS

The largest reuse application for construction is dust control; other applications include soil compaction, irrigation, and cement mixing. Treatment requirements vary with the method of application and potential contact by the public. Other considerations for treatment requirements are listed in Table 14–2 under unlisted types of direct reuse.

Other reuse applications include the flushing of toilets and urinals for office high-rises. The Irvine Ranch Water District in Orange County, California, has operated reuse facilities for over 30 years. Most of the recycled water is used for urban, landscape, and crop irrigation; however, because of water shortages, several new office high-rises were dual-plumbed for toilet flushing using recycled water.

14–6 GROUNDWATER RECHARGE AND POTABLE SUPPLY

Groundwater recharge may result from surface infiltration or direct injection. The state of Florida distinguishes between groundwater injection, discharge to surface waters used for potable water supplies, and rapid infiltration, as listed in Table 14–4. Discharge to surface waters used for potable water

TABLE 14–4

Florida Groundwater Recharge and Rapid-Rate Land Application for Indirect Potable Water Reuse

REUSE SYSTEM TYPE	REQUIREMENT FOR TREATMENT AND WATER QUALITY
Injection to Groundwater (TDS < 3000 mg/l)	Secondary treatment and filtration TOC 3 mg/l avg, 5 mg/l max TOX 0.2 mg/l avg, 0.3 mg/l max Meet primary and secondary drinking water standards Drinking water disinfection Total nitrogen <10 mg/l avg An evaluation of removal of enterovirus, *Cryptosporidium, Giardia*, and helminths is required to show pathogen-free treatment (less than detection) Water-supply wells within one mile of the injection well require reverse-osmosis treatment as a multiple barrier for organic compounds and pathogens
Injection to Groundwater (TDS > 3000 mg/l)	Secondary treatment and filtration Meet primary and secondary drinking water standards Drinking water disinfection Total nitrogen <10 mg/l avg
Rapid Infiltration Basins	Secondary treatment and basic disinfection levels prior to spreading Nitrate concentration <12 mg/l Discharge <10 mg/l total suspended solids, unless the absorption field and application/distribution system is designed for specific flexibility and reliability Because of the somewhat limited ability to renovate reclaimed water, the proposed treatment must meet groundwater criteria at the edge of the zone of discharge. Adsorption field designed and operated to preclude saturated conditions at the ground surface Loading rates must be set based on percolation tests and are limited to an annual average of 3 in./day or 1.9 gpd/sq ft. Rates must not exceed 9 in./day (5.6 gpd/sq ft). Hydraulic loading periods of 1–7 days with resting periods of 5–14 days to dry the cell bottom and enable scarification or removal of deposited solids are required.
Rapid Infiltration Basins with Unfavorable Hydrogeologic Conditions	Secondary treatment and filtration High-level disinfection Meet primary and secondary drinking water standards Total nitrogen <10 mg/l avg
Discharge to Surface Waters used for Potable Water Supply	Secondary treatment and filtration Meet primary and secondary drinking water standards Drinking water disinfection Total nitrogen <10 mg/l avg

(Adapted from *Florida Administrative Code, 2001*, Part V, groundwater recharge and indirect potable water reuse) supplies is a topic that continues to be addressed by many states.

Rapid infiltration operates on a fill-and-drain basis, in which water is added to a depth of 3 to 5 ft (1 to 2 m) and left to percolate into the ground. Beds are left to dry, reducing nuisance issues such as odor and mosquito problems. Infiltration beds are not used for crops and are periodically disked to break up any surface solids that reduce the infiltration rate. Key removal mechanisms include volatilization, chemical and biological reactions, precipitation and adsorption of metals, and sorption within the soil and sand matrix. Most of the treatment and removal takes place in the top 6 ft (2 m) of the soil profile (vadose zone). The degree of treatment depends on the type of application to the soil, soil formation and chemistry, depth to groundwater, dilution available, and residence time to the point of extraction. Physical filtration in the infiltration basin removes bacteria and algae. Biological reactions break down organics and create conditions for nitrification and denitrification in the unsaturated zone and upper level of the aquifer. Precipitation, cation exchange, and adsorption remove phosphorus, metals, and sodium. Virus and pathogen survival depends on soil conditions and infiltration rates. Proper operation, including drying, resting time, and disking, minimizes clogging caused by the surface straining of bacteria and algae.

The U.S. Geological Survey, Water Resources of California[8] studied the colloid transport beneath the recharge basins near Whittier Narrows in south-central Los Angeles County. Large infiltration basins (spreading fields) have been in operation since 1961 to recharge the groundwater. Because low flows during the summer result in a stream that is effluent dominated, treatment up to recycled water quality is required prior to discharge. Recycled water commingles with the stream flow and infiltrates into the groundwater under saturated conditions. The purpose of the study at the Whittier Narrows spreading fields was to find tracers in reclaimed water that might serve as a quantitative measure of the distribution of reclaimed water in the aquifer. Tritium/helium-3 and Freon analyses explain the wide range in water-quality data observed in deep production wells located very close to the infiltration basins. The recharge extends to a depth of only a few hundred feet, then extends horizontally due to the rapid groundwater movement. High levels of chlorofluorocarbons in wells with tritium/helium-3 suggest lesser treatment or more lax environmental practices in past years. Freon analysis has been suggested as an indicator of the reclaimed waterfront as it advances through groundwater. Additional research suggested that a 7-log subsurface removal of viruses occurs in less than 150 ft (50 m) of groundwater travel.

Groundwater quality may be divided by TDS level. The national secondary drinking water standard for TDS is between 0 and 500 mg/l. Groundwater with TDS > 3000 mg/l and up to 10,000 mg/l can be treated to reduce TDS to drinkable levels. The treatment required prior to potable water distribution is in addition to treatment required prior to injection.

Direct injection into groundwater containing TDS levels < 3000 mg/l must meet or exceed drinking water standards with the highest degree of wastewater treatment. To achieve drinking water quality, treatment requires biological, chemical, and physical processes to reclaim the water and remove pathogens and organic compounds. It includes removal of inorganic salts to reduce the total dissolved solids concentration, reducing salinity. Advanced treatment processes used in Water Factory 21 in southern California are diagrammed in Figure 13–9. Microbial pathogens such as parasites, bacteria, and viruses and chemical substances such as organics, pharmaceuticals, and inorganic compounds are present in untreated wastewater, but their presence and concentrations in treated wastewater varies greatly. Virus testing requires the concentration of large volumes, up to 100 to 400 l, and viruses are difficult to find in highly treated water. Treatment testing is based on wastewater seeded with attenuated viruses so that effluent virus concentrations and removal can be determined. Indicator organisms and chemicals are used to represent the degree of treatment. Fecal and total coliform measures are used to determine the effectiveness of secondary and tertiary treatment. Using total coliform is more conservative than using fecal coliform. Although the number of these organisms may not correlate to pathogen risk or virus concentrations, coliforms are an indication of the degree of treatment, since bacteria and viruses cannot be found in reclaimed water. Total organic carbon (TOC) is a measure of the gross amount of carbon from organic sources and includes natural and synthetic organic compounds. Total organic halide (TOX) and TOC are used as a measure for removal of organic compounds.

The California Department of Health requires groundwater recharge projects to include a maximum average of 50 percent reclaimed water extraction over 5 years, a minimum of 20 ft (6.5 m) of vertical distance between the surface and groundwater tables, a minimum of 300 ft (100 m) of horizontal distance between the recharge basin and the extraction wells, and a minimum retention time of 12 months.

The National Research Council (NRC) formed a committee in 1996 to evaluate the issues associated with the augmentation of potable water supplies with reclaimed water. A report published in 1998 was intended to help communities make decisions that will protect the public water supply.[9] Augmentation of a drinking water supply includes the direct injection of reclaimed water into the drinking water aquifer or discharge to the surface-water impoundment with limited dilution. These applications remain indirect applications because of the commingling of natural water and separation of time and space between discharge and potable water intake. Direct potable reuse is the immediate distribution of reclaimed water into the potable water distribution system. Direct potable reuse is not practiced in the United States.

The NRC report states that the best available water source should be used for drinking water. In some parts of the United States where quality water sources are becoming scarce, the committee reported that planned indirect potable reuse may be viable, but only after careful, thorough contaminant monitoring, health and safety testing, and addressing of treatment reliability issues. Issues include chemical constituents, microbial contaminants, and evaluation of potential health effects. Planned indirect reuse within a potable water supply should be considered only after implementation of water conservation, a review of other water sources, and evaluation of nonpotable reuse to replace existing potable water use.

14–7 DESIGN OF IRRIGATION SYSTEMS

The recommended design for an irrigation system is the iterative process diagrammed in Figure 14–2. The initial step in design is to define the characteristics of the wastewater, effluent quality, and site conditions. Regulatory limits on effluent quality are established to protect groundwater, surface water, and public exposure. The iterative procedure for determining the field area needed for irrigation involves the interdependence of the hydraulic loading rate, nitrogen loading rate, water storage volume, and crop selection. System monitoring, the method of water distribution, discharge control, and agricultural management are final considerations.

The field area required for irrigation is determined by either the hydraulic loading or nitrogen loading. The water balance is calculated by the relationship

$$\text{Water loading} + \text{precipitation} =$$
$$\text{evapotranspiration} + \text{percolation} + \text{runoff} \quad \textbf{(14–1)}$$

The value for precipitation is the calculated maximum for a 10-year return period. Evapotranspiration is estimated based on climatic conditions and crops, and percolation is determined by soil conditions. This relationship is used to calculate water balances for any desired length of time, usually by the week, month, and year.

The inorganic nitrogen in reclaimed water is readily oxidized to the nitrate form in storage ponds or during spray irrigation. In a properly operated system, the majority of the nitrogen is removed from the percolating water by crop synthesis. Under anoxic conditions created in wet soils, some of the nitrate may be denitrified to gaseous nitrogen and lost to the atmosphere. Nitrate ions can also be carried through the soil profile by percolating water and eventually appear in the groundwater. The relationship for nitrogen balance is

$$\text{Applied nitrogen} = \text{crop uptake}$$
$$+ \text{loss by dentrification}$$
$$+ \text{loss by percolation} \quad \textbf{(14–2)}$$

Figure 14–2

Iterative design process for an irrigation system.

(From *Process Design Manual for Land Treatment of Municipal Wastewater*, U.S. Environmental Protection Agency, EPA 625/1-77-008, 1977.)

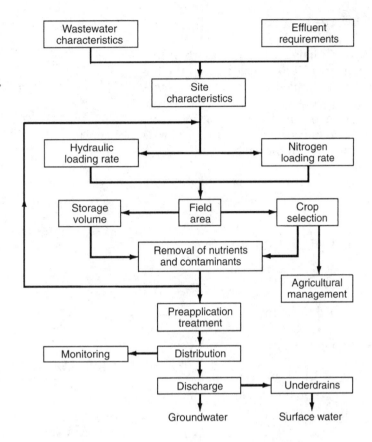

The greatest nitrogen uptake is in the approximate range of 200 to 400 lb/ac/yr (220 to 450 kg/ha · y) by forage crops such as alfalfa, clover, and grasses. Much less nitrogen, in the range of 70 to 170 lb/ac/yr (80 to 190 kg/ha · y) during one growing season, is taken up by field crops like corn, soybeans, and cotton. Denitrification is difficult to determine under field conditions; therefore, the nitrogen loss is generally assumed to be 15 to 25 percent of the applied nitrogen. The loss by percolation should be limited to 10 mg/l, which is the maximum contaminant level for drinking water.

Water storage is required in irrigation systems because of the imbalance between the reclaimed water supply and the application rate determined by crop growth and climatic conditions. In a cold climate, the operation of an irrigation site is stopped for several months each year when the ground is frozen. The length of time is determined by the growing season of either annual crops or perennial grasses. In a warm climate, the storage required is calculated by determining the difference between the application rate established by either hydraulic or nitrogen loading and the water supply. Calculating the surplus or deficit of water each month and then algebraically adding these values determines the storage needed to hold the maximum amount of surplus water.

Irrigated vegetation removes nutrients (primarily nitrogen and phosphorus) from the wastewater, maintains soil permeability, and reduces erosion of the soil. Crop selection is determined by the reclaimed water quality, required nitrogen uptake, and profitability. Grasses can be irrigated with a poorer quality of water, have a higher nutrient uptake over a longer growing season, and require less maintenance and skill to grow. Nevertheless, field crops like corn are generally preferred in large irrigation systems because of the greater monetary gain, even though watering must be scheduled more carefully to prevent the application of either too much water or nitrogen.

■ EXAMPLE 14–1

Calculate the water balance for the month of July for an irrigation site of 120 acres based on the following data. The reclaimed water supply averages 1.0 mgd, and precipitation is 2.1 in./mon. Evapotranspiration for the crop is 6.0 in./mon, and the allowable percolation is 10 in./mon with a nitrogen concentration in the percolate of 10 mg/l.

Solution

Water supply:

$$1.0 \text{ mil gal/day} \cdot 31\text{days/mon}$$

$$\cdot \frac{36.8 \text{ ac-in./mil gal}}{120 \text{ ac}}$$

$$= 9.5 \text{ in./mon}$$

Substituting into Eq. 14–1, assuming zero runoff,

$$\text{Water loading} + 2.1 \text{ in./mon}$$
$$= 6.0 \text{ in./mon} + 10 \text{ in./mon}$$
$$\text{Water loading} = 16.0 - 2.1$$
$$= 13.9 \text{ in./mon}$$

Water drawn from storage:

$$\frac{(13.9 - 9.5) \text{ in./mon} \cdot 120 \text{ ac}}{31 \text{ days/mon} \cdot 36.8 \text{ ac-in./mil gal}} = 0.46 \text{ mgd}$$

■ ■ ■

■ EXAMPLE 14–2

Calculate the nitrogen crop uptake required for the water balance data given in Example 14–1. Assume a concentration of 10 mg/l in the percolate and a denitrification loss of 20 percent. The average nitrogen content of the reclaimed water is 25 mg/l.

Solution

Applied nitrogen

$$= \frac{1.0 \text{ mil gal/day} \cdot 31 \text{ days/mon} \cdot 25\text{mg/l} \times 8.34}{120 \text{ ac}}$$

$$= 54 \text{ lb/ac/mon}$$

$$\text{Loss by denitrification} = 0.20 \cdot 54$$
$$= 11 \text{ lb/ac/mon}$$

Loss by percolation

$$= \frac{10 \text{ in./mon} \cdot 1.0 \text{ ac} \cdot 10 \text{ mg/l} \cdot 834}{36.8 \text{ ac-in./mil gal}}$$

$$= 23 \text{ lb/ac/mon}$$

Substituting into Eq. 14–2,

$$54 \text{ lb/ac/mon} = \text{crop uptake} + 11 \text{ lb/ac/mon}$$
$$+ 23 \text{ lb/ac/mon}$$

Required crop uptake $= 20 \text{ lb/ac/mon}$

■ ■ ■

14–8 DESIGN OF URBAN DISTRIBUTION SYSTEMS

Planning requirements for urban systems vary from individual irrigation applications to complete distribution systems with multiple application sites. The design of recycled water distribution systems requires coordination between supply and demand, water storage, and distribution system piping and pumping.

Coordination of Supply and Demand

All reuse applications require balancing supply and demand and coordinating disposal when the recycled water supply exceeds the available storage or adding supplementary water sources when demands exceed supply. Demand-side management strategies include limiting demand and optional demand applications where recycled water can be used as available. Supply-side management strategies include supplemental sources of supply such as groundwater, surface water, or potable water. For example, the city of Phoenix, Arizona, uses recycled water to recharge groundwater through constructed wetlands. In the winter, excess water flows to recharge areas, wetlands, and ephemeral waterways. In the summer, when less water is available and demand is at its peak, recharge areas, wetlands, and waterways receive little or no water.

A seasonal water balance is based on annual supplies and demands. Four conditions determine water availability, when precipitation creates years graded as wet, normal, dry, and drought. Planning for supply and demand is bounded by designing for dry-weather demands and wet-weather water supplies. If the design is based on limiting the water demand to the dry-year irrigation demand, then there will be plenty of recycled water during dry, normal, and wet years. The recycled water supply will be "drought resistant," except during successive dry years. During normal and wet years, excess water must be disposed of by another means of discharge (surface discharge, groundwater injection, or impoundment). Disposal costs may be higher than the cost of recycled water. All recycled water designs must include some method of disposal for excess water or treated water not meeting recycled water-quality standards. Water rationing may still be required during drought conditions. If the design is based on using all of the recycled water supply during a wet year, then the demand must be higher and there will be a shortage of recycled water during normal and dry years. The water shortfall must be met from another water supply, impoundments of surface runoffs, or stored or diverted potable water. Disposal would only be required in an emergency. Between drought-resistant and zero discharge, recycled water systems must be designed with provisions for both disposal of excess water and supplemental supplies for periods of supply shortfall. To support the development of recycled water use, some states allow recycled water supplies to be supplemented with surface waters, groundwater, treated storm water, and potable water. Supplemental water supplies are required to make the greatest use of recycled water possible and limit the disposal of excess recycled water.

Storage

Wastewater, the source of recycled water, varies in quantity from day to day and from month to month. Daily peak flows tend to be in the morning and evening, and peak irrigation demands tend to be at night during low flows. Daily storage scattered throughout the distribution system is used to equalize daily supply and demand, with the tanks tending to be between $\frac{1}{2}$ and 5 million gallons and similar to those used for potable water storage. All treatment plants tend to experience higher flows during the winter months as a result of infiltration and flow of storm water into the sewer system. At the same time, irrigation is prohibited during precipitation and generally for a period of 24 hr following a storm event. Irrigation is also limited by crop uptake during the growing season.

Winter water flows may be discharged or held in seasonal storage for use as recycled water in the irrigation season. Storage in ponds or reservoirs may be combined with decorative, recreational, or water resource uses. Many golf courses incorporate recycled water storage into their water features. Recycled water storage units are lined with clay, concrete, or hypalon to limit water loss and migration of nitrates or other constituents into the underlying groundwater. Groundwater monitoring is typically required to determine if seepage is impacting groundwater quality. Design must address control of algae, mosquitoes, and weeds. Many impoundments are fenced to limit access by the public and animals. Landscape impoundments are created for aesthetic purposes only, and recreational uses such as boating, swimming, and wading are prohibited. Recreational impoundments provide boating and fishing, but swimming, waterskiing, and other full-body activities are prohibited to limit the potential for ingesting water. Aquifer storage may be limited to groundwaters that cannot be used for public water supplies. Recycled water is injected into a subsurface formation with recovery for beneficial purposes at a later time.

Recycled water storage reservoirs are classified based on the degree of treatment and exposure to the public. Many states have adopted the following storage designations based on treatment level and use:

- A landscape impoundment is for recycled water storage, landscaping, or aesthetic purposes. Recreational activities are prohibited and public access is limited using fencing or natural barriers.
- Restricted recreational impoundment is for recycled water storage where boating and fishing

is an intended use (non-body-contact water recreational activities). Swimming and other full-body recreational activities such as waterskiing are prohibited to reduce the probability that water will be ingested.

- Nonrestricted recreational impoundments are designated in California for recycled water storage in which no limitations are imposed on body-contact or water recreational activities. The designation requires secondary, coagulated, filtration, and disinfection treatment to 2.2 MPN/100 ml with additional testing to monitor the presence of pathogenic organisms (*Giardia*, enteric viruses, and *Cryptosporidium*).

Under the appropriate conditions, recycled water may be stored in underground aquifers. An aquifer storage and recovery system allows storage of excess recycled water in a confined aquifer for use during subsequent dry seasons. Groundwater storage is less efficient that storage in an aboveground reservoir because water may migrate away from extraction wells or mix with poorer quality groundwater and become unusable. Use of underground storage requires characterization of the aquifer and study of changes to recycled water quality after storage and retrieval.

Distribution System Design

The distribution of recycled water is similar to water distribution as discussed in Chapter 6. Most ordinances require recycled water to be distributed in purple pipe or pipe wrapped with purple tape for identification and to avoid cross-connection with potable water. Recycled water lines must be 3 ft away from (horizontal) or 18 in. below potable water lines. Pipe size is typically smaller and based strictly on recycled water demand rather than the water supply that is required to meet fire flow demands. Distribution systems are analyzed for pipe size, pressure, velocity, and variations in demand using a computer modeling software such as EPA NET.[10]

The design of treatment and distribution systems must allow operating flexibility to provide the highest possible degree of treatment to be maintained under varying circumstances. Alarm devices are required for various unit processes, treatment, and power monitoring. Where short-term retention is used as a reliability feature, storage of partially treated wastewater must be sized for at least a 24-hr period. All of the equipment other than the pump-back equipment shall be either independent of the normal power supply or provided with a standby power source. Automatically actuated storage includes sensors, instruments, valves, and other devices to enable fully automatic diversion of partially treated wastewater to storage in the event of failure of a treatment process and a manual reset to prevent automatic restart until the failure is corrected.

Large storage reservoirs may adversely affect recycled water quality because of algae growth and nutrient peaks caused by seasonal turnover. Algae will grow in the nutrient-rich clear recycled water held in any open tank or reservoir. Although the water meets recycled water treatment requirements and may be distributed as is, algae and other solids substantially increase system maintenance by clogging small orifices, nozzles, and drip emitters. Their growth may be controlled by nutrient removal during wastewater treatment or by chlorination and filtration at the reservoir. Reservoir turnover occurs when warmer water near the surface mixes with cooler layers of water caused by thermal stratification. The turnover tends to bring solids and nutrients from the bottom of the reservoir to the top, where increased biological activity depletes the oxygen supply leading to odor and an increase in solids. Air diffusers and compressors may be used to break up thermal stratification and increase dissolved oxygen levels, thus improving recycled water quality.

Pressure

System operating pressures for recycled water distribution tend to be slightly higher than for water distribution systems. Higher pressures improve sprinkler coverage and are not limited by the water pressure limit of most household appliances (120 psi). The standard design pressure range is from 50 to 100 psi, but system pressures may be up to 150 psi, with individual pressure reduction where pressures exceed 80 to 90 psi. Whenever possible, water storage should be located at an elevation where pressure can be maintained by gravity. Unless topography allows convenient locations for elevated storage, recycled water systems tend to be pressurized by continuous pumping. Variable-speed drives on the pumps reduce power requirements by matching supply and demand.

Cross-Connections

Raw water and treated potable water may be used as backup for the recycled water supply. In addition, for some applications such as fire protection and irrigation, recycled water may back up the potable water supply. Potable water supplies must be protected from contamination by isolating and protecting cross-connections between recycled and potable water. Without protective devices, contamination may occur because of back-siphonage or reduced pressure. A back-siphon is a reduced or negative pressure in the potable water distribution system caused by very large demands or water line repair at a location lower than the service point. Connected systems will contaminate the potable water system when the potable water system is at a lower pressure because of low pressures in the distribution system or greater pressures in the interconnected piping system.

Backflow protection must be used to prevent contamination of a potable water supply. Where the public water system is used to supplement the recycled supply, an air gap must be used to prevent contamination of the water supply. An air-gap system requires a tank with a float-operated valve in which the water supply discharges above the high water level and free-falls into the tank. Pumps and a hydropneumatic tank are required to repressurize water to its point of use. Where recycled water is piped separately and there is not interconnection with the potable water system, a reduced-pressure backflow preventer may be used on the water supply. Under some circumstances, such as the use of recycled water for landscape irrigation as part of a dual-plumbed use area, a double-check assembly may be used with the approval of the utility agency and health services.

Agricultural Irrigation at Tallahassee, Florida

The application of wastewater effluent to agricultural land has been done in Tallahassee since 1966. A 16-acre agricultural farm was established as an

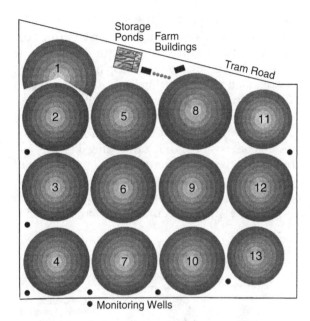

Figure 14–3

Site plan of farm for wastewater irrigation, Southeast Farm Wastewater Reuse Facility, Tallahassee, FL.

alternative to additional treatment for nutrient removal to avoid pollution of a downstream lake caused by the large contribution of wastewater. The farm was expanded several times over a 20-year period to over 2000 acres using 16 center-pivot irrigators with a capacity of 23 mgd of treated wastewater. Figure 14–3 shows a layout of the irrigated fields, wastewater storage ponds, farm buildings, and monitoring wells. Additional expansion of irrigated land is currently occurring to the east of this site. The figure does not show pivots 14–16 installed to the right of pivot 11. All of the center pivots are controlled by a computer system that tracks rainfall, temperature, wastewater flow, and storage pond water elevation. Aerosol drift is controlled through the use of low-drift nozzles on the irrigation equipment. Crops include canola (used to make cooking oil), corn and sorghum (sold as cattle feed), and various grasses (used as cattle feed or harvested as hay).

Wastewater is treated at the T.P. Smith and Lake Bradford Road plants, stored in ponds, and pumped for irrigation. The T.P. Smith facility is the largest, with a design treatment capacity of 27.5 mgd; the Lake Bradford Road plant has a capacity of 4.5 mgd. The T.P. Smith Water Reclamation facility was expanded by constructing parallel treatment plants (see Figure 14–4). Plant I started as a single-stage trickling filter plant and was expanded using rotating biological contactors and sand filtration to provide additional BOD and suspended solids removal. Plants II and III were constructed using an activated-sludge process. Plant IV consists of anoxic tanks ahead of the aeration basins to provide denitrification. The reduction in nitrogen allows water to be applied at higher rates without being nitrogen limited. Treated water is stored on-site in ponds with a

total volume of 102 mil gal. An additional 48 mil gal of storage for the treated water from the T.P. Smith and Lake Bradford plants is located at the Southeast Farm.

Crop irrigation is maintained year-round, avoiding the need for a second method of wastewater disposal. The ponds are sized to hold wastewater during brief interruptions caused during planting, harvesting, weather events, and equipment maintenance. Continuous irrigation is maintained through crop management that includes planting half of a circular field with one crop and half with another (see Figure 14–5). The

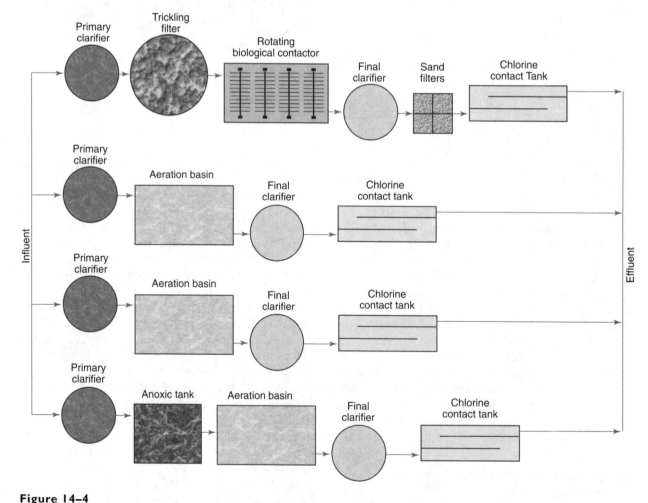

Figure 14–4

Schematic diagram of the T.P. Smith Water Reclamation Facility, Tallahassee, FL. The treatment facility consists of four parallel plants followed by storage ponds and pumping for crop irrigation.

Figure 14–5

Aerial photo of irrigation fields. The photo shows operating center-pivot irrigator on a field planted with half grass and half corn. The area of the irrigated circle is 134 acres (54 ha).

(Courtesy of City of Tallahassee Water & Sewer Department, Tallahassee, FL.)

center pivot irrigates half of the circle while workers are harvesting or planting crops in the other. Crop rotation reduces problems associated with weed and pest control.

Field observations showed that excessive water leads to problems with weed control and drives nutrients into the groundwater. Water application rates averaged 2.4 in./week (6 cm/week) and provided sufficient nitrogen, phosphorus, and potassium for rye and bermuda grass. The permitted application rate is 3.1 in./week (8 cm/week). Studies documented the long-term trends in soil parameters and found that two-thirds of the nutrients were consumed by the plants, with the balance unaccounted for. Soil comparisons in 1980, 1984, 1988, and 1993 indicate that pH, organic matter concentration, and cation exchange capacity have stabilized in the upper 3 ft. Phosphorus continues to increase in concentration due to fixation by iron and aluminum compounds in the soil. The soil is mainly a fine sand; therefore, an extensive groundwater monitoring program was developed. The groundwater contains a slightly elevated level of nutrients that migrate with the aquifer toward the Gulf of Mexico.

Frontyard and Backyard Residential Irrigation at El Dorado Hills, California

El Dorado Irrigation District, El Dorado County, California, provides potable water and operates two recycled water treatment plants and a dual-plumbed residential distribution system. The recycled water systems from each plant are connected and operate under a single Master Reclamation Permit. A schematic of the treatment plants and distribution system is shown in Figure 14–6.

Reclamation of wastewater at the El Dorado Hills Wastewater Treatment Plant began in 1979 when the State of California Regional Water Quality Control Board (RWQCB) changed the discharge requirements of the plant. Rather than make plant improvements so that these discharge requirements could be met, the district opted for the construction of a 66-mg reservoir and a pumping system so that recycled water could be disposed of at the El Dorado Hills Golf Course and a local lumber mill. Part of the reasoning for this was that the district had just been through the 1976–1977 drought.

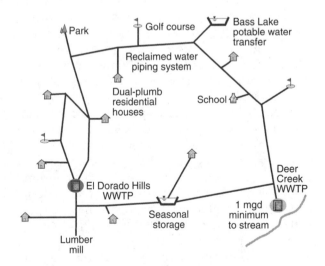

Figure 14–6

El Dorado Irrigation District wastewater treatment plants, recycled water distribution system uses, and storage locations. Recycled water use includes dual-plumb residential irrigation (front- and backyards), golf courses, parks, schools, industry, and construction (dust control). A minimum of 1 mgd is required for discharge to Deer Creek to maintain the aquatic environment.

In 1989, when the district declared a water emergency, Serrano (a local developer) approached the district with the concept of reclaiming wastewater so that their development, which included a golf course and a 3500-home subdivision, could move forward. The district approved the concept of upgrading the Deer Creek Waste Water Treatment Plant in 1990 to meet unrestricted urban reuse requirements. Treatment improvements made in 1996 included primary sedimentation, nitrifying activated sludge, water storage, filtration, and disinfection to meet California Title 22 requirements for disinfected tertiary recycled water with a treatment capacity of 3.6 mgd. The recycled water conveyance system has a peak capacity of approximately 5.0 mgd. A complicating factor was the mandate from the State of California Regional Board for the release of 1.0 mgd into Deer Creek year-round to maintain the wildlife habitat dependent on continuous stream flow. This significantly reduced the ability of the plant to deliver recycled water.

Front- and backyard recycling increases the use of recycled water and offsets additional potable water for other uses. Other operators of urban recycled water systems use frontyard recycling only to keep the watering timing and maintenance under the control of the utility. For front- and backyard irrigation, the homeowner is provided with a user's manual and is designated as the "Recycled Water Supervisor." Responsibilities include all liability and responsibility for operation, maintenance, and all phases of construction (design, review, and inspection) for modification to the irrigation system. Piping is wrapped with purple tape or colored purple and aboveground components are tagged or colored purple to designate recycled water use. Irrigation times for spray systems are restricted to 9:00 P.M. to 6:00 A.M. to limit the opportunity for public contact, while drip irrigation systems may be operated at any time. Dual-plumbed parcels cannot be converted from recycled water to potable water for landscape irrigation. Potable water may only be used indoors and through hose bibs outside the house for swimming pools and spas. Annual inspection and cross-connection control is conducted annually by the district.

As the recycled water system developed, the distribution systems of the Deer Creek and El Dorado Hills plants were connected and included daily storage tanks, pumping stations, and the development of pressure zones. Additional recycled water uses include street median irrigation, parks, and construction dust control. Considerations critical to expansion of the recycled water system include success of the existing system, continued interest among developers, and pressure from the RWQCB to increase treatment requirements for discharge to the creek. To increase the amount of recycled water available, the Recycled Water Master Plan included construction of additional treatment capacity for year-round treatment to Title 22 water quality and seasonal storage to allow the storage of winter wastewater flows for use during the summer irrigation season. Expansion of the recycled water distribution system includes additional daily storage to meet peak flows.

As an emergency backup to the recycled water system, a pumping station constructed at Bass Lake pumps potable water into the recycled water distribution system. Bass Lake can be used as a backup supply when the plant does not produce Title 22 water, the pumping and storage system cannot meet peak irrigation demands, or the total daily demand exceeds the capacity of the distribution system.

Urban Reuse at St. Petersburg, Florida

The city of St. Petersburg is located on a peninsula between Tampa Bay and the Gulf of Mexico. Effluent regulations enacted by the state of Florida in 1972 required limits of 5 mg/l of BOD, 5 mg/l of suspended solids, 3 mg/l of total nitrogen, and 1 mg/l total phosphorus. Potable water is imported from well fields up to 60 miles away, and increasing population growth has resulted in competition for available groundwater supplies. To avoid the high cost of nutrient removal and to offset the irrigation water demands, the city chose to upgrade its wastewater treatment plants to allow unrestricted urban irrigation. The city is the largest community in the United States to achieve complete reuse with a goal of zero discharge. When irrigation is not available, excess recycled water is injected into a deep-well saltwater aquifer for disposal. Reject water that does not meet reuse standards is also injected for disposal, resulting in zero discharge to surface waters. To meet the state's requirements for unrestricted reuse, St. Petersburg upgraded its four wastewater treatment plants with effluent granular-media filtration, chlorination with extended contact time for pathogen removal, and irrigation water storage. The treatment plants have a total capacity of 68.4 mgd. The reclaimed water system began operation in 1978.

Figure 14–7 is a diagram of the water reuse system showing the location of the four treatment plants, reuse piping, and key irrigation areas. Irrigation areas include parks, schools, golf courses, commercial customers, and connections for residential irrigation. The reuse mains are 12- to 36-in.-diameter ductile-iron pipe with secondary mains of 4- to 12-in.-diameter purple-colored PVC plastic pipe, for a total of 300 miles of pipe. Main pressures are kept between 50 to 85 psi to allow sufficient pressure for sprinkler systems. Separation of the piping systems and backflow prevention devices installed in the potable system isolate the public drinking water supply from the reuse system.

During dry months, usually March to June, the typical residential lawn requires 1.5 in./week or about 30,000 gal/mo., while the average wastewater discharge per household is 6000 gal month. This water imbalance suggests that five residential connections are required to produce enough water for one reclaimed water connection. Therefore,

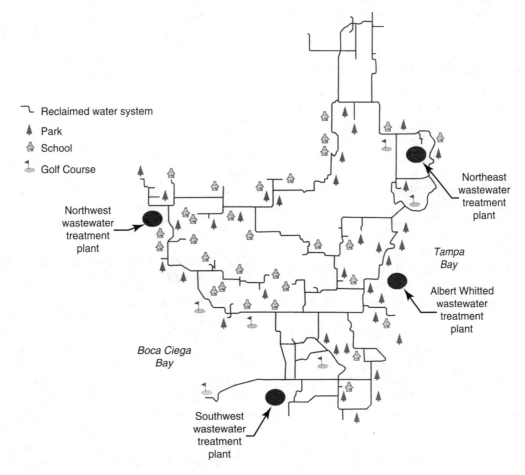

Figure 14–7

Map outline of the City of St. Petersburg, FL, located on a peninsula with Tampa Bay on the east and Boca Ciega Bay on the west. Reclaimed water system showing four tertiary treatment plants, arterial pipe network, and locations of irrigated parks, schoolyards, and golf courses.
(Courtesy of the City of St. Petersburg)

only 20 percent of the 80,000 potential residential customers are able to receive reclaimed water. Due to seasonal variations and irrigation demand, about 50 percent of the annually available reclaimed water is applied for irrigation; the balance is injected into the deep-well system.

The reclaimed water-quality requirements established by the state of Florida for urban water reuse include a maximum of 5 mg/l suspended solids (in any sample), a minimum of 1 mg/l chlorine residual after 15 min, and fecal coliform concentrations that are below detection in 75 percent of the samples (no sample greater than 25/100 ml). All four plants use effluent filtration and reduce suspended solids in the range of 1 to 3 mg/l; fecal coliforms are rarely detected. A minimum chlorine residual of 4.0 mg/l at the outlet of the storage tank is being used as an operating standard.

Industrial Reuse for Power Generation

The Delta Diablo Wastewater Treatment plant provides recycled water to the Los Medanos Energy Center in Pittsburg, CA. Power plant uses include feedwater for the heat-recovery steam generator make-up, and water for the evaporative cooling system.

The Delta Diablo plant receives wastewater from the cities of Antioch and Pittsburg and from unincorporated areas. Treatment facilities consist of screening, grit removal, primary clarification, tower trickling filters, activated sludge, secondary clarification, disinfection, and dechlorination. The design capacity is 16.5 mgd with excess treated effluent discharged into New York Slough via a deep water outfall.

The treatment process for industrial use is designed for 1200 gpm of total treatment and 500 gpm of reverse osmosis followed by demineralization. Treatment includes dual-media filtration, ultrafiltration, ultraviolet disinfection, and granular activated-carbon filtration for TOC removal. An antiscalant chemical is added to the portion of the flow used for evaporative cooling. The steam generator make-up water requires additional treatment using reverse osmosis for demineralization. The system includes a 1-micron cartridge filter, high-pressure pumps, reverse-osmosis trains, a forced-draft decarbonator system, mixed-bed ion exchange demineralizers, and a reverse-osmosis

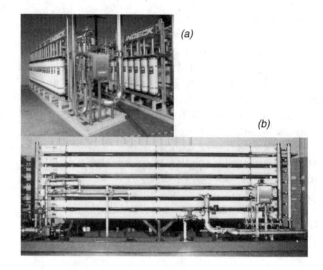

Figure 14–8

Industrial treatment of reuse water for steam generator feed and evaporative cooling make-up water. *(a)* Ultrafiltration modules for solids reduction and reverse-osmosis pretreatment. *(b)* Reverse-osmosis modules. (Courtesy of Indeck, Wheeling, IL.)

cleaning system. Ultrafiltration and reverse-osmosis modules are pictured in Figure 14–8*a* and *b*, respectively. Ancillary systems include waste neutralization, acid and caustic feed, and regeneration systems. The treatment is equivalent to the water reclamation system in Figure 13–9*b*.

Water Reclamation for Groundwater Recharge and Injection

Since 1976, the Orange County Water District, Fountain Valley, CA, has operated Water Factory 21 to reclaim wastewater for injection into groundwater. Due to overdraft from irrigation and well water for potable uses, saltwater had encroached as far as 5 miles inland. The objective of Water Factory 21 was to replenish the groundwater supply for 75 percent of the nearly 2 million residents and build a hydraulic mound to protect inland areas from seawater intrusion. Reclaimed water was blended with imported water prior to injection.

The treatment process consists of lime precipitation, pH adjustment, granular filtration, granular activated-carbon adsorption, reverse osmosis, decarbonation, and disinfection prior to blending, as depicted in Figure 13–9*a*, with a capacity of 15 mgd (57,000 m³/d) through carbon adsorption and 5 mgd (19,000 m³/d) through reverse osmosis. Lime precipitation operates at a pH of 11 to disinfect the wastewater and precipitate solids as a pretreatment for reverse osmosis. To neutralize pH, carbon dioxide is bubbled into a tank containing the settling tank effluent. Granular activated-carbon columns operate in a downflow direction with a bed contact time of approximately 30 min. Adsorption of a wide variety of organic compounds results in a COD reduction of between 50 and 60 percent. Heavy metals such as chromium and copper are partially removed. One-third of the activated-carbon effluent is processed by reverse osmosis to remove dissolved solids. After pressurizing to 550 psi by high-pressure feed pumps, sulfuric acid is injected to adjust the pH to 5.5 and minimize membrane hydrolysis and scale formation, primarily calcium carbonate and calcium sulfate. The three-stage reverse-osmosis process results in a 90 percent reduction of dissolved solids to a concentration of about 100 mg/l with a recovery of about 85 percent. The reject brine is discharged to an outfall sewer for ocean disposal.

Through the 1990s, the Orange County Water District conducted pilot and demonstration scale tests using microfiltration to reduce fouling and extend the runtime of the reverse-osmosis unit, as depicted in Figure 13–9b. Pretreatment using lime, as shown in Figure 13–9a, resulted in a reverse-osmosis operating rate of 10 gpd/sq ft and required cleaning approximately every 4 to 5 weeks to remove fouling materials. Replacing pretreatment, as shown in Figure 13–9b, produced an increase in reverse-osmosis output to 12 gpd/sq ft and an increase in operating time to 14 to 22 weeks. Microfiltration requires significantly less space, eliminates chemical treatment, is easily automated, and requires less maintenance. Ultrafiltration membranes operate up to 4 weeks without chemical cleaning and obtain 85 to 95 percent throughputs.

Orange County, in conjunction with Orange County Sanitation District, is planning a groundwater replenishment system to replace Water Factory 21, expand the existing seawater intrusion barrier, and develop a groundwater replenishment program to supplement water in the Santa Ana River. Distribution will include a pipeline to the groundwater recharge basins in Anaheim and expansion of the recharge basins (see Figure 14–9).

The system is to be constructed in phases starting from a flow rate of 71,000 acre-ft (87.6 mil m³) to more than 130,000 acre-ft (160 mil m³) per year as wastewater flows and recycled water demands increase. A 72-in. steel-cement mortar-lined pipeline is being considered to connect the treatment facility with the groundwater recharge basins. Lime will be added to prevent corrosion of the mortar lining. Pipe pressures range from 170 psi (1,172 kPa) at the pump discharge to 4 psi (27.5 kPa) at the basins. The implementation of groundwater recharge increases the reliability of the water supply during drought conditions and reduces dependence on imported water. The existing spreading fields have an estimated capacity of 65 mgd (246,025 m³/d) based on the use of imported water. The recycled water is lower in dissolved salts and will reduce the overall salinity of groundwater supplies by replacing imported Colorado River water.

The new facility will receive secondary treated wastewater from Orange County Sanitation District Plant No. 1. About 80 percent of the wastewater is treated using activated sludge and 20 percent from a

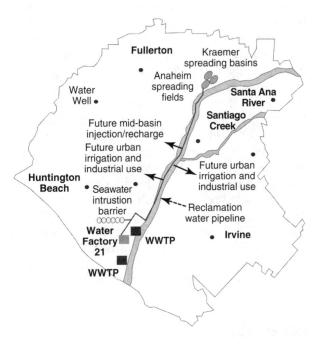

Figure 14–9

Orange County Water District groundwater replenishment system plan for expansion of Water Factory 21 with expansion of the seawater intrusion barrier and pipeline to the spreading fields along the Santa Ana River.

trickling filter process resulting in a feed containing 20 mg/l BOD and suspended solids. The facility will consist of microfiltration, reverse osmosis, and ultraviolet disinfection, similar to Figure 13–9*b*. Microfiltration is used to remove suspended solids, protozoans, and bacteria. The result is a wastewater with no significant particulate materials (turbidity less than 0.2 NTU). As a pretreatment, microfiltration will reduce fouling and extend the runtime and recovery of the reverse-osmosis process. Water recovery is expected to be between 80 and 97 percent. Reverse osmosis removes dissolved organic and inorganic compounds to produce a reclaimed water low in total dissolved solids and low total organic carbon. Phase 1 includes 15 trains (1 standby) with 150 pressure vessels, containing 7 membrane elements, with a flux of 12 gpd/sq ft (0.5 $m^3/d/m^2$) for a total of 5 mgd (18,925 m^3/d). Ultraviolet disinfection is an additional barrier for the inactivation of viruses, removing 99.99 percent of viruses. Combining reverse osmosis and disinfection, the total virus removal is 99.9999 percent or a 6-log removal.

The indirect reuse in the drinking water supply from the spreading fields or from groundwater injection cannot be avoided. The principal source of water for Orange County, CA, is the Santa Ana River, which, during dry periods every summer, is dominated by wastewater discharges from upstream communities. Water in the river percolates into the underlying groundwater and is withdrawn for drinking water. The spreading fields enhance percolation and groundwater recharge. A computer model was used to determine the optimal injection-well capacities and locations. The objective of the barrier is to maintain a mound of fresh water that is higher than the seawater level, thus preventing seawater from migrating into the aquifer. Demands for freshwater vary between winter and summer. Well depths range between 120 ft (37 m) and 690 ft (207 m) with capacities between 1 to 2.4 mgd (3,785 to 9,084 m^3/d).

REFERENCES

1. *Estimated Use of Water in the United States in 2000*, Water Resources of the United States, U.S. Geological Survey, 2000. http://water.usgs.gov/watuse/

2. *Water Facts No. 23: Water Recycling*, California Department of Water Resources, 2002. http://www.owue.water.ca.gov/recycle/docs/WaterFact23.pdf

3. *2005 Reuse Inventory Report*, Florida Department of Environmental Protection, 2005. http://www.dep.state.fl.us/water/reuse/inventory.htm

4. *Arizona Administrative Code, 2001*, Title 18, Chapter 11, Article 3. Reclaimed Water Quality Standards, State of Arizona, 2001.

5. *California Code of Regulations, 2001*, Title 22, Division 4, Chapter 3, Water Recycling Criteria, State of California, 2001.

6. *Florida Administrative Code, 2001*, Chapter 62–610. Reuse of Reclaimed Water and Land Application, State of Florida, 2001.

7. *Guidelines for the Treatment and Use of Reclaimed Water*, Hawaii State Department of Health, November 22, 1993.

8. *Denitrification, attenuation of organic compounds, and colloid transport beneath a reclaimed-water artificial recharge basin*, U.S. Geological Survey, USGS Water Resources of California, 2000. http://ca.water.usgs.gov/projects00/ca498.html

9. *Issues in Potable Reuse: The Viability of Augmenting Drinking Water Supplies with Reclaimed Water*. National Academy Press, 1998.

10. EPANET 2.0, Water Supply & Water Resources, U.S. Environmental Protection Agency, 2002. http://www.epa.gov/ORD/NRMRL/wswrd/epanet.html

Additional Web Site of Interest

WaterReuse Association http://www.watereuse.org/

PROBLEMS

14–1 How is water reuse different from a disposal and water resource planning perspective?

14–2 What are the general uses for recycled water? How are uses in California and Florida similar and different?

14–3 How are water-quality requirements for reuse governed? Include numeric criteria and treatment process examples for unrestricted reuse requirements.

14–4 How do the water-quality requirements for unrestricted reuse vary from the requirements for irrigation of nonfood crops?

14–5 What are the elements of a preliminary design report? Under project evaluation, what is the importance of a pretreatment program? (Refer to Section 13–1 for additional information to be used in your answer.)

14–6 What are the design criteria required for filtration and disinfection in California?

14–7 The following data were collected from a wastewater treatment plant providing recycled water to irrigate a golf course, road medians, and a green belt in front of a shopping center. The plant has 6 filters that are each 30 ft in diameter, a chlorine contact basin of 5.1 million cu ft, and a chlorine dose of 8 mg/l. What problems can you find?

14–8 Compare the treatment and water-quality requirements for restricted agricultural irrigation versus unrestricted urban irrigation.

14–9 A city wants to develop land for agricultural irrigation. The effluent flow of 2.3 mgd contains 24 mg/l of nitrate nitrogen. Crops being considered have nitrogen uptakes of 320 lb/ac/yr and 380 lb/ac/yr. The state limits application rates to a maximum of 4.0 in./week and gives no credit for denitrification in the soil. How many acres are required for recycled water application?

14–10 Another city wants to develop land for agricultural irrigation. The effluent flow of 2.3 mgd contains 24 mg/l of nitrate nitrogen. Crop uptake is 400 lb/ac/yr. Annual rainfall is 2.3 in./mo. The state has no application limits. Evapotransporation of 5.9 in./ac/mo and percolation of 14 in./mo were measured on a test plot. If a portion of the flow is denitrified, resulting in < 10 mg/l for 1.0 mgd, how many acres are necessary? Calculate the area required for recycled water application.

14–11 The design for urban irrigation is based on the data in the following table. Determine the area required to distribute wastewater and the storage requirements. (*Hint*: Size the storage reservoir to be empty on October 1.)

14–12 What quality standards and treatment requirements apply to golf course irrigation? Discuss

Problem 14–7

EFFLUENT FLOW (mgd)	FILTER TURBIDITY (NTU)	TOTAL COLIFORM (No./100 ml)	EFFLUENT FLOW (mgd)	FILTER TURBIDITY (NTU)	TOTAL COLIFORM (No./100 ml)
15.3	1	<2	21.2	15	280
15.4	1	<2	18.3	1.8	4
17.2	1	<2	18.1	1.4	<2
22.3	1.4	2.2	15.4	1.2	<2
31.5	1.8	8	15.3	1.2	<2
13.4	4	28	15.6	1.3	<2

Problem 14–11

MONTH	WASTEWATER FLOW (mgd)	PRECIPITATION (IN./MONTH)	EVAPORATION (IN./MONTH)	IRRIGATION REQUIREMENTS (in./ac/wk)
Jan	3.05	4	2	0
Feb	4.5	6	1.8	0
Mar	3.15	4	2.3	0
April	2.46	2	6	2.5
May	2.51	1	8	3
June	1.87	0	10	4
July	1.76	0	12	4
Aug	1.75	0	12	4
Sept	1.7	0	12	4
Oct	1.62	2	7	3
Nov	1.64	2	6	0
Dec	2.25	3	3	0

in terms of restricted and unrestricted requirements. What are the considerations for buffer area, fencing, and public access?

14–13 What are the differences between gray water and recycled water?

14–14 How are water-quality requirements for groundwater injection protective of public health? How do treatment requirements differ from rapid infiltration? What assumptions are made for additional treatment in the soil? (*Hint*: Vadose zone)

14–15 What cautions were suggested by the NRC for augmentation of potable water supplies?

14–16 The design of an urban irrigation system is based on the supply-and-demand data listed in the following table. Plot supply and demand on the same graph. Determine the storage requirements. (*Hint*: Assume storage is empty in November.)

14–17 Reuse water is used for golf course irrigation; however, insufficient water is available during July and August and the supply requires

augmentation from potable water supplies. What protections are necessary to guard against cross-connection?

14–18 How has Tallahassee solved its water discharge problem? How does the agency provide zero discharge?

14–19 What is the recycled use for the El Dorado Irrigation District, and what are the instructions to the users?

14–20 What treatment is required for industrial reuse for power supply and cooling? (*Note*: Treatment requirements in Hawaii.)

14–21 What degree of treatment is used for groundwater recharge?

14–22 Refer to the inside cover. Assuming that potable water supplies may come from rivers (surface water), lakes, and groundwater, explain the importance of wastewater, advanced wastewater, recycled water treatment, and water reclamation. Why are water-quality goals critical to an overall water use/reuse cycle? Think about the downstream users.

Problem 14–16

Month	Supply (mgd)	Demand (ac-ft/mo)	Month	Supply (mgd)	Demand (ac-ft/mo)
Jan	6.38	0	July	4.19	143.61
Feb	7.71	0	Aug	4.09	148.40
Mar	6.03	17.28	Sept	3.77	143.61
April	5.32	35.90	Oct	3.59	107.71
May	4.95	71.81	Nov	3.89	0
June	4.13	107.71	Dec	4.31	0

APPENDIX

COMMON ABBREVIATIONS USED IN THE TEXT

ac	acres
BOD	biochemical oxygen demand (5-day)
°C	degrees Celsius
cfm	cubic feet per minute
cm	centimeters
cu ft	cubic feet
d	day
dia.	diameter
DO	dissolved oxygen
°F	degrees Fahrenheit
F/M	food-to-microorganism ratio
fps	feet per second
ft	feet
g	grams
gal	gallons
gpcd	gallons per capita per day
gpm	gallons per minute
gr	grains
h, hr	hour
ha	hectare
hp	horsepower
in.	inches
kg	kilogram
km	kilometer
kN	kilonewton
kPa	kilopascal
kW	kilowatt
l	liters
lb	pounds
m	meters
MCL	maximum contaminant level
meq	milliequivalents
meq/l	milliequivalents per liter

mg	milligrams		ppm	parts per million
mg/l	milligrams per liter		psi	pounds per square inch
mgad	million gallons per acre per day		s, sec	second
mgd	million gallons per day		sour	specific oxygen uptake rate
mi	miles		sq ft	square feet
mil gal	million gallons		SS	suspended soils
min	minutes		THM	trihalomethane
mm	millimeters		TOC	total organic carbon
mon	months		VSS	volatile suspended solids
MPN	most probable number		y, yr	year

COMMON CONVERSION FACTORS

Multiply	By	To Obtain	Multiply	By	To Obtain
acres	43,560	square feet	grains per gallon	142.9	pounds per million gallons
acres	0.001 56	square miles	grams	0.035 27	ounces
acre-feet	43,560	cubic feet	grams	0.002 21	pounds
acre-feet	325,850	gallons	inches	2.540	centimeters
atmospheres	76.0	centimeters of mercury	kilograms	2.205	pounds
atmospheres	29.92	inches of mercury	kilograms per cubic meter	0.062 43	pounds per cubic foot
atmospheres	33.90	feet of water	kilograms per square centimeter	14.22	pounds per square inch
atmospheres	14.7	pounds per square inch			
centimeters	0.3937	inches	kilometers	0.6214	miles
cubic centimeters	0.061 02	cubic inches	liters	1000	cubic centimeters
cubic feet	1728	cubic inches	liters	0.035 31	cubic feet
cubic feet	7.481	gallons	liters	0.2642	gallons
cubic feet	28.32	liters	liters per second	15.85	gallons per minute
cubic feet per second	0.6463	million gallons per day	$\log_{10}$	2.303	$\log_e$
cubic feet per second	448.8	gallons per minute	$\log_e$ (e = 2.7183)	0.4343	$\log_{10}$
cubic meters	35.31	cubic feet	meters	100	centimeters
cubic meters	264.2	gallons	meters	3.281	feet
cubic meters	1000	liters	meters	39.37	inches
cubic meters per second	2,120	cubic feet per minute	meters per second	3.281	feet per second
			microns	0.001	millimeters
day	1440	minutes	milligrams per liter	1	parts per million
day	86,400	seconds	milligrams per liter	0.0584	grains per gallon
feet	30.48	centimeters	milligrams per liter	8.345	pounds per million gallons
feet	12	inches			
feet	0.3048	meters	pounds	7000	grains
feet of water	0.4335	pounds per square inch	pounds	453.6	grams
			pounds per square foot	0.006 95	pounds per square inch
gallons	3785	cubic centimeters	square centimeters	0.1550	square inches
gallons	0.1337	cubic feet	square meters	10.76	square feet
gallons	231	cubic inches	square miles	640	acres
gallons	3.785	liters	temperature		
gallons of water	8.345	pounds of water	°C	9/5	+32 = °F
gallons per minute	0.002 23	cubic feet per second	°F − 32	5/9	°C
gallons per minute	8.021	cubic feet per hour	tons	2000	pounds
grains per gallon	17.12	milligrams per liter			

SELECTED SI UNITS (Système International d'Unités)

Unit Type	Quantity	Unit	Symbol	Expression in Other Units
Base Units	Length	meter	m	
	Mass	kilogram[a]	kg	
	Time	second	s	
	Thermodynamic temperature	kelvin[b]	K	
	Amount of substance	mole	mol	
Supplementary Unit	Plane Angle	Radian	Rad	
Derived Units Having Special Names	Force	newton	N	$kg \cdot m/s^2$
	Pressure	pascal	Pa	N/m^2
	Energy, work	joule	J	$N \cdot m$
	Power	watt	W	J/s
	Area	hectare	ha	$10^4 \, m^2$
	Volume	liter	l	$10^{-3} \, m^3$
Derived Units	Area	square meter		m^2
	Volume	cubic meter		m^3
	Linear velocity	meter per second		m/s
	Linear acceleration	meter per second squared		m/s^2
	Density	kilogram per cubic meter		kg/m^3

[a]The kilogram is the only base unit with a prefix.
[b]The Celsius temperature scale (previously called centigrade) is commonly used for measurements even though it is not part of the SI system. A difference of 1 degree on the Celsius scale (°C) equals 1 kelvin (K), and zero on the thermodynamic scale is 273.15 kelvin below 0 degrees Celsius.

SI UNIT PREFIXES

Amount	Multiples and Sub-multiples	Prefixes	Symbols	Amount	Multiples and Sub-multiples	Prefixes	Symbols
1 000 000 000 000	10^{12}	tera	T	0.1	10^{-1}	deci	d
1 000 000 000	10^9	giga	G	0.01	10^{-2}	centi	c
1 000 000	10^6	mega	M	0.001	10^{-3}	milli	m
1 000	10^3	kilo	k	0.000 001	10^{-6}	micro	μ
100	10^2	hecto	h	0.000 000 001	10^{-9}	nano	n
10	10	deka	da	0.000 000 000 001	10^{-12}	pico	P
				0.000 000 000 000 001	10^{-15}	femto	f
				0.000 000 000 000 000 001	10^{-18}	atto	a

CONVERSION FACTORS ENGLISH TO SI METRIC

Multiply	By	To Obtain	Multiply	By	To Obtain
Length			*Density*		
mile, mi	1.609	kilometer, km	pound/cubic foot, lb/cu ft	16.02	kilogram/cubic meter, kg/m^3
yard, yd	0.914 4	meter, m	*BOD Loading Rate*		
foot, ft	0.304 8	meter, m			
inch, in.	25.40	millimeter, mm	pound/1000 cubic foot/day, lb/1000 cu ft/day	16.02	gram/cubic meter · day, g/m^3 · d
Area			pound/acre/day, lb/ac/day	0.1121	gram/square meter · day, g/m^2 · d
square mile, sq mi	2.590	square kilometer, km^2		1.121	kilogram/hectare · day, kg/ha · d
acre, ac	0.404 7	hectare, ha	*Solids Loading Rate*		
	4047.	square meter, m^2			
square yard, sq yd	0.836 1	square meter, m^2	pound/cubic foot/day, lb/cu ft/day	16.02	kilogram/cubic meter · day, kg/m^3 · d
square foot, sq ft	0.092 9	square meter, m^2	pound/square foot/day, lb/sq ft/day	4.883	kilogram/square meter · day, kg/m^2 · d
square inch, sq in.	645.2	square millimeter, mm^2	*Hydraulic Loading Rate*		
Volume					
acre foot, ac ft	1234.	cubic meter, m^3	million US gallons/acre/day, mgad	0.935 3	cubic meter/square meter · day, m^3/m^2 · d
cubic yard, cu yd	0.764 9	cubic meter, m^3		9353	cubic meter/hectare · day, m^3/ha · d
cubic foot, cu ft	0.028 32	cubic meter, m^3	US gallon/square foot/day, gal/sq ft/day	0.040 75	cubic meter/square meter · day, m^3/m^2 · d
	28.32	liter, l	US gallon/square foot/minute, gpm/sq ft	0.679 0	liter/square meter · second, l/m^2 · s
US gallon, gal	3.785	liter, l	US gallon/cubic foot/day, gal/cu ft/day	0.133 7	cubic meter/cubic meter · day, m^3/m^3 · d
Velocity					
foot/second, ft/sec	0.304 8	meter/second, m/s	*Concentration*		
foot/minute, ft/min	0.005 08	meter/second, m/s	pound/million US gallons, lb/mil gal	0.119 8	milligram/liter, mg/l
Flow			*Force*		
million US gallons/day, mgd	3785.	cubic meter/day, m^3/d	pound force, lb	4.448	newton, N
	43.81	liter/second, l/s			
US gallons/minute, gpm	5.450	cubic meter/day, m^3/d			
	0.063 09	liter/second, l/s			
cubic foot/second, cu ft/sec	0.028 32	cubic meter/second, m^3/s			
Mass					
ton (2000 lb)	0.907 2	tonne (1000 kg), t			
	907.2	kilogram, kg			
pound, lb	0.453 6	kilogram, kg			
	453.6	gram, g			

CONVERSION FACTORS ENGLISH TO SI METRIC (continued)

Multiply	By	To Obtain	Multiply	By	To Obtain
PRESSURE			*AERATION*		
pound/square foot, psf	0.047 88	kilopascal, kPa	cubic foot (air)/pound (BOD), cu ft/lb	0.062 43	cubic meter/ kilogram, m³/kg
pound/square inch, psi	6.895	kilopascal, kPa			
POWER					
horsepower, hp	0.7457	kilowatt, kW			

CONVERSION FACTORS SI METRIC TO ENGLISH

Multiply	By	To Obtain	Multiply	By	To Obtain
LENGTH			cubic meter/second, m³/s	35.31	cubic foot/second, cu ft/sec
kilometer, km	0.621 4	mile, mi	liter/second, l/s	15.85	US gallon/minute, gpm
meter, m	3.281	foot, ft		0.022 83	million US gallons/ day, mgd
	1.094	yard, yd			
millimeter, mm	0.039 37	inch, in.			
AREA			*MASS*		
square kilometer, km²	0.386 1	square mile, sq mi	tonne (1000 kg), t	1.102	ton (2000 lb)
hectare, ha	2.471	acre, ac	kilogram, kg	2.205	pound, lb
square meter, m²	10.76	square foot, sq ft	*DENSITY*		
	1.196	square yard, sq yd	kilogram/cubic meter, kg/m³	0.624 2	pound/cubic foot, lb/cu ft
square millimeter, mm²	0.001 55	square inch, sq in.			
VOLUME			*BOD LOADING RATE*		
cubic meter, m³	264.2	US gallon, gal	gram/cubic meter · day, g/m³ · d	0.062 43	pound/1000 cubic foot/day, lb/1000 cu ft/day
	35.31	cubic foot, cu ft			
	1.308	cubic yard, cu yd	gram/square meter · day, g/m² · d	8.921	pound/acre/day, lb/ac/day
liter, l	0.264 2	US gallon, gal	kilogram/hectare · day, kg/ha · d	0.892 1	pound/acre/day, lb/ac/day
	0.035 31	cubic foot, cu ft			
VELOCITY			*SOLIDS LOADING RATE*		
meter/second, m/s	3.281	foot/second, ft/sec	kilogram/cubic meter · day, kg/m³ · d	0.062 43	pound/cubic foot/day, lb/cu ft/day
	196.8	foot/minute, ft/min			
FLOW			kilogram/square meter · day, kg/m² · d	0.204 8	pound/square foot/day, lb/sq ft/day
cubic meter/day, m³/d	0.183 5	US gallon/minute, gpm			

CONVERSION FACTORS SI METRIC TO ENGLISH (continued)

MULTIPLY	BY	TO OBTAIN	MULTIPLY	BY	TO OBTAIN
HYDRAULIC LOADING RATE			*FORCE*		
cubic meter/square meter · day, $m^3/m^2 \cdot d$	24.54	US gallon/square foot/day, gal/sq ft/day	newton, N	0.224 8	pound force, lb
			PRESSURE		
	1.069	million US gallons/ acre/day, mgad	kilopascal, kPa	0.145 0	pound/square inch, psi
cubic meter/cubic meter · day, $m^3/m^3 \cdot d$	7.481	US gallon/cubic foot/day, gal/cu ft/day	*POWER*		
			kilowatt, kW	1.341	horsepower, hp
liter/square meter · second, $l/m^2 \cdot s$	1.473	US gallon/square foot/minute, gpm/sq ft	*AERATION*		
			cubic meter (air)/ kilogram (BOD), m^3/kg	16.02	cubic foot/pound, cu ft/lb
CONCENTRATION					
milligram/liter, mg/l	8.345	pound/million US gallons, lb/mil gal			

TABLE OF CHEMICAL ELEMENTS

NAME	SYMBOL	ATOMIC NUMBER	ATOMIC WEIGHT	NAME	SYMBOL	ATOMIC NUMBER	ATOMIC WEIGHT
Actinium	Ac	89	—	Cobalt	Co	27	58.9332
Aluminum	Al	13	26.9815	Copper	Cu	29	63.546
Americium	Am	95	—	Curium	Cm	96	—
Antimony	Sb	51	121.75	Dysprosium	Dy	66	162.50
Argon	Ar	18	39.948	Einsteinium	Es	99	—
Arsenic	As	33	74.9216	Erbium	Er	68	167.26
Astatine	At	85	—	Europium	Eu	63	151.96
Barium	Ba	56	137.34	Fermium	Fm	100	—
Berkelium	Bk	97	—	Fluorine	F	9	18.9984
Beryllium	Be	4	9.0122	Francium	Fr	87	—
Bismuth	Bi	83	208.980	Gadolinium	Gd	64	157.25
Boron	B	5	10.811	Gallium	Ga	31	69.72
Bromine	Br	35	79.904	Germanium	Ge	32	72.59
Cadmium	Cd	48	112.40	Gold	Au	79	196.967
Calcium	Ca	20	40.08	Hafnium	Hf	72	178.49
Californium	Cf	98	—	Helium	He	2	4.0026
Carbon	C	6	12.01115	Holmium	Ho	67	164.930
Cerium	Ce	58	140.12	Hydrogen	H	1	1.00797
Cesium	Cs	55	132.905	Indium	In	49	114.82
Chlorine	Cl	17	35.453	Iodine	I	53	126.9044
Chromium	Cr	24	51.996	Iridium	Ir	77	192.2

TABLE OF CHEMICAL ELEMENTS (continued)

Name	Symbol	Atomic Number	Atomic Weight	Name	Symbol	Atomic Number	Atomic Weight
Iron	Fe	26	55.847	Radon	Rn	86	—
Krypton	Kr	36	83.80	Rhenium	Re	75	186.2
Lanthanum	La	57	138.91	Rhodium	Rh	45	102.905
Lead	Pb	82	207.19	Rubidium	Rb	37	85.47
Lithium	Li	3	6.939	Ruthenium	Ru	44	101.07
Lutetium	Lu	71	174.97	Samarium	Sm	62	150.35
Magnesium	Mg	12	24.312	Scandium	Sc	21	44.956
Manganese	Mn	25	54.9380	Selenium	Se	34	78.96
Mendelevium	Md	101	—	Silicon	Si	14	28.086
Mercury	Hg	80	200.59	Silver	Ag	47	107.868
Molybdenum	Mo	42	95.94	Sodium	Na	11	22.998
Neodymium	Nd	60	144.24	Strontium	Sr	38	87.62
Neon	Ne	10	20.183	Sulfur	S	16	32.064
Neptunium	Np	93	—	Tantalum	Ta	73	189.948
Nickel	Ni	28	58.71	Technetium	Tc	43	—
Niobium	Nb	41	92.906	Tellurium	Te	52	127.60
Nitrogen	N	7	14.0067	Terbium	Tb	65	158.924
Nobelium	No	102	—	Thallium	Tl	81	204.37
Osmium	Os	76	190.2	Thorium	Th	90	232.038
Oxygen	O	8	15.9994	Thulium	Tm	69	168.934
Palladium	Pd	46	106.4	Tin	Sn	50	118.69
Phosphorus	P	15	30.9738	Titanium	Ti	22	47.90
Platinum	Pt	78	195.09	Tungsten	W	74	183.85
Plutonium	Pu	94	—	Uranium	U	92	238.03
Polonium	Po	84	—	Vanadium	V	23	50.942
Potassium	K	19	39.102	Xenon	Xe	54	131.30
Praseodymium	Pr	59	140.907	Ytterbium	Yb	70	173.04
Promethium	Pm	61	—	Yttrium	Y	39	88.905
Protactinium	Pa	91	—	Zinc	Zn	30	65.37
Radium	Ra	88	—	Zirconium	Zr	40	91.22

GEOMETRIC FORMULAS FOR BASINS OF VARIOUS SHAPES ($\pi = 3.1416$)

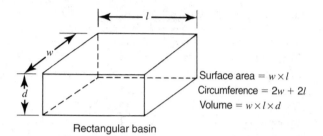

Surface area $= w \times l$
Circumference $= 2w + 2l$
Volume $= w \times l \times d$

Rectangular basin

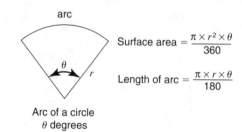

Surface area $= \dfrac{\pi \times r^2 \times \theta}{360}$

Length of arc $= \dfrac{\pi \times r \times \theta}{180}$

Arc of a circle
θ degrees

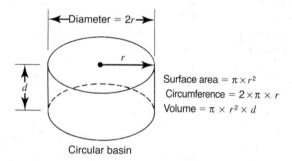

Surface area $= \pi \times r^2$
Circumference $= 2 \times \pi \times r$
Volume $= \pi \times r^2 \times d$

Circular basin

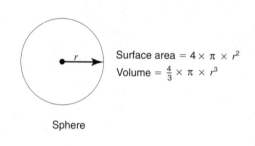

Surface area $= 4 \times \pi \times r^2$
Volume $= \dfrac{4}{3} \times \pi \times r^3$

Sphere

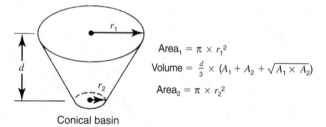

Area$_1 = \pi \times r_1^2$
Volume $= \dfrac{d}{3} \times (A_1 + A_2 + \sqrt{A_1 \times A_2})$
Area$_2 = \pi \times r_2^2$

Conical basin

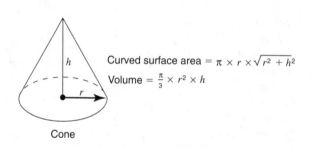

Curved surface area $= \pi \times r \times \sqrt{r^2 + h^2}$
Volume $= \dfrac{\pi}{3} \times r^2 \times h$

Cone

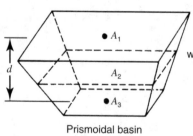

Volume $= \dfrac{d}{6} \times (A_1 + 4A_2 + A_3)$

where $A_1 =$ surface area
$A_2 =$ area of midsection
$A_3 =$ bottom area
$d =$ depth

Prismoidal basin

Volume $= \dfrac{a \times b \times h}{3}$

Area of side $= \dfrac{a \times l}{2}$
$= \dfrac{b \times l}{2}$

Area of base $= a \times b$

Pyramid

PERIODIC TABLE OF THE ELEMENTS

1a	2a	3b	4b	5b	6b	7b	8			1b	2b	3a	4a	5a	6a	7a	0
1 +1 **H** 1.0079																	**2** 0 **He** 4.0026
3 +1 **Li** 6.94	**4** +2 **Be** 9.01218											**5** +3 **B** 10.81	**6** +4 +3 **C** 12.011	**7** +5 **N** 14.0067 −3	**8** **O** 15.9994 −2	**9** **F** 18.998403 −1	**10** 0 **Ne** 20.17
11 +1 **Na** 22.98977	**12** +2 **Mg** 24.305											**13** +3 **Al** 26.98154	**14** +4 **Si** 28.0855	**15** +5 +3 **P** 30.97376 −3	**16** +6 +4 **S** 32.06 −2	**17** **Cl** 35.453 −1	**18** 0 **Ar** 39.948
19 +1 **K** 39.0983	**20** +2 **Ca** 40.08	**21** **Sc** 44.9559	**22** **Ti** 47.9	**23** **V** 50.941	**24** +6 +3 **Cr** 51.996	**25** +7 +4 +2 **Mn** 54.938	**26** +3 +2 **Fe** 55.847	**27** +3 +2 **Co** 58.9332	**28** +3 +2 **Ni** 58.71	**29** +2 +1 **Cu** 63.546	**30** +2 **Zn** 65.38	**31** +3 **Ga** 69.72	**32** +4 +2 **Ge** 72.59	**33** +5 +3 **As** 74.9216	**34** +6 +4 **Se** 78.96 −2	**35** **Br** 79.904 −1	**36** 0 **Kr** 83.8
37 +1 **Rb** 85.467	**38** +2 **Sr** 87.62	**39** **Y** 88.9059	**40** **Zr** 91.22	**41** **Nb** 92.9064	**42** **Mo** 95.94	**43** **Tc** 98.9062	**44** **Ru** 101.07	**45** **Rh** 102.9055	**46** **Pd** 106.4	**47** +1 **Ag** 107.868	**48** +2 **Cd** 112.41	**49** +3 **In** 114.82	**50** +4 +2 **Sn** 118.69	**51** +5 +3 **Sb** 121.75	**52** +6 +4 **Te** 127.6 −2	**53** **I** 126.9045 −1	**54** 0 **Xe** 131.3
55 +1 **Cs** 132.9054	**56** +2 **Ba** 137.33	**57** **La** 138.9055	**72** **Hf** 178.49	**73** **Ta** 180.947	**74** **W** 183.85	**75** **Re** 186.2	**76** **Os** 190.2	**77** **Ir** 192.22	**78** **Pt** 195.09	**79** **Au** 196.9665	**80** +2 +1 **Hg** 200.59	**81** +3 +1 **Ti** 204.37	**82** +4 +2 **Pb** 207.2	**83** +5 +3 **Bi** 208.9808	**84** +6 +4 **Po** (210)	**85** **At** (210)	**86** 0 **Rn** (222)
87 +1 **Fr** (223)	**88** +2 **Ra** 226.0254	**89** **Ac** (227)	**104** **Ku*** (261)	**105** **Ha*** (262)													

Atomic Number → **50** +2 ← Oxidation States
+4

Symbol → **Sn**

Atomic Weight → 118.69

Lanthanum Series

58	59	60	61	62	63	64	65	66	67	68	69	70	71
Ce 140.12	**Pr** 140.9	**Nd** 144.24	**Pm** (145)	**Sm** 150.4	**Eu** 151.96	**Gd** 157.2	**Tb** 158.93	**Dy** 162.50	**Ho** 164.93	**Er** 167.26	**Tm** 168.93	**Yb** 173.04	**Lu** 174.97

Actinium Series

90	91	92	93	94	95	96	97	98	99	100	101	102	103
Th 232.0	**Pa** 231.0	**U** 238.0	**Np** 237.0	**Pu** (244)	**Am** (243)	**Cm** (247)	**Bk** (247)	**Cf** (251)	**Es** (254)	**Fm** (257)	**Md** (256)	**No** (254)	**Lr** (257)

Mass numbers of the most stable or most abundant isotopes are shown in parentheses.

The elements to the right of the bold lines are called the non metals and the elements to the left of the bold line are called the metals.

Common oxidation number are given for the representative elements and some transition elements.

*Kurchatovium and Hahnium are tentative names for these elements.

TWO-CYCLE LOGARITHMIC PROBABILITY PAPER FOR USE IN SOLVING PROBLEMS 4–52 AND 4–53

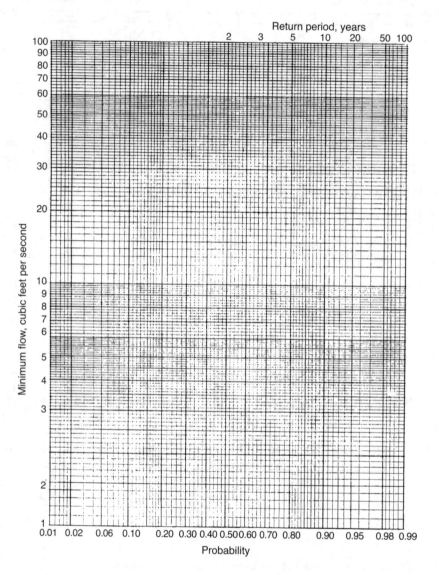

INDEX